Advances in Experimental Medicine and Biology

Advances in Microbiology, Infectious Diseases and Public Health

Volume 1435

Paola Mastrantonio • Maja Rupnik
Editors

Updates on *Clostridioides difficile* in Europe

Advances in Microbiology, Infectious Diseases and Public Health
Volume 18

Second Edition

Springer

Editors
Paola Mastrantonio
Former Research Director in the
Department of Infectious Diseases
National Institute of Health (ISS)
Rome, Italy

Maja Rupnik
Department for Microbiological Research
National Laboratory of Health,
Environment and Food (NLZOH)
Maribor, Slovenia

ISSN 0065-2598 ISSN 2214-8019 (electronic)
Advances in Experimental Medicine and Biology
ISSN 2365-2675 ISSN 2365-2683 (electronic)
Advances in Microbiology, Infectious Diseases and Public Health
ISBN 978-3-031-42107-5 ISBN 978-3-031-42108-2 (eBook)
https://doi.org/10.1007/978-3-031-42108-2

This is a second edition of Volume 8 in this series

This Springer imprint is published by the registered company Springer Nature Switzerland AG
The registered company address is: Gewerbestrasse 11, 6330 Cham, Switzerland

Paper in this product is recyclable.

This book is dedicated to the memory of our dear friend and colleague Professor Anne Collignon who passed away in 2022. She contributed significantly to a better knowledge of C. difficile *and motivated and mentored many young researchers during their first steps in this intriguing research area.*

Preface

The AIMI Volume 8 *Updates on Clostridium difficile in Europe* has been the most successful thematic volume of the AIMI Series in the last 5 years. As editors we have received the invitation to prepare a second edition with updated chapters.

In the meantime, *Clostridium difficile* was renamed to *Clostridioides difficile*, but the main clinical, diagnostic, and research challenges remained unchanged.

All but one chapter from the previous volume were updated, some of them in a rather substantial way. The chapters of this book were planned to cover the most important issues to be addressed in the study of infections due to *C. difficile*. Two new chapters, not included in the first edition, were also added, one on sporulation and the other on membrane vesicles.

C. difficile is a microorganism still feared not only as the cause of nosocomial diarrhea related to protracted antibiotic administration but more and more frequently of diarrheal diseases unrelated to the hospital environment, including those affecting animals. In the last few decades, a growing number of clinicians, microbiologists, and epidemiologists have investigated this topic, as evidenced by the large number of scientific publications still in an upward trend over the years.

In particular, this book has been focused on the clinical and experimental activities carried out in Europe for a better knowledge of this pathogen and its molecular characteristics, associated pathologies, and possible transmission routes, as well as to build preventive and diagnostic strategies, and efficacious therapeutic approaches for the treatment of *Clostridioides difficile* infection (CDI).

Thanks also to the foundation of the European Study Group on *C.difficile* (ESGCD) in 2000, in the framework of the European Society of Clinical Microbiology and Infectious Diseases (ESCMID), European microbiologists, clinicians, and epidemiologists, together with other experts from all over the world, were able to consolidate the already existing positive collaboration that led in recent years to the establishment of a European network for the epidemiological surveillance, the molecular characterization, and the evaluation of the antibiotic resistance profile of the clinical isolates, with obvious advantages for the continuous updating of the treatment strategies of CDI. To emphasize the positive role of the ESCMID in the fight against *C. difficile* infection, an updated chapter written by the current and past presidents of ESGCD has been included in this volume.

We are grateful to all the authors for their contributions to the book. In our view and intention, they also ideally represent the work of many other European experts in this field who did not get involved on this occasion for obvious limits of space.

Rome, Italy Paola Mastrantonio
Maribor, Slovenia Maja Rupnik

Contents

Economic Burden of *Clostridioides difficile* Infection in European Countries

Elena Reigadas, Silvia Vázquez-Cuesta, and Emilio Bouza

Abstract

Clostridioides difficile infection (CDI) remains a considerable challenge to healthcare systems worldwide. Although CDI represents a significant burden on healthcare systems in Europe, few studies have attempted to estimate the consumption of resources associated with CDI in Europe. The reported extra costs attributable to CDI vary widely according to the definitions, design, and methodologies used, making comparisons difficult to perform. In this chapter, the economic burden of healthcare facility-associated CDI in Europe will be assessed, as will other less explored areas such as the economic burden of recurrent CDI, community-acquired CDI, pediatric CDI, and CDI in outbreaks.

E. Reigadas (✉)
Department of Clinical Microbiology and Infectious Diseases, Hospital General Universitario Gregorio Marañón, Madrid, Spain

Medicine Department, School of Medicine, Universidad Complutense de Madrid (UCM), Madrid, Spain

Instituto de Investigación Sanitaria Gregorio Marañón, Madrid, Spain

S. Vázquez-Cuesta
Instituto de Investigación Sanitaria Gregorio Marañón, Madrid, Spain

E. Bouza (✉)
Department of Clinical Microbiology and Infectious Diseases, Hospital General Universitario Gregorio Marañón, Madrid, Spain

Medicine Department, School of Medicine, Universidad Complutense de Madrid (UCM), Madrid, Spain

Instituto de Investigación Sanitaria Gregorio Marañón, Madrid, Spain

CIBER de Enfermedades Respiratorias (CIBERES CB06/06/0058), Madrid, Spain

1 Introduction

In this chapter, the economic burden of healthcare facility-associated *Clostridioides difficile* infection (CDI) in Europe will be assessed, as will other less explored areas such as recurrent CDI (R-CDI), community-acquired CDI, pediatric CDI, and CDI in outbreaks.

Despite advances in the diagnosis and treatment of CDI and prevention efforts to reduce the incidence of CDI, the disease remains a significant challenge to healthcare systems worldwide (Dubberke and Olsen 2012; Bouza 2012). From an economic point of view, CDI increases patient healthcare costs as a result of extended length of hospital stay (LOS), re-admission, laboratory tests, and medication (Wiegand et al. 2012; Gabriel and Beriot-Mathiot 2014; Nanwa et al. 2015). *C. difficile* infection is costly, not only to third-party payers and hospitals but also to society as a whole (Mcglone et al. 2012).

Most of the existing literature is from the United States, where an in silico economic

P. Mastrantonio, M. Rupnik (eds.), *Updates on Clostridioides difficile in Europe*, Advances in Experimental Medicine and Biology 1435, https://doi.org/10.1007/978-3-031-42108-2_1

model suggested that the annual US economic burden of CDI would be $496 million from a hospital perspective, $547 million from a third-party payer perspective, and $796 million from a societal perspective (Mcglone et al. 2012). Another study including both healthcare setting and the community estimated that the annual economic cost of all CDI in the United States was $5.4 billion, with $4.7 billion of the costs incurred in healthcare settings (Desai et al. 2016). However, few published studies have attempted to estimate the consumption of resources associated with CDI in Europe (Wiegand et al. 2012; Wingen-Heimann et al. 2022). It has been estimated that the annual cost of CDI in Europe is €3 billion per year (Jones et al. 2013); consequently, approaches that can reduce CDI-associated resource use and costs are of interest.

Although antibiotics are a key component of therapy for CDI, they currently represent a minimal cost in the overall budget for CDI management, and the main extra associated cost reported in most studies is the extended LOS attributable to CDI (Wiegand et al. 2012; Asensio et al. 2013, 2015; Wilcox et al. 1996; Hubner et al. 2015).

CDI-related costs are also likely to increase as the population ages. In a systematic European meta-analysis on clinical and economic burden, the authors reported that the incremental cost of CDI may have increased by £1857 to £4266 (27–93%) over a 12-year period (Wiegand et al. 2012). In a review by Kuijper et al., the potential cost of CDI was estimated to be €3 billion/year and is expected to almost double over the next four decades, assuming a European Union population of 457 million inhabitants (Kuijper et al. 2006).

The reported extra costs attributable to CDI vary widely according to the definitions, design, and methodologies used (Ghantoji et al. 2010; Wiegand et al. 2012). Most studies do not separate the costs of resources due to CDI from those generated by the underlying disease. Therefore, comparisons need to be made with caution and limited to results obtained in a similar manner.

A clearer understanding of the healthcare and economic burden of CDI is of value to hospital administrators, infection prevention teams, and persons involved in antimicrobial stewardship programs, who can use this key information to determine the appropriate degree of investment in infection control measures and in other priority areas.

Future studies should follow standard methodology, include other indirect cost perspectives such as societal and patient perspectives, and examine poorly explored populations, such as individuals with community-acquired CDI.

2 Economic Burden of Hospital-Acquired CDI in European Countries

A wide range of CDI costs in Europe have been reported, ranging from €5798 to €12,867 per case (Wiegand et al. 2012; Braae et al. 2020). Detailed data are only available from eight European countries (Ireland, England, Wales, Denmark, Germany, Poland, Spain, and Italy). Table 1 summarizes CDI costs by study and country. Recently, a retrospective analysis within the Combatting Bacterial Resistance in Europe CDI (COMBACTE-CDI) point-prevalence study project was conducted based on resource costs for inpatient treatment and productivity costs (Wingen-Heimann et al. 2022). This study analyzed 430 hospitalized patients from 12 European countries, reporting a cost of €15,242 per CDI primary episodes.

2.1 Primary Episodes

The economic burden of primary episodes in Europe is reviewed below by region. The most abundant literature comes from Northern, Western, and Southern Europe.

2.1.1 Northern Europe

A large study conducted in Denmark, including 12,768 patients with healthcare-related CDI and 23,272 matched controls, revealed that the total healthcare cost was significantly larger for CDI cases than controls throughout all periods (Braae et al. 2020). During the index admission period, cost was €12,867 per CDI case compared to

Table 1 Summary of European *Clostridioides difficile* infection costs by study and country

Reference Author (year)	Country or region	CDI cases examined	Study population	Study period	Cost
Braae et al. (2020)	Denmark	$N = 12{,}768$	Healthcare CDI	2011–2014	€12,867/CDI
Ryan et al. (2017)	Ireland	$N = 13$	Healthcare CDI	August 2015	€5820/CDI
Al-Eidan et al. (2000)	Ireland	$N = 87$	Healthcare CDI	1994–1995	£2860/CDI
Wilcox et al. (1996)	England	$N = 50$	Healthcare CDI	1994–1995	£4107/CDI
Tresman and Goldenberg (2018)	England	$N = 90$ (45 CDI, 45 recurrent CDI)	Healthcare CDI	2014–2017	£12,710 (CDI) £31,121 (recurrent CDI)
Wilcox et al. (2017)	United Kingdom	$N = 128$	Healthcare CDI, recurrences	2012–2014	£6294/CDI £7539/recurrent CDI
Vonberg et al. (2008)	Germany	$N = 116$	Healthcare CDI	2006	€7147/CDI
Hubner et al. (2015)	Germany	$N = 43$	Healthcare CDI	2010	€5262.96/CDI
Grube et al. (2015)	Germany	$N = 2767$	Healthcare CDI, recurrences	2011	€4132/CDI as primary diagnosis €19,381/CDI as secondary diagnosis €20,755/recurrent CDI
Sierocka et al. (2021)	Poland	$N = 53$	Healthcare CDI	2018	€1664 cost per CDI person-day
Le Monnier et al. (2015)	France	$N = 1097$	Healthcare CDI, recurrences	2011	€9575/CDI (€6056 CDI as primary diagnosis/€11,251 CDI as secondary diagnosis) €9625/recurrent CDI
Asensio et al. (2013)	Spain	$N = 7601$	Healthcare CDI, recurrences	2012	€3901 CDI €4875 first recurrent CDI €5916 second recurrent CDI
Asensio et al. (2015)	Spain and Italy	$N = 232$ (Spain) $N = 145$ (Italy)	Healthcare CDI, recurrences, children	2011–2013 (adults) 2006–2012 (pediatrics)	€4265/CDI case (Spain) €14,936/adult CDI case (Italy) €17,714/recurrent CDI case (Italy) €3545/pediatric CDI case (Italy)
Bouza et al. (2021)	Spain	$N = 282$	Recurrent CDI	2010–2018	€10,877 recurrent CDI case

Ref reference, *CDI Clostridioides difficile* infection

€4522 ($p < 0.001$) for controls, with a CDI incremental cost with respect to matched controls in Year 1 of €11,876 (Braae et al. 2020).

A recent study conducted in a tertiary referral hospital in Ireland during August 2015 showed that the total incremental cost of CDI was €75,680, with a mean cost of €5820 per patient (Ryan et al. 2017).

Another study conducted in Ireland established the mean cost per treated case of CDI in terms of bed occupancy, laboratory requests, and treatment to be £4577 (2010 GBP) (Al-Eidan et al. 2000).

It has been estimated that the cost for CDI is €5000–€15,000 per case in England (Kuijper et al. 2006). The earliest data on economic burden from England were communicated by Wilcox et al. (1996), who performed a study in geriatric wards. Cases and controls were matched for age, sex, and distribution of the main diagnoses. The total identifiable increased cost of CDI was £6986 in 2010 GBP.

A more recent study in England, comparing first CDI episodes and recurrent CDI episodes, revealed a cost of £12,710 for primary episodes (Tresman and Goldenberg 2018).

2.1.2 Western Europe

A retrospective multicenter study analyzed a sample of 12 large, public, acute-care hospitals in France representing 5.82% of the cumulative annual number of patient-days spent in public acute-care hospitals in France in 2011 (Le Monnier et al. 2015). The costs of CDI incurred by public insurance and by the hospital itself (euros) were based on full unit cost per diagnosis-related group in hospitals at 2010 values. The annual incidence of CDI based on laboratory reporting was estimated at 3.74 cases per 10,000 patient-days. In cases where CDI was the primary diagnosis, the mean cost per stay was €6056 (median €4410), and the cumulative cost for the whole set of stays observed in 2011 for the 12 hospitals was €823,656. In patients where CDI was considered a secondary diagnosis, the mean extra cost adjusted for age-, sex-, and diagnosis-related groups in cases without CDI was €11,251 (median: €8822) per stay (Le Monnier et al. 2015). The extrapolated annual nationwide cost of CDI in 2011 in France was €163.1 million.

A single-center retrospective analysis of data from patients with nosocomial CDI carried out over a 1-year period at a teaching hospital in Germany showed an additional cost of €5262 per case (Hubner et al. 2015).

Another single-center German study showed that costs for CDI patients were significantly higher than for their matched controls (median: €7147) (Vonberg et al. 2008). A large multicenter study conducted in 37 German hospitals based on data from the German DRG system analyzed 2767 CDI cases grouped according to whether CDI was a primary or secondary diagnosis (Grube et al. 2015). For comparison, non-CDI cases from the same hospitals during the same year were matched using propensity score matching.

Patients from the primary diagnosis group ($n = 817$) showed a mean cost per case of €4132(€536 more than controls), while the secondary diagnosis group ($n = 1840$) had costs of €19,381 (€13,082 for controls) (Grube et al. 2015). The authors extrapolated their data and declared that CDI generates a yearly cost burden of €464 million for the German healthcare system.

2.1.3 Southern Europe

Evidence regarding the impact of CDI on healthcare resources in southern Europe is generally scarce. In the case of Spain, few studies have assessed the economic burden of CDI. An economic model analysis performed in 2012 by Asensio et al. (2013) assessed the cost of CDI in adult patients (≥18 years) treated for 1 year with metronidazole or vancomycin from the perspective of the Spanish National Health System Service. The resources used in clinical practice were obtained through a Delphi panel of Spanish clinicians with expertise in CDI. Unit costs (€2012) were obtained from Spanish sources.

This study estimated that 7601 episodes of CDI occur annually in Spain (incidence of 17.1 episodes/year/10,000 hospital discharges) with an annual cost to the Spanish National Health System Service of €32,157,093. The cost per episode of CDI was €3901 for initial or primary CDI episodes. More recently, another study assessed the impact of CDI on hospital resources and costs in both Spain and Italy (Asensio et al. 2015). Each patient was matched with two randomly selected uninfected controls in the same institution. Data were collected for 232 adult infected patients and 426 matched non-infected patients in Spain ($n = 106$) and Italy ($n = 126$). CDI-associated costs were due to excess hospitalization. The difference in LOS between the two countries resulted in a significant variation in costs.

Hospitalization costs attributable to CDI in Spain were €4265 per patient for all patients, €2882 per patient for patients aged ≤65 years, and €4885 for those aged >65 years (Asensio et al. 2015).

For Italy, the total cost attributable to CDI was €14,023 per patient for all patients. The cost was €15,668 for those aged ≤65 years and €13,862 for those aged >65 years, with the difference in cost being due to differences in LOS (21 vs. 19 days, respectively). The authors estimated a cost of CDI in Italy of €32,371 per 10,000 patient-days (Asensio et al. 2015).

A multicenter Italian cost analysis study performed in hospitalized patients from the hospital's perspective (Poli et al. 2015). This study showed that the mean total incremental cost for a patient with CDI was €3270 per case.

2.2 Recurrent Episodes

One of the first studies to assess the cost of recurrent CDI in an European country was a Spanish study in which the cost of the initial CDI episode was estimated to be €3901, the cost of the first recurrence was €4875 and that of the second recurrence was €5916 (Asensio et al. 2013). More recently, data from another multicenter Spanish study estimated the mean cost for R-CDI episode which was €10,877 (Bouza et al. 2021).

In an Italian multicenter study including recurrences (Asensio et al. 2015), the cost attributable to recurrent CDI was €17,714 per patient, while for patients with a single episode of CDI, the cost was €14,936. In this study, a total of 34 adult patients (12.5%) and 2 pediatric patients (10.5%) experienced a first recurrence of CDI. Three of the 34 adult patients and 1 of the 2 pediatric patients had an additional recurrence.

A French multicenter study estimated the median extra cost per stay with CDI to be €7514, i.e., approximately €9.5 million in 2011 for the 12 facilities included. The fraction of that total cost attributable to recurrences was 12.5% (Le Monnier et al. 2015). Recurrences occurring in acute-care settings were present in 12.0% of hospital stays with CDI. In addition, 9.3% (11/118) of recurrences were coded as the primary diagnosis and led to readmission of the patients, which resulted in prolonged LOS and additional medical costs.

Data from 37 German hospitals revealed high costs for recurrent CDI of €20,755 vs. €13,101 for matched controls from the same hospitals during the same year (Grube et al. 2015).

Wilcox et al. recently analyzed the impact of recurrent CDI in terms of hospital resource use and health-related quality of life associated with hospitalizations for recurrent CDI in six UK acute-care hospitals (Wilcox et al. 2017). The median cost per patient during a 28-day post-index period was £7539 for recurrent CDI and £6294 for first CDI episodes (Wilcox et al. 2017). A recent micro-costing study conducted in London determined the health resource utilization of patients with R-CDI versus first episode CDI; in this study the mean total costs (variable and fixed costs) of R-CDI were £31,121, which exceeded by more than double the cost of the first episodes CDI episodes (£12,710) (Tresman and Goldenberg 2018).

The retrospective analysis within the COMBACTE-CDI project was recently conducted based on resource costs for inpatient treatment and productivity costs. This study included 430 hospitalized patients from 12 European countries; mean overall costs per patient between the CDI case group and recurrence group were €15,242 and €52,024, respectively (Wingen-Heimann et al. 2022).

2.3 Length of Stay

In their review, Wiegand et al. (2012) estimated the average LOS in Europe to be 15 days. When examined by country, they found that Switzerland had the lowest LOS (12 days), followed by Belgium, France, Ireland (17 days), and Spain (18 days), while the highest LOS were observed for the Netherlands (21 days), Germany (27 days), and the UK (37 days) (Wiegand et al. 2012). In the recent multicenter study including 12 European countries, the overall median LOS was 22 days (95% CI 17–27 days) (Wingen-Heimann et al. 2022).

Even though LOS values are more reproducible between studies than costs, data on the excess LOS attributable to CDI are limited. Not many studies assess the attributable LOS, reporting only total LOS. It was recently suggested that, compared with newer statistical models, models that were previously used to determine the LOS attributable to CDI overestimated the additional LOS (Mitchell and Gardner 2012). Therefore, future studies must take this into account. Table 2 shows the European studies reporting LOS attributable to CDI; the mean incremental LOS attributable to CDI ranged from 4.2 days to 20 days (Ryan et al. 2017).

As for recurrent CDI, the mean incremental LOS in Europe is 9.1 days to 26 days (Asensio et al. 2013, 2015). Although data may vary, most studies agree that recurrent CDI presents longer LOS than primary episodes. In a study conducted in England, Wilcox et al. observed a median LOS of 21 days for recurrent CDI in contrast to 15.5 days for first episodes (Wilcox et al. 2017). More recently, in the European COMBACTE-CDI study, the median overall LOS for recurrent CDI was 55 vs. 22 for primary CDI episodes (Wingen-Heimann et al. 2022).

Few studies have assessed differences in extra costs between mild to moderate CDI cases and severe CDI cases. A study conducted by van Kleef et al. in a large English-teaching hospital showed that severe cases had an average excess LOS which was twice that of the nonsevere cases (11.6 days (95% CI: 3.6–19.6) vs. approximately 5 days (95% CI: 1.1–9.5)) (Van Kleef et al. 2014).

2.4 Distribution of Costs

The expense associated with CDI stems mainly from extended LOS. Various studies in Europe place the additional cost of LOS at 43.2–95.6% of the total extra costs of the CDI episode (Ryan et al. 2017; Asensio et al. 2013, 2015; Wilcox et al. 1996; Poli et al. 2015). Figure 1 represents the distribution of CDI costs of the abovementioned studies.

In contrast, cost for CDI antibiotics account for a low percentage of the total cost, ranging from 0.43% to 13.3% (Asensio et al. 2013, 2015; Wilcox et al. 1996; Ryan et al. 2017; Poli et al. 2015). Figure 2 illustrates the distribution of costs in patients with CDI antibiotics as percent of total cost per country. Most of these studies only include vancomycin and metronidazole as treatment for CDI, probably because they were conducted before fidaxomicin was licensed in those countries; only one recent study conducted in Ireland included fidaxomicin as treatment for CDI. In a recent pan-European study, one of the most important variables associated with increased overall costs were change escalation in CDI medication (OR 3.735), which included increased dosage or change of CDI active agent and treatment in an intensive care unit (ICU) (OR 5.454) (Wingen-Heimann et al. 2022). In this study, treatment with fidaxomicin as first-line therapy, age >65 years, evidence of CDI related colitis were found to have no influence on the overall costs(Wingen-Heimann et al. 2022).

Regarding distribution of costs for R-CDI, a recent study conducted in England observed that the cost of hospital admissions and emergency department visits accounted for more than 85%, similar to first-case CDI. The median cost for CDI-specific drugs was higher in R-CDI patients (£376 per patient) than first-case CDI (£46 per patient) (Wilcox et al. 2017).

0

3 Economic Burden of Community-Acquired CDI

Community-acquired CDI is a growing problem, and additional data are needed to accurately quantify the contribution of this subpopulation to the overall burden of CDI. Few studies provide insight on this understudied patient group (Kuntz et al. 2012; Sammons et al. 2013; Nanwa et al. 2017), and none have been performed in European patients. In addition, across studies, the case definition of community-acquired CDI may differ depending on the time between a previous hospital admission and whether the case of CDI was an incident case (Kutty et al. 2010; Freeman et al. 2010).

The most recent and largest study is a population-based matched cohort study including, between 2005–2006 and 2014–2015, 33,909 new cases of *C. difficile* infection in Ontario, Canada. In this study, Pereira et al.

Table 2 Cost of length of stay (LOS) attributable to *Clostridioides difficile* by study and country

Reference Author (year)	Country	CDI cases examined	Study population	Study period	LOS attributable to CDI (days)
Eckmann et al. (2013)	Netherlands	$N = 270$	Healthcare CDI	2008–2009	Mean 12.58
Ryan et al. (2017)	Ireland	$N = 13$	Healthcare CDI	August 2015	Mean 4.2
Al-Eidan et al. (2000)	Ireland	$N = 87$	Healthcare CDI	1994–1995	Mean 13
Eckmann et al. (2013)	England	$N = 10{,}602$	Healthcare CDI	2007–2009	Mean 16.09
van Kleef et al. (2014)	England	$N = 157$	Healthcare CDI	2012	Mean 7.2 (all CDI) Mean 11.6 (severe CDI) Mean 5.3 (nonsevere CDI)
Tresman and Goldenberg (2018)	England	$N = 90$ (45 CDI, 45 recurrent CDI)	Healthcare CDI	2014–2017	Mean overall LOS 17 (primary CDI) Mean overall LOS 33 (recurrent CDI)
Vonberg et al. (2008)	Germany	$N = 116$	Healthcare CDI	2006	Median 7
Eckmann et al. (2013)	Germany	$N = 109{,}526$	Healthcare CDI	2008–2010	Mean 15.47
Hubner et al. (2015)	Germany	$N = 43$	Healthcare CDI	2010	Mean 11.4
Sierocka et al. (2021)	Poland	$N = 53$	Healthcare CDI	2018	Mean 11.95
Le Monnier et al. (2015)	France	$N = 1097$	Healthcare CDI, recurrences	2011	Mean 8.9
Eckmann et al. (2013)	Spain	$N = 830$	Healthcare CDI	2008–2010	Mean 13.56
Asensio et al. (2013)	Spain	$N = 7601$	Healthcare CDI, recurrences	2012	Mean 7.4 (CDI) Mean 9.1 (first recurrent CDI) Mean 10.8 (second recurrent CDI)
Asensio et al. (2015)	Spain and Italy	$N = 232$ (Spain) $N = 145$ (Italy)	Healthcare CDI, recurrences, children	2011–2013 (adults) 2006–2012 (pediatrics)	Median 6.4 (Madrid) Median 20.0 (Barcelona) Median 20.0 (Rome) Median 26.0 for first recurrent CDI case (Spain and Italy) Median 5.0 for pediatric case (Naples)
Bouza et al. (2021)	Spain	$N = 282$	Recurrent CDI	2010–2018	Median 17.18 for recurrent CDI

Ref reference, *CDI Clostridioides difficile* infection, *LOS* length of stay

studied 7216 (21.3%) subjects with community-associated/community-onset CDI and 7098 (20.9%) with healthcare-associated/community-onset infection. Community-associated cases increased by 36.3% between 2005–2006 (6.09 cases per 100,000 person-years) and 2014–2015 (9.56 cases per 100,000 person-years). Median costs attributable to *C. difficile* infection were $13,249 for community-associated infection and $11,917 for healthcare-associated/community-onset infection (Pereira et al. 2020).

Sammons et al. examined a cohort of children and performed a subanalysis on community-onset and hospital-onset CDI (Sammons et al. 2013).

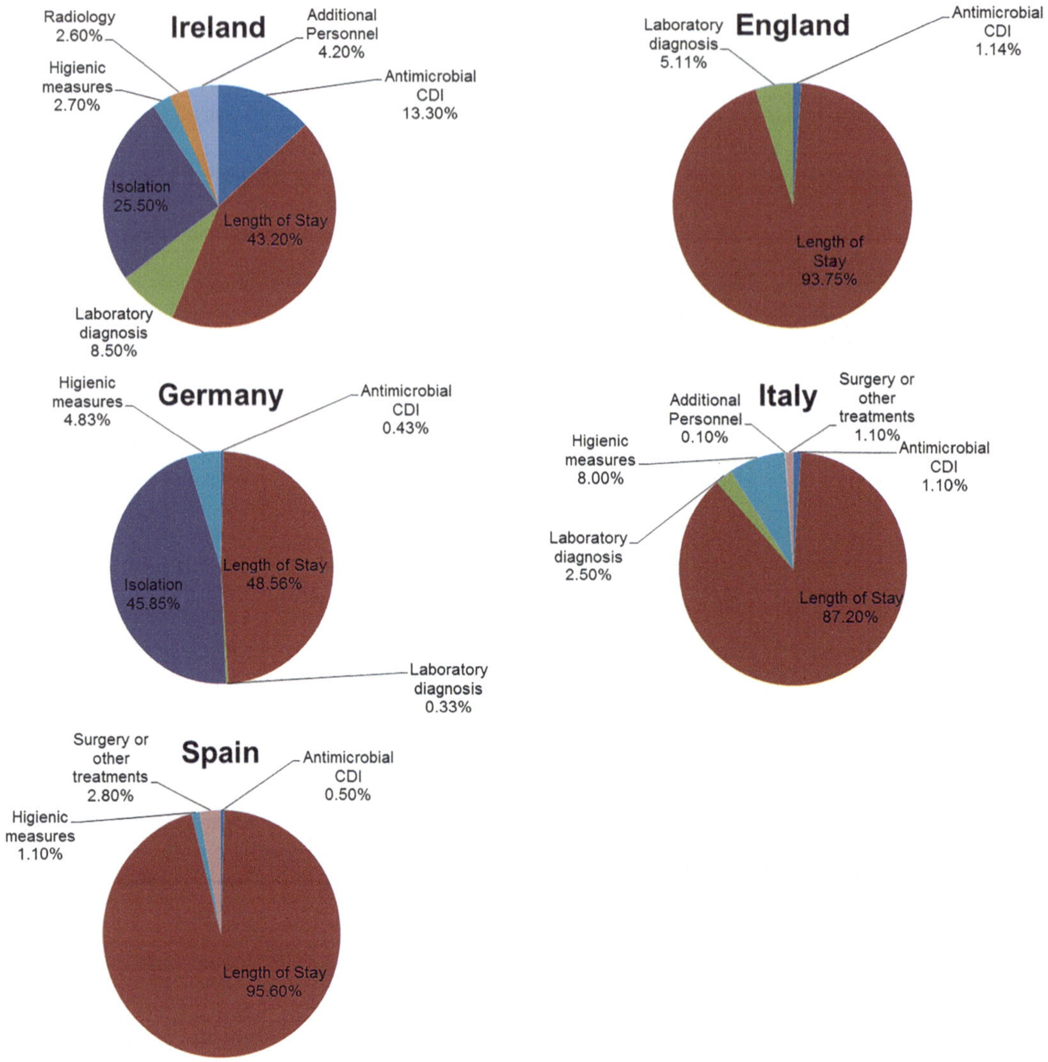

Fig. 1 Distribution of costs of *Clostridioides difficile* infection

They found that patients with community-onset CDI comprised 54% of cases (2414 cases). Patients with hospital-onset CDI had significantly higher mortality rates and longer LOS than those with community-onset CDI, and mean differences in LOS and total standardized costs were 21.60 days and $93,600 for hospital-onset CDI and 5.55 days and $18,900 for community-onset CDI. Although mortality rates did not differ between those with community-onset CDI and matched unexposed subjects, community-onset CDI patients had significantly longer LOS and total hospital costs (Sammons et al. 2013).

Kuntz et al. performed a population-based study in which they identified 3067 CDIs and classified CDI by whether it was identified in the outpatient or inpatient healthcare setting (Kuntz et al. 2012). A total of 1712 (56%) were identified in the outpatient setting. These patients tended to be younger, with fewer comorbid conditions than patients with CDI identified in the inpatient setting. Eleven percent of patients with outpatient-identified CDI were hospitalized with a CDI-related diagnosis code during the follow-up period. These hospitalizations occurred, on average, 27 days after outpatient

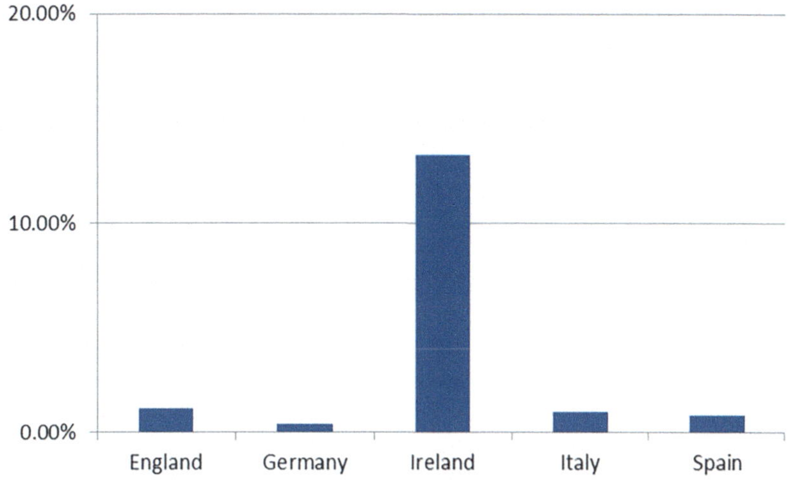

Fig. 2 Costs in *Clostridioides difficile* infection antibiotics as percent of total cost

identification of CDI and lasted an average of 10 days.

As expected, the impact of CDI on healthcare utilization and cost was most notable in the setting in which the patient's infection had been identified. Outpatient care costs were higher among persons with CDI identified in the outpatient setting, with drugs representing the greatest percentage of these costs in both groups. Similarly, patients with inpatient-identified CDI had higher inpatient costs than patients with outpatient-identified CDI ($10,708.40 vs. $837.40). Total costs for community-onset CDI were $1697 vs. $11,315 of hospital-onset CDI (in US$2009 per patient) (Kuntz et al. 2012).

4 Pediatric Population

Data on the burden of CDI in children are very limited—as with adults—and most of the literature on this topic comes from studies performed in the United States. In Italy, Asensio et al. (2015) reported separate data on the economic burden of CDI in children, although they found that the number of patients included was low ($n = 19$). Most cases of CDI in children were community-acquired as opposed to nosocomial. Disease characteristics were generally comparable to those of adults, although the incidence of ulceration and bowel wall thickening was higher than in adults. The authors found that the median LOS attributable to CDI was lower than in adults (5 vs. 19 days in Rome), as was the frequency of isolation and admission to the ICU, probably because most cases were community-acquired. Therefore, although daily costs of care are higher for children than adults, the overall burden of CDI in the pediatric population in Italy is lower than in adults. The total cost attributable to CDI in pediatric patients in Naples was €3545 per patient (Asensio et al. 2015).

The only data on the economic burden of pediatric CDI in larger populations are from American studies. In their multicenter cohort study, Sammons et al. found that CDI was associated with worse outcomes among hospitalized children who were otherwise similar in the main demographic and clinical characteristics, although the difference was most pronounced in children with hospital-onset disease. The presence of CDI was associated with >6-fold higher mortality rates among those with healthcare-onset CDI and resulted in significantly longer LOS and increased total hospital costs, corresponding to a mean difference in total standardized costs of $48,500 between matched exposed and unexposed patients (Sammons et al. 2013).

In another study performed in acute care hospitals in the Michigan Health and Hospital Association, children younger than 5 years of age had mean charges of $148,525, compared with $56,796 for discharges of patients who were aged ≥65 years, probably because of longer LOS: children younger than 5 years of age were hospitalized for a mean of more than 25 days per discharge vs. 14.2 days for the remaining age groups reported (Verlee et al. 2012).

A large propensity score-matching analysis in 313,664 patients aged 1–18 years was performed to evaluate the influence of CDI on mortality, LOS, and costs in hospitalized surgical pediatric patients. The authors observed that after propensity score matching, the mean excess LOS and costs attributable to CDI were 5.8 days and $12,801 ($p < 0.001$), accounting for 8295 days spent in the hospital and $18.4 million (2012 USD) in annual expenditure (Kulaylat et al. 2017).

5 Economic Costs of CDI Outbreaks

Few data have been published on the costs derived from outbreaks. One of the few studies to assess this situation was that conducted in Ireland by Ryan et al. (Ryan et al. 2017). The authors collected data on LOS, diagnosis, diagnosis-related group codes at discharge, time in isolation because of CDI, additional measures because of CDI (medications, consultations, investigations, and procedures), unit costs (laboratory testing, personal protective equipment, single-room accommodation, and cleaning/decontamination), and personnel time.

This study covered only a 1-month period (August 2015), during which they observed that the CDI outbreak resulted in additional costs of €46,967. The outbreak resulted in 58 bed days lost due to bed closures on the outbreak ward, with an estimated value of €34,585. Five outbreak control meetings were held, each with a mean duration of 47 min and supported by 15 h of administrative input. All meetings involved a consultant microbiologist, a senior laboratory scientist, a senior antimicrobial specialist pharmacist, an assistant director of nursing, multiple clinical nursing managers, and a number of other staff members. The mean personnel cost per meeting was €546, and the aggregate cost was €2728. The cost of outbreak-related cleaning/decontamination during August was €9654 (Ryan et al. 2017).

For the patients involved in the CDI outbreak, excluding the value of the 58 bed days lost (€34,585), costs were 30% higher (€7589 per patient) than those not involved in the outbreak during the same period (Ryan et al. 2017).

Van Beurden et al. assessed the costs of an outbreak of *C. difficile* ribotype 027 at the VU University medical center, a 750-bed tertiary care center in the Netherlands, from May 2013 to May 2014 (Van Beurden et al. 2017). Several control measures were implemented, such as reinforcement of infection control, the introduction of hydrogen peroxide as disinfectant, extra cleaning, optimization of CDI diagnosis, optimization of CDI treatment, and antibiotic stewardship. Twelve meetings of the outbreak management team (consisting of five medical specialists, one infection prevention specialist, one care manager, and two co-workers from facility management) were held during the study period. Several beds had to be closed to ensure that every patient with suspected CDI was placed in contact isolation in a single room. After the implementation of these control measures, the incidence of CDI decreased to around 1.5 cases per 10,000 patient days in early 2014.

Missed revenue due to prolonged LOS among CDI patients, costs of the outbreak meetings, extra surveillance, contact isolation material (compared with the same period 1 year earlier and 1 year later), and additional microbiological diagnostics (compared with the same period 1 year earlier) were calculated directly from available data for the entire outbreak. Overall costs for additional cleaning, contact isolation, and missed revenue due to closed beds were extrapolated from the costs incurred during the previous 3 months of the outbreak. Attributable costs per item (in 2014 euros) were assessed over a 365-day period.

The total identifiable costs of this *C. difficile* outbreak were €1,222,376. Most costs (36%) stemmed from the loss of revenue resulting from decreased hospital capacity because of the increased LOS of CDI patients and the closure of multiple beds to ensure contact isolation of a single CDI patient. Twenty-five percent of the costs were from extra surveillance and the work of the department of infection control, 24% were for extra cleaning of the affected wards, 6% for extra microbiological diagnostic procedures, 3% for the outbreak meetings, and 3% for the use of extra gloves and aprons. Extra antibiotic treatment of CDI patients counted for 2% of the total costs (Van Beurden et al. 2017).

As can be seen in both studies, the cost of one missed hospital admission due to closed beds or prolonged LOS is a major cost. The economic and healthcare impact of loss of revenue is very difficult to determine, and closed beds prevent inpatient accommodation, with the resultant morbidity and mortality (Singer et al. 2011). In addition, increased bed usage by medical specialties is associated with cancelled elective surgeries (Robb et al. 2004; Nasr et al. 2004).

Outbreak control generates extra work, which often relies on staff already overburdened with administrative tasks from patient care activities. Extra cleaning measures and multidisciplinary infection control teams are key elements for outbreak control (Barbut et al. 2011; Barbut 2015). Healthcare facilities should be able to assess the economic impact of an outbreak, and knowing the costs of additional measures will make it possible to establish a cost-efficient program for outbreak control, with adequate resource allocation.

References

Al-Eidan FA, McElnay JC, Scott MG, Kearney MP (2000) Clostridium difficile-associated diarrhoea in hospitalised patients. J Clin Pharm Ther 25:101–109

Asensio A, Bouza E, Grau S, Rubio-Rodriguez D, Rubio-Terres C (2013) Cost of Clostridium difficile associated diarrhea in Spain. Rev Esp Salud Publica 87:25–33

Asensio A, Di Bella S, Lo Vecchio A, Grau S, Hart WM, Isidoro B, Scotto R, Petrosillo N, Watt M, Nazir J (2015) The impact of Clostridium difficile infection on resource use and costs in hospitals in Spain and Italy: a matched cohort study. Int J Infect Dis 36:31–38

Barbut F (2015) How to eradicate Clostridium difficile from the environment. J Hosp Infect 89:287–295

Barbut F, Jones G, Eckert C (2011) Epidemiology and control of Clostridium difficile infections in healthcare settings: an update. Curr Opin Infect Dis 24:370–376

Bouza E (2012) Consequences of Clostridium difficile infection: understanding the healthcare burden. Clin Microbiol Infect 18(Suppl 6):5–12

Bouza E, Cobo J, Rodríguez-Hernández MJ, Salavert M, Horcajada JP, Iribarren JA, Obi E, Lozano V, Maratia S, Cuesta M, Uría E, Limón E (2021) Economic burden of recurrent Clostridioides difficile infection in adults admitted to Spanish hospitals. A multicentre retrospective observational study. Rev Esp Quimioter 34:126–135

Braae UC, Møller FT, Ibsen R, Ethelberg S, Kjellberg J, Mølbak K (2020) The economic burden of Clostridioides difficile in Denmark: a retrospective cohort study. Front Public Health 8:562957

Desai K, Gupta SB, Dubberke ER, Prabhu VS, Browne C, Mast TC (2016) Epidemiological and economic burden of Clostridium difficile in the United States: estimates from a modeling approach. BMC Infect Dis 16:303

Dubberke ER, Olsen MA (2012) Burden of Clostridium difficile on the healthcare system. Clin Infect Dis 55 (Suppl 2):S88–S92

Eckmann C, Wasserman M, Latif F, Roberts G, Beriot-Mathiot A (2013) Increased hospital length of stay attributable to Clostridium difficile infection in patients with four co-morbidities: an analysis of hospital episode statistics in four European countries. Eur J Health Econ

Freeman J, Bauer MP, Baines SD, Corver J, Fawley WN, Goorhuis B, Kuijper EJ, Wilcox MH (2010) The changing epidemiology of Clostridium difficile infections. Clin Microbiol Rev 23:529–549

Gabriel L, Beriot-Mathiot A (2014) Hospitalization stay and costs attributable to Clostridium difficile infection: a critical review. J Hosp Infect 88:12–21

Ghantoji SS, Sail K, Lairson DR, Dupont HL, Garey KW (2010) Economic healthcare costs of Clostridium difficile infection: a systematic review. J Hosp Infect 74: 309–318

Grube RF, Heinlein W, Scheffer H, Rathmayer M, Schepp W, Lohse AW, Stallmach A, Wilke MH, Lerch MM (2015) Economic burden of Clostridium difficile enterocolitis in German hospitals based on routine DRG data. Z Gastroenterol 53:391–397

Hubner C, Hubner NO, Muhr M, Claus F, Leesch H, Kramer A, Flessa S (2015) Cost analysis of hospitalized Clostridium difficile-associated diarrhea (CDAD). GMS Hyg Infect Control 10:Doc13

Jones AM, Kuijper EJ, Wilcox MH (2013) Clostridium difficile: a European perspective. J Infect 66:115–128

Kuijper EJ, Coignard B, Tull P (2006) Emergence of Clostridium difficile-associated disease in North America and Europe. Clin Microbiol Infect 12(Suppl 6):2–18

Kulaylat AN, Rocourt DV, Podany AB, Engbrecht BW, Twilley M, Santos MC, Cilley RE, Hollenbeak CS, Dillon PW (2017) Costs of Clostridium difficile infection in pediatric operations: a propensity score-matching analysis. Surgery 161:1376–1386

Kuntz JL, Johnson ES, Raebel MA, Petrik AF, Yang X, Thorp ML, Spindel SJ, Neil N, Smith DH (2012) Epidemiology and healthcare costs of incident Clostridium difficile infections identified in the outpatient healthcare setting. Infect Control Hosp Epidemiol 33: 1031–1038

Kutty PK, Woods CW, Sena AC, Benoit SR, Naggie S, Frederick J, Evans S, Engel J, Mcdonald LC (2010) Risk factors for and estimated incidence of community-associated Clostridium difficile infection, North Carolina, USA. Emerg Infect Dis 16:197–204

Le Monnier A, Duburcq A, Zahar JR, Corvec S, Guillard T, Cattoir V, Woerther PL, Fihman V, Lalande V, Jacquier H, Mizrahi A, Farfour E, Morand P, Marcade G, Coulomb S, Torreton E, Fagnani F, Barbut F (2015) Hospital cost of Clostridium difficile infection including the contribution of recurrences in French acute-care hospitals. J Hosp Infect 91:117–122

Mcglone SM, Bailey RR, Zimmer SM, Popovich MJ, Tian Y, Ufberg P, Muder RR, Lee BY (2012) The economic burden of Clostridium difficile. Clin Microbiol Infect 18:282–289

Mitchell BG, Gardner A (2012) Prolongation of length of stay and Clostridium difficile infection: a review of the methods used to examine length of stay due to healthcare associated infections. Antimicrob Resist Infect Control 1:14

Nanwa N, Kendzerska T, Krahn M, Kwong JC, Daneman N, Witteman W, Mittmann N, Cadarette SM, Rosella L, Sander B (2015) The economic impact of Clostridium difficile infection: a systematic review. Am J Gastroenterol 110:511–519

Nanwa N, Sander B, Krahn M, Daneman N, Lu H, Austin PC, Govindarajan A, Rosella LC, Cadarette SM, Kwong JC (2017) A population-based matched cohort study examining the mortality and costs of patients with community-onset Clostridium difficile infection identified using emergency department visits and hospital admissions. PLoS One 12:e0172410

Nasr A, Reichardt K, Fitzgerald K, Arumugusamy M, Keeling P, Walsh TN (2004) Impact of emergency admissions on elective surgical workload. Ir J Med Sci 173:133–135

Pereira JA, Mcgeer A, Tomovici A, Selmani A, Chit A (2020) Incidence and economic burden of Clostridioides difficile infection in Ontario: a retrospective population-based study. CMAJ Open 8:E16–E25

Poli A, Di Matteo S, Bruno GM, Fornai E, Valentino MC, Colombo GL (2015) Economic burden of Clostridium difficile in five hospitals of the Florence health care system in Italy. Risk Manag Healthc Policy 8:207–213

Robb WB, O'sullivan MJ, Brannigan AE, Bouchier-Hayes DJ (2004) Are elective surgical operations cancelled due to increasing medical admissions? Ir J Med Sci 173:129–132

Ryan P, Skally M, Duffy F, Farrelly M, Gaughan L, Flood P, Mcfadden E, Fitzpatrick F (2017) Evaluation of fixed and variable hospital costs due to Clostridium difficile infection: institutional incentives and directions for future research. J Hosp Infect 95:415–420

Sammons JS, Localio R, Xiao R, Coffin SE, Zaoutis T (2013) Clostridium difficile infection is associated with increased risk of death and prolonged hospitalization in children. Clin Infect Dis 57:1–8

Sierocka A, Kiersnowska Z, Lemiech-Mirowska E, Marczak M (2021) Costs associated with the treatment of Clostridioides difficile infections. Int J Environ Res Public Health 18

Singer AJ, Thode HC Jr, Viccellio P, Pines JM (2011) The association between length of emergency department boarding and mortality. Acad Emerg Med 18:1324–1329

Tresman R, Goldenberg SD (2018) Healthcare resource use and attributable cost of Clostridium difficile infection: a micro-costing analysis comparing first and recurrent episodes. J Antimicrob Chemother 73: 2851–2855

Van Beurden YH, Bomers MK, Van Der Werff SD, Pompe EA, Spiering S, Vandenbroucke-Grauls CM, Mulder CJ (2017) Cost analysis of an outbreak of Clostridium difficile infection ribotype 027 in a Dutch tertiary care centre. J Hosp Infect 95:421–425

Van Kleef E, Green N, Goldenberg SD, Robotham JV, Cookson B, Jit M, Edmunds WJ, Deeny SR (2014) Excess length of stay and mortality due to Clostridium difficile infection: a multi-state modelling approach. J Hosp Infect 88:213–217

Verlee KE, Finks JL, Wilkins MJ, Wells EV (2012) Michigan Clostridium difficile hospital discharges: frequency, mortality, and charges, 2002-2008. Public Health Rep 127:62–71

Vonberg RP, Reichardt C, Behnke M, Schwab F, Zindler S, Gastmeier P (2008) Costs of nosocomial Clostridium difficile-associated diarrhoea. J Hosp Infect 70:15–20

Wiegand PN, Nathwani D, Wilcox MH, Stephens J, Shelbaya A, Haider S (2012) Clinical and economic burden of Clostridium difficile infection in Europe: a systematic review of healthcare-facility-acquired infection. J Hosp Infect 81:1–14

Wilcox MH, Ahir H, Coia JE, Dodgson A, Hopkins S, Llewelyn MJ, Settle C, Mclain-Smith S, Marcella SW (2017) Impact of recurrent Clostridium difficile infection: hospitalization and patient quality of life. J Antimicrob Chemother

Wilcox MH, Cunniffe JG, Trundle C, Redpath C (1996) Financial burden of hospital-acquired Clostridium difficile infection. J Hosp Infect 34:23–30

Wingen-Heimann SM, Davies K, Viprey VF, Davis G, Wilcox MH, Vehreschild M, Lurienne L, Bandinelli PA, Cornely OA, Vilken T, Hopff SM, Vehreschild JJ (2022) Clostridioides difficile infection (CDI): A pan-European multi-center cost and resource utilization study, results from the Combatting Bacterial Resistance in Europe CDI (COMBACTE-CDI). Clin Microbiol Infect

The Need for European Surveillance of CDI

Camilla Wiuff Coia, A-Lan Banks, Laura Cottom, and Fidelma Fitzpatrick

For surveillance systems to be useful, they must adapt to the changing environment in which they operate and accommodate emerging public health requirements that were not conceived previously.
Joseph S. Lombardo and David L. Buckeridge

Abstract

Since the turn of the millennium, the epidemiology of *Clostridioides difficile* infection (CDI) has continued to challenge. Changes in clinical presentation, severity of disease, descriptions of new risk factors and the occurrence of outbreaks all emphasised the importance of early diagnosis and standardised surveillance systems. However, a lack of consensus on case definitions, clinical guidelines and optimal laboratory diagnostics across Europe has led to the underestimation of CDI and impeded comparison between countries. These inconsistencies have prevented the true burden of disease from being appreciated.

Acceptance that a multi-country CDI surveillance program and optimised diagnostic strategies are required has built the foundations for a more robust, unified surveillance. The concerted efforts of the European Centre for Disease Prevention and Control (ECDC) CDI networks led to the development of the European surveillance protocol and an over-arching long-term CDI surveillance strategy for 2014–2020, which has been followed by the development of surveillance systems in at least 20 European countries. However, surveillance activities in individual countries have slowed during the COVID-19 pandemic as resources were diverted to the global health crisis. A renewed and strengthened focus on CDI surveillance and prevention is therefore urgently needed post COVID-19.

C. W. Coia (✉)
Department of Bacteria, Parasites and Fungi, Statens Serum Institut, Copenhagen, Denmark
e-mail: cmwi@ssi.dk

A.-L. Banks
St. Helens & Knowsley Teaching Hospitals NHS Trust Whiston Hospital, Prescot, Merseyside, UK
e-mail: a-lan.banks@nhs.net

L. Cottom
Department of Clinical Microbiology, Glasgow Royal Infirmary, Greater Glasgow & Clyde, Glasgow, UK
e-mail: Laura.Cottom@ggc.scot.nhs.uk

F. Fitzpatrick
Departments of Clinical Microbiology, The Royal College of Surgeons in Ireland, and Beaumont Hospital, Dublin, Ireland
e-mail: fidelmafitzpatrick@rcsi.ie

P. Mastrantonio, M. Rupnik (eds.), *Updates on Clostridioides difficile in Europe*, Advances in Experimental Medicine and Biology 1435, https://doi.org/10.1007/978-3-031-42108-2_2

1 Epidemiology of CDI in Europe

Clostridioides difficile is a leading infectious cause of antimicrobial-associated diarrhoea, with symptoms ranging from mild diarrhoea to pseudomembranous colitis (PMC). *C. difficile* infection (CDI) has been classified as an urgent public health threat by the Centres for Disease Control and Prevention and has an attributed healthcare cost of over $1 billion per year (CDC 2019). Until the end of the millennium interest in this pathogen was primarily in relation to healthcare and its impact on morbidity and mortality in the elderly. However, since 2000 significant changes in the clinical presentation of CDI have been reported, including more severe disease, CDI in the community in patients without traditional risk factors (such as antimicrobial treatment and recent hospitalisation) and the occurrence of outbreaks (Bauer et al. 2009; Freeman et al. 2010; Wilcox et al. 2008). The changes in the epidemiology of CDI leading to several outbreaks in North America and Europe correlated with the emergence of a new hypervirulent strain PCR ribotype 027 (Kuijper et al. 2006a), and to a lesser extent, PCR ribotype 078 (Goorhuis et al. 2008). Ribotype 027 was associated with more severe disease, higher mortality, increased risk of relapse and higher colectomy rates (Kuijper et al. 2006a; Ricciardi et al. 2007; Warny et al. 2005). However, other ribotypes of *C. difficile* were also linked to outbreaks and contributed to the spread of this infection not only in Europe but worldwide (Bauer et al. 2011).

The reasons for the emergence and rapid global spread of *C. difficile* ribotype 027 remained unexplained until the genomes of a global collection of *C. difficile* ribotype 027 isolates from hospital patients between 1985 and 2010 were sequenced. Phylogenetic analysis showed that two separate lineages of ribotype 027, FQR1 and FQ2, had emerged in North America within a short period of time, after acquiring the same fluoroquinolone resistance mutation, of which one spread throughout the United States (USA), South Korea and Switzerland, and the other spread more widely across continents throughout Europe and Australia (see Fig. 1) (He et al. 2013). Isolates obtained prior to the emergence of these two lineages were not associated with hospital outbreaks, suggesting that they represented pre-epidemic lineages of ribotype 027. These findings highlighted the important role of selective pressure from fluoroquinolone use in the evolution and spread of these two lineages in healthcare settings and highlighted the interconnectedness of the global healthcare systems due to human travel.

In the 2011/2012, the European Centre for Disease Prevention and Control (ECDC) acute hospital point prevalence survey of hospital-acquired infection (HAI), *C. difficile* was the most frequently reported pathogen associated with healthcare-associated gastrointestinal disease in European hospitals (accounting for 48% of all gastrointestinal disease) (ECDC 2013). Based on this data, it was estimated that 152,905 new cases of CDI occur every year in Europe with an incidence of 30 cases per 100,000 population. Moreover, CDI was associated with considerable short- and long-term disability, with a reported 8382 attributable deaths per year (Cassini et al. 2016).

ECDC commenced coordination of CDI surveillance in acute care hospitals in EU/EEA countries in 2016 (ECDC 2015). In 2016–2017, the crude incidence of CDI in 23 European countries was 3.48 cases per 10,000 patient days, the majority healthcare-associated CDI (HA-CDI) (60.9%) with 32.7% community-associated CDI (CA-CDI) or unknown association (ECDC 2022). Among the healthcare-associated cases (HA-CDI), the crude incidence density was 2.12 cases per 10,000 patient-days. The highest national annual crude CDI incidence densities were reported in Estonia, Lithuania and Poland (at 5.92–7.51 cases per 10,000 patient-days) in 2016. The patterns of high and low national incidence densities were mirrored among the HA-CDI cases in both years (Fig. 2). Ten countries also reported ribotype data, of

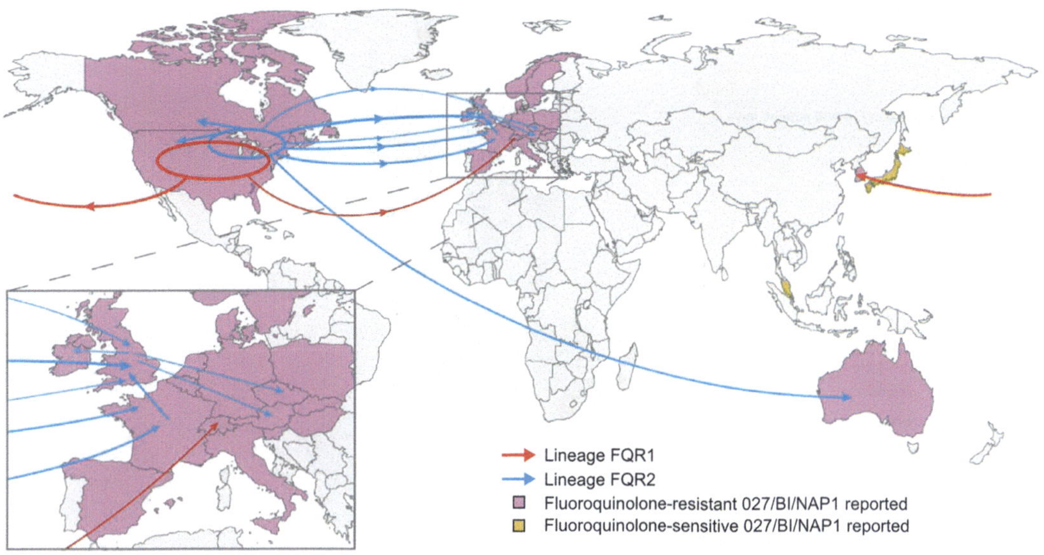

Fig. 1 Global transmission events of *C. difficile* PCR ribotype 027 (with permission from authors). Arrows indicate individual introductory transmission events of FQR1 and FQR2 (He et al. 2013)

which 81% of the data originated from three countries (Belgium, the Netherlands and the UK (Wales)). The latest ribotype data are therefore not representative of the overall distribution in Europe. Despite its limited coverage of ribotype data, the ECDC CDI report provides the largest harmonised epidemiological data set from Europe. This report of CDI data collected from countries over a 2-year period used the ECDC surveillance protocol and to a high degree the same diagnostic algorithm as defined in the protocol.

A systematic literature review that included data from France, Germany, Italy, Poland, Spain and the UK (Scotland) reported an overall incidence of 3.5 cases per 10,000 patient days with the UK at 1.99 and Poland at 6.18 cases per 10,000 patient days (Finn et al. 2021). Although, these reports arrive at the same overall incidence in Europe, the ECDC report covers a larger number of countries and contains standardised and comparable data from a larger number of European hospitals using the same case definitions, diagnostic testing and reporting methods as defined in the ECDC surveillance protocol. The development of incidence over time (i.e. trends) is rarely reported for European countries in the scientific literature (Finn et al. 2021) but can be found in national reports (e.g. the UK (England), the UK (Scotland) and Ireland).

2 Developments in Approaches to Monitoring the Epidemiology of CDI in European Countries

In 2002, the European Study Group on *C. difficile* (ESGCD) conducted a survey of 212 laboratories in eight countries (Belgium, Denmark, Finland, the Netherlands, Germany, Italy, Spain, the UK) to obtain an overview of diagnostic methods used and to estimate the average incidence of CDI across Europe (Barbut et al. 2003). The survey revealed an inconsistent approach to diagnosing and typing CDI, including variation in the criteria for testing, laboratory methodology and strategy for testing and possible bias in the study (by inclusion of only the most responsive laboratories). These factors raised concern of under-ascertainment due to undiagnosed and misdiagnosed cases and inaccurate estimates of

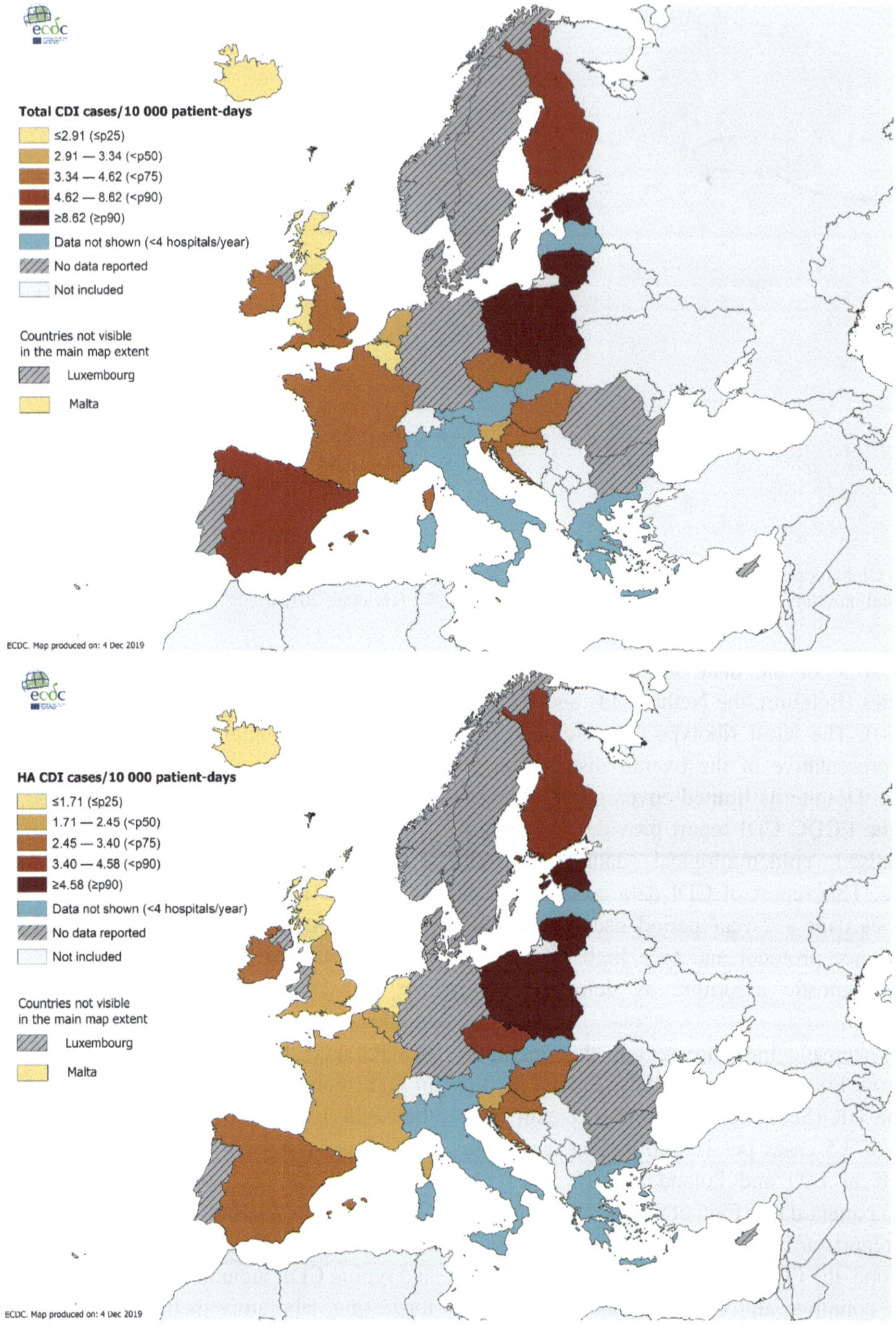

Fig. 2 ECDC European surveillance 2016–2017: (**a**) crude CDI incidence density of all cases per 10,000 patient-days, (**b**) incidence density of healthcare associated CDI (HA-CDI) per 10,000 patient-days (reproduced from European Centre for Disease Prevention and Control. *Clostridioides* (*Clostridium*) *difficile* infections. From Annual epidemiological report for 2016–2017, Stockholm: ECDC; 2022) (ECDC 2022)

the overall burden of disease and highlighted the need for international guidance.

The comprehensive ESCGD review of the emergence of CDI in North America and Europe (Kuijper et al. 2006b) specified for the first time a case definition for CDI (including healthcare and community association), provided advice on optimal diagnostic testing and recommended that each member state should develop systematic and comprehensive surveillance systems in order to detect, monitor and respond to changes in the epidemiology of CDI (in particular ribotype 027) at both national and European levels. Following 2006, national surveillance systems were developed or expanded in countries across Europe.

In 2011, the European *C. difficile* Infection Surveillance Network (ECDIS-net) surveyed the national surveillance systems through a web-based questionnaire and reviewed extant surveillance protocols at the time. Of the 31 countries, 14 reported that 18 surveillance systems were in place with some countries reporting more than one CDI data collection system (Kola et al. 2016). The majority were continuous and prospective CDI surveillance; 11 countries used mandatory reporting, while 7 used voluntary reporting. Key features of the surveillance systems varied widely with considerable variation in case definitions, data collection methods, reporting and availability of reference typing. Of note only 12 countries used the ECDC/ Centres for Disease Control and Prevention (CDC) case definition of CDI. More detailed case definitions for community-associated/community-onset and healthcare-onset/healthcare-associated CDI were used in nine systems, seven of which were consistent with ECDC definitions, while the remainder had different cut-off time points for healthcare association. For 13 systems a definition for severe disease was included, while 11 had a definition for recurrence, but both definitions varied between countries, and not all were consistent with ECDC definitions. Despite the increasingly recognised role of CDI in community settings, few countries engaged general practitioners in their surveillance systems. Descriptive-enhanced patient data were only collected in six systems and death within 30 days in five. Reference typing was performed routinely in 13/14 countries using various different criteria for submission including the presence of severe CDI, outbreaks or a more systematic periodic collection of a representative sample of cases. Finally, the reporting of the CDI burden varied widely with the use of a non-standardised denominators and stratification by geographical region, healthcare facility or laboratory making comparisons over time and between regions and facilities difficult (Kola et al. 2016).

In 2017, surveillance systems in 33 European countries were surveyed and reviewed again by ESGCD (Krutova et al. 2018). Of the 33 countries, 20 reported having 24 national surveillance systems for CDI, an increase from 2011 (see above). Many of the current national surveillance systems still reported data only on healthcare-associated CDI despite the growing evidence of CDI occurring in community settings. Multiple strategies for CDI surveillance have developed across Europe. Most commonly, as part of a national surveillance system for healthcare-associated or hospital-acquired infection (Belgium, Denmark, Finland, Germany, Hungary, Latvia, Lithuania, Norway, the UK (England, Ireland, Northern Ireland, Scotland, Wales)). Other approaches include laboratory-based surveillance systems (Poland, Sweden) sentinel surveillance (Luxembourg, the Netherlands), early warning and response systems (France, Slovakia), national infectious disease and discharge registers (Finland), notifiable surveillance (Ireland, Slovenia) and enhanced surveillance (Ireland, UK (Northern Ireland)). Reporting of data included most commonly healthcare-associated CDI per patient days, hospital admissions and/or population size. Annual epidemiology reports were issued in 13 of 20 countries with national surveillance systems (and in one country, Switzerland, based on national reference laboratory data). Fewer countries reported continuously on the burden of community-acquired CDI (Hungary, Lithuania, Luxembourg, Slovakia, Northern Ireland and Scotland). Moreover, seven countries routinely reported on severe CDI (Belgium, Finland,

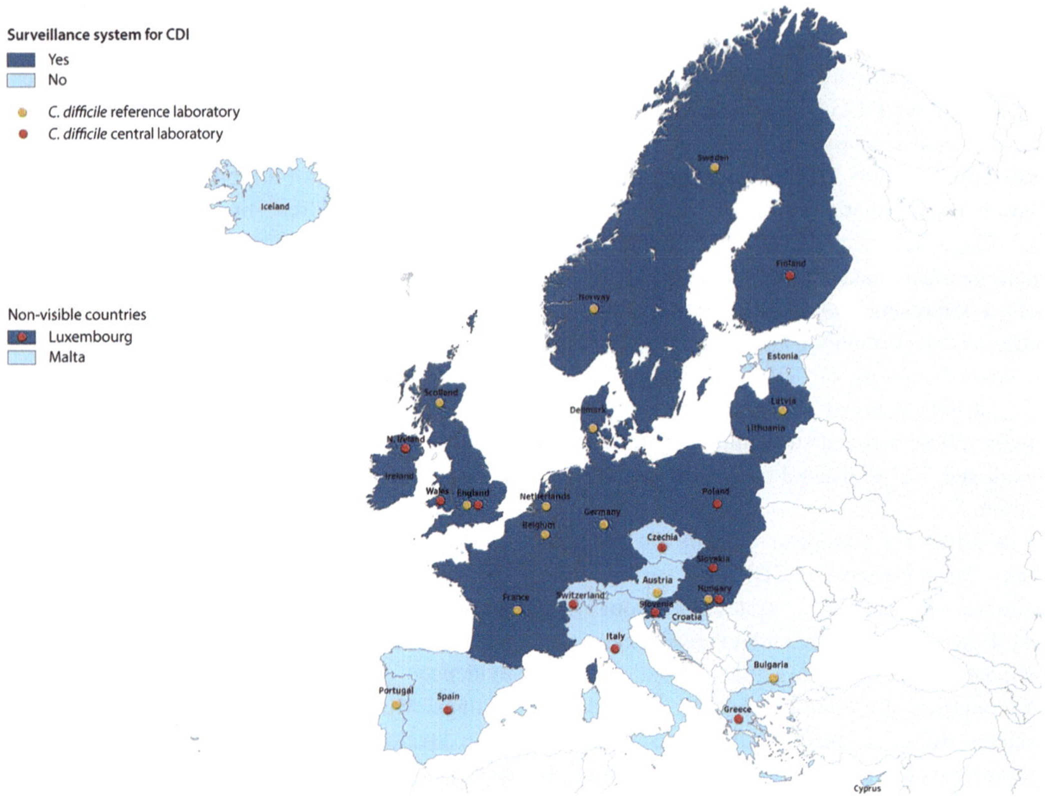

Fig. 3 Surveillance and NRLs in Europe. Reproduced from Krutova et al. (2018) (permission from Elsevier pending)

France, Germany, Ireland, the Netherlands, UK (England)), and some included data on ICU admission, surgery related to CDI and/or death or death within 30 days of CDI.

In the past decade, the national reference testing capacity has expanded through the establishment of national reference laboratories. In 2017, 26 of 33 countries indicated having a national reference laboratory (NRL) (the UK (England) and Hungary having two laboratories) or central laboratory for typing and further investigation of *C. difficile* isolates in support of the enhanced surveillance option in the ECDC CDI surveillance protocol (an increase from 13 NRLs in 2011). NRLs for *C. difficile* have been established in some countries without national surveillance systems for CDI. Reference typing support provided by NRL included PCR-ribotyping in all countries but 1 (Denmark used tandem repeat sequence typing (TRST) instead) and multiple locus variable-number tandem repeat analysis (MLVA) typing in 14 countries, antimicrobial susceptibility testing in 5 countries, toxino-typing in 1 country (Slovenia) and whole-genome sequencing (WGS) in 5 countries as outlined below (Germany, the Netherlands, Portugal, Sweden and Switzerland). Criteria for NRL investigation most commonly included the presentation of severe disease and suspected outbreaks. The overall combined capacity for CDI surveillance and NRL typing support is higher in Northern and Eastern parts of Europe than in the Southern parts (see Fig. 3).

The most recent survey of compliance with international guidance on testing and surveillance for CDI was conducted in 2018–2019 in 12 European countries by the Combatting Bacterial Resistance in Europe CDI consortium (COMBACTE-CDI) (Viprey et al. 2023). Of these countries, 11 had national surveillance programmes for CDI; 7 countries had continuous monitoring in hospitals, 2 countries had periodic monitoring, and 1 country had a temporary programme only. Less than half of the countries however (Belgium, Italy, Poland, Slovakia and the UK) participated in the ECDC surveillance. Most hospitals in the national surveillance systems monitored hospital-diagnosed CDI (40/47), while only a third of hospitals reported cases diagnosed in other settings.

3 Clinical Suspicion and Diagnostic Testing: A Prerequisite for Surveillance

In the 1990s, a large number of diagnostic tests for *C. difficile* became commercially available, including faecal culture on selective media, detection of GDH (glutamate dehydrogenase) a non-specific antigen, direct detection of toxin A and B from stool using enzyme immune assay or cytotoxicity assay (Delmee 2001), but system-wide or national surveillance programmes remained rare.

Stool assays for toxin A and B became quickly the main clinical test for diagnosing CDI, while stool cultures were used mainly for epidemiological investigations (Kelly and LaMont 1998). However, the majority of the available testing methods were associated with either low sensitivity or specificity, or both (see also Chap. 4), and some required culture facilities. Moreover, at that time, there was no consensus across Europe in terms of diagnostic testing and surveillance due to the lack of guidance.

The attention given to diagnostic procedures and surveillance of CDI varied widely between countries. In 2008, with the support of ECDC, a Europe-wide survey (involving 106 laboratories in 34 countries) assessed the epidemic preparedness and current CDI epidemiology aiming to ultimately build capacity for diagnosis and surveillance of CDI in each country (Bauer et al. 2011). The frequency of testing varied between countries from 3 to 141 CDI tests conducted per 10,000 patient days, and a correlation between testing rate and CDI incidence was identified resulting in North European countries reporting the highest incidence rates.

Optimum laboratory diagnosis of CDI depends on testing patients at the correct time using appropriate testing methodology and strategy. A point prevalence study in a multi-centre setting in Spain evaluated 988 unformed stools (from 897 patients) and found 66% of CDI episodes were undiagnosed or misdiagnosed due to lack of clinical suspicion (48%) or due to using a non-sensitive test (19%) (Alcala et al. 2012). In the Europe-wide point prevalence study (EUCLID) conducted in 2012 and 2013, 7297 unformed stools (from 482 hospitals across 20 countries) were tested at a central laboratory using the recommended two-step diagnostic algorithm. In total, only 63% of unformed stools were tested for *C. difficile* at the participating hospitals and 23% of patients with positive samples were misdiagnosed due to using an inadequate laboratory test. It was estimated that on a single day on average 74 patients with diarrhoea due to *C. difficile* in hospitals across Europe were not diagnosed due to the lack of suspicion (Davies et al. 2014). When using the optimised diagnostic method (including a method to detect toxins in faecal samples/recommend in European guidance) to examine the isolates, the mean incidence increased 2.4- to 2.9-fold relative to the reported rates in both study measurements (e.g. from 7.3 cases to 17.2 cases per 10,000 patient bed days in the second measurement). In addition, only 32% of participating hospitals used the optimum diagnostic method at the first study measurement (in 2011–2012), whereas this had improved at the second study measurement (48% in 2012–2013).

In a more recent study, 3163 diarrhoeal samples obtained, on 2 selected days, from 119 sites in 12 European countries (1 site per 3 million population) were tested, cultured and typed centrally. Testing rates varied between regions of Europe and between hospital and community settings. The testing (i.e. examination) rates of diarrhoeal samples was 74.9% at hospitals and only 29.6% in community settings.

Among samples testing positive for *C. difficile* at the central laboratory, *C. difficile* was not detected by local sites in 16% of hospitals and 55% of community samples. Reduced sampling and testing rates were most pronounced in Eastern European countries, which also had the highest positivity rates (13.1% of all stools) and highest prevalence of epidemic toxinotype IIIb strains (ribotypes 027, 181, and 176), suggesting that lack of suspicion and underdiagnosis lead to outbreaks (K. Davies et al. 2020b).

Information collected in the ECDC surveillance programme in 2016–2017 suggests that large variation in stool testing frequency remains in European hospitals—the mean being 96.1 stools tested per 10,000 patient days and the median being only 38.6 stool tested per 10,000 patient days, as many hospitals tested infrequently for CDI (ECDC 2022). The impact of testing frequencies has been studied in COMBACTE-CDI in which very low levels of testing were found to mask true CDI incidence rates. When adjusting for variation in testing rates, the true CDI incidence can be estimated (K. Davies et al. 2020a). The highest reported testing rates were observed in the UK in all healthcare settings, and patients were tested for CDI significantly earlier than those in other countries with a mean 3 days between admission and testing (compared to up to 9 days in other European countries). The CDI incidence in the UK has reduced significantly over the past 15 years during which testing frequency was consistently high and patients were tested early on after their admission.

The European Society of Clinical Microbiology and Infectious Diseases (ESCMID) recommended diagnostic testing algorithms were used in 76.8% of hospital surveillance periods in 2016–2017 (ECDC 2022). Consistent with these findings, the ESCMID-recommended testing methodologies were used in 82% (86/105) of hospitals across Europe (Viprey et al. 2023). Moreover, non-recommended test methodologies were used in countries reporting the highest incidence of CDI. One-step testing setups using molecular tests is less time-consuming than the recommended two-steps algorithms and is available commercially as automated cartridge systems. However, careful clinical evaluation of patients who are positive by PCR alone is required as relying solely on molecular (PCR-based) diagnostics, without using a diagnostic algorithm, risks overdiagnosis, overcounting cases and overtreatment. In a study of hospitalised patients with suspected CDI, complications and deaths only occurred in toxin-positive patients, while patients with a combined positive PCR-test and negative-toxin immunoassay test had clinical outcomes comparable with those of non-infected patients (Polage et al. 2015). Germany has a relatively high incidence of CDI compared with that of the UK and France, which has been suggested to be related to a very frequent use of non-toxin methods in Germany (in 78.9% of surveyed hospitals) (K. Davies et al. 2020a).

4 Whole-Genome Sequencing-Based Typing as a Tool for In-Depth Analysis of the Epidemiology of CDI

In addition to applying appropriate testing strategies and using optimum laboratory diagnosis of CDI, using an optimum method for typing of isolates is instrumental in detecting changes in the epidemiology, transmissions and outbreaks and evaluating efforts to control spread of disease. PCR-ribotyping (often in combination with multilocus variable-number tandem repeat analysis (MLVA) typing for phylogenetic analysis) has been the most-used reference testing method at European NRLs for investigating epidemiological questions since the mid-2000s. Capillary-electrophoresis (CE) PCR ribotyping, allowing comparison of PCR-fragments of known ribotypes stored in a central database, is currently considered gold standard for typing of *C. difficile*. A standardised protocol for CE PCR-ribotyping allowing comparison of typing data between laboratories and transfer of data across laboratories nationally and internationally was developed and validated in 2015 (Fawley et al. 2015).

However, PCR ribotyping even in combination with MLVA-typing still lacks sufficient discriminatory power to distinguish between closely related strains needed to investigate transmission events and outbreaks and is unable to characterize virulence and antimicrobial resistance genes, determine relatedness to international clones and microevolutionary events in epidemic and hypervirulent strains (Baktash et al. 2022; Janezic and Rupnik 2019).

The utility and superiority of WGS-based typing of *C. difficile* as a novel surveillance and investigative epidemiological tool have been demonstrated in a number of studies on transmission, outbreaks and recurrent CDI (Janezic and Rupnik 2019; Lim et al. 2020). Using WGS-based surveillance over a 3-year period in four hospitals, only 35% of isolates from hospital patients were found to be related to strains of previous hospital patients, while 45% of isolates from patients were completely unrelated to isolates of previous patients, which suggested that other sources and reservoirs of *C. difficile* play a role in the local epidemiology (Eyre et al. 2013a). Lower transmission rates (of 7–24%) were observed in six hospitals 3 years later, which was explained by the decrease in 027, as this ribotype had the highest proportion (57%) of related isolates (transmission events) among all ribotypes (Eyre et al. 2017).

WGS-based analysis was used to investigate elevated incidence rates of CDI in Northern Wales. Despite variation in transmission rates between three hospitals (11–27%), only 17% of CDI could have been plausibly acquired from a previous patient (transmission rates being within the same range as the six English hospitals), which could not explain the nearly double incidence rates in North Wales compared to England. Other predictors of transmission were examined in risk factor analysis to explain the higher incidence in North Wales, of which cephalosporin exposure, healthcare exposure in the last 12 weeks and infection with ribotype 027 were implicated (Eyre et al. 2019).

As many laboratories and NRLs now are switching to WGS as their primary methodology, WGS-based analysis has the potential to replace CE-PCR ribotyping as the main tool for in-depth investigations of the epidemiology of *C. difficile*.

However, as the investigations from North Wales showed, collection of epidemiological information about the cases, medicines use and healthcare facilities is essential to interpreting the WGS data.

5 Benefits of National Surveillance Programmes: Experiences from The UK and Ireland

5.1 *C. difficile* Infection in Ireland

In Ireland, CDI surveillance is coordinated by the national public health surveillance centre (Health Protection Surveillance Centre). New cases of CDI have been a notifiable infectious disease since May 2008, with recurrent CDI notifiable since January 2012. A voluntary national-enhanced CDI programme has been in place since 2009 capturing data on CDI origin, onset and severity, with 97% of all tertiary and general hospitals taking part since quarter 1 of 2012. Hospital-acquired CDI rates per 10,000 bed days used (BDU) are a national key performance indictor (KPI) since April 2014.

Aside from 2019, when a number of hospitals reported hospital-associated outbreaks due to ribotype (RT) 002, there has been a trend in HA-CDI reduction with a concurrent rise in CA-CDI. In 2021, the CDI national crude incidence rate for new and recurrent CDI per 100,000 population was higher than that reported in 2020 (32.8 vs. 30.7; and lower than 39.0, the annual mean of 2015–2019) (HSE 2022). As in previous years, the majority of CDI was reported in people over 65 years (65%). In the voluntary-enhanced CDI, surveillance scheme HA-CDI represented 54% of all cases, equating to a national incidence rate for new and recurrent HA-CDI, that originated within the participating hospital, of 2.1 per 10,000 bed days used (BDU), which was lower than that of 2020 (2.4), and of the 2015–2019 annual mean (2.4). Information on the patient's location at CDI symptom onset showed 46% of patients were in the community and 11% were reported as healthcare onset in a long-term care facility. Of community-acquired CDI, 93% people experienced onset of symptoms in the community outside of a healthcare facility and without discharge from a healthcare facility in the previous 12 weeks.

Until recently, Ireland lacked a national *C. difficile* reference laboratory; hence, limited national information on the epidemiology and clinical consequences of circulating *C. difficile* ribotypes was available. Only 22% of CDI cases reported in 2021 had associated ribotyping data, with 078 (16%), 014 (9%), 002 (9%), 020 (8%) and 005 (7%) being the most common, similar to recent years. Notably, the increase in ribotype 002 which had peaked at 33% of ribotyped cases in 2019 has fallen back to historical levels of 9% in 2021. A slow increase in the proportion of cases with ribotype 020 year on year is evident, which is at 8% of ribotyped cases in 2021 (6% in 2020, 4% in 2019). No cases of the virulent ribotype 027 were detected in Ireland in 2021.

5.2 *C. difficile* Infection in the UK

Prior to the year 2000, data on *C. difficile* was collected on a voluntary basis. In the UK, a steady increase in laboratory reports was observed during the 1990s (DH-HPA 2008). In England, this was suggested to reflect a failure to implement guidelines published in 1994, as well as the result of increased testing and awareness of CDI, and an increase in community-associated CDI (DH-HPA 2008).

The increasing CDI rates and emergence of ribotype 027 precipitated the implementation of mandatory national surveillance of CDI by England, Wales and Northern Ireland in 2004 and by Scotland in 2006. Initially, the surveillance programmes included only those aged 65 years and above but have since expanded to include all ages except the very young (HPS 2017; Pearson 2009). Between 2003 and 2007, several large hospital outbreaks of CDI occurred, involving ribotype 027 (two in England and one in Scotland), which brought CDI to the public attention (Careinspectorate 2014; HC 2006). Among the many key findings and recommendations contained within the critical reports that followed was an acknowledgement of a lack of appropriate surveillance mechanisms, both locally and nationally, that could have identified an outbreak, and the need for formal communication channels to be in place to allow information on CDI numbers and severity to be quickly disseminated. These major incidents were quickly followed by the setting of national targets within the UK to reduce CDI rates by 30% (Duerden 2011; SG 2012).

Around the same time as the UK was implementing national surveillance schemes, the ECDC and the US CDC produced recommendations for surveillance of CDI (Kuijper et al. 2006a; McDonald et al. 2007). The publication of these documents enabled a standardised surveillance case definition to be developed as well as definitions for severe CDI, recurrence, outbreaks and origin of infection that could be used as necessary within a surveillance programme. Shortly thereafter, evidence-based recommendations for infection prevention and control of CDI were published (Vonberg et al. 2008), with strong recommendations for the implementation of routine surveillance of CDI, including the setting of thresholds to identify outbreaks, emphasis on the importance of early diagnosis and awareness of changes in incidence or severity of disease. The foundations were laid for the development of a range of tools and strategies to deal with the CDI epidemic (DH-HPA 2008; HPS 2009).

Continuous and prospective surveillance at the national level in healthcare and community settings was mandated by governments in England and Scotland and real-time 'local surveillance' (by ward, unit or facility) to monitor the number of cases, disease severity, surgery and mortality rates with a duty for the multidisciplinary clinical and infection prevention control team to investigate the root cause of any anomalies or 'exceedances' identified at local level in order rectify deficiencies in patient care and/or infection control (DH-HPA 2008; HPS 2014). The heightened focus on local surveillance was a result of recommendations that emerged from investigations of previous hospital outbreaks (Careinspectorate 2014; HC 2006).

CDI incidence rates in the UK peaked during 2007/2008 and then rapidly declined over the next few years (HPS 2014; McDonald et al. 2007; Vonberg et al. 2008). Between 2007 and 2010, significant reductions in the incidence rate of CDI were observed in England (from 120 to 35 per 100,000 bed days, a decrease of 71%), while there was a 78% decrease in the Scottish incidence rate between 2007 and 2012 (from 150 to 29 per 100,000 bed days) (Duerden 2011; HPS 2013). Following the large reductions observed in the early stages of the UK surveillance programmes, the trend in CDI incidence rates from 2013 onwards in the UK levelled off and were relatively stable up to 2020, during which the first wave of the COVID-19 pandemic resulted in a decrease in the total number of reported cases, though hospital-onset incidence rates increased due to large decreases in the hospital patient population (UKHSA 2022).

The most recent data from England for the 2021/2022 financial year shows reported CDI cases to be at a 9-year high (14,248 cases), with hospital-onset cases increasing over three consecutive financial years since 2018/2019 (from 12 to 16 cases per 100,000 bed days) (UKHSA 2022). Similarly, Scotland reported an increase in hospital-onset cases between 2020 and 2021 although the annual trend in hospital-onset incidence rates does not appear to be increasing (15.7 compared to 15.1 per 100,000 bed days in 2020 and 2021, respectively) (ARHAI 2022, 2023). Interpretation of these more recent trends is difficult as they occurred during the COVID-19 pandemic, and it is not yet certain what impact the pandemic has had, and whether any changes reflect long-term changes in the epidemiology of CDI. Despite the observed increases in England, the case numbers and incidence rates are still far lower than the 2008 peak. Current data from England and Scotland does not suggest any underlying changes in ribotype distribution, and the explanation for the more recent trends may lie in differences in infection prevention and control, antibiotic prescribing and CDI case ascertainment during a period where healthcare services are still adjusting/recovering from changes introduced during the pandemic (ARHAI 2022; UKHSA 2022).

In order to respond to the public health need and to provide more detailed epidemiological information on circulating strains of *C. difficile*, a network of reference laboratories was established in England (the *Clostridium difficile* ribotyping network, CDRN) with collaborative links to a single-reference laboratory in Scotland. Investigations and isolate typing criteria focussed on severe cases of CDI, clusters of cases and unexplained increases in incidence in both countries. In the first 3 years after establishing these laboratory services, the prevalence of ribotype 027 decreased markedly in England (from 55% to 21%). This change in distribution of ribotypes in England coincided with a 61% reduction in reports of CDI cases (from 36,095 in 2008–2009 to 21,698 in 2010–2011) and a decrease in reports of complications, including mortality (Wilcox et al. 2012). Likewise, the three major epidemic ribotypes 027, 001 and 106 were gradually replaced with other less prevalent ribotypes, while rates of CDI were reducing in Scotland between 2008 and 2013 (Wiuff et al. 2011, 2014). In parallel with the relatively stable CDI incidence rates since 2013, prevalence rates for individual ribotypes have also shown little fluctuation with ribotypes 002, 005, 014, 015, 020, 023 and 078 now predominating in the UK (ARHAI 2022; PHE 2019). The timely provision of ribotype information to infection prevention and control teams may have facilitated the targeting of interventions and resources to high-incidence settings (PHE 2019; Wilcox et al. 2012). However, this also needs to be viewed in the context of a heightened awareness and an improved understanding of the need for clinical vigilance and aggressive interventions at a time when CDI incidence and mortality rates in hospitals were much higher.

The overall decrease in CDI has been attributed to a multidisciplinary approach including evidence-based guidance for the treatment and management of CDI patients, restrictive antimicrobial stewardship policies and, arguably, due to the government targets for reducing CDI (Duerden 2011; Lawes et al. 2017; Nathwani et al. 2011). Newer evidence suggests a stronger role for antimicrobial stewardship within CDI control programmes, with a 2017 study by Dingle et al. showing a correlation between restrictions of fluoroquinolone use in hospitals and the community and the significant declines in CDI incidence rates observed in the early period of surveillance in England (Dingle et al. 2017). WGS demonstrated that the development of fluoroquinolone resistance occurred prior to the appearance of the most prevalent genotypes of that period. These resistant types then experienced the most significant declines compared to fluoroquinolone-susceptible isolates, the latter of which did not appear to be markedly affected by improved infection control policies. A prominent role for restricting fluroquinolones on CDI incidence (as opposed to infection control measures) was also highlighted in a Scottish study by Lawes et al. (Lawes et al. 2017), which found that reducing fluroquinolones in conjunction with other 4C antibiotics (cephalosporins, clindamycin and co-amoxiclav) reduced *C. difficile* prevalence rate by 68% in hospitals and 45% in the community.

The establishment of mandatory surveillance systems across the UK driven by government policy was instrumental to the development of standardised, evidence-based diagnostic testing and expansion of national reference laboratory services. The success of the UK surveillance programmes has undoubtedly been due to the rapid and joined up development of diagnostic and surveillance capability and capacity, with coverage of all healthcare settings.

Standardised national surveillance programmes are crucial to enable the monitoring of trends within and between countries, as well as facilitating the monitoring of interventions for improving care and outcomes of CDI patients. Central to all of this has been the adoption within the UK national surveillance programmes of standardised protocols for sampling, testing, typing of isolates, reporting and feeding back data in management structures. This has resulted in more solid reporting and accountability structures that lead to rapid responses to increases in CDI.

6 Epidemiological Sentinel Surveillance

A ribotyping-based sentinel surveillance programme was developed in Scotland to monitor the circulating strains of *C. difficile* in healthcare and community settings to complement the national mandatory surveillance of CDI cases and reference typing of isolates from severe cases and suspected outbreaks (Banks et al. 2016). The Scottish sentinel programme, which is now integral to the national UK surveillance programme, has helped understand changes in the epidemiology of endemic and hypervirulent strain types and flagged potential clusters by ribotyping a representative number of isolates from each healthcare region on a quarterly basis.

In Denmark, surveillance efforts began following a number of smaller and larger outbreaks of ribotype 027 in 2006–2009 (Bacci et al. 2009; Soes et al. 2009). However, surveillance efforts were concentrated on detecting ribotype 027 and other binary toxin-positive strains due to guidance by the Danish Health Authority in 2008 asking the clinical laboratories only to submit isolates to the NRL if consistent with (a) binary toxin-positive or (b) presence of severe clinical manifestations or (c) part of an outbreak. As a result of these national requirements and a healthcare system with autonomous regions, an NRL-typing-based sentinel surveillance became the main national monitoring system for CDI alongside access to a national available database, Healthcare-Associated Infections Database (HAIBA), holding laboratory results from all laboratories in the country. The European recommended diagnostic testing algorithms, case and surveillance definitions were not implemented at a national level.

In 2016, a typing-based sentinel surveillance was developed centrally by the Danish NRL (at Statens Serum Institut) to monitor all clinically relevant strain types. In the sentinel surveillance, all laboratories are assigned 2 months a year (one in spring and one in autumn), where they submit all toxigenic isolates detected locally to the NRL (Persson et al. 2022). Initially, sequence types were inferred from tandem repeat sequence typing (TRST), but from 2018 and onwards, all isolates have been analysed by WGS using core genome multi-locus sequence typing (cgMLST). The sentinel surveillance system has allowed temporal analysis by investigation of 15% of all national cases of CDI and has detected regional and temporal differences. Over a 4-year period, binary toxin-negative strains have gradually replaced binary toxin strains and increased from 70% to 79.5% of all strains, and the overall diversity of strain has increased. Moreover, outbreaks and transmission events among all strain types are now investigated routinely. As in many other European countries, incidence rates have declined in Denmark during the past 8 years as a result of improved infection prevention and control measures and restriction of antimicrobial use, but the epidemiology is monitored nationally by a typing-based sentinel surveillance system rather than case-based epidemiological surveillance.

7 European CDI Surveillance at the ECDC

A European pilot study of surveillance of CDI carried out by ECDIS-Net in 2013, including 37 hospitals in 14 countries, demonstrated the feasibility of coordinated and standardised European CDI surveillance (van Dorp et al. 2016). Participating hospitals could choose between three options for CDI surveillance from a 'minimal' (aggregated numerator and denominator hospital data), 'light' (including individual patient data for CDI cases and aggregated denominator data) to an 'enhanced' option (including collection data on patient comorbidity and characterization of isolates).

Following on from this, ECDC developed the European Surveillance of *Clostridium difficile* infections surveillance protocol 2.1 (2015) addressing disease specific aspects, case definitions, criteria for inclusion and exclusion of cases and specifics of the three optional surveillance systems. The first European protocol has been superseded by subsequent versions, the most recent version being 2.4 (ECDC 2019).

The ECDC CDI protocol is aimed at 'providing a tool for hospitals and countries to estimate the incidence of CDI; to assess the burden of adverse outcomes of CDI, including morbidity and mortality; and to describe the epidemiology of *C. difficile* at the local, national and European level'. The protocol specified three surveillance options (minimal, light or enhanced monitoring): the minimal option collecting only hospital-level aggregate numerators and denominators, the light option also collecting case-based numerators including mortality and the enhanced option that links epidemiological and microbiological data on at least the first five cases with case-based data (current version 2.4) (ECDC 2019). The linkage of individual epidemiological case-data with microbiological typing data in the enhanced surveillance option of the ECDC protocol will potentially permit faster identification of new highly virulent strains. Regardless of the surveillance option used, ECDC recommends continuous incidence surveillance of CDI for a period of 12 months (ECDC 2022).

To date, data from the European CDI surveillance system has been reported and published only for the years 2016–2017 (ECDC 2022) in which 23 countries/administrations provided data suitable for analysis (including separate data sets from devolved UK administrations in Wales, England and Scotland), although only 14 countries contributed data in both years. Key data from this report are also summarized above (see Sect. 1: 'Epidemiology of CDI in Europe').

Comparison of the countries participating in the European surveillance programme (ECDC 2022) with the participants of the 2017-surveillance capacity survey (Krutova et al. 2018) suggests divergent priorities and lack of harmonization of CDI surveillance efforts within countries and across Europe. A gap in publicly available European epidemiological CDI data has developed since the issuing of the first ECDC CDI report in 2020 possibly as a result of competing priorities under the COVID-19 pandemic. It is also unknown how the pandemic has affected CDI surveillance efforts and protocols in the individual European countries and the international collaborations aiming at obtaining comparable epidemiological data.

8 The Need for European Surveillance of CDI

The suite of guidance documents on CDI diagnostics, infection prevention and control and treatment developed by ESGCD and supported by ESCMID has provided the evidence platform for the development of European surveillance of CDI now undertaken and coordinated by ECDC.

Suboptimal laboratory diagnostics, a continued lack of consensus on optimal testing methodology for CDI and availability of typing across Europe have led to underdiagnosis and impeded comparison between countries. Underestimation of CDI has also resulted from a deficiency in uniformity of case definitions, clinical algorithms and recognition among clinicians of when to suspect CDI. These inconsistencies have prevented the true burden of disease from being appreciated. The international surveys reviewed above highlighted again variation in awareness and capability and capacity to diagnose, sub-type, report, collect patient risk factor data and monitor CDI across Europe. Although the overall capability and capacity for monitoring and investigating CDI isolates has increased tremendously across Europe over the last 20 years, there is still scope for improvement and standardisation of diagnostic and surveillance setups in many countries. This will enable countries to monitor their national situation, compare trends over time and with other countries and prevent new global epidemics of hypervirulent types of *C. difficile*. The ESCMID guidance and ECDC surveillance protocol provides a comprehensive evidence base for diagnosing and monitoring CDI in European countries and in Europe as a whole. Key steps to optimize and standardise CDI surveillance are listed in Table 1.

The future collection of a European standardised data set on CDI (annually or 3–5 yearly) will strengthen and consolidate national surveillance systems and prevent deterioration of established surveillance in the individual countries and declining participation in the European surveillance programme through provision of a standardised approach. European guidance currently addresses nearly all aspects of national surveillance for CDI except for public health governance details (e.g. mandatory surveillance, quality improvement indicators and targets for reduction of CDI) (see Table 1).

In countries where large reductions in CDI incidence have been achieved, comprehensive national surveillance programmes have been a key driver in the standardisation of diagnostic approach, sampling and reporting practices and in developing coordinated approaches and resources to infection prevention and management of CDI by highlighting the evolving epidemic of CDI.

Additional benefits could be achieved from the introduction of WGS to investigate clusters, cross-transmission routes, emergence of new hypervirulent strains and global epidemiology (Eyre et al. 2013b; Eyre and Walker 2013; Fawley et al. 2011).

Table 1 Steps in optimising and standardising surveillance of CDI and availability of European guidance

Steps in optimising and standardising CDI surveillance	Recommendations for standardization by ECDC
Criteria for testing patients (including diarrhoea and faecal samples)	Yes
Use of recommended two-step diagnostic *C. difficile*, a two-step testing algorithm including a toxin-based method	Yes
Inclusion of healthcare facilities in monitoring (primary, secondary and tertiary sectors)	Yes
Options for use of CDI case data (laboratory data combined with patient information): (a) Minimal surveillance (b) Light surveillance (c) Enhanced surveillance	Yes
Mandatory vs. voluntary reporting of CDI data	No
Public health targets for improvement	No
Prospective and continuous surveillance of CDI	Yes
Reporting of healthcare-associated CDI according to definition (including community-onset and healthcare-onset cases)	Yes
Reporting of community-associated CDI according to definition	Yes
Reporting of severe disease according to definition	Yes
Reporting of recurrent disease according to definition	Yes
Reporting of complicated course of CDI according to definition (including admission to ICU, surgery for CDI, admission to healthcare facility for treatment CDI)	Yes
Reporting of death according to definition	Yes
National targets for driving the reduction of CDI	No
Use of national reference laboratory ribotyping (and MLVA) typing data	Yes
Use of WGS-based typing data for national and international comparisons	Yes (in other document)[a]
Use of sentinel surveillance for monitoring circulating strains	No

[a]Covered in European Centre for Disease Prevention and Control. ECDC strategic framework for the integration of molecular and genomic typing into European surveillance and multi-country outbreak investigations—2019–2021. Stockholm: ECDC; 2019

9 Conclusion

Significant reductions in CDI have been reported in countries across Europe, but surveillance activities in individual countries have slowed down during the COVID-19 pandemic as resources were diverted to the global health crisis. A renewed and strengthened focus on CDI surveillance and prevention is therefore urgently needed post COVID-19.

References

Alcala L, Martin A, Marin M, Sanchez-Somolinos M, Catalan P, Pelaez T, Bouza E, Spanish Clostridium difficile Study, G (2012) The undiagnosed cases of Clostridium difficile infection in a whole nation: where is the problem? Clin Microbiol Infect *18*(7): E204–E213. https://doi.org/10.1111/j.1469-0691.2012.03883.x

ARHAI (2022) Antimicrobial Resistance and Healthcare Associated Infection Scotland. ARHAI Scotland 2021 Annual Report. ARHAI Scotland. https://www.nss.nhs.scot/publications/arhai-scotland-2021-annual-report/

ARHAI (2023) Antimicrobial Resistance and Healthcare Associated Infection Scotland. Quarterly epidemiological data on Clostridioides difficile infection, Escherichia coli bacteraemia, Staphylococcus aureus bacteraemia and Surgical Site Infection in Scotland. July to September (Q3) 2022 | National Services Scotland (nhs.scot). https://www.nss.nhs.scot/publications/quarterly-epidemiological-data-on-clostridioides-difficile-infection-escherichia-coli-bacteraemia-staphylococcus-aureus-bacteraemia-and-surgical-site-infection-in-scotland-july-to-september-q3-2022/

Bacci S, St-Martin G, Olesen B, Bruun B, Olsen KE, Nielsen EM, Molbak K (2009) Outbreak of

Clostridium difficile 027 in North Zealand, Denmark, 2008-2009. Euro Surveill 14(16) https://www.ncbi.nlm.nih.gov/pubmed/19389341

Baktash A, Corver J, Harmanus C, Smits WK, Fawley W, Wilcox MH, Kumar N, Eyre DW, Indra A, Mellmann A, Kuijper EJ (2022) Comparison of whole-genome sequence-based methods and PCR ribotyping for subtyping of Clostridioides difficile. J Clin Microbiol 60(2):e0173721. https://doi.org/10.1128/JCM.01737-21

Banks A, Brown DJ, Mather H, Coia JE, Wiuff C (2016) Sentinel community Clostridium difficile infection (CDI) surveillance in Scotland, April 2013 to March 2014. Anaerobe 37:49–53. https://doi.org/10.1016/j.anaerobe.2015.12.008

Barbut F, Delmee M, Brazier JS, Petit JC, Poxton IR, Rupnik M, Lalande V, Schneider C, Mastrantonio P, Alonso R, Kuipjer E, Tvede M, difficile, E. S. G. o. C. (2003) A European survey of diagnostic methods and testing protocols for Clostridium difficile. Clin Microbiol Infect 9(10):989–996. https://doi.org/10.1046/j.1469-0691.2003.00683.x

Bauer MP, Notermans DW, van Benthem BH, Brazier JS, Wilcox MH, Rupnik M, Monnet DL, van Dissel JT, Kuijper EJ (2011) Clostridium difficile infection in Europe: a hospital-based survey. Lancet 377(9759): 63–73. https://doi.org/10.1016/s0140-6736(10)61266-4

Bauer MP, Veenendaal D, Verhoef L, Bloembergen P, van Dissel JT, Kuijper EJ (2009) Clinical and microbiological characteristics of community-onset Clostridium difficile infection in The Netherlands. Clin Microbiol Infect 15(12):1087–1092. https://doi.org/10.1111/j.1469-0691.2009.02853.x

Careinspectorate (2014) The Vale of Leven Hospital Inquiry Report. https://hub.careinspectorate.com/media/1415/vale-of-leven-hospital-inquiry-report.pdf

Cassini A, Plachouras D, Eckmanns T, Abu Sin M, Blank HP, Ducomble T, Haller S, Harder T, Klingeberg A, Sixtensson M, Velasco E, Weiss B, Kramarz P, Monnet DL, Kretzschmar ME, Suetens C (2016) Burden of six healthcare-associated infections on European population health: estimating incidence-based disability-adjusted life years through a population prevalence-based modelling study. PLoS Med 13(10):e1002150. https://doi.org/10.1371/journal.pmed.1002150

CDC (2019) Centers for Disease Control and Prevention. Antibiotic resistance threats in the United States, 2019. https://stacks.cdc.gov/view/cdc/82532

Davies K, Davis G, Barbut F, Eckert C, Petrosillo N, Pisapia R, Gartner B, Berger FK, Reigadas E, Bouza E, Demont C, Wilcox MH (2020a) Factors affecting reported Clostridioides difficile infection rates; the more you look the more you find, but should you believe what you see? Anaerobe 62:102178. https://doi.org/10.1016/j.anaerobe.2020.102178

Davies K, Viprey V, Ewin D, Spittal W, Vernon J, Tkalec V, Fawley W, Benson A, Davis G, Rupnik M, Wilcox M (2020b) Burden of Clostridium difficile infection (CDI) across whole healthcare economies and European borders; COMBACTE-CDI results: the Sixth Decennial International Conference on Healthcare-Associated Infections Abstracts, March 2020: Global Solutions to Antibiotic Resistance in Healthcare. Infect Control Hosp Epidemiol 41(S1): S24–S25. https://www.cambridge.org/core/journals/infection-control-and-hospital-epidemiology/article/burden-of-clostridium-difficile-infection-cdi-across-whole-healthcare-economies-and-european-borders-combactecdi-results/CFE7440B7DD8597C68164B4D986F20FB

Davies KA, Longshaw CM, Davis GL, Bouza E, Barbut F, Barna Z, Delmee M, Fitzpatrick F, Ivanova K, Kuijper E, Macovei IS, Mentula S, Mastrantonio P, von Muller L, Oleastro M, Petinaki E, Pituch H, Noren T, Novakova E et al (2014) Underdiagnosis of Clostridium difficile across Europe: the European, multicentre, prospective, biannual, point-prevalence study of Clostridium difficile infection in hospitalised patients with diarrhoea (EUCLID). Lancet Infect Dis 14(12):1208–1219. https://doi.org/10.1016/S1473-3099(14)70991-0

Delmee M (2001) Laboratory diagnosis of Clostridium difficile disease. Clin Microbiol Infect 7(8):411–416. https://www.ncbi.nlm.nih.gov/pubmed/11591203

DH-HPA (2008) Clostridium difficile infection: how to deal with the problem. Department of Health – Health Protection Agency. https://www.gov.uk/government/publications/clostridium-difficile-infection-how-to-deal-with-the-problem

Dingle KE, Didelot X, Quan TP, Eyre DW, Stoesser N, Golubchik T, Harding RM, Wilson DJ, Griffiths D, Vaughan A, Finney JM, Wyllie DH, Oakley SJ, Fawley WN, Freeman J, Morris K, Martin J, Howard P, Gorbach S et al (2017) Effects of control interventions on Clostridium difficile infection in England: an observational study. Lancet Infect Dis 17(4):411–421. https://doi.org/10.1016/S1473-3099(16)30514-X

Duerden BI (2011) Contribution of a government target to controlling Clostridium difficile in the NHS in England. Anaerobe 17(4):175–179. https://doi.org/10.1016/j.anaerobe.2010.12.004

ECDC (2013) Point prevalence survey of healthcare associated infections and antimicrobial use in European acute care hospitals. European Centre for Disease Prevention and Control. https://www.ecdc.europa.eu/en/publications-data/point-prevalence-survey-healthcare-associated-infections-and-antimicrobial-use-0

ECDC (2015) European surveillance of Clostridium difficile infections Surveillance protocol version 2.2. https://www.ecdc.europa.eu/en/publications-data/european-surveillance-clostridium-difficile-infections-surveillance-protocol-0

ECDC (2019) European surveillance of Clostridioides (Clostridium) difficile infections – surveillance protocol version 2.4. European Centre for Disease Prevention and Control. https://www.ecdc.europa.eu/en/publications-data/european-surveillance-clostridium-difficile-infections-surveillance-protocol-2

ECDC (2022) Clostridioides (Clostridium) difficile infections. Annual epidemiological report for 2016–2017. European Centre for Disease Prevention and Control. https://www.ecdc.europa.eu/en/publications-data/clostridiodes-difficile-infections-annual-epidemiological-report-2016-2017

Eyre DW, Cule ML, Wilson DJ, Griffiths D, Vaughan A, O'Connor L, Ip CLC, Golubchik T, Batty EM, Finney JM, Wyllie DH, Didelot X, Piazza P, Bowden R, Dingle KE, Harding RM, Crook DW, Wilcox MH, Peto TEA, Walker AS (2013a) Diverse sources of C. difficile infection identified on whole-genome sequencing. N Engl J Med 369(13):1195–1205. https://doi.org/10.1056/NEJMoa1216064

Eyre DW, Fawley WN, Best EL, Griffiths D, Stoesser NE, Crook DW, Peto TE, Walker AS, Wilcox MH (2013b) Comparison of multilocus variable-number tandem-repeat analysis and whole-genome sequencing for investigation of Clostridium difficile transmission. J Clin Microbiol 51(12):4141–4149. https://doi.org/10.1128/JCM.01095-13

Eyre DW, Fawley WN, Rajgopal A, Settle C, Mortimer K, Goldenberg SD, Dawson S, Crook DW, Peto TEA, Walker AS, Wilcox MH (2017) Comparison of control of Clostridium difficile infection in six English hospitals using whole-genome sequencing. Clin Infect Dis 65(3):433–441. https://doi.org/10.1093/cid/cix338

Eyre DW, Shaw R, Adams H, Cooper T, Crook DW, Griffin RM, Mannion P, Morgan M, Morris T, Perry M, Jones S, Peto TEA, Sutton J, Walker AS, Williams D, Craine N (2019) WGS to determine the extent of Clostridioides difficile transmission in a high incidence setting in North Wales in 2015. J Antimicrob Chemother 74(4):1092–1100. https://doi.org/10.1093/jac/dky523

Eyre DW, Walker AS (2013) Clostridium difficile surveillance: harnessing new technologies to control transmission. Expert Rev Anti Infect Ther 11(11):1193–1205. https://doi.org/10.1586/14787210.2013.845987

Fawley WN, Knetsch CW, MacCannell DR, Harmanus C, Du T, Mulvey MR, Paulick A, Anderson L, Kuijper EJ, Wilcox MH (2015) Development and validation of an internationally-standardized, high-resolution capillary gel-based electrophoresis PCR-ribotyping protocol for Clostridium difficile. PLoS One 10(2):e0118150. https://doi.org/10.1371/journal.pone.0118150

Fawley WN, Wilcox MH, Clostridium difficile Ribotyping Network for, E, Northern, I (2011) An enhanced DNA fingerprinting service to investigate potential Clostridium difficile infection case clusters sharing the same PCR ribotype. J Clin Microbiol 49(12):4333–4337. https://doi.org/10.1128/JCM.05873-11

Finn E, Andersson FL, Madin-Warburton M (2021) Burden of Clostridioides difficile infection (CDI) – a systematic review of the epidemiology of primary and recurrent CDI. BMC Infect Dis 21(1):456. https://doi.org/10.1186/s12879-021-06147-y

Freeman J, Bauer MP, Baines SD, Corver J, Fawley WN, Goorhuis B, Kuijper EJ, Wilcox MH (2010) The changing epidemiology of Clostridium difficile infections. Clin Microbiol Rev 23(3):529–549. https://doi.org/10.1128/CMR.00082-09

Goorhuis A, Bakker D, Corver J, Debast SB, Harmanus C, Notermans DW, Bergwerff AA, Dekker FW, Kuijper EJ (2008) Emergence of Clostridium difficile infection due to a new hypervirulent strain, polymerase chain reaction ribotype 078. Clin Infect Dis 47(9):1162–1170. https://doi.org/10.1086/592257

HC (2006) Healthcare Commission. Investigation into outbreaks of Clostridium difficile at Stoke Mandeville Hospital, Buckinghamshire Hospitals NHS Trust

He M, Miyajima F, Roberts P, Ellison L, Pickard DJ, Martin MJ, Connor TR, Harris SR, Fairley D, Bamford KB, D'Arc S, Brazier J, Brown D, Coia JE, Douce G, Gerding D, Kim HJ, Koh TH, Kato H et al (2013) Emergence and global spread of epidemic healthcare-associated Clostridium difficile. Nat Genet 45(1):109–113. https://doi.org/10.1038/ng.2478

HPS (2009) Health Protection Scotland. Guidance on Prevention and Control of Clostridium difficile Infection (CDI) in Healthcare Settings in Scotland. Health Protection Network Scottish Guidance 6. https://www.processcleaningsolutions.com/pdf/clostridium-difficile-infection-guidelines%20Scotland.pdf

HPS (2013) Health Protection Scotland. Healthcare Associated Infection Annual Report 2012. https://www.cne-siar.gov.uk/media/committeearchive/oldcommitteedocs/healthandsocialcare/agendas/september2013/J%20Item%2010C%20Appendix%202%20HAI%20Annual%20Report%20(H&SCC%2005.09.13).pdf

HPS (2014) Health Protection Scotland. Guidance on Prevention and Control of Clostridium difficile Infection (CDI) in Care Settings in Scotland. Health Protection Network Scottish Guidance 6. Version 2.1. https://www.nhsggc.org.uk/media/239962/guidance-on-prevention-and-control-of-cdi-in-care-settings-oct-2014-v21-hpn.pdf

HPS (2017) Health Protection Scotland. Protocol for the Scottish Surveillance Programme for Clostridium difficile infection. User manual. https://www.nss.nhs.scot/media/3445/cdi-surveillance-protocol-2017.pdf

HSE (2022) HSE Health Protection Surveillance Centre. Clostridioides difficile infection in Ireland 2021. https://www.hpsc.ie/a-z/microbiologyantimicrobialresistance/clostridioidesdifficile/cdifficiledataandreports/annualreports/AER_CDI_2021_v1.1.pdf

Janezic S, Rupnik M (2019) Development and implementation of whole genome sequencing-based typing schemes for Clostridioides difficile. Front Public

Health 7:309. https://doi.org/10.3389/fpubh.2019.00309
Kelly CP, LaMont JT (1998) Clostridium difficile infection. Annu Rev Med 49:375–390. https://doi.org/10.1146/annurev.med.49.1.375
Kola A, Wiuff C, Akerlund T, van Benthem BH, Coignard B, Lyytikainen O, Weitzel-Kage D, Suetens C, Wilcox MH, Kuijper EJ, Gastmeier P, members of, E-N (2016) Survey of Clostridium difficile infection surveillance systems in Europe, 2011. Euro Surveill 21(29). https://doi.org/10.2807/1560-7917.ES.2016.21.29.30291
Krutova M, Kinross P, Barbut F, Hajdu A, Wilcox MH, Kuijper EJ, survey, c (2018) How to: surveillance of Clostridium difficile infections. Clin Microbiol Infect 24(5):469–475. https://doi.org/10.1016/j.cmi.2017.12.008
Kuijper EJ, Coignard B, Tull P (2006a) Emergence of Clostridium difficile-associated disease in North America and Europe. Clin Microbiol Infect 12(Suppl 6):2–18. https://doi.org/10.1111/j.1469-0691.2006.01580.x
Kuijper EJ, Coignard B, Tull P, difficile, E. S. G. f. C., States, E. U. M., European Centre for Disease, P., & Control (2006b) Emergence of Clostridium difficile-associated disease in North America and Europe. Clin Microbiol Infect 12(Suppl 6):2–18. https://doi.org/10.1111/j.1469-0691.2006.01580.x
Lawes T, Lopez-Lozano JM, Nebot CA, Macartney G, Subbarao-Sharma R, Wares KD, Sinclair C, Gould IM (2017) Effect of a national 4C antibiotic stewardship intervention on the clinical and molecular epidemiology of Clostridium difficile infections in a region of Scotland: a non-linear time-series analysis. Lancet Infect Dis 17(2):194–206. https://doi.org/10.1016/S1473-3099(16)30397-8
Lim SC, Knight DR, Riley TV (2020) Clostridium difficile and One Health. Clin Microbiol Infect 26(7):857–863. https://doi.org/10.1016/j.cmi.2019.10.023
McDonald LC, Coignard B, Dubberke E, Song X, Horan T, Kutty PK, Ad Hoc Clostridium difficile Surveillance Working, G (2007) Recommendations for surveillance of Clostridium difficile-associated disease. Infect Control Hosp Epidemiol 28(2):140–145. https://doi.org/10.1086/511798
Nathwani D, Sneddon J, Malcolm W, Wiuff C, Patton A, Hurding S, Eastaway A, Seaton RA, Watson E, Gillies E, Davey P, Bennie M, Scottish Antimicrobial Prescribing, G (2011) Scottish Antimicrobial Prescribing Group (SAPG): development and impact of the Scottish National Antimicrobial Stewardship Programme. Int J Antimicrob Agents 38(1):16–26. https://doi.org/10.1016/j.ijantimicag.2011.02.005
Pearson A (2009) Historical and changing epidemiology of healthcare-associated infections. J Hosp Infect 73(4):296–304. https://doi.org/10.1016/j.jhin.2009.08.016
Persson S, Nielsen HL, Coia JE, Engberg J, Olesen BS, Engsbro AL, Petersen AM, Holt HM, Lemming L, Marmolin ES, Sondergaard TS, Andersen LP, Jensen MBF, Wiuff C, Sorensen G, Nielsen SH, Nielsen EM (2022) Sentinel surveillance and epidemiology of Clostridioides difficile in Denmark, 2016 to 2019. Euro Surveill 27(49). https://doi.org/10.2807/1560-7917.ES.2022.27.49.2200244
PHE (2019) Public Health England. Clostridium difficile Ribotyping Network (CDRN) for England and Northern Ireland 2015-2018. https://www.gov.uk/government/publications/clostridium-difficile-ribotyping-network-cdrn-report
Polage CR, Gyorke CE, Kennedy MA, Leslie JL, Chin DL, Wang S, Nguyen HH, Huang B, Tang YW, Lee LW, Kim K, Taylor S, Romano PS, Panacek EA, Goodell PB, Solnick JV, Cohen SH (2015) Overdiagnosis of Clostridium difficile infection in the molecular test era. JAMA Intern Med 175(11):1792–1801. https://doi.org/10.1001/jamainternmed.2015.4114
Ricciardi R, Rothenberger DA, Madoff RD, Baxter NN (2007) Increasing prevalence and severity of Clostridium difficile colitis in hospitalized patients in the United States. Arch Surg 142(7):624-631; discussion 631. https://doi.org/10.1001/archsurg.142.7.624
SG (2012) Scottish Government. NHS Scotland Local Delivery Plan Guidance 2013/14. http://www.gov.scot/Publications/2012/12/8405/4
Soes L, Molbak K, Strobaek S, Truberg Jensen K, Torpdahl M, Persson S, Kemp M, Olsen KE (2009) The emergence of Clostridium difficile PCR ribotype 027 in Denmark–a possible link with the increased consumption of fluoroquinolones and cephalosporins? Euro Surveill 14(15) https://www.ncbi.nlm.nih.gov/pubmed/19371514
UKHSA (2022) Annual epidemiological commentary: gram-negative, MRSA, MSSA bacteraemia and C. difficile infections, up to and including financial year 2021 to 2022. UK Health Security Agency. https://www.gov.uk/government/statistics/mrsa-mssa-and-e-coli-bacteraemia-and-c-difficile-infection-annual-epidemiological-commentary/annual-epidemiological-commentary-gram-negative-mrsa-mssa-bacteraemia-and-c-difficile-infections-up-to-and-including-financial-year-2021-to-2022
van Dorp SM, Kinross P, Gastmeier P, Behnke M, Kola A, Delmee M, Pavelkovich A, Mentula S, Barbut F, Hajdu A, Ingebretsen A, Pituch H, Macovei IS, Jovanovic M, Wiuff C, Schmid D, Olsen KE, Wilcox MH, Suetens C et al (2016) Standardised surveillance of Clostridium difficile infection in European acute care hospitals: a pilot study, 2013. Euro Surveill 21(29). https://doi.org/10.2807/1560-7917.ES.2016.21.29.30293
Viprey VF, Granata G, Vendrik KEW, Davis GL, Petrosillo N, Kuijper EJ, Vilken T, Lammens C, Schotsman JJ, Benson AD, Cataldo MA, van der Kooi TII, Wilcox MH, Davies KA, consortium, C-C (2023) European survey on the current surveillance practices, management guidelines, treatment pathways and heterogeneity of testing of Clostridioides difficile,

2018-2019: results from The Combatting Bacterial Resistance in Europe CDI (COMBACTE-CDI). J Hosp Infect 131:213–220. https://doi.org/10.1016/j.jhin.2022.11.011
Vonberg RP, Kuijper EJ, Wilcox MH, Barbut F, Tull P, Gastmeier P, European, C d-I C G, European Centre for Disease, P, Control, van den Broek PJ, Colville A, Coignard B, Daha T, Debast S, Duerden BI, van den Hof S, van der Kooi T, Maarleveld HJ, Nagy E et al (2008) Infection control measures to limit the spread of Clostridium difficile. Clin Microbiol Infect 14(Suppl 5):2–20. https://doi.org/10.1111/j.1469-0691.2008.01992.x
Warny M, Pepin J, Fang A, Killgore G, Thompson A, Brazier J, Frost E, McDonald LC (2005) Toxin production by an emerging strain of Clostridium difficile associated with outbreaks of severe disease in North America and Europe. Lancet 366(9491):1079–1084. https://doi.org/10.1016/S0140-6736(05)67420-X
Wilcox MH, Mooney L, Bendall R, Settle CD, Fawley WN (2008) A case-control study of community-associated Clostridium difficile infection. J Antimicrob Chemother 62(2):388–396. https://doi.org/10.1093/jac/dkn163
Wilcox MH, Shetty N, Fawley WN, Shemko M, Coen P, Birtles A, Cairns M, Curran MD, Dodgson KJ, Green SM, Hardy KJ, Hawkey PM, Magee JG, Sails AD, Wren MW (2012) Changing epidemiology of Clostridium difficile infection following the introduction of a national ribotyping-based surveillance scheme in England. Clin Infect Dis 55(8):1056–1063. https://doi.org/10.1093/cid/cis614
Wiuff C, Banks A, Brown DJ, Henderson D, Coia JE, Inkster T (2014) The changing epidemiology of Clostridium difficile ribotypes in Scotland between 2009-2013 European Congress on Clinical Microbiology and Infectious Diseases, 2014, Barcelona, Spain
Wiuff C, Brown DJ, Mather H, Banks AL, Eastaway A, Coia JE (2011) The epidemiology of Clostridium difficile in Scotland. J Infect 62(4):271–279. https://doi.org/10.1016/j.jinf.2011.01.015

Diagnostic Guidance for *C. difficile* Infections

Joffrey van Prehn, Monique J. T. Crobach, Amoe Baktash, Nikolas Duszenko, and Ed J. Kuijper

Abstract

Diagnosis of *Clostridioides difficile* infection (CDI) can be challenging. First of all, there has been debate on which of the two reference assays, cell cytotoxicity neutralization assay (CCNA) or toxigenic culture (TC), should be considered the gold standard for CDI detection. Although the CCNA suffers most from suboptimal storage conditions and subsequent toxin degradation, TC is reported to falsely increase CDI detection rates as it cannot differentiate CDI patients from patients asymptomatically colonised by toxigenic *C. difficile*. Several rapid assays are available for CDI detection and fall into three broad categories: (1) enzyme immunoassays for glutamate dehydrogenase, (2) enzyme immunoassays or single-molecule array assays for toxins A/B and (3) nucleic acid amplification tests detecting toxin genes. All three categories have their own limitations, being suboptimal specificity and/or sensitivity or the inability to discern colonised patients from CDI patients. In light of these limitations, multi-step algorithmic testing has been advocated by international guidelines (IDSA/SHEA and ESCMID) in order to optimize diagnostic accuracy. As a result, a survey performed in 2018–2019 in Europe revealed that most of all hospital sites reported using more than one test to diagnose CDI. CDI incidence rates are also influenced by sample selection criteria, as several studies have shown that if not all unformed stool samples are tested for CDI, many cases may be missed due to an absence of clinical suspicion. Since methods for diagnosing CDI remain imperfect, there has been a growing interest in alternative testing strategies like faecal microbiota biomarkers, immune modulating interleukins, cytokines and imaging methods. At the moment, these alternative methods might play an adjunctive role, but they are not suitable to replace conventional CDI testing strategies.

J. van Prehn (✉) · E. J. Kuijper
Department of Medical Microbiology, Leiden University Centre for Infectious Diseases (LU-CID), Leiden University Medical Centre, Leiden, The Netherlands

ESCMID Study Group for C. difficile (ESGCD) and Study Group for Host and Microbiota Interaction (ESGHAMI), Basel, Switzerland
e-mail: J.van_Prehn@lumc.nl

M. J. T. Crobach · A. Baktash · N. Duszenko
Department of Medical Microbiology, Leiden University Centre for Infectious Diseases (LU-CID), Leiden University Medical Centre, Leiden, The Netherlands

1 Introduction

Diagnosis of *Clostridioides difficile* infection (CDI) is challenging, as there is no optimal laboratory assay and even no universal reference test. Due to imperfect assays, combinations of assays

P. Mastrantonio, M. Rupnik (eds.), *Updates on Clostridioides difficile in Europe*, Advances in Experimental Medicine and Biology 1435, https://doi.org/10.1007/978-3-031-42108-2_3

Table 1 Available assays for CDI detection. *CDI Clostridium difficile* infection

Type of assay	Target of detection	Detected condition
Culture	*C. difficile*	*C. difficile* colonisation, can be CDI
Glutamate dehydrogenase enzyme immunoassay (GDH EIA)	Glutamate dehydrogenase	*C. difficile* colonisation, can be CDI
Toxins A/B enzyme immunoassay (Tox A/B EIA)	Toxins A and B	CDI (above 1 year of age, after exclusion of other causes)
Nucleic acid amplification test (NAAT)	*TcdB* and/or *TcdA* genes, sometimes *cdt* and deletion in *tcdC*	Toxigenic *C. difficile* colonisation, can be CDI
Cell cytotoxicity neutralization assay (CCNA)	Toxin B	CDI (above 1 year of age, after exclusion of other causes)
Toxigenic culture (TC)	*C. difficile* and thereafter in vitro toxin production by Tox A/B EIA, NAAT or CCNA	Toxigenic *C. difficile* colonisation, can be CDI

to optimize their performance have been proposed. However, diverse testing strategies are applied across laboratories. These diverse testing strategies may impact CDI incidence rates. In addition to the conventional testing methods (Table 1), alternative methods are sometimes applied either to diagnose CDI or as an aid to predict severity. Here, we will describe the diverse testing strategies with their advantages and limitations and clinical relevance.

2 Reference Tests

The diagnosis of CDI relies on one of two approaches: demonstrate the presence of toxins responsible for the clinical manifestations of CDI or demonstrate the presence of *C. difficile* which is capable of producing toxins, the so-called toxigenic *C. difficile* (Planche and Wilcox 2011). The reference test for detection of toxins in stools is the cell cytotoxicity neutralisation assay (CCNA) (Burnham and Carroll 2013; Planche and Wilcox 2011). For CCNA, stool sample filtrate is inoculated onto an in vitro cell monolayer, using cell lines such as Vero cells, HeLa cells, human foreskin fibroblast cells or Hep-2 cells. At 24- and 48-h intervals, these cultures are evaluated for the characteristic rounding effect engendered mainly by toxin B. Reversal of this effect by toxins A and B antitoxin (either *C. sordellii* or *C. difficile* antiserum) demonstrates the roles of toxins in inducing the cytopathic effects observed, and thus its presence (Burnham and Carroll 2013; Delmee 2001). Although toxin B is primarily detected in this assay, toxin A is also detected to some extent. The reference test for detection of toxigenic *C. difficile* is toxigenic culture (TC) (Burnham and Carroll 2013; Planche and Wilcox 2011). For TC, stool samples are inoculated onto selective media and incubated for at least 48 h (Hink et al. 2013). Colonies suspected of being *C. difficile*, by, e.g. gramme staining, colony morphology, odour or more sophisticated techniques, are isolated and identified by matrix-assisted laser desorption ionization time of flight mass spectrometry (MALDI-TOF MS). MALDOI-TOF has also been tested a rapid method to diagnose the so-called hypervirulent *C. difficile* PCR ribotype 027 by specific peptides (Corver et al. 2019; Flores-Trevino et al. 2019). The toxigenic potential is assessed by testing for in vitro toxin production via the aforementioned CCNA, by enzyme immunoassays (EIA) for toxins A/B or by testing for toxin-producing genes via nucleic acid amplification tests (NAAT) (Burnham and Carroll 2013; Persson et al. 2008).

During the last years, there has been debate on which of these two reference tests represents true disease, as the CCNA detects in vivo toxins, while TC detects in vitro toxin production (T. Planche and Wilcox 2011). There is a growing body of evidence demonstrating that toxigenic strains are often carried asymptomatically (Kyne et al. 2000; Loo et al. 2011). TC is not able to make a distinction between asymptomatic carriage of toxigenic *C. difficile* strains and true infection. Studies have shown that patients with positive CCNA or Tox A/B EIA have a worse prognosis than patients who test only positive in TC, indicating that this latter category may actually be colonised patients instead of patients with true CDI (Alonso et al. 2022; Planche et al. 2013; Polage et al. 2015). Although CCNA may therefore better reflect true CDI, it is this reference test that suffers most from lack of standardization and suboptimal storage or collection conditions of faeces, thereby possibly generating false-negative results. Both reference tests are laborious and expensive and require trained personnel. Therefore, easy-to perform rapid assays have been developed. These include enzyme immunoassays for GDH, enzyme immunoassays for toxins A/B, and during the last decade, NAATs for toxin genes have become available. Given their ease of use and rapid turnaround time, these rapid tests have become the mainstays of CDI diagnosis in a clinical setting.

3 Rapid Assays

Reference methods are accurate, but the lengthy, specific requirements of faeces collection and storage (degradation), laborious nature of such testing precludes its application in a clinical setting. Rapid tests are ideally suited for clinical use, but each suffers from its own shortcomings. Tox A/B EIAs directly detect free toxins in stools and are therefore believed to correlate to clinical symptoms (Polage et al. 2015). They are relatively cheap and easy to use. However, sensitivity of Tox A/B EIAs is suboptimal, also strongly dependent on storage time and storage temperature. Compared to CCNA, pooled sensitivity of Tox A/B EIA was 83%. In comparison to toxigenic culture, pooled sensitivity of Tox A/B EIA was as low as 57%. Pooled specificity of Tox A/B EIAs was however reported to be as high as 99%, both compared to CCNA and TC (Crobach et al. 2016).

In 2018, a new highly sensitive 'single-molecule array assays (SIMOA)' for detection of *C. difficile* toxin in stool samples was assessed relative to positive glutamate dehydrogenase (GDH) screen and cell cytotoxicity neutralizing assay (CCNA) (Banz et al. 2018). The SIMOA toxin A and toxin B assays showed very low limits of detection of 0.6 and 2.9 pg/ml, respectively, and detected toxins in 24% more samples than the high-performing toxin EIA. However, another study performed in 2019 concluded that the assay was not specific enough to diagnose CDI and did not differentiate an individual with CDI from one with asymptomatic carriage (Pollock et al. 2019). Interestingly, when the test was only used for patients with proven CDI, stool concentration correlated with severe baseline disease, severe CDI-attributable outcomes and recurrence (Alonso et al. 2022; Sandora et al. 2023). Though test performance may improve by setting a new threshold for a positive test result, new studies are lacking and the test has not been marketed.

GDH EIAs are relatively easy to perform and cheap. They detect glutamate dehydrogenase, an enzyme that is produced by both toxigenic and non-toxigenic *C. difficile* strains. GDH is a metabolic enzyme that converts glutamate to α-ketoglutarate and is commonly presents in many eukaryotes and microbes including *C. difficile* and other *Clostridium* species. GDH EIAs are sensitive (pooled sensitivity compared to CCNA and TC 94% and 96%, respectively) (Crobach et al. 2016). However, they cannot make a distinction between the presence of toxigenic or non-toxigenic strains and are thus less specific to detect true disease. This was demonstrated by a specificity of only 90% in comparison to CCNA (Crobach et al. 2016). Rapid GDH assays can be used to rule out CCDI, with subsequent reduction in patient isolation time (Doolan et al. 2023; Vogelzang et al. 2020); however, other infectious causes of diarrhoea may also require isolation.

NAATs include PCR assays, helicase-dependent amplification assays and loop-mediated isothermal amplification assays. Most of these assays target conserved regions within the gene for toxin B (*tcdB*), although some target a highly conserved sequence of the toxin A gene (*tcdA*). Assays that detect the ribotype 027/NAP1 strain (and related ribotypes) are also available; these detect the genes for binary toxin (*cdt*) and the deletion at nucleotide 117 on the regulatory gene *tcdC*.

NAATs are sensitive (sensitivity compared to CCNA and TC 96% and 95%, respectively) (Crobach et al. 2016). As they only detect toxigenic strains instead of all *C. difficile*, they are more specific than GDH EIA (specificity compared to CCNA and TC 94% and 98%, respectively) (Crobach et al. 2016). However, NAATs only detect the presence of toxin genes and hence the toxin producing capacity of *C. difficile*. Therefore, a major drawback of NAAT is that in addition to CDI cases, it will also detect asymptomatic carriers of toxigenic *C. difficile* (Crobach et al. 2018b). In the past years, there has been a focus on the predictive value of the PCR-cycle threshold. Low-cycle threshold correlates with the presence of free toxin (Crobach et al. 2018a; Davies et al. 2018; Senchyna et al. 2017). PCR-cycle threshold has also been proposed to correlate with clinical course of patients with CDI, i.e. lower threshold may be associated with severe disease and poor outcome (Davies et al. 2018; Jazmati et al. 2016; Reigadas et al. 2016). It is important to realize that thresholds may differ per PCR platform and test (Doolan et al. 2021). Reporting PCR results with an interpretation (or predicted presence of free toxin) based on cycle threshold may help clinicians to identify patients at higher risk for CDI-related complications and reduce overtreatment of CDI (Hitchcock et al. 2019). So there is a role for PCR cycle to help establish the diagnosis of CDI, though the relative low sensitivity and specificity for prediction of free toxin status remain problematic and ROC-AUC for the prediction of mortality is insufficient (0.568) (Crobach et al. 2018a; Davies et al. 2018). Therefore, when using NAAT only, cycle threshold may help to establish the diagnosis of CDI, but clinical judgment remains essential (Doolan et al. 2021).

Since 2019, various commercially available NAATs have been developed and implemented in routine diagnostics of CDI, often without sufficient information on targets and included primer sets. Since some *C. difficile* belong to 'cryptic clades' and contain divergent pathogenicity locus (PaLoc) sequences, diagnostic tests may fail, as demonstrated in a patient with severe pseudomembranous colitis and negative Cepheid Xpert *C. difficile* BT (XCBT) assay due to *C. difficile* PCR ribotype 151 (cryptic clade C-2) (Ducarmon et al. 2022). Therefore, vigilance towards *C. difficile* infection as a result of cryptic clade isolates and regular critical evaluation of NAAT testing results is warranted.

4 Recommended Testing Algorithms

Although it would be the easiest to use one of the rapid assays for CDI detection in daily practice, this will falsely impact CDI detection rates. First of all, GDH EIA and NAAT results do not directly correlate with clinical symptoms possibly leading to overdiagnosis of CDI. Second, all of these three tests, even the very specific Tox A/B EIAs, are not specific enough to be used as a stand-alone test (Crobach et al. 2016). Namely, most of the samples submitted for CDI testing will not have the disease. Assuming a CDI prevalence rate of 5% among submitted samples, positive predictive values of the most specific assays (Tox A/B EIA) range from 69 to 81%, indicating that 19–31% of samples with a positive test result do not have the disease (Crobach et al. 2016).

In light of these limitations of the rapid assays, common guidelines for CDI diagnosis put forth by the European Society of Clinical Microbiology and Infectious Diseases (ESCMID) and the Society for Healthcare Epidemiology of America/the Infectious Diseases Society of America (SHEA/IDSA) recommend the use of multistep algorithmic testing to maximize diagnostic accuracy (Fig. 1) (Crobach et al. 2016; McDonald et al. 2018). The premise of this strategy is sequential testing that most efficiently uses molecular tests' different strengths. First, stool samples are screened by a sensitive test. According to both guidelines, this could either be either be GDH EIA or NAAT (Crobach et al. 2016; McDonald et al. 2018). The high sensitivity of these tests provides them a high negative predictive value

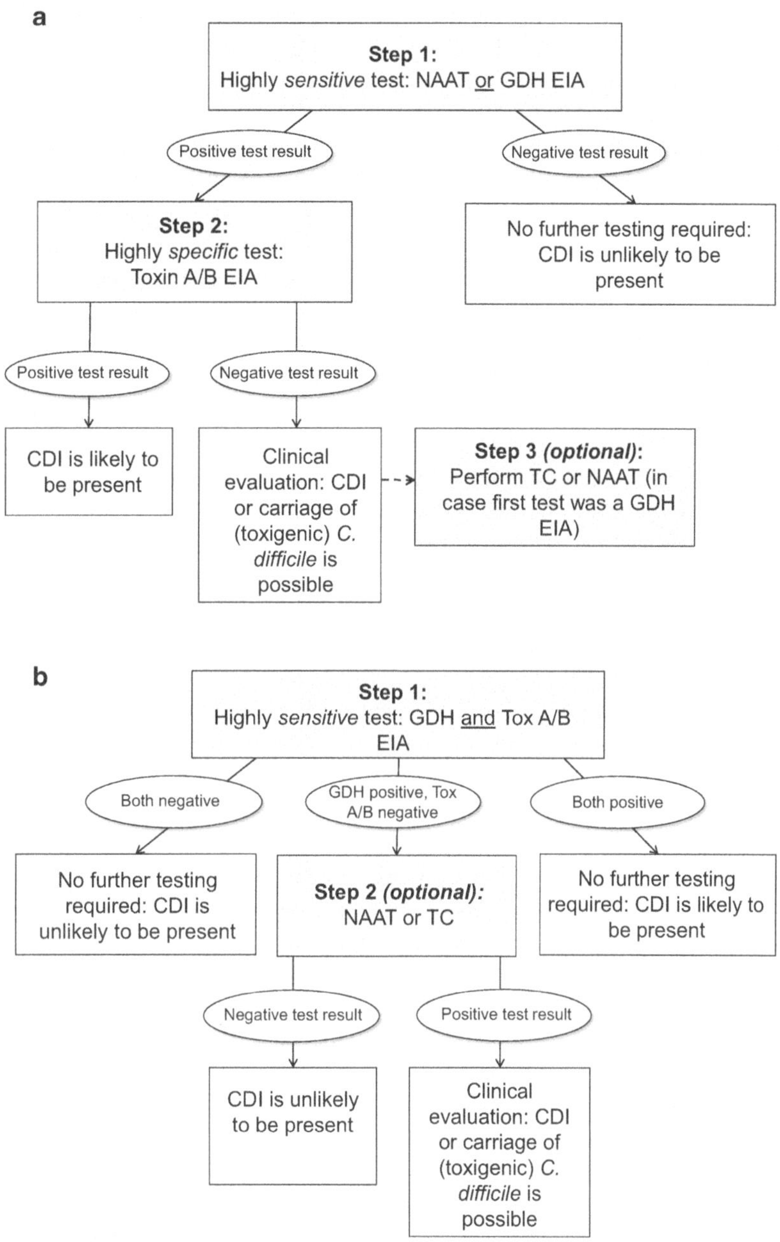

Fig. 1 Algorithms for CDI testing as recommended by ESCMID guidelines. (**a**) GDH or NAAT-Tox A/B algorithm, (**b**) GDH and Tox A/B-NAAT/TC algorithm. CDI *Clostridium difficile* infection, *GDH* glutamate dehydrogenase, *NAAT* nucleic acid amplification test, *TC* toxigenic culture, *Tox A/B* toxin A/B, *EIA* enzyme immunoassay. Figure reprinted from Crobach et al., CMI 2016;22:S63, https://doi.org/10.1016/j.cmi.2016.03.010, available under a Creative Commons Attribution-NonCommercial-NoDerivates License (CC BY NC ND), https://creativecommons.org/licenses/by-nc-nd/3.0/legalcode

(NPV) with which to be reasonably confident that a negative test is in fact indicative of no CDI. In this manner, a large proportion of diarrhoeal cases can be quickly ruled out for CDI. If the first test is positive, reflex testing occurs by Tox A/B EIA (Crobach et al. 2016; McDonald et al. 2018), a test of high specificity with a correspondingly high positive predictive value (PPV) as it is now used in selected samples with a higher pre-test probability of CDI. Thus, a positive result on this second test is likely indicative of CDI. In the event of a positive first test and a negative second, the result is considered an ambiguous one in need of resolution by clinical evaluation or further testing, e.g. via TC. In the ESCMID guidelines, an alternative algorithm starting with both GDH and Tox A/B EIA in the first step, optionally followed by TC or NAAT in the case of ambiguous results, is mentioned as a suitable equivalent (Crobach et al. 2016). The 2021 update of the American College of Gastroenterology guideline now has adopted the ESCMID recommendation for a two-step testing algorithm (Kelly et al. 2022). The 2017 IDSA/SHEA guideline accepts NAAT testing only, provided there are pre-agreed institutional criteria for patient stool submission (McDonald et al. 2018). The 2021 update of the ESCMID treatment guidance document acknowledges in the definition of CDI that some laboratories use NAAT only but warns against overdiagnosis of CDI (van Prehn et al. 2021).

The gains in diagnostic accuracy achieved by such algorithmic testing are substantial. It was calculated that in a typical endemic setting of 5% CDI prevalence among submitted samples, PPV and NPV of the most accurate algorithm, NAAT followed by Tox A/B EIA, are 98.5 and 98.9%, respectively. In comparison, PPV and NPV of standalone NAAT are 45.7 and 99.8%, respectively; PPV and NPV of standalone Tox A/B EIA are 81.4 and 99.1%, respectively (Crobach et al. 2016).

Algorithmic testing does have its own drawback: increased turnaround time. While patients with a negative result can quickly be ruled out for CDI, actually establishing a CDI diagnosis requires two positive tests, inevitably requiring more time, especially if CCNA is used as the second test as recommended by IDSA/SHEA guidelines. This is a non-trivial drawback, as it has been shown that decreasing the time to diagnosis positively affects patient outcomes (Barbut et al. 2014). Numerous studies have found an association between low CT values and toxin presence or outcome (Chung and Lee 2017; Dionne et al. 2013; Jazmati et al. 2016; Kaltsas et al. 2012; Leslie et al. 2012; Reigadas et al. 2016). Efforts have been made to address the longer turnaround time of algorithms by examining whether quantitation of NAAT results by cycle threshold (CT), the point during a PCR when product begins being fluorescently detectable that serves as an indirect measure of the starting number of DNA copies in a sample, can be used by itself to establish a CDI diagnosis (Senchyna et al. 2017; Crobach et al. 2018a). Although studies indicate that NAAT CT values can be used to predict the toxin status, the relationship between the two is not strong enough to negate the need for toxin testing by a second test at his moment (Senchyna et al. 2017; Crobach et al. 2018a). For now, the increased turnaround time of algorithms must be accepted, as algorithms seem to represent the most accurate, clinically implementable testing strategy for CDI diagnosis. A 2018–2019 European survey showed that a two-step testing algorithm is increasingly being implemented, as the percentage of hospital sites that used an ESCMID-recommended diagnostic algorithm had increased to >80% (Viprey et al. 2023). Yet, low compliance with diagnostic testing guidelines continued to be reported in some countries.

Although TC is not an efficient method for screening large numbers of diarrhoeal samples for potential CDI, it nevertheless remains an important technique for laboratories to be able to carry out. Isolating *C. difficile* by TC serves several post-diagnostic purposes. These include antimicrobial susceptibility testing and molecular typing of isolates. For molecular typing, pulsed-field gel electrophoresis (PFGE) was considered the standard method in North America, with the resulting banding patterns described as 'North American pulse-field' (NAP) types (Killgore et al. 2008; Kristjansson et al. 1994). In Europe PCR ribotyping is most commonly applied, with the resulting patterns described as PCR ribotypes (Bidet et al. 1999; Stubbs et al. 1999). Reference laboratories in Canada and the USA have also applied PCR ribotyping, using a standardized protocol for capillary-electrophoresis PCR ribotyping (Fawley et al. 2015). While PFGE and PCR ribotyping have been adopted as the choice for surveillance purposes, additional typing methods like multilocus variable-number of tandem-repeat analysis (MLVA), multilocus sequence typing (MLST) and whole-genome sequencing (WGS) are of use for outbreak investigations (Knetsch et al. 2013; Maiden et al. 1998; van den Berg et al. 2007). Core genome MLST is a WGS-based approached that has the potential to be a future alternative to ribotyping for surveillance purposes (Janezic and Rupnik 2019; Baktash et al. 2022). Furthermore, TC may be needed to resolve discrepant results of algorithmic testing where *C. difficile* is detected by GDH EIA or NAAT but toxin is not. A positive TC result rules out a false-positive GDH EIA/NAAT result in these patients. In that case, clinical evaluation is needed; these patients can either be CDI patients with a false-negative Tox A/B EIA result due to low toxin levels or degradation of toxins, or *C. difficile* carriers.

5 Selection of Stool Samples

Testing for CDI should only be performed on unformed stools as the presence of clinical symptoms is a prerequisite to diagnose CDI (Cohen et al. 2010; Crobach et al. 2016; Surawicz et al. 2013). However, it can be difficult to assess which unformed stools should be tested. A large study in 482 hospitals across 20 European countries showed that 23% of samples positive for CDI were not diagnosed by the local hospital because of an absence of clinical suspicion (Davies et al. 2014). It was reported that mostly younger patients and patients who are not hospitalized or have been hospitalized for <3 days are inadvertently not tested for CDI (Alcala et al. 2012; Davies et al. 2014). In general practice, CDI is also often missed due to lack of suspicion, as was shown in a study among 12,714 unformed stool samples (Hensgens et al. 2014). In this study, general practitioners requested CDI testing in 7% of unformed stool samples, thereby detecting only 40% of all CDI cases (Hensgens et al. 2014). In light of these problems, testing of all submitted unformed stool samples is now endorsed by the ESCMID guidelines (Crobach et al. 2016). This approach has been shown to increase the diagnostic yield (Davies et al. 2014; Reigadas et al. 2015). Restricting CDI testing to liquid samples instead of all unformed samples seems to be too stringent and may cause the diagnosis of CDI to be missed (Berrington and Settle 2007).

A special situation exists for patients with ileus due to CDI. In this case, formed stools or rectal swabs can be tested for CDI (McFarland et al. 1987; Rogers et al. 2013). Although perirectal swabs have also been proposed as suitable alternatives, their use may depend on the presence of faecal staining on the swab (Kundrapu et al. 2012; Rogers et al. 2013).

Asymptomatic *C. difficile* intestinal carriage occurs most commonly in neonates and toddlers (Ferraris et al. 2019). At 1 month of age, the carriage rate is on average 37%, declining to 10% at >1 year, compared with 1–3% in adults (Jangi and Lamont 2010). In these pooled data (n = 928), 13% were carriers of toxigenic and 17% of nontoxigenic *C. difficile* strains. As such, routine testing for *C. difficile* is not recommended in this group (McDonald et al. 2018). Here, colonization frequently occurs without clinical symptoms of diarrhoea, even if the faeces contains detectable levels of *C. difficile* toxins, a criterion recommended to define CDI in adults (Crobach et al. 2016). On the other hand, the incidence of CDI among hospitalized children has been increasing (Schutze and Willoughby 2013). CDI testing is therefore burdensome in young children and should always include clinical evaluation. Routine testing for CDI in children <1 year should be avoided, according to guidelines launched by the American Academy of Pediatrics (Schutze and Willoughby 2013). For children between 1 and 3 years of age with diarrhoea (IDSA/SHEA between 1 and 2 years), CDI testing can be considered, but testing for other causes, particularly viral infections, is recommended first (McDonald et al. 2018; Schutze and Willoughby 2013). For children above 3 years of age, normal testing procedures can be followed (Crobach et al. 2016; Schutze and Willoughby 2013).

There is still a paucity of data on community-based CDI and its recognition. A European point prevalence study performed in 2018 showed CDI positivity rate was 4.4% in hospital samples and 1.3% in community samples (Viprey et al. 2022). Half of community CDI cases were undiagnosed because of the absence of clinical suspicion. This conclusion is in line with studies conducted in general practice populations in France and the Netherlands, in which 48 and 60% of cases would have been missed based on the requested test by the general practitioner (Barbut et al. 2019; Hensgens et al. 2014). These data illustrate the need for improved awareness for diagnosing CDI in patients presenting with diarrhoea in the community.

6 Repeat Testing

Before the introduction of algorithms, lack of confidence in Tox A/B EIAs led to the submission of multiple stool samples during one diarrhoeal episode. There was a common misconception that three sequential tests were necessary to 'rule out' CDI. This policy resulted in a decrease of the positive predictive value of each subsequent test and subsequently an increase in the likelihood of false-positive results. Several studies sought to determine the yield of such repeat testing. Diagnostic yield can either be expressed in the percentage of first test negative samples converting to positive in a repeat test, or the percentage of positive samples that is detected by repeat testing. After a first negative Tox A/B EIA result, it was reported that 0.9–2.5% of samples test positive in a repeat sample submitted within 7 days (Aichinger et al. 2008; van Prehn et al. 2015). These samples constitute around 9% of all positive samples (Aichinger et al. 2008; van Prehn et al. 2015). Although the former studies were performed in endemic situations, a study performed during an outbreak situation demonstrated that there was a definite diagnostic yield of retesting in such a situation; of all samples submitted for repeat Tox A/B EIA testing, 8.2% tested positive. These samples constituted 5% of all positive CDI samples (Debast et al. 2008).

The utility of repeat NAAT testing has been evaluated in several studies, too. The percentages of samples that were positive within 7 days after a negative test range from 0.9 to 2.9% (Aichinger et al. 2008; Green et al. 2014; Khanna et al. 2012; Luo and Banaei 2010; van Prehn et al. 2015). The number of CDI cases detected by a repeat test range from 1.7 to 4.5% (Aichinger et al. 2008; van Prehn et al. 2015). The chance of turning positive was lower in the first 7 days after a negative NAAT result than in the 7–14 days period after the negative test result (Khanna et al. 2012; Luo and Banaei 2010). In one study, a history of CDI seemed to increase the risk of a positive repeat NAAT result within 7 days after a first negative test (Green et al. 2014).

The general consensus is that in a non-epidemic situation, the diagnostic yield of repeat testing by both Tox A/B EIA and NAAT is too low, and therefore, repeat testing within 7 days should be discouraged (Cohen et al. 2010; Crobach et al. 2016; Surawicz et al. 2013). If an algorithm is used instead of stand-alone NAAT or Tox A/B EIA, the even higher predictive values make repeat testing redundant. However, in epidemic situations, or in patients with very high clinical suspicion, repeat testing may be of value (Crobach et al. 2016). This conclusion is also included in the updated IDSA/SHEA guideline (2017), but is not supported by a meta-analysis published in 2019 (Kraft et al. 2019). This meta-analysis concluded that there was insufficient evidence to recommend against repeat testing of the sample using NAAT after an initial negative result due to a lack of evidence of harm, which is a remarkable approach. The ASM recommendation that patients suspected of having CDI, NAAT-only testing is a recommended practice for detection of the C. *difficile toxin* gene, is also deviant from the IDSA/SHEA and ESCMID guideline but is understandable since the ASM-supported systematic review analysed NAAT algorithm and not NAAT followed by toxin testing.

How to implement this algorithm in daily routine? A computerized clinical decision support (CDS) tool incorporated in the electronic medical record system has been tested in a 1250-bed tertiary care hospital in St. Louis, Missouri (Kwon et al. 2019). A hard-stop intervention limited repeat *C. difficile* toxin enzyme immunoassay testing within 96 h of a previous negative test. The testing rate and number of admissions with repeat tests decreased significantly post-intervention ($p < 0.01$ for both), whereas the percentage of positive tests was unchanged.

Sometimes, repeat samples are taken after CDI treatment as a test of cure. However, after resolution of diarrhoea, patients can still test positive for toxins (Wenisch et al. 1996). Furthermore, patients can become asymptomatic carriers after treatment for CDI: one small study showed that 1–4 weeks after treatment, 29/56 (56%) of patients were found to be asymptomatic carriers of *C. difficile* (Sethi et al. 2010). Testing for cure is therefore not recommended in current guidelines (Cohen et al. 2010; Crobach et al. 2016; Schutze and Willoughby 2013).

7 Consequences of Testing Strategy on CDI Incidence/Reporting Rates

Despite the common recommendations of ESCMID and SHEA/IDSA advocating the use of algorithmic testing in CDI diagnosis, testing methods between hospitals vary widely. A large study across 60 European hospitals found that only 64% of hospitals use a recommended testing algorithm for CDI testing (K. Davies et al. 2016). However, ESCMID-recommended CDI testing methodologies were used by 82% (86/105) of hospital in a survey performed in hospital sites of 11 countries in 2018–2019, indicating a better implementation (Viprey et al. 2023). Almost all hospital sites across Europe (95%) reported using more than one test to diagnose CDI demonstrating that the 2016 ESCMID guidance on not relying on a single assay was being followed (Crobach et al. 2016). The 2017 update of Clinical Practice Guidelines for *Clostridium difficile* Infection in Adults and Children by IDSA and SHEA also recommends a multistep algorithm but left more room to use a NAAT alone (McDonald et al. 2018). An interesting study performed a retrospective analysis of positive *C. difficile* cases over 2 years spanning a year preceding and following transition from PCR to two-step testing (Dbeibo et al. 2023). A reduction of CDI-specific antibiotic use was found without restricting clinician diagnostic ordering, indicating that the two-step algorithm has also important implications for antibiotic stewardship. In the study across 60 European hospitals, a 2.5-fold higher CDI positivity rate was demonstrated when stand-alone or GDH/NAAT were used instead of a recommended algorithm. This was reflected in the subsequent incidence rates; hospitals relying on NAAT or GDH/NAAT reported a mean incidence rate of 5.2 per 10,000 patient-days, while hospitals relying on an algorithm reported a lower mean incidence rate of 2.0 per 10,000 patient-days, despite similar testing frequencies (Davies et al. 2016). Interestingly, a European study performed in 2018–2019 showed that countries reporting the highest incidence of CDI used non-ESCMID-recommended single test to diagnose CDI (Viprey et al. 2023).

These observations hold true when the same samples are concomitantly tested with both stand-alone NAAT and an algorithm. In 1 study of 1321 stool samples, the CDI positivity rate by NAAT was 6.4%, while the CDI positivity rate by a GDH and Tox A/B EIA—CCNA algorithm on the same samples—was 4.2%. The overall incidence rates were 8.9 and 5.8 per 10,000 patient-days for stand-alone NAAT and the algorithm, respectively (Longtin et al. 2013). When stand-alone NAAT was compared to stand-alone Tox A/B EIA, higher CDI positivity rates and higher CDI incidence rates for NAAT compared to Tox A/B EIA were reported, too (Grein et al. 2014). Even so, hospitals that switch from non-molecular tests to stand-alone NAAT testing are reported to experience an increase in their CDI incidence rates (Moehring et al. 2013).

The implications of testing method-dependent CDI incidence rates are consequential. Besides the obvious effect of interfering with attempts to accurately monitor CDI for surveillance purposes, financially tangible effects also result. For instance, UK hospitals can be assessed financial penalties for excessive numbers of hospital-acquired CDI cases (Davies et al. 2016). Similarly, in the USA, the Centres for Medicare and Medicaid Services (CMS) value-based purchasing program are affected by reported incidence rates (Marra et al. 2017). In the latter's case, an attempt to normalize rates by factoring in testing method has been made, although the study demonstrated the inadequacy of such normalization and stressed the need for refinement.

In the European COMBACTE-CDI study, all diarrhoeal stool samples sent to the laboratories were tested for CDI, irrespective of the tests requested. A model was constructed to predict the incidence in participating countries, based on national-reported incidence rates and sampling and testing rates observed in the study (Agnew et al. 2023). Antimicrobial usage rates, national sampling and testing rates and community prevalence of CDI influenced CDI incidence. Notably, countries with the smallest difference between known and true incidences are the countries with the highest levels of sampling and testing.

In conclusion, CDI incidence is clearly affected by testing method. Given the heterogeneity of such methods between institutions, and the importance of correctly ascertaining CDI incidence, it is necessary to somehow normalize incidence rates in a way that takes into consideration testing method.

8 Non-microbiological Diagnostic Tests and Procedures

The diagnosis of CDI is established by a combination of clinical findings in combination with positive microbiological evidence (van Prehn et al. 2021). The following diagnostic modalities can support the diagnosis of CDI.

8.1 CT Imaging

CT imaging can be useful in diagnosing fulminant CDI and pseudomembranous colitis (PMC). Several features are suggestive of advanced PMC such as colonic-wall thickening, pericolonic stranding, the accordion sign, the double-halo sign and ascites (Kirkpatrick and Greenberg 2001; Bartlett and Gerding 2008). The radiography is usually normal in the absence of ileus or toxic megacolon. Kirkpatrick et al. evaluated whether diagnosis of *C. difficile* colitis could be made with CT. They included 110 patients of which 54 had a positive stool assay and 56 patients a negative stool assay. The sensitivity at their centre was 52% and the specificity 93%, and the positive and negative predictive values were, respectively, 88 and 67%. CT imaging is less sensitive when compared with NAAT or stool toxin tests but can be useful when there is a need for quick results (Bartlett and Gerding 2008).

8.2 Endoscopy

Nearly all cases of PMC are caused by CDI (Tang et al. 2016), though other causes are sometimes found, such as chemotherapy, toxin producing *Staphylococcus aureus* and cytomegalovirus infection (Sundar and Chan 2003; Pressly et al. 2016). PMC is not very common and not all CDI will develop PMC (Bartlett 2002). Therefore, endoscopy is a relatively insensitive procedure. Furthermore, in one-third of the patients, PMC is missed by sigmoidoscopy because of involvement of the right colon; making colonoscopy the preferred endoscopic procedure. Endoscopy is an invasive procedure with perforation risks and is often expensive (Bartlett 2002; Bartlett and Gerding 2008). Endoscopy is therefore not recommended to diagnose CDI, though it may be used to establish an alternative diagnosis.

8.3 Histopathology

CDI is more likely when pseudomembranes are detected histologically. Pseudomembrane lesions are microscopically visualized as ‘mushroom’ like and consist of pus, mucin and fibrin. Their reported sensitivity is 44% and the specificity is 89% (Wang et al. 2013). Biopsy is not needed for the diagnosis of CDI. However, histologic findings of pseudomembranes may suggest CDI and should stimulate stool testing.

9 Alternative Testing Strategies

Methods for diagnosing CDI remain imperfect, which naturally has spurned an interest in alternative testing strategies. Alternative testing strategies cannot only possibly aid in the diagnosis of CDI but might also be able to predict severity or prognosis of CDI. These testing methods include faecal biomarkers, immune modulating interleukins and cytokines and more recently intestinal microbiota analysis. Their role is discussed below.

9.1 Calprotectin

Calprotectin, a calcium- and zinc-binding protein, is found predominantly in the cytosol of neutrophils (Popiel et al. 2015; Usacheva et al. 2016; Whitehead et al. 2014). In vitro studies have shown that it has bacteriostatic and fungostatic properties (Peretz et al. 2016). It is a marker of inflammation due to release into the gut lumen by neutrophils during infiltration and can be measured in stool (Popiel et al. 2015). However, infection cannot be differentiated from inflammation by this marker, since both give a rise in faecal calprotectin (FCP) levels (Usacheva et al. 2016). The role of calprotectin in evaluating disease severity has been well studied in IBD (Vrabie and Kane 2014). Several studies evaluated the role of FCP in CDI testing (Table 2). First, the usefulness of FCP testing to diagnose CDI was evaluated in several studies. In most studies, median FCP levels were found to be significantly higher in CDI patients than in diarrhoeal patients who tested negative for CDI and in non-diarrhoeal controls (Barbut et al. 2017; Darkoh et al. 2014; Kim et al. 2017; Popiel et al. 2015; Swale et al. 2014; Whitehead et al. 2014). A study in cancer patients also found higher FCP values in toxin-positive samples compared to toxin-negative samples (He et al. 2018). Yet, faecal calprotectin did not had higher concentrations in CDI cases compared with asymptomatic carriers in a more recent study (Villafuerte Galvez et al. 2023).

Studies that calculated optimal FCP cut-off points for distinguishing CDI from non-CDI samples reported sensitivities ranging from 77 to 88% and specificities ranging from 75 to 79% (Kim et al. 2017; Popiel et al. 2015; Swale et al. 2014; Whitehead et al. 2014). However, in two of these studies, the discriminative power of FCP might have been attenuated as the group of CDI patients might have included CD carriers due to testing for CDI by NAAT only (Kim et al. 2017; Popiel et al. 2015). On the other hand, the use of healthy controls instead of patients suspected of CDI might have falsely increased the specificity in one study (Kim et al. 2017). Overall, the suboptimal sensitivity and specificity demonstrated in these observational studies, of which several with limitations or small sample sizes, does not provide enough evidence for the use of FCP to detect CDI.

Interestingly, besides the expected suboptimal specificity of FCP, sensitivity is also moderate. One study reported that in 20% of CDI patients, FCP levels were lower than in hospitalised patients without diarrhoea (Darkoh et al. 2014). Another study reported that from 120 CDI subjects, only 5 had normal FCP levels (<50 μg/g) and speculated that these cases might represent mild disease (Whitehead et al. 2014). The correlation between FCP levels and CDI severity has also been evaluated, but results are conflicting (He et al. 2018; Kim et al. 2017; Peretz et al. 2016; Swale et al. 2014). A correlation between CDI due to ribotype 027 and FCP levels was also evaluated in two studies (Peretz et al. 2016; Swale et al. 2014). Significantly higher FCP levels compared to non-027 CDI were found in one small study comprising seven 027 cases and 22 non-027 cases (Peretz et al. 2016); the same trend was shown in a somewhat larger study, but results were not significant (Swale et al. 2014). In conclusion, there is also insufficient evidence for the use of FCP levels to predict severity or presence of ribotype 027.

Table 2 Overview of relevant studies evaluating the role of FCP in patients with CDI. *AAD* antibiotic-associated diarrhoea, *CCNA* cell cytotoxicity neutralization assay, *CDI Clostridium difficile* infection, *FCP* faecal calprotectin, *NAAT* nucleic acid amplification test

Study	Type of study	Detection of CDI	Number of cases/controls	Results
Kim et al., Ann Lab Med 2017	Retrosp. cohort study	NAAT for toxin gene	30 severe CDI, (group 1), 50 mild CDI (group 2) and 71 negative CDI healthy controls (group 3)	*CDI diagnosis* Median levels of FCP were significantly higher in group 1 than in group 2 and group 3, 1391.5 μg/g (170.0–2088.1 μg/g) vs 188.2 μg/g (41.4–188.2 μg/g) and 35.0 μg/g (10.7–108.9 μg/g), respectively Optimal cut-off value for CDI diagnosis 112.5 μg/g ROC curve AUC 0.821 Sens 75% and spec. 79% *CDI severity* Median levels of FCP were significantly higher in group 1 than in group 2, 1391.5 μg/g (173.5–2075.9 μg/g) vs 188.2 μg/g (41.4–591.6 μg/g), respectively Optimal cut-off value for differentiating mild from severe CDI 729.8 μg/g ROC curve AUC 0.746 Sens 70% and spec. 80%
Peretz et al., BMC Infect Dis 2016	Retrosp. cohort study	NAAT for toxin gene and identification 027 strains	29 pts with CDI 7 CDI ribotype 027, 22 other ribotype	Overall mean levels of FCP 331.4 μg/g (21–932 μg/g) Mean levels of FCP were significantly higher in 027 positive group than in 027 negative group, 331.4 μg/g (21–932 μg/g) vs 249 μg/g (155–498 μg/g), respectively A trend was found between higher FCP levels and higher *Clostridium* severity score
Popiel et al., JCM 2015	Prosp. exploratory observ. study	NAAT for toxin gene	44 CD-PCR positive vs 20 CD-PCR negative	Median levels of higher-range assay of FCP (assay range, 100–1800 μg/g) were significantly higher in CD-PCR+ than in CD-PCR- 983 μg/g (351 to >1800 μg/g) vs <100 μg/g (<100–194 μg/g) and also in the lower-range assay of FCP (assay range, 30–300 μg/g) >300 μg/g (>300 to >300 μg/g) vs 77.5 μg/g (30–238 μg/g) Optimal cut-off value 135 μg/g High-range FCP ROC curve AUC 0.82 Sens. 88.6% and spec. 75%
Whitehead et al., J Med Microbiol 2014	Prosp. cohort study	Phase 1: Toxin EIA ($N = 75$) Phase 2: GDH EIA + NAAT for toxin gene ($N = 45$) Change of departmental *C. difficile* testing methodology during evaluation	Group 1: 75 cases in toxin EIA positive Group 2: 45 cases in GDI NAAT positive Group 3: 99 cases in CDI negative	Median levels of FCP were significantly higher in group 1 than in group 2, 336 μg/g (208–536 μg/g) vs 249 μg/g (155–498 μg/g), respectively Both were significantly higher than in group 3, 106 (46–176 μg/g) Optimal cut-off value 176 μg/g and 169 μg/g, ROC curve AUC 0.84 and 0.80 Sens 81% and 73%, spec. 77% and 77% for group 1 and 2, respectively
Swale et al., PLOS One 2014	Prosp. cohort study	NAAT for toxin gene and toxin EIA	164 CDI cases vs 52 AAD controls	*CDI diagnosis* Median levels of FCP were significantly higher in CDI cases vs AAD, 684.8 μg/g (203.7–1581.0 μg/g) vs 66.5 μg/g (23.1–145.7 μg/g), respectively Optimal cut-off value 148 μg/g ROC curve AUC 0.864

(continued)

Table 2 (continued)

Study	Type of study	Detection of CDI	Number of cases/controls	Results
				Sens 81.8% and spec. 76.9% PPV 91.5%, NPV 57.4%
			8 severe CDI cases vs 116 non-severe CDI cases *C. difficile* isolates recovered from 149 CDI cases 72 cases with ribotype 027 vs 77 non-ribotype 027	Sub-group analyses *CDI severity* Median levels of FCP were not significantly higher in severe CDI cases vs non-severe CDI cases, 969.3 μg/g vs 512.7 μg/g), respectively *Ribotype 027* Median levels of FCP were not significantly higher in ribotype 027 cases vs non-ribotype 027 cases, 1011 μg/g vs 658 μg/g), respectively
Darkoh et al., Clin Vaccine Immunol 2014	Prosp. cohort study	AAD stools: CCNA, NAAT for toxin genes and toxin EIAcontrol stools: NAAT for toxin gene and toxin EIA	CDI-positive stools ($N = 50$), CDI-negative stools ($N = 50$), hospitalized patients without diarrhoea ($N = 45$)	FCP concentration in CDI positive stools, 18 μg/g (2.8–70.2 μg/g) was threefold higher than in CDI-negative stools, 6.5 μg/g (2.0–31.0 μg/g), and twofold higher than of hospitalized pts without diarrhoea, 8.7 μg/g (1.8–33.2 μg/g) FCP levels of 80% of the CDI-positive stools and 30% of the CDI-negative stools higher than hospitalized pts without diarrhoea
Barbut et al., Eur J Clin Microbiol 2017	Prosp cohort study	Stool cytotoxicity and/or toxigenic culture	135 CDI cases and 135 controls	The FCP values were 218.0 μg/g and 111.5 μg/g, respectively Among patients with CDI, faecal calprotectin levels were higher in those with free toxins in their stools (274.0 vs 166.0 μg/g, $p = 0.051$), respectively
He et al., Eur J Clin Microbiol Infect Dis 2018	Cohort study in cancer patients	NAAT for toxin gene (GeneXpert)	117 PCR+ samples and 115 PCR- samples	Median FCP 183.6 μg/g and 145.6 μg/g, respectively, $p = 0.006$
		GDH/toxin EIA	24 toxin positive and 86 toxin negative	toxin positive 200.2 μg/g vs. 182.8 μg/g, toxin negative. $p = 0.044$
			95 mild/ moderate CDI and 22 severe/ severe complicated CDI	182.1 μg/g and 218.5 μg/g, respectively $p = 0.014$
Villafuerte Galvez et al., Clin Infect Dis 2023	Prosp. cohort study	NAAT for toxin gene	43 CDI 42 asymptomatic carriers	Non-significant difference FCP values ($p = 0.36$) in CDI versus asymptomatic cases: Median FCP 185 μg/g (IQR 60–851.5) vs 168.4 μg/g (75.3–406.8)
			26 non-CDI diarrhoea	75.8 μg/g (22.8–169.5)
			28 hospital controls, no diarrhoea, no colonization	171 μg/g (75.5–400.5)

9.2 Lactoferrin

Lactoferrin is a glycoprotein and resides in neutrophils. It is released upon neutrophil activation. The faecal lactoferrin (FL) levels can be measured in stool and correlate with the number of infiltrated neutrophils. Multiple studies have proven that it can be an accurate marker of intestinal inflammation and useful in diagnosis of inflammatory diarrhoea (Usacheva et al. 2016).

The usefulness of FL to detect CDI was evaluated in a handful studies (Table 3). All studies report higher median FL levels in CDI samples than in control samples (either diarrhoeal samples without CDI or non-diarrhoeal samples) (Barbut et al. 2017; Boone et al. 2014; Darkoh et al. 2014; LaSala et al. 2013; Swale et al. 2014). However, a substantial proportion of CDI-negative patients have elevated FL levels, too (Boone et al. 2014; Darkoh et al. 2014). This was also reflected in the suboptimal specificity of 77% that was found when an optimal cut-off point to distinguish CDI from patients with non-CDI antibiotic-associated diarrhoea was determined (Swale et al. 2014). In a small retrospective study, 59 patients with positive PCR and diarrhoea were compared with a group of 59 PCR positive patients without diarrhoea; lactoferrin was not capable to classify patients with or without diarrhoea (Anikst et al. 2016). Interestingly, also no differences were found between organism burden and toxin concentrations, questioning whether patients with CDI were appropriate diagnosed.

Whether FL could be used as a marker for severe CDI was also evaluated in some studies. Severe CDI was found to be associated with higher median FL levels in two small studies (Archbald-Pannone 2014; Boone et al. 2013). In addition, higher FL levels were associated with a higher white blood cell count and decreased serum albumin (Boone et al. 2013), but no association with mortality was demonstrated (Archbald-Pannone 2014), possibly due to small cohorts. Furthermore, it was demonstrated that patients with CDI due to ribotype 27 and positive stool toxin had significantly higher FL levels and WBC counts than non-027 CDI patients (Boone et al. 2013, 2014). In patients with CDI due to ribotype 027, patients with positive stool toxin and elevated FL had a higher mortality risk (Boone et al. 2014).

To conclude, all of the studies report an association between elevated FL and CDI. However, the reported specificity is insufficient for implementing it in the diagnosis of CDI. Furthermore, as the studies report different median FL levels, this would reduce predictive accuracy. Some parts may be ascribed to variation in disease severity, while other parts are due to laboratory handling and the volume of diluent. Another problem is that FL can be elevated due to comorbidities, such as ulcerative colitis and Crohn's disease (Shi et al. 2022).

Some studies also report an association between elevated FL and CDI severity (Archbald-Pannone 2014; Boone et al. 2013). However these studies had small sample sizes. To our knowledge, there are no studies that observed that FL on its own is a predictor of severity or mortality. Therefore, more research is needed to understand the role of LF.

9.3 Faecal Leukocyte Test

The faecal leukocyte test is performed on stool specimens, which are smeared on slides and Wright stained. The test takes approximately 1 h and samples are positive when >1 WBC/highfield are observed (Reddymasu et al. 2006). However, in a study evaluating 263 stool samples from patients suspected of CDI for the diagnosis of CDI, the faecal leukocyte test showed a sensitivity and specificity of 30 and 74.9%, respectively, compared to toxin EIA (Reddymasu et al. 2006). A larger study ($n = 797$ stool samples) reported a sensitivity and specificity of 14 and 90%, respectively (Savola et al. 2001). Thus, faecal leukocyte testing is not a good test for CDI and a poor predictor of the toxin assay result.

Table 3 Overview of relevant studies evaluating the role of FL in patients with CDI. *AAD* antibiotic-associated diarrhoea, *CCNA* cell cytotoxicity neutralization assay, *CDI Clostridium difficile* infection, *EIA* enzyme immunoassay, *FL* faecal lactoferrin, *GDH* glutamate dehydrogenase, *NAAT* nucleic acid amplification test, *TC* toxigenic culture

Study	Study type	Detection of CDI	Number of cases/controls	Results
Darkoh et al., Clin Vaccine Immunol 2014	Prosp. cohort study	AAD stools: CCNA, NAAT and toxin EIA Control stools: NAAT and toxin EIA	CDI-positive stools ($N = 50$), CDI-negative stools ($N = 50$), hospitalized pts without diarrhoea ($N = 45$)	FL concentration in CDI-positive stools, 31.4 μg/g (3.0–155.2 μg/g) was significantly different and was fivefold higher than in CDI-negative stools, 6.3 μg/g (0.6–140.3 μg/g), and sixfold higher than of hospitalized pts without diarrhoea, 5.6 μg/g (0.5–35.0 μg/g) FL levels of 88% of the CDI-positive stools and 44% of the CDI-negative stools higher than hospitalized pts without diarrhoea
Swale et al., PLOS One 2014	Prosp. cohort study	Toxin EIA	164 CDI cases vs 52 AAD controls	Median levels of FL were significantly higher in CDI cases 57.9 μg/ml (11.4–177.5 μg/ml) vs AAD 2.7 μg/ml (0.7–7.8 μg/ml) Optimal cut-off value 8.06 ng/ml ROC curve AUC 0.859, Sens 81.7% Spec 76.9%, PPV 91.8%, NPV 57.1%
			8 severe CDI cases vs 116 non-severe CDI cases	Sub-group analysis CDI severity Median levels of FL were significantly higher in severe CDI cases vs non-severe CDI cases, 104.6 μg/ml vs 40.1 μg/ml), respectively
			C. difficile isolates recovered from 149 CDI cases 72 cases with ribotype 027 vs 77 non-ribotype 027	Ribotype 027 Median levels of FL were not significantly higher in ribotype 027 cases vs non-ribotype 027 cases, 83.2 μg/ml vs 51.0 μg/ml), respectively
Archbald-Pannone, J Geriatr Paliat Care 2014	Prosp. cohort study	Not described	$N = 79$ 41 severe CDI vs 38 non-severe CDI	Overall mean concentration of FL in the cohort was 388.8 μg/ml Mean levels of LF in severe CDI 580 μg/ml (SD 989.0 μg/ml) were significantly higher than in non-severe CDI 181.7 μg/ml (SD 244.2 μg/ml)
Boone et al., Eur J Clin Microbiol Infect Dis 2014	Prosp. cohort study	NAAT and TC	$N = 210$ 129 TC +&CCNA + (group 1), 62 TC +&CCNA− (group 2) and 19 TC &CCNA (group 3)	FL concentration in group 1, 90 μg/g was significantly higher than in group 2, 24 μg/g and group 3, 20 μg/g
Boone et al., Eur J Clin Microbiol	Prosp. cohort study	GDH membrane-based EIA and toxin EIA	$N = 98$ (85 toxigenic strains, 6 non-toxigenic, 6 negative for *C. difficile*, 1 mixed infection)	96% of ptn with positive toxin stool had elevated LF and 59% of pts negative stool toxin had elevated levels

(continued)

Table 3 (continued)

Study	Study type	Detection of CDI	Number of cases/controls	Results
Infect Dis 2013			85 toxigenic (21 severe CDI, 57 moderate, 7 milds)	Mean levels of severe CDI (961 μg/g (SE 303 μg/g) were significantly higher than in moderate CDI, 292 μg/g (SE 42 μg/g) and mild CDI 73 μg/g (SE 52 μg/g)
			38 ptns had a 027 infection (45%)	There is a significant difference for elevated LF between ptn with 027 and non-027
LaSala et al., J Clin Microbiol 2013	Retrosp. cohort study	GDH EIA, toxin EIA and NAAT	$N = 112$ 43 GDH negative (group 1) 14 GDH positive/toxin neg/PCR negative (group 2) 25 GDH and toxin positive (group 3) 30 GDH positive/toxin neg/PCR positive (group 4)	Median levels of LF were sign. higher in group 3, 80 μg/ml (3–124 μg/ml) than in group 1, 13 μg/ml (3–143 μg/ml), group 2, 18 μg/ml (4–78 μg/ml) and group 4, 24 μg/ml (4–160 μg/ml)
Anikst et al., Diagn Microbiol Infect Dis 2016	Retrosp. electronic chart review	NAAT and quantitative culture *of C. difficile*, stool toxin EIA	59 PCR positive patients with diarrhoea and 59 PCR positive patients without diarrhoea	Lactoferrin concentrations were significantly increased in patients with clinically significant diarrhoea (median, 99.0 vs 55.1 μg/ml, $p = 0.05$) but could not sufficiently classify patients with and without clinically significant diarrhoea
Barbut et al. Eur J Clin Microbiol 2017	Prosp. cohort study	Stool cytotoxicity and/or toxigenic culture	135 CDI cases and 135 controls	The median lactoferrin values were 26.8 μg/g and 8.0 μg/g in CDI patients and control group, respectively Among patients with CDI, faecal lactoferrin levels were higher in those with free toxins in their stools (39.2 vs 10.2 μg/g, $p = 0.003$)

9.4 Interleukins and Chemokines

Interleukins and chemokines mediate inflammatory responses, so it is no surprise that this is an active field of CDI research. Since the previous version of this diagnostic chapter, several new markers have been suggested to correlate with CDI disease and severity. We here focus on clinical studies in humans with either blood or faecal biomarkers that are of most interest. However, more biomarkers then described below have been suggested to correlate with CDI.

9.4.1 Interleukins and Chemokines to Discriminate CDI Patients

Interleukine-8 (IL-8) is a chemoattractant and recruits neutrophils to sites of infection. Activated dendritic cells and macrophages produce IL-23. This interleukin is involved in host defence against bacterial infections and the development of chronic inflammation. Darkoh and colleagues tested CDI stools, diarrhoeal non-CDI stools and non-diarrhoeal stools for interleukins both by a cytokine assay and by a quantitative EIA (Darkoh et al. 2014). Both IL-8 and IL-23 were detected in more CDI-positive stools than CDI-negative stools. The cytokine assay showed that the relative amount of IL-8 was higher in the 50 CDI-positive stools, compared to 50 CDI-negative stools. This is in contrast with IL-23 where the relative amount was higher in the CDI-negative stools. When the findings were confirmed by EIA, they found that CDI-positive stools showed a significantly higher amount of IL-8 (mean 318.2 pg/ml) in stools compared to the CDI-negative stools (mean 84.7 pg/ml) and hospitalized patients without diarrhoea (mean 79.8 ppg/ml). In contrast, IL-23 was significantly higher in CDI-negative stools and hospitalized patients without diarrhoea than in the CDI positive stools, 946.7 pg/ml (185.5–2016 pg/ml), 1617 pg/ml (489.0–6810 pg/ml) and 722 pg/ml (110.0–7069 pg/ml), respectively. This study shows that IL-8 plays a role in CDI and that increased levels are associated with more severe forms of CDI. Interestingly, Czepiel et al. found that the presence of an IL-8 genetic polymorphism was associated with severe disease in 65 CDI patients, while IL-1β (a stimulator of IL-8) polymorphism was not (Czepiel et al. 2018). In contrast, IL-23 amounts during CDI may be inadequate to sustain sufficient cellular immunity. Therefore, lower concentrations of IL-23 may show a lack of immunological response in a proportion of CDI patients and may explain also recurrence (Darkoh et al. 2014).

Wang and colleagues found that IL-27 was significantly elevated in serum and stool of 76 CDI patients as compared with 72 CDI-negative patients and 7 healthy volunteers (Wang et al. 2018). Interestingly, in their mouse model, IL-27 receptor-deficient mice had enhanced colonic histology damage, less *C. difficile* clearance and decreased survival compared to controls during CDI.

A recent cohort study conducted in the USA compared 120 CDI patients with groups of asymptomatic carriers, non-CDI (NAAT negative) diarrhoea-hospitalized patients and a (NAAT negative) control group without diarrhoea (Galvez et al. 2022). Stool markers that differed significantly between the CDI cohort and the remaining cohorts included IL-1β, IL-6, IL-8, IL-15, tumour necrosis factor alpha (TNF-α) and granulocyte-colony-stimulating factor (G-CSF). Median IL-1β stool concentrations were significantly higher (20-fold) in patients with severe CDI-attributed outcomes compared to those without them, but were not associated with initial disease severity. In patients with CDI median stool IL-1β concentrations were >40-fold higher than in patients with non-CDI diarrhoea and asymptomatic carriers. IL-1β seems a promising diagnostic stool marker to differentiate true CDI from control groups, as areas under the receiver operating characteristic curve (ROC-AUCs) ranged from 0.83 to 0.88. A similar study from this group found that median serum levels of IL-4, IL-6, IL-8, IL-10, IL-15, G-CSF, TNF-α and monocyte chemoattractant protein-1 (MCP-1 were significantly higher in CDI patients compared with all other groups) (Kelly et al. 2020). Concentration distributions for IL-6, G-CSF and TNF-α separated CDI patients from the other groups. Serum G-CSF seemed to be most promising with an ROC-AUC of 0.844 to discriminate CDI from the other groups combined.

9.4.2 Interleukins and Chemokines to Discriminate CDI Severity and Predict Outcome

CXCL-5 is a CXC chemokine and recruits and activates neutrophils. El Feghaly and colleagues studied the correlation between intestinal inflammation and disease severity in hospitalized patients with symptomatic CDI (El Feghaly et al. 2013). They found that faecal CXCL-5 mRNA and IL-8 mRNA were associated with diarrhoeal persistence and longer time to diarrhoea resolution. The levels were also higher in patients with CDI in the prior 90 days than in patients with no history of CDI (El Feghaly et al. 2013).

Abhyankar et al. analysed cytokines plasma level in 341 CDI inpatients. Increased IL-6, IL-8, IL-15, TNF-α, C-C motif chemokine ligand 5 (CCL-5) and suppression of tumorigenicity 2 receptor (sST-2) predicted mortality by univariate analysis (Abhyankar et al. 2020). A subsequent logistic regression risk prediction model had an ROC-AUC of 0.91 for 90-day mortality and 0.77 for 90-day recurrence. Likewise, Dieterle et al. constructed prediction models based on 156 CDI cases and validated this on a cohort of 272 cases (Dieterle et al. 2020). The best model for 30-day mortality included interleukin-8 (IL-8), PCT, CXCL-5, IP-10 and IL-2Rα with an AUC of 0.89. In this study, the best model for prediction of disease-related complication included IL-8, procalcitonin, hepatocyte growth factor (HGF) and IL-2Rα with an AUC of 0.84.

In conclusion, many markers of inflammation have been suggested to play a role in CDI and may correlate to disease severity. However as stated in the 2017 IDSA/SHEA guideline, at this point, no recommendations for their routine use can be made (McDonald et al. 2018). Several prediction models of interest have been published since 2018 (Abhyankar et al. 2020; Dieterle et al. 2020; Galvez et al. 2022; C. P. Kelly et al. 2020). More prospective research and validation of markers on external cohorts is needed to confirm these associations and models.

9.5 Microbiota-Based Markers for Prediction or Diagnostic of *Clostridioides difficile* Infection

Patients suffering from CDI harbour a disrupted intestinal microbiota characterized by reduced diversity and elevated levels of *Proteobacteria*, yeasts and *Enterococcus* species alongside reduced levels of members of the *Bacteroidetes* phylum, the *Lachnospiraceae* and *Ruminococcaceae* families. Gut dysbiosis is probably dependent on the previous use of antibiotics, but microbiome-mediated diagnosis of CDI remains understudied. The gut microbiota difference between patients with *C. difficile* colonization and infection is difficult to assess, though the relative abundance of *Bacteroides* and *Veillonella* has been reported (Crobach et al. 2020). However, analysis of microbial composition before the use of antibiotics can perhaps provide identification of microbial markers predictive of the risk of CDI development. One prospective study has investigated microbial composition as a potential predictor of CDI and concluded that the decrease of *Clostridiales*, namely, members of Clostridiales Incertae Sedis XI, in the intestinal microbiota was associated with an increased risk of CDI (Vincent et al. 2013). In a multicentre, observational, prospective study, the intestinal microbiota was determined utilizing 16S rRNA gene profiling of hospitalized patients aged 50 years and above in 34 hospitals across 6 European countries prior to antibiotic therapy with the aim of identifying robust microbial markers predictive of CDI (Berkell et al. 2021). It was concluded that a distinct microbiota enriched in *Enterococcus* and depleted of *Ruminococcus*, *Blautia*, *Prevotella* and *Bifidobacterium* identified patients at risk for CDI development before antibiotic treatment was started. Findings were validated on an external Canadian cohort. In combination with clinical and microbiological characteristics, carbapenem treatment (hazard ratio (95% CI): 5.3 (1.7–16.6)), toxigenic *C. difficile* rectal carriage (10.3 (3.2–33.1)) and high relative abundance of *Enterococcus* spp. vs low relative abundance of *Ruminococcus* spp. or *Alistipes* spp. and low Shannon alpha diversity index as determined by 16S rRNA gene profiling (9.7 (3.2–29.7)) predicted an increased CDI risk (van Werkhoven et al. 2021).

In conclusion, microbiome analysis may be used to identify patients at risk for the development of CDI. This may, for example, help to tailor antibiotic therapy or initiate pre-emptive CDI therapy. Whether microbiome analyses can be useful for diagnosis of CDI remains to be seen. For widespread application, cost-effective standardized (high-throughput) analyses with low turn-around times will be necessary.

References

Abhyankar MM, Ma JZ, Scully KW, Nafziger AJ, Frisbee AL, Saleh MM et al (2020) Immune profiling to predict outcome of Clostridioides difficile infection. mBio 11(3). https://doi.org/10.1128/mBio.00905-20

Agnew E, Davies KA, Viprey VF, Evans S, Davis GL, Hope R et al (2023) Impact of testing on Clostridioides difficile infection in hospitals across Europe: a mathematical model. Clin Microbiol Infect. https://doi.org/10.1016/j.cmi.2023.02.004

Aichinger E, Schleck CD, Harmsen WS, Nyre LM, Patel R (2008) Nonutility of repeat laboratory testing for detection of Clostridium difficile by use of PCR or enzyme immunoassay. J Clin Microbiol 46(11):3795–3797. https://doi.org/10.1128/jcm.00684-08

Alcala L, Martin A, Marin M, Sanchez-Somolinos M, Catalan P, Pelaez T, Bouza E (2012) The undiagnosed cases of Clostridium difficile infection in a whole nation: where is the problem? Clin Microbiol Infect 18(7):E204–E213. https://doi.org/10.1111/j.1469-0691.2012.03883.x

Alonso CD, Kelly CP, Garey KW, Gonzales-Luna AJ, Williams D, Daugherty K et al (2022) Ultrasensitive and quantitative toxin measurement correlates with baseline severity, severe outcomes, and recurrence among hospitalized patients with Clostridioides difficile infection. Clin Infect Dis 74(12):2142–2149. https://doi.org/10.1093/cid/ciab826

Anikst VE, Gaur RL, Schroeder LF, Banaei N (2016) Organism burden, toxin concentration, and lactoferrin concentration do not distinguish between clinically significant and nonsignificant diarrhea in patients with Clostridium difficile. Diagn Microbiol Infect Dis 84(4):343–346. https://doi.org/10.1016/j.diagmicrobio.2015.11.022

Archbald-Pannone LR (2014) Quantitative fecal lactoferrin as a biomarker for severe Clostridium difficile infection in hospitalized patients. J Geriatr Palliat Care 2(1):3. https://doi.org/10.13188/2373-1133.1000006

Baktash A, Corver J, Harmanus C, Smits WK, Fawley W, Wilcox MH et al (2022) Comparison of whole-genome sequence-based methods and PCR ribotyping for subtyping of Clostridioides difficile. J Clin Microbiol 60(2):e0173721. https://doi.org/10.1128/JCM.01737-21

Banz A, Lantz A, Riou B, Foussadier A, Miller M, Davies K, Wilcox M (2018) Sensitivity of single-molecule array assays for detection of Clostridium difficile toxins in comparison to conventional laboratory testing algorithms. J Clin Microbiol 56(8). https://doi.org/10.1128/JCM.00452-18

Barbut F, Day N, Bouee S, Youssouf A, Grandvoinnet L, Lalande V et al (2019) Toxigenic Clostridium difficile carriage in general practice: results of a laboratory-based cohort study. Clin Microbiol Infect 25(5):588–594. https://doi.org/10.1016/j.cmi.2018.12.024

Barbut F, Gouot C, Lapidus N, Suzon L, Syed-Zaidi R, Lalande V, Eckert C (2017) Faecal lactoferrin and calprotectin in patients with Clostridium difficile infection: a case control study. Eur J Clin Microbiol Infect Dis 36(12):2423–2430. https://doi.org/10.1007/s10096-017-3080-y

Barbut F, Surgers L, Eckert C, Visseaux B, Cuingnet M, Mesquita C et al (2014) Does a rapid diagnosis of Clostridium difficile infection impact on quality of patient management? Clin Microbiol Infect 20(2):136–144. https://doi.org/10.1111/1469-0691.12221

Bartlett JG (2002) Clinical practice. Antibiotic-associated diarrhea N Engl J Med 346(5):334–339. https://doi.org/10.1056/NEJMcp011603

Bartlett JG, Gerding DN (2008) Clinical recognition and diagnosis of Clostridium difficile infection. Clin Infect Dis 46(Suppl 1):S12–S18. https://doi.org/10.1086/521863

Berkell M, Mysara M, Xavier BB, van Werkhoven CH, Monsieurs P, Lammens C et al (2021) Microbiota-based markers predictive of development of Clostridioides difficile infection. Nat Commun 12(1):2241. https://doi.org/10.1038/s41467-021-22302-0

Berrington A, Settle CD (2007) Which specimens should be tested for Clostridium difficile toxin? J Hosp Infect 65(3):280–282. https://doi.org/10.1016/j.jhin.2006.12.011

Bidet P, Barbut F, Lalande V, Burghoffer B, Petit JC (1999) Development of a new PCR-ribotyping method for Clostridium difficile based on ribosomal RNA gene sequencing. FEMS Microbiol Lett 175(2):261–266

Boone JH, Archbald-Pannone LR, Wickham KN, Carman RJ, Guerrant RL, Franck CT, Lyerly DM (2014) Ribotype 027 Clostridium difficile infections with measurable stool toxin have increased lactoferrin and are associated with a higher mortality. Eur J Clin Microbiol Infect Dis 33(6):1045–1051. https://doi.org/10.1007/s10096-013-2043-1

Boone JH, DiPersio JR, Tan MJ, Salstrom SJ, Wickham KN, Carman RJ et al (2013) Elevated lactoferrin is associated with moderate to severe Clostridium difficile disease, stool toxin, and 027 infection. Eur J Clin Microbiol Infect Dis 32(12):1517–1523. https://doi.org/10.1007/s10096-013-1905-x

Burnham CA, Carroll KC (2013) Diagnosis of Clostridium difficile infection: an ongoing conundrum for clinicians and for clinical laboratories. Clin Microbiol Rev 26(3):604–630. https://doi.org/10.1128/cmr.00016-13

Chung HS, Lee M (2017) Evaluation of the performance of C. DIFF QUIK CHEK COMPLETE and its usefulness in a hospital setting with a high prevalence of Clostridium difficile infection. J Investig Med 65(1):88–92. https://doi.org/10.1136/jim-2016-000231

Cohen SH, Gerding DN, Johnson S, Kelly CP, Loo VG, McDonald LC et al (2010) Clinical practice guidelines for Clostridium difficile infection in adults: 2010 update by the society for healthcare epidemiology of America (SHEA) and the infectious diseases society of America (IDSA). Infect Control Hosp Epidemiol 31(5):431–455. https://doi.org/10.1086/651706

Corver J, Sen J, Hornung BVH, Mertens BJ, Berssenbrugge EKL, Harmanus C et al (2019) Identification and validation of two peptide markers for the recognition of Clostridioides difficile MLST-1 and MLST-11 by MALDI-MS. Clin Microbiol Infect 25(7):904 e901–904 e907. https://doi.org/10.1016/j.cmi.2018.10.008

Crobach MJ, Planche T, Eckert C, Barbut F, Terveer EM, Dekkers OM et al (2016) European Society of Clinical Microbiology and Infectious Diseases: update of the diagnostic guidance document for Clostridium difficile infection. Clin Microbiol Infect 22(Suppl 4):S63–S81. https://doi.org/10.1016/j.cmi.2016.03.010

Crobach MJT, Ducarmon QR, Terveer EM, Harmanus C, Sanders I, Verduin KM et al (2020) The bacterial gut microbiota of adult patients infected, colonized or noncolonized by Clostridioides difficile. Microorganisms 8(5). https://doi.org/10.3390/microorganisms8050677

Crobach MJT, Duszenko N, Terveer EM, Verduin CM, Kuijper EJ (2018a) Nucleic acid amplification test quantitation as predictor of toxin presence in Clostridium difficile infection. J Clin Microbiol 56(3). https://doi.org/10.1128/JCM.01316-17

Crobach MJT, Vernon JJ, Loo VG, Kong LY, Pechine S, Wilcox MH, Kuijper EJ (2018b) Understanding Clostridium difficile colonization. Clin Microbiol Rev 31(2). https://doi.org/10.1128/CMR.00021-17

Czepiel J, Biesiada G, Drozdz M, Gdula-Argasinska J, Zuranska J, Marchewka J et al (2018) The presence of IL-8 +781 T/C polymorphism is associated with the parameters of severe Clostridium difficile infection. Microb Pathog 114:281–285. https://doi.org/10.1016/j.micpath.2017.11.066

Darkoh C, Turnwald BP, Koo HL, Garey KW, Jiang ZD, Aitken SL, DuPont HL (2014) Colonic immunopathogenesis of Clostridium difficile infections. Clin Vaccine Immunol 21(4):509–517. https://doi.org/10.1128/cvi.00770-13

Davies K, Davis G, Barbut F, Eckert C, Petrosillo N, Wilcox MH (2016) Variability in testing policies and impact on reported Clostridium difficile infection rates: results from the pilot Longitudinal European Clostridium difficile Infection Diagnosis surveillance study (LuCID). Eur J Clin Microbiol Infect Dis 35(12): 1949–1956. https://doi.org/10.1007/s10096-016-2746-1

Davies KA, Longshaw CM, Davis GL, Bouza E, Barbut F, Barna Z et al (2014) Underdiagnosis of Clostridium difficile across Europe: the European, multicentre, prospective, biannual, point-prevalence study of Clostridium difficile infection in hospitalised patients with diarrhoea (EUCLID). Lancet Infect Dis 14(12): 1208–1219. https://doi.org/10.1016/s1473-3099(14)70991-0

Davies KA, Planche T, Wilcox MH (2018) The predictive value of quantitative nucleic acid amplification detection of Clostridium difficile toxin gene for faecal sample toxin status and patient outcome. PLoS One 13(12): e0205941. https://doi.org/10.1371/journal.pone.0205941

Dbeibo L, Lucky CW, Fadel WF, Sadowski J, Beeler C, Kelly K et al (2023) Two-step algorithm-based Clostridioides difficile testing as a tool for antibiotic stewardship. Clin Microbiol Infect, in press. https://doi.org/10.1016/j.cmi.2023.02.008

Debast SB, van Kregten E, Oskam KM, van den Berg T, Van den Berg RJ, Kuijper EJ (2008) Effect on diagnostic yield of repeated stool testing during outbreaks of Clostridium difficile-associated disease. Clin Microbiol Infect 14(6):622–624. https://doi.org/10.1111/j.1469-0691.2008.01999.x

Delmee M (2001) Laboratory diagnosis of Clostridium difficile disease. Clin Microbiol Infect 7(8):411–416

Dieterle MG, Putler R, Perry DA, Menon A, Abernathy-Close L, Perlman NS et al (2020) Systemic inflammatory mediators are effective biomarkers for predicting adverse outcomes in clostridioides difficile infection. mBio 11(3). https://doi.org/10.1128/mBio.00180-20

Dionne LL, Raymond F, Corbeil J, Longtin J, Gervais P, Longtin Y (2013) Correlation between Clostridium difficile bacterial load, commercial real-time PCR cycle thresholds, and results of diagnostic tests based on enzyme immunoassay and cell culture cytotoxicity assay. J Clin Microbiol 51(11):3624–3630. https://doi.org/10.1128/JCM.01444-13

Doolan CP, Louie T, Lata C, Larios OE, Stokes W, Kim J et al (2021) Latent class analysis for the diagnosis of Clostridioides difficile infection. Clin Infect Dis 73(9): e2673–e2679. https://doi.org/10.1093/cid/ciaa1553

Doolan CP, Sahragard B, Leal J, Sharma A, Kim J, Spackman E et al (2023) Clostridioides difficile near patient testing versus centralized testing: a pragmatic cluster randomized cross-over trial. Clin Infect Dis. https://doi.org/10.1093/cid/ciad046

Ducarmon QR, van der Bruggen T, Harmanus C, Sanders I, Daenen LGM, Fluit AC et al (2022) Clostridioides difficile infection with isolates of cryptic clade C-II: a genomic analysis of polymerase chain reaction ribotype 151. Clin Microbiol Infect. https://doi.org/10.1016/j.cmi.2022.12.003

El Feghaly RE, Stauber JL, Deych E, Gonzalez C, Tarr PI, Haslam DB (2013) Markers of intestinal inflammation, not bacterial burden, correlate with clinical outcomes in Clostridium difficile infection. Clin Infect Dis 56(12):1713–1721. https://doi.org/10.1093/cid/cit147

Fawley WN, Knetsch CW, MacCannell DR, Harmanus C, Du T, Mulvey MR et al (2015) Development and validation of an internationally-standardized, high-resolution capillary gel-based electrophoresis PCR-ribotyping protocol for Clostridium difficile. PLoS One 10(2):e0118150. https://doi.org/10.1371/journal.pone.0118150

Ferraris L, Couturier J, Eckert C, Delannoy J, Barbut F, Butel MJ, Aires J (2019) Carriage and colonization of C. difficile in preterm neonates: a longitudinal prospective study. PLoS One 14(2):e0212568. https://doi.org/10.1371/journal.pone.0212568

Flores-Trevino S, Garza-Gonzalez E, Mendoza-Olazaran-S, Morfin-Otero R, Camacho-Ortiz A, Rodriguez-Noriega E et al (2019) Screening of biomarkers of drug resistance or virulence in ESCAPE pathogens by MALDI-TOF mass spectrometry. Sci Rep 9(1):18945. https://doi.org/10.1038/s41598-019-55430-1

Galvez JAV, Pollock NR, Alonso CD, Chen X, Xu H, Wang L et al (2022) Stool interleukin-1beta differentiates Clostridioides difficile infection from asymptomatic carriage and non-C.difficile infection diarrhea. Clin Infect Dis. https://doi.org/10.1093/cid/ciac624

Green DA, Stotler B, Jackman D, Whittier S, Della-Latta P (2014) Clinical characteristics of patients who test positive for Clostridium difficile by repeat PCR. J Clin Microbiol 52(11):3853–3855. https://doi.org/10.1128/JCM.01659-14

Grein JD, Ochner M, Hoang H, Jin A, Morgan MA, Murthy AR (2014) Comparison of testing approaches for Clostridium difficile infection at a large community hospital. Clin Microbiol Infect 20(1):65–69. https://doi.org/10.1111/1469-0691.12198

He T, Kaplan SE, Gomez LA, Lu X, Ramanathan LV, Kamboj M, Tang YW (2018) Fecal calprotectin concentrations in cancer patients with Clostridium difficile infection. Eur J Clin Microbiol Infect Dis 37(12):2341–2346. https://doi.org/10.1007/s10096-018-3381-9

Hensgens MP, Dekkers OM, Demeulemeester A, Buiting AG, Bloembergen P, van Benthem BH et al (2014) Diarrhoea in general practice: when should a Clostridium difficile infection be considered? Results of a nested case-control study. Clin Microbiol Infect 20(12):O1067–O1074. https://doi.org/10.1111/1469-0691.12758

Hink T, Burnham CA, Dubberke ER (2013) A systematic evaluation of methods to optimize culture-based recovery of Clostridium difficile from stool specimens. Anaerobe 19:39–43. https://doi.org/10.1016/j.anaerobe.2012.12.001

Hitchcock MM, Holubar M, Hogan CA, Tompkins LS, Banaei N (2019) Dual reporting of Clostridioides difficile PCR and predicted toxin result based on PCR cycle threshold reduces treatment of toxin-negative patients without increases in adverse outcomes. J Clin Microbiol 57(11). https://doi.org/10.1128/JCM.01288-19

Janezic S, Rupnik M (2019) Development and implementation of whole genome sequencing-based typing schemes for Clostridioides difficile. Front Public Health 7:309. https://doi.org/10.3389/fpubh.2019.00309

Jangi S, Lamont JT (2010) Asymptomatic colonization by Clostridium difficile in infants: implications for disease in later life. J Pediatr Gastroenterol Nutr 51(1):2–7. https://doi.org/10.1097/MPG.0b013e3181d29767

Jazmati N, Hellmich M, Licanin B, Plum G, Kaasch AJ (2016) PCR cycle threshold value predicts the course of Clostridium difficile infection. Clin Microbiol Infect 22(2):e7–e8. https://doi.org/10.1016/j.cmi.2015.09.012

Kaltsas A, Simon M, Unruh LH, Son C, Wroblewski D, Musser KA et al (2012) Clinical and laboratory characteristics of Clostridium difficile infection in patients with discordant diagnostic test results. J Clin Microbiol 50(4):1303–1307. https://doi.org/10.1128/jcm.05711-11

Kelly CP, Chen X, Williams D, Xu H, Cuddemi CA, Daugherty K et al (2020) Host immune markers distinguish clostridioides difficile infection from asymptomatic carriage and non-C. difficile diarrhea. Clin Infect Dis 70(6):1083–1093. https://doi.org/10.1093/cid/ciz330

Kelly CR, Fischer M, Allegretti JR, LaPlante K, Stewart DB, Limketkai BN, Stollman NH (2022) Correction to: ACG Clinical Guidelines: Prevention, diagnosis, and treatment of Clostridioides difficile infections. Am J Gastroenterol 117(2):358. https://doi.org/10.14309/ajg.0000000000001529

Khanna S, Pardi DS, Rosenblatt JE, Patel R, Kammer PP, Baddour LM (2012) An evaluation of repeat stool testing for Clostridium difficile infection by polymerase chain reaction. J Clin Gastroenterol 46(10):846–849. https://doi.org/10.1097/MCG.0b013e3182432273

Killgore G, Thompson A, Johnson S, Brazier J, Kuijper E, Pepin J et al (2008) Comparison of seven techniques for typing international epidemic strains of Clostridium difficile: restriction endonuclease analysis, pulsed-field gel electrophoresis, PCR-ribotyping, multilocus sequence typing, multilocus variable-number tandem-repeat analysis, amplified fragment length polymorphism, and surface layer protein A gene sequence typing. J Clin Microbiol 46(2):431–437. https://doi.org/10.1128/jcm.01484-07

Kim J, Kim H, Oh HJ, Kim HS, Hwang YJ, Yong D et al (2017) Fecal calprotectin level reflects the severity of Clostridium difficile infection. Ann Lab Med 37(1):53–57. https://doi.org/10.3343/alm.2017.37.1.53

Knetsch CW, Lawley TD, Hensgens MP, Corver J, Wilcox MW, Kuijper EJ (2013) Current application and future perspectives of molecular typing methods to study Clostridium difficile infections. Euro Surveill 18(4):20381

Kirkpatrick ID, Greenberg HM (2001) Evaluating the CT diagnosis of Clostridium difficile colitis: should CT guide therapy? AJR Am J Roentgenol 176(3):635–639. https://doi.org/10.2214/ajr.176.3.1760635

Kraft CS, Parrott JS, Cornish NE, Rubinstein ML, Weissfeld AS, McNult P et al (2019) A laboratory medicine best practices systematic review and meta-analysis of nucleic acid amplification tests (NAATs) and algorithms including NAATs for the diagnosis of clostridioides (Clostridium) difficile in adults. Clin Microbiol Rev 32(3). https://doi.org/10.1128/CMR.00032-18

Kristjansson M, Samore MH, Gerding DN, DeGirolami PC, Bettin KM, Karchmer AW, Arbeit RD (1994) Comparison of restriction endonuclease analysis, ribotyping, and pulsed-field gel electrophoresis for molecular differentiation of Clostridium difficile strains. J Clin Microbiol 32(8):1963–1969

Kundrapu S, Sunkesula VC, Jury LA, Sethi AK, Donskey CJ (2012) Utility of perirectal swab specimens for diagnosis of Clostridium difficile infection. Clin Infect Dis 55(11):1527–1530. https://doi.org/10.1093/cid/cis707

Kwon JH, Reske KA, Hink T, Jackups R, Burnham CD, Dubberke ER (2019) Impact of an electronic hard-stop clinical decision support tool to limit repeat Clostridioides difficile toxin enzyme immunoassay testing on test utilization. Infect Control Hosp Epidemiol 40(12):1423–1426. https://doi.org/10.1017/ice.2019.275

Kyne L, Warny M, Qamar A, Kelly CP (2000) Asymptomatic carriage of Clostridium difficile and serum levels of IgG antibody against toxin A. N Engl J Med 342(6):390–397. https://doi.org/10.1056/nejm200002103420604

LaSala PR, Ekhmimi T, Hill AK, Farooqi I, Perrotta PL (2013) Quantitative fecal lactoferrin in toxin-positive and toxin-negative Clostridium difficile specimens. J Clin Microbiol 51(1):311–313. https://doi.org/10.1128/jcm.02735-12

Leslie JL, Cohen SH, Solnick JV, Polage CR (2012) Role of fecal Clostridium difficile load in discrepancies between toxin tests and PCR: is quantitation the next step in C. difficile testing? Eur J Clin Microbiol Infect Dis 31(12):3295–3299. https://doi.org/10.1007/s10096-012-1695-6

Longtin Y, Trottier S, Brochu G, Paquet-Bolduc B, Garenc C, Loungnarath V et al (2013) Impact of the type of diagnostic assay on Clostridium difficile infection and complication rates in a mandatory reporting program. Clin Infect Dis 56(1):67–73. https://doi.org/10.1093/cid/cis840

Loo VG, Bourgault AM, Poirier L, Lamothe F, Michaud S, Turgeon N et al (2011) Host and pathogen factors for Clostridium difficile infection and colonization. N Engl J Med 365(18):1693–1703. https://doi.org/10.1056/NEJMoa1012413

Luo RF, Banaei N (2010) Is repeat PCR needed for diagnosis of Clostridium difficile infection? J Clin Microbiol 48(10):3738–3741. https://doi.org/10.1128/jcm.00722-10

Maiden MC, Bygraves JA, Feil E, Morelli G, Russell JE, Urwin R et al (1998) Multilocus sequence typing: a portable approach to the identification of clones within populations of pathogenic microorganisms. Proc Natl Acad Sci U S A 95(6):3140–3145

Marra AR, Edmond MB, Ford BA, Herwaldt LA, Algwizani AR, Diekema DJ (2017) Failure of risk-adjustment by test method for C. difficile laboratory-identified event reporting. Infect Control Hosp Epidemiol 38(1):109–111. https://doi.org/10.1017/ice.2016.227

McDonald LC, Gerding DN, Johnson S, Bakken JS, Carroll KC, Coffin SE et al (2018) Clinical practice guidelines for Clostridium difficile infection in adults and children: 2017 Update by the Infectious Diseases Society of America (IDSA) and Society for Healthcare Epidemiology of America (SHEA). Clin Infect Dis 66(7):e1–e48. https://doi.org/10.1093/cid/cix1085

McFarland LV, Coyle MB, Kremer WH, Stamm WE (1987) Rectal swab cultures for Clostridium difficile surveillance studies. J Clin Microbiol 25(11): 2241–2242

Moehring RW, Lofgren ET, Anderson DJ (2013) Impact of change to molecular testing for Clostridium difficile infection on healthcare facility-associated incidence rates. Infect Control Hosp Epidemiol 34(10): 1055–1061. https://doi.org/10.1086/673144

Peretz A, Tkhawkho L, Pastukh N, Brodsky D, Halevi CN, Nitzan O (2016) Correlation between fecal calprotectin levels, disease severity and the hypervirulent ribotype 027 strain in patients with Clostridium difficile infection. BMC Infect Dis 16:309. https://doi.org/10.1186/s12879-016-1618-8

Persson S, Torpdahl M, Olsen KE (2008) New multiplex PCR method for the detection of Clostridium difficile toxin A (tcdA) and toxin B (tcdB) and the binary toxin (cdtA/cdtB) genes applied to a Danish strain collection. Clin Microbiol Infect 14(11):1057–1064. https://doi.org/10.1111/j.1469-0691.2008.02092.x

Planche T, Wilcox M (2011) Reference assays for Clostridium difficile infection: one or two gold standards? J Clin Pathol 64(1):1–5. https://doi.org/10.1136/jcp.2010.080135

Planche TD, Davies KA, Coen PG, Finney JM, Monahan IM, Morris KA et al (2013) Differences in outcome according to Clostridium difficile testing method: a prospective multicentre diagnostic validation study of C difficile infection. Lancet Infect Dis 13(11): 936–945. https://doi.org/10.1016/s1473-3099(13)70200-7

Polage CR, Gyorke CE, Kennedy MA, Leslie JL, Chin DL, Wang S et al (2015) Overdiagnosis of Clostridium difficile Infection in the Molecular Test Era. JAMA Intern Med 175(11):1792–1801. https://doi.org/10.1001/jamainternmed.2015.4114

Pollock NR, Banz A, Chen X, Williams D, Xu H, Cuddemi CA et al (2019) Comparison of Clostridioides difficile stool toxin concentrations in adults with symptomatic infection and asymptomatic carriage using an ultrasensitive quantitative immunoassay. Clin Infect Dis 68(1):78–86. https://doi.org/10.1093/cid/ciy415

Popiel KY, Gheorghe R, Eastmond J, Miller MA (2015) Usefulness of adjunctive fecal calprotectin and serum procalcitonin in individuals positive for Clostridium difficile toxin gene by PCR assay. J Clin Microbiol 53(11):3667–3669. https://doi.org/10.1128/jcm.02230-15

Pressly KB, Hill E, Shah KJ (2016) Pseudomembranous colitis secondary to methicillin-resistant Staphylococcus aureus (MRSA). BMJ Case Rep 2016. https://doi.org/10.1136/bcr-2016-215225

Reddymasu S, Sheth A, Banks DE (2006) Is Fecal Leukocyte Test a good predictor of Clostridium difficile associated diarrhea? Ann Clin Microbiol Antimicrob 5:9. https://doi.org/10.1186/1476-0711-5-9

Reigadas E, Alcala L, Marin M, Burillo A, Munoz P, Bouza E (2015) Missed diagnosis of Clostridium difficile infection; a prospective evaluation of unselected stool samples. J Infect 70(3):264–272. https://doi.org/10.1016/j.jinf.2014.10.013

Reigadas E, Alcala L, Valerio M, Marin M, Martin A, Bouza E (2016) Toxin B PCR cycle threshold as a predictor of poor outcome of Clostridium difficile infection: a derivation and validation cohort study. J Antimicrob Chemother 71(5):1380–1385. https://doi.org/10.1093/jac/dkv497

Rogers DS, Kundrapu S, Sunkesula VC, Donskey CJ (2013) Comparison of perirectal versus rectal swabs for detection of asymptomatic carriers of toxigenic Clostridium difficile. J Clin Microbiol 51(10):3421–3422. https://doi.org/10.1128/JCM.01418-13

Sandora TJ, Kociolek LK, Williams DN, Daugherty K, Geer C, Cuddemi C et al (2023) Baseline stool toxin concentration is associated with risk of recurrence in children with Clostridioides difficile infection. Infect Control Hosp Epidemiol 1-7. https://doi.org/10.1017/ice.2022.310

Savola KL, Baron EJ, Tompkins LS, Passaro DJ (2001) Fecal leukocyte stain has diagnostic value for outpatients but not inpatients. J Clin Microbiol 39(1):266–269. https://doi.org/10.1128/jcm.39.1.266-269.2001

Schutze GE, Willoughby RE (2013) Clostridium difficile infection in infants and children. Pediatrics 131(1):196–200. https://doi.org/10.1542/peds.2012-2992

Senchyna F, Gaur RL, Gombar S, Truong CY, Schroeder LF, Banaei N (2017) Clostridium difficile PCR cycle threshold predicts free toxin. J Clin Microbiol 55(9):2651–2660. https://doi.org/10.1128/JCM.00563-17

Sethi AK, Al-Nassir WN, Nerandzic MM, Bobulsky GS, Donskey CJ (2010) Persistence of skin contamination and environmental shedding of Clostridium difficile during and after treatment of C. difficile infection. Infect Control Hosp Epidemiol 31(1):21–27. https://doi.org/10.1086/649016

Shi JT, Zhang Y, She Y, Goyal H, Wu ZQ, Xu HG (2022) Diagnostic utility of non-invasive tests for inflammatory bowel disease: an umbrella review. Front Med (Lausanne) 9:920732. https://doi.org/10.3389/fmed.2022.920732

Stubbs SL, Brazier JS, O'Neill GL, Duerden BI (1999) PCR targeted to the 16S-23S rRNA gene intergenic spacer region of Clostridium difficile and construction of a library consisting of 116 different PCR ribotypes. J Clin Microbiol 37(2):461–463

Sundar S, Chan SY (2003) Cholestatic jaundice and pseudomembranous colitis following combination therapy with doxorubicin and docetaxel. Anti-Cancer Drugs 14(4):327–329. https://doi.org/10.1097/01.cad.0000065044.82984.ae

Surawicz CM, Brandt LJ, Binion DG, Ananthakrishnan AN, Curry SR, Gilligan PH et al (2013) Guidelines for diagnosis, treatment, and prevention of Clostridium difficile infections. Am J Gastroenterol 108(4):478–498; quiz 499. https://doi.org/10.1038/ajg.2013.4

Swale A, Miyajima F, Roberts P, Hall A, Little M, Beadsworth MB et al (2014) Calprotectin and lactoferrin faecal levels in patients with Clostridium difficile infection (CDI): a prospective cohort study. PLoS One 9(8):e106118. https://doi.org/10.1371/journal.pone.0106118

Tang DM, Urrunaga NH, von Rosenvinge EC (2016) Pseudomembranous colitis: not always Clostridium difficile. Cleve Clin J Med 83(5):361–366. https://doi.org/10.3949/ccjm.83a.14183

Usacheva EA, Jin JP, Peterson LR (2016) Host response to Clostridium difficile infection: diagnostics and detection. J Glob Antimicrob Resist 7:93–101. https://doi.org/10.1016/j.jgar.2016.08.002

van den Berg RJ, Schaap I, Templeton KE, Klaassen CH, Kuijper EJ (2007) Typing and subtyping of Clostridium difficile isolates by using multiple-locus variable-number tandem-repeat analysis. J Clin Microbiol 45(3):1024–1028. https://doi.org/10.1128/jcm.02023-06

van Prehn J, Reigadas E, Vogelzang EH, Bouza E, Hristea A, Guery B et al (2021) European Society of Clinical Microbiology and Infectious Diseases: 2021 update on the treatment guidance document for Clostridioides difficile infection in adults. Clin Microbiol Infect 27(Suppl 2):S1–S21. https://doi.org/10.1016/j.cmi.2021.09.038

van Prehn J, Vandenbroucke-Grauls CM, van Beurden YH, van Houdt R, Vainio S, Ang CW (2015) Diagnostic yield of repeat sampling with immunoassay, real-time PCR, and toxigenic culture for the detection of toxigenic Clostridium difficile in an epidemic and a non-epidemic setting. Eur J Clin Microbiol Infect Dis 34(12):2325–2330. https://doi.org/10.1007/s10096-015-2484-9

van Werkhoven CH, Ducher A, Berkell M, Mysara M, Lammens C, Torre-Cisneros J et al (2021) Incidence and predictive biomarkers of Clostridioides difficile infection in hospitalized patients receiving broad-spectrum antibiotics. Nat Commun 12(1):2240. https://doi.org/10.1038/s41467-021-22269-y

Villafuerte Galvez JA, Pollock NR, Alonso CD, Chen X, Xu H, Wang L et al (2023) Stool interleukin-1beta differentiates Clostridioides difficile infection (CDI) from asymptomatic carriage and non-CDI diarrhea. Clin Infect Dis 76(3):e1467–e1475. https://doi.org/10.1093/cid/ciac624

Vincent C, Stephens DA, Loo VG, Edens TJ, Behr MA, Dewar K, Manges AR (2013) Reductions in intestinal Clostridiales precede the development of nosocomial Clostridium difficile infection. Microbiome 1(1):18. https://doi.org/10.1186/2049 2618 1 18

Viprey VF, Davis GL, Benson AD, Ewin D, Spittal W, Vernon JJ et al (2022) A point-prevalence study on community and inpatient Clostridioides difficile infections (CDI): results from Combatting Bacterial Resistance in Europe CDI (COMBACTE-CDI), July to November 2018. Euro Surveill 27(26). https://doi.org/10.2807/1560-7917.ES.2022.27.26.2100704

Viprey VF, Granata G, Vendrik KEW, Davis GL, Petrosillo N, Kuijper EJ et al (2023) European survey on the current surveillance practices, management guidelines, treatment pathways and heterogeneity of testing of Clostridioides difficile, 2018-2019: results from The Combatting Bacterial Resistance in Europe CDI (COMBACTE-CDI). J Hosp Infect 131:213–220. https://doi.org/10.1016/j.jhin.2022.11.011

Vogelzang EH, Lankelma JM, van Mansfeld R, van Prehn J, van Houdt R (2020) Implementing a Clostridium difficile testing algorithm and its effect on isolation duration and treatment initiation: a pre- and post-implementation study. Eur J Clin Microbiol Infect Dis 39(6):1071–1076. https://doi.org/10.1007/s10096-020-03823-w

Vrabie R, Kane S (2014) Noninvasive markers of disease activity in inflammatory bowel disease. Gastroenterol Hepatol (NY) 10(9):576–584

Wang T, Matukas L, Streutker CJ (2013) Histologic findings and clinical characteristics in acutely symptomatic ulcerative colitis patients with superimposed Clostridium difficile infection. Am J Clin Pathol 140 (6):831–837. https://doi.org/10.1309/ajcp2lbrttjbf3kd

Wang L, Cao J, Li C, Zhang L (2018) IL-27/IL-27 receptor signaling provides protection in Clostridium difficile-induced colitis. J Infect Dis 217(2):198–207. https://doi.org/10.1093/infdis/jix581

Wenisch C, Parschalk B, Hasenhundl M, Hirschl AM, Graninger W (1996) Comparison of vancomycin, teicoplanin, metronidazole, and fusidic acid for the treatment of Clostridium difficile-associated diarrhea. Clin Infect Dis 22(5):813–818

Whitehead SJ, Shipman KE, Cooper M, Ford C, Gama R (2014) Is there any value in measuring faecal calprotectin in Clostridium difficile positive faecal samples? J Med Microbiol 63(Pt 4):590–593. https://doi.org/10.1099/jmm.0.067389-0

European Practice for CDI Treatment

Fidelma Fitzpatrick, Robert Brennan, Joffrey van Prehn, Mairead Skally, Melissa Brady, Karen Burns, Christopher Rooney, and Mark H. Wilcox

Abstract

Clostridioides difficile infection (CDI) remains a significant cause of morbidity and mortality worldwide. Historically, two antibiotics (metronidazole and vancomycin) and a recent third (fidaxomicin) have been used for CDI treatment; convincing data are now available showing that metronidazole is the least efficacious agent. The European Society of Clinical Microbiology and Infectious Diseases (ESCMID) management guidance for CDI were updated in 2021. This guidance document outlines the treatment options for a variety of CDI clinical scenarios and for non-antimicrobial management (e.g., faecal microbiota transplantation, FMT). One of the main changes is that metronidazole is no longer recommended as first-line CDI treatment. Rather, fidaxomicin is preferred on the basis of reduced recurrence rates with vancomycin as an acceptable alternative. Recommended options for recurrent CDI now include bezlotoxumab as well as FMT.

A 2017 survey of 20 European countries highlighted variation internationally in CDI management strategies. A variety of restrictions were in place in 65% countries prior to use of new anti-CDI treatments, including committee/infection specialist approval or economic review/restrictions. This survey was repeated in November 2022 to assess the current landscape of CDI

F. Fitzpatrick (✉) · M. Skally
Department of Clinical Microbiology, The Royal College of Surgeons in Ireland, Dublin, Ireland

Department of Clinical Microbiology, Beaumont Hospital, Dublin, Ireland
e-mail: fidelmafitzpatrick@rcsi.ie; maireadskally@beaumont.ie

R. Brennan
Department of Clinical Microbiology, The Royal College of Surgeons in Ireland, Dublin, Ireland
e-mail: rbrennan@rcsi.com

J. van Prehn
Department of Medical Microbiology, Leiden University Medical Center, Leiden, the Netherlands
e-mail: J.van_Prehn@lumc.nl

M. Brady
Health Protection Surveillance Centre (HPSC), Dublin, Ireland
e-mail: melissa.brady1@hse.ie

K. Burns
Department of Clinical Microbiology, Beaumont Hospital, Dublin, Ireland
e-mail: karenburns@beaumont.ie

C. Rooney
Microbiology, Leeds Teaching Hospitals, Leeds, UK

University of Leeds, Leeds, UK
e-mail: christopherrooney@nhs.net

M. H. Wilcox (✉)
University of Leeds, Leeds, UK

Leeds Teaching Hospitals and Leeds Regional Public Health Laboratory, UK Health Security Agency (UKHSA), Leeds, UK
e-mail: mark.wilcox@nhs.net

P. Mastrantonio, M. Rupnik (eds.), *Updates on Clostridioides difficile in Europe*, Advances in Experimental Medicine and Biology 1435, https://doi.org/10.1007/978-3-031-42108-2_4

management practices in Europe. Of 64 respondents from 17 countries, national CDI guidelines existed in 14 countries, and 11 have already/plan to incorporate the ESCMID 2021 CDI guidance, though implementation has not been surveyed in 6. Vancomycin is the most commonly used first-line agent for the treatment of CDI ($n = 42$, 66%), followed by fidaxomicin ($n = 30$, 47%). Six (9%) respondents use metronidazole as first-line agent for CDI treatment, whereas 22 (34%) only in selected low-risk patient groups. Fidaxomicin is more likely to be used in high-risk patient groups. Availability of anti-CDI therapy influenced prescribing in six respondents (9%). Approval pre-prescription was required before vancomycin ($n = 3$, 5%), fidaxomicin ($n = 10$, 6%), bezlotoxumab ($n = 11$, 17%) and FMT ($n = 10$, 6%). Implementation of CDI guidelines is rarely audited.

Novel anti-CDI agents are being evaluated; it is not yet clear what will be the roles of these agents. The treatment of recurrent CDI is particularly troublesome, and several different live biotherapeutics are being developed, in addition to FMT.

1 Introduction

CDI is a leading cause of healthcare-associated (HA) diarrhoea ranging from 1.1 to 631.8 per 100,000 population globally (Finn et al. 2021). In Europe, HA-CDI infects 1-in-20 patients with HA infection and is responsible for 48% of all HA gastrointestinal infections (European Centre for Disease Control, Point prevalence survey 2013). The European Society of Clinical Microbiology and Infectious Diseases (ESCMID) first published guidelines for CDI treatment in 2009, which were revised in 2014 (Debast et al. 2014), and most recently in 2021 (van Prehn et al. 2021). Many European countries have published their own national CDI treatment guidelines, which are broadly similar to the ESCMID guidelines, though contextualised to the local setting (ECDC 2017). The Grades of Recommendation Assessment, Development and Evaluation (GRADE) system (Guyatt et al. 2008) was used to grade the strength of recommendations and the quality of the evidence in the 2021 document. One of the main changes in the 2021 update is that metronidazole is no longer recommended as first-line CDI treatment. Rather, fidaxomicin is preferred on the basis of reduced recurrence rates, with vancomycin as an acceptable alternative. Recommended options for recurrent CDI now include bezlotoxumab and faecal microbiota transplantation (FMT). Of note, recommendations for CDI diagnosis and infection prevention and control are described in separate ESCMID guidance documents (Crobach et al. 2016; Tschudin-Sutter et al. 2018).

In 2021, updated CDI treatment guidelines were also published by the Infectious Diseases Society of America (IDSA) and Society for Healthcare Epidemiology of America (SHEA) (Johnson et al. 2021) and the American College of Gastroenterology (ACG) (Kelly et al. 2021). The IDSA/SHEA guidelines also recommend fidaxomicin as the preferred option for the first episode of non-severe CDI and the first CDI recurrence and recommend metronidazole only when fidaxomicin or vancomycin is unavailable. In contrast, the ACG guidelines continue to recommend metronidazole for the first episode of non-severe CDI but in younger low-risk patients with minimal comorbidities. Both sets of guidelines recommend fidaxomicin is an equal alternate to vancomycin for severe CDI.

When discussing European practice for CDI treatment, variability between countries is inevitable for a number of reasons. Treatment of patients with CDI begins with making the diagnosis, specifically having a high index of clinical suspicion if a patient has a combination of signs and symptoms and/or CDI risk factors and thereafter confirmation by microbiological testing or colonoscopic/histopathological findings. Clinician awareness of CDI as part of the differential diagnosis and access to timely laboratory diagnostics is therefore crucial for appropriate patient management. However, there remains considerable variability across countries with an estimated 40,000 inpatients potentially undiagnosed annually in European hospitals (Davies et al. 2014). Mnemonic checklists can be useful tools to reduce clinician error and promote awareness (Chew et al. 2016). Albeit potentially more useful when English is the commonly spoken language, the SIGHT mnemonic is a useful aide memoire for clinicians when managing patients with suspected potentially infectious diarrhoea (Fig. 1) (Public Health England 2013).

Once CDI is diagnosed, variability in anti-CDI treatment practices may be due to individual judgement and/or knowledge, individualised patient factors and national regulatory or economic issues, e.g. the availability of newer (more expensive) anti-CDI agents. Lastly, the ESCMID (and national) guidelines recommend a number of potential treatment options for similar CDI clinical scenarios, so individual clinician preference will likely be a potential cause of variability. This variability in anti-CDI treatment preferences has previously been described in Ireland (Prior et al. 2017). In the United States (USA), almost half of patients with severe CDI were treated with metronidazole, despite vancomycin being recommended in national guidelines at that time (Stevens et al. 2017).

In 2017 a survey of 20 European countries and their implementation of CDI guidelines found that, while the majority ($n = 14$) have national CDI guidelines, the guidelines provide a range of recommendations for CDI treatment and only 5 countries had audited guideline implementation. A variety of restrictions were in place in 13 (65%) countries prior to use of new anti-CDI treatments, including committee/infection specialist approval or economic review/restrictions (Fitzpatrick et al. 2018). Interestingly, when the impact of the revised IDSA/SHEA guidelines were subsequently reviewed, prescriptions of oral vancomycin and fidaxomicin had increased, and metronidazole had decreased significantly in the USA (Clancy et al. 2021).

In this chapter, we summarise the updated 2021 ESCMID CDI guideline recommendations and present the findings of a revised 2022 European survey of CDI guidelines and their implementation and lastly look to the future as we summarise promising new therapies for CDI treatment.

2 ESCMID Guidelines for CDI Treatment

A number of CDI scenarios are considered including the initial management of CDI in addition to the management of recurrent and severe CDI (Table 1). For all scenarios the timely implementation of appropriate infection prevention and control measures to prevent further cross-infection is highlighted, in addition to the discontinuation of antimicrobial therapy (if clinically

Fig. 1 SIGHT mnemonic protocol

S	Suspect that a case may be infective where there is no clear alternative cause for diarrhoea.
I	Isolate the patient/resident. Consult with the infection prevention and control team where available while determining the cause of the diarrhoea.
G	Gloves and aprons must be used for all contacts with the patient/resident and their environment.
H	Hand washing with soap and water should be carried out after each contact with the patient/resident and the patient/resident's environment.
T	Test the stool for *C. difficile* toxin, by sending a specimen immediately.

Adapted with permission from SIGHT Mnemonic UK protocol (DH. and HPA. 2008)

Table 1 Summary of definitions used in the updated European Society of Clinical Microbiology and Infectious Diseases (ESCMID) management guidelines for *Clostridioides difficile* infection (CDI) (van Prehn et al. 2021)

	Clinical parameters		Laboratory diagnostics
CDI	Clinical findings compatible with CDI *without* reasonable evidence of another cause of diarrhoea	*and*	Microbiological evidence of *C. difficile*-free toxins by enzyme immunoassay
	Clinical picture compatible with CDI	*and*	Positive nucleic acid amplification test (preferably with a low cycle threshold value) *or* positive toxigenic *C. difficile* culture
	Pseudomembranous colitis diagnosed during endoscopy *or* after colectomy *or* on autopsy	*and*	Positive test for the presence of toxigenic *C. difficile*
Severe CDI	Fever >38.5°C	*or*	Leucocyte count $>15 \times 10^9$/L *or* Rise in serum creatinine >50% above baseline
Severe-complicated/ fulminant CDI	One of the following attributed to CDI: hypotension, septic shock, ileus, toxic megacolon, bowel perforation, any fulminant course of disease (i.e. rapid deterioration of the patient)	*or*	One of the following needs to be attributed to CDI the following: – Elevated serum lactate
Refractory CDI	No response after 3–5 days of therapy		
Recurrent CDI	CDI recurs within 8 weeks after a previous episode, provided the symptoms from the previous episode resolved after completion of initial treatment		

indicated), fluid and electrolyte replacement, review of proton pump inhibitor use and avoidance of anti-motility medications. Management of CDI in paediatric patients is not covered, though was subsequently reviewed in a best practice summary (Krutova et al. 2022a). Specific management of CDI patients prescribed concomitant antibiotics was also recently reviewed (Fitzpatrick et al. 2022).

2.1 Non-severe CDI

In contrast to previous guidance, oral metronidazole is no longer recommended as first-line therapy for the initial episode of CDI. Rather, metronidazole is recommended only when vancomycin and fidaxomicin are not available or feasible. Oral metronidazole administration achieves very low stool concentrations, especially as mucosal inflammation resolves, and interaction with faecal microbiota reduces its antimicrobial bioactivity (Krutova et al. 2022b). Increased metronidazole minimum inhibitory concentrations in epidemic *C. difficile* ribotypes and the emergence of plasmid-mediated resistance also contribute to clinical failure. If metronidazole is used, it is important that antimicrobial susceptibility is performed on agar containing heme (Boekhoud et al. 2021).

The updated ESCMID guidance recommends fidaxomicin preferentially over vancomycin for initial CDI (strong recommendation, moderate level of evidence), because of reduced recurrence rates: 101 fewer per 1000 (95% CI; 138 fewer versus 49 fewer) (Table 2). Fidaxomicin also has a narrower spectrum of activity than vancomycin and is less detrimental to the gut microbiome (Louie et al. 2009; Tannock et al. 2010). CDI treatment choice may also have implications for healthcare facility infection prevention and control and CDI cross-infection risk. In a small randomised controlled trial of 31 patients, fidaxomicin and vancomycin therapy reduced *C. difficile* shedding an environmental contamination in comparison with metronidazole (Turner et al. 2022). A larger study in four English hospitals of 244 CDI patients ($n = 83$ fidaxomicin, $n = 102$ vancomycin, $n = 70$ metronidazole) found that *C difficile* environmental contamination of patient rooms was lower during fidaxomicin therapy in comparison to

Table 2 Overview of the European Society of Clinical Microbiology and Infectious Diseases (ESCMID) management guidelines for *Clostridioides difficile* infection (CDI) (van Prehn et al. 2021)

		Initial CDI	First recurrence	Two or more recurrences
Standard of care (SoC)	First	Fidaxomicin[a] 200 mg BD × 10 days	SoC + bezlotoxumab	FMT
	Second	Vancomycin 125 mg QDS × 10 days	Fidaxomicin[b] 200 mg BD × 10 days	FMT *or* oC + bezlotoxumab
High risk of recurrence[b]	First	Fidaxomicin[b] 200 mg BD × 10 days		
	Second	SoC + Bezlotoxumab		
Preferred options not available	Metronidazole 500 mg TDS × 10 days		Vancomycin taper and pulse: 125 mg QDS × 2 weeks, then BD × 1 week, then daily × 1 week, then every 48 h × 1 week. Finally every 72 h × 1 week	
Severe CDI	Vancomycin or fidaxomicin Oral administration not possible: local (rectal or nasoduodenal delivery), +/− adjunctive IV metronidazole or IV tigecycline			
Severe-complicated and refractory severe CDI	Vancomycin or fidaxomicin Multidisciplinary approach with surgical consultation Consider IV tigecycline and FMT when refractory			

[a]Risk stratification for CDI recurrence risk can be applied for selective use of fidaxomicin in cases of limited access/ resources

[b]Consider extended fidaxomicin 200 mg BD day 1–5 and then 200 mg q48h d7–25. Most important risk for recurrence is age >65/70 years. Additional risk factors to consider are healthcare-associated CDI, hospitalisation in previous 3 months, prior CDI, concomitant non-CDI antibiotics and PPI started during/after CDI diagnosis

metronidazole or vancomycin therapy from approximately day 4 of anti-CDI therapy, though it was similar after therapy was completed (Davies et al. 2020). In addition, a significant reduction in environmental contamination rates of CDI patient rooms was reported for patients treated with fidaxomicin compared with those receiving metronidazole or vancomycin at multiple times after anti-CDI therapy commenced.

Previously, fidaxomicin-treated hospital inpatients were reported less likely to contaminate their environment than patients treated with metronidazole or vancomycin (Biswas et al. 2015).

When access to fidaxomicin is limited, a risk stratification for selected use is recommended whenever the clinicians deem the risk of recurrence high. Alternatively, vancomycin is a suitable alternative. Elements of risk stratification to determine CDI recurrence risk include age over 65 years along with the presence of one or more additional risk factor(s). These include healthcare-associated CDI, hospitalization in the previous 3 months, use of concomitant antibiotics, PPIs started during/after CDI diagnosis and a prior CDI episode (van Rossen et al. 2021). The risk of recurrence is assumed to be higher with more risk factors present.

Two other options are included in the ESCMID guidance, and both are 'considerations' rather than recommendations namely bezlotoxumab and extended fidaxomicin. The addition of bezlotoxumab to standard of care oral CDI treatment is discussed only in the context of CDI with increased risk of recurrence when fidaxomicin is not available or feasible. This is because of the higher acquisition costs and no clear benefit when compared with fidaxomicin. As the (high-risk) population of interest was studied in a *post hoc* analysis, the quality of evidence was graded moderate. Lastly, caution is advised in patients with a history of congestive heart failure. Extended (off label) fidaxomicin regimens receive a weak recommendation (low level of evidence) and only for an episode of CDI with increased risk of recurrence, especially in elderly hospitalized patients.

The updated guidance also recommends that the diagnosis of CDI should be reconsidered if patients do not respond to either fidaxomicin or vancomycin, provided that the patient is stable (i.e. not deteriorating) and not progressing to complicated CDI. This is based on the rarity of resistance to vancomycin and fidaxomicin in Europe (Freeman et al. 2015a). In these patients an alternative diagnosis should be sought as the patient may be colonised with *C. difficile* and their symptoms due to a different pathology. It is also important to ensure that patients are adhering to their treatment regimens especially in an outpatient setting.

2.2 Definition and Treatment of Severe CDI

Classification of CDI by severity can be problematic, as patients with severe ileus may not have diarrhoea. In practice, the clinical spectrum of severe CDI varies considerably, and the diagnosis is usually reached using a combination of findings. The most important risk factors for severe CDI are older age (>65 years) and presence of multiple comorbidities (van Rossen et al. 2021). The ESCMID guidelines summarise the range of patient, laboratory, endoscopic and radiological factors associated with severity of CDI colitis (Table 1). Factors associated with severe-complicated CDI include hypotension, septic shock, elevated serum lactate, ileus, toxic megacolon, bowel perforation or any rapid deterioration of the patient that is attributed to CDI.

In the previous ESCMID guidelines, the recommended treatment of choice for severe CDI in the ESCMID guidelines was oral vancomycin (Grade A-I) which achieves high intracolonic concentrations with minimal systemic adverse effects (Debast et al. 2014). Intravenous metronidazole combined with vancomycin retention enema or oral/NG vancomycin at the higher 500 mg dose was provided as an alternative. In the 2021 guidelines, both vancomycin and fidaxomicin are recommended options for severe and severe-complicated CDI (good practice statement, Table 2). There is no data supporting the superiority of one over the other. As with non-severe CDI, routine addition of metronidazole to oral standard of care therapy is not recommended. One option that may be considered on a case-by-case basis for deteriorating patients is addition of intravenous (IV) tigecycline (weak recommendation, very low level of evidence). In patients when oral therapy is not possible, intraluminal delivery of vancomycin or fidaxomicin is recommended (good practice statement) in addition to adjuvant IV therapy with metronidazole or tigecycline (weak, very low). The rationale behind the recommendation for intraluminal anti-CDI therapy is that standard-of-care anti-CDI treatment is based on achieving high intraluminal concentrations that are minimally absorbed in the gastrointestinal tract. The evidence base however is very limited with case series for intraluminal vancomycin and no data on intraluminal fidaxomicin delivery. Addition of an intravenous antibiotic might be beneficial on a theoretical basis when low intraluminal concentrations of oral CDI agents are expected.

The precise role of surgical management in severe CDI remains a topic of debate (Fitzpatrick 2008). As in previous versions, the updated ESCMID guidelines recommend surgical review for patients with severe-complicated CDI (good practice statement). There are no clear guidelines or protocols to guide the timing of surgical intervention. Certainly, the decision that surgical management is required for CDI should be taken by the multidisciplinary team, surgeons consulted at an 'early' stage (though there is no clear definition as to when this is) and an interdisciplinary risk/benefit analysis of surgery individualised for that patient.

2.3 Recurrent CDI

Recurrent CDI itself is a significant risk factor with the risk of recurrence increasing significantly with each episode of recurrence. The updated ESCMID guidelines report that a variety of factors increase the risk for recurrent CDI including older age (>65 years) and prior CDI (both strong, moderate), healthcare-associated CDI and hospitalization in the previous 3 months (weak, low), concomitant non-CDI antibiotic use after

CDI and new proton pump inhibitor (PPI) use during/after CDI diagnosis (both weak, very low) (van Rossen et al. 2021). Predicting which patients will develop recurrent CDI would enable clinicians to minimise recurrence risk (e.g. avoid concomitant antimicrobials) and also by heightening awareness, facilitates prompt diagnosis and treatment of recurrences (Hu et al. 2009).

The ESCMID guideline recommendation for the first recurrence of non-severe CDI is if the initial CDI episode was treated with vancomycin or metronidazole, then fidaxomicin is recommended (strong, low, Table 2). Alternatively, the addition of bezlotoxumab (when available and feasible) to an oral standard of care regimen is recommended if the initial CDI episode was treated with fidaxomicin (good practice statement). Aside from availability, concerns around high cost of bezlotoxumab have limited its use in some centres. There is also an unexplained observation of poor outcome in some bezlotoxumab recipients who have congestive heart failure. When other options for treatment of a first (or second) recurrent CDI episode are not available (i.e. fidaxomicin, bezlotoxumab and FMT), a vancomycin taper and pulse regimen may be considered (weak, very low).

Options for the prevention of recurrent CDI include follow on rifaximin (after standard of care treatment), though rates of rifampicin resistance in circulating strains are concerning. Yet, no formal recommendation on its use has been given by ESCMID. In the future preventative options may likely include non-toxigenic *C. difficile*, bacterial spores, bacterial consortia or other live biotherapeutic products. In November 2022, the United Stated Food and Drug Administration for the first time approved a faecal microbiota product (US Food and Drug Administration 2022). In Europe however, there is currently no agreed EU approach in relation to the classification of FMT products (European Medicines Agency 2022).

Options for patients with a second or further CDI recurrence include FMT after standard-of-care antibiotic pretreatment or bezlotoxumab in addition to standard-of-care antibiotic treatment (weak, moderate (FMT)/low (bezlotoxumab). The updated guidelines acknowledge that local regulations, availability and feasibility will likely play a role in the choice between either. For FMT an adequate multidisciplinary risk assessment is essential, and FMT products should standardized and be screened appropriately. In the future it is likely that more targeted (and potentially safer) approaches will play a role in the management of recurrent CDI (Kampouri et al. 2021).

2.4 CDI Prophylaxis

The updated ESCMID guidelines contain a new section on the role of anti-CDI antibiotic prophylaxis for patients on systemic antibiotic treatment. Neither routine administration of probiotics (strong, low) nor routine prophylaxis with anti-CDI antibiotics (good practice statement) is recommended. The guidelines however include a good practice statement for selected patients with a history of multiple recurrent CDI precipitated by systemic antibiotic use. In these patients, prophylaxis with microbiota-sparing anti-CDI antibiotics may be considered but only after careful consideration and consultation with a clinical microbiologist or infectious diseases specialist.

CDI prophylactic treatment strategies and their implementation in clinical practice were reviewed subsequently (Reigadas et al. 2021). While there are no evidenced-based prophylaxis options for primary CDI, FMT can be an option as secondary prevention for patients with multiple recurrences. Bezlotoxumab can be added to standard of care CDI treatment for patients at high risk for recurrent CDI (Gerding et al. 2018).

3 Updated Survey of European CDI Experts on CDI Treatment

In 2017 an international online survey of CDI treatment guideline recommendations and their implementation was circulated to colleagues that are involved in CDI treatment in 20 European countries (Fitzpatrick et al. 2018). This survey was repeated in November 2022 to assess the current landscape of CDI management practices in Europe. An online survey of CDI management practices was designed using SurveyMonkey® (Table 3). The original questionnaire was used and updated with additional questions regarding the following:

- Whether national and/or local CDI treatment guidelines have been adapted with the publication of updated 2021 ESCMID guidance.
- If when managing a patient with severe CDI, do you distinguish severe and severe-complicated CDI?
- Is metronidazole still an option for CDI treatment?
- What is the first choice of agent for an initial episode?

The survey was circulated by email to members of the ESCMID *C. difficile* study group and European infection societies. Data was analysed using an Excel® database (Microsoft Corp., Redmond, WA, USA). The survey was closed once the point of data saturation was reached.

Of 64 respondents from 17 countries which included clinical microbiologists (consultants $n = 31$, 48% and trainees $n = 4$, 6%), infectious diseases physicians (consultants $n = 14$, 22%, and trainees $n = 1$, 2%) and other physicians/healthcare workers ($n = 14$, 22%), 45 were based in tertiary referral/university hospitals. Subjective questions were analysed based on majority opinion in cases of discrepancy among respondents from the same country. Questions with definitive answers (e.g. date of guideline publication) were fact-checked by the authors, and the accurate response was included in the analysis.

National CDI guidelines existed in 14 of 17 countries; 11 had already/planned to incorporate(d) the ESCMID 2021 CDI guidelines. Of the three countries that did not have national guidelines, guidance was sought from the ESCMID CDI guidelines ($n = 1$) or local guidelines ($n = 2$). National guidelines were revised or first implemented more than 5 years ago ($n = 3$), during the last 5 years ($n = 3$), 1 year ($n = 3$), or presently under revision ($n = 4$). Revisions had not been undertaken in one country, with these guidelines published in 2020. Guideline implementation has not been surveyed in eight (57%) countries. One country had audited all aspects of guideline implementation, six countries have audited some aspects of guideline implementation, and one country is currently in the progress of auditing their guideline implementation.

In total, 40 (62.5%) respondents define severe/severe-complicated CDI as per ESCMID guidance. Thirty-two (50%) make a distinction between severe and severe-complicated as per ESCMID definition, 4 (6.3%) use local guidelines to make the distinction and 28 (43.8%) do not make any distinction. The three most cited markers for severe CDI were leucocytosis ($n = 46$), raised creatinine ($n = 45$) and fever ($n = 32$). A variety of anti-CDI regimens were recommended as summarised in Table 3. Vancomycin is the most commonly used first-line agent, prescribed by respondents, for the treatment of CDI ($n = 42$, 66%), followed by fidaxomicin ($n = 30$, 47%). Six (9%) respondents use metronidazole as a first-line agent for CDI treatment in a normal cohort, whereas 22 (34%) would prescribe metronidazole only in selected low-risk patient groups. Fidaxomicin is more likely to be used in high-risk patient groups.

In addition, several other factors were reported to influence the choice of the recommended anti-CDI therapy including the following:

- Patient factors
  - Risk factors for recurrence ($n = 6$)
  - Patient tolerance/ability to take oral medications/response to treatment ($n = 3$)
  - Patient suitability for surgery ($n = 1$)
  - Allergies ($n = 1$)
  - Whether they are taking any additional antimicrobials ($n = 2$)
- Fidaxomicin use
  - Economic considerations because of high cost ($n = 2$)
  - Availability in community care settings ($n = 4$)
- FMT
  - Availability of facilities for a FMT service ($n = 1$)

Table 3 Survey of CDI management practices in Europe, November 2022. Questions highlighted with an asterisk represent new questions that were not included in the 2017 survey

Question	*Answer choices*
Has your country issued national guidelines for managing patients with CDI?	Yes/no
What year were these guidelines first published?	
What year were these guidelines last revised?	A. Currently being revised, B. Revised in the last 5 years, C. More than 5 years since last revision, D. No revisions have been undertaken
Do these guidelines provide recommendations on:	A. Prevention of CDI, B. Surveillance of CDI, C. Surveillance of CDI, D. Laboratory diagnosis of CDI, E. Treatment of patients with CDI, F. Management of outbreaks and clusters of CDI, G. CDI key performance indicators (KPIs), H. Audit of guideline implementation, I. Other (please specify)
Has your national CDI treatment guideline been adapted with the publication of new ESCMID guidance? *	A. Yes, has been adapted, B. Yes, will be adapted/update planned, C. Don't have a national guideline, D. Don't know, E. No
Has your local CDI treatment guideline been adapted with the publication of new ESCMID guidance? *	A. Yes, has been adapted, B. Yes, will be adapted/update planned, C. Don't have a national guideline, D. Don't know, E. No
Have you surveyed/audited the implementation of these national/local CDI guidelines?	A. Yes all aspects of guidelines surveyed/audited, B. Yes, some aspects of guidelines surveyed/audited, C. Survey of guidelines in progress, D. No
When was this survey conducted?	Surveyed in the last year
	Surveyed in the last 5 years
	More than 5 years since last survey
Did you include CDI treatment as part of this survey?	Yes/no
Which facilities were included in the survey/audit? (tick all that apply)	A. Primary care/general practice, B. Hospitals, C. Nursing homes, D. Long-term care facilities, E. Other (please specify)
Does your agency/institution recommend using the European Society of Clinical Microbiology and Infectious Diseases (ESCMID) guidelines for managing patients with CDI?	A. Yes, B. No—recommend local guidelines, C. No—no recommendations
How do you define a severe case of CDI in your country? (tick all that apply)	A. Fevers, B. Raised lactate, C. Shock/hypotension, D. Rigors, E. Abdominal pain, F. Leucocytosis of ≥15,000 cells/μL, G. Serum creatinine of ≥50% above baseline, H. Serum creatinine >133 μmol/L, I. Pseudomembranous colitis on endoscopy, J. Evidence of colitis or ascites on CT imaging, K. Other (please specify)
When managing a patient with severe CDI—do you distinguish severe and severe-complicated CDI for patient management? *	A. Yes—using the ESCMID definition of severe-complicated CDI, B. No, C. If you use another definition, please specify
Which of the following CDI treatments are available in your country? (please tick all apply)	A. Metronidazole, B. Vancomycin, C. Fidaxomicin, D. Extended fidaxomicin, E. Bezlotoxumab, F. Tapering vancomycin regimen, G. Immunoglobulin therapy, H. Faecal microbiota transplantation, I. Other (please specify)
Is metronidazole still an option for CDI treatment? *	A. No, I no longer use metronidazole for CDI management, B. Yes, part of 1st choice, C. Yes, only in a selected low risk* patient group (*risk refers to adverse outcome/recurrence)

(continued)

Table 3 (continued)

Question	*Answer choices*
What is first choice for an initial episode? *	A. Metronidazole, B. Vancomycin, C. Fidaxomicin, D. Fidaxomicin in selected high-risk* patient group (*risk refers to adverse outcome/recurrence)
Which of the following factors influence the choice of treatment for CDI? (please select all relevant factors that influence choice of CDI treatment)	A. N/A, B. New CDI, C. Recurrent CDI (first episode), D. Recurrent CDI (second episode), E. Recurrent CDI (≥ third episode), F. History of CDI, G. Severe CDI, H. Host factors, e.g. serum albumin, age, I. Other factors (please state)
If other factors influence your choice of treatment for CDI, what are they?	
Are there restrictions in your country for clinical use of new/novel anti-CDI therapies, e.g. monoclonal antibodies, fidaxomicin?	Yes/no
Which restrictions apply? (tick all that apply)	A. Health technology assessment, B. Pharmacoeconomic review, C. National committee approval, e.g. drugs and therapeutics, D. Local committee approval, E. CEO approval, F. Regulatory authority, G. I don't know, H. Other (please specify)

  - Not used if patient's unable to swallow oral FMT ($n = 1$)
  - Use as an option for severe CDI when surgery is not possible ($n = 1$)
- Bezlotoxumab
  - Availability ($n = 3$)
  - Often only used in high-risk cohorts ($n = 1$)
- Immunoglobulin therapy
  - Used as an option for severe CDI when surgery is not possible ($n = 1$)

Approval pre-prescription was required before vancomycin ($n = 3$, 5%), fidaxomicin ($n = 10$, 6%), bezlotoxumab ($n = 11$, 17%) and FMT ($n = 10$, 6%).

Only six countries were represented in both surveys. The comparison of data from these countries reveals a decrease in metronidazole utilisation as a first-line treatment option for first episode CDI among respondents from all six countries (82.4% in 2017 vs 34.8% in 2022) and first CDI recurrence (23.5% in 2017 vs 8.7% in 2012). This decrease coincides with a rise in the use of fidaxomicin, from 17.6% in 2017 to 43.5% in 2022 (first episode CDI) and from 47.1% to 82.6% (first CDI recurrence). Though there was little change in prescription of vancomycin for first episode CDI between the two surveys (70.6% in 2017 vs 73.9% in 2022), vancomycin use decreased for first CDI recurrence (88.2% in 2017 to 34.8% in 2022). However, the accessibility of anti-CDI agents improved between the two surveys. Accessibility to the following agents increased between the two time periods: fidaxomicin (52.2% in 2017 vs 91.3% in 2022), immunoglobulin therapy (34.8% vs 43.5%) and FMT (47.8% vs 65.2%). The number of respondents who reported requiring approval prior to the prescription of certain anti-CDI agents has decreased from 55.8% in 2017 to 34.8% in 2022 (Tables 4 and 5).

4 *Clostridioides difficile* Pipeline Prophylactic and Therapeutic Agents

The four current approved therapeutic agents for CDI vary markedly in efficacy. While metronidazole has historically been the most commonly used option for treating CDI, as previously discussed, it is now known that this antibiotic is inferior to vancomycin (Johnson et al. 2014; Nelson et al. 2017). Concern regarding treatment failures with metronidazole remains (Vardakas et al. 2012). Metronidazole achieves poor intraluminal colonic concentrations, especially as mucosal inflammation subsides, such that the

Table 4 Recommendations for CDI treatment in 16 European countries with national ($n = 14$) or local ($n = 2$) CDI guidelines

	MTZ	Vancomycin	Fidaxomicin	Extended Fidaxomicin	Bezlotoxumab	Tapering vancomycin regimen	Immunoglobulin therapy	FMT
	Number (*n*) and percentage (%) of countries surveyed							
New CDI, *n* (%)	10 (62.5)	15 (93.8)	8 (50)	3 (18.8)	1 (6.3)	0 (0)	0 (0)	0 (0)
Recurrence (first), *n* (%)	2 (12.5)	11 (68.8)	10 (62.4)	7 (43.8)	5 (32.3)	5 (32.3)	1 (6.3)	5 (32.3)
Recurrence (second), *n* (%)	1 (6.3)	9 (56.3)	7 (43.8)	7 (43.8)	8 (50)	11 (68.8)	2 (12.5)	8 (50)
Three or more recurrences, *n* (%)	1 (6.3)	6 (37.5)	6 (37.5)	5 (32.3)	7 (43.8)	9 (56.3)	4 (25)	10 (62.5)
Severe CDI, *n* (%)	6 (37.5)	12 (75)	9 (56.3)	3 (18.8)	2 (12.5)	2 (12.5)	4 (25)	4 (25)
Other, *n* (%)	5 (31.3)	1 (6.3)	5 (32.3)	2 (12.5)	3 (18.8)	1 (6.3)	2 (12.5)	3 (18.8)

MTZ metronidazole, *FMT* faecal microbiota transplantation

Table 5 Recommendations for CDI management in 14 European countries with national CDI guidelines. Two countries with local guidelines were not included as applicable data was not available for these countries. Comparison made to 2018 survey

Recommendation	Included in guideline Number (*n*) and percentage (%) of countries 2017	2023
Surveillance of CDI, *n* (%)	11 (79)	8 (57)
Laboratory diagnosis of CDI, *n* (%)	12 (86)	13 (93)
Treatment of patients with CDI, *n* (%)	13 (93)	14 (100)
Management of outbreaks and clusters of CDI, *n* (%)	11 (79)	8 (57)
CDI key performance indicators (KPIs), *n* (%)	3 (21)	6 (43)
Audit of guideline implementation, *n* (%)	3 (21)	4 (29)
Other recommendations, *n* (%)	5 (36)	0 (0)

antibiotic may be undetectable as diarrhoea resolves. Also some *C. difficile* isolates show reduced susceptibility to metronidazole, which may be relevant given the suboptimal pharmacokinetics for this antibiotic in CDI. Laboratory detection of reduced metronidazole susceptibility is itself problematic with variations in methodology and MIC interpretation limiting analysis of trends and comparisons with published data (Moura et al. 2013).

Fidaxomicin and bezlotoxumab, a monoclonal anti-toxin B have been shown to reduce the risk of recurrent CDI by 40–50% in comparison with vancomycin alone (Wilcox et al. 2017; Cornely et al. 2012; Crook et al. 2012). High-acquisition cost of fidaxomicin has inhibited uptake in some settings and was observed in our survey of European countries as outlined above. However, a real-world study suggested a reduction in mortality associated with fidaxomicin use and that this was therapy was cost-effective (Goldenberg et al. 2016). In the phase 3 trials, bezlotoxumab was associated with a significant reduction in CDI readmissions.

The ideal antimicrobial agent for CDI should reduce vegetative *C. difficile* cells, toxins and spores in the host gut lumen without perturbation of the host microbiota, both to avoid creating an environment that is conducive to *C. difficile* expansion or to select for resistant potential pathogens (e.g. vancomycin resistant enterococci (VRE) or multiresistant Gram-negative bacilli) (Chang et al. 2008). This is a very challenging profile for an antibiotic, and indeed recent 'failures' of two antimicrobial agents in late-stage clinical trials emphasise how difficult it is to improve on current CDI therapies.

Emerging treatment options for CDI have been recently reviewed (Gonzales-Luna et al. 2023). In the following section, we update the data on options for CDI management that were considered emerging at the time of publication of the previous version of this chapter. Some of these options have been discontinued; others are now included in CDI guidelines, and others like ridinilazole, live biotherapeutic products and toxoid vaccine look promising. Ibezapolstat (previously ACX-362E) is a Gram-positive antimicrobial that inhibits bacterial DNA polymerase IIIC. This polymerase is present in Gram-positive bacteria including *C. difficile* though absent in *Actinobacteria* and Gram-negative host microbiota. It is an effective CDI therapy in animal models achieving high colonic and low systemic concentrations (van Eijk et al. 2019). In a recent phase 2a study (Garey et al. 2022), sustained clinical cure was reported in ten (of ten) CDI patients. Ibezapolstat was well tolerated, demonstrated high colonic and low systemic concentrations with beneficial microbiome and bile acid results (Tables 4, 5 and 6).

4.1 Surotomycin and Cadazolid

Surotomycin, an oral lipopeptide derivative of daptomycin, was examined in two phase 3 trials (NCT01598311 and NCT01597505) but did not

Table 6 Anti-CDI agents in the pipeline agents that have completed at least a phase 2 clinical trial for treatment or prevention of CDI

Clinical trial phase	Drug/product (developer)	Indication notes
Phase III	*C. difficile* vaccine (Sanofi Pasteur)	Primary prevention of CDI. NCT01887912: efficacy of vaccine (three doses) containing toxin A and B toxoids Last update posted: March 28, 2022 Terminated (the Independent Data Monitoring Committee (IDMC) concluded that the probability that the study will meet its primary objective is low)
	C. difficile vaccine (Pfizer)	Primary prevention of CDI Vaccine containing toxoids of toxin A and B. Three doses NCT03090191: efficacy of vaccine (three doses) containing toxin A and B toxoids Update: The full results are yet to be published though preliminary results are available on the company's website (Pfizer 2022)
	SER-109 (Seres)	Treatment of recurrent CDI Oral microbiome therapeutic (mixture of bacterial spores) tested in a single-arm, open-label clinical trial NCT03183128: Is SER-109 superior vs placebo to reduce recurrence of CDI? Update: Ended early due to COVID-19 pandemic. The study found that SER-109 was superior to placebo in reducing the risk of recurrent CDI (12% in the SER-109 group vs 40% placebo group), and the safety profile of SER-109 was similar to that of placebo
	Ridinilazole	Treatment of CDI NCT03595553: A global phase 3 trial evaluated ridinilazone (200 mg BD) versus vancomycin (125 QDS) for 10 days Status: Completed
	RBX2660 (Rebiotix)	NCT03244644: PUNCHCD3 is a prospective, multicentre, randomized, double-blinded, placebo-controlled phase 3 study of a microbiota suspension of intestinal microbes Status: Completed, enrolling for PUNCH CD3-OLS (NCT03931941)
Phase 2	Ridinilazole (SMT 19969, Summit)	Treatment of CDI Ridinilazole is a novel, small molecule, highly selective antibiotic. Successful phase 2 trial completed; phase 3 initiation expected 2018
	RBX2660 (Rebiotix)	Treatment of recurrent CDI Microbiota suspension. Three completed phase 2 trials Expected to enter phase 3 in 2017/2018
	SYN-004 (Synthetic Biologics)	Prevention of CDI. SYN-004 is a class A b-lactamase Successful phase 2 trial completed Phase 2b trial: A study of SYN-004 for the prevention of *C. difficile* in patients with a LRTI completed (NCT02563106) SYN-004 safety and tolerability in Allo-HCT subjects recruiting (NCT04692181)
	VLA84 (Valneva)	Primary prevention of CDI Vaccine consisting of a fusion protein with portions of toxins A and B Successful phase 2 trial completed in 2016 Phase 3 trial not planned
	Non-toxigenic *C. difficile* (Viropharma)	Prevention of recurrent CDI Biological therapy. Completed successful phase 2 trial in 2013 Phase 3 trial not planned
	Ramoplanin (Nanotherapeutics)	Treatment of CDI No new clinical efficacy data published since a phase 2 study was completed in 2004. Development plans/potential is therefore unclear. No clinical studies listed in clinicaltrials.gov

demonstrate non-inferiority compared with vancomycin (Boix et al. 2017). Notably, surotomycin dosing caused an overgrowth of Gram-negative bacilli in both in mice and in a gut model of CDI that is highly predictive of human disease; recurrent CDI was also seen in the latter model (Deshpande et al. 2016; Chilton et al. 2014b). A more recent meta-analysis (surotomycin versus vancomycin) again noted no significant difference in clinical cure and CDI recurrence overall, though lower rates of recurrence with surotomycin if NAP1/BI/027 (Muhammad et al. 2019). No further studies have been performed since 2017, and it appears that its production was discontinued due to its non-superiority to current therapies.

Cadazolid (Actelion), a novel hybrid oxazolidinone-fluroquinolone antibiotic that inhibits *C. difficile* protein synthesis and, to a lesser extent, DNA synthesis, did not meet its primary endpoint in comparison with vancomycin in one of two phase 3 trials (Actelion Ltd. 2017; Gehin et al. 2015; Chilton et al. 2014a; Baldoni et al. 2014). This may relate to the activity of cadazolid on the gut microbiome in vivo and/or persistence of *C. difficile* spores (Chilton et al. 2014a). In 2018, Johnson and Johnson announced the cessation of its clinical development program for cadazolid for CDI (Daley et al. 2017).

4.2 Ridinilazole

Ridinilazole (SMT19969) is a novel, non-absorbable, very-narrow-spectrum antimicrobial with minimal activity against host gut microbiota (Goldstein et al. 2013). It appears to act through classical antibiotic pathways, such as inhibition of cell wall, protein, lipid, RNA or DNA synthesis (Vickers et al. 2016). Basseres et al. described the effects of ridinilazole on *C. difficile* cell morphology, as visualised by scanning electron microscopy and confocal microscopy (Basseres et al. 2016). Following exposure to sub-lethal concentrations of ridinilazole, bacterial cell division was halted, and there was an absence of septum formation; this resulted in marked cell elongation. It has not been confirmed whether these observations are a direct effect of ridinilazole or a downstream response to the antibiotic. Ridinilazole has good activity against some but not all clostridia; it is 7- to 17-fold more active in vitro than metronidazole and vancomycin and has similar potency to fidaxomicin against *C. difficile* (Baines et al. 2015; Weiss et al. 2014; Sattar et al. 2015; Corbett et al. 2015). Notably, in vitro, in vivo and gut model data confirm that ridinilazole has little antimicrobial activity against indigenous gut microflora groups, except selected clostridia (Freeman et al. 2015b; Goldstein et al. 2013; Baines et al. 2015; Corbett et al. 2015; Chang et al. 2016b).

Safety and tolerability of ridinilazole was established in healthy subjects and in a recently reported phase II randomised double-blind trial (CoDIFy) (Vickers et al. 2015, 2017). CoDIFy was designed as a non-inferiority study and compared 10 days therapy of either oral ridinilazole 200-mg BD or oral vancomycin 125 mg QDS. Sustained clinical response rates were 67% and 42%, respectively ($n = 69$ mITT population); CDI recurrence occurred in 14% of ridinilazole recipients compared with 35% of vancomycin subjects; this difference meant that ridinilazole achieved a sustained response rate of 66.7% vs 42.4% for vancomycin, which met pre-set statistical superiority criteria (Vickers et al. 2017). Microbiome analyses of faecal samples from subjects in this phase 2 study showed that vancomycin recipients had a marked loss of diversity and replacement of the predominant phyla of healthy stool (*Bacteroides* and *Firmicutes*) by *Enterobacteriaceae*. These disruptions were still present 2 weeks after the end of treatment, even in subjects who had not had a recurrence at that point. By contrast, ridinilazole had a minimal effect on gut microbiota (Chang et al. 2016a).

Treatment with ridinilazone was recently reported to significantly decrease the rate of recurrent CDI compared with vancomycin (Okhuysen et al. 2022). A global phase 3 trial evaluated ridinilazone (200 mg BD) versus vancomycin (125 QDS) for 10 days. Of 745 patients in the modified intention-to-treat analysis (ridinilazone $n = 370$; vancomycin $n = 375$), there was no difference in sustained clinical response as defined by clinical response and no recurrent CDI though 30 days post end of treatment. However, patients treated with ridinilazole had significantly reduced rates of recurrent CDI compared with those in the vancomycin group (8.1% vs 17.3%, respectively; $P = .0002$). This was more significant among patients who were not receiving additional antibiotics, with CDI recurrence rates of 6.7% and 16.5% observed among those in the ridinilazone and vancomycin groups, respectively ($P = .0005$). Of note, increased microbiome diversity and reduced abundance and concentrations of faecal secondary bile acids at treatment completion was noted in patients who received ridinilazole.

A clinical trial (clinicaltrials.gov, NCT04802837) comparing ridinilazole versus vancomycin for CDI treatment is currently underway in adolescents aged 12–17 years.

A longitudinal study comparing ridinilazole versus vancomycin indicated that ridinilazole maintains an intestinal bile acid profile associated with a lowered risk of CDI recurrence (Qian et al. 2020). In this study, the ratio of conjugated to secondary bile acids in patients treated with vancomycin increased almost 100-fold increase, whereas ridinilazole had little impact. *Bacteroidales* and *Clostridiales* spp. were depleted in the vancomycin group but preserved at near-baseline levels in the ridinilazole group. Bile acid ratios at the end of CDI treatment were significantly different between in those with CDI recurrence.

4.3 CDI Prophylaxis

As outlined in Sect. 2.4 previously, the updated ESCMID guidelines contain a new section on the role of anti-CDI antibiotic prophylaxis for patients on systemic antibiotic treatment. The most hopeful strategies are those aimed at reducing changes in intestinal microbiota and development of non-toxin-based vaccines.

4.3.1 Ribaxamase

Ribaxamase (SYN-004, synthetic biologics) is a recombinant beta-lactamase that has been formulated to be administered orally in patient receiving beta-lactam antibiotic therapy (Kaleko et al. 2016; Connelly et al. 2015). Ribaxamase degrades unmetabolised antibiotic in the colon to reduce the deleterious effects on the gut microbiota (Roberts et al. 2016). Animal studies have demonstrated safety and notably no reduction in the systemic concentration of co-administered ceftriaxone (Connelly et al. 2015). A phase 2 double-blind placebo-controlled study has examined the potential of ribaxamase to prevent CDI, antibiotic-associated diarrhoea and the emergence of antimicrobial-resistant potential pathogens in patients hospitalized with a lower respiratory tract infection treated with IV ceftriaxone (Synthetic Biologics 2017). Patients who received ribaxamase had a 71.4% relative risk reduction for CDI ($P = 0.045$). There was also a significant reduction in new colonisation by VRE in ribaxamase versus placebo recipients ($P = 0.0002$). Adverse events were similar in active and placebo patients.

A subsequent phase 2b proof-of-concept study of 412 hospitalized patients reported that the co-administration of ribaxamase with ceftriaxone resulted in a 2.4% reduction in CDI occurrence, although the lower limit of the 95% confidence interval did fall below zero (−0.6), indicating the possibility of no effect. Microbiome analysis demonstrated reduction of ceftriaxone-induced changes in patients treated with ribaxamase that recovered more quickly than placebo. A subsequent analysis reported reduced changes to the gut resistome subsequent to ceftriaxone administration in patients also treated with ribaxamase (Kokai-Kun et al. 2019, 2020).

4.3.2 DAV132

Another novel approach to CDI prophylaxis is DAV132 (DaVolterra), which is an activated charcoal based product that is administered as an enteric coated capsule. DAV132 irreversibly captures antibiotics in the intestine while avoiding interruption of antibiotic absorption.

DAV132 has been examined in a proof-of-concept study involving 18 healthy subjects who had received DAV132, uncoated formulated activated charcoal (FAC) or water 16 and 8 h before, alongside the probe drugs, and 8 h thereafter. The AUC0-96 h of amoxicillin was reduced by more than 70% when it was taken with FAC, but was not adversely affected when taken with water or DAV132. By contrast, the AUC0-96 h of sulfapyridine was reduced by >90% when administered with either FAC or DAV132 in comparison with water. Hence, DAV132 can selectively adsorb drugs in the proximal colon, without interfering with their absorption.

In healthy volunteers treated with moxifloxacin, DAV-132 was shown to be effective to protect the gut microbiome (de Gunzburg et al. 2018). In this study, one group received moxifloxacin with DAV132 coadministration ($n = 14$), while another group received moxifloxacin alone ($n = 14$). Additionally, two control groups of eight volunteers each were included, one receiving DAV132 alone and the other receiving a nonactive substitute.

When DAV132 was co-administered with moxifloxacin, free moxifloxacin faecal concentrations decreased by 99%, with plasmatic levels remaining largely unaffected. Shotgun quantitative metagenomics were used to analyse the composition of the intestinal microbiota, which was mostly preserved in subjects who were co-treated with DAV132. Furthermore, no adverse effects were observed. Ex vivo experiments also demonstrated that DAV132 efficiently adsorbed a broad range of clinically relevant antibiotics.

A recent randomized control trial (Vehreschild et al. 2022), designed to assess safety and efficacy of DAV132 in 243 hospitalized patients receiving fluroquinolones (123 of whom also received DAV132), reported no significant difference in adverse effects: 18 (14.8%) DAV132 vs 13 (10.8%) No-DAV132 patients (difference 3.9%; 95% CI: −4.7 to 12.6). DAV132 was associated with a >98% reduction in faecal fluroquinolone levels (Day 4 to end of treatment; $P < 0.001$), less impaired microbiota diversity (Shannon index; $P = 0.003$), increased ex vivo resistance to *C. difficile* colonization ($P = 0.0003$) and less frequent FQ-induced VRE acquisition ($P = 0.01$).

4.4 Active *C. difficile* Immunisation

Vaccination to boost host antibody-mediated immunity is an attractive strategy to prevent CDI. The relative importance of *C. difficile* toxins A and B to human infection remains controversial, but host immune response to these toxins likely influences the likelihood of infection, clinical severity and outcome of CDI (Solomon et al. 2013; Kuehne et al. 2010). Higher serum IgG levels to toxin A have been shown in patients with asymptomatic colonisation compared with those with CDI, and recurrent infection is associated with poor IgG and IgM responses (Kyne et al. 2000, 2001). Interestingly, the effectiveness of the anti-toxin B monoclonal antibody bezlotoxumab at reducing the risk of CDI recurrence was not enhanced by the addition of an anti-toxin A monoclonal antibody, actoxumab; also, actoxumab alone was not efficacious at preventing recurrence. Nevertheless, it remains logical to design a vaccine around the augmentation of the host response to both toxins A and B (Kuehne et al. 2010). Other *C. difficile* antigens may also be important, noting, for example, that antibodies to surface proteins are greater in colonised versus infected patients (Pechine et al. 2005). To date, two vaccine candidates have completed phase 3 trials (PF06425090 (Kitchin et al. 2020) and a Sanofi Pasteur vaccine candidate that has yet to be named (de Bruyn et al. 2021)) with two additional vaccine candidates in early clinical trials (GSK2904545A (clinicaltrials.gov, NCT04026009) and VLA84 (clinicaltrials.gov, NCT02316470)).

Three vaccines that use *C. difficile* toxin targets have progressed to phase 2 or 3 clinical development. The first to reach a phase 3 clinical trial is a formalin-inactivated toxoid-based vaccine developed by Sanofi Pasteur (Foglia et al. 2012). Following vaccination, seroconversion to toxin A was more pronounced than to toxin B (but took up to 70 days) and notably was less common in elderly subjects; three vaccine doses were required to achieve an adequate neutralising-antibody response (Foglia et al. 2012; Kotloff et al. 2001). A 100-μg dose (given with an AlOH adjuvant) was found to yield the best immunogenic response, and a phase 3 trial of this vaccine in the prevention of primary CDI in at-risk subjects aged >50 years commenced in 2013 (NCT01887912). Unfortunately it was discontinued as it was not shown to prevent CDI (de Bruyn et al. 2021; Reigadas et al. 2021).

PF06425090 (Pfizer Inc.) is a formalin-inactivated toxoid-based vaccine, but with alterations in both toxins A and B to reduce toxigenicity, has recently completed a phase 3 primary CDI prevention trial (clinicaltrials.gov, NCT03090191), also based on a three-dose strategy (Donald et al. 2013; Sheldon et al. 2016). The phase 3 trial was conducted in people over 50 years who had recent antibiotic therapy (previous 12 weeks) or were likely to have future contact with the healthcare system. The full results are yet to be published though preliminary results are available on the company's website (Pfizer 2022). Of 8766 people who received PF06425090 (placebo 8769), vaccine efficacy was 28.6% (96.4% CI −28.4 to 61.0%) after dose two and 31% (96.4% CI −38.7 to 66.6%) after three doses. The vaccine was very well tolerated and showed a favourable safety profile.

A third *C. difficile* vaccine candidate (VLA84, Valneva) has completed a phase 2 trial with 500 subjects (Valneva 2016). To date this candidate has not progressed to phase 3 trials. VLA84 uses a different antigen approach to either of the two toxoid-based vaccines that are currently undergoing phase 3 evaluation. VLA84 is a single recombinant fusion protein consisting of portions of the C-terminal cell-binding domains of toxins A and B. The developers claim that production and characterization of VLA84 could be simpler and less costly compared with toxoid-based vaccines. The phase 2 study of VLA84 met its primary endpoint in terms of identifying the dose and formulation with the highest seroconversion rate against both toxins A and B (subjects were followed up to day 210) and confirmed the favourable safety profile that was seen in phase 1.

In a recent paper, the authors Wang et al. (2022) describe a genetically modified, nontoxigenic *C. difficile* strain expressing immunodominant fragments of toxins A and B. Oral immunization of mice provided effective protection against infection with a hypervirulent strain of *C. difficile*. This may represent a candidate for a novel mucosal vaccine against CDI by targeting both toxins and colonization of pathogenic *C. difficile*.

4.5 Microbiome-Based Therapeutics

In recent years, the central role of the microbiome in a person's risk for CDI and subsequence recurrences has driven renewed investigation of microbiome-based therapies. The updated 2021 ECCMID CDI treatment guidelines have reiterated the role of FMT in multiply recurrent CDI and provided some guidance regarding its role in other CDIs (e.g. severe infection). Live biotherapeutic products (LBP) are a promising option as a strategy to prevent CDI recurrence by restoration of dysbiosis. As they are regulated as 'drugs' (FDA 2016), they are much more regulated than, e.g. probiotics, and will therefore require efficacy as well as safety data before approval.

4.5.1 Faecal Microbiota Transplantation

The evidence base concerning the effectiveness of FMT continues to grow, but it remains a non-regulated product, with many different versions reported. FMT comprises the administration of a complex live faeces-derived mixture of microorganisms, including some of uncertain significance (some beneficial, others possibly harmful or neither) and so (particularly longer term) safety remains unproven. Of particular concern here is the increasing use of FMT when licensed CDI therapeutics has not been tried. Hence, different regulatory authorities have taken varied stances on FMT to safeguard patient interests. Requirements for consenting subjects, screening of donors and recipients, faecal material preparation and delivery via either rectal or nasogastro-/duodenal routes, mean that there are intensive endeavours to develop alternatives to FMT that can still harness the restorative and protective effectiveness of specific components of the gut microbiota, but possibly with greater reassurance on safety.

The first randomised (sham procedure controlled) trial of FMT to treat recurrent CDI demonstrated an intention-to-treat (ITT) efficacy rate of 81% to prevent further recurrences; notably, however, the study contained only 16 patients in the FMT arm (van Nood et al. 2013). In a randomised but non-blinded clinical trial, 39 subjects with recurrent CDI were given FMT (preceded by vancomycin 125 mg QDS for 3 days), comprising at least 1 infusion of faeces via colonoscopy or vancomycin 125 mg QDS for 10 days and then 125–500 mg/day every 2–3 days for at least 3 weeks. The primary endpoint was the resolution of diarrhoea related to CDI at week 10; surprisingly, a positive *C. difficile* test was not required to define recurrence post-study treatment (Cammarota et al. 2015). The study was stopped after a 1-year interim analysis, at which point 18/20 (90%) vs 5/19 (26%) patients in the FMT vs vancomycin treatment groups, respectively, had resolution of *C. difficile* diarrhoea ($P < 0.0001$). There were no significant adverse events in either of the study groups.

Adults with recurrent or refractory CDI were enrolled in a randomised, double-blind, non-inferiority study in six Canadian centres of freeze-thawed ($n = 114$) vs fresh ($n = 118$) FMT via enema. Clinical resolution without recurrence up to 13 weeks did not differ significantly in the per-protocol (83.5% vs 85.1%) and mITT (75.0% vs 70.3%) populations (Lee et al. 2016). These results suggest that using freeze-thawed faecal material is a practicable alternative to fresh donor material. All patients received suppressive antibiotics for the most recent episode of CDI, and these were discontinued 24–48 h before FMT; this probably explains why only 38% of the subjects were positive for toxin or toxin gene immediately prior to FMT administration. Notably, about one third of FMT recipients in both groups, who were ultimately, classified as resolved, required two FMTs, which is a relatively common observation. A non-blinded, non-randomised study of encapsulated (and freeze-thawed) faeces was performed in 20 subjects with at least 3 episodes of mild-to-moderate CDI and failure of 6 to 8 weeks of vancomycin therapy, or $\geq$2 episodes of severe CDI requiring hospitalization (Youngster et al. 2014). Diarrhoea resolution occurred in 14 patients (70%; 95% CI, 47–85%) after a single capsule-based FMT; 4/6 retreated nonresponders had resolution of diarrhoea, giving an overall 90% (95% CI, 68–98%) response rate. No serious adverse events were attributed to FMT.

The six randomised controlled trials of FMT have been recently reviewed, three that compared FMT to antibiotic management; the remainder compared FMT to various 'types' of FMT in terms of preparation, source and delivery (Johnson and Gerding 2017). It is important to note that, unlike prior uncontrolled studies that reported FMT efficacy rates of at least 90%, efficacy (for one FMT) in these RCTs was 44–91%, with four recording success rates of $\leq$65%. These include a randomized controlled trial of FMT versus a 6-week vancomycin tapering regimen (VAN-TP) (Hota et al. 2017). VAN-TP was stopped early for futility; 56% of patients randomized to FMT by enema developed recurrent CDI, compared with 42% VAN-TP recipients.

There are many important factors for European clinicians to consider when establishing or using a FMT service. Factors that should be taken into account at an institutional level when commencing an FMT service are the national regulatory frameworks that FMT falls under (i.e. as a drug or biological material), donor selection and screening practices, stool preparation techniques and long-term safety of microbiome manipulation in these patients. Concerns regarding the long-term safety of FMT are not unfounded, especially in patients with inflammatory bowel disease. Reports of peripheral neuropathy, Sjögren syndrome, idiopathic thrombocytopenic purpura, microscopic colitis, contact dermatitis, rheumatoid arthritis, obesity, bacteraemia and ulcerative colitis flare after FMT (Tariq et al. 2016; De Leon et al. 2013; Quera et al. 2014; Alang and Kelly 2015). Institutions need to ensure they are working within their national and European frameworks and regulations. Where national regulations are absent, comparisons should be made to international standards to ensure the highest level of safety. In Europe, the regulation of FMT is currently at the discretion of the EU member states, though in many countries no such national regulation exists. Future planned EU regulation of FMT donor material may hinder its widespread use, depending on whether it is regulated as a drug or bodily tissue. A European consensus paper provided recommendations on a number of areas pertinent to FMT implementation, including regulatory, administrative and laboratory guidelines (Cammarota et al. 2017).

In addition to the management of recurrent CDI, the 2021 ESCMID guidelines (van Prehn et al. 2021) note that FMT may also have a role in severe-complicated refractory CDI especially in patients that are not considered appropriate for surgical management and/or as an alternate to surgical management. However, surgical consultation must always be sought first, and clinicians would require ready access to standardized, screened FMT products, and a case-by-case risk assessment with patient consent should be carried out prior to a decision regarding FMT. Most recently, the EarlyFMT trial (Baunwall et al. 2022) examined the role of FMT in patients with first or second episode of CDI. The trial was terminated early as FMT was superior in achieving sustained CDI resolution at 8 weeks. Surprisingly, the investigators used a positive PCR test alone to determine CDI, though international guidance recommends a two-step testing protocol (van Prehn et al. 2023). Interestingly, the low sustained response rate (33%) in the placebo arm was comparable to previous reports in patients with multiple recurrent CDI.

4.5.2 Live Biotherapeutic Microbiota Preparations

4.5.2.1 RBX2660

RBX2660 is a live bio therapeutic microbiota suspension that aims to harness the effectiveness of FMT, but within a standardised, regulated product, for the treatment of recurrent CDI. It has been studied in three phase 2 clinical trials. PUNCH CD (NCT01925417) was a safety-focussed, prospective multicentre, open-label study; 34 subjects (with ≥2 recurrent CDI episodes or ≥2 severe episodes resulting in hospitalization) received at ≥1 dose of RBX2660 and 31 completed 6-month follow-up (Orenstein et al. 2016). Following a 10–14-day course of anti-CDI antibiotics and a 24–48 h washout period, RBX2660 was administered as a single dose via enema. Further recurrent CDI occurred in 48% of subjects after one dose of RBX2660, with 15/31 patients receiving a second enema; of these, 78.6% were considered to be treatment successes, contributing to an overall success rate of 27/31 (87.1%). No serious adverse events were related to RBX2660.

PUNCH CD 2 (NCT02299570) was a phase 2b multicentre randomized double-blind, placebo-controlled trial with 2-year follow-up (Dubberke et al. 2016). The primary efficacy objective was assessment of response (defined as no CDI recurrence) to RBX2660 versus placebo at 8 weeks. A total of 127 patients formed the ITT population (enrolled at 21 sites in the USA and Canada); patients were randomized into three treatment arms: two doses of RBX2660 (Group A, $n = 41$), two doses of placebo (Group B, $n = 44$) or one dose of RBX2660 and one dose of placebo (Group C, $n = 42$) via enema with doses 7 days apart. Efficacy for Group A was 61% vs 45.5% for Group B, $P = 0.152$. Efficacy for Group C was 66.7% compared with Group B (45.5%), $P = 0.048$; efficacy of Group A and C (63.9%) vs B (45.5%), $P = 0.046$. For subjects who developed recurrent CDI after receipt of study drug, open-label treatment success was Group A (68.8%, 11/16); Group B (87.5%, 21/24); Group C (71.4%; 10/14) for an overall open label success rate of 77.8%. Adverse events at 56 days were primarily gastrointestinal, with no significant difference in the proportion of adverse or serious adverse events among the treatment groups. As the two doses of RBX2660 treatment arm was not superior to two doses of placebo, the primary efficacy endpoint was not met.

The third phase 2 study, PUNCH Open Label (NCT02589847), had 31 active treatment sites and 4 control sites in the USA and Canada. One hundred thirty-two RBX2660 and 110 historical control subjects were included; follow-up results at 8 weeks have been reported, although there is a 2-year assessment point also (Rebiotix Inc 2017). RBX2660 met its primary efficacy endpoint at 8 weeks, preventing CDI recurrence, with a success rate of 78.8% compared with 51.8% in historical controls treated with antibiotics alone ($P < 0.0001$). No new safety concerns were identified. Analyses of faecal microbiomes shows that these became more diverse and aligned to a 'healthy' microbiome after treatment with RBX2660 (Blount et al. 2017; Ray et al. 2017). 16S rRNA sequencing was also performed on stool samples collected from 42 subjects treated with RBX2660 treatment arm and for 19 RBX2660 drug lots. The RBX2660 microbial profiles had similar taxonomic distributions, with a group mean that was highly divergent and significantly different from those of patients at baseline. However, after RBX2660 treatment, patients' microbiomes progressively resembled those of RBX2660.

The phase 3 randomized, double-blind, placebo-controlled trial (PUNCH CD3) of RBX2660 (Khanna et al. 2022a) was recently conducted in adults who had one or more CDI recurrences and were previously treated with antibiotics. Participants were randomly assigned to receive either RBX2660 ($n = 180$) or a placebo enema ($n = 87$), with the primary endpoint being the absence of CDI diarrhoea within 8 weeks of treatment. Because of difficulties recruiting patients because of the COVID-19 pandemic, the use of a Bayesian analysis to assess the primary endpoint that incorporated patients receiving one dose of RBX2660 from the phase 2 PUNCH CD2 trial and patients from the PUNCH CD3 trial was permitted. This demonstrated a treatment success rate of 70.6% with RBX2660 vs 57.5% with placebo, with an estimated treatment effect of 13.1%. The sustained response rate was high for both groups after 6 months. RBX2660 was generally well-tolerated but had a higher incidence of mild gastrointestinal events compared to placebo. Overall, the study concluded that RBX2660 is a safe and effective treatment for reducing recurrent CDI following standard-of-care antibiotics. Updated interim analysis of PUNCH CD3-OLS (Kraft et al. 2021) in a patient cohort with broad eligibility criteria (including inflammatory bowel disease and irritable bowel syndrome) reported that RBX2660 consistently reduced CDI recurrence and was well-tolerated. Across all five trials, treatment success has ranged from 50 to 79% (Bancke and Su 2021).

4.5.2.2 SER-109

SER-109 (Seres) is also a live biotheraputic that comprises an encapsulated mixture of purified *Firmicutes* spores, obtained from the faeces of healthy humans, which were effective at preventing CDI in animal models. The resilience of the spores means that an ethanol-based purification process can be applied to reduce the risk that transmissible infectious agents contaminate the therapeutic product. Also, resistance to gastric acid facilitates oral dosing. Two phase 2 studies of SER-109 have been completed. The first was a non-comparative study in patients with ≥3 CDI episodes during 12 months (Khanna et al. 2016). Following standard-of-care CDI antibiotic treatment, patients received SER-109 either on 2 consecutive days (geometric mean dose, 1.7×10^9 spores) or on 1 day (geometric mean dose, 1.1×10^8 spores). The primary endpoint was absence of *C. difficile*-positive diarrhoea during 8 weeks of follow-up. In total, 26/30 patients (86.7%) across the 2 dosing groups met the primary efficacy endpoint. Three patients with early, self-limiting *C. difficile*-positive diarrhoea did not require antibiotic treatment and were *C. difficile*-negative on re-testing at 8 weeks; thus, 29/30 (96.7%) were considered to have achieved clinical resolution. Notably, gut microbiome analyses showed that baseline loss of microbiota diversity was rapidly reversed after receipt of SER-109, with persistence of *Firmicutes* spores. There were no safety concerns in the study.

A phase 2 (ECOSPORE) study of SER-109 enrolled 89 subjects with ≥3 recurrences who were randomized (2:1 ratio) in a placebo-controlled, double-blind, 24-week trial (Trucksis et al. 2017). SER-109 was administered orally as a single dose (1×10^8 bacterial spores), after CDI antibiotic treatment. Recurrence was defined as diarrhoea for ≥2 consecutive days, a positive CDI test and the need for antibiotic treatment. The study's primary endpoint of reducing the relative risk of CDI recurrence at 8 weeks was not achieved, despite a (nonsignificant) reduction in the relative risk of CDI recurrence. In the ITT population, recurrence occurred in 44% (26/59) vs 53% (16/30) of subjects who received SER-109 vs placebo, respectively. A pre-specified sub-group analysis showed that the lack of efficacy of SER-109 to prevent recurrence occurred in subjects aged <65 years old. However, in subjects aged ≥65 years old, CDI recurrence occurred in 45% of SER-109 (14 of 31) recipients, and in 80% of those who received placebo (12 of 15). A re-analysis showed that the disappointing results may be because cases were included and recurrences diagnosed without the most stringent requirement for free faecal toxin to be present. Also, while SER-109 was biologically active, a higher dose may be necessary.

The phase 3, double-blind, randomized, placebo-controlled ECOSPOR III trial (Feuerstadt et al. 2022) included the requirement for a positive *C. difficile* toxin assay for eligibility and used an increased dose of four SER109 capsules given once daily for 3 days. Patients who had three or more CDI episodes received SER-109 or placebo after standard-of-care antibiotic treatment. The primary objective was to show the superiority of SER-109 in reducing the CDI recurrence risk up to 8 weeks after treatment. Because of the COVID-19 pandemic, the trial ended early with 182 enrolled patients. The study found that SER-109 was superior to placebo in reducing the risk of recurrent CDI (12% in the SER-109 group vs 40% placebo group), and the safety profile of SER-109 was similar to that of placebo. SER-109 dose species were detected as early as week 1 and were associated with bile acid profiles that inhibit *C. difficile* spore germination. Most recently, SER-109 was shown to be well tolerated in patients with recurrent CDI and comorbidities (Sims et al. 2023). In this study of 263 patients with a history of recurrent CDI, the rate of recurrent CDI was low at 8.7% regardless of the number of prior recurrences or diagnostic approach.

4.5.2.3 Non-toxigenic *C. difficile*

Non-toxigenic *C difficile* (NTCD) strains are avirulent. Theoretically, it may be possible to displace toxigenic strains in colonised (or infected) individuals. A randomized, double-blind, placebo-controlled, dose-ranging study examined the efficacy of a NTCD strain to prevent recurrent CDI in patients with either primary (>80%) or recurrent CDI who had completed treatment with metronidazole, vancomycin or both (Gerding et al. 2015). Approximately two thirds (69%) of recipients became colonised by NTCD. CDI recurrence rates were 2% in colonized subjects, compared with 31% (similar to placebo) in those not colonised ($P < 0.001$), highlighting the correlation between engraftment and clinical efficacy. Interestingly, no subjects who were colonised at week 6 remained so at week 26. It remains unclear whether this successful proof of concept phase 2 clinical trial will lead to commercial development of NTCD.

5 Summary

In summary, there are varied approaches in advanced clinical trials for the primary prevention, treatment and/or secondary prevention of CDI. Unfortunately, however, recent experience shows us that developing new management options for CDI is very challenging. Well-designed trials with clearly defined patient populations are key to delivering new therapeutic and preventative options. Research gaps outlined in the 2021 ESCMID guidelines include delineating optimal CDI treatment and treatment algorithms in large-scale trials independent from pharmaceutical industry, investigation of the exact mechanism of FMT for CDI treatment and health-economic studies in different settings and population for selection of CDI treatments. The updated European survey among clinical microbiologist and infectious disease specialists indicates that significant variation remains among CDI management practices in Europe. Implementation of CDI guidelines is not routinely audited. Access to anti-CDI agents still impacts treatment practices in some countries.

Acknowledgements We wish to thank colleagues who took the time to complete the online survey of CDI treatment in Europe in November 2022.

Conflicts of Interest FF has received research grant support and been a recipient of a consultancy fee from Tillots Pharma.

JvP has received research grant support by Merck Sharp & Dohme.

MW has received consulting fees from the following: AiCuris, Bayer, Crestone, Da Volterra, Deinove, EnteroBiotix, the European Tissue Symposium, Ferring, GSK, Menarini, Merck, Nestlé, Paion, Paratek, Pfizer, Phico Therapeutics, Qpex Biopharma, Seres, Surface Skins, Summit, Tillotts, Vaxxilon/Idorsia and Vedanta; Lecture fees from GSK, Merck, Pfizer, Seres & Tillotts; and Grant support from Almirall, Da Volterra, EnteroBiotix, GSK, Merck, MicroPharm, Nabriva, Paratek, Pfizer, Seres, Summit, the European Tissue Symposium and Tillotts.

References

A randomized, double-blind, active controlled study to evaluate the safety and tolerability of ridinilazole compared with vancomycin and to assess the pharmacokinetics of ridinilazole in adolescent subjects (aged 12 to <18 years) with Clostridioides ClinicalTrials.gov identifier: NCT04802837. Updated October 13, 2022. https://ClinicalTrials.gov/show/NCT04802837. Accessed 22 Mar 2023

A phase 3, placebo-controlled, randomized, observer-blinded study to evaluate the efficacy, safety, and tolerability of a Clostridium difficile vaccine in adults 50 years of age and older. ClinicalTrials.gov identifier: NCT03090191. Updated February 13, 2023. https://ClinicalTrials.gov/show/NCT03090191. Accessed 16 Mar 2023

ActelionLtd (2017) Actelion provides an update on the Phase III IMPACT program with cadazolid in CDAD

Alang N, Kelly CR (2015) Weight gain after fecal microbiota transplantation. Open Forum Infect Dis 2(1):ofv004. https://doi.org/10.1093/ofid/ofv004

Baines SD, Crowther GS, Freeman J, Todhunter S, Vickers R, Wilcox MH (2015) SMT19969 as a treatment for *Clostridium difficile* infection: an assessment of antimicrobial activity using conventional susceptibility testing and an *in vitro* gut model. J Antimicrob Chemother 70(1):182–189. https://doi.org/10.1093/jac/dku324

Baldoni D, Gutierrez M, Timmer W, Dingemanse J (2014) Cadazolid, a novel antibiotic with potent activity against *Clostridium difficile:* safety, tolerability and pharmacokinetics in healthy subjects following single and multiple oral doses. J Antimicrob Chemother 69(3):706–714. https://doi.org/10.1093/jac/dkt401

Bancke L, Su X (2021) Efficacy of investigational microbiota-based live biotherapeutic RBX2660 in individuals with recurrent Clostridioides difficile infection: data from five prospective clinical studies. Open Forum Infect Dis 8(Supplement_1):S100–1

Basseres E, Endres BT, Khaleduzzaman M, Miraftabi F, Alam MJ, Vickers RJ, Garey KW (2016) Impact on toxin production and cell morphology in *Clostridium difficile* by ridinilazole (SMT19969), a novel treatment for *C. difficile* infection. J Antimicrob Chemother 71(5):1245–1251. https://doi.org/10.1093/jac/dkv498

Baunwall SMD, Andreasen SE, Hansen MM, Kelsen J, Høyer KL, Rågård N, Eriksen LL, Støy S, Rubak T, Damsgaard EMS, Mikkelsen S, Erikstrup C, Dahlerup JF, Hvas CL (2022) Faecal microbiota transplantation for first or second Clostridioides difficile infection (EarlyFMT): a randomised, double-blind, placebo-controlled trial. Lancet Gastroenterol Hepatol 7(12):1083–1091. https://doi.org/10.1016/S2468-1253(22)00276-X

Biswas J, Patel A, Otter J, Van Kleef E, Goldenberg S (2015) Contamination of the hospital environment from potential Clostridium difficile excretors without active infection. Infect Control Hosp Epidemiol 36(8):975–977. https://doi.org/10.1017/ice.2015.79

Blount K, Jones C, Shannon W, Carter S (2017) Changing the microbiome: patients with a successful outcome following microbiota-based RBX2660 treatment trend toward human microbiome project healthy subjects' profile. Paper presented at the ASM Microbe, New Orleans, USA. Abstract 212

Boekhoud IM, Sidorov I, Nooij S, Harmanus C, Bos-Sanders IMJG, Viprey V, Spittal W, Clark E, Davies K, Freeman J, Kuijper EJ, Smits WK, COMBACTE-CDI Consortium (2021) Haem is crucial for medium-dependent metronidazole resistance in clinical isolates of Clostridioides difficile. J Antimicrob Chemother 76(7):1731–1740. https://doi.org/10.1093/jac/dkab097

Boix V, Fedorak RN, Mullane KM, Pesant Y, Stoutenburgh U, Jin M, Adedoyin A, Chesnel L, Guris D, Larson KB, Murata Y (2017) Primary outcomes from a Phase 3, randomized, double-blind, active-controlled trial of surotomycin in subjects with *Clostridium difficile* infection. Open Forum Infect Dis 4(1):ofw275. https://doi.org/10.1093/ofid/ofw275

Cammarota G, Ianiro G, Tilg H, Rajilic-Stojanovic M, Kump P, Satokari R, Sokol H, Arkkila P, Pintus C, Hart A, Segal J, Aloi M, Masucci L, Molinaro A, Scaldaferri F, Gasbarrini G, Lopez-Sanroman A, Link A, de Groot P, de Vos WM, Hogenauer C, Malfertheiner P, Mattila E, Milosavljevic T, Nieuwdorp M, Sanguinetti M, Simren M, Gasbarrini

A (2017) European consensus conference on faecal microbiota transplantation in clinical practice. Gut 66(4):569–580. https://doi.org/10.1136/gutjnl-2016-313017

Cammarota G, Masucci L, Ianiro G, Bibbo S, Dinoi G, Costamagna G, Sanguinetti M, Gasbarrini A (2015) Randomised clinical trial: faecal microbiota transplantation by colonoscopy vs. vancomycin for the treatment of recurrent *Clostridium difficile* infection. Aliment Pharmacol Therapeut 41(9):835–843. https://doi.org/10.1111/apt.13144

Chang J, Kane A, McDermott L, Vickers R, Snydman D, Thorpe C (2016a) Ridinilazole preserves major components of the intestinal microbiota during treatment of *Clostridium difficile* infection. Paper presented at the ECCMID Amsterdam, Netherlands. Abstract LB-116

Chang J, Kane A, Snydman D (2016b) Ridinilazole preserves major components of the intestinal microbiota during treatment of *Clostridium difficile* infection. Paper presented at the American Society for Microbiology, Boston, MA, 16–20 June 2016. Abstr LB-116

Chang JY, Antonopoulos DA, Kalra A, Tonelli A, Khalife WT, Schmidt TM, Young VB (2008) Decreased diversity of the fecal Microbiome in recurrent *Clostridium difficile*-associated diarrhea. J Infect Dis 197(3): 435–438. https://doi.org/10.1086/525047

Chew KS, van Merrienboer J, Durning SJ (2016) A portable mnemonic to facilitate checking for cognitive errors. BMC Res Notes 9(1):445. https://doi.org/10.1186/s13104-016-2249-2

Chilton CH, Crowther GS, Baines SD, Todhunter SL, Freeman J, Locher HH, Athanasiou A, Wilcox MH (2014b) *In vitro* activity of cadazolid against clinically relevant *Clostridium difficile* isolates and in an *in vitro* gut model of *C. difficile* infection. J Antimicrob Chemother 69(3):697–705. https://doi.org/10.1093/jac/dkt411

Chilton CH, Crowther GS, Todhunter SL, Nicholson S, Freeman J, Chesnel L, Wilcox MH (2014a) Efficacy of surotomycin in an in vitro gut model of *Clostridium difficile* infection. The Journal of antimicrobial chemotherapy 69(9):2426–2433. https://doi.org/10.1093/jac/dku141

Clancy CJ, Buehrle D, Vu M, Wagener MM, Nguyen MH (2021) Impact of Revised Infectious Diseases Society of America and Society for Healthcare Epidemiology of America Clinical Practice Guidelines on the Treatment of Clostridium difficile infections in the United States. Clin Infect Dis 72(11):1944–1949. https://doi.org/10.1093/cid/ciaa484

Connelly S, Widmer G, Mukherjee J, Huynh K, Bristol JA, Hubert S, Sliman J, Tzipori S, Kaleko M (2015) Tu2054 SYN-004, a clinical stage oral beta-lactamase therapy, protects the intestinal microflora from antibiotic-mediated damage in humanized pigs. Gastroenterology 148(4):S-1195. https://doi.org/10.1016/S0016-5085(15)34082-8

Corbett D, Wise A, Birchall S, Warn P, Baines SD, Crowther G, Freeman J, Chilton CH, Vernon J, Wilcox MH, Vickers RJ (2015) *In vitro* susceptibility of *Clostridium difficile* to SMT19969 and comparators, as well as the killing kinetics and post-antibiotic effects of SMT19969 and comparators against *C. difficile*. J Antimicrob Chemother 70(6):1751–1756. https://doi.org/10.1093/jac/dkv006

Cornely OA, Crook DW, Esposito R, Poirier A, Somero MS, Weiss K, Sears P, Gorbach S (2012) Fidaxomicin versus vancomycin for infection with *Clostridium difficile* in Europe, Canada, and the USA: a double-blind, non-inferiority, randomised controlled trial. Lancet Infect Dis 12(4):281–289. https://doi.org/10.1016/s1473-3099(11)70374-7

Crobach MJ, Planche T, Eckert C, Barbut F, Terveer EM, Dekkers OM, Wilcox MH, Kuijper EJ (2016) European Society of Clinical Microbiology and Infectious Diseases: update of the diagnostic guidance document for Clostridium difficile infection. Clin Microbiol Infect Suppl 4:s63–s81. https://doi.org/10.1016/j.cmi.2016.03.010

Crook DW, Walker AS, Kean Y, Weiss K, Cornely OA, Miller MA, Esposito R, Louie TJ, Stoesser NE, Young BC, Angus BJ, Gorbach SL, Peto TE (2012) Fidaxomicin versus vancomycin for *Clostridium difficile* infection: meta-analysis of pivotal randomized controlled trials. Clin Infect Dis 55(Suppl 2):S93–S103. https://doi.org/10.1093/cid/cis499

Daley P, Louie T, Lutz JE, Khanna S, Stoutenburgh U, Jin M, Adedoyin A, Chesnel L, Guris D, Larson KB, Murata Y (2017) Surotomycin versus vancomycin in adults with Clostridium difficile infection: primary clinical outcomes from the second pivotal, randomized, double-blind, Phase 3 trial. J Antimicrob Chemother 72(12):3462–3470. https://doi.org/10.1093/jac/dkx299

Davies KA, Longshaw CM, Davis GL, Bouza E, Barbut F, Barna Z, Delmee M, Fitzpatrick F, Ivanova K, Kuijper E, Macovei IS, Mentula S, Mastrantonio P, von Muller L, Oleastro M, Petinaki E, Pituch H, Noren T, Novakova E, Nyc O, Rupnik M, Schmid D, Wilcox MH (2014) Underdiagnosis of *Clostridium difficile* across Europe: the European, multicentre, prospective, biannual, point-prevalence study of *Clostridium difficile* infection in hospitalised patients with diarrhoea (EUCLID). Lancet Infect Dis 14(12): 1208–1219. https://doi.org/10.1016/s1473-3099(14)70991-0

Davies K, Mawer D, Walker AS, Berry C, Planche T, Stanley P, Goldenberg S, Sandoe J, Wilcox MH (2020) An analysis of Clostridium difficile environmental contamination during and after treatment for C difficile infection. Open Forum Infect Dis 7(11): ofaa362. https://doi.org/10.1093/ofid/ofaa362

de Bruyn G, Gordon DL, Steiner T, Tambyah P, Cosgrove C, Martens M, Bassily E, Chan ES, Patel D, Chen J, Torre-Cisneros J, De Magalhães F, Francesconi C, Gesser R, Jeanfreau R, Launay O,

Laot T, Morfin-Otero R, Oviedo-Orta E, Park YS, Piazza FM et al (2021) Safety, immunogenicity, and efficacy of a Clostridioides difficile toxoid vaccine candidate: a phase 3 multicentre, observer-blind, randomised, controlled trial. Lancet Infect Dis 21(2): 252–262. https://doi.org/10.1016/S1473-3099(20)30331-5

de Gunzburg J, Ghozlane A, Ducher A, Le Chatelier E, Duval X, Ruppé E, Armand-Lefevre L, Sablier-Gallis-F, Burdet C, Alavoine L, Chachaty E, Augustin V, Varastet M, Levenez F, Kennedy S, Pons N, Mentré F, Andremont A (2018) Protection of the human gut microbiome from antibiotics. J Infect Dis 217(4):628–636. https://doi.org/10.1093/infdis/jix604

De Leon LM, Watson JB, Kelly CR (2013) Transient flare of ulcerative colitis after fecal microbiota transplantation for recurrent *Clostridium difficile* infection. Clin Gastroenterol Hepatol 11(8):1036–1038. https://doi.org/10.1016/j.cgh.2013.04.045

Debast SB, Bauer MP, Kuijper EJ (2014) European Society of Clinical Microbiology and Infectious Diseases: update of the treatment guidance document for *Clostridium difficile* infection. Clin Microbiol Infect 20 (Suppl 2):1–26. https://doi.org/10.1111/1469-0691.12418

Deshpande A, Hurless K, Cadnum JL, Chesnel L, Gao L, Chan L, Kundrapu S, Polinkovsky A, Donskey CJ (2016) Effect of surotomycin, a novel cyclic lipopeptide antibiotic, on intestinal colonization with vancomycin-resistant enterococci and *Klebsiella pneumoniae* in mice. Antimicrob Agents Chemother 60(6):3333–3339. https://doi.org/10.1128/aac.02904-15

Donald RG, Flint M, Kalyan N, Johnson E, Witko SE, Kotash C, Zhao P, Megati S, Yurgelonis I, Lee PK, Matsuka YV, Severina E, Deatly A, Sidhu M, Jansen KU, Minton NP, Anderson AS (2013) A novel approach to generate a recombinant toxoid vaccine against *Clostridium difficile*. Microbiology (Reading, England) 159(Pt 7):1254–1266. https://doi.org/10.1099/mic.0.066712-0

Dose-confirmation, immunogenicity and safety study of the Clostridium difficile vaccine candidate VLA84 in healthy adults aged 50 years and older. Phase II study. ClinicalTrials.gov identifier: NCT02316470. Updated June 8, 2017. https://clinicaltrials.gov/ct2/show/NCT02316470. Accessed 16 Mar 2023

Dubberke E, Lee C, Orenstein R, Khanna S, Hecht G, Fraiz J (2016) Efficacy and safety of RBX2660 for the prevention of recurrent clostridium difficile infection: results of the PUNCH CD 2 trial. Paper presented at the IDWeek 2016. Abstract 1341

ECDC (2017) *Clostridium difficile*. https://ecdc.europa.eu/en/publications-data/directory-guidance-prevention-and-control/clostridium-difficile. Accessed 30 June 2017

European Centre for Disease Prevention and Control (2013) Healthcare-associated infections in acute care hospitals. https://www.ecdc.europa.eu/en/healthcare-associated-infections-acute-care-hospitals

European Medicines Agency (2022, June) Faecal Microbiota Transplantation EU-IN Horizon Scanning Report. Retrieved from https://www.ema.europa.eu/en/documents/report/faecal-microbiota-transplantation-eu-horizonscanning-report_en.pdf

Feuerstadt P, Louie TJ, Lashner B, Wang EEL, Diao L, Bryant JA, Sims M, Kraft CS, Cohen SH, Berenson CS, Korman LY, Ford CB, Litcofsky KD, Lombardo MJ, Wortman JR, Wu H, Auniņš JG, McChalicher CWJ, Winkler JA, McGovern BH et al (2022) SER-109, an oral microbiome therapy for recurrent Clostridioides difficile infection. New Engl J Med 386(3):220–229. https://doi.org/10.1056/NEJMoa2106516

Finn E, Andersson FL, Madin-Warburton M (2021) Burden of Clostridioides difficile infection (CDI) - a systematic review of the epidemiology of primary and recurrent CDI. BMC Infect Dis 21(1):456. https://doi.org/10.1186/s12879-021-06147-y

Fitzpatrick F (2008) Management of Clostridium difficile infection–medical or surgical? Surgeon 6(6):325–328. https://doi.org/10.1016/s1479-666x(08)80002-9

Fitzpatrick F, Safdar N, van Prehn J, Tschudin-Sutter S (2022) How can patients with Clostridioides difficile infection on concomitant antibiotic treatment be best managed? Lancet Infect Dis 22(11):336–340. https://doi.org/10.1016/S1473-3099(22)00274-2

Fitzpatrick F, Skally M, Brady M, Burns K, Rooney C, Wilcox MH (2018) European practice for CDI treatment. Adv Exp Med Biol 1050:117–135. https://doi.org/10.1007/978-3-319-72799-8_8

Foglia G, Shah S, Luxemburger C, Pietrobon PJ (2012) *Clostridium difficile*: development of a novel candidate vaccine. Vaccine 30(29):4307–4309. https://doi.org/10.1016/j.vaccine.2012.01.056

Food and Drug Administration (2016) Early clinical trials with live biotherapeutic products: chemistry, manufacturing, and control information. https://www.apa.org/news/press/releases/2020/11/anxiety-depression-treatment Accessed 16 Mar 2022

Freeman J, Vernon J, Morris K, Nicholson S, Todhunter S, Longshaw C, Wilcox MH, Pan-European Longitudinal Surveillance of Antibiotic Resistance among Prevalent Clostridium difficile Ribotypes' Study Group (2015b) Pan-European longitudinal surveillance of antibiotic resistance among prevalent Clostridium difficile ribotypes. Clin Microbiol Infect 21(3):248.e9–248.e16. https://doi.org/10.1016/j.cmi.2014.09.017

Freeman J, Vernon J, Vickers R, Wilcox MH (2015a) Susceptibility of *Clostridium difficile* isolates of varying antimicrobial resistance phenotypes to SMT19969 and 11 comparators. Antimicrob Agents Chemother 60(1):689–692. https://doi.org/10.1128/aac.02000-15

Garey KW, McPherson J, Dinh AQ, Hu C, Jo J, Wang W, Lancaster CK, Gonzales-Luna AJ, Loveall C, Begum K, Jahangir Alam M, Silverman MH, Hanson BM (2022) Efficacy, safety, pharmacokinetics, and microbiome changes of ibezapolstat in adults with Clostridioides difficile infection: a phase 2a multicenter clinical trial. Clin Infect Dis 75(7):1164–1170. https://doi.org/10.1093/cid/ciac096

Gehin M, Desnica B, Dingemanse J (2015) Minimal systemic and high faecal exposure to cadazolid in patients with severe *Clostridium difficile* infection. Int J Antimicrob Agents 46(5):576–581. https://doi.org/10.1016/j.ijantimicag.2015.07.015

Gerding DN, Kelly CP, Rahav G, Lee C, Dubberke ER, Kumar PN, Yacyshyn B, Kao D, Eves K, Ellison MC, Hanson ME, Guris D, Dorr MB (2018) Bezlotoxumab for prevention of recurrent Clostridium difficile infection in patients at increased risk for recurrence. Clin Infect Dis 67(5):649–656. https://doi.org/10.1093/cid/ciy171

Gerding DN, Meyer T, Lee C, Cohen SH, Murthy UK, Poirier A, Van Schooneveld TC, Pardi DS, Ramos A, Barron MA, Chen H, Villano S (2015) Administration of spores of nontoxigenic *Clostridium difficile* strain M3 for prevention of recurrent *C. difficile* infection: a randomized clinical trial. JAMA 313(17):1719–1727. https://doi.org/10.1001/jama.2015.3725

Guyatt GH, Oxman AD, Vist GE, Kunz R, Falck-Ytter Y, Alonso-Coello P, Schünemann HJ, GRADE Working Group (2008) GRADE: an emerging consensus on rating quality of evidence and strength of recommendations. BMJ 336(7650):924–926. https://doi.org/10.1136/bmj.39489.470347.AD

Goldenberg SD, Brown S, Edwards L, Gnanarajah D, Howard P, Jenkins D, Nayar D, Pasztor M, Oliver S, Planche T, Sandoe JA, Wade P, Whitney L (2016) The impact of the introduction of fidaxomicin on the management of *Clostridium difficile* infection in seven NHS secondary care hospitals in England: a series of local service evaluations. Eur J Clin Microbiol Infect Dis 35(2):251–259. https://doi.org/10.1007/s10096-015-2538-z

Goldstein EJ, Citron DM, Tyrrell KL, Merriam CV (2013) Comparative *in vitro* activities of SMT19969, a new antimicrobial agent, against *Clostridium difficile* and 350 gram-positive and gram-negative aerobic and anaerobic intestinal flora isolates. Antimicrob Agents Chemother 57(10):4872–4876. https://doi.org/10.1128/aac.01136-13

Gonzales-Luna AJ, Carlson TJ, Garey KW (2023) Emerging options for the prevention and management of Clostridioides difficile infection. Drugs 83(2):105–116. https://doi.org/10.1007/s40265-022-01832-x

Hota SS, Sales V, Tomlinson G, Salpeter MJ, McGeer A, Coburn B, Guttman DS, Low DE, Poutanen SM (2017) Oral vancomycin followed by fecal transplantation versus tapering oral vancomycin treatment for recurrent *Clostridium difficile* infection: an open-label, randomized controlled trial. Clin Infect Dis 64(3):265–271. https://doi.org/10.1093/cid/ciw731

Hu MY, Katchar K, Kyne L, Maroo S, Tummala S, Dreisbach V, Xu H, Leffler DA, Kelly CP (2009) Prospective derivation and validation of a clinical prediction rule for recurrent *Clostridium difficile* infection. Gastroenterology 136(4):1206–1214. https://doi.org/10.1053/j.gastro.2008.12.038

Johnson S, Gerding DN (2017) Fecal fixation: fecal microbiota transplantation for *Clostridium difficile* infection. Clin Infect Dis 64(3):272–274. https://doi.org/10.1093/cid/ciw735

Johnson S, Lavergne V, Skinner AM, Gonzales-Luna AJ, Garey KW, Kelly CP, Wilcox MH (2021) Clinical Practice Guideline by the Infectious Diseases Society of America (IDSA) and Society for Healthcare Epidemiology of America (SHEA): 2021 focused update guidelines on management of clostridioides difficile infection in adults. Clin Infect Dis 73(5):1029–1044. https://doi.org/10.1093/cid/ciab549

Johnson S, Louie TJ, Gerding DN, Cornely OA, Chasan-Taber S, Fitts D, Gelone SP, Broom C, Davidson DM (2014) Vancomycin, metronidazole, or tolevamer for *Clostridium difficile* infection: results from two multinational, randomized, controlled trials. Clin Infect Dis 59(3):345–354. https://doi.org/10.1093/cid/ciu313

Kaleko M, Bristol JA, Hubert S, Parsley T, Widmer G, Tzipori S, Subramanian P, Hasan N, Koski P, Kokai-Kun J, Sliman J, Jones A, Connelly S (2016) Development of SYN-004, an oral beta-lactamase treatment to protect the gut microbiome from antibiotic-mediated damage and prevent *Clostridium difficile* infection. Anaerobe 41:58–67. https://doi.org/10.1016/j.anaerobe.2016.05.015

Kampouri E, Croxatto A, Prod'hom G, Guery B (2021) Clostridioides difficile infection, still a long way to go. J Clin Med 10(3):389. https://doi.org/10.3390/jcm10030389

Kelly CR, Fischer M, Allegretti JR, LaPlante K, Stewart DB, Limketkai BN, Stollman NH (2021) ACG clinical guidelines: prevention, diagnosis, and treatment of Clostridioides difficile infections. Am J Gastroenterol 116(6):1124–1147. https://doi.org/10.14309/ajg.0000000000001278

Khanna S, Assi M, Lee C, Yoho D, Louie T, Knapple W, Aguilar H, Garcia-Diaz J, Wang GP, Berry SM, Marion J, Su X, Braun T, Bancke L, Feuerstadt P (2022b) Efficacy and safety of RBX2660 in PUNCH CD3, a Phase III, randomized, double-blind, placebo-controlled trial with a bayesian primary analysis for the prevention of recurrent Clostridioides difficile infection. Drugs 82(15):1527–1538. https://doi.org/10.1007/s40265-022-01797-x

Khanna S, Assi M, Feuerstadt P, Harvey A, Bancke L (2022a) S132. An interim analysis of a phase 3, open-label study indicates efficacy and safety of RBX2660 in patients with recurrent Clostridioides difficile infection. Am J Gastroenterol 117(10S):e96. https://doi.org/10.14309/01.ajg.0000857168.31201.9f

Khanna S, Pardi DS, Kelly CR, Kraft CS, Dhere T, Henn MR, Lombardo MJ, Vulic M, Ohsumi T, Winkler J, Pindar C, McGovern BH, Pomerantz RJ, Aunins JG, Cook DN, Hohmann EL (2016) A novel microbiome therapeutic increases gut microbial diversity and prevents recurrent *Clostridium difficile* infection. J Infect Dis 214(2):173–181. https://doi.org/10.1093/infdis/jiv766

Kitchin N, Remich SA, Peterson J, Peng Y, Gruber WC, Jansen KU, Pride MW, Anderson AS, Knirsch C, Webber C (2020) A phase 2 study evaluating the safety, tolerability, and immunogenicity of two 3-dose regimens of a Clostridium difficile vaccine in healthy US adults aged 65 to 85 years. Clin Infect Dis 70(1):1–10. https://doi.org/10.1093/cid/ciz153

Kokai-Kun JF, Roberts T, Coughlin O, Le C, Whalen H, Stevenson R, Wacher VJ, Sliman J (2019) Use of ribaxamase (SYN-004), a β-lactamase, to prevent Clostridium difficile infection in β-lactam-treated patients: a double-blind, phase 2b, randomised placebo-controlled trial. Lancet Infect Dis 19(5):487–496. https://doi.org/10.1016/S1473-3099(18)30731-X

Kokai-Kun JF, Le C, Trout K et al (2020) Ribaxamase, an orally administered beta-lactamase, diminishes changes to acquired antimicrobial resistance of the gut resistome in patients treated with ceftriaxone. Infect Drug Resist 13:2521–2535

Kotloff KL, Wasserman SS, Losonsky GA, Thomas W Jr, Nichols R, Edelman R, Bridwell M, Monath TP (2001) Safety and immunogenicity of increasing doses of a *Clostridium difficile* toxoid vaccine administered to healthy adults. Infect Immun 69(2):988–995. https://doi.org/10.1128/iai.69.2.988-995.2001

Kraft C, Khanna S, Assi M, Feuerstadt P, Harvey A, Bancke L (2021) Sa611 interim analysis of a phase 3 open-label study indicates safety and efficacy of RBX2660, an investigational live biotherapeutic, in a "Real-world" population of patients with recurrent Clostridioides difficile infection. Gastroenterology 160(6):S-573

Kuehne SA, Cartman ST, Heap JT, Kelly ML, Cockayne A, Minton NP (2010) The role of toxin A and toxin B in *Clostridium difficile* infection. Nature 467(7316):711–713. https://doi.org/10.1038/nature09397

Krutova M, de Meij TGJ, Fitzpatrick F, Drew RJ, Wilcox MH, Kuijper EJ (2022b) How to: Clostridioides difficile infection in children. Clin Microbiol Infect 28(8):1085–1090. https://doi.org/10.1016/j.cmi.2022.03.001

Krutova M, Wilcox M, Kuijper E (2022a) Clostridioides difficile infection: are the three currently used antibiotic treatment options equal from pharmacological and microbiological points of view? Int J Infect Dis 124:118–123. https://doi.org/10.1016/j.ijid.2022.09.013

Kyne L, Warny M, Qamar A, Kelly CP (2000) Asymptomatic carriage of *Clostridium difficile* and serum levels of IgG antibody against toxin A. New Engl J Med 342(6):390–397. 0.1056/nejm200002103420604

Kyne L, Warny M, Qamar A, Kelly CP (2001) Association between antibody response to toxin A and protection against recurrent *Clostridium difficile* diarrhoea. Lancet (London, England) 357(9251):189–193. 0.1016/s0140-6736(00)03592-3

Lee CH, Steiner T, Petrof EO, Smieja M, Roscoe D, Nematallah A, Weese JS, Collins S, Moayyedi P, Crowther M, Ropeleski MJ, Jayaratne P, Higgins D, Li Y, Rau NV, Kim PT (2016) Frozen vs fresh fecal microbiota transplantation and clinical resolution of diarrhea in patients with recurrent *Clostridium difficile* infection: a randomized clinical trial. JAMA 315(2):142–149. https://doi.org/10.1001/jama.2015.18098

Louie TJ, Emery J, Krulicki W, Byrne B, Mah M (2009) OPT-80 eliminates Clostridium difficile and is sparing of bacteroides species during treatment of C. difficile infection. Antimicrob Agents Chemother 53(1):261–263. https://doi.org/10.1128/AAC.01443-07

Moura I, Spigaglia P, Barbanti F, Mastrantonio P (2013) Analysis of metronidazole susceptibility in different *Clostridium difficile* PCR ribotypes. J Antimicrob Chemother 68(2):362–365. https://doi.org/10.1093/jac/dks420

Muhammad A, Madhav D, Rawish F, Viveksandeep TC, Albert E, Mollie J, Prateek S, Surotomycin (A Novel Cyclic Lipopeptide) vs. (2019) Vancomycin for the treatment of Clostridioides difficile infection: a systematic review and meta-analysis. Curr Clin Pharmacol 14(3):166–174. https://doi.org/10.2174/1574884714666190328162637. PMID: 30924421

Nelson RL, Suda KJ, Evans CT (2017) Antibiotic treatment for *Clostridium difficile*-associated diarrhoea in adults. Cochrane Database Syst Rev 3:Cd004610. https://doi.org/10.1002/14651858.CD004610.pub5

Orenstein R, Dubberke E, Hardi R, Ray A, Mullane K, Pardi DS, Ramesh MS (2016) Safety and durability of RBX2660 (microbiota suspension) for recurrent *Clostridium difficile* infection: results of the PUNCH CD study. Clin Infect Dis 62(5):596–602. https://doi.org/10.1093/cid/civ938

Okhuysen PC, Ramesh M, Garey KW, Louie TJ, Torre Cisneros J, Stychneuskaya A, Kiknadze N, Li J, Duperchy E, Wilcox MH, Montoya JG, Styles L, Clow F, James D, Dubberke ER, De Oliveira CM, Van Steenkiste C (2022) A phase 3, randomized, double-blind study to evaluate the efficacy and safety of ridinilazole compared with vancomycin for the treatment of clostridioides difficile infection. Open Forum Infect Dis 9(Suppl 2):ofac492.021. https://doi.org/10.1093/ofid/ofac492.021

Pechine S, Gleizes A, Janoir C, Gorges-Kergot R, Barc MC, Delmee M, Collignon A (2005) Immunological properties of surface proteins of *Clostridium difficile*. J Med Microbiol 54(Pt 2):193–196. https://doi.org/10.1099/jmm.0.45800-0

Pfizer (2022) Phase 3 CLOVER trial for Pfizer's investigational clostridioides difficile vaccine indicates strong potential effect in reducing duration and severity of disease based on secondary endpoints [Press release]. 01 Mar 2022. https://www.pfizer.com/news/press-release/press-release-detail/phase-3-clover-trial-pfizers-investigational-clostridioides

Prior AR, Kevans D, McDowell L, Cudmore S, Fitzpatrick F (2017) Treatment of *Clostridium difficile* infection: a national survey of clinician recommendations and the use of faecal microbiota transplantation. J Hosp Infect 95(4):438–441. https://doi.org/10.1016/j.jhin.2016.10.004

Public Health England (2013) Updated guidance on the management and treatment of *Clostridium difficile* infection

Qian X, Yanagi K, Kane AV, Alden N, Lei M, Snydman DR, Vickers RJ, Lee K, Thorpe CM (2020) Ridinilazole, a narrow spectrum antibiotic for treatment of Clostridioides difficile infection, enhances preservation of microbiota-dependent bile acids. Am J Physiol Gastrointest Liver Physiol 319(2):G227–G237. https://doi.org/10.1152/ajpgi.00046.2020

Quera R, Espinoza R, Estay C, Rivera D (2014) Bacteremia as an adverse event of fecal microbiota transplantation in a patient with Crohn's disease and recurrent *Clostridium difficile* infection. J Crohn's Colitis 8(3):252–253. https://doi.org/10.1016/j.crohns.2013.10.002

Ray A, Jones C, Shannon W, Carter S (2017) Resetting the microbial landscape: donor microbiome engraftment in patients treated with RBX2660 for multi-recurrent *Clostridium difficile* infection. Paper presented at the ASM Microbe, New Orleans, USA, Abstract 262

Rebiotix Inc (2017) Rebiotix reports positive top line data from open-label phase 2 trial of RBX2660 in recurrent *Clostridium difficile*

Reigadas E, van Prehn J, Falcone M, Fitzpatrick F, MJGT V, Kuijper EJ, Bouza E, European Society of Clinical Microbiology and Infectious Diseases Study Group on Clostridioides difficile (ESGCD) and Study Group for Host and Microbiota interaction (ESGHAMI) (2021) How to: prophylactic interventions for prevention of Clostridioides difficile infection. Clin Microbiol Infect 27(12):1777–1783. https://doi.org/10.1016/j.cmi.2021.06.037

Roberts T, Kokai-Kun JF, Coughlin O, Lopez BV, Whalen H, Bristol JA, Hubert S, Longstreth J, Lasseter K, Sliman J (2016) Tolerability and pharmacokinetics of SYN-004, an orally administered beta-lactamase for the prevention of *Clostridium difficile*-associated disease and antibiotic-associated diarrhea, in two phase 1 studies. Clini Drug Investig 36(9):725–734. https://doi.org/10.1007/s40261-016-0420-0

Safety and Immunogenicity Study of GSK's Clostridium Difficile Vaccine 2904545A When Administered in Healthy Adults Aged 18-45 Years and 50-70 Years. ClinicalTrials.gov identifier: NCT04026009. Updated June 10, 2022. https://ClinicalTrials.gov/show/NCT04026009. Accessed 21 Sept 2022

Sattar A, Thommes P, Payne L, Warn P, Vickers RJ (2015) SMT19969 for *Clostridium difficile* infection (CDI): in vivo efficacy compared with fidaxomicin and van comycin in the hamster model of CDI. J Antimicrob Chemother 70(6):1757–1762. https://doi.org/10.1093/jac/dkv005

Sheldon E, Kitchin N, Peng Y, Eiden J, Gruber W, Johnson E, Jansen KU, Pride MW, Pedneault L (2016) A phase 1, placebo-controlled, randomized study of the safety, tolerability, and immunogenicity of a *Clostridium difficile* vaccine administered with or without aluminum hydroxide in healthy adults. Vaccine 34(18):2082–2091. https://doi.org/10.1016/j.vaccine.2016.03.010

Sims MD, Khanna S, Feuerstadt P et al (2023) Safety and tolerability of SER-109 as an investigational microbiome therapeutic in adults with recurrent Clostridioides difficile infection: a phase 3, open-label, single-arm trial. JAMA Netw Open 6(2):e2255758. https://doi.org/10.1001/jamanetworkopen.2022.55758

Solomon K, Martin AJ, O'Donoghue C, Chen X, Fenelon L, Fanning S, Kelly CP, Kyne L (2013) Mortality in patients with *Clostridium difficile* infection correlates with host pro-inflammatory and humoral immune responses. J Med Microbiol 62(Pt 9):1453–1460. https://doi.org/10.1099/jmm.0.058479-0

Stevens VW, Nelson RE, Schwab-Daugherty EM, Khader K, Jones MM, Brown KA, Greene T, Croft LD, Neuhauser M, Glassman P, Goetz MB, Samore MH, Rubin MA (2017) Comparative effectiveness of vancomycin and metronidazole for the prevention of recurrence and death in patients with *Clostridium difficile* infection. JAMA Intern Med 177(4):546–553. https://doi.org/10.1001/jamainternmed.2016.9045

Synthetic Biologics I (2017) SYN-004 (Ribaxamase) receives breakthrough therapy designation from U.S. Food and Drug Administration for prevention of *Clostridium difficile* infection

Tannock GW, Munro K, Taylor C, Lawley B, Young W, Byrne B, Emery J, Louie T (2010) A new macrocyclic antibiotic, fidaxomicin (OPT-80), causes less alteration to the bowel microbiota of Clostridium difficile-infected patients than does vancomycin. Microbiology 156(11):3354–3359. https://doi.org/10.1099/mic.0.042010-0

Tariq R, Smyrk T, Pardi DS, Tremaine WJ, Khanna S (2016) New-onset microscopic colitis in an ulcerative colitis patient after fecal microbiota transplantation. Am J Gastroenterol 111(5):751–752. https://doi.org/10.1038/ajg.2016.67

Trucksis M, Baird I, Cornely O, Golan Y, Hecht G, Pardi D, Pullman J, Polage C, Wilcox M, Bernardo P, Ford C, O'Brien E, Vetro R, Wortman J, Weston J, Henn M (2017) An analysis of results from the first placebo-controlled trial of single-dose SER-109, an investigational oral microbiome therapeutic to reduce the recurrence of *Clostridium difficile infection* (CDI). Paper presented at the ECCMID, Vienna, Austria. Abstract OS0250C

Tschudin-Sutter S, Kuijper EJ, Durovic A, Vehreschild MJGT, Barbut F, Eckert C, Fitzpatrick F, Hell M, Norèn T, O'Driscoll J, Coia J, Gastmeier P, von Müller L, Wilcox MH, Widmer AF, Committee (2018) Guidance document for prevention of Clostridium difficile infection in acute healthcare settings. Clin

Microbiol Infect 24(10):1051–1054. https://doi.org/10.1016/j.cmi.2018.02.020

Turner NA, Warren BG, Gergen-Teague MF, Addison RM, Addison B, Rutala WA, Weber DJ, Sexton DJ, Anderson DJ (2022) Impact of oral metronidazole, vancomycin, and fidaxomicin on host shedding and environmental contamination with Clostridioides difficile. Clin Infect Dis 74(4):648–656. https://doi.org/10.1093/cid/ciab473

U.S. Food and Drug Administration (2022) FDA Approves First Fecal Microbiota Product. https://www.fda.gov/news-events/press-announcements/fda-approves-first-fecal-microbiota-product. Accessed 15 Jan 2023

Valneva S (2016) Valneva announces successful completion of phase II for *Clostridium difficile* vaccine candidate

van Eijk E, Boekhoud IM, Kuijper EJ, Bos-Sanders IMJG, Wright G, Smits WK (2019) Genome location dictates the transcriptional response to PolC inhibition in Clostridium difficile. Antimicrob Agents Chemother 63(2):e01363–e01318. https://doi.org/10.1128/AAC.01363-18

van Nood E, Vrieze A, Nieuwdorp M, Fuentes S, Zoetendal EG, de Vos WM, Visser CE, Kuijper EJ, Bartelsman JF, Tijssen JG, Speelman P, Dijkgraaf MG, Keller JJ (2013) Duodenal infusion of donor feces for recurrent *Clostridium difficile*. New Engl J Med 368(5):407–415. https://doi.org/10.1056/NEJMoa1205037

van Prehn J, Fitzpatrick F, Kuijper EJ, European Study Group for Clostridioides difficile and the European Study Group for Host and Microbiota Interaction (2023) Faecal microbiota transplantation for first and second episodes of Clostridioides difficile infection. Lancet Gastroenterol Hepatol 8(2):109. https://doi.org/10.1016/S2468-1253(22)00342-9

van Prehn J, Reigadas E, Vogelzang EH et al (2021) European Society of Clinical Microbiology and Infectious Diseases: 2021 update on the treatment guidance document for Clostridioides difficile infection in adults. Clin Microbiol Infect 27(Suppl 2):s1–s21. https://doi.org/10.1016/j.cmi.2021.09.038

van Rossen TM, Ooijevaar RE, Vandenbroucke-Grauls CMJE, Dekkers OM, Kuijper EJ, Keller JJ, van Prehn J (2021) Prognostic factors for severe and recurrent Clostridioides difficile infection: a systematic review. Clin Microbiol Infect 28(3):321–331. https://doi.org/10.1016/j.cmi.2021.09.026

Vardakas KZ, Polyzos KA, Patouni K, Rafailidis PI, Samonis G, Falagas ME (2012) Treatment failure and recurrence of *Clostridium difficile* infection following treatment with vancomycin or metronidazole: a systematic review of the evidence. Int J Antimicrob Agents 40(1):1–8. https://doi.org/10.1016/j.ijantimicag.2012.01.004

Vehreschild MJGT, Ducher A, Louie T, Cornely OA, Feger C, Dane A, Varastet M, Vitry F, de Gunzburg J, Andremont A, Mentré F, Wilcox MH (2022) An open randomized multicentre Phase 2 trial to assess the safety of DAV132 and its efficacy to protect gut microbiota diversity in hospitalized patients treated with fluoroquinolones. J Antimicrob Chemother 77(4):1155–1165. https://doi.org/10.1093/jac/dkab474

Vickers R, Robinson N, Best E, Echols R, Tillotson G, Wilcox M (2015) A randomised phase 1 study to investigate safety, pharmacokinetics and impact on gut microbiota following single and multiple oral doses in healthy male subjects of SMT19969, a novel agent for *Clostridium difficile* infections. BMC Infect Dis 15:91. https://doi.org/10.1186/s12879-015-0759-5

Vickers RJ, Tillotson G, Goldstein EJ, Citron DM, Garey KW, Wilcox MH (2016) Ridinilazole: a novel therapy for *Clostridium difficile* infection. Int J Antimicrob Agents 48(2):137–143. https://doi.org/10.1016/j.ijantimicag.2016.04.026

Vickers RJ, Tillotson GS, Nathan R, Hazan S, Pullman J, Lucasti C, Deck K, Yacyshyn B, Maliakkal B, Pesant Y, Tejura B, Roblin D, Gerding DN, Wilcox MH (2017) Efficacy and safety of ridinilazole compared with vancomycin for the treatment of *Clostridium difficile* infection: a phase 2, randomised, double-blind, active-controlled, non-inferiority study. Lancet Infect Dis 17(7):735–744. https://doi.org/10.1016/s1473-3099(17)30235-9

Wang S, Zhu D, Sun X (2022) Development of an effective nontoxigenic Clostridioides difficile-based oral vaccine against C. difficile infection. Microbiol Spectr 10(3):e0026322. https://doi.org/10.1128/spectrum.00263-22

Weiss W, Pulse M, Vickers R (2014) *In vivo* assessment of SMT19969 in a hamster model of *Clostridium difficile* infection. Antimicrob Agents Chemother 58(10):5714–5718. https://doi.org/10.1128/aac.02903-14

Wilcox MH, Gerding DN, Poxton IR, Kelly C, Nathan R, Birch T, Cornely OA, Rahav G, Bouza E, Lee C, Jenkin G, Jensen W, Kim YS, Yoshida J, Gabryelski L, Pedley A, Eves K, Tipping R, Guris D, Kartsonis N, Dorr MB (2017) Bezlotoxumab for prevention of recurrent *Clostridium difficile* infection. New Engl J Med 376(4):305–317. https://doi.org/10.1056/NEJMoa1602615

Youngster I, Russell GH, Pindar C, Ziv-Baran T, Sauk J, Hohmann EL (2014) Oral, capsulized, frozen fecal microbiota transplantation for relapsing *Clostridium difficile* infection. JAMA 312(17):1772–1778. https://doi.org/10.1001/jama.2014.13875

Fecal Microbiota Transplantation as Emerging Treatment in European Countries 2.0

Serena Porcari, Marcello Maida, Stefano Bibbò, James McIlroy, Gianluca Ianiro, and Giovanni Cammarota

Abstract

Clostridioides difficile infection (CDI) is one of the most common healthcare-associated infections and one of the leading causes of morbidity and mortality in hospitalized patients in the world. Although several antibiotics effectively treat CDI, some individuals may not respond to these drugs and may be cured by transplanting stool from healthy donors. FMT has demonstrated extraordinary cure rates for the cure of CDI recurrences.

Moreover, FMT has also been investigated in other disorders associated with the alteration of gut microbiota, such as inflammatory bowel disease (IBD), where the alterations of the gut microbiota ecology have been theorized to play a causative role. Although FMT is currently not recommended to cure IBD patients in clinical practice, several studies have been recently carried out with the ultimate goal to search new therapeutic options to patients.

This review summarizes data on the use of FMT for the treatment of both CDI and IBD, with a special attention to highlight studies conducted in European countries.

Author contributions: SP and MM are joint first authors. GC conceived the article and designed the structure of the article. SP, SB wrote the initial draft of the manuscript. SP and SB built the tables. All authors provided critical revision of the manuscript and approved its final version for submission.

S. Porcari · S. Bibbò · G. Ianiro · G. Cammarota (✉)
Department of Medical and Surgical Sciences, Digestive Disease Center, Fondazione Policlinico Universitario Agostino Gemelli IRCCS, Rome, Italy

Department of Translational Medicine and Surgery, Università Cattolica del Sacro Cuore, Rome, Italy
e-mail: giovanni.cammarota@unicatt.it

M. Maida
Gastroenterology and Endoscopy Unit, S. Elia-Raimondi Hospital, Caltanissetta, Italy

J. McIlroy
School of Medicine, Medical Sciences and Nutrition, University of Aberdeen, Aberdeen, UK

1 Introduction

The gut microbiome could be considered as a large community of microorganisms inhabiting the mucosa of the gastrointestinal tract, which consists of over 500 different species (Thursby and Juge 2017; Ianiro et al. 2014a, 2015). Gut microbiome harbors 10^{14} organisms and a number of microbial genes that are thousands times larger than human genomes (Ianiro et al. 2014b; Cani 2018). For this reason, gut microbiome could be considered a "super-organism" (Kramer and Bressan 2015).

Gut microbiome is involved in several functions, such as the digestion and absorption processes (El Kaoutari et al. 2013; Venema 2010; Ianiro et al. 2014a), the synthesis of nutrients

P. Mastrantonio, M. Rupnik (eds.), *Updates on Clostridioides difficile in Europe*, Advances in Experimental Medicine and Biology 1435, https://doi.org/10.1007/978-3-031-42108-2_5

(LeBlanc et al. 2013), the regulation of the gut barrier, and the facilitation of the innate and immune system development (Wu and Wu 2012; Cardinale et al. 2020; Bibbò et al. 2014). Moreover, gut microbiome is involved in promoting health, and its perturbation is recognized in many communicable and noncommunicable chronic disorders (Ianiro et al. 2020). For this reason, the manipulation of gut microbiota has been investigated as treatment option of several disorders.

Fecal microbiota transplantation (FMT) is defined as the transfer of feces from a healthy donor to a recipient to treat disorders directly or indirectly associated with a microbiome imbalance.

Nowadays, FMT is considered a highly effective treatment against *Clostridioides difficile* infection (CDI), with a success rate of nearly 90%; for this reason, FMT is recommended by international guidelines (McDonald et al. 2018; Van Prehn et al. 2021; Kelly et al. 2021; Cammarota et al. 2017) and it is considered a valid treatment option in clinical practice, for patients after a second or further episode of recurrent CDI and for patients with acute, severe, and/or fulminant CDI that is refractory to antibiotic therapy, especially when patients are poor surgical candidate (Kelly et al. 2021).

FMT has also been investigated in a lot of noncommunicable chronic disorders, including inflammatory bowel disease (IBD) (Cammarota et al. 2015a; Ianiro et al. 2015) with promising results (Sokol et al. 2020), but its efficacy rates are lower than for CDI; for these reasons, FMT in IBD is still considered an investigational treatment. FMT is, however, currently not recommended to cure IBD patients in clinical practice (Cammarota et al. 2017).

This review summarizes data on the use of FMT for the treatment of both CDI and IBD, with a special attention to highlight studies carried out in European countries.

2 Fecal Microbiota Transplantation for *Clostridioides difficile* Infection

2.1 The Burden of *Clostridioides difficile*

Clostridioides difficile (CD) is a spore-forming, anaerobic, gram-positive bacteria and is spread via the oral-fecal route. The infection generally implies two factors: the presence (endogenous infection) or acquisition (exogenous infection) of CD and an altered composition of gut microbiota. Different factors could facilitate CDI, including older age, hospitalization, recent use of antibiotics, long-term therapy with proton pump inhibitors, and chronic kidney disease (Asha et al. 2006; Mullane et al. 2013; Stevens et al. 2011).

Usually, the proliferation of *Clostridioides difficile* in the large intestine is encouraged by an impaired gut microbiota. The main bacterium virulence factors are toxin A (TcdA) and B (TcdB), which are responsible for mucosal inflammation and disruption of colonic epithelium. The most frequent symptoms of CDI are lower abdominal pain, fever, and diarrhea. Clinical pictures of CDI are variable and range widely from mild colitis to fulminant disease with toxic megacolon and death.

According to recent update of European guidelines (Van Prehn et al. 2021), diagnosis of CDI is defined as:

- clinical findings compatible with CDI and microbiological evidence of *Clostridioides difficile* free toxins by enzyme immunoassay without reasonable evidence of another cause of diarrhea
- a clinical picture compatible with CDI and a positive nucleic acid amplification test (NAAT) or positive toxigenic *Clostridioides difficile* culture

– the presence of pseudomembranous colitis, diagnosed during endoscopy, after colectomy or on autopsy, in combination with a positive test for the presence of toxigenic *Clostridioides difficile*.

A refractory CDI is defined as a not responding to recommended CDI antibiotic treatment, or related to a no response after 3–5 days of therapy.

Moreover, European guidelines reported the definition of recurrence as the presence of CDI evocative symptoms within 8 weeks after a previous episode, provided the symptoms from the previous episode resolved after completion of initial treatment.

In the last decades, CDI has been emerged as a leading cause of healthcare-associated infection in Western countries, with an incidence rate ranges from 1.1 to 631.8 per 100,000 population per year globally (Balsells et al. 2019).

This data has been confirmed, in a recent systematic review of 185 studies, reporting data about CDI incidence, rate and number of recurrences and risk factors, from different countries (France, Germany, Italy and Spain, UK, Poland, US, Canada, Australia, Japan, and China). The authors reported that the median CDI incidence per 10,000 patient days was 4.00 (0.30–74.4) (Finn et al. 2021).

Moreover in a recent meta-analysis, Balsells et al. (2019) evaluated the CDI incidence rate, for health care facility (HCF)-associated, hospital onset-health care facility-associated, medical or general intensive care unit (ICU), internal medicine (IM), long-term care facility (LTCF), and community-associated (CA), from 41 countries, and reported CDI rates more high among ICU and IM patients (11.08 and 10.80 per 1000 admissions/year, respectively) and for HCF patients (2.24 per 1000 admissions/year).

The health and economic burden of CDI is closely related to the recurrence of CDI. Recurrent CDI is directly associated with a rise of medical costs compared with primary episodes, because of an extension of the hospitalization length due to an increase of life-threatening complications. The most common rCDI complications are pseudomembranous colitis (PMC), bloodstream infection (BSI), toxic megacolon, shock, perforation, and death (Shields et al. 2015; Olsen et al. 2015; Falcone 2015.

Among Western countries, rCDI occurs in approximately 15–35% of all CDI cases and data suggest that second and subsequent recurrences are common among patients who experience a recurrent episode, with an increased incidence rate of 50–60% after the second recurrence (Singh et al. 2019).

Similar epidemiological data are observed in the Eastern countries. In a meta-analysis of 51 studies, similar incidence rates for CDI, between Asia and Western countries, particularly Europe and North America, were reported (Borren et al. 2017).

In recent years, these previous data have been confirmed in several studies, in which an increase in the spread of CDI, among low- and middle-income countries, such as India and Africa, has been related to the frequent use of antibiotics as empirical therapy (Ghia et al. 2021; Monaghan et al. 2022; Kulling et al. 2022; De Jager et al. 2021).

This raise in incidence and virulence of CD can be explained, at least in part, by the outbreaks of CDI in healthcare facilities, the spread of fluoroquinolone-resistant strains belonging to the PCR-ribotype 027 and by inappropriate antibiotic usage, which leads to compositional and functional changes in the gastrointestinal microbiome (Warny et al. 2005; McDonald et al. 2005; Imwattana et al. 2020).

Beside this, the epidemiological trend shows how the incidence of CDI has been increasing in the last decades, particularly in Western countries (Lessa et al. 2015).

Recently, in a retrospective analysis of hospitalized patients from 12 European countries, has been reported a higher mean hospital length and mean overall costs per patients, in those who experienced rCDI, with a mean hospital length 55 days (95% CI 17–94 days) and an overall cost of €52,024 (95% CI 715–103,334) (Wingen-Heimann et al. 2022).

2.2 Fecal Microbiota Transplantation and *Clostridioides difficile*

Nowadays, FMT is established as a highly effective treatment and a reliable therapeutic alternative to vancomycin and fidaxomicin, both in European and extra European countries (Hvas et al. 2019; Johnson et al. 2021; Van Prehn et al. 2021, 2023) in rCDI.

Over the years, the efficacy of FMT in the treatment of rCDI, compared to conventional therapy, has been investigated in several studies (MacConnachie et al. 2009; Garborg et al. 2010; Polak et al. 2011; Mattila et al. 2012; Jorup-Ronstrom et al. 2012; Van Nood et al. 2013; Cammarota et al. 2015c; Ianiro et al. 2017) and in different randomized clinical trials (RCTs), with an efficacy rate nearly 90% (Ianiro et al. 2018a. b; Hui et al. 2019).

FMT has also been an effective treatment in the management of severe and severe-complicated CDI (Cammarota et al. 2015b).

Moreover, the use of FMT is positively associated with an increase of overall survival in patients with rCDI (Ianiro et al. 2019), in a reduction in CDI-associated bloodstream infections (Ianiro et al. 2019) and in CDI-related surgery (Cammarota et al. 2015b).

Furthermore, CDI could be considered a very simple model of human gut microbial ecosystem alteration compared to complex chronic disorders in which the gut microbiome is only one among many pathways contributing to disease. For this reason, the high efficacy rate of FMT in rCDI seems to be weakly influenced by donor and recipient characteristics (Staley et al. 2017).

Indeed, one of the most important clinical efficacy predictors in CDI is related to the use of different route of delivery. In several studies, the administration of FMT by colonoscopy has proven to be more effective than other delivery approach (i.e., nasojejunal tube or enema) (Hagel et al. 2016; Li et al. 2016; Hamilton et al. 2012; Ianiro et al. 2017).

In a recent systematic review and meta-analysis, in which 26 studies have been included, by Ramai et al. (2021), different clinical outcomes depended on distinct FMT route of delivery, including colonoscopy, capsule, enema, and nasogastric tube, were compared. They found that the resolution rate of CDI symptoms after FMT via colonoscopy was comparable to the use of capsule (94.8% vs. 92.1%), though superior to that obtained by the delivery via enema and nasogastric tube (87.2% and 78.1%, respectively).

Similar results were confirmed in another meta-analysis of 45 studies and 3744 patients. The authors reported that among different explored routes of delivery (oral capsules, enemas, nasogastric tube) lower GI administration was the most effective, with a success rate of 81–96%, compared to 70–80% for upper administration and 26–84% for enemas. Similar success rate was obtained with the use of capsules, 75–90% vs 81–96% for colonoscopy (Baunwall et al. 2021).

These studies highlighted also the emerging role of capsuled FMT in the treatment of CDI that has been considered an effective route of delivery in the treatment of CDI, from 2015 with a cumulative resolution rate of 89% (Hirsh et al. 2015).

In contrast with these findings, in a pilot open-label randomized trial (Youngster et al. 2014) the efficacy rate of FMT in resolving CDI-associated diarrhea is similar among the administration by nasogastric tube and colonoscopy FMT. This study included a low number of patients, so more studies are advocated to confirm these data.

Despite these great results, the dissemination of FMT has been limited by different factors, such as safety issues. The lack of availability and standardization, that make FMT a low reproducible procedure, is in contrast with the potential expanding use of FMT in clinical practice (Table 1).

2.3 FMT Centers in Europe

In 2021, Baunwall et al. (2021) carried out a Europe-wide survey to describe the clinical use, conduct, and potential for FMT in Europe. Because of this survey emerged a partial spread

Table 1 Characteristics of main studies assessing FMT/LBP for *C. difficile* infection in European and extra European countries

Author	Study level	Single/ multiple center study	Area	Sample	Age (mean or median)	Route of delivery	Frozen/fresh material	Fecal dosage (g/ml)	Follow-up (weeks)	Overall resolution rate after minimum follow-up
Studies from European countries										
MacConnachie et al. (2009)	RCS	Single	UK	15	81.5 (68–95)	Nasogastric tube	Fresh	30 g/150 ml	16	11/15
Garborg et al. (2010)	RCS	Single	Norway	40	75 (53–94)	Gastroscopy and colonoscopy	Fresh	50–100 g/250 ml	11	33/40
Polak et al. (2011)	PCS	Single	Czech Republic	15	82 (NR)	Nasojejunal tube	Fresh	20–30/NR	12	13/15
Mattila et al. (2012)	RCS	Single	Finland	70	73 (22–90)	Colonoscopy	Fresh	20–30 ml feces/ 100–200 ml water	12	66/70
Jorup-Ronstrom et al. (2012)	RCS	Single	Sweden	32	75 (27–94)	Colonoscopy and enema	Fresh	NR/30 ml	104	22/32
Van Nood et al. (2013)	RCT	Single	Netherlands	16	73 (60–86)	Nasojejunal tube	Fresh	≥150/500 ml	10	15/16
Cammarota et al. (2015c)	RCT	Single	Italy	20	73 (29–89)	Colonoscopy	Fresh	152 ± 32 g/500 ml	10	18/20
Satokari et al. (2015)	RCS	Single	Finland	49	56.8 (20–88)	Colonoscopy or enema	Fresh and frozen	30 g/150 ml	12	47/49
Hagel et al. (2016)	RCS	Multiple	Germany	92	75 (59–81)	Gastroscopy, duodenal route, colonoscopy, capsule	Fresh, frozen (no separated efficacy data available)	NR	20	79/92
Ianiro et al. (2017)	PCS	Single	Italy	64	74 (29–94)	Colonoscopy	Fresh, frozen	120–180 g for fresh feces, 50 g for frozen feces/500 ml	8	62/64

(continued)

Table 1 (continued)

Author	Study level	Single/ multiple center study	Area	Sample	Age (mean or median)	Route of delivery	Frozen/fresh material	Fecal dosage (g/ml)	Follow-up (weeks)	Overall resolution rate after minimum follow-up
Ianiro et al. (2018b)	RCT	Single vs multiple	Italy	56	75 (59–91) vs 74 (49-93)	Colonoscopy	Fresh, frozen	50 g of feces for each sample	2	21/28 (single FMT) vs 28/28 (multiple FMT)
Hvas et al. (2019)	RCT	Single	Denmark	24	68 (22–90)	Colonoscopy or nasojejunal tube	Frozen	50 g	8	22/24
Baunwall et al. (2022)	RCT	Single	Denmark	42	59 years (25–89)	Capsules	Frozen	50 g	8	19/21
Studies from extra European countries										
Hamilton et al. (2012)	PCS	Single	USA	43	58 (39–68)	Colonoscopy	Frozen	50 g/250 cc	8	41/43
Youngster et al. (2014)	RCT	Single	USA	20	54.5 ± 24.2	Nasogastric tube and colonoscopy	Frozen	NR	8	18/20
Hirsch et al. (2015)	RCS	Single	USA	19	61 (26–92)	Capsule	Frozen	18–27 g/350 ml/8–12 capsules	12	17/19
Staley et al. (2017)	RCS	Single	USA	49	62.3 ± 17.1	Capsules	Freeze-dried	~1×1011 cells/capsule	8	43/49
Feuerstadt et al. (2022)	RCT (phase III)	Single	Canada	89	65.6 ± 16.5	Oral microbiome therapeutic (SER-109)	–	3×107 spore colony-forming units	8 after SER-109 + 16 weeks FU	Recurrence rate:12% in the SER-109
Khanna et al. (2022)	RCT (phase III)	Single	USA	180	63 (19–93)	Enema	LBP	Consortium of microbes	24	90%US

RCT randomized controlled trial, *PCS* prospective case series, *RCS* retrospective case series, *NR* not reported of FMT across different European countries, with the establishment of 31 FMT centers for a total amount of 1874 FMT procedures; of these, 1077 (57%) have been performed in clinical practice, compared with about 12,400 (6100–28,500) annual cases of multiple, recurrent CDI and indication for FMT in Europe, and 791 (42%) with experimental indications. Currently, European FMT activity covers approximately 10% of the patients who would benefit from the use of FMT.

2.4 Stool Banks

The first step in standardization of FMT was the introduction of frozen feces. The use of frozen preparations is associated with several advantages, such as the immediate availability of FMT, the possibility of administering FMT at centers that do not have an adequate laboratory for stool preparation, and a reduction in the number and frequency of donor screenings with a consequent reduction in costs.

Moreover, the use of frozen feces has also increased safety measures, allowing quarantine of stored feces (Vendrik et al. 2021) and expanded donor screening with the introduction of molecular testing (Ianiro et al. 2021), to prevent the transmission of multi-drug resistant organisms (MDRO) as indicated by Food and Drug Administration (FDA) (US Food and Drug Administration 2019).

Most recently, the donor screening has also been integrated to prevent the possible dissemination of Sars Cov 2 infection, with satisfactory outcomes (Ianiro et al. 2020).

In addition, in effort to address the increase in FMT requests and the need for adequate security measures, stool banks have been established to provide widespread and balanced access to FMT along with high security, quality, and traceability workflows (Cammarota et al. 2019).

Another important element of FMT standardization and quality control is represented by capsuled FMT, to alleviate the requirement for a structured endoscopy unit to provide fecal transplants.

2.5 FMT, New Perspectives

Next to capsulized FMT, live biotherapeutic products (LBPs) have been developed to improve the standardized procedures. LBPs differ in their approach toward product composition and delivery.

The efficacy of an oral microbiome therapeutic composed of live purified Firmicutes bacterial spores (SER-109) in preventing *Clostridioides difficile* infection recurrence in patients treated with standard-of-care antibiotics, was explored in a phase 3, double-blind, randomized, placebo-controlled trial (Feuerstadt et al. 2022). The authors reported that SER-109 is safe and more effective in reducing the risk of recurrent CDI than placebo, with a recurrence rate of 11% in the SER-109 group compared to 41% in the placebo group and the cure rate of 88%.

Recently, FDA approved a live biotherapeutic product consisting of a broad consortium of microbes prepared from human stool (REBYOTA). The commercial use of REBYOTA is limited to the prevention of rCDI (US Food and Drug Administration (FDA) 2022).

Although, a randomized, double-blind, placebo-controlled, phase III study, analyzed with a Bayesian hierarchical model formally incorporating data from a phase 2b trial, showed a treatment success rate of 70.6% vs. 57.5% with placebo (Khanna et al. 2022).

REBYOTA delivery is via enema, without the need for bowel preparation or colonoscopy and can be used in patients who are not able to take an oral product.

2.6 FMT as First-Line Therapy for CDI?

Beside the use of FMT in clinical practice, it has also been explored in the treatment of the first episode of CDI.

In a retrospective cohort, Hocquart et al. (2018) reported the efficacy of FMT in the improvement of survival in patients with severe CDI compared to medical treatment alone. These results have been confirmed recently in a randomized, double-blind, placebo-controlled trial. In this study, it was reported that early FMT was effective in 19 (90%; 95% CI 70–99) of 21 patients in the FMT group compared to 7 (33%, 95% CI 15–57) of 21 patients in the placebo group. The authors suggest that first-line FMT is highly effective and superior to the standard-of-care vancomycin alone in achieving sustained resolution from *Clostridioides difficile* (Baunwall et al. 2022).

Despite the increasing incidence of CDI and the growing evidence suggesting FMT as an highly effective treatment option against this communicable disorder, nowadays there is still a partial spread of FMT across different European countries. For these reasons future research focused on standardizing FMTis advocated, with the aim to ensure it's widespread across countries and to become an easily accessible therapy.

This, together with a rigorous monitoring by regulatory authorities, should be key to improving the efficacy and safety of FMT in Europe and beyond.

2.7 Fecal Microbiota Transplantation for Inflammatory Bowel Disease

The first successful reported use of FMT as a treatment intervention for inflammatory bowel disease (IBD) was published in 1989 by Bennet and Brinkman (1989). However, despite the encouraging results, in the following two decades little evidence has been published, mainly represented by case reports or small series (Borody et al. 1989, 2001, 2003, 2011a, b). The validity of these studies was limited by small patient numbers, vague methods of FMT preparation, and poorly defined and inconsistent results. Consequently, a systematic review in 2012 consisted of only nine retrospective studies, insufficient to perform a meta-analysis (Anderson et al. 2012). However, in 2013, Van Nood et al. (2013) published the first randomized trial on the efficacy of FMT in relapsing *Clostridioides difficile* infections. Hence, a great interest of researchers in the role of FMT in various gastrointestinal and non-gastrointestinal pathologies has arisen.

There are now several controlled and non-controlled studies on the role of FMT in the IBD subtypes of Crohn's disease (CD) and Ulcerative Colitis (UC) (Zhou et al., 2023; Wei et al. 2022).

In particular, the strongest evidence for FMT in IBD in European countries comes from a little number of randomized controlled trials (RCTs). The first European RCT was conducted in the Netherlands by Rossen et al. (2015), who randomized 50 adult patients suffering from active UC to undergo FMT from a healthy donor compared to autologous FMT as a placebo. The primary endpoint was clinical remission (simple clinical colitis activity index scores ≤ 2) combined with ≥ 1-point decrease in the Mayo endoscopic score at week 12. FMT was administered once through nasoduodenal tube at baseline and week 3. The authors reported a no statistical difference in the achievement of clinical and endoscopic remission, between the treatment and placebo arm. The authors reported that there was no statistically significant difference in clinical and endoscopic remission between the treatment arm and the autologous placebo arm of the study. In another RCT (Moayyedi et al. 2015), in which active UC patients were randomized to receive weekly frozen FMT or water enemas for 6 weeks, FMT appeared superior than placebo in the induction of combined remission (24% vs. 5%; $p = 0.03$). Interestingly, the authors suggest that donor characteristics may influence the efficacy of FMT in UC, which gives rise to the alluring prospect of matching donors to recipients.

Other pieces of evidence were reported in a large study by Paramsothy et al. (2017) that allocated 81 adult patients with active UC to receive FMT (from unrelated donors) or placebo (isotonic saline with added brown food colorant and odorant). Study participants were treated with a first colonoscopic infusion followed by self-administered enemas five times per week for 8 weeks (a total of 40 FMTs). The primary endpoint of steroid-free clinical remission together with endoscopic remission (total Mayo score ≤ 2 points) was met in 11 of 41 (27%) of patients receiving FMT vs. 3 of 40 (8%) of patients receiving placebo ($p = 0.02$). In this study, FMT was prepared using a mixture of fecal microbiota from unrelated donors, this approach was implemented in an attempt to maximize the microbial diversity of each FMT.

In the last 5 years, further evidence has emerged regarding the role of FMT in IBD, confirming encouraging results in UC but weak and conflicting evidence for CD. A randomized clinical trial by Costello et al. (2019) reported that 73 patients with mild to moderate UC received either FMT or placebo enemas for eight consecutive weeks. The clinical remission, defined as a Mayo score of ≤ 2 with no individual subscore >1, was achieved by 24 of 38 patients (63%) in the FMT group compared with 5 of 35 patients (14%) in the placebo group ($p < 0.001$). In addition, the FMT group had significant improvements in the Mayo score, endoscopic score, and fecal calprotectin levels compared to the placebo group. In another study by Moayyedi et al. (2015), 73 patients with active UC were randomized to receive either FMT or placebo via colonoscopy. The clinical remission, defined as a Mayo score of ≤ 2 with no individual subscore >1, was achieved by 29% in the FMT group, compared with 9% in the placebo group ($p = 0.03$). The FMT group also had significant improvements in the Mayo score, endoscopic score, and fecal calprotectin levels compared to the placebo group. Another randomized controlled trial by Rossen et al. (2015) investigated the efficacy of FMT in patients with active UC. A total of 73 patients were randomized to receive either FMT or placebo via colonoscopy. The primary endpoint was clinical remission, defined as a Mayo score of ≤ 2 with no individual subscore >1 at week 8. At week 8, 17 of 36 patients (47%) in the FMT group achieved clinical remission, compared with 8 of 37 patients (22%) in the placebo group ($p = 0.03$).

The first randomized controlled study on the role of FMT in maintaining remission in Crohn's disease came from France; Sokol et al. (2020) reported in 17 CD patients an efficacy of FMT in maintaining clinical remission without steroids at 10 and 24 weeks higher than sham transplant (87.5% and 50.0% vs 44.4% and 33.3%). The authors also showed an increase in CRP level 6 weeks after sham transplantation ($p = 0.008$) but not after FMT ($p = 0.5$) and a CDEIS that decreased significantly 6 weeks after FMT ($p = 0.03$). Furthermore, significantly results come from a RCT lead by a European group (van Lingen et al. 2023), on 113 IBD patients underwent FMT because of rCDI. Specifically, at 8-week post-FMT, 71% of patients showed a resolution of CDI, and a sustained cure after FMT occurred in 54 of 86 patients (62.8%) from 90 patients with a median of 784 days (402–1251). Moreover, at the moment of FMT, 54% of the enrolled patients had a concomitant active IBD. Follow up at 8 weeks, 63% of patients showed remission of the non communicable disorder , 34% showed a persistent activity of IBD, and 4% was operated for worsening of disease.

Finally, two meta-analyses have recently been published analyzing results from European and non-European studies.

The first meta-analysis (Zhou et al. 2023), including 11 cohort studies and one randomized controlled trial involving 228 patients, aimed to evaluate the efficacy of FMT in inducing remission in Crohn's disease; authors defined clinical remission as HBI 170 in patients with active CD and it was achieved in 57% (95% CI 49-64%) 2-4 weeks after FMT. In the second one (Wei et al. 2022), metadata from nine RCTs, with a total amount of 425 UC patients (213 FMT and 212 control) were analyzed. They found a positive association between the use of FMT and the achievement of clinical remission (40% vs 22%).

Beside the investigation of FMT as treatment option of IBD, FMT has also been investigated for the treatment of CDI in patients with underlying IBD; data reported that in these patients FMT may be considered an efficacy and safe treatment option (Table 2).

To date, data scaling the efficacy of FMT in IBD are still conflicting. Studies are often heterogeneous, and results are still weak to use FMT in clinical practice to treat IBD patients; studies are

Table 2 Characteristics of main studies assessing FMT for IBD

Study type	Author	FMT route	Patients	Frequency	Control/ comparison	Frozen vs fresh	Definition of clinical remission or primary endpoint (RCT)	Clinical remission	Definition of clinical response or primary endpoint (RCT)	Clinical response
RCT	Rossen et al. (2015)	Nasoduodenal	23 (25 placebo)	One at baseline, one at week 3	Autologous FMT	Fresh	Clinical remission and endoscopic improvement SCCAI ≤2 in combination with ≥1 point drop in combined Mayo endoscopic score	7/23 FMT (30%) vs 5/25 (20%) placebo	≥1.5 point reduction in SCCAI	11/23 (48%) FMT vs 13/25 (52%) placebo
RCT	Moayyedi et al. (2015)	Enema		Six enemas over 6 weeks	Saline	Frozen	Clinical and endoscopic remission Mayo < 3 with endoscopic Mayo 0	9/38 (24%) FMT vs 2/37 (5%) placebo	≥3 point reduction in Mayo score	15/38 FMT (39%) vs 9/37 (24%)
RCT	Paramsothy et al. (2017)	Colonoscopy at baseline followed by enemas	41 (40 placebo)	Colonoscopy at baseline followed by self-administered enemas for 8 weeks (n=40)	Isotonic saline.	Frozen	Total Mayo score ≤2, with all subscores ≤1, and ≥1 point reduction from baseline in endoscopy subscore	11/41 (27%) FMT vs 3/40 (8%) placebo	Steroid-free drop in combined Mayo subscore for bleeding & stool frequency of ≥ 3	22/41 (54%) FMT vs 9/40 (23%)Placebo
RCT	Costello et al. (2019)	Colonoscopy/ enema	38 (35 controls)	At baseline FMT colonoscopy followed by 2 enemas over 7 days	Autologous (A-FMT)	Frozen	Total Mayo score of ≤2 with an endoscopic Mayo score of 1 or less at week 8	12/38 (32%) in D-FMT group vs 3/35 (9%) in A-FMT group	A ≥3-point reduction in total Mayo score at week 8 and 12 months	Clinical response: At 8 weeks: 21/38 (55%) in D-FMT group vs 8/35 (23%) in A-FMT group; At 12 months: 18/29 (62%) in D-FMT group 9/20 (45%) in A-FMT group

RCT	Sokol et al. (2020)	Colonoscopy	8 (9 sham)	One FMT	Sham (S-FMT group)	Fresh	Primary endpoint: implantation of the donor microbiota at week 6 (Sorensen index >0.6)	Was obtained before FMT with corticosteroid	Steroid-free clinical remission at 10 and 24 weeks	At 10 weeks 44% (S-FMT group) vs 87.5% (D-FMT group); at 24 weeks was 33.3% (S-FMT group) vs 50.0% (D-FMT group)

RCT randomized controlled trial, *PCS* prospective case series, *RCS* retrospective case series, *NR* not reported often heterogeneous, and results are still weak for the use of FMT in clinical practice for the treatment of IBD patients; for these reasons, large cohort studies are encouraged with the aim of identifying the main predictors of FMT efficacy in these patients, with the hope of offering another possible treatment strategy in the future.

Conflict of Interest All authors declare no conflict of interest related to this publication.

References

Anderson JL, Edney RJ, Whelan K (2012) Systematic review: faecal microbiota transplantation in the management of inflammatory bowel disease. Aliment Pharmacol Ther 36:503–516

Asha NJ, Tompkins D, Wilcox MH (2006) Comparative analysis of prevalence, risk factors, and molecular epidemiology of antibiotic-associated diarrhea due to Clostridium difficile, Clostridium perfringens, and Staphylococcus aureus. J Clin Microbiol 44:2785–2791

Balsells E, Shi T, Leese C et al (2019) Global burden of Clostridium difficile infections: a systematic review and meta-analysis. J Glob Health 9(1):010407. https://doi.org/10.7189/jogh.09.010407

Baunwall SMD, Andreasen SE, Hansen MM et al (2022) Faecal microbiota transplantation for first or second Clostridioides difficile infection (EarlyFMT): a randomised, double-blind, placebo-controlled trial. Lancet Gastroenterol Hepatol 7(12):1083–1091. https://doi.org/10.1016/S2468-1253(22)00276-X

Baunwall SMD, Terveer EM, Dahlerup JF et al (2021) The use of Faecal Microbiota Transplantation (FMT) in Europe: a Europe-wide survey. Lancet Reg Health Europe 9:100181. https://doi.org/10.1016/j.lanepe.2021.100181

Bennet JD, Brinkman M (1989) Treatment of ulcerative colitis by implantation of normal colonic flora. Lancet 333(8630):164. https://doi.org/10.1016/S0140-6736(89)91183-5

Bibbò S, Lopetuso LR, Ianiro G et al (2014) Role of microbiota and innate immunity in recurrent Clostridium difficile infection. J Immunol Res 2014:462740. https://doi.org/10.1155/2014/462740

Borody TJ, Campbell J, Torres M et al (2011a) Reversal of idiopathic thrombocytopenic purpura [ITP] with fecal microbiota transplantation [FMT]. Am J Gastroenterol 106:S352

Borody TJ, George L, Andrews P et al (1989) Bowel-flora alteration: a potential cure for inflammatory bowel disease and irritable bowel syndrome? Med J Aust 150:604

Borody TJ, Leis S, McGrath K (2001) Treatment of chronic constipation and colitis using human probiotic infusions. Probiotics, Prebiotics and New Foods Conference, Universita Urbaniana, Rome, 2–4 September 2001

Borody TJ, Torres M, Campbell J et al (2011b) Reversal of inflammatory bowel disease (IBD) with recurrent faecal microbiota transplants (FMT). Am J Gastroenterol 106:S366–S366

Borody TJ, Warren EF, Leis S et al (2003) Treatment of ulcerative colitis using fecal bacteriotherapy. J Clin Gastroenterol 37:42–47

Borren NZ, Ghadermarzi S, Hutfless S et al (2017) The emergence of Clostridium difficile infection in Asia: a systematic review and meta-analysis of incidence and impact. PLoS One 12(5):e0176797. https://doi.org/10.1371/journal.pone.0176797

Cammarota G, Ianiro G, Cianci R et al (2015a) The involvement of gut microbiota in inflammatory bowel disease pathogenesis: potential for therapy. Pharmacol Therap 149:191–212. https://doi.org/10.1016/j.pharmthera.2014.12.006

Cammarota G, Ianiro G, Magalini S et al (2015b) Decrease in surgery for Clostridium difficile infection after starting a program to transplant fecal microbiota. Ann Intern Med 163:487–488. https://doi.org/10.7326/L15-5139

Cammarota G, Masucci L, Ianiro G et al (2015c) Randomised clinical trial: faecal microbiota transplantation by colonoscopy vs. vancomycin for the treatment of recurrent Clostridium difficile infection. Aliment Pharmacol Ther 41:835–843

Cammarota G, Ianiro G, Kelly CR et al (2019) International consensus conference on stool banking for faecal microbiota transplantation in clinical practice. Gut 68:2111–2121. https://doi.org/10.1136/GUTJNL-2019-319548

Cammarota G, Ianiro G, Tilg H et al (2017) European consensus conference on faecal microbiota transplantation in clinical practice. Gut 66(4):569–580. https://doi.org/10.1136/gutjnl-2016-313017

Cani PD (2018) Human gut microbiome: hopes, threats and promises. Gut 67(9):1716–1725. https://doi.org/10.1136/gutjnl-2018-316723

Cardinale V, Capurso G, Ianiro G et al (2020) Intestinal permeability changes with bacterial translocation as key events modulating systemic host immune response to SARS-CoV-2: a working hypothesis. Digest Liver Dis 52(12):1383–1389. https://doi.org/10.1016/j.dld.2020.09.009

Costello SP, Hughes PA, Waters O et al (2019) Effect of fecal microbiota transplantation on 8-week remission in patients with ulcerative colitis: a randomized clinical trial. JAMA 321(2):156–164

De Jager P, Smith O, Bolon S et al (2021) Epidemiology of Clostridioides difficile in South Africa. PLoS Onesss 16(11):e0259771. https://doi.org/10.1371/journal.pone.0259771

El Kaoutari A, Armougom F, Gordon JI et al (2013) The abundance and variety of carbohydrate-active enzymes in the human gut microbiota. Nat Rev Microbiol 11(7): 497–504. https://doi.org/10.1038/nrmicro3050

Falcone M, Russo A, Iraci F et al (2015) Risk factors and outcomes for bloodstream infections secondary to Clostridium difficile infection. Antimicrob Agents Chemother 60(1):252–257. https://doi.org/10.1128/AAC.01927-15

FDA (2019) Important safety alert regarding use of fecal microbiota for transplantation and risk of serious adverse reactions due to transmission of multi-drug resistant organisms. https://www.fda.gov/vaccines-blood-biologics/safety-availability-biologics/important-safety-alert-regarding-use-fecal-microbiota-transplantation-and-risk-serious-adverse

Feuerstadt P, Louie TJ, Lashner B et al (2022) SER-109, an oral microbiome therapy for recurrent Clostridioides difficile infection. New Engl J Med 386(3):220–229. https://doi.org/10.1056/NEJMoa2106516

Finn E, Andersson FL, Madin-Warburton M (2021) Burden of Clostridioides difficile infection (CDI) – a systematic review of the epidemiology of primary and recurrent CDI. BMC Infect Dis 21(1):456. https://doi.org/10.1186/s12879-021-06147-y

Garborg K, Waagsbø B, Stallemo A et al (2010) Results of faecal donor instillation therapy for recurrent Clostridium difficile-associated diarrhoea. Scand J Infect Dis 42:857–861

Ghia CJ, Waghela S, Rambhad GS (2021) Systematic literature review on burden of Clostridioides difficile infection in India. Clin Pathol 14: 2632010X211013816. https://doi.org/10.1177/2632010X211013816

Hagel S, Fischer A, Ehlermann P et al (2016) Fecal microbiota transplant in patients with recurrent Clostridium difficile infection. Deutsches Arzteblatt Int 113(35–36):583–589. https://doi.org/10.3238/arztebl.2016.0583

Hamilton MJ, Weingarden AR, Sadowsky MJ et al (2012) Standardized frozen preparation for transplantation of fecal microbiota for recurrent Clostridium difficile infection. Am J Gastroenterol 107:761–767

Hirsch BE, Saraiya N, Poeth K et al (2015) Effectiveness of fecal-derived microbiota transfer using orally administered capsules for recurrent Clostridium difficile infection. BMC Infect Dis 15:191

Hocquart M, Lagier JC, Cassir N et al (2018) Early fecal microbiota transplantation improves survival in severe Clostridium difficile infections. Clin Infect Dis 66(5): 645–650

Hui W, Li T, Liu W et al (2019) Fecal microbiota transplantation for treatment of recurrent C. difficile infection: an updated randomized controlled trial meta-analysis. PLoS One 14. https://doi.org/10.1371/JOURNAL.PONE.0210016

Hvas CL, Dahl Jørgensen SM, Jørgensen SP et al (2019) Fecal microbiota transplantation is superior to fidaxomicin for treatment of recurrent Clostridium difficile infection. Gastroenterology 156:1324–1332.e3. https://doi.org/10.1053/J.GASTRO.2018.12.019

Ianiro G, Valerio L, Masucci L et al (2017) Predictors of failure after single faecal microbiota transplantation in patients with recurrent Clostridium difficile infection: results from a 3-year, single-centre cohort study. Clin Microbiol Infect 23:337.e1–337.e3

Ianiro G, Bruno G, Lopetuso L et al (2014a) Role of yeasts in healthy and impaired gut microbiota: the gut mycome. Curr Pharm Design 20(28):4565–4569. https://doi.org/10.2174/13816128113196660723

Ianiro G, Bibbò S, Scaldaferri F et al (2014b) Fecal microbiota transplantation in inflammatory bowel disease: beyond the excitement. Medicine 93(19):e97. https://doi.org/10.1097/MD.0000000000000097

Ianiro G, Maida M, Burisch J et al (2018a) Efficacy of different faecal microbiota transplantation protocols for Clostridium difficile infection: a systematic review and meta-analysis. United Eur Gastroenterol J 6:1232–1244. https://doi.org/10.1177/2050640618780762

Ianiro G, Masucci L, Quaranta G et al (2018b) Randomised clinical trial: faecal microbiota transplantation by colonoscopy plus vancomycin for the treatment of severe refractory Clostridium difficile infection-single versus multiple infusions. Aliment Pharmacol Ther 48:152–159. https://doi.org/10.1111/APT.14816

Ianiro G, Molina-Infante J, Gasbarrini A (2015) Gastric microbiota. Helicobacter 20(Suppl 1):68–71. https://doi.org/10.1111/hel.12260

Ianiro G, Murri R, Sciumè GD et al (2019) Incidence of bloodstream infections, length of hospital stay, and survival in patients with recurrent Clostridioides difficile infection treated with fecal microbiota transplantation or antibiotics: a prospective cohort study. Ann Intern Med 171:695–702. https://doi.org/10.7326/M18-3635

Ianiro G, Porcari S, Bibbò S et al (2021) Donor program for fecal microbiota transplantation: a 3-year experience of a large-volume Italian stool bank. Dig Liver Dis 53:1428–1432. https://doi.org/10.1016/J.DLD.2021.04.009

Ianiro G, Segal JP, Mullish BH et al (2020) Fecal microbiota transplantation in gastrointestinal and extraintestinal disorders. Future Microbiol 15:1173–1183. https://doi.org/10.2217/fmb-2020-006

Imwattana K, Knight DR, Kullin B et al (2020) Antimicrobial resistance in Clostridium difficile ribotype 017. Expert Rev Anti-infect Ther 18(1):17–25. https://doi.org/10.1080/14787210.2020.1701436

Johnson S, Lavergne V, Skinner AM, Gonzales-Luna AJ, Garey KW, Kelly CP, Wilcox MH (2021) Clinical Practice Guideline by the Infectious Diseases Society of America (IDSA) and Society for Healthcare Epidemiology of America (SHEA): 2021 focused update guidelines on management of Clostridioides difficile infection in adults. Clin Infect Dis 73(5):1029–1044. https://doi.org/10.1093/cid/ciab549

Jorup-Ronstrom C, Hakanson A, Sandell S et al (2012) Fecal transplant against relapsing Clostridium difficile-associated diarrhea in 32 patients. Scand J Gastroenterol 47:548–552

Kelly CR, Fischer M, Allegretti JR et al (2021) ACG Clinical Guidelines: Prevention, diagnosis, and treatment of Clostridioides difficile infections. Am J Gastroenterol 116(6):1124–1147. https://doi.org/10.14309/ajg.0000000000001278

Khanna S, Assi M, Lee C et al (2022) Efficacy and safety of RBX2660 in PUNCH CD3, a phase III, randomized, double-blind, placebo-controlled trial with a Bayesian primary analysis for the prevention of recurrent Clostridioides difficile infection. Drugs 82(15): 1527–1538. https://doi.org/10.1007/s40265-022-01797-x

Kramer P, Bressan P (2015) Humans as superorganisms: how microbes, viruses, imprinted genes, and other selfish entities shape our behavior. Perspect Psychol Scis 10(4):464–481. https://doi.org/10.1177/1745691615583131

Kullin B, Abratt VR, Reid SJ et al (2022) Clostridioides difficile infection in Africa: a narrative review. Anaerobe 74:102549. https://doi.org/10.1016/j.anaerobe.2022.102549

LeBlanc JG, Milani C, de Giori GS et al (2013) Bacteria as vitamin suppliers to their host: a gut microbiota perspective. Curr Opin Biotechnol 24(2):160–168. https://doi.org/10.1016/j.copbio.2012.08.005

Lessa FC, Mu Y, Bamberg WM et al (2015) Burden of Clostridium difficile infection in the United States. N Engl J Med 372:825–834

Li YT, Cai HF, Wang ZH et al (2016) Systematic review with meta-analysis: long-term outcomes of faecal microbiota transplantation for Clostridium difficile infection. Aliment Pharmacol Ther 43:445–457

McDonald LC, Killgore GE, Thompson A et al (2005) An epidemic, toxin gene-variant strain of Clostridium difficile. N Engl J Med 353:2433–2441

McDonald LC, Gerding DN, Johnson S et al (2018) Clinical Practice Guidelines for Clostridium difficile Infection in Adults and Children: 2017 Update by the Infectious Diseases Society of America (IDSA) and Society for Healthcare Epidemiology of America (SHEA). Clin Infect Dis 66

MacConnachie AA, Fox R, Kennedy DR et al (2009) Faecal transplant for recurrent Clostridium difficile-associated diarrhoea: a UK case series. QJM 102: 781–784

Mattila E, Uusitalo-Seppala R, Wuorela M et al (2012) Fecal transplantation, through colonoscopy, is effective therapy for recurrent Clostridium difficile infection. Gastroenterology 142:490–496

Moayyedi P, Surette MG, Kim PT et al (2015) Fecal microbiota transplantation induces remission in patients with active ulcerative colitis in a randomized controlled trial. Gastroenterology 149:102–109

Monaghan TM, Biswas R, Satav A et al (2022) Clostridioides difficile epidemiology in India. Anaerobe 74:102517. https://doi.org/10.1016/j.anaerobe.2022.102517

Mullane KM, Cornely OA, Crook DW et al (2013) Renal impairment and clinical outcomes of Clostridium difficile infection in two randomized trials [published correction appears in Am J Nephrol. 38(3):266]. Am J Nephrol 38(1):1–11. https://doi.org/10.1159/000351757

Olsen MA, Yan Y, Reske KA et al (2015) Recurrent Clostridium difficile infection is associated with increased mortality. Clin Microbiol Infect 21(2): 164–170. https://doi.org/10.1016/j.cmi.2014.08.017

Paramsothy S, Kamm MA, Kaakoush NO et al (2017) Multidonor intensive faecal microbiota transplantation for active ulcerative colitis: a randomised placebo-controlled trial. Lancet 389:1218–1228

Polak P, Freibergerova M, Jurankova J et al (2011) First experiences with faecal bacteriotherapy in the treatment of relapsing pseudomembranous colitis due to Clostridium difficile. Klin Mikrobiol Infekc Lek 17: 214–217

Ramai D, Zakhia K, Fields PJ et al (2021) Fecal microbiota transplantation (FMT) with colonoscopy is superior to enema and nasogastric tube while comparable to capsule for the treatment of recurrent Clostridioides difficile infection: a systematic review and meta-analysis. Digest Dis Sci 66(2):369–380. https://doi.org/10.1007/s10620-020-06185-7

Rossen NG, Fuentes S, van der Spek MJ et al (2015) Findings from a randomized controlled trial of fecal transplantation for patients with ulcerative colitis. Gastroenterology 149:110–118

Satokari R, Mattila E, Kainulainen V et al (2015) Simple faecal preparation and efficacy of frozen inoculum in faecal microbiota transplantation for recurrent *Clostridium difficile* infection—an observational cohort study. Aliment Pharmacol Ther 41:46–53

Shields K, Araujo-Castillo RV, Theethira TG et al (2015) Recurrent Clostridium difficile infection: from colonization to cure. Anaerobe 34:59–73. https://doi.org/10.1016/j.anaerobe.2015.04.012

Singh T, Bedi P, Bumrah K et al (2019) Updates in treatment of recurrent Clostridium difficile infection. J Clin Med Res 11(7):465–471. https://doi.org/10.14740/jocmr3854

Sokol H, Landman C, Seksik P et al (2020) Fecal microbiota transplantation to maintain remission in Crohn's disease: a pilot randomized controlled study. Microbiome 8(1):12. https://doi.org/10.1186/s40168-020-0792-5

Staley C, Hamilton MJ, Vaughn BP, Graiziger CT, Newman KM, Kabage AJ, Sadowsky MJ, Khoruts A (2017) Successful resolution of recurrent Clostridium difficile infection using freeze-dried encapsulated fecal microbiota; pragmatic cohort study. Am J Gastroenterol 112(6):940–947. https://doi.org/10.1038/ajg.2017.6

Stevens V, Dumyati G, Fine LS et al (2011) Cumulative antibiotic exposures over time and the risk of

Clostridium difficile infection. Clin Infect Dis 53(1): 42–48. https://doi.org/10.1093/cid/cir301

Thursby E, Juge N (2017) Introduction to the human gut microbiota. Biochem J 474(11):1823–1836. https://doi.org/10.1042/BCJ20160510

US Food and Drug Administration (FDA) (2022) REBYOTA. https://www.fda.gov/vaccines-blood-biologics/vaccines/rebyota

Van Lingen EE, Baunwall SSMD, Lieberknecht SSC et al (2023) Short- and long-term follow-up after fecal microbiota transplantation as treatment for recurrent Clostridioides difficile infection in patients with inflammatory bowel disease. Therap Adv Gastroenterol 16:17562848231156285. https://doi.org/10.1177/17562848231156285

Van Nood E, Vrieze A, Nieuwdorp M et al (2013) Duodenal infusion of donor faeces for recurrent Clostridium difficile. N Engl J Med 368:407–415

Van Prehn J, Reigadas E, Vogelzang EH et al (2021) European Society of Clinical Microbiology and Infectious Diseases: 2021 update on the treatment guidance document for Clostridioides difficile infection in adults. Clin Microbiol Infect 27(Suppl 2):S1–S21. https://doi.org/10.1016/j.cmi.2021.09.038

Van Prehn J, Fitzpatrick F, Kuijper EJ, European Study Group for Clostridioides difficile and the European Study Group for Host and Microbiota Interaction (2023) Faecal microbiota transplantation for first and second episodes of Clostridioides difficile infection. Lancet Gastroenterol Hepatol 8(2):109. https://doi.org/10.1016/S2468-1253(22)00342-9

Vendrik KEW, Terveer EM, Kuijper EJ et al (2021) Periodic screening of donor faeces with a quarantine period to prevent transmission of multidrug-resistant organisms during faecal microbiota transplantation: a retrospective cohort study. Lancet Infect Dis 21:711–721. https://doi.org/10.1016/S1473-3099(20)30473-4

Venema K (2010) Role of gut microbiota in the control of energy and carbohydrate metabolism. Curr Opin Clin Nutr Metab Care 13(4):432–438. https://doi.org/10.1097/MCO.0b013e32833a8b60

Warny M, Pepin J, Fang A et al (2005) Toxin production by an emerging strain of Clostridium difficile associated with outbreaks of severe disease in North America and Europe. Lancet 366:1079–1084

Wei ZJ, Dong HB, Ren YT et al (2022) Efficacy and safety of fecal microbiota transplantation for the induction of remission in active ulcerative colitis: a systematic review and meta-analysis of randomized controlled trials. Ann Transl Med 10(14):802. https://doi.org/10.21037/atm-22-3236

Wingen-Heimann SM, Davies K, Viprey VF et al (2022) Clostridioides difficile infection (CDI): a pan-European multi-center cost and resource utilization study, results from the Combatting Bacterial Resistance in Europe CDI (COMBACTE-CDI). Clin Microbiol Infect. S1198-743X(22)00643-7. https://doi.org/10.1016/j.cmi.2022.12.019

Wu HJ, Wu E (2012) The role of gut microbiota in immune homeostasis and autoimmunity. Gut Microbes 3(1):4–14. https://doi.org/10.4161/gmic.19320

Youngster I, Sauk J, Pindar C et al (2014) Fecal microbiota transplant for relapsing Clostridium difficile infection using a frozen inoculum from unrelated donors: a randomized, open-label, controlled pilot study. Clin Infect Dis 58(11):1515–1522. https://doi.org/10.1093/cid/ciu135

Zhou S, Cui Y, Zhang Y et al (2023) Fecal microbiota transplantation for induction of remission in Crohn's disease: a systematic review and meta-analysis. Int J Colorect Dis 38(1):62. https://doi.org/10.1007/s00384-023-04354-4

Probiotics for Prevention and Treatment of *Clostridium difficile* Infection

Lorena Valdés-Varela, Miguel Gueimonde, and Patricia Ruas-Madiedo

Abstract

Probiotics have been claimed as a valuable tool to restore the balance in the intestinal microbiota following a dysbiosis caused by, among other factors, antibiotic therapy. This perturbed environment could favor the overgrowth of *Clostridium difficile*, and in fact, the occurrence of *C. difficile*-associated infections (CDI) is increasing in recent years. In spite of the high number of probiotics able to in vitro inhibit the growth and/or toxicity of this pathogen, its application for treatment or prevention of CDI is still scarce since there are not enough well-defined clinical studies supporting efficacy. Only a few strains, such as *Lactobacillus rhamnosus* GG and *Saccharomyces boulardii*, have been studied in more extent. The increasing knowledge about the probiotic mechanisms of action against *C. difficile*, some of them reviewed here, makes promising the application of these live biotherapeutic agents against CDI. Nevertheless, more effort must be paid to standardize the clinical studies conducted to evaluate probiotic products, in combination with antibiotics, in order to select the best candidate for *C. difficile* infections.

1 Introduction

The gut microbiota is a complex and diverse microbial community that has coevolved with humans in a commensal way (Donaldson et al. 2016). In a healthy state, this collection of microorganisms protects the host by inhibiting colonization and growth of pathogens. However, antibiotic exposure strongly perturbs the intestinal microbiota, producing a decrease in microbial abundance and species diversity, as well as a suppression of the innate immune system disrupting the gut barrier and frequently causing antibiotic-associated diarrhea. In some cases, the intestinal dysbiosis followed after antibiotic treatment allows the overgrowth of *Clostridium difficile* given that this perturbed environment has a low abundance of short chain fatty acids, a high abundance of primary bile acids, a high carbohydrate availability, and an immunosuppressed host

Note: This chapter is an exact replication of the corresponding chapter in the previous edition: Valdés-Varela, L., Gueimonde, M., Ruas-Madiedo, P. (2018). Probiotics for Prevention and Treatment of *Clostridium difficile* Infection. In Mastrantonio, P., Rupnik, M. (Eds.) Updates on *Clostridium difficile* in Europe. Advances in Experimental Medicine and Biology(), Vol 1050. Springer, Cham. https://doi.org/10.1007/978-3-319-72799-8_10

L. Valdés-Varela · M. Gueimonde · P. Ruas-Madiedo (✉)
Department of Microbiology and Biochemistry of Dairy Products, Instituto de Productos Lacteos de Asturias – Consejo Superior de Investigaciones Cientıficas (IPLA-CSIC), Villaviciosa, Asturias, Spain
e-mail: lvaldes@ipla.csic.es; mgueimonde@ipla.csic.es; ruas-madiedo@ipla.csic.es

P. Mastrantonio, M. Rupnik (eds.), *Updates on Clostridioides difficile in Europe*, Advances in Experimental Medicine and Biology 1435, https://doi.org/10.1007/978-3-031-42108-2_6

in the absence of microbial competitors in the gut (Lawley and Walker 2013).

C. difficile can be found in the gut microbiota of both healthy infants and adults, the occurrence being higher in infant (70%) than in the adult (17%) population (Ozaki et al. 2004; Jangi and Lamont 2010). In these healthy carriers, the presence of this microorganism does not seem to cause any disease. However, at the same time, *C. difficile* is the main causative agent of antibiotic-associated diarrhea in nosocomial environments (Leffler and Lamont 2015). As previously indicated, the antimicrobial therapy affects the endogenous gut microbiota diminishing colonization resistance, allowing the overgrowth of this pathogen and causing *C. difficile*-associated diarrhea (CDAD). This problem has been traditionally linked to elderly and institutionalized/hospitalized persons under antibiotic therapy (Rupnik et al. 2009); however, the occurrence of *C. difficile*-associated infections (CDI) seems to be increasing also in traditionally considered low-risk populations (Carter et al. 2012). This change in the epidemiology of CDI has been related to the worldwide distribution of hypervirulent strains (Yakob et al. 2015); besides, foods and animals have been found to act as carriers of this pathogen pointing at *C. difficile* as a zoonotic agent and suggesting potential foodborne transmission (Rodriguez et al. 2016). A range of virulent factors are the cause of colitis during CDI course, the main ones being several toxins, encoded in pathogenicity loci, and the flagella, which are factors allowing mobility and adherence of the pathogen (Abt et al. 2016). Pathogenesis was initially attributed to the production of toxins A (TcdA) and B (TcdB), belonging to the large clostridial toxin (LCT) family, which act as intracellular glycosyltransferases that inactivate Rho family GTPases, thus blocking downstream cellular events (Carter et al. 2012). More recently, strains producing a third toxin, the binary toxin (CDT), have been associated with an increase in the CDI severity; this toxin has two components the CDTa, which acts as an ADP-ribosyltransferase targeting actin, and CDTb that is able to bind to the cell and translocate the first component to the cytosol (Gerding et al. 2014). In spite of recent advances in the identification of processes involved on receptor binding and entry into mammalian cells, the mode of action of clostridial toxins remains to be totally elucidated (Orrell et al. 2017).

The standard treatment for *C. difficile* infection is the administration of antibiotics, mainly metronidazole, vancomycin, or fidaxomicin, but unfortunately, the recurrence rate of the disease is very high and this treatment becomes less effective. Indeed, it has been described that some *C. difficile* subpopulations (ribotypes) have a reduced susceptibility to metronidazole (Moura et al. 2013). In case of multiple recurrent CDI, fecal microbiota transplantation (FMT) is being more frequently used as the ultimate therapy, although the selection of the appropriate donor is a critical issue (Woodworth et al. 2017). These facts have prompted researchers to look for alternative therapeutic options (Fig. 1) which have been recently reviewed by different authors (Mathur et al. 2014; Hussack and Tanha 2016; Kachrimanidou et al. 2016; Kociolek and Gerding 2016; Martin and Wilcox 2016; McFarland 2016; Ofosu 2016; Padua and Pothoulakis 2016; Unal and Steinert 2016). Among them, probiotics have been proposed as a potential tool for preventing the dysbiosis of microbiota, caused by the administration of antibiotics, and for assisting the microbiota restoration after antibiotics or infection (Reid et al. 2011); thus, they have also been evaluated for prevention and treatment of CDI (Na and Kelly 2011).

Probiotics were defined in 2001 by a group of experts joined by FAO/WHO as "live microorganisms that, when administered in adequate amounts, confer a health benefit on the host"; this definition was recently revised, and accepted after minor grammatical modifications, by members of the International Scientific Association for Probiotics and Prebiotics (ISAPP) which also proposes an overall framework for use of this term, encompassing diverse end uses (Hill et al. 2014). In next sections, we will review the current available data about the efficacy of probiotics in prevention and therapy for CDI, as

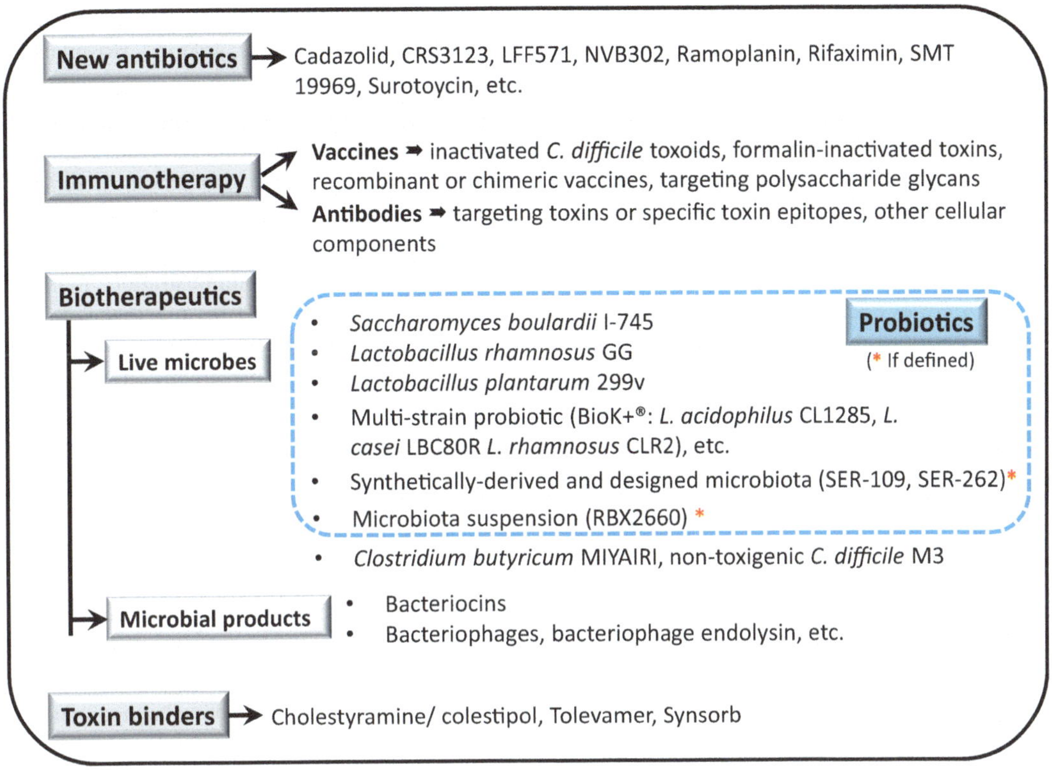

Fig. 1 Some therapeutic options currently under study for the prevention and treatment of *Clostridium difficile* infection

well as some putative mechanisms involved in this anti-*C. difficile* effect.

2 Clinical Studies Evaluating Probiotic Efficacy

The ability of probiotics for inhibiting the growth of *C. difficile* has been characterized by using different experimental approaches (Auclair et al. 2015; Forssten et al. 2015; Valdes-Varela et al. 2016b; Fredua-Agyeman et al. 2017). This use of probiotic microorganisms has long been considered a potential option to combat CDI. However, despite the large number of in vitro studies performed for the selection of probiotic strains with activity against *C. difficile* and for their use for CDI prevention or treatment, the evidence from human clinical trials is still limited. Different probiotic strains have been reported to increase the colonization resistance against *C. difficile* (Hopkins and Macfarlane 2003; Kondepudi et al. 2014; Auclair et al. 2015; Forssten et al. 2015). Certain strains of bifidobacteria and lactobacilli have been found to reduce the adhesion of *C. difficile* to intestinal epithelial cells or intestinal mucus (Collado et al. 2005; Banerjee et al. 2009) or to be able to inhibit its growth (Lee et al. 2013; Schoster et al. 2013; Valdes-Varela et al. 2016b). Moreover, animal studies seem to confirm a potential benefit of probiotics on the inhibition of *C. difficile* colonization (Mansour et al. 2017). Nevertheless, to date most of the clinical studies have focused on prevention, and there is a lack of data on the potential use of probiotics on the treatment of *C. difficile* infection.

During the last couple of decades, several studies have evaluated the usefulness of different probiotic strains in the prevention of CDAD.

However, in spite of the large number of strains screened in vitro, most of the evidence from clinical trials regards only a few bacterial strains, and most often, the studies have focused on the prevention of antibiotic-associated diarrhea, without further confirmation of *C. difficile* etiology. Among the assessed strains, the effect of *Lactobacillus rhamnosus* strain GG (Arvola et al. 1999; Vanderhoof et al. 1999), or the yeast species *Saccharomyces boulardii* (Kotowska et al. 2005; Can et al. 2006), in the prevention of antibiotic associated diarrhea has been widely recognized. Although not so extensively studied, other probiotic strains and probiotic mixes have also been evaluated around the world with positive results (Wullt et al. 2003; Maziade et al. 2015). The availability of a large number of clinical studies focusing on antibiotic-associated diarrhea has provided enough data for carrying out systematic reviews and meta-analysis studies, either considering probiotics as a group, which shows important limitations due to interstrain and/or interproduct variability, or meta-analyses focused on specific strains. The meta-analysis studies on the general use of probiotics for the prevention of antibiotic-associated diarrhea have consistently provided evidence for a beneficial role, especially in children (Cremonini et al. 2002; D'Souza et al. 2002; Sazawal et al. 2006; Johnston et al. 2007; Hempel et al. 2012; Goldenberg et al. 2015). Moreover, meta-analyses conducted for some specific probiotics, such as *S. boulardii* or *L. rhamnosus* GG, have further confirmed the beneficial effect of these strains in the prevention of antibiotic-associated diarrhea (McFarland 2006; Szajewska et al. 2007a, b). This has resulted in recommendations issued by the ESPGHAN (European Society for Paediatric Gastroenterology Hepatology and Nutrition) with regard to the use of probiotics for the prevention of antibiotic-associated diarrhea in children (Szajewska et al. 2016).

Furthermore, some studies have specifically focused on confirmed *C. difficile*-associated diarrhea, and these have also provided positive results for primary prevention (Wullt et al. 2003; Gao et al. 2010; Sampalis et al. 2010; Allen et al. 2013; Dietrich et al. 2014; Maziade et al. 2015). Some practical examples exist as well, such as that of the "Pierre-Le Gardeur" Hospital in Canada, which after a *C. difficile* outbreak began to administer a probiotic mix (BioK+®) together with any antibiotic prescriptions, achieving a significant reduction on the number of *C. difficile* disease cases (Maziade et al. 2015). Recent meta-analyses and systematic reviews have assessed the effects of probiotic administration, most of them administering the strains together with the antibiotic treatment, on the primary prevention of CDAD in different population groups (Table 1). In general the data support a beneficial effect of probiotics on the primary prevention of CDAD. However, the high heterogeneity among the available clinical studies makes difficult defining the best probiotic to be used, its dose, and the administration regime.

Regarding the prevention of the recurrence of the disease, the available data are more limited than in the case of primary prevention. Some clinical intervention studies have been conducted with variable results (McFarland et al. 1994; Surawicz et al. 2000), with reviews and meta-analyses indicating that there is only limited evidence on the benefit of probiotics in secondary prevention of CDI (Allen et al. 2013; O'Horo et al. 2014; McFarland 2015). The limited data available on secondary prevention underlines the need for more clinical intervention trials to be conducted in this topic.

To sum up, the available evidence strongly suggests that probiotics are helpful for primary prevention with only moderate evidence of a role in avoiding disease relapse. However, the potential role of probiotics in the treatment during the active phase of the disease remains largely unknown. Perhaps the major criticism that can be done to the available data is that there has not been a serious standardization effort for the probiotic products, doses, antibiotics, and therapeutic protocols to be used. Moreover, analyses of the cost-effectiveness of probiotic use on the prevention of *C. difficile* disease have not been performed until recently, with variable results, indicating the need for further studies conducted under different healthcare systems (Leal et al. 2016; Starn et al. 2016).

Table 1 Recent meta-analyses and systematic reviews on the use of probiotics in primary prevention of *C. difficile* infection

Target population	Probiotic	N° eligible RCTs	N° volunteers included	Conclusion	References
Elderly	Any	5	>3400	No significant effect	Vernaya et al. (2017)
Adults	Any	19	>6200	Significant reduction	Shen et al. (2017)
Adults	*Lactobacillus* (any)	10	>4800	Inconclusive evidence	Sinclair et al. (2016)
Adults and children	Any	26	>7900	Significant reduction	Lau and Chamberlain (2016)
Adults and children	Any (and by species)	21	>3700	Significant reduction	McFarland (2015)
Adults and children	Any	31	>4200	Significant reduction	Goldenberg et al. (2013)

RCT randomized controlled trial

3 Models to Study Probiotics Against *C. difficile*

Different experimental models have been developed in order to study the interaction of *C. difficile* with the host (recently reviewed by Young (2017)); additionally, these models can be used in the search for new therapeutic alternatives and adjuvant strategies for preventing or treating CDI (Table 2). Investigations using in vitro models of bacterial cultures are valuable systems for the screening of potential probiotics against *C. difficile*, but as disadvantage, they have the lack of feedback mechanisms with host and/or host-microbe interactions (Best et al. 2012). However, these microbial culturing models can be combined with cell culture systems to better mimic the interaction *C. difficile*—probiotic—host (Venema and van den Abbeele 2013). Co-cultures of toxigenic *C. difficile* strains with probiotic candidates have been carried out to determine the potential of the latter for reducing the germination of spores and outgrowth into vegetative toxin-producing cells of the pathogen (Table 2). Models of gut microbiota have been assayed to in vitro evaluate the potential of probiotic candidates for decreasing the growth of *C. difficile* in this complex microbial ecosystem. These models range from simple batch fermentations to complex multi-compartmental continuous systems (Venema and van den Abbeele 2013). Static batch cultures, containing fecal suspensions, have been used to observe the influence of probiotics on the survival of *C. difficile* (Tejero-Sarinena et al. 2013). Continuous culture systems (human "colonic" model) allow the study of the pathogen in an environment closer to the reality, over considerably longer periods than in static batch cultures (Best et al. 2012; Le Lay et al. 2015). Currently, most of the colonic simulators consists of four different units (glass vessels) continuously connected, having different pH and flow rates, thus representing the ascending, transverse, descending, and distal colon (Forssten et al. 2015).

Several in vitro studies investigated the effect of probiotic treatment on the interaction of *C. difficile* with components of the intestinal mucosa, such as mucus or epithelial cells (Table 2). The cytotoxicity of clostridial cell-free supernatants (obtained from co-cultures of probiotic vs. *C. difficile*) or of caecum contents (collected from animals infected with *C. difficile* and treated with potential probiotics) has been evaluated upon cell lines using classic label-based, endpoint methods (Banerjee et al. 2009; Trejo et al. 2010, 2013; Valdes-Varela et al. 2016a). However, label-free technologies are currently been available and being used in drug development processes, which are noninvasive techniques that allow the continuous (real-time)

Table 2 Summary of some in vitro models used to study potential probiotics against *Clostridium difficile*

In vitro experimental models			References
Microbial cultivation	*Vs.* probiotic	Co-cultures of *C. difficile* with probiotic candidates	Trejo et al. (2010), Best et al. (2012), Kolling et al. (2012), Lee et al. (2013), Schoster et al. (2013), Kondepudi et al. (2014), Yun et al. (2014), Ambalam et al. (2015), Andersen et al. (2016), Spinler et al. (2016), and Rätsep et al. (2017)
	Vs. microbiota/ probiotic	Static batch system	Tejero-Sarinena et al. (2013)
		Semicontinuous system	Le Lay et al. (2015)
		"Colonic" model	Forssten et al. (2015)
Intestinal cell lines	Adhesion/ exclusion	HT29-MTX cell	Zivkovic et al. (2015)
		Immobilized intestinal mucus	Collado et al. (2005), Banerjee et al. (2009), and Ferreira et al. (2011)
	Cytotoxicity	Label-based endpoint methods	Banerjee et al. (2009), Trejo et al. (2010, 2013), and Valdes-Varela et al. (2016a)
		Label-free, RTCA (real-time cell analyzer) method	Valdes et al. (2015) and Valdes-Varela et al. (2016a, b)

monitoring of the status of live cells (Xi et al. 2008). Indeed the label-free, impedance-based RTCA (real-time cell analyzer) technology has been applied to develop methods allowing the clinical diagnosis of toxigenic *C. difficile* in different biological samples (Yu et al. 2015). Recently, this RTCA technology was also used in our group to develop a model to test the cytotoxicity of *C. difficile* supernatants upon the intestinal epithelial cell lines HT29 and Caco-2 (Valdes et al. 2015). Moreover, this model was used to search for potential probiotic strains able to counteract the toxic effect of *C. difficile* supernatants upon HT29 (Valdes-Varela et al. 2016a) as well as to evaluate the toxicity of *C. difficile* co-cultured with some of these probiotics (Valdes-Varela et al. 2016b).

On the other hand, several models have been used to assess the ability of probiotic candidates to modify the adhesion of *C. difficile* to the intestinal mucosa, such as those using immobilized (human) intestinal mucus which showed a good correlation with data obtained with a enterocyte-like (Caco-2) model (Collado et al. 2005; Banerjee et al. 2009; Ferreira et al. 2011). The ability of potential probiotic strains to inhibit the adhesion of *C. difficile* has also been evaluated using intestinal cell lines, such as HT29-MTX which is a derivative from HT29 (adapted to methotrexate) thus synthesizing higher amounts of mucus (Zivkovic et al. 2015). A study has suggested that this cell model may be more suitable for studying cell-pathogen interactions, as well as effectiveness of antimicrobial treatments, as compared to Caco-2 or HT29 models which do not have goblet cells or do not constitutively secrete mucus, respectively (Gagnon et al. 2013).

In a step forward, several authors have evaluated the protective effect of selected probiotic candidates against CDI in animal models (Best et al. 2012; Kolling et al. 2012; Trejo et al. 2013; Kondepudi et al. 2014; Yun et al. 2014; Andersen et al. 2016; Arruda et al. 2016; Spinler et al. 2016; Rätsep et al. 2017). This infection has been studied in different models, including mice, hamsters, rats, rabbits, hares, guinea pigs, prairie dogs, quails, foals, piglets, and monkeys. Moreover, zebrafish embryos have been described as suitable models for identification of in vivo targets of *C. difficile* toxins and evaluation of novel candidate therapeutics; zebrafish possess many of the major organs present in humans, and due to the transparency of the embryo, damage by toxins can be visualized by standard light microscopy (Best et al. 2012). Each of the *C. difficile* animal models has inherent advantages and disadvantages. The hamster model has been widely used to study pseudomembranous colitis in human because of extreme sensitivity to infection following antibiotic administration, using clindamycin as agent of choice; however, this model does not represent the usual course and spectrum of CDI in humans. Recently, new mouse and piglet CDI models have been developed which appear to mimic many of the disease symptoms observed in humans (Sun et al. 2011; Best et al. 2012; Hutton et al. 2014).

4 Mechanisms of Probiotic Action

As pointed in previous sections, probiotics are gaining more and more interest as preventive and co-adjuvant therapies for treatment of antibiotic-associated dysbiosis. However, their modes of action are poorly understood and vary between probiotic microorganisms. Indeed, the effects of any probiotic are strain-specific, and therefore, beneficial effects cannot be extrapolated to other species or strains (Hickson 2011). It has been described that probiotics could have diverse positive actions on the host by (1) modulating the intestinal microbiota and inhibiting pathogenic microorganisms at the intestinal luminal environment, (2) enhancing of intestinal barrier function at the intestinal epithelium, and (3) modulating the immune response, among others (Ng et al. 2009). Several mechanisms have been proposed for explaining the potential role of probiotics against *C. difficile*. Some of these effects, such as the production of antimicrobial factors (Corr et al. 2007), competitive inhibition of the pathogen (Collado et al. 2005), and the ability to degrade and to reduce the toxicity of *C. difficile* (Castagliuolo et al. 1999; Valdes-Varela et al. 2016a), could be of help not only in the prevention but also in the treatment of CDI.

4.1 Microbial Antagonism: Interaction Probiotics vs. *C. difficile*

The restoration of intestinal microbiota after dysbiosis, caused by any etiological agent, is the main way of action of any treatment against intestinal pathogens including *C. difficile* (Gareau et al. 2010; Reid et al. 2011). This was evidenced, for example, in an in vivo study with a murine CDI model of antibiotic-induced dysbiosis, in which the gut microbiota was restored after treatment with a multi-strain probiotic supplement (*Lactobacillus plantarum* F44, *Lactobacillus paracasei* F8, *Bifidobacterium breve* 46, *Bifidobacterium animalis* subsp. *lactis* 8:8) (Kondepudi et al. 2014). There are several mechanisms by which probiotics can help the restoration of the intestinal microbiota, some of them being related to typical bacterial antagonism (Ng et al. 2009); however, little is known about those mechanisms acting specifically in the context of CDI (Parkes et al. 2009; Ollech et al. 2016).

Some probiotic strains are able to compete with pathogenic bacteria for the adhesion sites, that is, competitive exclusion, thus providing a "physical" barrier that increases the colonization resistance (Fig. 2a). In vitro studies showed the ability of selected *Bifidobacterium* and *Lactobacillus* strains to modify the adhesion of *C. difficile* to intestinal epithelial cells or intestinal mucus, the effect being strain-dependent (Collado et al. 2005; Zivkovic et al. 2015). A reduction from 60% to 3% in the adhesion of *C. difficile* to gingival epithelial cell cultures (obtained from healthy horses) was reported when *Lactobacillus reuteri* Lr1 was added; additionally, it was detected that this strain was able to co-aggregate with the pathogen (Dicks et al. 2015). In this regard, it has been suggested that the aggregation capability between lactobacilli and *C. difficile* could be a way to reduce the adhesion of the pathogen to the intestinal mucosa (Ferreira et al. 2011). *S. boulardii* is also able to reduce the adhesion of *C. difficile* to epithelial cells, and the same effect was detected using extracts obtained from the cell wall of this yeast (Tasteyre et al. 2002). Similarly, it has been proved that cell-free supernatants obtained from *Lactobacillus delbrueckii* ssp. *bulgaricus* B-30892 (Banerjee et al. 2009) and different bifidobacterial strains (Trejo et al. 2006) were able to reduce the adhesion of *C. difficile* to intestinal epithelial Caco-2 cells. Different treatments of the bifidobacterial supernatants showed that the factors related to the anti-clostridial adhesion

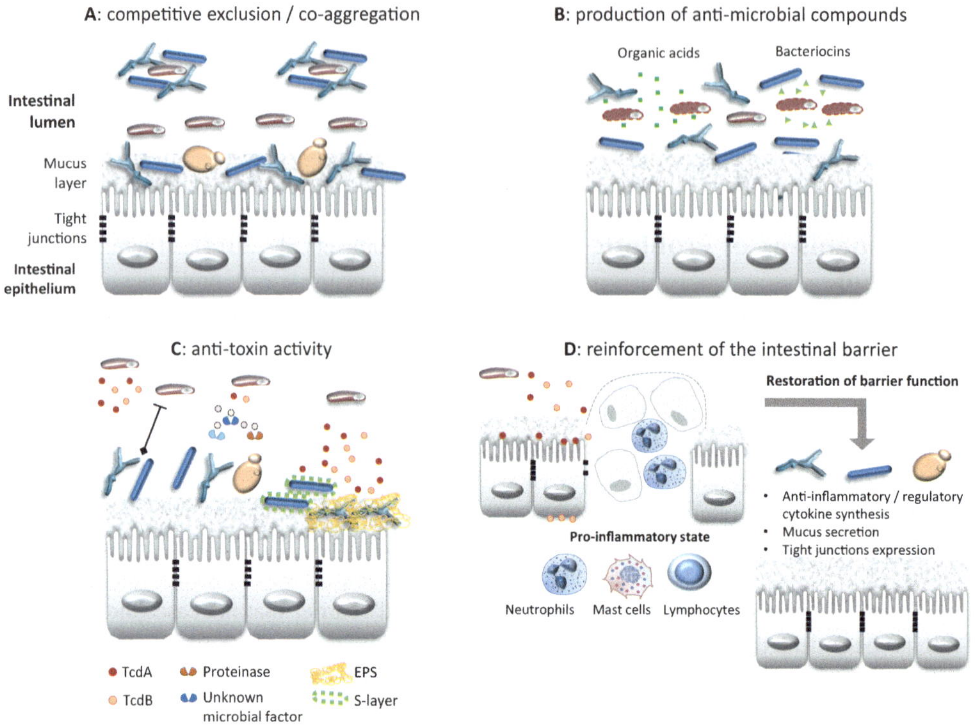

Fig. 2 Potential mechanisms of action proposed for probiotics against *C. difficile*. (**a**) Competitive exclusion/co-aggregation. (**b**) Production of antimicrobial compounds. (**c**) Anti-toxin activity. (**d**) Reinforcement of the intestinal barrier

were not heat-resistant and nonrelated with acids (active at neutral pH) and were not affected by proteinases, but its nature remains unknown (Trejo et al. 2006). Indirect evidence suggests that exopolysaccharides covering the surface of some probiotics could be involved in the inhibition of the binding capability of some pathogens, including *C. difficile*, by probiotics (Ruas-Madiedo et al. 2006). Thus, altogether, these studies suggest that different surface molecules and/or secreted factors might be implicated in the interference of probiotics against *C. difficile* adhesion to the intestinal mucosa.

Another mechanism of probiotic action is the inhibition of the pathogen growth through the competition for the limiting nutritional sources and/or by the production of antimicrobial factors, such as organic acids and bacteriocins (Fig. 2b). In a study carried out with a CDI animal model, it was shown that mice treated with *Streptococcus thermophilus* LMD-9 exhibited less pathology and lower detectable toxin levels in cecal contents, compared with untreated controls; an inverse correlation was observed between the levels of luminal lactate and the abundance of *C. difficile*, suggesting that the anti-clostridial effect was due to the production of this organic acid (Kolling et al. 2012). Similarly, the lactic acid synthesized by *Lactobacillus acidophilus* GP1B had an inhibitory effect on *C. difficile* growth in a CDI mouse model, which may be related to a reduction in pH as a result of organic acids produced by the probiotic bacterium (Yun et al. 2014). Several in vitro studies have investigated the activity of probiotics to inhibit *C. difficile* growth; using a fecal, pH-controlled (between 6.7 and 6.9), anaerobic batch model, it was found that *Lactobacillus casei* NCIMB30185 and *B. breve* NCIMB30180 were able to reduce the numbers of *C. difficile* in this complex microbial ecosystem (Tejero-Sarinena et al. 2013). Co-cultivation of *C. difficile* with cell-free supernatants from different commercial probiotics highlighted that the mechanism of inhibition was pH-dependent; thus, the production of organic acids, mainly lactic and acetic acids, is the inhibition factor controlling the growth of *C. difficile* (Schoster et al. 2013). In another in vitro study, the co-incubation of *C. difficile* with *L. rhamnosus* LR5, *Lactococcus lactis* SL3, *B. breve* BR3, and *B. animalis* subsp. *lactis* BL3 demonstrated their potential to decrease *C. difficile* numbers, mainly mediated by the organic acid production. However, among those strains, SL3 appeared to have the strongest activity which seems to be pH-independent and likely could be mediated through the action of a bacteriocin (Lee et al. 2013). Similar pH-dependent and pH-independent effects against *C. difficile* were also reported using cell-free supernatants from other commercially available probiotics (Fredua-Agyeman et al. 2017). With respect to the competition for nutrients, some studies have been carried out using "synbiotic" combinations, which are mixtures of probiotics and prebiotic substrates that (theoretically) will improve the performance of probiotics or other beneficial microbes in the gut. In a mice (C57Bl/6) model of CDI, the feeding with a synbiotic formulation, consisting of four strains (*L. plantarum* F44, *L. paracasei* F8, *B. breve* 46, *B. animalis* subsp. *lactis* 8:8) and three prebiotics (galacto-oligosaccharides, isomalto-oligosaccharides, and resistant starch), conferred protection against this pathogen (Kondepudi et al. 2014). Some studies have suggested that the growth inhibition of *C. difficile* by probiotics is strain but also carbon source specific. Ambalam et al. reported the ability of cell-free supernatants from *L. paracasei* F8 and *L. plantarum* F44 to inhibit the growth of *C. difficile* strains when they grew on glucose, due to the production of organic acids and heat-stable antimicrobial proteins, while the effect was only pH-dependent when growing on prebiotics (Ambalam et al. 2015). Our workgroup recently analyzed the influence of carbon sources upon *C. difficile* growth and toxicity when co-cultured with *Bifidobacterium longum* IPLA20022 or *B. breve* IPLA20006 in the presence of short-chain fructo-oligosaccharides (scFOS) or inulin. The use of scFOS reduced the growth of the pathogen, as well as the toxicity of the co-culture supernatants, which was not observed with inulin (Valdes-Varela et al. 2016b).

4.2 Probiotics Against *C. difficile* Toxin Activity

The toxins produced by *C. difficile* are responsible for the clinical profile of the CDI. Therefore, therapeutic agents that reduce toxin-induced damage could be valuable tools to alleviate the severity of symptoms and to improve the course of the disease. Some authors have reported that probiotics are able to reduce the activity of *C. difficile* toxins but, in most cases, the specific mechanisms of action by which probiotics exert the protective effect in this infection are unknown (Fig. 2c). In a hamster model of enterocolitis induced by *C. difficile*, *Bifidobacterium bifidum* CIDCA5310 protected the animals, and avoided mortality, when compared with the control (infected) group; besides, the supernatants obtained from caecum contents were less toxics upon Vero (cells from monkey's kidney) cultures in animals fed with the bifidobacteria, suggesting that this strain is able to in vivo counteract the effect of clostridial toxins (Trejo et al. 2013).

Co-culture of toxigenic strains of *C. difficile* with different strains of bifidobacteria and lactobacilli leads to a reduction of the cytotoxic effects of spent culture supernatants on cultured Vero cells, which correlates with a diminution of clostridial toxins present in these supernatants (Trejo et al. 2010). However, the growth of clostridial strains in BHI medium with different concentrations of cell-free supernatants from bifidobacteria or lactobacilli cultures did not decrease the toxic effect of pathogens; taking into account these results, authors hypothesized that co-culture of clostridia with lactobacilli or bifidobacteria leads to the modification of the environment, thus leading to the repression of toxin synthesis/secretion pathway. Similarly, a cell extract from *L. acidophilus* GP1B was able to decrease the pathogenicity of *C. difficile* by inhibiting quorum sensing signaling, probably by lowering the expression of quorum sensing-regulated toxin genes (Yun et al. 2014).

On the other hand, it was observed that some microorganisms release metabolites that are able to inhibit the harmful effects of toxins. A bacterial cell-free supernatant obtained from *L. delbrueckii* subsp. *bulgaricus* LDB B-30892 reduced cytotoxic effects of *C. difficile* ATCC9689 upon the human intestinal epithelial cell line Caco-2- (Banerjee et al. 2009). Banerjee et al. (2009) suggested that bioactive components, of unknown nature, were released by this strain which were the probable causative agents of inhibition of the clostridial toxins. Similarly, bacterial cell-free supernatants obtained from *L. lactis* CIDCA8221 contained heat-sensitive metabolites, higher than 10 kDa, that were not affected by treatment with different proteases or protease inhibitors, which were able to inhibit cytotoxic effects of *C. difficile* toxins upon epithelial Vero cells (Bolla et al. 2013). These results suggest that the protective effect of *L. lactis* CIDCA8221 supernatant could be owing to a non-covalent interaction between molecules present in the lactococcal supernatant and toxins. In this regard, surface components of the bacterial cell envelope, such as exopolysaccharides which can be released to the environment, have been proposed to in vitro inhibit the adverse effect of pathogenic toxins (Ruas-Madiedo et al. 2010). A study showed the ability of the outermost (proteinaceous) S-layer from *Lactobacillus kefir* strains to inhibit the damage induced by supernatants obtained from *C. difficile* upon Vero cells; the protective effect was not affected by inhibitors of proteases or heat treatment, while pre-incubation with specific anti-S-layer antibodies reduced the inhibitory effect of these proteins (Carasi et al. 2012). From this study, it was concluded that the capability for reducing the toxigenic effect of *C. difficile* could be attributed to an interaction between its toxins and the *L. kefir* S-layer protein (Carasi et al. 2012). Recently, our workgroup analyzed the capability of *Bifidobacterium* and *Lactobacillus* strains to reduce the toxic effect of supernatants obtained from *C. difficile* LMG21717 ($TcdA^+$, $TcdB^+$) culture upon the human intestinal epithelial cell line HT29. For this purpose, the probiotic candidates were incubated together with a toxigenic supernatant of *C. difficile*, and the analyzed strains from *B. longum* and *B. breve* species were able to reduce the toxic effect of the pathogen; more specifically, the strain *B. longum* IPLA20022, in a viable state, showed the highest ability to reduce the levels of both clostridial toxins and to counteract the cytotoxic effect upon HT29 (Valdes-Varela et al. 2016a). Furthermore, the incubation of supernatant from *B. longum* IPLA20022 with the toxigenic *C. difficile* supernatant showed similar effect on the cell line than that obtained with the bifidobacterial biomass. The treatment of the clostridial supernatant with this probiotic strain prevented the rounding of HT29 cells, detected in cells treated only with *C. difficile* supernatant, thus keeping a monolayer structure resembling that of the control (nontreated HT29) (Fig. 3). Taking into account these results, we hypothesize that the adsorption of toxins to the bifidobacterial surface and the secretion of molecules able to reduce the cytotoxic effect by degrading the toxins are both probable mechanisms of action (Valdes-Varela et al. 2016a). In this regard, 20 years ago, it had been reported that *S. boulardii* inhibited *C. difficile* TcdA effects in the rat ileum by releasing a 54kDa serine protease which hydrolyzed toxin A and its intestinal receptor (Castagliuolo et al. 1996); this could be the mechanism behind the effectiveness of this yeast in both the prevention and the treatment of antibiotic-associated colitis in humans (Castagliuolo et al. 1999). More recently, it was observed that a protease secreted by *Bacillus clausii* O/C is able to inhibit the cytotoxic effect of *C. difficile*; thus this enzyme could be involved in the protective effect of this bacilli in antibiotic-associated diarrhea (Ripert et al. 2016). A similar phenomenon may be taking place with the abovementioned *Bifidobacterium* strains (Valdes-Varela et al. 2016a).

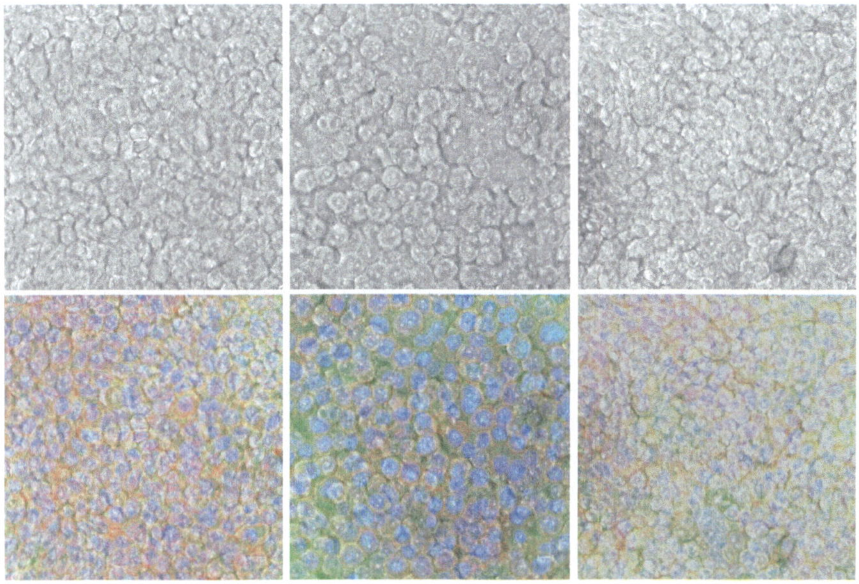

Fig. 3 CSLM (Leica TCSAOBS SP8 X confocal microscopy) images obtained, after 20 h incubation, for HT29 cells submitted to different treatments. (**a**) Panel shows transmission (visible) images and (**b**) panel shows Z-projection snapshots resulting from a combination of the transmission image with the "blue" image, captured with the violet laser diode (excited at 405 nm, showing DAPI-stained nucleus); the "red" image, captured with the white laser (excited at 578 nm, showing phalloidind-alexa-fluor-568-stained F-actin); and the "green" image resulting from the autofluorescence emitted by the intracellular components of HT29. The 63×/1.4 oil objective was used; bars 10 µm. Individual images of stained nucleus and/or F-actin were included in the reference Valdes-Varela et al. (2016a)

4.3 Other Mechanisms of Action

The intestinal barrier function given, among other factors, by the presence of an intact intestinal epithelium enabling the absorption of nutrients and the exclusion of harmful substances can be compromised by the activity of enteric pathogens including *C. difficile* (Barreau and Hugot 2014). In fact, internalized clostridial toxins induce changes in the F-actin cytoskeleton and a breakdown of the tight junctions, thus contributing to the disruption of the epithelial barrier function; the increase in the permeability of this barrier ends with an inflammatory process due to the infiltration of neutrophils, production of chemokines and pro-inflammatory cytokines, and activation of mast cells and lymphocytes, among other events (Voth and Ballard 2005; Rupnik et al. 2009; Abt et al. 2016). Thus some probiotics have been claimed to be able to reinforce the intestinal barrier function, although there is not much information in the context of CDI (Fig. 2d). In a hamster model of CDI, the oral administration of live *S. boulardii* five days before the infection significantly reduced cecal tissue damage, NF-κB phosphorylation, and TNFα protein expression caused by different *C. difficile* ribotypes, thus indicating that this probiotic can prevent intestinal damage and inflammation (Koon et al. 2016). In fact, after a literature search conducted by Stier and Bischoff (2016), they found that mechanisms of *S. boulardii* action involve not only a direct effect on the pathogen or its toxins but also impact on the innate and adaptive immune response of the host induced after CDI. Regarding probiotic bacteria, it has been shown that *L. rhamnosus* L34 and *L. casei* L39 are able to modulate, by different ways, the inflammation caused by *C. difficile*, thus making suitable the use of these vancomycin-resistant lactobacilli for treating CDI (Boonma et al. 2014). In our research group, we have detected that lactobacilli strains are able to increase the synthesis of interleukin (IL)-8 and mucins by HT29-MTX monolayers challenged with *C. difficile*, thus helping to the reinforcement of the innate immune defense (Zivkovic et al. 2015). More recently, a combination of *Lactobacillus helveticus* BGRA43, *Lactobacillus fermentum* BGHI14, and *S. thermophilus* BGVLJ1–44 was in vitro tested against *C. difficile* in a Caco-2 model, and results showed an increase in the release of transforming growth factor (TGF)-β, thus resulting in a promising probiotic candidate to be further evaluated against CDI (Golic et al. 2017).

Finally, recombinant lactobacilli, although they cannot be considered as probiotics, could be suitable vehicles for the in situ production and delivery of therapeutic molecules in the intestine. In a recent study, the basis for an oral antitoxin strategy based on engineered *Lactobacillus* strains expressing TcdB-neutralizing antibody fragments in the gastrointestinal tract was explored; the results showed that only lactobacilli displaying the anti-TcdB variable domain of the heavy chain antibody can inhibit the cytotoxic effect of TcdB in the gastrointestinal tract of a hamster model (Andersen et al. 2016).

5 Conclusion and Future Trends

The search for probiotics with anti-*C. difficile* activity has been an active area of research for more than two decades. However, in spite of the abundance of in vitro studies, the in vivo evidence is less conclusive. The role of probiotics in preventing antibiotic-associated diarrhea is well established by several clinical intervention studies and meta-analyses. Good evidence is also available regarding the benefit of certain probiotics in the prevention of specific *C. difficile* diarrhea, being still necessary to define the best conditions for maximizing the efficacy. However, the studies on the use of probiotics in the treatment of CDI are still scarce; this is in spite of the several potential mechanisms of action that would be of interest in the case of *C. difficile* infection. Among them, the ability of certain strains to inhibit the growth of *C. difficile*, or to promote the restoration of the normal gut microbiota, represents two very direct potentially beneficial mechanisms of action. Moreover, specific probiotic strains have been found to be able to reduce the toxicity of this pathogen and/or to degrade the produced toxins. This inhibition of *C. difficile* toxicity may constitute an interesting strategy for the treatment of CDI by probiotics: first by eliminating the toxins from the intestine and second by the promotion of the microbiota restoration by the use of selected probiotic strains with both properties.

The existing clinical interest of CDI together with the successful application of FMT allows foreseeing that the interest in the use for probiotic therapies, likely using defined combinations of strains, will continue rising during the next years. In this regard, the development of products, based on the combination of strains with different properties and anti-*C. difficile* mechanisms of action, promises to allow the development of highly efficacy products for both prevention and treatment of CDI.

Acknowledgments The funds supporting this research topic in our group are given by the Spanish Ministry of Economy and Competitiveness (current project AGL2015-64901-R) partially co-funded by FEDER (European Union) grants.

References

Abt MC, McKenney PT, Pamer EG (2016) *Clostridium difficile* colitis: pathogenesis and host defense. Nat Rev Microbiol 14:609–620

Allen SJ, Wareham K, Wang D et al (2013) Lactobacilli and bifidobacteria in the prevention of antibiotic-associated diarrhoea and *Clostridium difficile* diarrhoea in older inpatients (PLACIDE): a randomised, double-blind, placebo-controlled, multicentre trial. Lancet 382:1249–1257

Ambalam P, Kondepudi KK, Balusupati P et al (2015) Prebiotic preferences of human lactobacilli strains in co-culture with bifidobacteria and antimicrobial activity against *Clostridium difficile*. J Appl Microbiol 119: 1672–1682

Andersen KK, Strokappe NM, Hultberg A et al (2016) Neutralization of *Clostridium difficile* toxin B mediated by engineered lactobacilli that produce single-domain antibodies. Infect Immun 84:395–406

Arruda PHE, Madson DM, Ramirez A et al (2016) Bacterial probiotics as an aid in the control of *Clostridium difficile* disease in neonatal pigs. Can Vet J 57:183–188

Arvola T, Laiho K, Torkkeli S et al (1999) Prophylactic *Lactobacillus* GG reduces antibiotic-associated diarrhea in children with respiratory infections: a randomized study. Pediatrics 104:64

Auclair J, Frappier M, Millette M (2015) *Lactobacillus acidophilus* CL1285, *Lactobacillus casei* LBC80R, and *Lactobacillus rhamnosus* CLR2 (Bio-K+): characterization, manufacture, mechanisms of action, and quality control of a specific probiotic combination for primary prevention of *Clostridium difficile* infection. Clin Infect Dis 60:S135–S143

Banerjee P, Merkel GJ, Bhunia AK (2009) *Lactobacillus delbrueckii* ssp. *bulgaricus* B-30892 can inhibit cytotoxic effects and adhesion of pathogenic *Clostridium difficile* to Caco-2 cells. Gut Pathog 1:8

Barreau F, Hugot JP (2014) Intestinal barrier dysfunction triggered by invasive bacteria. Curr Opin Microbiol 17: 91–98

Best EL, Freeman J, Wilcox MH (2012) Models for the study of *Clostridium difficile* infection. Gut Microbes 3:145–167

Bolla PA, Carasi P, Serradell MA et al (2013) Kefir-isolated *Lactococcus lactis* subsp. lactis inhibits the cytotoxic effect of *Clostridium difficile* in vitro. J Dairy Res 80:96–102

Boonma P, Spinler JK, Venable SF et al (2014) *Lactobacillus rhamnosus* L34 and *Lactobacillus casei* L39 suppress *Clostridium difficile*-induced IL-8 production by colonic epithelial cells. BMC Microbiol 14:177

Can M, Besirbellioglu BA, Avci IY et al (2006) Prophylactic *Saccharomyces boulardii* in the prevention of antibiotic associated diarrhea: a prospective study. Med Sci Monit 12:19–22

Carasi P, Trejo FM, Perez PF et al (2012) Surface proteins from *Lactobacillus kefir* antagonize in vitro cytotoxic effect of *Clostridium difficile* toxins. Anaerobe 18: 135–142

Carter GP, Rood JI, Lyras D (2012) The role of toxin A and toxin B in the virulence of *Clostridium difficile*. Trends Microbiol 20:21–29

Castagliuolo I, Lamont JT, Nikulasson ST (1996) *Saccharomyces boulardii* protease inhibits *Clostridium difficile* Toxin A effects in the rat ileum. Infect Immun 64: 5225–5232

Castagliuolo I, Rieger MF, Valenick L et al (1999) *Saccharomyces boulardii* protease inhibits the effect of *Clostridium difficile* toxins A and B in human colonic mucosa. Infect Immun 67:302–307

Collado MC, Gueimonde M, Hernandez M et al (2005) Adhesion of selected *Bifidobacterium* strains to human intestinal mucus and its role in enteropathogen exclusion. J Food Protect 68:2672–2678

Corr SC, Li Y, Riedel CU et al (2007) Bacteriocin production as a mechanism for the anti-infective activity of *Lactobacillus salivarius* UCC118. Proc Natl Acad Sci U S A 104:7617–7621

Cremonini F, Caro S, Nista EC et al (2002) Meta-analysis: the effects of probiotic administration on antibiotic associated diarrhoea. Aliment Pharmacol Ther 16:1461

D'Souza AL, Rajkumar C, Cooke J et al (2002) Probiotics in the prevention of antibiotic associated diarrhoea: meta-analysis. Br Med J 324:1361

Dicks LMT, Botha M, Loos B et al (2015) Adhesion of *Lactobacillus reuteri* strain Lr1 to equine epithelial cells and competitive exclusion of *Clostridium difficile* from the gastro-intestinal tract of horses. Ann Microbiol 65:1087–1096

Dietrich CG, Kottmann T, Alavi M (2014) Commercially available probiotic drinks containing *Lactobacillus casei* DN-114001 reduce antibiotic-associated diarrhea. World J Gastroenterol 20:15837–15844

Donaldson GP, Lee SM, Mazmanian SK (2016) Gut biogeography of the bacterial microbiota. Nat Rev Microbiol 14:20–32

Ferreira CL, Grzeskowiak Ł, Collado MC et al (2011) In vitro evaluation of *Lactobacillus gasseri* strains of infant origin on adhesion and aggregation of specific pathogens. J Food Prot 74:1482–1487

Forssten SD, Roytio H, Hibberd AA et al (2015) The effect of polydextrose and probiotic lactobacilli in a *Clostridium difficile*-infected human colonic model. Microb Ecol Health Dis 26:27988

Fredua-Agyeman M, Stapleton P, Basit AW et al (2017) In vitro inhibition of *Clostridium difficile* by commercial probiotics: a microcalorimetric study. Int J Pharm 517: 96–103

Gagnon M, Zihler Berner A, Chervet N et al (2013) Comparison of the Caco-2, HT29 and the mucus-secreting HT29-MTX intestinal cell models to investigate *Salmonella* adhesion and invasion. J Microbiol Methods 94:274–279

Gao XW, Mubasher M, Fang CY et al (2010) Dose-response efficacy of a proprietary probiotic formula of *Lactobacillus acidophilus* CL1285 and *Lactobacillus casei* LBC80R for antibiotic associated diarrhea and *Clostridium difficile*-associated diarrhea prophylaxis in adults patients. Am J Gastroenterol 105: 1636–1641

Gareau MG, Sherman PM, Walker WA (2010) Probiotics and the gut microbiota in intestinal health and disease. Nat Rev Gastroentol Hepatol 7:503–514

Gerding DN, JohnsonS RM et al (2014) *Clostridium difficile* binary toxin CDT: mechanism, epidemiology, and potential clinical importance. Gut Microb 5:15–27

Goldenberg JZ, Ma SSY, Saxton JD et al (2013) Probiotics for the prevention of *Clostridium difficile*-associated diarrhea in adults and children. Cochrane Database Syst Rev 5:CD006095

Goldenberg JZ, Lytvyn L, Steurich J et al (2015) Probiotics for prevention of pediatric antibiotic-associated diarrhea. Cochrane Database Syst Rev 12: CD004827

Golic N, Veljovic K, Popovic N et al (2017) In vitro and in vivo antagonistic activity of new probiotic culture against *Clostridium difficile* and *Clostridium perfringens*. BMC Microbiol 17:108

Hempel S, Newberry SJ, Maher AR et al (2012) Probiotics for the prevention and treatment of antibiotic-associated diarrhea: a systematic review and meta-analysis. JAMA 307:1959–1969

Hickson M (2011) Probiotics in the prevention of antibiotic-associated diarrhea and *Clostridium difficile* infection. Ther Adv Gastroenterol 4:185–197

Hill C, Guarner F, Reid G et al (2014) The International Scientific Association for Probiotics and Prebiotics consensus statement on the scope and appropriate use of the term probiotic. Nat Rev Gastroenterol Hepatol 11:506–514

Hopkins MJ, Macfarlane GT (2003) Nondigestible oligosaccharides enhance bacterial colonization resistance against *Clostridium difficile* in vitro. Appl Environ Microbiol 69:1920–1927

Hussack G, Tanha J (2016) An update on antibody-based immunotherapies for *Clostridium difficile* infection. Clin Exp Gastroenterol 9:209–224

Hutton ML, Mackin KE, Chakravorty A et al (2014) Small animal models for the study of *Clostridium difficile* disease pathogenesis. FEMS Microbiol Lett 352:140–149

Jangi S, Lamont JT (2010) Asymptomatic colonization by *Clostridium difficile* in infants: implications for dis-ease in later life. J Pediatr Gastroenterol Nutr 51: 2–7

Johnston B, Supina A, Ospina M et al (2007) Probiotics for the prevention of pediatric antibiotic-associated diarrhea. Cochrane Database Syst Rev 18:CD004827

Kachrimanidou M, Sarmourli T, Skoura L et al (2016) *Clostridium difficile* infection: new insights into therapeutic options. Crit Rev Microbiol 42:773–779

Kociolek LK, Gerding DN (2016) Breakthroughs in the treatment and prevention of *Clostridium difficile* infection. Nat Rev Gastroenterol Heptol 13:150–160

Kolling GL, Wu M, Warren CA et al (2012) Lactic acid production by *Streptococcus thermophilus* alters *Clostridium difficile* infection and in vitro toxin A production. Gut Microb 3:523–529

Kondepudi KK, Ambalam P, Karagin PH et al (2014) A novel multi-strain probiotic and synbiotic supplement for prevention of *Clostridium difficile* infection in a murine model. Microbiol Immunol 58:552–558

Koon HW, Su B, Xu C et al (2016) Probiotic *Saccharomyces boulardii* CNCM I-745 prevents outbreak-associated *Clostridium difficile*-associated cecal inflammation in hamsters. Am J Physiol Gastroenterol Liver Physiol 311:G610–G623

Kotowska M, Albrecht P, Szajewska H (2005) *Saccharomyces boulardii* in the prevention of antibiotic-associated diarrhea in children: a randomized double-blind placebo-controlled trial. Aliment Pharmacol Ther 21:583–590

Lau CSM, Chamberlaim RS (2016) Probiotics are effective at preventing *Clostridium difficile*-associated diarrhea: a systematic review and meta-analysis. Int J Gen Med 9:27–37

Lawley TD, Walker AW (2013) Intestinal colonization resistance. Immunology 138:1–11

Le Lay C, Fernandez B, Hammami R et al (2015) On *Lactococcus lactis* UL719 competitivity and nisin (Nisaplin®) capacity to inhibit *Clostridium difficile* in a model of human colon. Front Microbiol 6:1020

Leal JR, Heitman SJ, Conly JM et al (2016) Cost-effectiveness analysis of the use of probiotics for the prevention of *Clostridium difficile*-associated diarrhea in a provincial health care system. Infect Cont Hosp Epidemiol 37:1079–1086

Lee JS, Chung MJ, Seo JG (2013) In vitro evaluation of antimicrobial activity of lactic acid bacteria against *Clostridium difficile*. Toxicol Res 29:99–106

Leffler DA, Lamont JT (2015) *Clostridium difficile* infection. N Engl J Med 372:1539–1548

Mansour NM, Elkhatib WF, Aboshnad KM et al (2017) Inhibition of *Clostridium difficile* in mice using a mixture of potential probiotic *strains Enterococcus faecalis* NM815, *E. faecalis* NM915, and *E. faecium* NM1015: novel candidates to control *C. difficile* infection (CDI). Probiotics Antimicrob Prot. https://doi.org/10.1007/s12602-017-9285-7. [E-pub ahead of print]

Martin J, Wilcox M (2016) New and emerging therapies for *Clostridium difficile* infection. Curr Opin Infect Dis 29:546–554

Mathur H, Rea MC, Cotter PD et al (2014) The potential for emerging therapeutic options for *Clostridium difficile* infection. Gut Microb 5:696–710

Maziade PJ, Pereira P, Goldstein EJC (2015) A decade of experience in primary prevention of *Clostridium difficile* infection at a community hospital using the probiotic combination *Lactobacillus acidophilus* CL1285, *Lactobacillus casei* LBC80R, and *Lactobacillus rhamnosus* CRL2 (Bio-K+). Clin Infect Dis 60(Suppl 2):S144–S147

McFarland LV (2006) Meta-analysis of probiotics for the prevention of antibiotic associated diarrhea and the treatment of *Clostridium difficile* disease. Am J Gastroenterol 101:812–822

McFarland LV (2015) Probiotics for the primary and secondary prevention of *C. difficile* infections. A meta-analysis and systematic review. Antibiotics 4:160–178

McFarland LV (2016) Therapies on the horizon for *Clostridium difficile* infections. Expert Opin Investig Drugs 25:541–555

McFarland LV, Surawicz CM, Greenberg RN et al (1994) A randomized placebo-controlled trial of *Saccharomyces boulardii* in combination with standard antibiotics for *Clostridium difficile* disease. J Am Med Assoc 271: 1913

Moura I, Spigaglia P, Barbanti F et al (2013) Analysis of metronidazole susceptibility in different *Clostridium difficile* PCR ribotypes. J Antimicrob Chemother 68: 362–365

Na X, Kelly C (2011) Probiotics in *Clostridium difficile* infection. J Clin Gastroenterol 45:S154–S158

Ng SC, Hart AL, Kamm MA et al (2009) Mechanisms of action of probiotics: recent advances. Inflamm Bowel Dis 15:300–310

O'Horo JC, Jindai K, Kunzer B et al (2014) Treatment of recurrent *Clostridium difficile* infection: a systematic review. Infection 42:43–59

Ofosu A (2016) *Clostridium difficile* infection: a review of current and emerging therapies. Ann Gastroenterol 29: 147–154

Ollech JE, Shen NT, Crawford CV et al (2016) Use of probiotics in prevention and treatment of patients with *Clostridium difficile* infection. Best Pract Res Clin Gastroenterol 30:111–118

Orrell KE, Zhanga Z, Sugiman-Marangosa SN et al (2017) *Clostridium difficile* toxins A and B: receptors, pores, and translocation into cells. Crit Rev Biochem Mol Biol 52:461–473

Ozaki E, Kato H, Kita H et al (2004) *Clostridium difficile* colonization in healthy adults: transient colonization and correlation with enterococcal colonization. J Med Microbiol 53:167–172

Padua D, Pothoulakis C (2016) Novel approaches to treating *Clostridium difficile*-associated colitis. Expert Opin Investig Drugs 10:193–204

Parkes GC, Sanderson JD, Whelan K (2009) The mechanisms and efficacy of probiotics in the prevention of *Clostridium difficile*-associated diarrhoea. Lancet Infect Dis 9:237–244

Rätsep M, Köljalg S, Sepp E et al (2017) A combination of the probiotic and prebiotic product can prevent the germination of *Clostridium difficile* spores and infection. Anaerobe 47:94–103

Reid G, Younes JA, Van der Mei HC et al (2011) Microbiota restoration: natural and supplemented recovery of human microbial communities. Nat Rev Microbiol 9:27–38

Ripert G, Racedo SM, Elie AM et al (2016) Secreted compounds of the probiotic *Bacillus clausii* strain O/C inhibit the cytotoxic effects induced by *Clostridium difficile* and *Bacillus cereus* toxins. Antimicrob Agents Chemother 60:3445–3454

Rodriguez C, Taminiau B, Van Broeck J et al (2016) *Clostridium difficile* in food and animals: a comprehensive review. Adv Exp Med Biol 4:65–92

Ruas-Madiedo P, Gueimonde M, Margolles A et al (2006) Exopolysaccharides produced by probiotic strains modify the adhesion of probiotics and enteropathogens to human intestinal mucus. J Food Prot 69:2011–2015

Ruas-Madiedo P, Medrano M, Salazar N et al (2010) Exopolysaccharides produced by *Lactobacillus* and *Bifidobacterium* strains abrogate in vitro the cytotoxic effect of bacterial toxins on eukaryotic cells. J Appl Microbiol 109:2079–2086

Rupnik M, Wilcox MH, Gerding DN (2009) *Clostridium difficile* infection: new developments in epidemiology and pathogenesis. Nat Rev Microbiol 7:526–536

Sampalis J, Psaradellis E, Ranpakakis E (2010) Efficacy of BIO K+ CL1285 in the reduction of antibiotic associated diarrhea – a placebo controlled double blind randomized, multi-centre study. Arch Med Sci 6:56–64

Sazawal S, Hiremath G, Dhingra U et al (2006) Efficacy of probiotics in prevention of acute diarrhoea. A meta-analysis of masked, randomized, placebo-controlled trials. Lancet Infect Dis 6:374–382

Schoster A, Kokotovic B, Permin A (2013) In vitro inhibition of *Clostridium difficile* and *Clostridium perfringens* by commercial probiotic strains. Anaerobe 20:36–41

Shen NT, Maw A, Tmanova LL et al (2017) Timely use of probiotics in hospitalized adults prevents *Clostridium difficile* infection: a systematic review with meta-regression analysis. Gastroenterology 52:1889–1900

Sinclair A, Xie X, Saab L et al (2016) *Lactobacillus* probiotics in the prevention of diarrhea associated with *Clostridium difficile*: a systematic review and bayesian hierarchical meta-analysis. CMAJ Open 4: E706–E718

Spinler JK, Brown A, Ross CL et al (2016) Administration of probiotic kefir to mice with *Clostridium difficile* infection exacerbates disease. Anaerobe 40:54–57

Starn ES, Hampe H, Cline T (2016) The cost-efficiency and care effectiveness of probiotic administration with antibiotics to prevent hospital-acquired *Clostridium difficile* infection. Qual Manag Health Care 25:238–243

Stier H, Bischoff SC (2016) Influence of *Saccharomyces boulardii* CNCM I-745 on the gut-associated immune system. Clin Exp Gastroenterol 9:269–279

Sun X, Wang H, Zhang Y et al (2011) Mouse relapse model of *Clostridium difficile* infection. Infect Immun 79:2856–2864

Surawicz CM, McFarland LV, Greenberg RN et al (2000) The search for a better treatment for recurrent *Clostridium difficile* disease: use of high dose vancomycin combined with *Saccharomyces boulardii*. Clin Infect Dis 31:1012–1017

Szajewska H, Skorka A, Dylag M (2007a) Meta-analysis: *Saccharomyces boulardii* for treating acute diarrhoea in children. AP&T 25:257–264

Szajewska H, Skorka A, Ruszczynski M et al (2007b) Meta-analysis: *Lactobacillus* GG for treating acute diarrhoea in children. AP&T 25:871–881

Szajewska H, Canani RB, Guarino A et al (2016) Probiotics for the prevention of antibiotic-associated diarrhea in children. J Pediatr Gastroenterol Nutr 62: 495–506

Tasteyre A, Barc MC, Karjalainen T et al (2002) Inhibition of in vitro cell adherence of *Clostridium difficile* by *Saccharomyces boulardii*. Microb Pathog 32:219–225

Tejero-Sarinena S, Barlow J, Costabile A et al (2013) Antipathogenic activity of probiotics against *Salmonella* Typhimurium and *Clostridium difficile* in anaerobic batch culture systems: is it due to synergies in

probiotic mixtures or the specificity of single strains? Anaerobe 24:60–65
Trejo FM, Minnaard J, Pereza PF et al (2006) Inhibition of *Clostridium difficile* growth and adhesion to enterocytes by *Bifidobacterium* supernatants. Anaerobe 12:186–193
Trejo FM, Perez PF, De Antoni GL (2010) Co-culture with potentially probiotic microorganisms antagonises virulence factors of *Clostridium difficile* in vitro. Antonie Van Leeuwenhoek 98:19–29
Trejo FM, De Antoni GL, Perez PF (2013) Protective effect of bifidobacteria in an experimental model of *Clostridium difficile* associated colitis. J Dairy Res 80:263–269
Unal CM, Steinert M (2016) Novel therapeutic strategies for *Clostridium difficile* infections. Expert Opin Investig Drugs 20:269–285
Valdes L, Gueimonde M, Ruas-Madiedo P (2015) Monitoring in real time the cytotoxic effect of *Clostridium difficile* upon the intestinal epithelial cell line HT29. J Microbiol Method 119:66–73
Valdes-Varela L, Alonso-Guervos M, Garcıa-Suärez O et al (2016a) Selection of bifidobacteria and lactobacilli able to antagonise the cytotoxic effect of *Clostridium difficile* upon intestinal epithelial HT29 monolayer. Front Microbiol 7:577
Valdes-Varela L, Hernandez-Barranco AM, Ruas-Madiedo P et al (2016b) Effect of *Bifidobacterium* upon *Clostridium difficile* growth and toxicity when co-cultured in different prebiotic substrates. Front Microbiol 7:738
Vanderhoof J, Whitney D, Antonson D et al (1999) *Lactobacillus* GG in the prevention of antibiotic associated diarrhea in children. J Pediatr 135:564–568
Venema K, van den Abbeele P (2013) Experimental models of the gut microbiome. Best Pract Res Clin Gastroenterol 27:115–126
Vernaya M, McAdam J, Hamptom MD (2017) Effectiveness of probiotics in reducing the incidence of *Clostridium difficile*-associated diarrhea in elderly patients: a systematic review. JBI Database Syst Rev Impl Rep 15:140–164
Voth DE, Ballard JD (2005) *Clostridium difficile* toxins: mechanism of action and role in disease. Clin Microbiol Rev 18:247–263
Woodworth MH, Carpentieri C, Sitchenko KL et al (2017) Challenges in fecal donor selection and screening for fecal microbiota transplantation: a review. Gut Microb 8:225–237
Wullt M, Hagslatt ML, Odenholt I (2003) *Lactobacillus plantarum* 299v for the treatment of recurrent *Clostridium difficile*-associated diarrhoea: a double-blind, placebo-controlled trial. Scand J Infect Dis 35:365–367
Xi B, Yu N, Wang X et al (2008) The application of cell-based label-free technology in drug discovery. Biotechnol J 3:484–495
Yakob L, Riley TV, Paterson DL et al (2015) Mechanisms of hypervirulent *Clostridium difficile* ribotype 027 dis-placement of endemic strains: an epidemiological model. Sci Rep 5:12666
Young VB (2017) Old and new models for studying host-microbe interactions in health and disease: *C. difficile* as an example. Am J Physiol Gastrointest Liver Physiol 312:G623–G627
Yu H, Chen K, Wu J et al (2015) Identification of toxemia in patients with *Clostridium difficile* infection. PLoS One 10:e0124235
Yun B, Oh S, Griffiths MW (2014) *Lactobacillus acidophilus* modulates the virulence of *Clostridium difficile*. J Dairy Sci 97:4745–4758
Zivkovic M, Hidalgo-Cantabrana C, Kojic M et al (2015) Capability of exopolysaccharide-producing *Lactobacillus paraplantarum* BGCG11 and its non-producing isogenic strain NB1, to counteract the effect of enteropathogens upon the epithelial cell line HT29-MTX. Food Res Int 74:199–207

Immunization Strategies Against *Clostridioides difficile*

Camille Campidelli, Jean-François Bruxelle, Anne Collignon, and Severine Péchiné

Abstract

Clostridioides difficile (*C. difficile*) infection (CDI) is an important healthcare but also a community-associated disease. CDI is considered a public health threat and an economic burden. A major problem is the high rate of recurrences. Besides classical antibiotic treatments, new therapeutic strategies are needed to prevent infection, to treat patients, and to prevent recurrences. If fecal transplantation has been recommended to treat recurrences, another key approach is to elicit immunity against *C. difficile* and its virulence factors. Here, after a summary concerning the virulence factors, the host immune response against *C. difficile*, and its role in the outcome of disease, we review the different approaches of passive immunotherapies and vaccines developed against CDI. Passive immunization strategies are designed in function of the target antigen, the antibody-based product, and its administration route. Similarly, for active immunization strategies, vaccine antigens can target toxins or surface proteins, and immunization can be performed by parenteral or mucosal routes. For passive immunization and vaccination as well, we first present immunization assays performed in animal models and second in humans and associated clinical trials. The different studies are presented according to the mode of administration either parenteral or mucosal and the target antigens and either toxins or colonization factors.

Anne Collignon has died before the publication of this book.

C. Campidelli · S. Péchiné (✉)
Université Paris-Saclay, INRAE, AgroParisTech, Micalis Institute, Jouy-en-Josas, France
e-mail: severine.pechine@universite-paris-saclay.fr

J.-F. Bruxelle
CIRI-Centre International de Recherche en Infectiologie, Université de Lyon, Université Claude Bernard Lyon 1, Inserm U1111, CNRS UMR5308, ENS Lyon, Lyon, France

1 Introduction

Clostridioides difficile is an anaerobic spore-forming intestinal pathogen responsible for post-antibiotic diarrhea and pseudomembranous colitis (PMC) (Lawson et al. 2016). *C. difficile* infection (CDI) is characterized by a large spectrum of clinical signs from asymptomatic carriage to fulminant colitis. CDI is an important healthcare—but also community-associated disease causing almost half a million infections each year in the USA (Finn et al. 2021). Increased morbidity and mortality have been associated with the emergence of hypervirulent epidemic strains such as BI/NAP1/027 strains. Even if a decrease in prevalence of these 027 strains in some European countries has been observed, CDI remains poorly

P. Mastrantonio, M. Rupnik (eds.), *Updates on Clostridioides difficile in Europe*, Advances in Experimental Medicine and Biology 1435, https://doi.org/10.1007/978-3-031-42108-2_7

controlled, and 027 and other epidemic strains are still prevalent (van Dorp et al. 2016). A major problem is the high rate of recurrences, 20–30% after a first episode and up to 60% after a first recurrence (Shields et al. 2015). Thus, CDI is considered a public health threat and an economic burden.

CDI is most commonly triggered by disruption of the intestinal microbiota by antibiotics and subsequent intestinal colonization. *C. difficile* highly resistant spores serve the transmission agent. After contamination of the host, spores germinate in response to bile acids and glycine and resume vegetative growth. Then, vegetative forms colonize the gut thanks to several colonization factors. Finally, the toxins are released and led to diarrhea and colitis. Initial colonization is influenced by the intestinal microbiota, and *C. difficile* persistence in the gut is dependent on the microbiota and the host immune response (Péchiné and Collignon 2016). Persistence of spores in the gut associated with an altered microbiota and a poor immune response could be responsible for recurrences.

Guidelines for CDI treatment have been recently updated in America and in Europe (van Prehn et al. 2021). The treatment of a first episode is well defined and is based on antibiotherapy, such as vancomycin or fidaxomicin depending on the severity of the episode. In case of recurrences, variable guidelines have been recommended and there is no firm consensus on optimal treatment. Novel therapeutic strategies are needed to prevent infection, to treat patients and prevent recurrences. If fecal transplantation has been recommended to treat recurrences, another key approach is to elicit immunity against *C. difficile* and its virulence factors.

C. difficile studies are mainly performed in vivo in two different animal models, the hamster and mouse models. Hamsters are extremely susceptible to *C. difficile* and are used as virulence and protection model. In mice, several models have been described either in germ-free or conventional animals. Mouse models are used to monitor intestinal colonization by *C. difficile* and also in virulence and protection assays (Best et al. 2012).

Here, after a summary concerning the virulence factors, the host immune response against *C. difficile* and its role in the outcome of disease, we review the different approaches of passive immunotherapies and vaccines developed to treat and prevent CDI.

2 Virulence Factors and Host Immune Response

The main *C. difficile* virulence factors are the toxins, especially TcdA and TcdB. However, surface proteins involved in the colonization process participate also in the pathogenesis (Janoir 2016).

2.1 Surface Proteins and Colonization Factors

The first interaction between *C. difficile* and the host involves bacterial surface components. Some have been identified in *C. difficile* and shown to be involved in the colonization process (Bruxelle 2017).

2.1.1 Cell Wall Proteins (Cwp)

Many bacteria have an outer layer called the S-layer which gives them significant immunogenic potential. S-layer proteins (SLPs) were detected in all *C. difficile* strains. Unlike most bacteria, where the S-layer is composed of a particular protein species, the *C. difficile*'s S-layer is composed of two protein subunits called high molecular weight surface layer protein (HMW-SLP) consisting of a 47 kDa protein and another 36 kDa protein called low molecular weight surface layer protein (LMW-SLP). These forms are obtained following proteolytic cleavage mediated by the Cwp84 protease of the polypeptide precursor SlpA, encoded by the slpA gene (Janoir 2016). The two S-layer proteins (SLPs) are the main components of the bacterial surface and form a crystalline array over the entire cell surface. The low molecular weight (LMW)-SLP is surface exposed, involved in cell adherence, and highly variable between strains (Eidhin et al. 2006). The high molecular weight

(HMW)-SLP is anchored in the cell wall, is involved in adherence to intestinal tissue and extra cellular matrix proteins, and is conserved between strains (Karjalainen et al. 2001; Calabi et al. 2002). In addition, SLPs interact with the Toll-like receptor 4 (TLR 4), expressed on the surface of the host cell. S-layer binding to dendritic cells initiates downstream signaling of nuclear transcription factor kappa β (NF-kB) and interferon regulatory factor 3, resulting in production of inflammatory cytokines and activation of immune cells (Ryan et al. 2011).

The Cwp66 protein also serves adhesin function (Waligora et al. 2001) with its surface-exposed, highly variable, and highly immunogenic C-terminal domain. The cysteine protease Cwp84 is also surface exposed (Janoir et al. 2007; Chapetón Montes et al. 2013) and anchored in the cell wall through its C-terminal domain. The N-terminal domain contains the proteolytic site responsible for the cleavage of the SlpA precursor into the two SLPs (Kirby et al. 2009; Dang et al. 2010) and for the degradation of extracellular matrix components facilitating therefore bacterial spread (Janoir et al. 2007). Interestingly, the Cwp84 protease is conserved among *C. difficile* strains and has been shown to be immunogenic in humans (Péchiné et al. 2005b).

2.1.2 Flagellar Proteins

Flagellated and non-flagellated *C. difficile* strains have been described. Flagella are involved in motility, adherence to host cells and host signaling, and promotes toxin-mediated gut inflammation by interacting with the immune Toll-like receptor 5 (TLR5) to activate NF-kB and mitogen-activated protein kinase (MAPK) signaling pathways (Stevenson et al. 2015; Batah et al. 2016).

Dapa et al. have shown that flagella are involved in the biofilm formation process (Dapa and Unnikrishnan 2013). Tasteyre et al. have shown that naturally occurring nonflagellated strains are less adherent to mouse caecum than flagellated strains and that the flagellin FliC and cap protein FliD are able to bind murin mucus (Tasteyre et al. 2000, 2001). More recently, it has been shown that *fliC* and *fliD* mutants in the 630*Δerm* strain displayed increased adherence to Caco2 cells compared to the parenteral strain. Thus, in 630*Δerm* genetic background, flagella do not seem to play a role in adherence (Dingle et al. 2011). In contrast, *fliC* and *fliD* mutants in *C. difficile* 027 strain R20291 displayed decreased adherence to Caco2 cells and mouse caeca suggesting a role of flagella in cell adherence and colonization (Baban et al. 2013). *C. difficile* flagellin FliC has been shown to activate an innate immune response via its interaction with TLR5 and activation of NF-κB signaling (Yoshino et al. 2013; Batah et al. 2016). Interestingly, Batah et al. demonstrated, in an animal model, a synergic effect of flagella and toxins in eliciting an inflammatory mucosal response (Batah et al. 2017). In addition, toxin and flagellar genes are co-regulated in strains such as the 630 and not in others such as the R20291 (Baban et al. 2013). More recently, Chebly et al. demonstrated for the first time that FliC from *C. difficile* is also internalized in Caco-2/TC7 cells and triggers the activation of the NLRC4 inflammasome, resulting in the cleavage of pro-caspase-1 thereby contributing to the inflammatory process of *C. difficile* infection through the release of inflammasome-bound cytokines IL-18 and IL-33 (Chebly et al. 2022).

Taken together, those results suggest that the contribution of flagella to the pathogenic process is complex and could be different according to the genetic background.

2.1.3 Other Surface Components

Other colonization factors have been characterized. The surface-exposed fibronectin-binding protein FbpA (Hennequin et al. 2003; Barketi-Klai et al. 2011), highly conserved between *C. difficile* isolates, interacts with fibronectin in host tissues. The heat shock protein GroEL, highly conserved, also serves adhesin function (Hennequin et al. 2001). The collagen-binding protein CbpA with a N-terminal collagen-binding domain is surface-localized (Tulli et al. 2013). Recently, Arato et al. characterized and demonstrated that CD2831 is a collagen-binding protein capable of binding immobilized collagen types I, III, and V as well as native collagen produced by human fibroblasts. Overexpression of this protein increased the ability to form a biofilm in *C. difficile*. This protein is believed to have a dual role in adhesion to collagen-rich tissues and host immune evasion by binding to the collagen-like domain of human complement component C1q (Arato et al. 2019).

The lipoprotein CD0873, part of an ABC transporter, is surface-associated, displays significant adhesive properties, and is immunogenic in patients (Kovacs-Simon et al. 2014). Recently, Bradshaw et al. demonstrated that the lipoprotein CD0873 plays a role in intestinal colonization by *C. difficile* and is part of an import system for tyrosine, a key amino acid in *C. difficile* infection (Bradshaw et al. 2019). Zmp1 and CD2830 metalloproteases are able to cleave several host proteins such as IgA2, fibrinogen, or fibronectin (Cafardi et al. 2013; Hensbergen et al. 2014).

Other surface components include polysaccharides (PS) such as PS-I and PS-II. However, only PS-II is common to all strains of *C. difficile* (Ganeshapillai et al. 2008).

There are also immunogenic proteins located on the outer layers of *C. difficile* spores such as the BclA3 protein and the *C. difficile* exosporium cysteine-rich protein CD1067 (CdeC). The collagen-like BclA3 exosporium protein is common to most *C. difficile* strains. It is composed of an N-terminal domain, possibly oriented inward; a collagen-like domain formed by GXX repeats, strongly glycosylated (Strong et al. 2014); and a C-terminal domain that probably faces outward from the exosporium (Pizarro-Guajardo et al. 2014).

Recently, Aubry et al. showed that several glycosylated peptides from the collagen region of BclA3 were able to induce a humoral immune response in mice (Aubry et al. 2020). CdeC is expressed under sporulation conditions and localizes in the exosporium-like layer to the *C. difficile* spore, is accessible to IgGs, and is involved in resistance to lysozyme, ethanol, and heat. In addition, Barra-Carrasco et al. show that CdeC is essential for exosporium morphogenesis and the correct assembly of the spore coat of *C. difficile* (Barra-Carrasco et al. 2013). Ghose et al. found that this spore protein is immunogenic in mice and is able to protect them against challenge with *C. difficile* UK1, a clinically relevant 027/B1/NAP1 strain. CdeC is also able to afford high levels of protection against challenge with *C. difficile 630Δerm* in golden Syrian hamsters (Ghose et al. 2016a). Pizarro-Guajardo et al. also confirm that CdeC of the epidemic strain R20291 (B1/NAP1/027) is immunogenic in mice (027/B1/NAP1) (Pizarro-Guajardo et al. 2018).

2.2 Toxins

2.2.1 TcdA and TcdB

Both toxins have the same ABCD domain structure: the binding, cutting, and delivery domains acting sequentially to deliver the N-terminal glucosyltransferase domain (GTD) in the cytosol of enterocytes (Jank and Aktories 2008). This N-terminal domain glucosylates and inactivates the Rho-GTPases leading to actin cytoskeleton disruption, cell death, and epithelial barrier disruption (Voth and Ballard 2005; Popoff and Geny 2011). The receptor-binding C-terminal domain (RBD) is composed of combined repetitive oligopeptides (CROPs) that are responsible for binding cell receptors (Dingle et al. 2008).

TcdA and TcdB, despite their similar structure, are immunologically distinct. Antibodies directed to TcdA are able to neutralize TcdA but fail to neutralize TcdB, and the opposite is true for antibodies directed against TcdB (Libby and Wilkins 1982). The two toxins display high variability especially in the C-terminal domain (Leuzzi et al. 2013).

Different recombinant fragments derived from TcdA and TcdB have been identified for the generation of neutralizing antibodies (Leuzzi et al. 2013; Maynard-Smith et al. 2014). The RBD of both TcdA and TcdB was first identified as an important antigenic motif (Lyerly et al. 1990; Sauerborn et al. 1997; Belyi and Varfolomeeva 2003). In contrast, the TcdA GTD induces low antibody responses (Leuzzi et al. 2013; Maynard-Smith et al. 2014). Several regions of TcdB induce neutralizing antibodies: the central region domains (Maynard-Smith et al. 2014), the RBD

(Kink and Williams 1998), and the GTD (Libby and Wilkins 1982; Leuzzi et al. 2013).

The respective role of TcdA and TcdB in pathogenesis is a key question. One group concluded that TcdB is essential for virulence (Lyras et al. 2009; Carter et al. 2015). For the other group, both toxins are responsible for disease. Interestingly, the full virulence of *tcdB* mutant was restored when it expressed the binary toxin in addition to TcdA (Kuehne et al. 2010, 2014). It seems wise to take into account both toxins TcdA and TcdB for immunization strategies.

2.2.2 Binary Toxin

An additional toxin, the binary toxin or *C. difficile* transferase (CDT), is produced by some strains (Perelle et al. 1997) such as the epidemic/hypervirulent BI/NAP1/027 strains.

The lipolysis-stimulated lipoprotein receptor (LSR) has been identified as the host cell receptor for CDT (Papatheodorou et al. 2011). The component CDTb (99 kDa) induces LSR clustering and accumulation in lipid rafts, and CDTb N-terminal domain serves as a binding site for the component CDTa (48 kDa), thus triggering endocytosis of this complex in cells. At low pH, the endosomes seem to induce the insertion of CDTb into the membrane and allow the formation of a transmembrane β-barrel channel to deliver CDTa into the cytosol (Sheedlo et al. 2020). Subsequently, CDTa ADP-ribosylates G-actin and leads to complete depolymerization of the actin cytoskeleton, thereby causing changes in cell morphology and tight junctions (Stieglitz et al. 2021). In addition, formation of microtubule-based protrusions leads to enhanced adherence (Schwan et al. 2009, 2014; Papatheodorou et al. 2011). The role of CDT is not fully understood, but it appears to enhance the disruption of host protective mechanisms stimulated by toxins A and B, increase the virulence of B1/NAP1/027 strains in animal models, would activate the NF-κB pathway, and induce the production of pro-inflammatory cytokine (Nibbering et al. 2021).

Of note, there are naturally occurring TcdA− TcdB−CDT+ strains, which can be responsible for diarrhea in humans (Eckert et al. 2015).

2.3 Host Antibody Response Against *C. difficile*

Several studies assessed the antibody response to *C. difficile* surface components and toxins (Hernández Del Pino et al. 2021).

Regarding surface proteins, SLPs are highly immunogenic. The LMW-SLP is an immunodominant antigen, as demonstrated by the presence of antibodies against this protein in sera of patients infected by *C. difficile* (Cerquetti et al. 1992; Wright et al. 2008). Drudy et al. found that antibody levels to SLPs were similar in patients with CDI, asymptomatic carriers, and controls. However, patients with recurrences failed to mount an efficient IgM immune response to SLPs compared to patients with a single episode of CDI (Drudy et al. 2004). Nevertheless, in recent studies, Mizrahi et al. showed that during the early response further to infection, hospitalized patients with a single episode of CDI had a significant higher level of IgG against SlpA precursor compared to patients with recurrence or control group (Mizrahi et al. 2018). Taken together, these results suggest that SLPs play a role in the early antibody response to *C. difficile* and then might become tolerogenic as described for other TLR-inducer commensal bacterial antigens (Valentini et al. 2014).

The adhesin Cwp66, the protease Cwp84, the flagellar proteins FliC and FliD, and the Fbp protein were found to be expressed during the course of infection and to be immunogenic. Most patients with CDI developed antibodies to FliC, FliD, Cwp84, and Cwp66 C-terminal domain, confirming the expression of these surface proteins during the course of the disease (Péchiné et al. 2005a). In another study, serum antibody levels were compared in a CDI patient group with a control group. Regarding the adhesins Cwp66 and FbpA, the protease Cwp84, and the FliC and FliD flagellar proteins, the total antibody levels in blood were statistically lower in the CDI group than in the control group suggesting a role of these specific antibodies in CDI occurrence (Péchiné et al. 2005b).

Concerning *C. difficile* PS, two studies in CDI patients have reported an antibody response specific to *C. difficile* PS. Oberli et al. detected PS-II-specific IgA in CDI patient stool, and Martin et al. detected PS-I-specific IgA and IgG in CDI patient stool and sera, respectively (Oberli et al. 2011; Martin et al. 2013).

Regarding toxins, TcdA and TcdB have been shown to be immunogenic. In some studies, a correlation was found between anti-TcdB antibodies and asymptomatic carriage or absence of recurrence. Whereas in other studies, anti-TcdA antibody levels were shown to be more significant. Viscidi et al. found that antibody levels to TcdB were higher in sera of convalescent CDI patients than in sera of controls (Viscidi et al. 1983). Another study showed a correlation between clinical recovery without relapse, high TcdB IgG titers, and/or neutralizing antibodies (Aronsson et al. 1985). Kyne et al. monitored antibody response to *C. difficile* toxins and non-toxin antigens over time in hospitalized patients (Kyne et al. 2000). Although 15–31% of high-risk hospitalized patients were colonized with *C. difficile*, only a minority developed symptomatic infection. The asymptomatic carriers had significantly higher serum IgG antibody levels to TcdA within three days of colonization than those who developed diarrhea. So after contamination by *C. difficile*, a rise in IgG antibody to TcdA resulted in asymptomatic colonization rather than symptomatic infection. Interestingly, serum IgG levels against TcdB and nontoxin antigens were also higher in asymptomatic carriers, but the difference was not statistically significant. The same group also observed that patients with a single episode of CDI had significantly higher levels of IgM against TcdA, TcdB, and nontoxin antigens by day 3 of illness. These patients had also significantly higher levels of circulating IgG against TcdA by day 12, compared to patients who later developed recurrent CDI. After adjusting for other risk factors, patients with CDI and a low level of serum IgG against TcdA had a 48-fold greater risk of recurrence (Kyne et al. 2001). In recent studies in humans, De Roo et al. have shown that rCDI is associated with low serum antibody titers against TcdA and TcdB (De Roo and Regenbogen 2020).

Besides circulating antibodies, neutralizing anti-TcdA IgA in stool have also been detected (Kelly et al. 1992). Warny et al. showed that fecal anti-TcdA IgA titers were significantly higher in patients who suffered a single episode compared to those relapsing (Warny et al. 1994). Jonhson et al. found that anti-TcdA sIgA titers were higher in the intestinal secretions of CDI convalescent patients compared to noncarrier subjects (Johnson et al. 1992). Anti-TcdA secretory IgA (sIgA) could inhibit toxin binding to intestinal receptors (Kelly et al. 1992; Warny et al. 1994). Low levels of fecal IgA and reduction in colonic IgA-producing cells associated with the gut mucosa have been shown to be associated with prolonged CDI and recurrences of infection (Johal et al. 2004). For Islam et al. in the early course of CDI (<72 h), low specific sIgA titers against TcdB but not TcdA were associated with susceptibility to disease suggesting that the mucosal immunity to TcdB may be particularly important in the early stages of infection (Islam et al. 2014).

All these results demonstrated that the adaptative host immune response plays a role in disease presentation and outcome.

A better knowledge of *C. difficile* pathogenesis and the host response has paved the way to the development of several antibody-based products (APs), and passive and active immunization strategies have been developed for the prevention and/or treatment of CDI (Mizrahi et al. 2014).

3 Passive Immunization Strategies with Antibody-Based Products

Passive immunization strategies are designed in function of the target antigen (*C. difficile* toxins or surface proteins), the antibody-based product (AP), and its administration route (oral or parenteral) (Bruxelle 2017).

Toxins, as key virulence factors, represent the first studied target for passive immunization.

However, such a strategy does not act on the bacterial clearance and consequently neither on colonization or dissemination of *C. difficile* in the environment. Another rational strategy is to target the whole bacterium or its surface proteins.

In passive immunization strategies, AP must be present in the intestinal lumen in order to act directly against *C. difficile*. If administered via a parenteral route, the AP should have a low immunogenicity and a good bioavailability and should be transferred from the systemic circulation to the intestinal lumen. If directly administered via the oral route, the AP faces digestion process. In both cases, the main issue of passive immunization strategy resides in pharmacokinetic properties of the AP.

Currently, the progress on antibody engineering enables to design a variety of APs ranging from polyclonal antibodies through monoclonal antibodies and various antibody fragments such as heavy chain single-domain antibodies (V_HHs). Therefore, the concomitant choices of the target, the administration route, and the variety of APs explain the diversity of studies dealing with passive immunization strategies against *C. difficile*.

3.1 Assays in Animal Models

3.1.1 Parenteral Administration of Antibody-Based Products in Animal Models

Polyclonal Antibodies Against Toxins

First, polyclonal antibodies were used in passive immunotherapy against *C. difficile*. In 1982, Libby and Wilkins were the first to demonstrate that passive immunization of mice with specific rabbit antiserum against toxins protected mice against the homologous toxin but not the heterologous toxin (Libby and Wilkins 1982). In another animal model, Giannasca et al. showed that intraperitoneally (i.p.) injection of mouse antitoxin antibodies before challenge protected hamsters in a dose-dependent manner against *C. difficile* (Giannasca et al. 1999). Robert et al. produced polyvalent antitoxin antibodies in sheep; the i.p. administration of this antiserum to hamsters after challenge with different strains of *C. difficile* was protective in a dose-dependent manner (Roberts et al. 2012).

Monoclonal Antibodies Against Toxins

Then, monoclonal antibodies (MAbs) have been produced and tested in different models. Corthier et al. developed a passive immunotherapy with MAbs targeting specifically *C. difficile* toxins in monoxenic mice (Corthier et al. 1991). Intravenous (i.v.) administration of a mouse monoclonal IgG targeting TcdA C-terminal repeating units was able to protect mice against lethal *C. difficile* infection. After administration, the titer of MAbs remained high for at least eight days, and mice were fully protected against *C. difficile* while no mouse survives in the control group. Of note, no impact on *C. difficile* colonization in passively immunized mice was observed.

In a second generation of AP, fully humanized monoclonal antibodies (HuMAbs) directed against either the RBD of TcdA or TcdB were produced and assessed by Babcock et al. (2006). After characterization, anti-TcdA CDA1 and anti-TcdB MDX1388 were selected for protection assays in animal models. In a classic infection model after a primary challenge, hamsters were treated i.p. with CDA1, MDX1388 alone, or in combination for four days before challenge with *C. difficile* spores. In a relapse model, treatment with CDA1 and/or MDX1388 was associated with a vancomycin treatment. CDA1 alone led to early partial protection compared to controls without treatment. However, this protection did not persist. MDX1388 administered alone did not lead to protection. Interestingly, compared to controls, combination therapy with CDA1 and MDX1388, leading to neutralization of RBD of both toxins, provided better and prolonged protection in both models. These antibodies were shown to neutralize the toxin effects of diverse and clinically relevant strains of *C. difficile*, including multiple isolates of the BI/NAP1/027 and BK/NAP7/078 strains (Hernandez et al. 2015). In addition, Babcock et al. reported that levels of circulating HuMAbs in hamsters were much lower than anticipated and that 10% of the hamsters had no detectable circulating antibodies after a total of 200 mg of antibody i.p.-administered. They suggest that it could be due to the inefficient transport of human antibodies from the peritoneum into the bloodstream or that some hamsters develop an immune response to the human antibody leading to their rapid clearance.

Then, others produced HuMAbs targeting toxins based on the sequence of CDA1 and MDX1388. For instance, Davies et al. have developed a mixture of three humanized IgG1 MAbs (UCB MAbs), of which one neutralize TcdA and two TcdB (Davies et al. 2013). The UCB MAbs showed high potency in a variety of in vitro binding and neutralization assays. Compared to CDA1 and MDX1388, UCB MAbs led to higher levels of protection in their hamster model of CDI and displayed higher valencies of toxin binding. Interestingly, pharmacokinetic and biodistribution assays of i.p. administered humanized IgG1 in noninfected hamsters showed that antibody half-life in serum was of about six days. MAbs were detectable in healthy hamster colon (about 28 ng/ml/cm of mucosa ±17) seven days after i.p. administration of about 2 mg of humanized IgG1. This persistence may likely explain the levels of protection provided by these UCB MAbs. Qiu et al. developed anti-TcdA and anti-TcdB HuMAbs from murine MAb candidates. Administered parenterally, they were able to protect animals in a dose-dependent manner against lethal challenge (85% of hamster survival after *C. difficile* challenge) and to reduce the severity and duration of diarrhea associated with several *C. difficile* clinical strains (Qiu et al. 2016).

These RBD-specific MAbs block toxin activity by inhibiting receptor binding and subsequently internalization in epithelial cells. Another way to block toxin activity is to target the N-terminal domain either the translocation domain (TD) or the glucosyltransferase domain (GTD). Indeed, these domains are more conserved between *C. difficile* strains and therefore represent targets for AP against a broader range of clinical strains. For Anosova et al., the combination of three fully HuMAbs, one specific to the RBD of TcdA and two specific for the GTD of TcdB, protected hamsters from CDI (Anosova et al. 2015).

These studies showed that treatments with AP targeting both toxins could increase protection against CDI. In these animal models, MAbs were administered with several doses a few days before challenge, therefore mimicking either a prophylactic strategy or an established circulating antibody response against toxins.

Monoclonal Antibody Subunits Against Toxins

With the development of antibody engineering, various MAb fragments have been produced and tested (Péchiné et al. 2017), in particular V_HH fragments or nanobodies, which correspond to the N-terminal region of a single variable (VH) domain from camel heavy chain antibody.

Yang et al. and Schmidt et al. developed two neutralizing, tetravalent antibodies composed of V_HHs targeting both TcdA and TcdB (designated ABA and VNA2-Tcd) (Yang et al. 2014; Schmidt et al. 2016). ABA and VNA2-Tcd, two chimeric multivalent APs, were composed of two V_HHs recognizing the GTD and translocation domain (TD) of TcdA, respectively, and two V_HHs recognizing the GTD of TcdB. After i.p. administration, both V_HHs protected against CDI in mice and gnotobiotic piglets but not in hamsters. Moreover, ABA was able to neutralize toxins from a panel of genotypically diverse $TcdA^+$ $TcdB^+$ clinical isolates, including some BI/NAP1/027 strains. However, to increase protective efficacy of parenterally administered V_HH, its serum half-life has been improved by developing a replication-deficient recombinant adenovirus expressing the heteromultimeric V_HH-based agents (ABA and VNA2-Tcd). This strategy, optimizing the AP delivery, has shown its efficacy to neutralize toxins and prevent CDI.

Of note, V_HHs against the two fragments of CDT have also been designed and seem to efficiently neutralize the toxin in vitro (Unger et al. 2015).

To develop novel *C. difficile* APs, Hussack et al. isolated several single-domain antibodies (V_HHs) capable of toxin A neutralization through recognition of the extreme C-terminal combined repetitive oligopeptide (CROP) domain. However, they did not succeed at identifying neutralizing V_HHs that bound a similar region on the toxin B (Hussack et al. 2011). In a more recent study, they reported the isolation of a panel of 29 V_HHs targeting at least seven unique epitopes on a toxin B immunogen composed of a portion of the central delivery domain and the entire CROP domain. Unfortunately, none of the V_HHs tested neutralized the toxin B. However, toxin B inhibition was observed with a chimeric form of the V_HH fused to a human Fc domain (V_HH-Fc fusions), reaching the neutralizing potency of the recently approved antitoxin B monoclonal antibody bezlotoxumab in in vitro assays (Hussack et al. 2018).

The advantage of such antibodies is the possibility of their genetic manipulation to increase their efficiency. Sulea et al. (2018) considered an affinity maturation platform to construct mutant antibodies neutralizing TcdA. These results supported the role of mutation in enhancing the affinity of antibodies. In this regard, the development of double-mutant T56R and T103R neutralized TcdA cytotoxicity with a half maximal inhibitory concentration (IC50) of 12 nM and enhanced sdAb affinity to toxin A (Sulea et al. 2018).

Polyclonal and Fragments Antibodies Against Surface Proteins

Another approach is to target colonization factors, which may help to eliminate colonizing bacteria. Few studies have tested parenteral passive immunization targeting colonization factors with polyclonal antibodies. For instance, Malderelli et al. targeted *C. difficile* pilin with anti-PilW serum but did not obtain protection in mice against *C. difficile* (Maldarelli et al. 2016). Ghose et al. produced polyclonal antibodies targeting the *C. difficile* flagellin FliC (Ghose et al. 2016b). Passive immunization of mice via i.p. route with anti-FliC hyperimmune serum was able to protect 80% of treated mice against *C. difficile* after lethal challenge. Since FliC plays a key role in the pathogenesis ranging from bacterial colonization through immunomodulatory effects and gene regulation, protection elicited by anti-FliC antibody may involve various mechanisms. Kandalaft et al. described the production of V_HHs targeting the SLPs; in vitro test showed that a combination of three V_HHs targeting the LMW-SLP inhibited *C. difficile*'s motility (Kandalaft et al. 2015).

3.1.2 Mucosal Administration of Antibody-Based Products in Animal Models

Lyerly et al. used a bovine immunoglobulin G (IgG) concentrate (BIC) from gestating cow's colostrum vaccinated with *C. difficile* formalin-inactivated culture filtrate to orally passively immunize hamsters (Lyerly et al. 1991). BIC contained high levels of neutralizing IgG specific to both toxins and probably to other antigens. Treated hamsters were completely protected from the disease during the treatment compared to controls. However, treated hamsters developed diarrhea and died after the end of treatment. These results showed for the first time that passive immunization by oral route against *C. difficile* targeting mainly toxins can protect against the toxin-mediated virulence.

Van Dissel et al. used in the hamster model an immune whey protein concentrate (immune WPC-40, Mucomilk) containing high concentration of sIgA antibodies against the whole bacterial cell as well as TcdA and TcdB (van Dissel et al. 2005). Immune WPC-40 conferred 80–90% protection in hamsters challenged with a toxigenic *C. difficile* strain. In contrast to Lyerly et al., the protection was maintained in surviving hamsters after treatment cessation for at least 28 days. These authors suggest that sIgA directed against the whole bacterial cell may reduce *C. difficile* gut colonization and promote bacterial clearance.

In another study, Kink et al. tested therapeutic or prophylactic strategy, neutralizing avian antitoxin antibodies (IgY) directed against the C-terminal domain of TcdA or TcdB administered orally to hamsters (Kink and Williams 1998). Prophylactic treatment before challenge with anti-TcdA alone was efficient to protect hamsters from CDI. However, for therapeutic treatment, co-administration of anti-TcdA and anti-TcdB after *C. difficile* challenge was necessary to fully protect hamsters.

The main issue of passive immunization by oral route is the AP stability in the digestive environment. Immunoglobulins (Igs) have to resist to acidity in the stomach and to pancreatic enzymes in the small intestine. However, antibody sensitivity to digestive enzymes depends on antibody isotype. For instance, IgG1 are more susceptible to hydrolysis by pepsin than IgG2. Trypsin preferentially digests bovine IgG1 and IgG2 than IgM, whereas chymotrypsin preferentially hydrolyzes IgM than IgG (de Rham and Isliker 1977; Brock et al. 1977). Notably, sIgA are more resistant than IgG to degradation in the stomach and intestine (Fagarasan and Honjo 2003). This emphasizes the importance of isotype selection for oral passive immunization. Nevertheless, to improve orally administrated AP half-life, a specific formulation and/or vectorization may be recommended to maintain activity and to target *C. difficile* in the colonic infection site.

To extend life time of orally AP against *C. difficile* toxins, Andersen et al. engineered a *Lactobacillus* strain in order to express cell wall-anchored TcdB-neutralizing antibody fragments (V_HHs) (Andersen et al. 2015). In a prophylactic treatment in a hamster model, oral administration of a combination of two *L. paracasei* strains expressing two different V_HHs conferred a partial (50%) protection against lethal *C. difficile* challenge. Hamsters showed either no damage or limited inflammation of the colonic mucosa after four days of *C. difficile* infection although they were colonized by *C. difficile*.

Bovine antibodies from hyperimmune colostrum milk is considered a powerful orally administered drug candidate that is currently in clinical development. A pregnant dairy cow was repeatedly immunized with recombinant mutants of toxins A and B produced by *C. difficile*, and the resultant hyperimmune bovine colostrum (HBC) was evaluated for therapeutic efficacy in gnotobiotic piglets with diarrhea due to CDI. As a result, nonimmune colostrum-treated piglets developed moderate to severe diarrhea and colitis. In contrast, HBC-treated piglets had mild or no diarrhea and mild or no colitis (Sponseller et al. 2014).

Few years later, Hutton et al. worked on pregnant cows that were immunized intramuscularly to generate HBC containing antibodies that target essential *C. difficile* virulence components, specifically spores, vegetative cells, and toxin B (TcdB) and SLP. Mouse infection and relapse models were used to compare the capacity of HBC to prevent or treat primary CDI as well as prevent recurrence. Administration by the oral route of TcdB-specific colostrum alone, or in combination with spore or vegetative cell-targeted colostrum, prevents and treats *C. difficile* disease in mice and reduces disease recurrence by 67% (Hutton et al. 2017).

In another study with the same approach, cows were repeatedly immunized to establish specific immunoglobulin G and A titers against toxins A (TcdA) and B (TcdB) and against *C. difficile* cells in mature milk or colostrum. The effect of three different concentrations of anti-*C. difficile* whey protein isolates (anti-CD-WPI) and the standard of care antibiotic vancomycin were investigated in the hamster animal model of CDI. WPI obtained from the milk of exactly the same cows preimmunization and a vehicle group served as negative controls. The survival of hamsters receiving orally anti-CD-WPI was 50%, 80%, and 100% compared to 10% and 0% for the control groups, respectively. The surviving hamsters of the anti-CD-WPI groups survived the entire study period, although they were treated for only 75h. The specific antibodies not only inactivated the toxins for initial suppression of CDI but also prevented recurrence (Heidebrecht et al. 2019).

Roberts et al. developed a novel orally delivered ovine polyclonal antibody product targeting *C. difficile* toxins. This so-called OraCAb shows high antibody titers and was optimized with a formulation protecting the antibodies from gastrointestinal-mediated inactivation. The potential of OraCAb to prevent CDI was assessed in vivo in a hamster model. Results show a significant difference in animal survival for those treated with the optimized OraCAb formulation versus the untreated control group. Also, treatment with a combination of vancomycin and OraCAb prevented simulated CDI recurrence, unlike vancomycin therapy alone (Roberts et al. 2020).

Finally, Chiari et al. tested co-administration of human secretory IgA (sIgA) targeting TcdA and TcdB together with subtherapeutic vancomycin. They observed that this treatment enhanced survival in the CDI hamster model (Chiari et al. 2021).

Targeting the colonization factors may protect against early stage of *C. difficile* infection. O'Brien et al. showed that antiserum directed against SLP administered orally to hamsters was able to delay mortality after lethal challenge with *C. difficile* compared to untreated hamsters (O'Brien et al. 2005).

3.2 Assays in Humans and Clinical Trials

In humans, passive immunotherapy against *C. difficile* has been considered mainly to prevent recurrences.

3.2.1 Parenteral Administration of Antibody-Based Products in Humans

Treatment with Polyvalent Immunoglobulins

In 1991, passive immunotherapy administered to humans against CDI consisted of human polyvalent gamma globulins (Leung et al. 1991). Children with chronic recurrent CDI presented a deficient level of anti-TcdA IgG and IgA, and i.v. administration of gamma globulins (IVGG), every three weeks (400 mg/kg), was able to increase anti-TcdA IgG level in serum and was associated with resolution of clinical symptoms and clearance of toxins in stool. It was then confirmed in two adults with severe PMC receiving either 300 mg/kg or 200 mg/kg of IVGG; both patients presented a rapid resolution of diarrhea, pain, and fever within 36 h. The efficacy of IVGG was correlated with the presence of anti-TcdA-neutralizing IgG. The mechanism of action of IVGG in treatment for CDI is most likely explained by exudation of serum proteins across an already inflamed colonic mucosa. Then, IVGG were tested in many patients with CDI with various results (Diraviyam et al. 2016). Recently, Negm et al., in 17 CDI patients, observed a therapeutic response to polyvalent i.v. immunoglobulins (IVIg) in 41% (10/17) of the patients (Negm et al. 2017). In addition, they observed differences in TcdA-neutralizing efficacy between three commercial IVIg preparations as well as differences of level of specific IgG isotypes against *C. difficile* antigens.

Clinical Trials with Monoclonal Antibodies

After successful assays in animal models, CDA1 and MDX1388 targeting TcdA and TcdB RBD, respectively, were selected and tested in clinical trials.

A phase I clinical trial with CDA1 in healthy volunteers was completed (Taylor et al. 2008). Single injection of CDA1 at different doses did not lead to serious adverse events nor antihuman antibodies' production. The half-life of CDA1 ranged from 25 to 31 days. Two phase II clinical trials were performed. In first phase II clinical trial, CDA1 was tested in patients receiving standard-of-care (SOC) treatment for CDI and in a placebo group. The recurrence rate was not significantly different between the two groups (Leav et al. 2010). A second phase II trial (sponsored by University of Massachusetts and Medarex Inc.) tested the efficacy of the combination of CDA1 (actoxumab, MK-3415) and MDX1388 (bezlotoxumab, MK-6072) in patients with symptomatic CDI with SOC antibiotics (NCT00350298) (Lowy et al. 2010). The recurrence rate was significantly lower in the patient group treated with this combination (7%) compared to placebo group (25%). Two phase III trials have been completed under Merck license (MODIFY I NCT01241552 and MODIFY II NCT01513239). They were randomized, double-blind, placebo-controlled trials conducted at 322 sites in 30 countries involving 2655 adults receiving SOC antibiotics for primary or recurrent CDI. The primary endpoint was the recurrence of the infection. MODIFY I was a four-arm study: patients were randomly assigned to receive a single dose of bezlotoxumab (10 mg/kg), actoxumab (10 mg/kg), actoxumab plus bezlotoxumab (10 mg/kg each), and placebo. Interestingly, this clinical trial did not show a significant efficacy of actoxumab alone on recurrence rate. Consequently, actoxumab alone was discontinued after an interim analysis and not evaluated alone in MODIFY II. Thus MODIFY II is a three-arm study: bezlotoxumab (10 mg/kg), actoxumab plus bezlotoxumab (10 mg/kg each), and placebo. In both trials, the recurrence rate was significantly lower with bezlotoxumab alone or combined with actoxumab than with placebo. However, addition of actoxumab to bezlotoxumab did not improve efficacy, which may attest a major role of TcdB in the pathogenesis (Wilcox et al. 2017). The rate of adverse events was similar among the three groups; the most common events were diarrhea and nausea. The use of a single dose was supported by the long half-life of the MAbs (approximately 19 days).

Finally, in 2016, bezlotoxumab was approved by the Food and Drug Administration and the European Medicines Agency, for the prevention of rCDI in adult patients (≥18 years) (US Food and Drug Administration, 2016 (https://www.fda.gov/). Notably, bezlotoxumab can only reduce the rate of CDI relapse to ~40% compared to placebo and is unfavorable for treating acute CDI. Therefore, bezlotoxumab can be applied as an effective therapy for preventing rCDI. However, the clinical effectiveness of the drug should be assessed in further studies (Navalkele and Chopra 2018).

Since 2021, important changes compared with previous guideline have been suggested. This includes that metronidazole is no longer recommended for treatment of CDI when fidaxomicin and vancomycin are available, FMT or bezlotoxumab in addition to SOC antibiotics are preferred for treatment of a second or further recurrence of CDI, bezlotoxumab in addition to SOC is recommended for the first recurrence of CDI when fidaxomicin was used to manage the initial CDI episode, and bezlotoxumab is considered as an ancillary treatment to vancomycin for a CDI episode with high risk of recurrence when fidaxomicin is not available (van Prehn et al. 2021).

Among the proposed antibiotics, fidaxomicin is the only specific antibiotic for *C. difficile*; however, there is no study to date that has compared the cost-effectiveness of fidaxomicin with bezlotoxumab. The only cost-effectiveness analysis is related to the comparison of the effect of fidaxomicin with standard therapy plus bezlotoxumab as reported by Lam et al. focusing only on rCDI (Lam et al. 2018).

Additionally, it is proven that fidaxomicin plus bezlotoxumab has similar effect to other SOC antibiotics (i.e., vancomycin or metronidazole) plus bezlotoxumab. Notably, pharmacoeconomic analyses demonstrate that standard therapy plus bezlotoxumab could be cost-effective compared with standard therapy alone, especially in preventing rCDI episodes in those >65 years of age, those with severe CDI, and immunocompromised patients (Raeisi et al. 2022). Regarding the treatment of initial episode of CDI to prevent recurrence, fidaxomicin seems to be the most cost-effective regimen according to Chen et al.'s study. Bezlotoxumab-vancomycin was found to be better than fidaxomicin (Chen et al. 2021).

Reports on real-world experience on efficacy of bezlotoxumab has been done. Oksi et al. retrospectively studied the efficacy and safety of bezlotoxumab in preventing the recurrence of *C. difficile* infection in five university hospitals in Finland. Seventy-three percent of 46 patients remained free of recurrence in the following 3 months, and the performance remained at 71% protection among immunocompromised patients. In severe infections, bezlotoxumab prevented recurrence in 63% of cases (Oksi et al. 2019).

In a recent revue, authors interpret the most recent safety data and the clinical application of bezlotoxumab, highlighting specific high-risk patient populations. Overall, bezlotoxumab demonstrated a 40% relative reduction rate (Alonso and Mahoney 2018).

To conclude, the bezlotoxumab as adjunctive treatment has a high success rate at preventing rCDI in patients.

Systemically administered AP targeting the toxins can protect from recurrent CDI. However, it does not confer a locally oriented and a long-lasting protection against *C. difficile*. To date, no clinical trial has evaluated parenteral passive immunization targeting *C. difficile* colonization factor. Circulating antibodies are poorly transferred through a healthy intestinal mucosa and are most likely to encounter *C. difficile* after epithelium disruption. So investigators have assessed the protective efficacy of orally administered AP to target toxin and *C. difficile* colonization in the intestinal tract.

3.2.2 Mucosal Administration of Antibody-Based Products in Humans and Clinical Trials

Van Dissel et al. tested WPC-40 in 16 patients with confirmed CDI, among them 9 had a history of relapses (van Dissel et al. 2005). No adverse effect was observed during the follow-up. In addition, no toxin was detected in feces after treatment in 14 out of 15 patients, and *C. difficile* could no longer be cultured from the stools in 9 out of 15 subjects. Interestingly, none of the patients experienced another episode of CDI after treatment during the follow-up. In accordance with this study, a larger cohort was conducted in 101 patients with CDI (median age 74 years). After completion of at least 10 days of antibiotic treatment, patients received orally anti-CD-WPC for 2 weeks and were followed for 60 days. Interestingly, only 10% relapsed within the follow-up. A phase II clinical trial has been performed and completed but the results are not posted (NCT00177775).

After phase I trial, Mattila et al. performed a double-blind phase II study comparing another *C. difficile* immune whey IgG concentrate (CDIW) with metronidazole for recurrent CDI (Mattila et al. 2008). CDIW was produced from colostrum of cows immunized with formalin-inactivated *C. difficile*. Patients included in the study were adults who experienced at least two episodes of CDI. No significant differences were observed between the two treatment groups. At the end of the study, 8 patients out of 18 experienced a relapse in CDIW (44%) treated group and 9 out of 20 in metronidazole group (45%). These authors suggest that partial failure may probably be explained by a weak neutralization of colonization and toxin activity in vivo. These results emphasize the need of a well-engineered AP with high specificity and neutralizing activity in the infection site.

To conclude, the development of passive immunization strategies against *C. difficile* has led to many interesting APs targeting toxins but only a few targeting colonization factors. APs are developed mainly for the prevention of recurrences associated with SOC antibiotic treatment against CDI. Currently, the most efficient strategy to prevent recurrences is to target and neutralize toxins systemically. A combination of antibodies could increase the effectiveness of antibody therapy. The combination of antibodies directed against different antigen targets may have a synergistic effect, thus increasing the performance of antibody therapy. Even well tolerated, these treatments are costly and do not confer a long-lasting protection. In addition, passive immunization strategies targeting toxins alone do not seem to influence intestinal colonization nor prevent epidemic burden. To obtain a long-term protection, vaccine strategies are under development.

4 Active Immunization Strategies: Vaccines

Active immunization strategies are defined by the type of the induced antibody response and depend on the targeted antigen, the administration route, and the regimen.

Vaccine candidates must be immunogenic to induce an antibody response. Obviously, this antibody response must be protective against CDI. The targeted antigen must be specific to *C. difficile* and conserved among diverse clinical strains. The target must be biologically accessible to the elicited antibodies, which restricts vaccine candidates to exposed antigens. Antitoxin antibodies are associated with protection against CDI and recurrences. Thus, as in passive immunization strategies, toxins represent the first target studied for vaccination against *C. difficile*. However, to prevent colonization, and therefore to limit dissemination of bacteria in the environment, surface antigens represent an interesting alternative.

Protective antibody response to *C. difficile* can be either systemic or mucosal (Bruxelle 2017). The immune response depends on the administration route and the adjuvant (Zhang et al. 2015; Savelkoul et al. 2015). Here, we discuss the different vaccination strategies against *C. difficile* targeting toxins or colonization factors tested in animal models and in humans and clinical trials after vaccination by parenteral or mucosal routes.

4.1 Vaccines Targeting Toxins

4.1.1 In Animal Models

Parenteral Immunization in Animal Models

The respective role of TcdA and TcdB in the pathogenesis has been investigated in vaccination assays as well. Usually, both toxins are simultaneously used as vaccine antigens. The first generation of vaccine was composed of formalin-inactivated toxins (toxoids) from culture filtrates, and then toxoids were purified. Libby et al. and Fernie et al. obtained a full protection of hamsters against *C. difficile* after immunization with inactivated culture filtrate with Freund adjuvant (Libby et al. 1982; Fernie et al. 1983). In contrast, other authors (Kim et al. 1987) observed that toxoid A was sufficient to protect hamsters against *C. difficile*. However, differences in vaccination regimen, antigen purity, and *C. difficile* strain could explain the discrepancies between studies. Recently, a highly purified toxoid vaccine targeting TcdA and TcdB, adjuvanted with alum, has been developed and first tested in animal models (Anosova et al. 2013). Intramuscular (i.m.) immunization of hamsters was protective against mortality and disease in a dose-dependent manner, with 90% of protection with the highest dose tested. The protection was correlated with a neutralizing toxin-specific IgG response. These promising results in animal models using toxoids in parenteral vaccination against *C. difficile* have led to the development in clinical trials, which will be further discussed.

In a second generation of vaccine, alternatives to toxoids were investigated with vaccine based on nontoxic recombinant fragments of *C. difficile* toxins such as the antigenic C-terminal RBD. Sauerborn et al. first used recombinant *C. difficile* TcdA C-terminal domain in a subcutaneous (s.c.) immunization assay in mice with Freund's adjuvant (Sauerborn et al. 1997). Seven out of ten immunized mice were protected against a lethal dose of TcdA, correlated with production of anti-TcdA antibodies. Then, different combinations of recombinant toxin fragments have been used in parenteral immunizations. For instance, hamsters were immunized via i.p. route with a combination of fragments adjuvanted with MF29 and the RBD of TcdA associated either with TcdB GTD or TcdB RBD fragment (Leuzzi et al. 2013; Spencer et al. 2014). These combinations induced systemic IgGs, which neutralized both toxins and protected vaccinated hamsters from a lethal challenge of various *C. difficile* ribotypes. Karczewski et al. obtained full protection of hamsters after i.m. immunization with toxoids A and B adjuvanted with aluminum hydroxyphosphate sulfate and ISCOMATRIX™. Only partial protection was obtained with full-length toxoid A combined with different fragments of TcdB (the enzymatic domain + different fragments of the C-terminal domain) (Karczewski et al. 2014). To reduce the antigenic cocktail, fragments can be fused, and the larger fusion protein may benefit of an increased immunogenicity. Tian et al. constructed a recombinant fusion protein composed of TcdA and TcdB RBD fragments (Tian et al. 2012). This fusion protein was shown to be immunogenic in mice after i.m. immunization, and the produced antibodies were able to neutralize toxin cytotoxicity in vitro. In addition, full protection against *C. difficile* challenge was observed in the hamster model. Wang et al. constructed a chimeric atoxic toxin constituted of inactivated GTD and TD of TcdB and the RBD of TcdA (cTxAB) (Wang et al. 2012). A rapid and potent neutralizing antibody response against both toxins was induced after parenteral immunization of mice with cTxAB adjuvanted with alum. cTxAB parenteral immunizations protected mice from a primary infection and relapses as well, thus conferring a long-lasting protection against *C. difficile*.

With the emergence of hypervirulent strains such as BI/NAP1/027 strains producing the binary toxin, it could be of interest to broaden vaccine-protective efficacy. Secore et al. (2017) recently described a tetravalent vaccine composed of recombinant inactivated TcdA and TcdB and binary toxin components CDTa and CDTb adjuvanted with ISCOMATRIX i.m.-administered in hamsters. The addition of CDT to TcdA and TcdB significantly improved vaccine efficacy against BI/NAP1/027 strains. Interestingly, they observed that this tetravalent vaccine was able to elicit neutralizing antibodies against the three toxins in hamsters and in *Rhesus macaques*. Another group targeted TcdB variants, TcdA and CDT (Tian et al. 2017). They produced two fusion proteins, a trivalent one with CDTb/$TcdB_{(003)}$/TcdA (T-toxin) and a quadrivalent one with CTDb/$TcdB_{(003)}$/TcdA/$TcdB_{(027)}$ (Q-toxin). They i.m. immunized mice and hamsters with T-toxin or Q-toxin with alum as adjuvant and showed that this vaccination induced toxin-neutralizing antibodies to each of the toxins and a broad protection in hamsters against *C. difficile* 630 (ribotype 003) and *C. difficile* ribotype 027 strains. Taken together, these studies showed the interest of multivalent fusion proteins as vaccine antigens.

To increase antigen immunogenicity, polypeptides with immunoadjuvant properties can be fused to the targeted antigens. For instance, Ghose et al. constructed a fusion protein constituted of the *Salmonella enterica* serovar *Typhimurium flagellin* subunit D1 as an innate immune agonist and the RBDs of TcdA or TcdB as targeted antigen (Ghose et al. 2013). After i.p. immunization, mice displayed more anti-TcdA IgA and the same level of anti TcdB IgA in stool than mice immunized with unfused antigen adjuvanted with alum or heat-labile enterotoxin. Concerning the circulating IgG response, immunization with the fusion protein induced a higher anti-TcdA response than non-adjuvanted unfused antigens. However, the same level of anti-TcdA response was observed with the fusion protein and adjuvanted unfused antigens.

Recently, Liu et al. used biodegradable nanoparticles composed of poly-γ-glutamic acid (γ-PGA) and chitosan as an antigen carrier for their vaccine development. After construction of a recombinant rTcdB protein of the TcdB receptor-binding domain and its encapsulation in γ-PGA and chitosan nanoparticles, three cycles of intraperitoneal vaccination led to high antibody responses against TcdB and provided mice with complete protection against a lethal dose of *C. difficile* spores. They were able to observe that the protection was associated with high levels of toxin-neutralizing antibodies, and the rTcdBs encapsulated by the nanoparticles elicited a longer-lasting antibody titers than the antigen with the conventional adjuvant, aluminum hydroxide. Significant reductions in the level of pro-inflammatory cytokines and chemokines were observed in vaccinated mice. These results suggest that nanoparticle-based vaccine design may be useful in the development of vaccines against *C. difficile* infections (Liu et al. 2017).

Parenteral vaccination with nontoxic recombinant vaccines showed promising results in animal models. Of note, these fragment vaccines have several advantages compared to toxoid vaccine, such as elimination of the potential risk of incomplete toxoid inactivation, large-scale production in a cost-effective way, and decrease of batch-to-batch variations (Wang et al. 2015a).

In the third generation of vaccine, vectorized antigens and DNA vaccines targeting *C. difficile* were investigated. DNA vaccines are versatile, stable, and easy to produce. DNA vaccines facilitate antigen presentation and enable proper protein folding for correct epitope presentation. In addition, DNA vaccines have the ability to induce both humoral and cellular immune responses and a good immune priming (Saade and Petrovsky 2012). Gardiner et al. were the first to test DNA vaccination against *C. difficile* (Gardiner et al. 2009). A synthetic gene TxA-RBD optimized for expression in human cells was constructed. Mice were inoculated by electroporation with the TxA-RBD expressing plasmid and then challenged with a lethal dose of purified TcdA. Treated mice were fully protected and presented a strong serum anti-TcdA IgG antibody response. Jin et al. (2013) screened the immunogenicity of various toxin fragments by DNA vaccination. Antibody response was elicited by two DNA vaccines, one expressing fragment of the TcdA RBD (TcdA-C) and the other expressing the TcdA catalytic N-terminal domain (Tcd-N). Passive transfer in mice of immune serum elicited with both TcdA-C and TcdB-N fully protected mice against a lethal dose of *C. difficile* concentrated culture filtrate. Baliban et al. constructed an optimized DNA vaccine encoding the RBD of TcdA and TcdB and showed that i.m. followed by electroporation in mice and non-human primates was able to promote a strong serum IgG but not IgA response associated with neutralizing IgG antibodies to both toxins in blood and interestingly in stool (Baliban et al. 2014). In addition, immunized mice were significantly protected against *C. difficile*.

Zhang et al. constructed highly optimized plasmids to express the receptor binding domains of TcdA and TcdB from a single vector. This DNA vaccine has been evaluated in two animal models to test its immunogenicity and protective effects. They were able to observe high levels of serum antibodies against toxin A and/or B and demonstrated neutralizing activity in both in vitro and in vivo systems. In the hamster model, immunization with the DNA vaccine reduced the severity of infection and conferred significant protection against a lethal strain of *C. difficile* (Zhang et al. 2016).

Seregin et al. designed an adenovirus-based vaccine targeting TcdA (Seregin et al. 2012). This vaccine induced a rapid and strong antibody response and a T cell response against TcdA, which led to full protection in mice after *C. difficile* challenge.

Matchett et al. developed a novel single-cycle adenovirus (SC-Ad) vaccine against *C. difficile* expressing the RBD of TcdA and TcdB. Single immunization of mice generated sustained toxin-binding antibody responses and protected them from lethal toxin challenge for 38 weeks. Immunized Syrian hamsters produced significant toxin-neutralizing antibodies that increased over 36 weeks, and a single intramuscular immunization provided complete protection against lethal BI/NAP1/027 spore challenge 45 weeks later (Matchett et al. 2020).

All these three generations of parenteral vaccines targeting toxins aim to induce a systemic response. However, these studies did not report the induction of a parallel mucosal immune response. Even if antitoxin-circulating IgG antibodies against *C. difficile* can be protective, they are less likely effective locally on the early step of infection. In contrast, mucosal IgA response would be more likely able to rapidly act locally on the infection site.

Mucosal Immunization in Animal Models

Torres et al. aimed to identify the best route of immunization for a protective vaccine against *C. difficile* in hamsters. They compared the mucosal (intra-nasal i.n., intra-rectal i.r., intra-gastric i. g.), parenteral (s.c., i.p.), and a combination of mucosal and parenteral routes (i.n. and i.p.) with formalin-inactivated culture filtrate of toxigenic *C. difficile* with CT as adjuvant for mucosal immunizations and RIBI for parenteral immunizations (Torres et al. 1995). Immunizations via i.n., s.c., and i.p. routes led to full protection. Similarly, Giannasca et al. in a vaccination strategy targeting toxins tested several immunization routes. The optimal protection was obtained with combined i.m. and i.r. routes (Giannasca et al. 1999).

Inducing mucosal immunity via the mucosal route encounters many difficulties. Antigens have to cross the mucosal surface to be uptaken by the immune cells. In addition, the mucosal immune system is closely interacting with the intestinal microbiota resulting in an important regulation and immune tolerance (Chen and Cerutti 2010; Xiong and Hu 2015). Adjuvants and vectorization are key factors to modulate the mucosal immune system and develop a mucosal vaccine (Lavelle 2005).

Even if after i.n. immunization the intestinal mucosa can be stimulated through mucosal homing, it is not the optimal route to induce an intestinal antibody response compared to the oral route. Antigen vectorization is recommended for oral immunizations. Ryan et al. used a live attenuated bacterial vector for oral immunization (Ryan et al. 1997). A live attenuated *V. cholerae* strain was used, expressing a fusion protein consisting of the C-terminal RBD of TcdA fused to the secretion signal of *E. coli* hemolysin A as secretion system, co-administered with CT as adjuvant to orally immunize rabbits. Vaccination induced an anti-TcdA IgG response but did not significantly induce an IgA response. However, in an ileal loop challenge assay, this vaccination was protective against TcdA.

More recently, Winter et al. provided proof of concept by considering a locally invasive but highly attenuated *Salmonella enterica* serovar *Typhimurium* YS1646 vector expressing a fusion protein consisting of the *S. Typhimurium* secretion signals, and either toxin A or toxin B RBD could be even more effective in inducing local and systemic anti-RBD responses. When given in a five-day multimodal regimen (i.m. one time, p.o. three times), these candidate vaccines elicited high serum IgG titers and provided complete protection against lethal challenge in a mouse model. Although the amount of IgA present in the intestinal tissues after vaccination was relatively low, they found that the induction of an effective local immune response by these vaccines was strongly supported by the fact that oral vaccination alone provided substantial protection, despite the absence of detectable serum antibodies prior to challenge (Winter et al. 2019).

Permpoonpattana et al. used a *Bacillus subtilis* spores as vehicle to orally deliver the carboxy-terminal repeat domains of TcdA alone (amino acids 2388–2706) or with TcdB (amino acids 2137–2366) in mouse and hamster vaccination assays (Permpoonpattana et al. 2011). Such a strategy was able to induce a neutralizing and protective mucosal IgA and systemic IgG response. Of note, neutralizing sIgA antibodies to the TcdA repeat domain were shown to be cross-reactive with the analogous domain of TcdB. In this study, antibodies against TcdA provide protection against challenge with A+/B+ toxigenic *C. difficile* strains. Recently, Hong et al. expressed in *Bacillus subtilis* spores the same carboxy-terminal domain of TcdA (A26-39) and immunized hamsters by oral and sublingual routes (Hong et al. 2017). Hamsters were protected after challenge with *C. difficile* strain 630. In addition, protection was associated with the absence of *C. difficile* toxins and spores in fecal samples and high level of anti $TcdA_{26\text{-}39}$ fecal IgA and circulating IgG. The authors suggest that antibodies induced by $TcdA_{26\text{-}39}$ cross-react with seemingly unrelated proteins expressed on the vegetative cell surface or spore coat of *C. difficile*.

Wang et al. constructed a protein (mTcd138), composed of the glucosyltransferase and cysteine protease domains of TcdB and the RBD of TcdA to develop an oral vaccine that can target both *C. difficile* toxins and colonization/adhesion factors (Wang et al. 2015b). After expressing mTcd138 in nontoxigenic *C. difficile* (NTCD), oral immunization with NTCD_mTcd138 spores completely protected mice from the infection with the hypervirulent strain UK6 (ribotype 027). In addition, the immunization significantly protected hamster against a lethal dose of UK6 (Wang et al. 2018). More recently, they generated a new chimeric protein (Tcd169), composed of the GT and CP domains and the RBDs of TcdB and TcdA, and observed that after parenteral immunizations with this protein, mice were effectively protected from infection by *C. difficile R20291*. As before, they expressed Tcd169 in an NTCD to develop an oral vaccine that can target both *C. difficile* toxins and other *C. difficile* antigen. Oral immunizations with NTCD_Tcd169 spores induced systemic and mucosal antibody responses against both toxins but also against FliC and FliD. Interestingly, anti-Tcd169 sera showed significant cross-reactivity to FliC and FliD but also SlpA and Cwp2 (another S-layer protein). Oral immunizations with NTCD_Tcd169 spores provided mice with effective protection against *C. difficile R20291* infection and significantly reduced the number of spores in feces compared to NTCD or PBS immunized mice (Wang et al. 2022).

Another approach developed by Guo et al. used a *Lactococcus lactis* strain to express the TcdA and TcdB RBD (Guo et al. 2015). Live vaccines benefit from the ideal features of an effective mucosal delivery system such as antigen protection from degradation, antigen delivery to mucosal surface, antigen uptake by target cells, and expression of potent immunostimulatory molecules. After oral administration in mice, animals were partially protected from *C. difficile* challenge, and this protection was shown to be positively correlated with an IgG- and sIgA-specific response in immunized mice.

4.1.2 In Humans and Clinical Trials

The positive results of animal vaccination assays constitute a proof of concept for the development of human vaccines (Table 1).

Parenteral Immunization in Humans and Clinical Trials

Safety, immunogenicity, and dose response of the highly purified formalin-inactivated toxoid A and B vaccine, previously tested in hamsters, have been assessed in human volunteers. After i.m. injection with alum as adjuvant, vaccination was well tolerated, and volunteers developed neutralizing antitoxin antibodies in serum (Kotloff et al. 2001). Then, Sougioultzis et al. tested this vaccine in three patients with recurrent CDI (Sougioultzis et al. 2005). After vaccination, the patients discontinued treatment with oral vancomycin without any further recurrence.

Sanofi Pasteur had have started to develop a vaccine targeting TcdA and B (CDIFFENSE^{TM}). Unfortunately, results reported from a recent phase III multicenter *C. difficile* toxoid (TcdA and TcdB) vaccine trial (NCT01887912, Sanofi Pasteur, 9302 participants in 23 countries) were not encouraging (de Bruyn et al. 2021). Subjects included were adults ≥50 years old considered to be at increased risk of CDI. The candidate vaccine was unable to reduce the incidence of symptomatic CDI in the first efficacy analysis (34/6173 vs. 16/3085 cases of CDI in the vaccine and placebo groups, respectively). Clinical development of the vaccine candidate was stopped.

In parallel, Pfizer (USA) develops a genetically modified full-length TcdA and TcdB toxoid vaccine. A phase I clinical trial in healthy adults (50–85 years of age) has been completed and assessed safety and immunogenicity by testing a three-dose vaccination regimen by i.m. route with one of three dose levels of *C difficile* vaccine with or without alum as an adjuvant (NCT01706367). Only mild to moderate local reactions and systemic events were observed. No clear dose-level response pattern was detected. Vaccination induced potent antitoxin neutralizing immune response in all groups that lasted at least for 12 months (Sheldon et al. 2016). A phase II clinical trial assessed the safety and tolerability of the three-dose vaccine in healthy adults aged 50–85 years (NCT02117570). Among volunteers who were 65–85 years of age, redness around the vaccination area (injection site erythema) was the most common medical problem. Six volunteers, out of the total of 121 in the two vaccine groups, experienced severe redness. None of them withdrew from the study as a result of the reaction. However, following this finding, the sponsor decided not to carry out stage 3 of the study. Another phase II trial to study the safety, tolerability, and immune response of subjects to the vaccine has been completed. Two different vaccination schedules were compared (low dose and high dose). Each subject initially received three doses of vaccine or placebo, and then one year after the third dose, the subjects who had not received a placebo were randomized to receive a fourth dose. Subjects were followed for up to four years after their third vaccination (NCT02561195). The *C. difficile* vaccine has been shown to be safe, well tolerated, and immunogenic in healthy US adults ages 65–85.

Table 1 Vaccines in clinical development

Vaccine sponsor	Clinical trial	Mode of administration	Outcome measures	Results	References ClinicalTrials.gov
Genetically modified full-length TcdA and TcdB toxoids *Pfizer*	Phase I (completed) Healthy adult volunteers (50–85 years)	i.m. 3 dose injection different doses with/ without adjuvant (alum) versus placebo	Safety, and immunogenicity	Generally safe and well tolerated No clear dose response Good immunogenicity	NCT01706367 NCT02052726 Sheldon et al. (2016)
	Phase II (completed) Healthy adult volunteers (50–85 years)	i.m. three doses (days 1, 8, and 30) high dose, low dose versus placebo (three arms)	Safety, tolerability and immune response	Severe redness at the site of injection Sponsor decided not to carry out stage 3 of the study	NCT02117570
	Phase II (completed) Healthy adults 65–85 years	i.m. three doses on 1 of 2 schedule versus placebo (six arms)	Safety, tolerability, and immune response	Safe, well tolerated, and immunogenic Robust immune responses in the 200 μg monthly diet group	NCT02561195 Kitchin et al. (2020)
	Phase III (completed) Adults ≥50 years	Vaccine versus placebo (two arms)	Efficacy: CDI and recurrence	Reduce the CDI severity Vaccine 100% effective in preventing medically assisted CDI Did not meet its pre-specified primary endpoint of prevention of primary CDI	NCT03090191
	Phase III (completed) Healthy adults 65–85 years	Three vaccines lot versus placebo (four arms)	Assess the lot consistency, safety, and tolerability Immune response	Consistent lots and immunogenic Generally safe and tolerated	NCT03579459
Recombinant fusion protein consisting of truncated TcdA and TcdB VLA84 (formerly IC84 Intercell) *Valneva* Austria	Phase Ia/Ib (completed) Ia Healthy subjects 18–65 years Ib Elderly ≥65 years	i.m. four injections (days 0, 7, 28, and 56) of two different doses with or without adjuvant (alum)	Safety, immunogenicity, and dose response	Good safety and tolerability Highly immunogenic for TcdA and TcdB	NCT01296386 Bezay et al. (2016)
	Phase II (completed) 500 healthy adults ≥50 years	i.m. injections (days 0, 7, and 28) different doses with or without alum versus placebo	Dose confirmation, immunogenicity, and safety	High seroconversion for antibodies against toxins A and B and toxins alone	NCT02316470
F2 antigen (GSK2904545A) adjuvanted with	Phase I (completed) Healthy	i.m. injections according to a zero- and one-month	Generate data on safety, reactogenicity,	No results posted	NCT04026009

(continued)

Table 1 (continued)

Vaccine sponsor	Clinical trial	Mode of administration	Outcome measures	Results	References ClinicalTrials.gov
AS01B *GlaxoSmithKline*	adults 18–45 years and 50–70 years	schedule with or without adjuvant AS01B versus placebo The third dose is administered 15 months after the second dose	and immunogenicity		
CDVAX inactivated *Bacillus* spores expressing a toxoid antigen and a spore colonization factor *Royal Holloway Univ. Cutting S.M.*	Phase I (terminated) Healthy adults 18–50 years	Oral vaccine	Safety, mucosal, and systemic immunogenicity	No results posted	NCT02991417

SOC standard of care antibiotic treatment

Immune responses were particularly robust in the group monthly treated with 200 μg of vaccine candidate (Kitchin et al. 2020). In March 2022, initial analyses of two protocol-defined secondary endpoints from the phase III trial (NCT03090191) CLOVER (*CLOstridium difficile* Vaccine Efficacy TRial) indicated a highly favorable benefit in reducing the CDI severity and 100% efficacy of the vaccine in preventing medically assisted CDI. However, the trial did not meet its prespecified primary endpoint of prevention of primary CDI. Safety reviews indicated that the experimental vaccine was safe and well tolerated. Another trial (NCT03579459) to assess batch consistency of *C. difficile* vaccines in healthy adults 65–85 years of age has been completed in January 2023.

Valneva (Austria) announced positive phase I results for its *C. difficile* vaccine candidate VLA84 (formerly IC84), a recombinant fusion protein consisting of truncated forms of TcdA and TcdB. Phase Ia/phase Ib trials showed good safety and tolerability profile of the vaccine (Bézay et al. 2016). VLA84 was highly immunogenic and was able to induce similar immune responses to TcdA and TcdB in adults and elderly subjects (NCT01296386). A dose confirmation, immunogenicity, and safety study in 500 healthy adults (≥50 years) in a phase II trial has been completed (NCT02316470). Again, the results showed that it induced seroconversion in up to 78% of participants for antibodies against toxins A and B, up to 97% against toxin A alone, and up to 84% against toxin B alone.

GlaxoSmithKline (GSK) has initiated a phase I study aimed at generating data on safety, reactogenicity (assessment of any expected or unexpected side effects of the vaccine), and immunogenicity (ability to induce an immune response) for the development of a vaccine candidate targeting *C. difficile* composed of the F2 antigen (GSK2904545A) with or without AS01B adjuvant. This vaccine aims to protect against primary cases of CDI and recurrence (NCT04026009).

All these three vaccines aim to induce a systemic antibody response against both toxins and showed promising results and could elicit long-lasting protection. However today, no vaccine has been approved for clinical use.

Mucosal Immunization in Humans and Clinical Trials

Currently, only one trial was performed and aimed to elicit both mucosal and systemic immune responses to *C. difficile*. A phase I clinical trial under the supervision of the Cutting S.M. (Royal Holloway University) and funded by the European Union 7th Framework Program assessed the safety and immunogenicity of an oral vaccine against *C. difficile* in healthy adults (CDVAX). Their approach was a novel mucosal vaccine delivery system based on the use of inactivated *Bacillus subtilis* spores that express two different recombinant *C. difficile* antigens on their surface, a toxoid antigen and a unique spore colonization factor. Adverse events were monitored, and specific mucosal and systemic immunity (sIgA, circulating IgA and IgG) were evaluated in this trial (NCT02991417). Further clinical research is needed to test the efficacy and safety of CDVAX.

4.2 Vaccines Targeting Surface Components

Several studies rationalized the concept of targeting surface components to develop vaccines (Mizrahi et al. 2014). To induce a specific immune response against *C. difficile* and limit cross-reactivity, vaccines target specific *C. difficile* surface components conserved among strains. Presently, all these vaccines are in preclinical development in animal models.

Parenteral Immunization in Animal Models

Several antigen candidates have been investigated for parenteral vaccination targeting *C. difficile* surface components.

After s.c. mouse immunization with nontoxigenic *C. difficile* membrane fraction adjuvanted in an oil emulsion (TiterMax Gold Adjuvant), Senoh et al. induced a specific IgG and IgA response in sera and intestinal fluids, respectively. Interestingly, hyperimmune sera and intestinal fluids were able to inhibit *C. difficile* adhesion in vitro to human intestinal Caco2 cells (Senoh et al. 2015).

One approach aims to target *C. difficile* surface proteins involved in bacterial gut colonization and participating to *C. difficile* pathogenesis. For instance, Ní Eidhin et al. i.p. immunized hamsters with crude SLPs from a clinical *C. difficile* strain (PCR ribotype 001) with either alum or RIBI as adjuvant (Ní Eidhin et al. 2008). Hamsters immunized with alum mounted a strong IgG response whereas hamsters immunized with RIBI mounted a weak IgG response. However, all immunized hamsters developed diarrhea and died after *C. difficile* challenge. Another study successfully tested *C. difficile* flagellin FliC in i.p. vaccination assay (Ghose et al. 2016b). Ghose et al. i.p. immunized mice and hamsters with recombinant FliC adjuvanted with alum. As expected, immunization induced a high systemic anti-FliC IgG response in mice. In addition, immunized mice were fully protected against a clinical epidemic 027 strain (UK1) whereas immunized hamsters were partially protected against strain 630Δ*erm*. Surface spore proteins have also been tested as vaccine antigen. Indeed, spores may play an essential role in persistence of *C. difficile* in the intestinal tract. Ghose et al. tested several spore proteins and i.p. immunized mice with the exosporium CdeC or CdeM proteins with alum as adjuvant. They observed a full protection in mice against the 027, UK1 strain. This protection was correlated with circulating specific IgG and a significant decrease of the level of spore shedding compared to controls (Ghose et al. 2016a). In addition, these spore proteins were also able to afford an 80% protection against 630Δ*erm* in vaccinated hamsters.

Another approach is to target highly specific antigens abundantly present on *C. difficile* surface, such as cell wall polysaccharides (Monteiro et al. 2013; Monteiro 2016). Oberli et al. (2011) and Romano et al. (2014) demonstrated PS-II immunogenicity in mice with, respectively, diphtheria toxoid (CRM_{197}) or recombinant TcdA and TcdB fragments as carrier protein, to make PS immunogenic. PS-I was shown immunogenic in

CDI patients as anti-PS-I IgG in serum and anti-PS-I IgA were found in stool. Mice immunized with synthetic PS-I-CRM$_{197}$ conjugate adjuvanted with either alum or Freund's adjuvant produced specific anti PS-I IgG, IgM, and IgA in blood (Martin et al. 2013). In addition, Martin et al. identified a minimal epitope in PS-I, which is the disaccharide RhA(1-3)-Glc. Then, the same group (Broecker et al. 2016a) constructed a pentavalent glycoconjugate based on the PS-I minimal epitope with an increased antigenicity. Of note, this synthetic pentavalent vaccine candidate elicited a weak but highly specific IgG response to native PS-I glycan in mice, but vaccine efficacy was not tested against *C. difficile* colonization.

C. difficile LTA (also named PS-III) has been shown to be conserved in *C. difficile* strains. Cox et al. (2013) constructed different glycoconjugates, and immunization of rabbit and mice induced a specific serum IgG response to *C. difficile* live vegetative cells and spores. Broecker et al. (2016b) constructed a semi-synthetic LTA-CRM$_{197}$ glycoconjugate that elicited anti-LTA IgG in mice with or without alum adjuvant. Anti-LTA antibodies recognized *C. difficile* surface and significantly limited bacterial mouse gut colonization five days post challenge.

Mucosal Immunization in Animal Models

A mucosal immunization targeting surface proteins compared to parenteral immunization aims to induce locally an immune response against *C. difficile* correlated to a decrease of bacterial gut colonization. Several vaccine candidates have been tested via mucosal routes in animal models.

After i.r. immunization of human microbiota gnotobiotic mice with a *C. difficile* toxin-free cell wall extract adjuvanted with CT and after *C. difficile* challenge, a significant decrease of bacterial gut colonization in immunized mice compared to controls was observed (Péchiné et al. 2007). Pechiné et al. evaluated *C. difficile* surface proteins as mucosal vaccine candidates in this mouse model (Péchiné et al. 2007). First, the flagellar cap protein FliD was used to determine the best mucosal route of immunization between i.r., i.g., and i.n. The best immune response was induced with FliD and CT as adjuvant via the i.r. route, leading to systemic anti-FliD IgG and mucosal sIgA response. Then, to prevent *C. difficile* gut colonization, mice were i.r. immunized with CT as adjuvant with flagellar preparation containing FliC and FliD or an association of Cwp84 and FliD as antigens. All immunized groups showed a significant decrease of intestinal colonization from day 13 after challenge. In another study aiming to identify surface antigens in mucosal vaccination, Péchiné et al. i.r. immunized hamsters with a cell wall extract of a nontoxigenic *C. difficile* strain adjuvanted with CT (Péchiné et al. 2013). A partial protection of hamsters (33%) against a lethal dose of *C. difficile* was observed. Using a comparative proteomic analysis between sera from protected immunized animals and sera from the control group, three proteins have been identified as key factors leading to production of protective antibodies: the chaperon protein DnaK, the heat shock protein GroEL, and the S-layer protein precursor SlpA. GroEL was used as antigen to i.n. immunize mice with CT as adjuvant and was able to induce a systemic anti-GroEL IgG response associated with a significant decrease of bacterial colonization from day 8 after *C. difficile* challenge. We also tested the SlpA precursor as vaccine antigen (Bruxelle et al. 2016). Immunization (i.r.) of mice with SlpA as antigen and CT as adjuvant induced a systemic anti-SlpA IgG and a mucosal sIgA response. Furthermore, this regimen induced a trend in decrease of *C. difficile* intestinal colonization significant at day 10 after challenge. In the hamster model, this immunization regimen led to a partial and non-lasting protection against *C. difficile*. In comparison, Ní Eidhin et al. (2008) immunized hamsters with crude SLP extract. They tested different vaccination regimens with different antigen doses, different adjuvants, and different routes of immunization (i.p. prime with RIBI adjuvant, i.n. prime with CT, i.n. boost with CT). The best protection was obtained in hamsters immunized with the i.p. prime and i.n. boost combination; two hamsters out of three survived after a lethal challenge. In mice, this regimen induced a strong circulating anti-SLP IgG and IgA response.

Immunization (i.r., i.n., or s.c.) of hamsters with Cwp84 as antigen and CT or Freund complete as adjuvant has been performed (Péchiné et al. 2011). The best protection was obtained with the rectal route and CT as adjuvant (40% greater survival in the i.r. immunized group compared to a control group). Surprisingly, this protection was not correlated with circulating anti-Cwp84 antibodies. Then, oral immunization was performed in hamsters with Cwp84 encapsulated in pectin beads for colonic delivery. A similar partial protection (40%) was obtained, with no correlation to systemic antibody response. These results in the hamster model support the role of other mechanisms of protection in parallel of the circulating antibody response such as innate immunity and mucosal immune response.

C. difficile colonization is multifactorial, and combination of several surface components in vaccine is likely necessary to obtain a full protection against colonization (Crobach et al. 2018). A better knowledge of the colonization process and the mucosal immune response against *C. difficile* will improve mucosal vaccine development.

CD0873 is a surface-exposed lipoprotein and an adhesin of *C. difficile*. Immunization of conventional mice with recombinant CD0873 by intraperitoneal route resulted in a prevented long-term gut colonization and was correlated with a strong secretory IgA immune response (Bradshaw et al. 2019). Following the initial identification of CD0873 as a potential interesting vaccine candidate, this antigen was tested in a hamster model. CD0873 of *C. difficile* given orally in enteric capsules to hamsters induced local and serum-neutralizing antibody responses which afforded partial protection against infection with a hypervirulent strain. Following challenge with the R20291*ermB* *C. difficile* strain, the CD0873-immunized group showed a partial protection by a mean increase of 80% in time to experimental endpoint compared to naïve animals (Karyal et al. 2021a). The same authors tested oral delivery of the whole protein CD0873 displayed on the outer membrane of liposomal nanoparticles. They observed that this liposomal formulation induced a greater neutralizing antibody response than CD0873 given alone (Karyal et al. 2021b). Furthermore, another study investigated the interest of CD0873 as vaccine candidate. In nontoxigenic *C. difficile* strain T7, CD0873 was overexpressed. Vaccination of hamsters with spores of this recombinant strain T7-0873 administered by the oral route successfully induced intestinal antibodies. This immune response induced significantly reduced adhesion of toxigenic *C. difficile* to Caco-2 cells, and these responses were mirrored in sera. Unfortunately, no challenge with a toxigenic *C. difficile* strain has been performed after immunization regimen (Hughes et al. 2022).

Spores are major players of the infection and could be targets for vaccination. In this context, spore surface proteins of *C. difficile* could be considered as potential antigens. Maia et al. reported that the C-terminal domain of the spore surface protein BclA3 ($BclA3_{CTD}$) was identified as an antigenic epitope, overproduced in *E. coli* and tested as an immunogen in mice. To increase antigen stability and efficiency, $BclA3_{CTD}$ was also exposed on the surface of *B. subtilis* spores used as a mucosal vaccine delivery system. Mice were intranasally immunized. Administrations of the recombinant protein $BclA3_{CTD}$ induced antibody production and attenuated some *C. difficile* infection symptoms after a challenge with the toxigenic strain R20291 of *C. difficile*, while the spore-displayed antigen resulted less effective (Maia et al. 2020).

Interestingly, a multi-epitope vaccine was designed using computer methods. Two target proteins, CdeC, affecting spore germination, and FliD, affecting vegetative forms, were selected to construct a vaccine candidate so that it could simultaneously induce the immune response against two different forms of *C. difficile*. The antigenicity, toxicity, allergenicity, and other physicochemical properties of the vaccine were checked. The results of molecular docking and MD simulation showed that the vaccine could stably bind to TLRs and MHC molecules (Tan et al. 2022).

Another original in silico approach has been used for designing a multivalent chimeric vaccine consisting of several colonization factors including CotE, SlpA, and FliC proteins. The overall reliability of this candidate vaccine was validated in silico, and the molecular dynamic simulation verified the stability of the vaccine designed (Basak et al. 2021).

5 Conclusion

Highly specific and protective, the antibody response to *C. difficile* is closely associated with the outcome of the infection and the control of recurrences. Several strategies have been developed to prevent or treat CDI (Fig. 1).

One strategy is based on passive immunizations with various Aps; another is to develop vaccines. In both passive and active immunizations, the key virulence factors TcdA and TcdB were the first vaccine antigens. Then, surface components of the vegetative cells and the spores were also studied as promising candidates. The first assays have been performed in animal models and have paved the way to development in humans.

Today, bezlotoxumab (Zinplava ™) is the only AP against *C. difficile* i.v. administered approved for clinical use and indicated to reduce recurrence of CDI in patients who received SOC antibiotic treatment for CDI and are at high risk of recurrences. The main advantage of passive immunization with specific MAbs consists in its rapidity of protection capable to complement the poor host immune response. However, this strategy is expensive and does not confer a long-term protection. Vaccines display advantages such as long-term protection but depend on the faculty of the host to develop an immune response. It is well known that immune response is impaired in elderly who are particularly at risk of CDI.

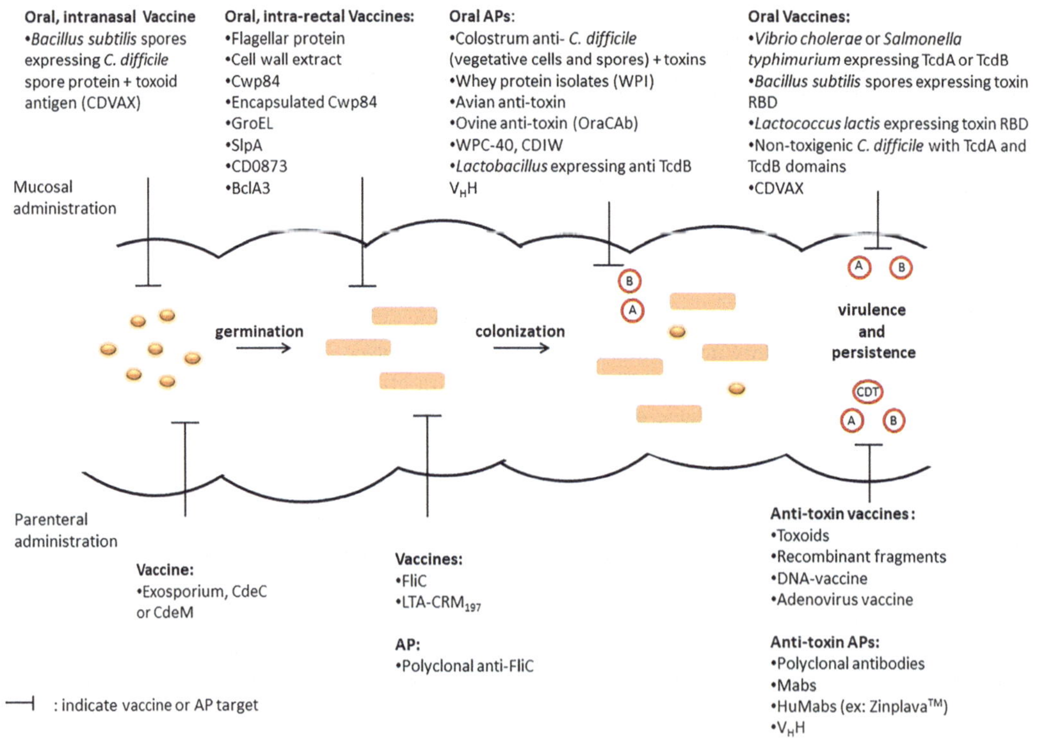

Fig. 1 Mucosal and parenteral immunization strategies against *C. difficile*

Several parenteral vaccines targeting both toxins TcdA and TcdB are tested in clinical trials (two in phase III). It is likely that parenteral toxin vaccines will be approved soon for human use.

The mucosal and systemic immune responses have both their role in the protection against CDI. Mucosal immunization displays advantages such as the local induction of an innate and adaptive immune response. Several assays have been performed in animal models either with toxin antigens or colonization factor antigens. A phase I clinical trial has been completed with bacillus spores expressing a toxin fragment and a spore protein. These promising assays should be confirmed. Surface components of the vegetative cells also led to promising results in animal models. However, a combination of various colonization factors seems necessary to reach full protection. In addition, the combination of toxin antigens with colonization factors antigens has the advantage to inhibit the two steps of the pathogenic process, colonization and toxin release. Future research should focus on development of novel immunologic strategies including systemic and mucosal vaccines targeting both virulence and colonization factors.

References

Alonso CD, Mahoney MV (2018) Bezlotoxumab for the prevention of *Clostridium difficile* infection: a review of current evidence and safety profile. Infect Drug Resist 12:1–9. https://doi.org/10.2147/IDR.S159957

Andersen KK, Strokappe NM, Hultberg A, Truusalu K, Smidt I, Mikelsaar R-H, Mikelsaar M, Verrips T, Hammarström L, Marcotte H (2015) Neutralization of *Clostridium difficile* toxin B mediated by engineered lactobacilli that produce single-domain antibodies. Infect Immun 84:395–406. https://doi.org/10.1128/IAI.00870-15

Anosova NG, Brown AM, Li L, Liu N, Cole LE, Zhang J, Mehta H, Kleanthous H (2013) Systemic antibody responses induced by a two-component *Clostridium difficile* toxoid vaccine protect against *C. difficile*-associated disease in hamsters. J Med Microbiol 62: 1394–1404. https://doi.org/10.1099/jmm.0.056796-0

Anosova NG, Cole LE, Li L, Zhang J, Brown AM, Mundle S, Zhang J, Ray S, Ma F, Garrone P, Bertraminelli N, Kleanthous H, Anderson SF (2015) A combination of three fully human toxin A- and toxin B-specific monoclonal antibodies protects against challenge with highly virulent epidemic strains of *Clostridium difficile* in the hamster model. Clin Vaccine Immunol CVI 22:711–725. https://doi.org/10.1128/CVI.00763-14

Arato V, Gasperini G, Giusti F, Ferlenghi I, Scarselli M, Leuzzi R (2019) Dual role of the colonization factor CD2831 in *Clostridium difficile* pathogenesis. Sci Rep 9:5554. https://doi.org/10.1038/s41598-019-42000-8

Aronsson B, Granström M, Möllby R, Nord CE (1985) Serum antibody response to *Clostridium difficile* toxins in patients with *Clostridium difficile* diarrhoea. Infection 13:97–101

Aubry A, Zou W, Vinogradov E, Williams D, Chen W, Harris G, Zhou H, Schur MJ, Gilbert M, Douce GR, Logan SM (2020) In vitro production and immunogenicity of a *Clostridium difficile* spore-specific BclA3 glycopeptide conjugate vaccine. Vaccines 8:73. https://doi.org/10.3390/vaccines8010073

Baban ST, Kuehne SA, Barketi-Klai A, Cartman ST, Kelly ML, Hardie KR, Kansau I, Collignon A, Minton NP (2013) The role of flagella in *Clostridium difficile* pathogenesis: comparison between a non-epidemic and an epidemic strain. PLoS One 8:e73026. https://doi.org/10.1371/journal.pone.0073026

Babcock GJ, Broering TJ, Hernandez HJ, Mandell RB, Donahue K, Boatright N, Stack AM, Lowy I, Graziano R, Molrine D, Ambrosino DM, Thomas WD (2006) Human monoclonal antibodies directed against toxins A and B prevent *Clostridium difficile*-induced mortality in hamsters. Infect Immun 74:6339–6347. https://doi.org/10.1128/IAI.00982-06

Baliban SM, Michael A, Shammassian B, Mudakha S, Khan AS, Cocklin S, Zentner I, Latimer BP, Bouillaut L, Hunter M, Marx P, Sardesai NY, Welles SL, Jacobson JM, Weiner DB, Kutzler MA (2014) An optimized, synthetic DNA vaccine encoding the toxin A and toxin B receptor binding domains of *Clostridium difficile* induces protective antibody responses in vivo. Infect Immun 82:4080–4091. https://doi.org/10.1128/IAI.01950-14

Barketi-Klai A, Hoys S, Lambert-Bordes S, Collignon A, Kansau I (2011) Role of fibronectin-binding protein A in *Clostridium difficile* intestinal colonization. J Med Microbiol 60:1155–1161. https://doi.org/10.1099/jmm.0.029553-0

Barra-Carrasco J, Olguin-Araneda V, Plaza-Garrido A, Miranda-Cardenas C, Cofre-Araneda G, Pizarro-Guajardo M, Sarker MR, Paredes-Sabja D (2013) The *Clostridium difficile* exosporium cysteine (CdeC)-rich protein is required for exosporium morphogenesis and coat assembly. J Bacteriol 195:3863–3875. https://doi.org/10.1128/JB.00369-13

Basak S, Deb D, Narsaria U, Kar T, Castiglione F, Sanyal I, Bade PD, Srivastava AP (2021) In silico designing of vaccine candidate against *Clostridium difficile*. Sci Rep 11:14215. https://doi.org/10.1038/s41598-021-93305-6

Batah J, Denève-Larrazet C, Jolivot P-A, Kuehne S, Collignon A, Marvaud J-C, Kansau I (2016)

Clostridium difficile flagella predominantly activate TLR5-linked NF-κB pathway in epithelial cells. Anaerobe 38:116–124. https://doi.org/10.1016/j.anaerobe.2016.01.002

Batah J, Kobeissy H, Pham PTB, Denève-Larrazet C, Kuehne S, Collignon A, Janoir-Jouveshomme C, Marvaud J-C, Kansau I (2017) *Clostridium difficile* flagella induce a pro-inflammatory response in intestinal epithelium of mice in cooperation with toxins. Sci Rep 7:3256. https://doi.org/10.1038/s41598-017-03621-z

Belyi IF, Varfolomeeva NA (2003) Construction of a fusion protein carrying antigenic determinants of enteric clostridial toxins. FEMS Microbiol Lett 225: 325–329

Best EL, Freeman J, Wilcox MH (2012) Models for the study of *Clostridium difficile* infection. Gut Microbes 3:145–167. https://doi.org/10.4161/gmic.19526

Bézay N, Ayad A, Dubischar K, Firbas C, Hochreiter R, Kiermayr S, Kiss I, Pinl F, Jilma B, Westritschnig K (2016) Safety, immunogenicity and dose response of VLA84, a new vaccine candidate against *Clostridium difficile*, in healthy volunteers. Vaccine 34:2585–2592. https://doi.org/10.1016/j.vaccine.2016.03.098

Bradshaw WJ, Bruxelle J-F, Kovacs-Simon A, Harmer NJ, Janoir C, Péchiné S, Acharya KR, Michell SL (2019) Molecular features of lipoprotein CD0873: a potential vaccine against the human pathogen *Clostridioides difficile*. J Biol Chem 294:15850–15861. https://doi.org/10.1074/jbc.RA119.010120

Brock JH, Arzabe R, Piñeiro A, Olivito AM (1977) The effect of trypsin and chymotrypsin on the bactericidal activity and specific antibody activity of bovine colostrum. Immunology 32:207–213

Broecker F, Hanske J, Martin CE, Baek JY, Wahlbrink A, Wojcik F, Hartmann L, Rademacher C, Anish C, Seeberger PH (2016a) Multivalent display of minimal *Clostridium difficile* glycan epitopes mimics antigenic properties of larger glycans. Nat Commun 7:11224. https://doi.org/10.1038/ncomms11224

Broecker F, Martin CE, Wegner E, Mattner J, Baek JY, Pereira CL, Anish C, Seeberger PH (2016b) Synthetic lipoteichoic acid glycans are potential vaccine candidates to protect from *Clostridium difficile* infections. Cell Chem Biol 23:1014–1022. https://doi.org/10.1016/j.chembiol.2016.07.009

Bruxelle J-F (2017) Développement d'une stratégie vaccinale par voie muqueuse ciblant les protéines de surface de *Clostridium difficile*. PhD thesis, Université Paris-Saclay, NNT:2017SACLS355. tel-01984391

Bruxelle J-F, Mizrahi A, Hoys S, Collignon A, Janoir C, Péchiné S (2016) Immunogenic properties of the surface layer precursor of *Clostridium difficile* and vaccination assays in animal models. Anaerobe 37:78–84. https://doi.org/10.1016/j.anaerobe.2015.10.010

Cafardi V, Biagini M, Martinelli M, Leuzzi R, Rubino JT, Cantini F, Norais N, Scarselli M, Serruto D, Unnikrishnan M (2013) Identification of a novel zinc metalloprotease through a global analysis of *Clostridium difficile* extracellular proteins. PLoS One 8:e81306. https://doi.org/10.1371/journal.pone.0081306

Calabi E, Calabi F, Phillips AD, Fairweather NF (2002) Binding of *Clostridium difficile* surface layer proteins to gastrointestinal tissues. Infect Immun 70:5770–5778

Carter GP, Chakravorty A, Pham Nguyen TA, Mileto S, Schreiber F, Li L, Howarth P, Clare S, Cunningham B, Sambol SP, Cheknis A, Figueroa I, Johnson S, Gerding D, Rood JI, Dougan G, Lawley TD, Lyras D (2015) Defining the roles of TcdA and TcdB in localized gastrointestinal disease, systemic organ damage, and the host response during *Clostridium difficile* infections. mBio 6:e00551. https://doi.org/10.1128/mBio.00551-15

Cerquetti M, Pantosti A, Stefanelli P, Mastrantonio P (1992) Purification and characterization of an immunodominant 36 kDa antigen present on the cell surface of *Clostridium difficile*. Microb Pathog 13: 271–279

Chapetón Montes D, Collignon A, Janoir C (2013) Influence of environmental conditions on the expression and the maturation process of the *Clostridium difficile* surface associated protease Cwp84. Anaerobe 19:79–82. https://doi.org/10.1016/j.anaerobe.2012.12.004

Chebly H, Marvaud J-C, Safa L, Elkak AK, Kobeissy PH, Kansau I, Larrazet C (2022) *Clostridioides difficile* flagellin activates the intracellular NLRC4 inflammasome. Int J Mol Sci 23:12366. https://doi.org/10.3390/ijms232012366

Chen J, Gong CL, Hitchcock MM, Holubar M, Deresinski S, Hay JW (2021) Cost-effectiveness of bezlotoxumab and fidaxomicin for initial *Clostridioides difficile* infection. Clin Microbiol Infect 27:1448–1454. https://doi.org/10.1016/j.cmi.2021.04.004

Chen K, Cerutti A (2010) Vaccination strategies to promote mucosal antibody responses. Immunity 33:479–491. https://doi.org/10.1016/j.immuni.2010.09.013

Chiari EF, Weiss W, Simon MR, Kiessig ST, Pulse M, Brown SC, Gerding HR, Mandago M, Gisch K, von Eichel-Streiber C (2021) Oral immunotherapy with human secretory immunoglobulin A improves survival in the hamster model of *Clostridioides difficile* infection. J Infect Dis 224:1394–1397. https://doi.org/10.1093/infdis/jiab087

Corthier G, Muller MC, Wilkins TD, Lyerly D, L'Haridon R (1991) Protection against experimental pseudomembranous colitis in gnotobiotic mice by use of monoclonal antibodies against *Clostridium difficile* toxin A. Infect Immun 59:1192–1195

Cox AD, St Michael F, Aubry A, Cairns CM, Strong PCR, Hayes AC, Logan SM (2013) Investigating the candidacy of a lipoteichoic acid-based glycoconjugate as a vaccine to combat *Clostridium difficile* infection. Glycoconj J 30:843–855. https://doi.org/10.1007/s10719-013-9489-3

Crobach MJT, Vernon JJ, Loo VG, Kong LY, Péchiné S, Wilcox MH, Kuijper EJ (2018) Understanding

Clostridium difficile colonization. Clin Microbiol Rev 31:e00021–e00017. https://doi.org/10.1128/CMR.00021-17

Dang THT, de la Riva L, Fagan RP, Storck EM, Heal WP, Janoir C, Fairweather NF, Tate EW (2010) Chemical probes of surface layer biogenesis in *Clostridium difficile*. ACS Chem Biol 5:279–285. https://doi.org/10.1021/cb9002859

Dapa T, Unnikrishnan M (2013) Biofilm formation by *Clostridium difficile*. Gut Microbes 4:397–402. https://doi.org/10.4161/gmic.25862

Davies NL, Compson JE, MacKenzie B, O'Dowd VL, Oxbrow AKF, Heads JT, Turner A, Sarkar K, Dugdale SL, Jairaj M, Christodoulou L, Knight DEO, Cross AS, Hervé KJM, Tyson KL, Hailu H, Doyle CB, Ellis M, Kriek M, Cox M, Page MJT, Moore AR, Lightwood DJ, Humphreys DP (2013) A mixture of functionally oligoclonal humanized monoclonal antibodies that neutralize *Clostridium difficile* TcdA and TcdB with high levels of in vitro potency shows in vivo protection in a hamster infection model. Clin Vaccine Immunol 20:377–390. https://doi.org/10.1128/CVI.00625-12

de Bruyn G, Gordon DL, Steiner T, Tambyah P, Cosgrove C, Martens M, Bassily E, Chan E-S, Patel D, Chen J, Torre-Cisneros J, Fernando De Magalhães Francesconi C, Gesser R, Jeanfreau R, Launay O, Laot T, Morfin-Otero R, Oviedo-Orta E, Park YS, Piazza FM, Rehm C, Rivas E, Self S, Gurunathan S (2021) Safety, immunogenicity, and efficacy of a *Clostridioides difficile* toxoid vaccine candidate: a phase 3 multicentre, observer-blind, randomised, controlled trial. Lancet Infect Dis 21:252–262. https://doi.org/10.1016/S1473-3099(20)30331-5

de Rham O, Isliker H (1977) Proteolysis of bovine immunoglobulins. Int Arch Allergy Appl Immunol 55:61–69

De Roo AC, Regenbogen SE (2020) *Clostridium difficile* infection: an epidemiology update. Clin Colon Rectal Surg 33:049–057. https://doi.org/10.1055/s-0040-1701229

Dingle T, Wee S, Mulvey GL, Greco A, Kitova EN, Sun J, Lin S, Klassen JS, Palcic MM, Ng KKS, Armstrong GD (2008) Functional properties of the carboxy-terminal host cell-binding domains of the two toxins, TcdA and TcdB, expressed by *Clostridium difficile*. Glycobiology 18:698–706. https://doi.org/10.1093/glycob/cwn048

Dingle TC, Mulvey GL, Armstrong GD (2011) Mutagenic analysis of the *Clostridium difficile* flagellar proteins, FliC and FliD, and their contribution to virulence in hamsters. Infect Immun 79:4061–4067. https://doi.org/10.1128/IAI.05305-11

Diraviyam T, He J-X, Chen C, Zhao B, Michael A, Zhang X (2016) Effect of passive immunotherapy against *Clostridium difficile* infection: a systematic review and meta-analysis. Immunotherapy 8:649–663. https://doi.org/10.2217/imt.16.8

Drudy D, Calabi E, Kyne L, Sougioultzis S, Kelly E, Fairweather N, Kelly CP (2004) Human antibody response to surface layer proteins in *Clostridium difficile* infection. FEMS Immunol Med Microbiol 41:237–242. https://doi.org/10.1016/j.femsim.2004.03.007

Eckert C, Emirian A, Le Monnier A, Cathala L, De Montclos H, Goret J, Berger P, Petit A, De Chevigny A, Jean-Pierre H, Nebbad B, Camiade S, Meckenstock R, Lalande V, Marchandin H, Barbut F (2015) Prevalence and pathogenicity of binary toxin-positive *Clostridium difficile* strains that do not produce toxins A and B. New Microbes New Infect 3:12–17. https://doi.org/10.1016/j.nmni.2014.10.003

Eidhin DN, Ryan AW, Doyle RM, Walsh JB, Kelleher D (2006) Sequence and phylogenetic analysis of the gene for surface layer protein, slpA, from 14 PCR ribotypes of *Clostridium difficile*. J Med Microbiol 55:69–83. https://doi.org/10.1099/jmm.0.46204-0

Fagarasan S, Honjo T (2003) Intestinal IgA synthesis: regulation of front-line body defences. Nat Rev Immunol 3:63–72. https://doi.org/10.1038/nri982

Fernie DS, Thomson RO, Batty I, Walker PD (1983) Active and passive immunization to protect against antibiotic associated caecitis in hamsters. Dev Biol Stand 53:325–332

Finn E, Andersson FL, Madin-Warburton M (2021) Burden of *Clostridioides difficile* infection (CDI) – a systematic review of the epidemiology of primary and recurrent CDI. BMC Infect Dis 21:456. https://doi.org/10.1186/s12879-021-06147-y

Ganeshapillai J, Vinogradov E, Rousseau J, Weese JS, Monteiro MA (2008) *Clostridium difficile* cell-surface polysaccharides composed of pentaglycosyl and hexaglycosyl phosphate repeating units. Carbohydr Res 343:703–710. https://doi.org/10.1016/j.carres.2008.01.002

Gardiner DF, Rosenberg T, Zaharatos J, Franco D, Ho DD (2009) A DNA vaccine targeting the receptor-binding domain of *Clostridium difficile* toxin A. Vaccine 27:3598–3604. https://doi.org/10.1016/j.vaccine.2009.03.058

Ghose C, Eugenis I, Edwards AN, Sun X, McBride SM, Ho DD (2016a) Immunogenicity and protective efficacy of *Clostridium difficile* spore proteins. Anaerobe 37:85–95. https://doi.org/10.1016/j.anaerobe.2015.12.001

Ghose C, Eugenis I, Sun X, Edwards AN, McBride SM, Pride DT, Kelly CP, Ho DD (2016b) Immunogenicity and protective efficacy of recombinant *Clostridium difficile* flagellar protein FliC. Emerg Microbes Infect 5:e8. https://doi.org/10.1038/emi.2016.8

Ghose C, Verhagen JM, Chen X, Yu J, Huang Y, Chenesseau O, Kelly CP, Ho DD (2013) Toll-like receptor 5-dependent immunogenicity and protective efficacy of a recombinant fusion protein vaccine containing the nontoxic domains of *Clostridium difficile* toxins A and B and *Salmonella enterica serovar typhimurium* flagellin in a mouse model of *Clostridium difficile* disease. Infect Immun 81:2190–2196. https://doi.org/10.1128/IAI.01074-12

Giannasca PJ, Zhang ZX, Lei WD, Boden JA, Giel MA, Monath TP, Thomas WD Jr (1999) Serum antitoxin antibodies mediate systemic and mucosal protection from *Clostridium difficile* disease in hamsters. Infect Immun 67:527–538

Guo S, Yan W, McDonough SP, Lin N, Wu KJ, He H, Xiang H, Yang M, Moreira MAS, Chang Y-F (2015) The recombinant *Lactococcus lactis* oral vaccine induces protection against *C. difficile* spore challenge in a mouse model. Vaccine 33:1586–1595. https://doi.org/10.1016/j.vaccine.2015.02.006

Heidebrecht H-J, Weiss WJ, Pulse M, Lange A, Gisch K, Kliem H, Mann S, Pfaffl MW, Kulozik U, von Eichel-Streiber C (2019) Treatment and prevention of recurrent *Clostridium difficile* infection with functionalized bovine antibody-enriched whey in a hamster primary infection model. Toxins 11:98. https://doi.org/10.3390/toxins11020098

Hennequin C, Janoir C, Barc M-C, Collignon A, Karjalainen T (2003) Identification and characterization of a fibronectin-binding protein from *Clostridium difficile*. Microbiol Read Engl 149:2779–2787

Hennequin C, Porcheray F, Waligora-Dupriet A, Collignon A, Barc M, Bourlioux P, Karjalainen T (2001) GroEL (Hsp60) of *Clostridium difficile* is involved in cell adherence. Microbiol Read Engl 147: 87–96

Hensbergen PJ, Klychnikov OI, Bakker D, van Winden VJ, Ras N, Kemp AC, Cordfunke RA, Dragan I, Deelder AM, Kuijper EJ, others (2014) A novel secreted metalloprotease (CD2830) from *Clostridium difficile* cleaves specific proline sequences in LPXTG cell surface proteins. Mol Cell Proteomics 13:1231–1244

Hernández Del Pino RE, Barbero AM, Español LÁ, Morro LS, Pasquinelli V (2021) The adaptive immune response to *Clostridioides difficile*: a tricky balance between immunoprotection and immunopathogenesis. J Leukoc Biol 109:195–210. https://doi.org/10.1002/JLB.4VMR0720-201R

Hernandez LD, Racine F, Xiao L, DiNunzio E, Hairston N, Sheth PR, Murgolo NJ, Therien AG (2015) Broad coverage of genetically diverse strains of *Clostridium difficile* by actoxumab and bezlotoxumab predicted by in vitro neutralization and epitope modeling. Antimicrob Agents Chemother 59: 1052–1060. https://doi.org/10.1128/AAC.04433-14

Hong HA, Hitri K, Hosseini S, Kotowicz N, Bryan D, Mawas F, Wilkinson AJ, van Broekhoven A, Kearsey J, Cutting SM (2017) Mucosal antibodies to the C terminus of toxin A prevent colonization of *Clostridium difficile*. Infect Immun 85. https://doi.org/10.1128/IAI.01060-16

Hughes J, Aston C, Kelly ML, Griffin R (2022) Towards development of a non-toxigenic *Clostridioides difficile* oral spore vaccine against toxigenic *C. difficile*. Pharmaceutics 14:1086. https://doi.org/10.3390/pharmaceutics14051086

Hussack G, Arbabi-Ghahroudi M, van Faassen H, Songer JG, Ng KK-S, MacKenzie R, Tanha J (2011) Neutralization of *Clostridium difficile* toxin A with single-domain antibodies targeting the cell receptor binding domain. J Biol Chem 286:8961–8976. https://doi.org/10.1074/jbc.M110.198754

Hussack G, Ryan S, van Faassen H, Rossotti M, MacKenzie CR, Tanha J (2018) Neutralization of *Clostridium difficile* toxin B with VHH-Fc fusions targeting the delivery and CROPs domains. PLoS One 13: e0208978. https://doi.org/10.1371/journal.pone.0208978

Hutton ML, Cunningham BA, Mackin KE, Lyon SA, James ML, Rood JI, Lyras D (2017) Bovine antibodies targeting primary and recurrent *Clostridium difficile* disease are a potent antibiotic alternative. Sci Rep 7: 3665. https://doi.org/10.1038/s41598-017-03982-5

Islam J, Taylor AL, Rao K, Huffnagle G, Young VB, Rajkumar C, Cohen J, Papatheodorou P, Aronoff DM, Llewelyn MJ (2014) The role of the humoral immune response to *Clostridium difficile* toxins A and B in susceptibility to *C. difficile infection*: a case-control study. Anaerobe 27:82–86. https://doi.org/10.1016/j.anaerobe.2014.03.011

Jank T, Aktories K (2008) Structure and mode of action of clostridial glucosylating toxins: the ABCD model. Trends Microbiol 16:222–229. https://doi.org/10.1016/j.tim.2008.01.011

Janoir C (2016) Virulence factors of *Clostridium difficile* and their role during infection. Anaerobe 37:13–24. https://doi.org/10.1016/j.anaerobe.2015.10.009

Janoir C, Péchiné S, Grosdidier C, Collignon A (2007) Cwp84, a surface-associated protein of *Clostridium difficile*, is a cysteine protease with degrading activity on extracellular matrix proteins. J Bacteriol 189:7174–7180. https://doi.org/10.1128/JB.00578-07

Jin K, Wang S, Zhang C, Xiao Y, Lu S, Huang Z (2013) Protective antibody responses against *Clostridium difficile* elicited by a DNA vaccine expressing the enzymatic domain of toxin B. Hum Vaccines Immunother 9:63–73. https://doi.org/10.4161/hv.22434

Johal SS, Lambert CP, Hammond J, James PD, Borriello SP, Mahida YR (2004) Colonic IgA producing cells and macrophages are reduced in recurrent and non-recurrent *Clostridium difficile* associated diarrhoea. J Clin Pathol 57:973–979. https://doi.org/10.1136/jcp.2003.015875

Johnson S, Gerding DN, Janoff EN (1992) Systemic and mucosal antibody responses to toxin A in patients infected with *Clostridium difficile*. J Infect Dis 166: 1287–1294

Kandalaft H, Hussack G, Aubry A, van Faassen H, Guan Y, Arbabi-Ghahroudi M, MacKenzie R, Logan SM, Tanha J (2015) Targeting surface-layer proteins with single-domain antibodies: a potential therapeutic approach against *Clostridium difficile*-associated disease. Appl Microbiol Biotechnol:1–14. https://doi.org/10.1007/s00253-015-6594-1

Karczewski J, Zorman J, Wang S, Miezeiewski M, Xie J, Soring K, Petrescu I, Rogers I, Thiriot DS, Cook JC, Chamberlin M, Xoconostle RF, Nahas DD, Joyce JG, Bodmer J-L, Heinrichs JH, Secore S (2014) Development of a recombinant toxin fragment vaccine for *Clostridium difficile* infection. Vaccine 32:2812–2818. https://doi.org/10.1016/j.vaccine.2014.02.026

Karjalainen T, Waligora-Dupriet AJ, Cerquetti M, Spigaglia P, Maggioni A, Mauri P, Mastrantonio P (2001) Molecular and genomic analysis of genes encoding surface-anchored proteins from *Clostridium difficile*. Infect Immun 69:3442–3446. https://doi.org/10.1128/IAI.69.5.3442-3446.2001

Karyal C, Hughes J, Kelly ML, Luckett JC, Kaye PV, Cockayne A, Minton NP, Griffin R (2021a) Colonisation factor CD0873, an attractive oral vaccine candidate against *Clostridioides difficile*. Microorganisms 9:306. https://doi.org/10.3390/microorganisms9020306

Karyal C, Palazi P, Hughes J, Griffiths RC, Persaud RR, Tighe PJ, Mitchell NJ, Griffin R (2021b) Mimicking native display of CD0873 on liposomes augments its potency as an oral vaccine against *Clostridioides difficile*. Vaccines 9:1453. https://doi.org/10.3390/vaccines9121453

Kelly CP, Pothoulakis C, Orellana J, LaMont JT (1992) Human colonic aspirates containing immunoglobulin A antibody to *Clostridium difficile* toxin A inhibit toxin A-receptor binding. Gastroenterology 102:35–40

Kim PH, Iaconis JP, Rolfe RD (1987) Immunization of adult hamsters against *Clostridium difficile*-associated ileocecitis and transfer of protection to infant hamsters. Infect Immun 55:2984–2992

Kink JA, Williams JA (1998) Antibodies to recombinant *Clostridium difficile* toxins A and B are an effective treatment and prevent relapse of *C. difficile*-associated disease in a hamster model of infection. Infect Immun 66:2018–2025

Kirby JM, Ahern H, Roberts AK, Kumar V, Freeman Z, Acharya KR, Shone CC (2009) Cwp84, a surface-associated cysteine protease, plays a role in the maturation of the surface layer of *Clostridium difficile*. J Biol Chem 284:34666–34673. https://doi.org/10.1074/jbc.M109.051177

Kitchin N, Remich SA, Peterson J, Peng Y, Gruber WC, Jansen KU, Pride MW, Anderson AS, Knirsch C, Webber C (2020) A phase 2 study evaluating the safety, tolerability, and immunogenicity of two 3-dose regimens of a *Clostridium difficile* vaccine in healthy US adults aged 65 to 85 years. Clin Infect Dis 70:1–10. https://doi.org/10.1093/cid/ciz153

Kotloff KL, Wasserman SS, Losonsky GA, Thomas W Jr, Nichols R, Edelman R, Bridwell M, Monath TP (2001) Safety and immunogenicity of increasing doses of a *Clostridium difficile* toxoid vaccine administered to healthy adults. Infect Immun 69:988–995. https://doi.org/10.1128/IAI.69.2.988-995.2001

Kovacs-Simon A, Leuzzi R, Kasendra M, Minton N, Titball RW, Michell SL (2014) Lipoprotein CD0873 is a novel adhesin of *Clostridium difficile*. J Infect Dis 210:274–284. https://doi.org/10.1093/infdis/jiu070

Kuehne SA, Cartman ST, Heap JT, Kelly ML, Cockayne A, Minton NP (2010) The role of toxin A and toxin B in *Clostridium difficile* infection. Nature 467:711–713. https://doi.org/10.1038/nature09397

Kuehne SA, Collery MM, Kelly ML, Cartman ST, Cockayne A, Minton NP (2014) Importance of toxin A, toxin B, and CDT in virulence of an epidemic *Clostridium difficile* strain. J Infect Dis 209:83–86. https://doi.org/10.1093/infdis/jit426

Kyne L, Warny M, Qamar A, Kelly CP (2000) Asymptomatic carriage of *Clostridium difficile* and serum levels of IgG antibody against toxin A. N Engl J Med 342:390–397. https://doi.org/10.1056/NEJM200002103420604

Kyne L, Warny M, Qamar A, Kelly CP (2001) Association between antibody response to toxin A and protection against recurrent *Clostridium difficile* diarrhoea. Lancet Lond Engl 357:189–193. https://doi.org/10.1016/S0140-6736(00)03592-3

Lam SW, Neuner EA, Fraser TG, Delgado D, Chalfin DB (2018) Cost-effectiveness of three different strategies for the treatment of first recurrent *Clostridium difficile* infection diagnosed in a community setting. Infect Control Hosp Epidemiol 39:924–930. https://doi.org/10.1017/ice.2018.139

Lavelle EC (2005) Generation of improved mucosal vaccines by induction of innate immunity. Cell Mol Life Sci CMLS 62:2750–2770. https://doi.org/10.1007/s00018-005-5290-1

Lawson PA, Citron DM, Tyrrell KL, Finegold SM (2016) Reclassification of *Clostridium difficile* as *Clostridioides difficile* (Hall and O'Toole 1935) Prévot 1938. Anaerobe 40:95–99. https://doi.org/10.1016/j.anaerobe.2016.06.008

Leav BA, Blair B, Leney M, Knauber M, Reilly C, Lowy I, Gerding DN, Kelly CP, Katchar K, Baxter R, Ambrosino D, Molrine D (2010) Serum anti-toxin B antibody correlates with protection from recurrent *Clostridium difficile* infection (CDI). Vaccine 28:965–969. https://doi.org/10.1016/j.vaccine.2009.10.144

Leung DYM, Kelly CP, Boguniewicz M, Pothoulakis C, LaMont JT, Flores A (1991) Treatment with intravenously administered gamma globulin of chronic relapsing colitis induced by *Clostridium difficile* toxin. J Pediatr 118:633–637. https://doi.org/10.1016/S0022-3476(05)83393-1

Leuzzi R, Spencer J, Buckley A, Brettoni C, Martinelli M, Tulli L, Marchi S, Luzzi E, Irvine J, Candlish D, Veggi D, Pansegrau W, Fiaschi L, Savino S, Swennen E, Cakici O, Oviedo-Orta E, Giraldi M, Baudner B, D'Urzo N, Maione D, Soriani M, Rappuoli R, Pizza M, Douce GR, Scarselli M (2013) Protective efficacy induced by recombinant

Clostridium difficile toxin fragments. Infect Immun 81: 2851–2860. https://doi.org/10.1128/IAI.01341-12
Libby JM, Jortner BS, Wilkins TD (1982) Effects of the two toxins of *Clostridium difficile* in antibiotic-associated cecitis in hamsters. Infect Immun 36:822–829
Libby JM, Wilkins TD (1982) Production of antitoxins to two toxins of *Clostridium difficile* and immunological comparison of the toxins by cross-neutralization studies. Infect Immun 35:374–376
Liu Y-W, Chen Y-H, Chen J-W, Tsai P-J, Huang I-H (2017) Immunization with recombinant TcdB-encapsulated nanocomplex induces protection against *Clostridium difficile* challenge in a mouse model. Front Microbiol 8:1411. https://doi.org/10.3389/fmicb.2017.01411
Lowy I, Molrine DC, Leav BA, Blair BM, Baxter R, Gerding DN, Nichol G, Thomas WD Jr, Leney M, Sloan S, Hay CA, Ambrosino DM (2010) Treatment with monoclonal antibodies against *Clostridium difficile* toxins. N Engl J Med 362:197–205. https://doi.org/10.1056/NEJMoa0907635
Lyerly DDM, Johnson JL, Frey SM, Wilkins TD (1990) Vaccination against lethal *Clostridium difficile* enterocolitis with a nontoxic recombinant peptide of toxin A. Curr Microbiol 21:29–32. https://doi.org/10.1007/BF02090096
Lyerly DM, Bostwick EF, Binion SB, Wilkins TD (1991) Passive immunization of hamsters against disease caused by *Clostridium difficile* by use of bovine immunoglobulin G concentrate. Infect Immun 59:2215–2218
Lyras D, O'Connor JR, Howarth PM, Sambol SP, Carter GP, Phumoonna T, Poon R, Adams V, Vedantam G, Johnson S, Gerding DN, Rood JI (2009) Toxin B is essential for virulence of *Clostridium difficile*. Nature 458:1176–1179. https://doi.org/10.1038/nature07822
Maia AR, Reyes-Ramírez R, Pizarro-Guajardo M, Saggese A, Ricca E, Baccigalupi L, Paredes-Sabja D (2020) Nasal immunization with the C-terminal domain of Bcla3 induced specific igg production and attenuated disease symptoms in mice infected with *Clostridioides difficile* spores. Int J Mol Sci 21:6696. https://doi.org/10.3390/ijms21186696
Maldarelli GA, Matz H, Gao S, Chen K, Hamza T, Yfantis HG, Feng H, Donnenberg MS (2016) Pilin vaccination stimulates weak antibody responses and provides no protection in a C57Bl/6 murine model of acute *Clostridium difficile* infection. J Vaccines Vaccin 7. https://doi.org/10.4172/2157-7560.1000321
Martin CE, Broecker F, Oberli MA, Komor J, Mattner J, Anish C, Seeberger PH (2013) Immunological evaluation of a synthetic *Clostridium difficile* oligosaccharide conjugate vaccine candidate and identification of a minimal epitope. J Am Chem Soc 135:9713–9722. https://doi.org/10.1021/ja401410y
Matchett WE, Anguiano-Zarate S, Malewana GBR, Mudrick H, Weldy M, Evert C, Khoruts A, Sadowsky M, Barry MA (2020) A replicating single-cycle adenovirus vaccine effective against *Clostridium difficile*. Vaccines 8:470. https://doi.org/10.3390/vaccines8030470
Mattila E, Anttila V-J, Broas M, Marttila H, Poukka P, Kuusisto K, Pusa L, Sammalkorpi K, Dabek J, Koivurova O-P, Vähätalo M, Moilanen V, Widenius T (2008) A randomized, double-blind study comparing *Clostridium difficile* immune whey and metronidazole for recurrent *Clostridium difficile*-associated diarrhoea: efficacy and safety data of a prematurely interrupted trial. Scand J Infect Dis 40:702–708. https://doi.org/10.1080/00365540801964960
Maynard-Smith M, Ahern H, McGlashan J, Nugent P, Ling R, Denton H, Coxon R, Landon J, Roberts A, Shone C (2014) Recombinant antigens based on toxins A and B of *Clostridium difficile* that evoke a potent toxin-neutralising immune response. Vaccine 32:700–705. https://doi.org/10.1016/j.vaccine.2013.11.099
Mizrahi A, Bruxelle J-F, Péchiné S, Le Monnier A (2018) Prospective evaluation of the adaptive immune response to SlpA in *Clostridium difficile* infection. Anaerobe 54:164–168. https://doi.org/10.1016/j.anaerobe.2018.09.008
Mizrahi A, Collignon A, Péchiné S (2014) Passive and active immunization strategies against *Clostridium difficile* infections: state of the art. Anaerobe 30:210–219. https://doi.org/10.1016/j.anaerobe.2014.07.006
Monteiro MA (2016) The design of a *Clostridium difficile* carbohydrate-based vaccine. Methods Mol Biol Clifton NJ 1403:397–408. https://doi.org/10.1007/978-1-4939-3387-7_21
Monteiro MA, Ma Z, Bertolo L, Jiao Y, Arroyo L, Hodgins D, Mallozzi M, Vedantam G, Sagermann M, Sundsmo J, Chow H (2013) Carbohydrate-based *Clostridium difficile* vaccines. Expert Rev Vaccines 12: 421–431. https://doi.org/10.1586/erv.13.9
Navalkele BD, Chopra T (2018) Bezlotoxumab: an emerging monoclonal antibody therapy for prevention of recurrent *Clostridium difficile* infection. Biol Targets Ther 12:11–21. https://doi.org/10.2147/BTT.S127099
Negm OH, MacKenzie B, Hamed MR, Ahmad OJ, Shone CC, Humphreys DP, Ravi Acharya K, Loscher CE, Marszalowska I, Lynch M, Wilcox MH, Monaghan TM (2017) Protective antibodies against *Clostridium difficile* are present in intravenous immunoglobulin and are retained in humans following its administration. Clin Exp Immunol. https://doi.org/10.1111/cei.12946
Ní Eidhin DB, O'Brien JB, McCabe MS, Athié-Morales V, Kelleher DP (2008) Active immunization of hamsters against *Clostridium difficile* infection using surface-layer protein. FEMS Immunol Med Microbiol 52: 207–218. https://doi.org/10.1111/j.1574-695X.2007.00363.x
Nibbering B, Gerding DN, Kuijper EJ, Zwittink RD, Smits WK (2021) Host immune responses to *Clostridioides difficile*: toxins and beyond. Front Microbiol 12: 804949. https://doi.org/10.3389/fmicb.2021.804949

Oberli MA, Hecht M-L, Bindschädler P, Adibekian A, Adam T, Seeberger PH (2011) A possible oligosaccharide-conjugate vaccine candidate for *Clostridium difficile* is antigenic and immunogenic. Chem Biol 18:580–588. https://doi.org/10.1016/j.chembiol.2011.03.009

O'Brien JB, McCabe MS, Athié-Morales V, McDonald GSA, Ní Eidhin DB, Kelleher DP (2005) Passive immunisation of hamsters against *Clostridium difficile* infection using antibodies to surface layer proteins. FEMS Microbiol Lett 246:199–205. https://doi.org/10.1016/j.femsle.2005.04.005

Oksi J, Aalto A, Säilä P, Partanen T, Anttila V-J, Mattila E (2019) Real-world efficacy of bezlotoxumab for prevention of recurrent *Clostridium difficile* infection: a retrospective study of 46 patients in five university hospitals in Finland. Eur J Clin Microbiol Infect Dis 38:1947–1952. https://doi.org/10.1007/s10096-019-03630-y

Papatheodorou P, Carette JE, Bell GW, Schwan C, Guttenberg G, Brummelkamp TR, Aktories K (2011) Lipolysis-stimulated lipoprotein receptor (LSR) is the host receptor for the binary toxin *Clostridium difficile* transferase (CDT). Proc Natl Acad Sci 108:16422–16427. https://doi.org/10.1073/pnas.1109772108

Péchiné S, Collignon A (2016) Immune responses induced by *Clostridium difficile*. Anaerobe 41:68–78. https://doi.org/10.1016/j.anaerobe.2016.04.014

Péchiné S, Denève C, Le Monnier A, Hoys S, Janoir C, Collignon A (2011) Immunization of hamsters against *Clostridium difficile* infection using the Cwp84 protease as an antigen. FEMS Immunol Med Microbiol 63:73–81. https://doi.org/10.1111/j.1574-695X.2011.00832.x

Péchiné S, Gleizes A, Janoir C, Gorges-Kergot R, Barc M-C, Delmée M, Collignon A (2005a) Immunological properties of surface proteins of *Clostridium difficile*. J Med Microbiol 54:193–196

Péchiné S, Hennequin C, Boursier C, Hoys S, Collignon A (2013) Immunization using GroEL decreases *Clostridium difficile* intestinal colonization. PLoS One 8:e81112. https://doi.org/10.1371/journal.pone.0081112

Péchiné S, Janoir C, Boureau H, Gleizes A, Tsapis N, Hoys S, Fattal E, Collignon A (2007) Diminished intestinal colonization by *Clostridium difficile* and immune response in mice after mucosal immunization with surface proteins of *Clostridium difficile*. Vaccine 25:3946–3954. https://doi.org/10.1016/j.vaccine.2007.02.055

Péchiné S, Janoir C, Collignon A (2005b) Variability of *Clostridium difficile* surface proteins and specific serum antibody response in patients with *Clostridium difficile*-associated disease. J Clin Microbiol 43:5018–5025. https://doi.org/10.1128/JCM.43.10.5018 5025.2005

Péchiné S, Janoir C, Collignon A (2017) Emerging monoclonal antibodies against *Clostridium difficile* infection. Expert Opin Biol Ther 17:415–427. https://doi.org/10.1080/14712598.2017.1300655

Perelle S, Gibert M, Bourlioux P, Corthier G, Popoff MR (1997) Production of a complete binary toxin (actin-specific ADP-ribosyltransferase) by *Clostridium difficile* CD196. Infect Immun 65:1402–1407

Permpoonpattana P, Hong HA, Phetcharaburanin J, Huang J-M, Cook J, Fairweather NF, Cutting SM (2011) Immunization with Bacillus spores expressing toxin A peptide repeats protects against infection with *Clostridium difficile* strains producing toxins A and B. Infect Immun 79:2295–2302. https://doi.org/10.1128/IAI.00130-11

Pizarro-Guajardo M, Olguín-Araneda V, Barra-Carrasco J, Brito-Silva C, Sarker MR, Paredes-Sabja D (2014) Characterization of the collagen-like exosporium protein, BclA1, of *Clostridium difficile* spores. Anaerobe 25:18–30. https://doi.org/10.1016/j.anaerobe.2013.11.003

Pizarro-Guajardo M, Ravanal MC, Paez MD, Callegari E, Paredes-Sabja D (2018) Identification of *Clostridium difficile* immunoreactive spore proteins of the epidemic strain R20291. PROTEOMICS – Clin Appl 12:1700182. https://doi.org/10.1002/prca.201700182

Popoff MR, Geny B (2011) Rho/Ras-GTPase-dependent and -independent activity of clostridial glucosylating toxins. J Med Microbiol 60:1057–1069. https://doi.org/10.1099/jmm.0.029314-0

Qiu H, Cassan R, Johnstone D, Han X, Joyee AG, McQuoid M, Masi A, Merluza J, Hrehorak B, Reid R, Kennedy K, Tighe B, Rak C, Leonhardt M, Dupas B, Saward L, Berry JD, Nykiforuk CL (2016) Novel *Clostridium difficile* anti-toxin (TcdA and TcdB) humanized monoclonal antibodies demonstrate in vitro neutralization across a broad spectrum of clinical strains and in vivo potency in a hamster spore challenge model. PLoS One 11:e0157970. https://doi.org/10.1371/journal.pone.0157970

Raeisi H, Azimirad M, Nabavi-Rad A, Asadzadeh Aghdaei H, Yadegar A, Zali MR (2022) Application of recombinant antibodies for treatment of *Clostridioides difficile* infection: current status and future perspective. Front Immunol 13:972930. https://doi.org/10.3389/fimmu.2022.972930

Roberts A, McGlashan J, Al-Abdulla I, Ling R, Denton H, Green S, Coxon R, Landon J, Shone C (2012) Development and evaluation of an ovine antibody-based platform for treatment of *Clostridium difficile* infection. Infect Immun 80:875–882. https://doi.org/10.1128/IAI.05684-11

Roberts AK, Harris HC, Smith M, Giles J, Polak O, Buckley AM, Clark E, Ewin D, Moura IB, Spitall W, Shone CC, Wilcox M, Chilton C, Donev R (2020) A novel, orally delivered antibody therapy and its potential to prevent *Clostridioides difficile* infection in pre-clinical models. Front Microbiol 11:578903. https://doi.org/10.3389/fmicb.2020.578903

Romano MR, Leuzzi R, Cappelletti E, Tontini M, Nilo A, Proietti D, Berti F, Costantino P, Adamo R, Scarselli M (2014) Recombinant *Clostridium difficile* toxin fragments as carrier protein for PSII surface polysaccharide preserve their neutralizing activity. Toxins 6:1385–1396. https://doi.org/10.3390/toxins6041385

Ryan A, Lynch M, Smith SM, Amu S, Nel HJ, McCoy CE, Dowling JK, Draper E, O'Reilly V, McCarthy C, O'Brien J, Ní Eidhin D, O'Connell MJ, Keogh B, Morton CO, Rogers TR, Fallon PG, O'Neill LA, Kelleher D, Loscher CE (2011) A role for TLR4 in *Clostridium difficile* infection and the recognition of surface layer proteins. PLoS Pathog 7:e1002076. https://doi.org/10.1371/journal.ppat.1002076

Ryan ET, Butterton JR, Smith RN, Carroll PA, Crean TI, Calderwood SB (1997) Protective immunity against *Clostridium difficile* toxin A induced by oral immunization with a live, attenuated *Vibrio cholerae* vector strain. Infect Immun 65:2941–2949

Saade F, Petrovsky N (2012) Technologies for enhanced efficacy of DNA vaccines. Expert Rev Vaccines 11: 189–209. https://doi.org/10.1586/erv.11.188

Sauerborn M, Leukel P, von Eichel-Streiber C (1997) The C-terminal ligand-binding domain of *Clostridium difficile* toxin A (TcdA) abrogates TcdA-specific binding to cells and prevents mouse lethality. FEMS Microbiol Lett 155:45–54

Savelkoul HFJ, Ferro VA, Strioga MM, Schijns VEJC (2015) Choice and design of adjuvants for parenteral and mucosal vaccines. Vaccines 3:148–171. https://doi.org/10.3390/vaccines3010148

Schmidt DJ, Beamer G, Tremblay JM, Steele JA, Kim HB, Wang Y, Debatis M, Sun X, Kashentseva EA, Dmitriev IP, Curiel DT, Shoemaker CB, Tzipori S (2016) A tetraspecific VHH-based neutralizing antibody modifies disease outcome in three animal models of *Clostridium difficile* infection. Clin Vaccine Immunol CVI 23:774–784. https://doi.org/10.1128/CVI.00730-15

Schwan C, Kruppke AS, Nölke T, Schumacher L, Koch-Nolte F, Kudryashev M, Stahlberg H, Aktories K (2014) *Clostridium difficile* toxin CDT hijacks microtubule organization and reroutes vesicle traffic to increase pathogen adherence. Proc Natl Acad Sci 111:2313–2318. https://doi.org/10.1073/pnas.1311589111

Schwan C, Stecher B, Tzivelekidis T, van Ham M, Rohde M, Hardt W-D, Wehland J, Aktories K (2009) *Clostridium difficile* toxin CDT induces formation of microtubule-based protrusions and increases adherence of bacteria. PLoS Pathog 5:e1000626. https://doi.org/10.1371/journal.ppat.1000626

Secore S, Wang S, Doughtry J, Xie J, Miezeiewski M, Rustandi RR, Horton M, Xoconostle R, Wang B, Lancaster C, Kristopeit A, Wang S-C, Christanti S, Vitelli S, Gentile M-P, Goerke A, Skinner J, Strable E, Thiriot DS, Bodmer J-L, Heinrichs JH (2017) Development of a novel vaccine containing binary toxin for the prevention of *Clostridium difficile* disease with enhanced efficacy against NAP1 strains. PLoS One 12:e0170640. https://doi.org/10.1371/journal.pone.0170640

Senoh M, Iwaki M, Yamamoto A, Kato H, Fukuda T, Shibayama K (2015) Inhibition of adhesion of *Clostridium difficile* to human intestinal cells after treatment with serum and intestinal fluid isolated from mice immunized with nontoxigenic *C. difficile* membrane fraction. Microb Pathog 81:1–5. https://doi.org/10.1016/j.micpath.2015.03.001

Seregin SS, Aldhamen YA, Rastall DPW, Godbehere S, Amalfitano A (2012) Adenovirus-based vaccination against *Clostridium difficile* toxin A allows for rapid humoral immunity and complete protection from toxin A lethal challenge in mice. Vaccine 30:1492–1501. https://doi.org/10.1016/j.vaccine.2011.12.064

Sheedlo MJ, Anderson DM, Thomas AK, Lacy DB (2020) Structural elucidation of the *Clostridioides difficile* transferase toxin reveals a single-site binding mode for the enzyme. Proc Natl Acad Sci 117:6139–6144. https://doi.org/10.1073/pnas.1920555117

Sheldon E, Kitchin N, Peng Y, Eiden J, Gruber W, Johnson E, Jansen KU, Pride MW, Pedneault L (2016) A phase 1, placebo-controlled, randomized study of the safety, tolerability, and immunogenicity of a *Clostridium difficile* vaccine administered with or without aluminum hydroxide in healthy adults. Vaccine 34:2082–2091. https://doi.org/10.1016/j.vaccine.2016.03.010

Shields K, Araujo-Castillo RV, Theethira TG, Alonso CD, Kelly CP (2015) Recurrent *Clostridium difficile* infection: from colonization to cure. Anaerobe 34:59–73. https://doi.org/10.1016/j.anaerobe.2015.04.012

Sougioultzis S, Kyne L, Drudy D, Keates S, Maroo S, Pothoulakis C, Giannasca PJ, Lee CK, Warny M, Monath TP, Kelly CP (2005) *Clostridium difficile* toxoid vaccine in recurrent *C. difficile*-associated diarrhea. Gastroenterology 128:764–770

Spencer J, Leuzzi R, Buckley A, Irvine J, Candlish D, Scarselli M, Douce GR (2014) Vaccination against *Clostridium difficile* using toxin fragments: observations and analysis in animal models. Gut Microbes 5:225–232. https://doi.org/10.4161/gmic.27712

Sponseller JK, Steele JA, Schmidt DJ, Kim HB, Beamer G, Sun X, Tzipori S (2014) Hyperimmune bovine colostrum as a novel therapy to combat *Clostridium difficile* infection. J Infect Dis jiu605. https://doi.org/10.1093/infdis/jiu605

Stevenson E, Minton NP, Kuehne SA (2015) The role of flagella in *Clostridium difficile* pathogenicity. Trends Microbiol 23:275–282. https://doi.org/10.1016/j.tim.2015.01.004

Stieglitz F, Gerhard R, Pich A (2021) The binary toxin of clostridioides difficile alters the proteome and phosphoproteome of HEp-2 cells. Front Microbiol 12:725612. https://doi.org/10.3389/fmicb.2021.725612

Strong PCR, Fulton KM, Aubry A, Foote S, Twine SM, Logan SM (2014) Identification and characterization of glycoproteins on the spore surface of *Clostridium difficile*. J Bacteriol 196:2627–2637. https://doi.org/10.1128/JB.01469-14

Sulea T, Hussack G, Ryan S, Tanha J, Purisima EO (2018) Application of Assisted Design of Antibody and Protein Therapeutics (ADAPT) improves efficacy of a *Clostridium difficile* toxin A single-domain antibody. Sci Rep 8:2260. https://doi.org/10.1038/s41598-018-20599-4

Tan C, Zhu F, Xiao Y, Wu Y, Meng X, Liu S, Liu T, Chen S, Zhou J, Li C, Wu A (2022) Immunoinformatics approach toward the introduction of a novel multi-epitope vaccine against *Clostridium difficile*. Front Immunol 13:887061. https://doi.org/10.3389/fimmu.2022.887061

Tasteyre A, Barc MC, Collignon A, Boureau H, Karjalainen T (2001) Role of FliC and FliD flagellar proteins of *Clostridium difficile* in adherence and gut colonization. Infect Immun 69:7937–7940. https://doi.org/10.1128/IAI.69.12.7937-7940.2001

Tasteyre A, Karjalainen T, Avesani V, Delmée M, Collignon A, Bourlioux P, Barc MC (2000) Phenotypic and genotypic diversity of the flagellin gene (fliC) among *Clostridium difficile* isolates from different serogroups. J Clin Microbiol 38:3179–3186

Taylor CP, Tummala S, Molrine D, Davidson L, Farrell RJ, Lembo A, Hibberd PL, Lowy I, Kelly CP (2008) Open-label, dose escalation phase I study in healthy volunteers to evaluate the safety and pharmacokinetics of a human monoclonal antibody to *Clostridium difficile* toxin A. Vaccine 26:3404–3409. https://doi.org/10.1016/j.vaccine.2008.04.042

Tian J-H, Fuhrmann SR, Kluepfel-Stahl S, Carman RJ, Ellingsworth L, Flyer DC (2012) A novel fusion protein containing the receptor binding domains of *C. difficile* toxin A and toxin B elicits protective immunity against lethal toxin and spore challenge in preclinical efficacy models. Vaccine 30:4249–4258. https://doi.org/10.1016/j.vaccine.2012.04.045

Tian J-H, Glenn G, Flyer D, Zhou B, Liu Y, Sullivan E, Wu H, Cummings JF, Ellllingsworth L, Smith G (2017) *Clostridium difficile* chimeric toxin receptor binding domain vaccine induced protection against different strains in active and passive challenge models. Vaccine. https://doi.org/10.1016/j.vaccine.2017.06.062

Torres JF, Lyerly DM, Hill JE, Monath TP (1995) Evaluation of formalin-inactivated *Clostridium difficile* vaccines administered by parenteral and mucosal routes of immunization in hamsters. Infect Immun 63:4619–4627

Tulli L, Marchi S, Petracca R, Shaw HA, Fairweather NF, Scarselli M, Soriani M, Leuzzi R (2013) CbpA: a novel surface exposed adhesin of *Clostridium difficile* targeting human collagen: Collagen binding protein of *Clostridium difficile*. Cell Microbiol. https://doi.org/10.1111/cmi.12139

Unger M, Eichhoff AM, Schumacher L, Strysio M, Menzel S, Schwan C, Alzogaray V, Zylberman V, Seman M, Brandner J, Rohde H, Zhu K, Haag F, Mittrücker H-W, Goldbaum F, Aktories K, Koch-Nolte F (2015) Selection of nanobodies that block the enzymatic and cytotoxic activities of the binary *Clostridium difficile* toxin CDT. Sci Rep 5:7850. https://doi.org/10.1038/srep07850

Valentini M, Piermattei A, Di Sante G, Migliara G, Delogu G, Ria F (2014) Immunomodulation by gut microbiota: role of toll-like receptor expressed by T cells. J Immunol Res 2014:1–8. https://doi.org/10.1155/2014/586939

van Dissel JT, de Groot N, Hensgens CM, Numan S, Kuijper EJ, Veldkamp P, van 't Wout J (2005) Bovine antibody-enriched whey to aid in the prevention of a relapse of *Clostridium difficile*-associated diarrhoea: preclinical and preliminary clinical data. J Med Microbiol 54:197–205

van Dorp SM, Kinross P, Gastmeier P, Behnke M, Kola A, Delmée M, Pavelkovich A, Mentula S, Barbut F, Hajdu A, Ingebretsen A, Pituch H, Macovei IS, Jovanović M, Wiuff C, Schmid D, Olsen KE, Wilcox MH, Suetens C, Kuijper EJ, European Clostridium difficile Infection Surveillance Network (ECDIS-Net) on behalf of all participants (2016) Standardised surveillance of *Clostridium difficile* infection in European acute care hospitals: a pilot study, 2013. Euro Surveill Bull Eur Sur Mal Transm Eur Commun Dis Bull 21. https://doi.org/10.2807/1560-7917.ES.2016.21.29.30293

van Prehn J, Reigadas E, Vogelzang EH, Bouza E, Hristea A, Guery B, Krutova M, Norén T, Allerberger F, Coia JE, Goorhuis A, van Rossen TM, Ooijevaar RE, Burns K, Scharvik Olesen BR, Tschudin-Sutter S, Wilcox MH, Vehreschild MJGT, Fitzpatrick F, Kuijper EJ (2021) European Society of Clinical Microbiology and Infectious Diseases: 2021 update on the treatment guidance document for *Clostridioides difficile* infection in adults. Clin Microbiol Infect 27:S1–S21. https://doi.org/10.1016/j.cmi.2021.09.038

Viscidi R, Laughon BE, Yolken R, Bo-Linn P, Moench T, Ryder RW, Bartlett JG (1983) Serum antibody response to toxins A and B of *Clostridium difficile*. J Infect Dis 148:93–100

Voth DE, Ballard JD (2005) *Clostridium difficile* toxins: mechanism of action and role in disease. Clin Microbiol Rev 18:247–263. https://doi.org/10.1128/CMR.18.2.247-263.2005

Waligora AJ, Hennequin C, Mullany P, Bourlioux P, Collignon A, Karjalainen T (2001) Characterization of a cell surface protein of *Clostridium difficile* with adhesive properties. Infect Immun 69:2144–2153. https://doi.org/10.1128/IAI.69.4.2144-2153.2001

Wang B, Wang S, Rustandi RR, Wang F, Mensch CD, Hong L, Kristopeit A, Secore S, Dornadula G, Kanavage A, Heinrichs JH, Mach H, Blue JT, Thiriot DS (2015a) Detecting and preventing reversion to toxicity for a formaldehyde-treated *C. difficile* toxin B mutant. Vaccine 33:252–259. https://doi.org/10.1016/j.vaccine.2014.06.032

Wang H, Sun X, Zhang Y, Li S, Chen K, Shi L, Nie W, Kumar R, Tzipori S, Wang J, Savidge T, Feng H (2012) A chimeric toxin vaccine protects against primary and recurrent *Clostridium difficile* Infection. Infect Immun 80:2678–2688. https://doi.org/10.1128/IAI.00215-12

Wang S, Zhu D, Sun X (2022) Development of an effective nontoxigenic *Clostridioides difficile*–based oral vaccine against *C. difficile* infection. Microbiol Spectr 10:e00263-22. https://doi.org/10.1128/spectrum.00263-22

Wang Y, Wang S, Bouillaut L, Li C, Duan Z, Zhang K, Ju X, Tzipori S, Sonenshein AL, Sun X (2018) Oral immunization with nontoxigenic *Clostridium difficile* strains expressing chimeric fragments of TcdA and TcdB elicits protective immunity against *C. difficile* infection in both mice and hamsters. Infect Immun 86:e00489–e00418. https://doi.org/10.1128/IAI.00489-18

Wang Y-K, Yan Y-X, Kim HB, Ju X, Zhao S, Zhang K, Tzipori S, Sun X (2015b) A chimeric protein comprising the glucosyltransferase and cysteine proteinase domains of toxin B and the receptor binding domain of toxin A induces protective immunity against *Clostridium difficile* infection in mice and hamsters. Hum Vaccines Immunother 11:2215–2222. https://doi.org/10.1080/21645515.2015.1052352

Warny M, Vaerman JP, Avesani V, Delmée M (1994) Human antibody response to *Clostridium difficile* toxin A in relation to clinical course of infection. Infect Immun 62:384–389

Wilcox MH, Gerding DN, Poxton IR, Kelly C, Nathan R, Birch T, Cornely OA, Rahav G, Bouza E, Lee C, Jenkin G, Jensen W, Kim Y-S, Yoshida J, Gabryelski L, Pedley A, Eves K, Tipping R, Guris D, Kartsonis N, Dorr M-B, MODIFY I and MODIFY II Investigators (2017) Bezlotoxumab for prevention of recurrent *Clostridium difficile* infection. N Engl J Med 376:305–317. https://doi.org/10.1056/NEJMoa1602615

Winter K, Xing L, Kassardjian A, Ward BJ (2019) Vaccination against *Clostridium difficile* by use of an attenuated *Salmonella enterica* serovar *Typhimurium* vector (YS1646) protects mice from lethal challenge. Infect Immun 87:e00089-19. https://doi.org/10.1128/IAI.00089-19

Wright A, Drudy D, Kyne L, Brown K, Fairweather NF (2008) Immunoreactive cell wall proteins of *Clostridium difficile* identified by human sera. J Med Microbiol 57:750–756. https://doi.org/10.1099/jmm.0.47532-0

Xiong N, Hu S (2015) Regulation of intestinal IgA responses. Cell Mol Life Sci CMLS 72:2645–2655. https://doi.org/10.1007/s00018-015-1892-4

Yang Z, Schmidt D, Liu W, Li S, Shi L, Sheng J, Chen K, Yu H, Tremblay JM, Chen X, Piepenbrink KH, Sundberg EJ, Kelly CP, Bai G, Shoemaker CB, Feng H (2014) A novel multivalent, single-domain antibody targeting TcdA and TcdB prevents fulminant *Clostridium difficile* infection in mice. J Infect Dis 210:964–972. https://doi.org/10.1093/infdis/jiu196

Yoshino Y, Kitazawa T, Ikeda M, Tatsuno K, Yanagimoto S, Okugawa S, Yotsuyanagi H, Ota Y (2013) *Clostridium difficile* flagellin stimulates toll-like receptor 5, and toxin B promotes flagellin-induced chemokine production via TLR5. Life Sci 92:211–217. https://doi.org/10.1016/j.lfs.2012.11.017

Zhang B-Z, Cai J, Yu B, Hua Y, Lau CC, Kao RY-TT, Sze K-H, Yuen K-Y, Huang J-D (2016) A DNA vaccine targeting TcdA and TcdB induces protective immunity against *Clostridium difficile*. BMC Infect Dis 16:596. https://doi.org/10.1186/s12879-016-1924-1

Zhang L, Wang W, Wang S (2015) Effect of vaccine administration modality on immunogenicity and efficacy. Expert Rev Vaccines 14:1509–1523. https://doi.org/10.1586/14760584.2015.1081067

Ribotypes and New Virulent Strains Across Europe

Jeanne Couturier, Kerrie Davies, and Frédéric Barbut

Abstract

Clostridioides (formerly *Clostridium*) *difficile* is a major bacterial cause of post-antibiotic diarrhoea. The epidemiology of *C. difficile* infections (CDIs) has dramatically changed since the early 2000s, with an increasing incidence and severity across Europe. This trend is partly due to the emergence and rapid worldwide spread of the hypervirulent and epidemic PCR ribotype 027. Profiles of patients with CDI have also evolved, with description of community-acquired (CA) infections in patients with no traditional risk factors for CDI. However, epidemiological studies indicated that some European countries have successfully controlled the dissemination of the 027 clone whereas other countries reported the emergence of other virulent or unusual strains. The aims of this review are to summarize the current European CDI epidemiology and to describe the new virulent *C. difficile* strains circulating in Europe, as well as other potential emerging strains described elsewhere. Standardized typing methods and surveillance programmes are mandatory for a better understanding and monitoring of CDI in Europe.

J. Couturier (✉)
National Reference Laboratory for C. difficile, Hôpital Saint-Antoine, Paris, France

Université Paris Cité, UMR INSERM 1139, Paris, France
e-mail: jeanne.couturier@aphp.fr

K. Davies
Healthcare Associated Infections Research Group, Leeds Teaching Hospitals NHS Trust and University of Leeds, Leeds, UK

European Society of Clinical Microbiology and Infectious Diseases (ESCMID) study group for Clostridioides difficile (ESGCD), Basel, Switzerland
e-mail: kerrie.davies@nhs.net

F. Barbut
National Reference Laboratory for C. difficile, Hôpital Saint-Antoine, Paris, France

Université Paris Cité, UMR INSERM 1139, Paris, France

European Society of Clinical Microbiology and Infectious Diseases (ESCMID) study group for Clostridioides difficile (ESGCD), Basel, Switzerland
e-mail: frederic.barbut@aphp.fr

1 Introduction

Clostridioides (formerly *Clostridium*) *difficile* is the main bacterial cause of hospital-acquired diarrhoea; it is responsible for 15–25% of post-antibiotic diarrhoea and for virtually all cases of pseudomembranous colitis (Bartlett and Gerding 2008). *C. difficile* infection (CDI) epidemiology has dramatically changed in Europe since the beginning of the 2000s (Fig. 1). The incidence has increased from 2.45 cases per 10,000 patient-days in 2005 (Barbut et al. 2007) to 4.1 in 2008 (Bauer et al. 2011) and 7.0 in 2012–2013 (Davies et al. 2014). The most recent data indicated a

P. Mastrantonio, M. Rupnik (eds.), *Updates on Clostridioides difficile in Europe*, Advances in Experimental Medicine and Biology 1435, https://doi.org/10.1007/978-3-031-42108-2_8

decrease of the incidence, suggesting an improvement in infection control measures. In 2016–2017, the ECDC-coordinated surveillance reported a crude incidence density of 3.48 cases per 10,000 patient-days among 1559 hospitals from 23 EU/EEA countries (ECDC 2022). In 2017–2018, the COMBACTE-CDI's study performed in 64 European healthcare facilities found a median incidence of 4.1 cases (IQR 2.7–6.4), ranging from 2.7 (France) to 39.7 (Romania) cases per 10,000 bed-days (Viprey et al. 2023) (Fig. 1). This variation is likely to result from a combination of several factors, including the level of awareness of CDI among physicians, the type of methods/algorithm for CDI diagnosis implemented in each country and the global spread of the PCR ribotype (RT) 027 clone. This study also showed that there is still a substantial underdiagnosis of CDI coupled with large disparities in testing policies among European countries (Viprey et al. 2023).

In Europe, the hypervirulent epidemic RT 027 strain (or REA-type BI/NAP1/toxinotype III) was first reported in England in 2005 (Smith 2005) and has since rapidly spread in other European countries. RT 027 is characterized by an 18 bp deletion and a deletion at position 117 in *tcdC* gene, resulting in the inactivation of the toxin repressor TcdC and higher amounts of toxin production (Warny et al. 2005), although the role of *tcdC* mutation in toxin overproduction is still debated (Murray et al. 2009; Cartman et al. 2012). Moreover, epidemic 027 strains also produce an additional toxin, the binary toxin, and are resistant to erythromycin and moxifloxacin, which may have conferred a selective advantage. The same combination of genetic and phenotypic features can be found in other rare RT, such as RT 176 (Krutova et al. 2015; Drabek et al. 2015). RT 027-related CDIs are associated with a higher rate of complications and recurrences (Sundram et al. 2009). The RT 027 has disseminated throughout Europe, with a clear shift in its regional repartition from the UK and Ireland in 2008 (Bauer et al. 2011) to Eastern Europe in 2012–2013 (Davies et al. 2016b). Some countries have successfully controlled its spread and decreased its prevalence (Hensgens et al. 2009; Fawley et al. 2016), while other countries were hit by large outbreaks (Bouza et al. 2017). In addition, other virulent or unusual PCR ribotypes are emerging.

2 *C. difficile* Typing Methods

2.1 PCR Ribotyping

PCR ribotyping is the reference method for *C. difficile* typing in Europe. It relies on the presence of several alleles of the rRNA operon in the *C. difficile* genome. The length polymorphism of the intergenic spacer region between 16S and 23S rRNA genes results in RT-specific patterns after genomic amplification and migration (Bidet et al. 1999). PCR ribotyping was first developed using agarose gel electrophoresis, but the capillary gel-based electrophoresis method has now been widely adopted. The latter enables better standardization and easier comparison between laboratories and is recommended as the reference technique in Europe (Fawley et al. 2015).

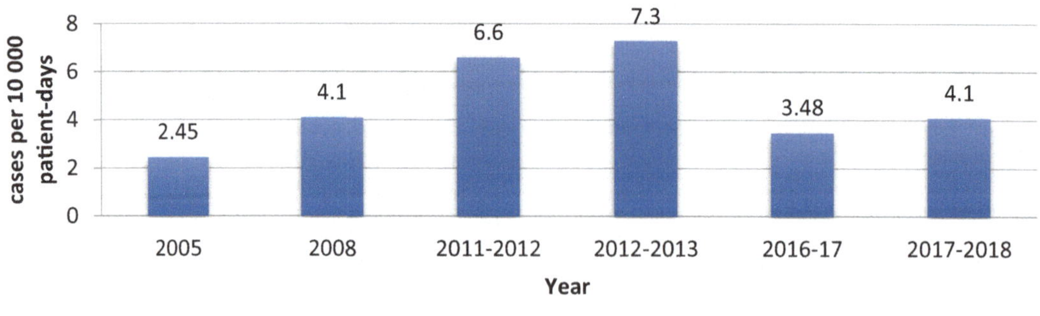

Fig. 1 Trend in CDI incidence in Europe (Bauer et al. 2011; Davies et al. 2014; ECDC 2022; Viprey et al. 2023)

Most European countries use a common nomenclature, but some laboratories developed their own local databases. An online database containing capillary electrophoresis RT profiles is available (WEBRIBO, https://webribo.ages.at/, (Indra et al. 2008). However, there is no standardized protocol since several primer sets were published (Stubbs et al. 1999; Bidet et al. 1999), some of them enabling direct PCR ribotyping from stool samples (Janezic et al. 2011). Harmonization of the PCR ribotyping method and nomenclature is therefore essential and needs to be improved in Europe, in order to detect emergence of new unreferenced RT in a timely manner.

2.2 Other Methods Used for *C. difficile* Typing

Toxins A and B, which are considered as the main virulence factors of *C. difficile* (Pruitt and Lacy 2012), are encoded by *tcdA* and *tcdB* genes located within a locus of pathogenicity (PaLoc). The PaLoc also contains *tcdR* (positive regulator of toxin expression), *tcdE* (holin required for toxin secretion) and *tcdC* (potential negative regulator). The genetic polymorphism of the PaLoc can be explored by toxinotyping, which is a PCR restriction-based method (Rupnik et al. 1998). Toxinotypes are defined according to differences in the PaLoc compared to the reference strain VPI 10463 (nonvariant toxinotype 0). To date, 34 toxinotypes have been described (Rupnik and Janezic 2016) (http://www.mf.um.si/mf/tox/profile.html). Toxinotyping and PCR ribotyping are well correlated since most of the strains in a given RT have similar changes in the PaLoc and thus belong to a single toxinotype. The analysis of 123 strains showed that in a few cases, PCR ribotyping can be more discriminatory than toxinotyping whereas RTs include several toxinotypes less frequently (Rupnik et al. 2001). To avoid ambiguities, a revised toxinotyping nomenclature was published (Rupnik and Janezic 2016).

PFGE (pulsed-field gel electrophoresis) is a genotype-based typing method developed in the 1980s and mostly used in North America. There is good concordance between results of PFGE and PCR ribotyping (Bidet et al. 2000). PFGE has a higher discriminatory power than PCR ribotyping (Killgore et al. 2008), but the interpretation of genetic relatedness is comparable between both typing methods. However, some strains are non-typeable with this method, and degradation of genomic DNA can hinder the analysis (Kristjánsson et al. 1994). PFGE is also very labour-intensive and the lack of standardisation makes inter-laboratory data comparison difficult.

The discriminatory power of PCR ribotyping is not sufficient to prove the nosocomial transmission of a strain, particularly when a RT is endemic at a regional or national level. In that case, another more discriminant typing method has to be used, such as multilocus variable-number tandem repeat (VNTR) analysis (MLVA). MLVA relies on the variability of the VNTR at different loci. The genetic relatedness of isolates is appreciated through the sum of tandem repeat number differences (STRD) (Marsh et al. 2006).

MLST is a typing method based on nucleotide sequence variation (defined as allele) of seven housekeeping gene fragments (*adk*, *dxr*, *glyA*, tpi, *recA*, *atpA* and *sodA*) compared to a reference strain (Griffiths et al. 2010). The combination of the different allele defines the sequence type (ST) (https://pubmlst.org/cdifficile/). This simple and standardized method allows comparison of large collections of *C. difficile* isolates. It has been used to build a phylogenetic tree of *C. difficile* which included five major clades. The hypervirulent ST027 belongs to clade 2 whereas the RT 078 belongs to clade 5. Interestingly some STs correspond to multiple ribotypes whereas a given ribotype may not belong to a single ST (Stabler et al. 2012).

Whole genome sequencing (WGS) can distinguish between strains at the single nucleotide level, highly increasing the discriminatory power over other typing schemes. Given the transferability of data and the diversity of potential applications, such as comparative genome analysis and lineage analysis, this method is increasingly being used for *C. difficile* typing (Knetsch 2013). Two methods can be considered to analyse genomic variations: one is based on the single nucleotide polymorphism (SNP) difference in the coding and non-coding region between the reference genome and the studied genome. The second is based on the gene-by-gene allelic profiling of the core genome (cgMLST) or the whole genome (wgMLST) (Bletz et al. 2018; Baktash et al. 2022). There are currently several cg/wgMLST schemes and software programmes available for *C. difficile* (SeqSphere software, Ridom GmbH, Germany; BioNumerics and bioMérieux, France; EnteroBase, University of Warwick, UK). These platforms are different according to the number of core genes or accessory genes included in their databases, and therefore the allelic cut-off threshold to distinguish clonal from non-clonal strains is not completely standardized. CgMLST and wgMLST have been successfully used for investigating *C. difficile* outbreaks and transmission and could become valuable tools in routine clinical practice (Werner et al. 2020; Barbar et al. 2022; Courbin et al. 2022).

3 Global Distribution of *C. difficile* PCR Ribotypes in Europe

Five major European surveys described the epidemiology of CDI at a European level, including incidence and RT distribution (Bauer et al. 2011; Davies et al. 2014; Freeman et al. 2020; ECDC 2022; Viprey et al. 2023) (Fig. 2).

The first pan-European study on *C. difficile* was performed in 2008 in 106 laboratories from 34 countries (Bauer et al. 2011). The incidence of CDI and the RT distribution varied greatly between hospitals, as well as the density testing for CDI. The authors could differentiate 65 RT among 389 *C. difficile* isolates. One of the main findings of this study was that RT 027 was not predominant in 2008, representing only 5% of the isolates. The most common RT were 014/020 (16%), 001 (9%) and 078 (8%). Some RT seemed to spread regionally, such as RT 106 mostly described in the UK and Ireland.

The EUCLID (European, multicentre, prospective, biannual, point-prevalence study of CDI in hospitalized patients with diarrhoea) study was conducted in 2012–2013 and included

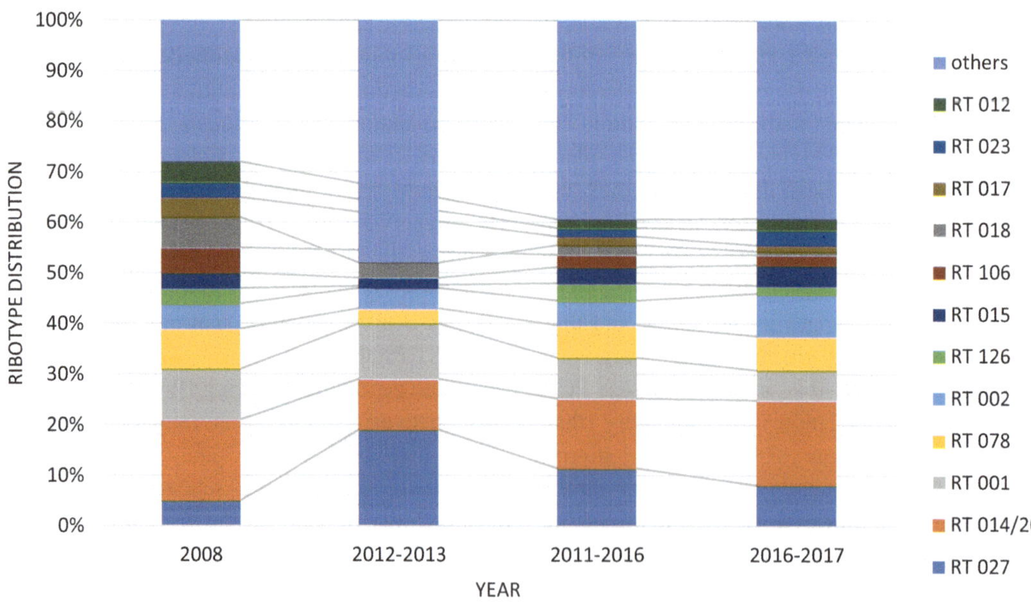

Fig. 2 PCR ribotype distribution during European surveillance (Bauer et al. 2011; Freeman et al. 2020; ECDC 2022)

482 hospitals from 19 European countries (Davies et al. 2016b). The objectives were to measure the underdiagnosis of CDI and to assess the diversity of RT repartition in Europe. During two sampling days (one in winter and one in summer), participating hospitals sent every diarrhoeal stool sample, irrespective of the request to test for *C. difficile* by the physician, to a national coordinating laboratory. The RT diversity was much higher than in the previous study, with 125 RT identified among 1196 isolates. Interestingly, the most common RT was 027 (19%), highlighting the rapid spread of this strain at a global scale. An inverse correlation was noted between the rate of testing and prevalence of ribotype 027 across north, south, east and west quadrants of Europe, which suggests that increased awareness of CDI and use of optimum testing methods and policies can reduce the dissemination of epidemic strains (Davies et al. 2014). The comparison with the 2008 data indicated a shift in the frequency of RT 027 from the UK and Ireland (decreasing prevalence) to Eastern Europe countries (increasing prevalence). RT 001/072 (11%) and 014/020 (10%) were the second and third most prevalent RT, consistent with the 2008 results; however, the prevalence of RT 078 dropped from 8% in 2008 to 3% in 2012–2013. The distribution of causative RT was country-specific as shown in Fig. 3 (Davies et al. 2016b).

The *Clos*ER study was initially designed to monitor antimicrobial susceptibility and geographical distribution of PCR ribotypes in 28 European countries from 2011 to 2016 (Freeman et al. 2020). Overall, 3499 isolates were characterized, resulting in 264 distinct RTs. RT prevalence and diversity scores varied markedly between countries and between each year of the study. However, for countries that submitted samples all five years, although fluctuations were apparent, the most prevalent RTs remained broadly consistent between 2011 and 2016.

During the ECDC-coordinated surveillance in 2016–2017, RT data were available for 4865 isolates (80.1% from Belgium, the Netherlands and the UK – Wales) (ECDC 2022). The most common RT were 014/020 (16.8%), 002 (8.4%), 027 (8.1%), 078 (6.8%), 001 (5.9%), 005 (4.7%), 015 (4%), 023 (3.1%), 012 (2.2%) and 106 (2%). Although the RT data were representative of strains in Belgium, Ireland and the Netherlands, they are not likely to be representative of the EU/EEA as a whole. Interestingly, Czech Republic, Hungary, Poland and Slovakia reported a high proportion of cases that had RT 027 and/or RT 027-like strains, confirming the trend observed in 2012–2013.

The COMBACTE-CDI study (including 119 sites in 12 European countries) found 67 different RTs among 198 *C. difficile* isolates from hospitals (Viprey et al. 2023). The five most common RTs were 027 (11%), 181 (12%), 014 (8%), 010 (5%) and 002 (5%). The highest prevalence of all toxinotype IIIb isolates (RTs 027, 181 and 176) was seen in Eastern Europe (55.9%) which also has the lowest testing rate in a hospital setting (correlation regional testing rate vs prevalence of toxinotype IIIb $r = -0.81$). The COMBACTE-CDI study also collected 82 *C. difficile* isolates from the community: 41 different RTs were identified, the most prevalent being 078 (9%), 039 (9%), 001 (6%), 020 (6%), 009 (5%), 010 (5%) and 181 (2%). Besides these five large epidemiological studies, several other European studies analysed RT distribution at a national level. The results of these national studies are summarized in Table 1.

A multicentre study characterized 3333 toxigenic strains isolated between 2010 and 2015 in 110 Belgian hospitals (Neely et al. 2017). RT 027 (4.2%) and RT 078 (7.0%) were associated with a higher rate of complications (unadjusted data) and higher levels of in vitro toxin production from cultured isolates.

A study compared epidemiological data for community-associated (CA)-CDI and healthcare-associated (HA)-CDI in 113 laboratories across England between 2011 and 2013 (Fawley et al. 2016). A total of 703 *C. difficile* toxin-positive faecal samples from CA-CDI cases were analysed, and the results were compared to HA-CDI data ($n = 10{,}754$) obtained from the *C. difficile* ribotyping network. RT distribution was similar in cases of CA- and HA-CDI, but RT 002 was

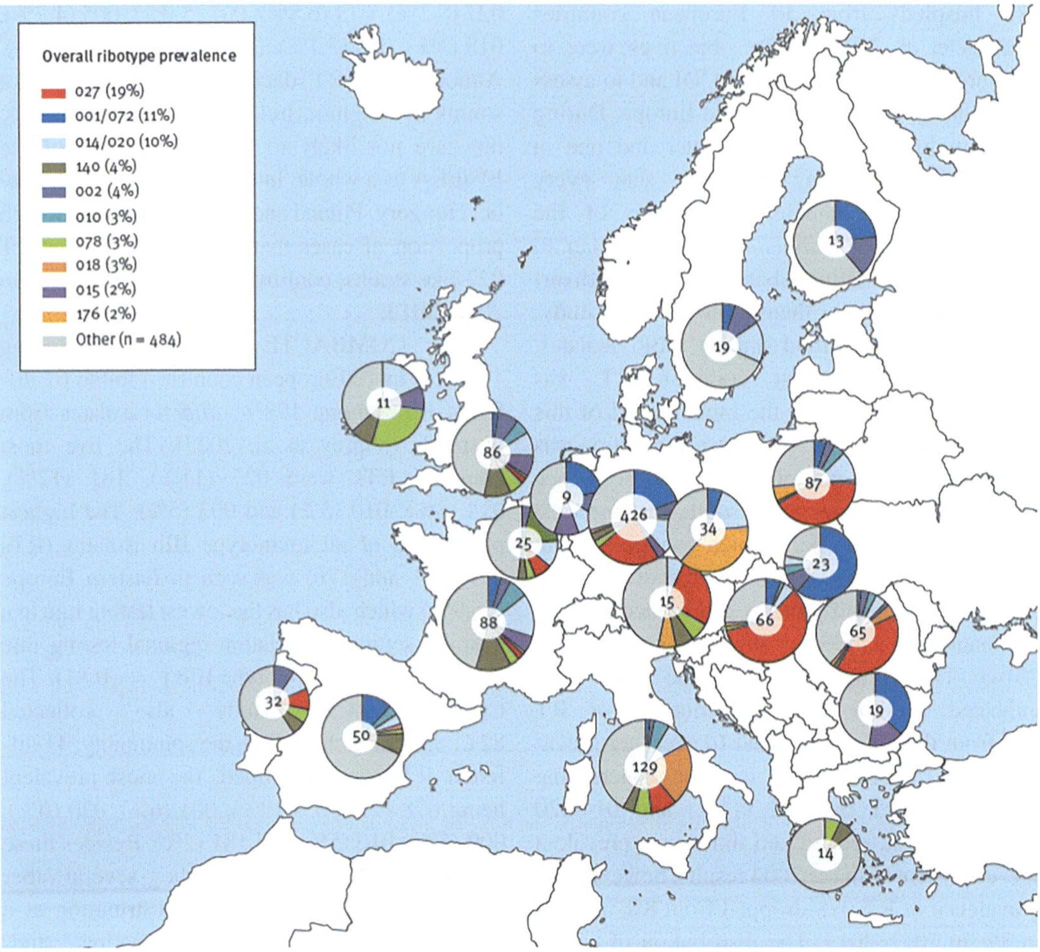

Fig. 3 Geographical distribution of *C. difficile* PCR ribotypes, by participating European country, EUCLID 2012–2013 and 2013 ($n = 1196$) (Davies et al. 2016b). Pie charts show the proportion of the most common ribotypes per country, and the number in the centre of the charts is the number of typed isolates in the country

more likely to cause CA-CDI, while RT 027 was more often associated with HA-CDI.

In Spain, Alcalá et al. performed *C. difficile* cultures on 807 unformed stool specimens sent to 118 Spanish microbiology laboratories on a single day, regardless of the prescription by the clinician (Alcalá et al. 2012). Among 42 toxigenic strains, RTs 014/020, 001 and 078/126 were the most prevalent (20.5%, 18.2% and 18.2%, respectively). RT 027 was not found.

The characterization of 498 clinical isolates from 20 hospitals in Portugal showed that RT 027 was predominant with 18.5% of all the strains and 19.6% of HA-associated CDI. RT 014 was the second most frequent overall (9.4%) and the most frequent among CA-CDI (12%). The prevalence of RTs 126 and 078 was low (3.8% and 2.8%, respectively) (Santos et al. 2016). The authors described a great heterogeneity of the RT distribution through the country with a higher diversity in the north, where RT 027 was not predominant.

The geographic distribution of *C. difficile* genotypes in Germany was assessed using 393 isolates sent to the national advisory laboratory for diagnostic reason between 2011 and 2013

Table 1 National epidemiological studies on *Clostridium difficile* PCR ribotype repartition

Country	N strains	PCR ribotyping method	Most prevalent RT (%)	Reference
Belgium	3333	Agarose gel electrophoresis	014 (11.6), 020 (8.5), 002 (7.6), 078 (7.0), 027 (4.2), 005 (3.5) and 106 (3.4)	Neely et al. (2017)
UK	11,457	Agarose gel electrophoresis	015 (10.2), 002 (9.1), 014 (9.1), 078 (8.0), 005 (7.4) and 027 (6.4)	Fawley et al. (2016)
Spain	42	Agarose gel electrophoresis	014/020 (20.5), 001 (18.2) and 078/126 (18.2)	Alcalá et al. (2012)
Portugal	498	Capillary electrophoresis	027 (18.5), 014 (9.4), 020 (5.6) and 017 (5.2)	Santos et al. (2016)
Germany	393	slpAST with assignment to RT	001 (35), 027 (26), 014/066 (9) and 078 (8)	von Müller et al. (2015)
France	224	Agarose gel electrophoresis	014/020/077 (18.7), 078/126 (12.1), 015 (8.5), 002 (8) and 005 (4.9)	Eckert et al. (2013)
Italy	881	Capillary electrophoresis	607 (20%), 018 (18%), 078 (5%), 126 (5%), 014 (5%), 027 (8%) and 010 (2%)	Barbanti and Spigaglia (2020)
Sweden	156	Capillary electrophoresis	014/020, 005, 002, 078, 023 and 070[a]	Enkirch et al. (2022)
Czech Republic	774	Capillary electrophoresis	176 (29) and 001 (24)	Krutova et al. (2016)

slpAST surface layer protein A sequence typing
[a]Frequency of each RT was not reported

(von Müller et al. 2015). The typing method used was surface layer protein A sequence typing, with strain assignment to RT for better comparison with international data. RTs 001 (35%) and 078 (8%) were prevalent nationwide; RTs 027 (26%) and 014/066 (9%) were detected in almost all regions. More recently, some healthcare facilities in North-Rhine Westphalia observed over the last 10 years a clear switch from RT 001 [18.75% in 2007 vs 3.75% in 2017 $P = 0.003$] to RT 027 [0% in 2007 vs 21.25% ($n = 17$) in 2017] (Piepenbrock et al. 2019).

In France, a multicentre study conducted in 2009 in 78 healthcare facilities showed that the most prevalent RTs were 014/020/077 (18.7%), followed by 078/126 (12.1%) (Eckert et al. 2013). The prevalence of RT 027 strains remained low (3.1%), and they were only isolated in Northern France, where RT 027 emergence was first described in 2006 (Coignard et al. 2006; Birgand et al. 2010). These results are consistent with the more recent LuCID (Longitudinal European *Clostridium difficile* Infection Diagnosis) surveillance study (Davies et al. 2016a), during which RTs 014/020/077 and 078/126 were the most prevalent in France (21.9% and 9.5%, respectively) (Eckert et al. 2015).

In Italy, the Istituto Superiore di Sanita Central Laboratory Service characterized 831 human and animal *C. difficile* strains isolated over a 10-year period (2006–2016) (Barbanti and Spigaglia 2020). Independently from the year of isolation, 42% of the clinical isolates belonged to the RT 018 lineage (including RT 018, RT 607, RT 541, PR07661 and PR14328), with RT 018 and RT 607 grouping the majority of isolates. This lineage was significantly associated with CDIs occurred in the General Medicine Units, Clinic Units or Long-Term Care Facilities, while it was rarely found in paediatric patients. Although the percentage of isolates positive for the binary toxin (CDT) was stable during the study (20%), several CDT-positive RTs emerged in 2012–2016, including RT 027.

In Sweden, the Public Health Agency investigated 122 CDI cases between October 2017 and March 2018, which were classified as CA (39%) (without previous hospital care or onset ≤2 days after admission or >12 weeks after discharge from hospital) or HA (61%) (onset >3 days after hospital admission or within

4 weeks after discharge). They found that RTs 005 (RR 3.1; 95% CI: 1.79–5.24) and 020 (RR 2.5; 95% CI: 1.31–4.63) were significantly associated with CA-CDI (Enkirch et al. 2022).

In conclusion, RTs 014/020 and 001/072 are endemic in almost all European countries while there is a national or regional specificity for other RTs. Moreover, the RT diversity is significantly increasing across Europe.

4 Emerging PCR Ribotypes

4.1 PCR Ribotype 176

RT 176 strains are closely related to RT 027 (Stabler et al. 2006). They belong to toxinotype III, produce the binary toxin and bear a deletion at position 117 of the *tcdC* gene, leading to a potential RT 027 misidentification with commonly used molecular assays such as Xpert® *C. difficile* (Cepheid). Moreover, their similar banding pattern (only one band difference) after gel electrophoresis can be confusing for RT attribution (Valiente et al. 2012). The first cases of RT 176-associated CDI were described in 2008 in Poland (Obuch-Woszczatyński et al. 2014), in 2009 in the Czech Republic (Nyč et al. 2011), in 2015 in Croatia (Rupnik et al. 2016) and in 2020 in Slovakia (Novakova et al. 2020). The first RT 176-related outbreak was described in France (Couturier et al. 2017). Four strains isolated in two geographically close hospitals, previously identified as RT 027 with the agarose gel method, were reassigned as RT 176 by capillary gel-based electrophoresis. MLVA showed that those four strains formed a clonal complex (STRD $\leq$ 2) and were genetically related to RT 027 strains (STRD $\leq$ 10). The results of the EUCLID study showed a regional specificity of RT 176, isolated mostly in the Czech Republic where it accounted for 38% of the strains (Davies et al. 2016b). In 2014, a study among 18 Czech hospitals showed that 29% of *C. difficile* isolates belonged to RT 176 and 24% to RT 001 (Krutova et al. 2016). Further typing analysis by MLVA indicated that both RTs formed clonal complexes in several hospitals, suggesting a rapid spread of these clones at a national level. Moreover, RT176 strains frequently exhibit a reduced susceptibility to moxifloxacin (Novakova et al. 2020).

These results suggest a rapid nosocomial spread of RT 176 strains through Europe, stressing the need for a common data base for PCR ribotyping.

4.2 PCR Ribotype 078

RT 078 strains can produce toxins A and B, as well as the binary toxin, and belong to toxinotype V. They are characterized by a 39 bp deletion in *tcdC*. RT 078 was reported as predominant in Greece in 2005 (Barbut et al. 2007) and was the third most common RT in the 2008 European study (Bauer et al. 2011). A study showed that RT 078 strains co-circulate with the hypervirulent 027 strains in Southern France (Cassir et al. 2017). In Southern Italy, RT 078 was the second most frequently isolated in 138 samples from the environment, animals, food and humans (14.5%), right after RT 126 (15.9%) (Romano et al. 2018). While RT 027 strains are mostly responsible for outbreaks of HA infections in the elderly, RT 078 strains are more frequently associated with CA infections in a younger population. CA-CDIs due to RT 078 strains were also described in England (Fawley et al. 2016) (see "*C. difficile* infection in the community" below). Finally, RT 078 strains are frequently resistant to fluoroquinolones and erythromycin, partly explaining this epidemiological success (Baldan et al. 2015).

4.3 PCR Ribotype 126

RTs 078 and 126 are highly related: they share similar banding patterns in agarose gel electrophoresis method and can only be differentiated with the capillary gel-based electrophoresis. Consequently, they are often reported together as RT 078/126. Like RT 078 strains, RT 126 strains belong to toxinotype V and are considered as "hypervirulent" (Knetsch et al. 2011). They also produce the binary toxin and are characterized by a 39 bp deletion in *tcdC*.

The prevalence of RT 126 strains in animals in Germany is high, suggesting the potential zoonotic spread of this RT (Schneeberg et al. 2013). MLVA showed that most of those strains are genetically related to RT 078 strains (STRD $\leq$ 10) and some of them belong to the same clonal complex (STRD $\leq$ 2). RT 126 strains are also frequently resistant to antibiotics, including erythromycin, moxifloxacin and tetracyclin (Álvarez-Pérez et al. 2017). A retrospective study showed that RT 078/126 was the most common RT (8.6%) in 206 hospitalized patients in Portugal, with a prevalence equal to that of RT 027 (Nazareth et al. 2022). In a large European study including 12 countries and investigating the *C. difficile* positivity rate on retail potatoes, RT 126 was the most frequently isolated, suggesting that food may be a source of contamination (Tkalec et al. 2022).

4.4 PCR Ribotype 033/Toxinotype XI

PCR ribotype 033 strains belong to toxinotype XI. They are characterized by the absence of TcdA and TcdB expression and therefore cannot be detected by EIA (enzyme immunoassay) methods for toxins. These strains were first described in 2001 (Rupnik et al. 2001). In 2014, six symptomatic CDI cases due to toxinotype XI strains were reported by the French National Reference Laboratory for *C. difficile* (Eckert et al. 2014). In four cases, the patient was successfully treated by oral metronidazole. These strains were characterized by PCR ribotyping; amplification of *tcdA*, *tcdB*, *cdtA* and *cdtB* genes; and toxinotyping. The six strains were defined as RT 033 (or 033-like) and were negative for TcdA and TcdB. The binary toxin genes were present and a 39 bp deletion was identified in the *tcdC* gene. The six strains were characterized by major deletions of the 5′ region of the PaLoc including *tcdB*, *tcdE* and *tcdR*; only a remnant part of *tcdA* (A2 and A3 fragments) and *tcdC* could be amplified.

In 2018, Bandelj et al. investigated the prevalence and transmission of *C. difficile* in calves from family dairy farms in Slovenia using MLVA (Bandelj et al. 2018). The most common RT in calves and in the environment was RT 033. RT 033 was also the second predominant RT (10/57 positive samples) in pigs in Czech Republic (Krutova et al. 2018).

The pathogenicity of toxinotype XI strains remains controversial. Studies on the role of the binary toxin as a virulence factor in animal models gave contradictory results. In the rabbit ileal loop model, an enterotoxic response was observed after inoculation of supernatants from culture of $A^-B^-CDT^+$ strains. However, despite colonization, no symptoms occurred in clindamycin-treated hamsters challenged with these strains (Geric et al. 2006). Although the prevalence of $A^-B^-CDT^+$ strains in Europe seems rather low (Barbut et al. 2007; Bauer et al. 2011), surveillance of this unusual strains is required. Indeed, the atypical genomic organization of the PaLoc can lead to a false-negative diagnosis, more particularly when methods relying on the presence of toxin A and/or toxin B only are used. However, the increasing use of the Xpert® *C. difficile* assay, which detects binary toxin genes, will possibly enable a better identification of toxinotype XI strains.

4.5 PCR Ribotype 018

RT 018 has been reported as an emerging RT responsible for outbreaks in Italy, where RT 126 was previously predominant (Spigaglia et al. 2010). The EUCLID study (Davies et al. 2016b) showed that prevalence of RT 018 was high in Italy (22%), as opposed to other European countries. In addition, Baldan et al. characterized 312 *C. difficile* isolates from a large Italian teaching hospital between 2009 and 2013 and observed that RT 018 was predominant. After epidemiological investigation of the outbreaks, RT 018 represented 42% of index CDI cases and virtually all secondary cases (due to nosocomial transmission). The transmission index (number of secondary cases divided by number of index cases) of RT 018 was significantly higher than that of RT 078 (0.640 and 0.0606, respectively) (Baldan et al. 2015). In Germany, Berger et al. investigated an outbreak that took place in 2015 and detected 9/82 CDI cases due to closely related RT 018 strains (Berger et al. 2019). The phenotypic analysis results showed a fluoroquinolone and macrolide resistance. It was the first description of a 018-related outbreak in this country. RT 018 was also responsible for a large outbreak in a geriatric unit in France, with 19 CDI cases (Gateau et al. 2019). MLVA indicated that 15/19 strains were included in two clonal complexes. Another study comparing RT 018, RT 126 and RT 078 demonstrated that RT 018 strains produced higher levels of toxins, showed increased adhesion to cells and became endemic in a short time (Barbanti and Spigaglia 2016). Moreover, RT 018 strains were all multidrug resistant (resistance to erythromycin, clindamycin and moxifloxacin). Together, these results suggest that RT 018 strains have phenotypic traits conferring an adaptive advantage and are able to spread widely. RT 018 strains were indeed reported in Southern Europe (Spain, Austria and Slovenia) and are associated with a higher rate of complicated infections (Bauer et al. 2011).

4.6 PCR Ribotype 017

RT 017 strains belong to toxinotype VIII and are part of *C. difficile* clade 4; they lack toxin A production and binary toxin genes (Cairns et al. 2012). The clinical relevance and the prevalence of this clone have been unclear for many years, since it was mainly found in asymptomatic infants (Depitre et al. 1993; Kato et al. 1998). However, it has now been established that RT 017 strains are predominant in Asian countries such as Korea, China and Japan (Collins et al. 2013) and that they have spread worldwide. A recent study showed that a key feature in this epidemiological success is the acquisition of macrolide resistance via *ermB*-positive transposon Tn*6194* (Imwattana et al. 2022). RT 017-related outbreaks have been reported in England (Cairns et al. 2015), the Netherlands (Kuijper et al. 2001), Poland (Pituch et al. 2001) and Ireland (Drudy et al. 2007). RT 017-related CA-CDIs appear to be more likely to affect younger patients (Fawley et al. 2016). Severe RT 017-related CDIs have been described in Germany, although RT 027 was the most prevalent strain in this study (Arvand et al. 2009).

4.7 Other Emerging PCR Ribotypes

RT 244 strains belong to the same hypervirulent clade as RT 027 (clade 2) (Lim et al. 2014). They produce binary toxin and bear a single nucleotide deletion at position 117 in *tcdC*. Severe CA-CDI and outbreaks due to RT 244 strains were reported in Australia and New Zealand, where it was previously uncommon (De Almeida et al. 2013; Eyre et al. 2015). Eyre et al. showed that a strain isolated in a patient recently returned from Australia to the UK was phylogenetically related to their outbreak, highlighting the potential rapid spread of RT 244 via international travel.

The previously quoted French multicentre survey showed that among 224 toxigenic strains, 19 (8.5%) belonged to RT 015 which was the third most frequent RT (Eckert et al. 2013). Fawley et al. showed that RT 015 was also predominant in England (Fawley et al. 2016). RT 015 strains were observed in wild rodents in and around food production buildings in the Netherlands (Krijger et al. 2019). Although RT 015 accounted for only 2% of the strains analysed in the EUCLID study, it seems that RT 015 strains can spread and become predominant at a national scale.

RT 106 strains represented 5% of all toxigenic isolates in the 2008 hospital-based European study, but their distribution showed a regional spread: among 20 strains, 13 were isolated in the UK and 5 in Ireland (Bauer et al. 2011). Although this strain had declined in prevalence, between 2012 and 2017, in France the prevalence of RT 106 strains increased from $<1\%$ to 4.65% (Colomb-Cotinat et al. 2019). In a Southern England healthcare facility, 38% of *C. difficile* isolates ($n = 97$) were identified as RT 106, the second most prevalent RT after 027 (45%) (Sundram et al. 2009). Almost all of these RT 106 strains were resistant to ciprofloxacin and erythromycin. Moreover, in the Belgian multicentre study (Neely et al. 2017), recurrences were more frequent with RT 106-related CDI.

Other data reported the emergence of RT 001 strains with reduced susceptibility to metronidazole, raising concerns about the potential spread of these strains due to this selective advantage (Baines et al. 2008). Studies showed that RT 001 is the most prevalent in CDI cases in Slovakia (Krehelova et al. 2019; Novakova et al. 2020) and in food samples (5/18) in Slovenia (Tkalec et al. 2020). In Southern Germany, the prevalence of RT 001 strains exhibiting resistance to erythromycin, ciprofloxacin and moxifloxacin is high in both in- and outpatients (Borgmann et al. 2008; Arvand et al. 2009). In 2021, the German National Reference Center for *C. difficile* analysed 1535 isolates from clinical samples: RT 001 was the second most frequent with 13.3% of the isolates, behind RT 027 (36.2%) (Abdrabou et al. 2021).

A new 027-like RT belonging to clade 2 and ST 1, RT 181, was described in Greece. In 2020, it was recognized as the cause of a large outbreak in a 180-bed rehabilitation clinic involving 15/19 CDI patients (Kachrimanidou et al. 2020). All RT 181 isolates were susceptible to vancomycin and metronidazole but resistant to fluoroquinolones and macrolides. All patients were successfully treated with a ten-day oral course of vancomycin, except for one case who suffered from a relapse. RT 181 was the most common RT between 2016 and 2019 in ten Greek healthcare facilities with 36% of *C. difficile* strains, and it was identified in seven out of the ten participating hospitals (Kachrimanidou et al. 2022). In the COMBACTE-CDI study, RT 181 was the most prevalent in *C. difficile* isolates from hospitals with 12% of the strains. It was also retrieved in samples from the community (Viprey et al. 2022). The highest frequency of toxinotype IIIb (027, 181 and 176) was observed in eastern European countries (56%, 43/77) where the testing rate was the lowest (58%, 164/281). These data suggest that RT 181 could rapidly spread at a national or a more global scale.

Other RTs such as RT826 (clade 5) and RT046 have been responsible for large outbreaks in the Netherlands and Sweden, respectively. These RTs have been associated with a higher mortality compared to other RTs and might display increased virulence (Crobach et al. 2018; Magnusson et al. 2022). Given their pathogenic and epidemic potential, the emergence of these RTs should be closely followed in European countries.

The genetic and epidemiological features of the emerging RT described above are summarized in Table 2.

4.8 Emerging Strains with a A+B− CDT− Unusual Profile

Three clinical strains with an atypical PaLoc structure were described in France (Monot et al. 2015), including the first variant strain producing only toxin A ($A^+B^-CDT^-$). Whole genome sequencing analysis of this strain showed that its PaLoc only contained *tcdA* and *tcdR*. None of the three strains belonged to any of the most frequent RTs. Moreover, the authors described variability in the sequence of the toxin genes, which may lead to potential false-negative results with the most commonly used diagnostic methods (immunoenzymatic or molecular assays).

5 *C. difficile* Infection in the Community

The epidemiology of CA-CDI is poorly known, since *C. difficile* testing is rarely requested in stool samples from community patients. However, according to recent epidemiological data, CA-CDIs account for 25% of CDI in Europe and Australia and 33% in the United States, and their incidence is rising (Chitnis et al. 2013; Moloney et al. 2021). In addition, CDIs were described among young patients from community settings without the traditional risk factors (antibiotic exposure, recent hospitalization,

Table 2 Characteristics of currently circulating and emerging PCR ribotypes in Europe

RT	Toxinotype	Toxins A and B	Binary toxin	Deletion in *tcdC*	Main circulation area
027	III	+/+	+	−18 bp/Δ117	Europe, mostly Eastern Europe (Davies et al. 2016b)
176	III	+/+	+	−18 bp/Δ117	Poland and Czech Republic (Nyč et al. 2011; Obuch-Woszczatyński et al. 2014)
078	V	+/+	+	−39 bp/A117T	Community-onset infections (Eckert et al. 2011; Fawley et al. 2016)
126	V	+/+	+	−39 bp/A117T	(Eckert et al. 2011)
033	XIa/XIb	−/−	+	−39 bp	Low prevalence in Europe (Eckert et al. 2014)
018	XIX	+/+	−	ND	Italy (Spigaglia et al. 2010; Rupnik and Janezic 2016)
017	VIII	−/+	−	ND	Asia (Collins et al. 2013), Ireland (Drudy et al. 2007), England (Cairns et al. 2015), the Netherlands (Kuijper et al. 2001), Poland (Pituch et al. 2001) and Germany
244	IXb	+/+	+	ND/Δ117	Australia (Lim et al. 2014; Rupnik and Janezic 2016)
015	NA	+/+	−	−18 bp or ND	France (Eckert et al. 2013)
106	NA	+/+	−	−18 bp or ND	UK and Ireland (Bauer et al. 2011)
001	XXIX	+/+	−	ND	Germany, multidrug-resistant strains (Borgmann et al. 2008; Rupnik and Janezic 2016)
181	NA	+/+	+	−18 bp/Δ117	Greece (Kachrimanidou et al. 2022)
826	NA	+/+	+	−39 bp	The Netherlands (Crobach et al. 2018)
046	NA	+/+	NA	NA	Sweden (Magnusson et al. 2022)

ND not deleted, *NA* not available

comorbidities) (Wilcox et al. 2008; Gupta and Khanna 2014).

Fawley et al. showed that RTs 002, 020 and 056 were largely responsible for CA-CDI whereas RT 027 was mostly associated with HA-CDI (Fawley et al. 2016). RT 078 strains have been reported in animals in the Netherlands (Goorhuis et al. 2008), and by using MLVA, Debast et al. showed that RT 078 strains found in animals and in humans were genetically highly related, suggesting a foodborne interspecies transmission of *C. difficile* (Debast et al. 2009). Similar results were obtained by Moloney et al., who showed using whole genome sequencing analysis a close overlap between 078 strains from humans and pigs isolated in Ireland (Moloney et al. 2021). In Canada, RT 078 epidemic strains (identified as pulsotype NAP7 by PFGE) were found in vegetables from grocery stores (Metcalf et al. 2010). RT 078 has also been described in the environment; it was the most frequently isolated RT in wastewater treatment plants in Switzerland (Romano et al. 2012). RT 078 was the commonest (19.0%) in 42 CA-CDI cases in a prospective study conducted in Scotland, followed by RT 014/020 (16.7%), 015 (14.3%) and 001 (11.9%) (Taori et al. 2014). However, in a US study of 984 CA-CDI cases, NAP1/RT 027 was the most frequent strain isolated (21.7%), while less than 7% of the isolates belonged to NAP7/RT 078 (Chitnis et al. 2013). In 2011, population- and laboratory-based surveillance for CDI was conducted in ten US areas (Lessa et al. 2015). A total of 1364 strains were characterized. The most common strains were NAP1/RT 027 (18.8% of CA-CDI and 30.7% of HA-CDI), NAP4/RT 020 (11.4% and 10.3%) and NAP11/RT 106 (10.7% and 10.0%). Less than 4% of the strains in both settings belonged to NAP7/RT 078. A recent Danish study revealed that the *C. difficile* prevalence in dog faeces harvested from public gardens was 4.9% (Bjöersdorff et al. 2021). RT 014/020 was the most frequent. CgMLST analysis showed a genetic relatedness between canine and human isolates, shedding light on the role of pets as potential community source of human CDI. These results show a large overlapping of the RT distribution in HA- and CA-CDI, suggesting the existence of common reservoirs and multiple transmission routes between community and hospital settings. In fact, the recognition of *C. difficile* as a ubiquitous bacteria is nowadays increasing, due to its detection in samples from parks, environmental waters, homes, food or commercial stores (Tkalec et al. 2020; Moloney et al. 2021).

6 Conclusion

In conclusion, there is a large diversity of RT across Europe, although some specific RTs are able to disseminate at a regional or national level. A national and European clinical surveillance system, associated with microbiological characterization of strains, is essential in order to monitor the constantly changing epidemiology of CDI. A common European data base of the circulating PCR ribotypes would be very helpful to detect emergence of new virulent clones in a timely manner.

References

Abdrabou AMM, Ul Habib Bajwa Z, Halfmann A et al (2021) Molecular epidemiology and antimicrobial resistance of *Clostridioides difficile* in Germany, 2014-2019. Int J Med Microbiol IJMM 311:151507. https://doi.org/10.1016/j.ijmm.2021.151507

Alcalá L, Martin A, Marin M et al (2012) The undiagnosed cases of *Clostridium difficile* infection in a whole nation: where is the problem? Clin Microbiol Infect 18:E204–E213. https://doi.org/10.1111/j.1469-0691.2012.03883.x

Álvarez-Pérez S, Blanco JL, Harmanus C et al (2017) Subtyping and antimicrobial susceptibility of *Clostridium difficile* PCR ribotype 078/126 isolates of human and animal origin. Vet Microbiol 199:15–22. https://doi.org/10.1016/j.vetmic.2016.12.001

Arvand M, Hauri AM, Zaiss NH et al (2009) *Clostridium difficile* ribotypes 001, 017, and 027 are associated with lethal *C. difficile* infection in Hesse, Germany. Euro Surveill Bull Eur Sur Mal Transm Eur Commun Dis Bull 14

Baines SD, O'Connor R, Freeman J et al (2008) Emergence of reduced susceptibility to metronidazole in *Clostridium difficile*. J Antimicrob Chemother 62: 1046–1052. https://doi.org/10.1093/jac/dkn313

Baktash A, Corver J, Harmanus C et al (2022) Comparison of whole-genome sequence-based methods and PCR ribotyping for subtyping of *Clostridioides difficile*. J Clin Microbiol 60:e0173721. https://doi.org/10.1128/JCM.01737-21

Baldan R, Trovato A, Bianchini V et al (2015) *Clostridium difficile* PCR Ribotype 018, a successful epidemic genotype. J Clin Microbiol 53:2575–2580. https://doi.org/10.1128/JCM.00533-15

Bandelj P, Harmanus C, Blagus R et al (2018) Quantification of *Clostridioides (Clostridium) difficile* in feces of calves of different age and determination of predominant *Clostridioides difficile* ribotype 033 relatedness and transmission between family dairy farms using multilocus variable-number tandem-repeat analysis. BMC Vet Res 14:298. https://doi.org/10.1186/s12917-018-1616-8

Barbanti F, Spigaglia P (2020) Microbiological characteristics of human and animal isolates of *Clostridioides difficile* in Italy: results of the Istituto Superiore di Sanità in the years 2006-2016. Anaerobe 61:102136. https://doi.org/10.1016/j.anaerobe.2019.102136

Barbanti F, Spigaglia P (2016) Characterization of *Clostridium difficile* PCR-ribotype 018: a problematic emerging type. Anaerobe 42:123–129. https://doi.org/10.1016/j.anaerobe.2016.10.003

Barbar R, Brazelton JN, Carroll KC et al (2022) Molecular epidemiology and genetic relatedness of *Clostridioides difficile* isolates in pediatric oncology and transplant patients using whole genome sequencing. Clin Infect Dis Off Publ Infect Dis Soc Am ciac459. https://doi.org/10.1093/cid/ciac459

Barbut F, Mastrantonio P, Delmee M et al (2007) Prospective study of *Clostridium difficile* infections in Europe with phenotypic and genotypic characterisation of the isolates. Clin Microbiol Infect 13:1048–1057

Bartlett JG, Gerding DN (2008) Clinical recognition and diagnosis of *Clostridium difficile* infection. Clin Infect Dis 46:S12–S18. https://doi.org/10.1086/521863

Bauer MP, Notermans DW, van Benthem BH et al (2011) *Clostridium difficile* infection in Europe: a hospital-based survey. Lancet 377:63–73. https://doi.org/10.1016/S0140-6736(10)61266-4

Berger FK, Gfrörer S, Becker SL et al (2019) Hospital outbreak due to *Clostridium difficile* ribotype 018 (RT018) in Southern Germany. Int J Med Microbiol IJMM 309:189–193. https://doi.org/10.1016/j.ijmm.2019.03.001

Bidet P, Barbut F, Lalande V et al (1999) Development of a new PCR-ribotyping method for *Clostridium difficile* based on ribosomal RNA gene sequencing. FEMS Microbiol Lett 175:261–266. https://doi.org/10.1111/j.1574-6968.1999.tb13629.x

Bidet P, Lalande V, Salauze B et al (2000) Comparison of PCR-ribotyping, arbitrarily primed PCR, and pulsed-field gel electrophoresis for typing *Clostridium difficile*. J Clin Microbiol 38:2484–2487

Birgand G, Blanckaert K, Carbonne A et al (2010) Investigation of a large outbreak of *Clostridium difficile* PCR-ribotype 027 infections in northern France, 2006-2007 and associated clusters in 2008-2009.

Euro Surveill Bull Eur Sur Mal Transm Eur Commun Dis Bull 15
Björsdorff OG, Lindberg S, Kiil K et al (2021) Dogs are carriers of *Clostridioides difficile* lineages associated with human community-acquired infections. Anaerobe 67:102317. https://doi.org/10.1016/j.anaerobe.2020.102317
Bletz S, Janezic S, Harmsen D et al (2018) Defining and evaluating a core genome multilocus sequence typing scheme for genome-wide typing of *Clostridium difficile*. J Clin Microbiol 56:e01987–e01917. https://doi.org/10.1128/JCM.01987-17
Borgmann S, Kist M, Jakobiak T et al (2008) Increased number of *Clostridium difficile* infections and prevalence of *Clostridium difficile* PCR ribotype 001 in southern Germany. Euro Surveill Bull Eur Sur Mal Transm Eur Commun Dis Bull 13
Bouza E, Alcalá L, Marín M et al (2017) An outbreak of *Clostridium difficile* PCR ribotype 027 in Spain: risk factors for recurrence and a novel treatment strategy. Eur J Clin Microbiol Infect Dis:1–10. https://doi.org/10.1007/s10096-017-2991-y
Cairns MD, Preston MD, Lawley TD et al (2015) Genomic epidemiology of a protracted hospital outbreak caused by a toxin A-negative *Clostridium difficile* sublineage PCR ribotype 017 strain in London, England. J Clin Microbiol 53:3141–3147. https://doi.org/10.1128/JCM.00648-15
Cairns MD, Stabler RA, Shetty N et al (2012) The continually evolving *Clostridium difficile* species. Future Microbiol 7:945–957. https://doi.org/10.2217/fmb.12.73
Cartman ST, Kelly ML, Heeg D et al (2012) Precise manipulation of the *Clostridium difficile* chromosome reveals a lack of association between the tcdC genotype and toxin production. Appl Environ Microbiol 78:4683–4690. https://doi.org/10.1128/AEM.00249-12
Cassir N, Fahsi N, Durand G et al (2017) Emergence of *Clostridium difficile* tcdC variant 078 in Marseille, France. Eur J Clin Microbiol Infect Dis:1–4. https://doi.org/10.1007/s10096-017-3022-8
Chitnis AS, Holzbauer SM, Belflower RM et al (2013) Epidemiology of community-associated *Clostridium difficile* infection, 2009 through 2011. JAMA Intern Med 173:1359–1367. https://doi.org/10.1001/jamainternmed.2013.7056
Coignard B, Barbut F, Blanckaert K et al (2006) Emergence of *Clostridium difficile* toxinotype III, PCR-ribotype 027-associated disease, France, 2006. Euro Surveill Bull Eur Sur Mal Transm Eur Commun Dis Bull 11:E060914.1
Collins DA, Hawkey PM, Riley TV (2013) Epidemiology of *Clostridium difficile* infection in Asia. Antimicrob Resist Infect Control 2:21. https://doi.org/10.1186/2047-2994-2-21
Colomb-Cotinat M, Assouvie L, Durand J et al (2019) Epidemiology of *Clostridioides difficile* infections, France, 2010 to 2017. Eurosurveillance 24:1800638. https://doi.org/10.2807/1560-7917.ES.2019.24.35.1800638
Courbin V, Le Neindre K, Couturier J et al (2022) Transmission patterns of *Clostridioides difficile* in a non-epidemic setting based on WGS analysis. Microbiol Res 13:530–538. https://doi.org/10.3390/microbiolres13030037
Couturier J, Eckert C, Barbut F (2017) Spatio-temporal variability of the epidemic 027 *Clostridium difficile* strains in France based on MLVA typing. Anaerobe. https://doi.org/10.1016/j.anaerobe.2017.08.007
Crobach MJT, Voor In't Holt AF, Knetsch CW et al (2018) An outbreak of *Clostridium difficile* infections due to new PCR ribotype 826: epidemiologic and microbiologic analyses. Clin Microbiol Infect Off Publ Eur Soc Clin Microbiol Infect Dis 24:309.e1–309.e4. https://doi.org/10.1016/j.cmi.2017.08.014
Davies K, Davis G, Barbut F et al (2016a) Variability in testing policies and impact on reported *Clostridium difficile* infection rates: results from the pilot Longitudinal European *Clostridium difficile* Infection Diagnosis surveillance study (LuCID). Eur J Clin Microbiol Infect Dis 35:1949–1956. https://doi.org/10.1007/s10096-016-2746-1
Davies KA, Ashwin H, Longshaw CM et al (2016b) Diversity of *Clostridium difficile* PCR ribotypes in Europe: results from the European, multicentre, prospective, biannual, point-prevalence study of *Clostridium difficile* infection in hospitalised patients with diarrhoea (EUCLID), 2012 and 2013. Eurosurveillance 21. https://doi.org/10.2807/1560-7917.ES.2016.21.29.30294
Davies KA, Longshaw CM, Davis GL et al (2014) Underdiagnosis of *Clostridium difficile* across Europe: the European, multicentre, prospective, biannual, point-prevalence study of *Clostridium difficile* infection in hospitalised patients with diarrhoea (EUCLID). Lancet Infect Dis 14:1208–1219. https://doi.org/10.1016/S1473-3099(14)70991-0
De Almeida MN, Heffernan H, Dervan A et al (2013) Severe *Clostridium difficile* infection in New Zealand associated with an emerging strain, PCR-ribotype 244. N Z Med J 126:9–14
Debast SB, Van Leengoed LAMG, Goorhuis A et al (2009) *Clostridium difficile* PCR ribotype 078 toxinotype V found in diarrhoeal pigs identical to isolates from affected humans. Environ Microbiol 11:505–511. https://doi.org/10.1111/j.1462-2920.2008.01790.x
Depitre C, Delmee M, Avesani V et al (1993) Serogroup F strains of *Clostridium difficile* produce toxin B but not toxin A. J Med Microbiol 38:434–441. https://doi.org/10.1099/00222615-38-6-434
Drabek J, Nyc O, Krutova M et al (2015) Clinical features and characteristics of *Clostridium difficile* PCR-ribotype 176 infection: results from a 1-year university hospital internal ward study. Ann Clin Microbiol Antimicrob 14. https://doi.org/10.1186/s12941-015-0114-0

Drudy D, Harnedy N, Fanning S et al (2007) Emergence and control of fluoroquinolone-resistant, toxin A–negative, toxin B–positive *Clostridium difficile*. Infect Control Amp Hosp Epidemiol 28:932–940. https://doi.org/10.1086/519181

ECDC (2022) European Centre for Disease Prevention and Control (ECDC). Healthcare-associated infections: *Clostridium difficile* infections – Annual Epidemiological Report for 2016-2017. ECDC, Stockholm

Eckert C, Bildan M-A, Quach S et al (2015) Caractérisation des souches de *Clostridium difficile* circulant en France en 2014 et 2015 : résultats d'une étude multicentrique (ref 350) - 35ème RICAI

Eckert C, Coignard B, Hebert M et al (2013) Clinical and microbiological features of *Clostridium difficile* infections in France: the ICD-RAISIN 2009 national survey. Médecine Mal Infect 43:67–74. https://doi.org/10.1016/j.medmal.2013.01.004

Eckert C, Emirian A, Le Monnier A et al (2014) Prevalence and pathogenicity of binary toxin–positive *Clostridium difficile* strains that do not produce toxins A and B. New Microbes New Infect 3:12–17. https://doi.org/10.1016/j.nmni.2014.10.003

Eckert C, Tessier C, Chassaing D et al (2011) Is deletion at 117 of the tcdC gene specific of PCR-ribotype 027 strains? Commun Affichée ICAAC 2011

Enkirch T, Mernelius S, Magnusson C et al (2022) Molecular epidemiology of community- and hospital-associated *Clostridioides difficile* infections in Jönköping, Sweden, October 2017 – March 2018. APMIS Acta Pathol Microbiol Immunol Scand 130:661–670. https://doi.org/10.1111/apm.13270

Eyre DW, Tracey L, Elliott B et al (2015) Emergence and spread of predominantly community-onset *Clostridium difficile* PCR ribotype 244 infection in Australia, 2010 to 2012. Euro Surveill Bull Eur Sur Mal Transm Eur Commun Dis Bull 20:21059

Fawley WN, Davies KA, Morris T et al (2016) Enhanced surveillance of *Clostridium difficile* infection occurring outside hospital, England, 2011 to 2013. Euro Surveill Bull Eur Sur Mal Transm Eur Commun Dis Bull 21. https://doi.org/10.2807/1560-7917.ES.2016.21.29.30295

Fawley WN, Knetsch CW, MacCannell DR et al (2015) Development and validation of an internationally-standardized, high-resolution capillary gel-based electrophoresis PCR-ribotyping protocol for *Clostridium difficile*. PLoS One 10. https://doi.org/10.1371/journal.pone.0118150

Freeman J, Vernon J, Pilling S et al (2020) Five-year Pan-European, longitudinal surveillance of *Clostridium difficile* ribotype prevalence and antimicrobial resistance: the extended ClosER study. Eur J Clin Microbiol Infect Dis 39:169–177. https://doi.org/10.1007/s10096-019-03708-7

Gateau C, Deboscker S, Couturier J et al (2019) Local outbreak of *Clostridioides difficile* PCR-Ribotype 018 investigated by multi locus variable number tandem repeat analysis, whole genome multi locus sequence typing and core genome single nucleotide polymorphism typing. Anaerobe 60:102087. https://doi.org/10.1016/j.anaerobe.2019.102087

Geric B, Carman RJ, Rupnik M et al (2006) Binary toxin–producing, large clostridial toxin–negative *Clostridium difficile* strains are enterotoxic but do not cause disease in hamsters. J Infect Dis 193:1143–1150. https://doi.org/10.1086/501368

Goorhuis A, Bakker D, Corver J et al (2008) Emergence of *Clostridium difficile* infection due to a new hypervirulent strain, polymerase chain reaction ribotype 078. Clin Infect Dis Off Publ Infect Dis Soc Am 47:1162–1170. https://doi.org/10.1086/592257

Griffiths D, Fawley W, Kachrimanidou M et al (2010) Multilocus sequence typing of *Clostridium difficile*. J Clin Microbiol 48:770–778. https://doi.org/10.1128/JCM.01796-09

Gupta A, Khanna S (2014) Community-acquired *Clostridium difficile* infection: an increasing public health threat. Infect Drug Resist 7:63–72. https://doi.org/10.2147/IDR.S46780

Hensgens MP, Goorhuis A, Notermans DW et al (2009) Decrease of hypervirulent *Clostridium difficile* PCR ribotype 027 in the Netherlands. Euro Surveill Bull Eur Sur Mal Transm Eur Commun Dis Bull 14

Imwattana K, Putsathit P, Collins DA et al (2022) Global evolutionary dynamics and resistome analysis of *Clostridioides difficile* ribotype 017. Microb Genom 8:000792. https://doi.org/10.1099/mgen.0.000792

Indra A, Huhulescu S, Schneeweis M et al (2008) Characterization of *Clostridium difficile* isolates using capillary gel electrophoresis-based PCR ribotyping. J Med Microbiol 57:1377–1382. https://doi.org/10.1099/jmm.0.47714-0

Janezic S, Strumbelj I, Rupnik M (2011) Use of modified PCR ribotyping for direct detection of *Clostridium difficile* ribotypes in stool samples. J Clin Microbiol 49:3024–3025. https://doi.org/10.1128/JCM.01013-11

Kachrimanidou M, Baktash A, Metallidis S et al (2020) An outbreak of *Clostridioides difficile* infections due to a 027-like PCR ribotype 181 in a rehabilitation centre: epidemiological and microbiological characteristics. Anaerobe 65:102252. https://doi.org/10.1016/j.anaerobe.2020.102252

Kachrimanidou M, Metallidis S, Tsachouridou O et al (2022) Predominance of *Clostridioides difficile* PCR ribotype 181 in northern Greece, 2016-2019. Anaerobe 76:102601. https://doi.org/10.1016/j.anaerobe.2022.102601

Kato H, Kato N, Watanabe K et al (1998) Identification of toxin A-negative, toxin B-positive *Clostridium difficile* by PCR. J Clin Microbiol 36:2178–2182

Killgore G, Thompson A, Johnson S et al (2008) Comparison of seven techniques for typing international epidemic strains of *Clostridium difficile*: restriction endonuclease analysis, pulsed-field gel electrophoresis, PCR-ribotyping, multilocus sequence typing, multilocus variable-number tandem-repeat analysis, amplified fragment length polymorphism, and surface layer protein A gene sequence typing. J Clin Microbiol 46:431–437. https://doi.org/10.1128/JCM.01484-07

Knetsch CW, Hensgens MPM, Harmanus C et al (2011) Genetic markers for *Clostridium difficile* lineages linked to hypervirulence. Microbiology 157:3113–3123. https://doi.org/10.1099/mic.0.051953-0

Knetsch EC for DP and C (ECDC)-HCU-E Editorial (2013) Current application and future perspectives of molecular typing methods to study *Clostridium difficile* infections. http://www.eurosurveillance.org/ViewArticle.aspx?ArticleId=20381. Accessed 2 Dec 2014

Krehelova M, Nyč O, Sinajová E et al (2019) The predominance and clustering of *Clostridioides (Clostridium) difficile* PCR ribotype 001 isolates in three hospitals in Eastern Slovakia, 2017. Folia Microbiol (Praha) 64: 49–54. https://doi.org/10.1007/s12223-018-0629-9

Krijger IM, Meerburg BG, Harmanus C et al (2019) *Clostridium difficile* in wild rodents and insectivores in the Netherlands. Lett Appl Microbiol 69:35–40. https://doi.org/10.1111/lam.13159

Kristjánsson M, Samore MH, Gerding DN et al (1994) Comparison of restriction endonuclease analysis, ribotyping, and pulsed-field gel electrophoresis for molecular differentiation of *Clostridium difficile* strains. J Clin Microbiol 32:1963–1969

Krutova M, Matejkova J, Kuijper EJ et al (2016) *Clostridium difficile* PCR ribotypes 001 and 176 – the common denominator of *C. difficile* infection epidemiology in the Czech Republic, 2014. Eurosurveillance 21. https://doi.org/10.2807/1560-7917.ES.2016.21.29.30296

Krutova M, Matejkova J, Tkadlec J et al (2015) Antibiotic profiling of *Clostridium difficile* ribotype 176 – a multidrug resistant relative to *C. difficile* ribotype 027. Anaerobe 36:88–90. https://doi.org/10.1016/j.anaerobe.2015.07.009

Krutova M, Zouharova M, Matejkova J et al (2018) The emergence of *Clostridium difficile* PCR ribotype 078 in piglets in the Czech Republic clusters with *Clostridium difficile* PCR ribotype 078 isolates from Germany, Japan and Taiwan. Int J Med Microbiol IJMM 308: 770–775. https://doi.org/10.1016/j.ijmm.2018.05.006

Kuijper E, Weerdt J, Kato H et al (2001) Nosocomial outbreak of *Clostridium difficile*-associated diarrhoea due to a clindamycin-resistant enterotoxin A-negative strain. Eur J Clin Microbiol Infect Dis 20:528–534. https://doi.org/10.1007/s100960100550

Lessa FC, Mu Y, Bamberg WM et al (2015) Burden of *Clostridium difficile* infection in the United States. N Engl J Med 372:825–834. https://doi.org/10.1056/NEJMoa1408913

Lim SK, Stuart RL, Mackin KE et al (2014) Emergence of a ribotype 244 strain of *Clostridium difficile* associated with severe disease and related to the epidemic ribotype 027 strain. Clin Infect Dis 58:1723–1730. https://doi.org/10.1093/cid/ciu203

Magnusson C, Mernelius S, Bengnér M et al (2022) Characterization of a *Clostridioides difficile* outbreak caused by PCR ribotype 046, associated with increased mortality. Emerg Microbes Infect 11:850–859. https://doi.org/10.1080/22221751.2022.2049981

Marsh JW, O'Leary MM, Shutt KA et al (2006) Multilocus variable-number tandem-repeat analysis for investigation of *Clostridium difficile* transmission in hospitals. J Clin Microbiol 44:2558–2566. https://doi.org/10.1128/JCM.02364-05

Metcalf DS, Costa MC, Dew WMV et al (2010) *Clostridium difficile* in vegetables, Canada. Lett Appl Microbiol 51:600–602. https://doi.org/10.1111/j.1472-765X.2010.02933.x

Moloney G, Eyre DW, Mac Aogáin M et al (2021) Human and porcine transmission of *Clostridioides difficile* ribotype 078, Europe. Emerg Infect Dis 27:2294–2300. https://doi.org/10.3201/eid2709.203468

Monot M, Eckert C, Lemire A et al (2015) *Clostridium difficile*: new insights into the evolution of the pathogenicity locus. Sci Rep 5. https://doi.org/10.1038/srep15023

Murray R, Boyd D, Levett PN et al (2009) Truncation in the tcdC region of the *Clostridium difficile* PathLoc of clinical isolates does not predict increased biological activity of Toxin B or Toxin A. BMC Infect Dis 9:103. https://doi.org/10.1186/1471-2334-9-103

Nazareth C, Leitão I, Reis E et al (2022) Epidemiology of *Clostridioides difficile* infection in Portugal: a retrospective, observational study of hospitalized patients. Acta Med Port 35:270–278. https://doi.org/10.20344/amp.15890

Neely F, Lambert M-L, Van Broeck J et al (2017) Clinical and laboratory features of the most common *Clostridium difficile* ribotypes isolated in Belgium. J Hosp Infect 95:394–399. https://doi.org/10.1016/j.jhin.2016.12.011

Novakova E, Stefkovicova M, Kopilec MG et al (2020) The emergence of *Clostridium difficile* ribotypes 027 and 176 with a predominance of the *Clostridium difficile* ribotype 001 recognized in Slovakia following the European standardized *Clostridium difficile* infection surveillance of 2016. Int J Infect Dis IJID Off Publ Int Soc Infect Dis 90:111–115. https://doi.org/10.1016/j.ijid.2019.10.038

Nyč O, Pituch H, Matějková J et al (2011) *Clostridium difficile* PCR ribotype 176 in the Czech Republic and Poland. Lancet 377:1407. https://doi.org/10.1016/S0140-6736(11)60575-8

Obuch-Woszczatyński P, Lachowicz D, Schneider A et al (2014) Occurrence of *Clostridium difficile* PCR-ribotype 027 and it's closely related PCR-ribotype 176 in hospitals in Poland in 2008-2010. Anaerobe 28:13–17. https://doi.org/10.1016/j.anaerobe.2014.04.007

Piepenbrock E, Stelzer Y, Berger F et al (2019) Changes in *Clostridium (Clostridioides) difficile* PCR-ribotype distribution and antimicrobial resistance in a German tertiary care hospital over the last 10 years. Curr Microbiol 76:520–526. https://doi.org/10.1007/s00284-019-01654-3

Pituch H, van den Braak N, van Leeuwen W et al (2001) Clonal dissemination of a toxin-A-negative/toxin-B-positive *Clostridium difficile* strain from patients with antibiotic-associated diarrhea in Poland. Clin Microbiol Infect 7:442–446. https://doi.org/10.1046/j.1198-743x.2001.00312.x

Pruitt RN, Lacy DB (2012) Toward a structural understanding of *Clostridium difficile* toxins A and B. Front Cell Infect Microbiol 2:28. https://doi.org/10.3389/fcimb.2012.00028

Romano V, Pasquale V, Krovacek K et al (2012) Toxigenic *Clostridium difficile* PCR ribotypes from wastewater treatment plants in Southern Switzerland. Appl Environ Microbiol 78:6643–6646. https://doi.org/10.1128/AEM.01379-12

Romano V, Pasquale V, Lemee L et al (2018) *Clostridioides difficile* in the environment, food, animals and humans in southern Italy: occurrence and genetic relatedness. Comp Immunol Microbiol Infect Dis 59:41–46. https://doi.org/10.1016/j.cimid.2018.08.006

Rupnik M, Avesani V, Janc M et al (1998) A novel toxinotyping scheme and correlation of toxinotypes with serogroups of *Clostridium difficile* isolates. J Clin Microbiol 36:2240–2247

Rupnik M, Brazier JS, Duerden BI et al (2001) Comparison of toxinotyping and PCR ribotyping of *Clostridium difficile* strains and description of novel toxinotypes. Microbiol Read Engl 147:439–447. https://doi.org/10.1099/00221287-147-2-439

Rupnik M, Janezic S (2016) An update on *Clostridium difficile* toxinotyping. J Clin Microbiol 54:13–18. https://doi.org/10.1128/JCM.02083-15

Rupnik M, Tambic Andrasevic A, Trajkovska Dokic E et al (2016) Distribution of *Clostridium difficile* PCR ribotypes and high proportion of 027 and 176 in some hospitals in four South Eastern European countries. Anaerobe 42:142–144. https://doi.org/10.1016/j.anaerobe.2016.10.005

Santos A, Isidro J, Silva C et al (2016) Molecular and epidemiologic study of *Clostridium difficile* reveals unusual heterogeneity in clinical strains circulating in different regions in Portugal. Clin Microbiol Infect 22: 695–700. https://doi.org/10.1016/j.cmi.2016.04.002

Schneeberg A, Neubauer H, Schmoock G et al (2013) *Clostridium difficile* genotypes in piglet populations in Germany. J Clin Microbiol 51:3796–3803. https://doi.org/10.1128/JCM.01440-13

Smith A (2005) Outbreak of *Clostridium difficile* infection in an English hospital linked to hypertoxin-producing strains in Canada and the US. Euro Surveill Bull Eur Sur Mal Transm Eur Commun Dis Bull 10:E050630.2

Spigaglia P, Barbanti F, Dionisi AM et al (2010) *Clostridium difficile* isolates resistant to fluoroquinolones in Italy: emergence of PCR ribotype 018. J Clin Microbiol 48:2892–2896. https://doi.org/10.1128/JCM.02482-09

Stabler RA, Dawson LF, Valiente E et al (2012) Macro and micro diversity of *Clostridium difficile* isolates from diverse sources and geographical locations. PLoS One 7:e31559. https://doi.org/10.1371/journal.pone.0031559

Stabler RA, Gerding DN, Songer JG et al (2006) Comparative phylogenomics of *Clostridium difficile* reveals clade specificity and microevolution of hypervirulent strains. J Bacteriol 188:7297–7305. https://doi.org/10.1128/JB.00664-06

Stubbs SL, Brazier JS, O'Neill GL et al (1999) PCR targeted to the 16S-23S rRNA gene intergenic spacer region of *Clostridium difficile* and construction of a library consisting of 116 different PCR ribotypes. J Clin Microbiol 37:461–463

Sundram F, Guyot A, Carboo I et al (2009) *Clostridium difficile* ribotypes 027 and 106: clinical outcomes and risk factors. J Hosp Infect 72:111–118. https://doi.org/10.1016/j.jhin.2009.02.020

Taori SK, Wroe A, Hardie A et al (2014) A prospective study of community-associated *Clostridium difficile* infections: the role of antibiotics and co-infections. J Infect 69:134–144. https://doi.org/10.1016/j.jinf.2014.04.002

Tkalec V, Jamnikar-Ciglenecki U, Rupnik M et al (2020) *Clostridioides difficile* in national food surveillance, Slovenia, 2015 to 2017. Euro Surveill Bull Eur Sur Mal Transm Eur Commun Dis Bull 25:1900479. https://doi.org/10.2807/1560-7917.ES.2020.25.16.1900479

Tkalec V, Viprey V, Davis G et al (2022) *Clostridioides difficile* positivity rate and PCR ribotype distribution on retail potatoes in 12 European countries, January to June 2018. Euro Surveill Bull Eur Sur Mal Transm Eur Commun Dis Bull 27:2100417. https://doi.org/10.2807/1560-7917.ES.2022.27.15.2100417

Valiente E, Dawson LF, Cairns MD et al (2012) Emergence of new PCR ribotypes from the hypervirulent *Clostridium difficile* 027 lineage. J Med Microbiol 61: 49–56. https://doi.org/10.1099/jmm.0.036194-0

Viprey VF, Davis GL, Benson AD et al (2022) A point-prevalence study on community and inpatient *Clostridioides difficile* infections (CDI): results from Combatting Bacterial Resistance in Europe CDI (COMBACTE-CDI), July to November 2018. Euro Surveill Bull Eur Sur Mal Transm Eur Commun Dis Bull 27:2100704. https://doi.org/10.2807/1560-7917.ES.2022.27.26.2100704

Viprey VF, Granata G, Vendrik KEW et al (2023) European survey on the current surveillance practices, management guidelines, treatment pathways and heterogeneity of testing of *Clostridioides difficile*, 2018-2019: results from The Combatting Bacterial Resistance in Europe CDI (COMBACTE-CDI). J Hosp Infect 131:213–220. https://doi.org/10.1016/j.jhin.2022.11.011

von Müller L, Mock M, Halfmann A et al (2015) Epidemiology of *Clostridium difficile* in Germany based on a single center long-term surveillance and German-wide genotyping of recent isolates provided to the advisory laboratory for diagnostic reasons. Int J Med Microbiol 305:807–813. https://doi.org/10.1016/j.ijmm.2015.08.035

Warny M, Pepin J, Fang A et al (2005) Toxin production by an emerging strain of *Clostridium difficile* associated with outbreaks of severe disease in North America and Europe. Lancet 366:1079–1084. https://doi.org/10.1016/S0140-6736(05)67420-X

Werner A, Mölling P, Fagerström A et al (2020) Whole genome sequencing of *Clostridioides difficile* PCR ribotype 046 suggests transmission between pigs and humans. PLoS One 15:e0244227. https://doi.org/10.1371/journal.pone.0244227

Wilcox MH, Mooney L, Bendall R et al (2008) A case-control study of community-associated *Clostridium difficile* infection. J Antimicrob Chemother 62:388–396. https://doi.org/10.1093/jac/dkn163

Antibiotic Resistances of *Clostridioides difficile*

Patrizia Spigaglia, Paola Mastrantonio, and Fabrizio Barbanti

Abstract

The rapid evolution of antibiotic resistance in *Clostridioides difficile* and the consequent effects on prevention and treatment of *C. difficile* infections (CDIs) are a matter of concern for public health. Antibiotic resistance plays an important role in driving *C. difficile* epidemiology. Emergence of new types is often associated with the emergence of new resistances, and most of the epidemic *C. difficile* clinical isolates is currently resistant to multiple antibiotics. In particular, it is to worth to note the recent identification of strains with reduced susceptibility to the first-line antibiotics for CDI treatment and/or for relapsing infections. Antibiotic resistance in *C. difficile* has a multifactorial nature. Acquisition of genetic elements and alterations of the antibiotic target sites, as well as other factors, such as variations in the metabolic pathways or biofilm production, contribute to the survival of this pathogen in the presence of antibiotics. Different transfer mechanisms facilitate the spread of mobile elements among *C. difficile* strains and between *C. difficile* and other species. Furthermore, data indicate that both genetic elements and alterations in the antibiotic targets can be maintained in *C. difficile* regardless of the burden imposed on fitness, and therefore resistances may persist in *C. difficile* population in absence of antibiotic selective pressure.

Abbreviations

AD	agar dilution
AI	agar incorporation
Ala	alanine
AMP	ampicillin
AMR	antibiotic resistance
Arg	arginine
Asn	asparagine
Asp	aspartic acid
BP	breakpoint
CCCP	carbonyl cyanide m-chlorophenyl hydrazone
CDI	*Clostridioides difficile* infection
CDS	coding DNA sequence
CEF	cefepime
CFs	cephalosporins
CHL	chloramphenicol
CIP	ciprofloxacin
CLI	clindamycin
CLSI	Clinical and Laboratory Standards Institute
CRO	ceftriaxone
CTT	cefotetan
CTX	cefotaxime

P. Spigaglia (✉) · P. Mastrantonio · F. Barbanti
Department of Infectious Diseases, Istituto Superiore di Sanità, Rome, Italy
e-mail: patrizia.spigaglia@iss.it

P. Mastrantonio, M. Rupnik (eds.), *Updates on Clostridioides difficile in Europe*, Advances in Experimental Medicine and Biology 1435, https://doi.org/10.1007/978-3-031-42108-2_9

DOX	doxycycline
ECOFF	epidemiological cutoff
ERY	erythromycin
EUCAST	European Committee on Antimicrobial Susceptibility Testing
FDX	fidaxomicin
FQs	fluoroquinolones
FUS	fusidic acid
GAT	gatifloxacin
Glu	glutamic acid
Gly	glycine
Gyr	gyrase
His	histidine
Ile	isoleucine
IMP	imipenem
Leu	leucine
LEV	levofloxacin
LNZ	linezolid
Lys	lysine
MATE	multidrug and toxic compound extrusion
MDR	multidrug resistance
Met	methionine
MGEs	mobile genetic elements
MIC	minimum inhibitory concentration
MLSB	macrolide-lincosamide-streptogramin B
MRP	meropenem
MTZ	metronidazole
MXF	noxifloxacin
NAP	North American pulsed field gel electrophoresis
PBPs	penicillin-binding proteins
Phe	phenylalanine
PhLOPSA	phenicols, lincosamides, oxazolidinones, pleuromutilins, and streptogramin A
PIP-TAZ	piperacillin/tazobactam
Pro	proline
QRDR	quinolone resistance-determining region
RFs	rifamycins
RFX	rifaximin
RIF	rifampicin
RNAP	RNA polymerase
RpoB	beta subunit of RNA polymerase
rRNA	ribosomal RNA
RT	PCR ribotype
Ser	serine
TB	antituberculosis
TET	tetracycline
TGC	tigecycline
Thr	threonine
Tn	transposon
Tyr	tyrosine
Val	valine
VAN	vancomycin

1 Introduction

Clostridioides difficile is recognized as the major cause of healthcare antibiotic-associated diarrhea (Lessa et al. 2012; European Centre for Disease Prevention and Control 2013). Potentially, all antibiotic classes may promote *C. difficile* infection (CDI) by disrupting intestinal microflora and allowing *C. difficile*, ingested or resident, to proliferate, colonize the gastrointestinal tract, and infect the host. Therefore, resistance to multiple agents represents a selective advantage for *C. difficile* strains to enhance their survival and spread.

An alarming increase in incidence of CDI has been observed worldwide over the last 15 years, with a significant financial burden on the healthcare system (Redelings et al. 2007; Burckhardt et al. 2008; Bauer et al. 2011; Gravel et al. 2009; Miller et al. 2011a; Dubberke and Olsen 2012; Lessa et al. 2015). The increased number of infections has been mainly associated with the emergence of highly virulent *C. difficile* strains. In particular, strains PCR ribotype (RT) 027/North American pulsed field gel electrophoresis type I (NAPI)/restriction endonuclease analysis group B1 have been recognized responsible for severe CDI, characterized by high rate of recurrences, mortality, and refractory to traditional therapy (Pépin et al. 2004, 2005a, b; McDonald et al. 2005; Muto et al. 2005; Goorhuis et al. 2007; Clements et al. 2010).

C. difficile RT population is in expansion with some dominant types that are flanked by new emerging RTs over time. Besides RT 027, other important epidemic types, such as RT 014/020,

RT 001/072, and RT 078, are endemic in many countries, whereas others, such as RT 018, have a local spread (Freeman et al. 2015a; Knight et al. 2019; Brajerova et al. 2022). Recent studies report the emergence of highly virulent RTs. In particular, a new type, denominated RT 181, genetically related to RT 027, has reported as the cause of several outbreaks in Europe (Kachrimanidou et al. 2020; Kachrimanidou et al. 2022; Viprey et al. 2022). Another type, phylogenetic related to RT 027 and denominated RT 244, has emerged in New Zealand (De Almeida et al. 2013), while the RT 251 has recently caused severe infections in Australia (Wehrhahn et al. 2019).

Antibiotics have a central role in driving the emergence of new *C. difficile* types. The global spread of *C. difficile* RT 027 has been associated with the massive use of fluoroquinolones (FQs) and the acquisition of resistance to these antibiotics by strains of this type (He et al. 2013). Actually, multidrug resistance (MDR) is widely diffused among the majority of epidemic and emergent strains, RT 027 or not (Spigaglia et al. 2011; Byun et al. 2019; Imwattana et al. 2021a, b, 2022; Aptekorz et al. 2022; Gargis et al. 2022; Aguilar-Zamora et al. 2022).

Genetic analyses have demonstrated that *C. difficile* has a versatile genome content, with a wide range of mobile elements, many of them encoding for predicted antibiotic resistances (Sebaihia et al. 2006; He et al. 2010, 2013). Besides horizontal gene transfer, other mechanisms may contribute to promote antibiotic resistance in *C. difficile*, which appears to be a multifactorial phenomenon.

In this chapter, antibiotic resistances of *C. difficile* will be discussed taking in consideration the most recent published data.

2 *C. difficile* Antibiotic Susceptibility

CDI is induced by exposure to antibiotics that is associated with a 60% increased risk of infection (Slimings and Riley 2014). *C. difficile* susceptibility is usually evaluated for antibiotics known to be significantly associated with CDI or used for CDI treatment. Among the first group, clindamycin (CLI) and cephalosporins (CFs) are historically recognized as high-risk agents for CDI (Bartlett et al. 1977; Bignardi 1998). Although a decreased number of infections have been observed in the hospitals that have curtailed the use of these antibiotics (de Lalla et al. 1989; Khan and Cheesbrough 2003; Wiström et al. 2001), the risk of hospital acquired CDI remains high after CLI or CF therapy, so their importance as promoting agents should not be minimized. In the last decades, a rise in the FQ-associated CDI has been observed in concomitant with the increasing incidence of *C. difficile* RT 027. Current strains RT 027 show high-level resistance to FQs, never observed in historical isolates of the same type (McDonald et al. 2005). Infection control procedures and antimicrobial stewardship have led to a significant reduction in the incidence of infections caused by RT 027, but this type is still globally widespread, in particular in the Eastern Europe where outbreaks due to RT 027 resistant to several antibiotics, including moxifloxacin (MXF), erythromycin (ERY), clindamycin (CLI), imipenem (IMP), and rifampicin (RIF), have been reported (Lessa et al. 2015; Freeman et al. 2015a; Kabała et al. 2021; Aptekorz et al. 2022). Resistance to FQs has become very common also in strains belonging to other epidemic types, including RT 001, RT 017, RT 018, and RT 078, with prevalence values up to 94% (Barbanti and Spigaglia 2020; Krutova et al. 2020; Aguilar-Zamora et al. 2022; Imwattana et al. 2022).

Although metronidazole (MTZ) and vancomycin (VAN) have usually been considered as standard CDI therapies for mild and severe CDI, respectively (Debast et al. 2014; Jarrad et al. 2015; Lyras and Cooper 2015), international guidelines recently recommend the use of oral metronidazole only when other agents are unavailable (Bishop and Tiruvoipati 2022). CDI therapies also include rifamycins (RFs), in particular rifaximin (RFX), that have been prosed as "chaser therapy" for treatment of relapsing CDI (Iv et al. 2014), while fidaxomicin (FDX), a bactericidal new narrow-spectrum macrocyclic antibiotic, is used for the management of CDI with high risk for recurrences (Chaparro-Rojas and Mullane 2013). Recent papers report that *C. difficile* strains isolated from human show a similar weighted pooled resistance for MTZ and VAN of 1.0% (95% CI 0–3% and 0–2%), respectively, while in the *C. difficile* isolates from animals or environment, the weighted pooled resistance increased to 1.9% (95% CI 0.5–3.6%) for MTZ and to 2.1% (95% CI 0–5.1%) for VAN (Saha et al. 2019; Sholeh et al. 2020). Only few *C. difficile* isolates with MICs from 1 to 64 mg/L for FDX have recently been detected (Goldstein et al. 2011; Peng et al. 2017; Schwanbeck et al. 2019; Freeman et al. 2020; Karlowsky et al. 2020).

2.1 Antibiotics Associated with CDI

Rates of antibiotic resistance varies considerably depending on geographic areas and local/national antibiotic policy (Table 1). In general, *C. difficile* strains have higher rates of resistance to older generation of antibiotics than never. In fact, resistance to CFs of second generation is more commonly observed compared to resistance to CFs of third generation (95% vs 38%) (Dong et al. 2013; Norman et al. 2014; Oka et al. 2012, Karlowsky et al. 2012, Büchler et al. 2014, Kuwata et al. 2015, Knight et al. 2015a, 2016). Similarly, resistance to ciprofloxacin (CIP), a FQ of second generation, is very common in *C. difficile* (99%) (Rodríguez-Pardo et al. 2013; Lee et al. 2014; Norman et al. 2014; Lachowicz et al. 2015; Kuwata et al. 2015; Shayganmehr et al. 2015), and resistance to FQs of fourth generation such as moxifloxacin (MXF) and gatifloxacin (GAT) has been detected in a percentage between 36% and 68% of the strains analyzed, respectively (Karlowsky et al. 2020; Tenover et al. 2012; Eckert et al. 2013; Rodríguez-Pardo et al. 2013; Lee et al. 2014; Kim et al. 2012; Liao et al. 2012; Terhes et al. 2014; Weber et al. 2013; Varshney et al. 2014; Freeman et al. 2015a; Senoh et al. 2015; Adler et al. 2015; Kociolek et al. 2016; Putsathit et al. 2021; Gao et al. 2016; Santos et al. 2016; Knight et al. 2015a, b; Kullin et al. 2017).

Data extrapolated from studies recently published indicate that *C. difficile* strains resistant to CLI are widely diffused in different countries (Table 1), with a percentage of resistance ranging between 41.8% (USA) and 91.5% (Japan). Resistance to MXF has been reported in different geographic areas with a percentage comprised between 21.7% (Canada) and 94% (Mexico), although in some area it is less frequently observed, such as in Australia (3.5%) (Putsathit et al. 2021; Du et al. 2022; Aguilar-Zamora et al. 2022; Jiménez et al. 2018).

The percentage of *C. difficile* strains resistant to meropenem (MRP) and RIF shows great variability in the different countries, ranging from 0.1% to 54% for MRP and from 9.3% to 79% for RIF, while a lower percentage of *C. difficile* strains is resistant to tetracycline (TET) (from 6% to 14.4%) (Table 1).

2.2 Antibiotics for CDI Treatment

Metronidazole Although percentage of *C. difficile* strains resistant to MTZ is still low (Table 1), several studies have reported high rate of treatment failures in patients who received this antibiotic (Musher et al. 2005; Pépin et al. 2005a, b; Vardakas et al. 2012; Zar et al. 2007). Recent data indicate the isolation of strains with minimum inhibitory concentrations (MICs) >2 mg/L or ≥32 mg/L, the European Committee on Antimicrobial Susceptibility Testing (EUCAST) epidemiological cutoff (ECOFF) (http://www.eucast.org/clinical_breakpoints), and the Clinical and Laboratory Standards Institute (CLSI) (Clinical and Laboratory Standards Institute 2012) breakpoint for MTZ, respectively (Tables 1 and 2). Besides RT 027, reduced susceptibility to MTZ has been observed also in other important *C. difficile*-epidemic RTs (Adler et al. 2015; Kouzegaran et al. 2016; Baines and Wilcox 2015; Spigaglia 2016). An association between a MIC breakpoint of ≥1 mg/L for MTZ and an increased risk of initial clinical failure has recently been reported (Gonzales-Luna et al. 2021). Furthermore, a link between prolonged/repeated use of MTZ and neurotoxicity has been

Table 1 Antibiotic susceptibility and multiresistance of *C. difficile* clinical isolates from different Countries

Country	Reference	Period	N. of strains analysed	Antibiotic[a]	N. of resistant	Strains (%)	% MDR	MDR patterns	Breakpoint in mg/L	Susceptibility[a] method
Australia	Putsathit et al. (2021)	2015–2018	1091				1.7	CLI, CFX, MXF. VAN, CLI, CFX. CLI, CFX, MRP, MXF. VAN, CLI, CFX, MXF.		AI
				CLI	929	(85.2)			≥8	
				MXF	38	(3.5)			≥8	
				MRP	1	(0.1)			≥16	
				RFX	0	0			≥32	
				FDX	0	0			>1	
				MTZ	0	0			>2	
				VAN	62	(5.7)			>2	
Canada	Karlowsky et al. (2020)	2013–2017	2099				12	NR[b]		AD
	Du et al. (2022)	2015–2019	1887	ERY	586	(31.0)			≥8	
				MXF	410	(21.7)			≥8	
				RIF	35	(1.9)			≥4	
				TGC	0	0			≥16	
				MTZ	0	0			≥32	
				VAN	0	0			≥32	
China	Jin et al. (2017)	2012–2015	411				74.5	NR		Etest
				CLI	257	(62.5)			≥8	
				ERY	266	(64.7)			≥8	
				CIP	334	(81.3)			≥8	
				MXF	104	(25.3)			≥8	
				LEV	308	(74.9)			≥8	
				GAT	133	(32.4)			≥8	
				RIF	38	(9.3)			≥4	
				PIP-TAZ	5	(1.2)			128-4	
				FUS	290	(70.6)			≥2	
				TET	149	(36.3)			≥16	
				MTZ	64	(15.6)			≥32	

(continued)

Table 1 (continued)

Country	Reference	Period	N. of strains analysed	Antibiotic[a]	N. of resistant	Strains (%)	% MDR	MDR patterns	Breakpoint in mg/L	Susceptibility[a] method
Europe	Spigaglia et al. (2011)	2005	148				55	ERY, CLI, TET. ERY, CLI, MXF. ERY, CLI, MXF, RIF. ERY, CLI, MXF, TET. ERY, CLI, RIF, TET. ERY, CLI, MXF, RIF, CHL. ERY, CLI, MXF, RIF, TET, CHL.		AI
	Freeman et al. (2018)	2011–2014	2698	CLI	1527	(56.6)			≥8	
				MXF	966	(35.8)			≥8	
				IMP	108	(4.0)			≥16	
				RIF	351	(13.0)			≥16	
				CHL	97	(3.6)			≥32	
				FDX	0	0			>1	
				MTZ	6	(0.2)			≥8	
				VAN	3	(0.1)			≥8	
				TGC	0	0			≥16	
Japan	Aoki et al. (2019)	2014–2016	188				NR	NR		AD
				CLI	172	(91.5)			≥8	
				MXF	124	(65.9)			≥8	
				FDX	0	0			≥1	
				MTZ	0	0			≥32	
				VAN	0	0			≥4	

Kenya	Mutai et al. (2021)	2016–2018	71				85.9	NR		Etest
				CLI	57	(80.3)				
				ERY	63	(88.7)				
				CIP	59	(83.1)				
				CRO	36	(50.7)				
				RIF	65	(91.5)				
				TET	7	(9.9)				
				MTZ	3	(4.2)				
				VAN	0	0				
Mexico	Aguilar-Zamora et al. (2022)	2014–2018	94				65	CLI, CIP, MRP. CLI, LEV, CIP, RIF, LNZ. CLI, LEV, CIP, RIF, MRP. CLI, LEV, CIP, RIF, MRP, TET.		Etest
				CLI	83	(88.2)			>8	
				MXF	88	(93.6)			>8	
				LEV	84	(89.3)			>8	
				CIP	93	(98.9)			>8	
				MRP	65	(69.1)			>16	
				RIF	74	(78.7)			>16	
				LNZ	57	(60.6)			>4	
				FUS	11	(11.7)			0.5	
				TET	7	(7.4)			>16	
				MTZ	0	0			>32	
				VAN	0	0			>2	
South Korea	Byun et al. (2019)	2017	106 (AMP, CTT, CLI, CHL, TET, MXF; IMP) / 257 (VAN, MTZ, RIF)				47	CTT, CLI, IPM, MXF. AMP, CTT, IPM, MXF. RIF, AMP, CLI, MXF. CTT, CLI, IPM, TET, MOX. AMP, CTT, CLI, IPM, MXF. RIF, AMP, CTT, CLI, IPM, MOX.		AD

(continued)

Table 1 (continued)

Country	Reference	Period	N. of strains analysed	Antibiotic[a]	N. of resistant	Strains (%)	% MDR	MDR patterns	Breakpoint in mg/L	Susceptibility[a] method
				CLI	68	(64.1)			≥8	
				MXF	55	(51.8)			≥8	
				CTT	49	(46.2)			≥64	
				IMP	57	(53.7)			≥16	
				RIF	62	(24.1)			≥32	
				AMP	51	(48.1)			≥2	
				CHL	0	0			≥32	
				TET	6	(5.6)			≥16	
				MTZ	0	0			>2	
				VAN	0	0			>2	
Thailand	Imwattana et al. (2021a, b)	2017–2018	321				9	NR		AI
				CLI	263	82			≥8	
				ERY	109	34			>8	
				MXF	76	24			≥8	
				MRP	2	1			≥16	
				RFX	31	10			>32	
				FDX	0 (0)	0			>1	
				MTZ	0 (0)	0			>2	
				VAN	0 (0)	0			>2	
USA	Tickler et al. (2019)	2011–2017	940 (CLI, MXF, RIF, MTZ) / 938 (TET) / 774 (VAN)				NR	NR		Etest
				CLI	393	41.8			≥8	
				MXF	269	28.6			≥8	
				RIF	102	10.8			≥32	
				TET	136	14.4			≥16	
				MTZ	0 (0)	0			≥32	
				VAN	81	10.4			>2	

[a]See section Abbreviations
[b]*NR* not reported

Table 2 Susceptibility to metronidazole and vancomycin of *C. difficile* clinical isolates

Review of reference	Period of time	N. of studies included for each antibiotic	Percentage (resistant/all strains) of *C. difficile*strains resistant to		
			MTZ[a]	VAN	FDX
Dilnessa et al. (2022)	2007–2020	15 (MTZ and VAN)[a]	4.9 (137/2753) BP[a] not known	4.1 (114/2755) BP not known	
Sholeh et al. (2020)	1992–2019	32 (MTZ) 58 (VAN) 6 (FDX)	3.2 (190/5900) BP EUCAST 2 mg/L	3.7 (416/11188) BP EUCAST 2 mg/L	0.08 (1/1188) BP ≥ 8mg/L
		69 (MTZ) 18 (VAN)	1 (129/13207) BP CLSI 32 mg/L	0.6 (13/2307) BP CLSI 32 mg/L	
		1 (VAN)		0.4 (10/2296) BP 16 mg/L	
		8 (VAN)		0.6 (7/1107) BP 4 mg/L	
Saha et al. (2019)	1982–2017	55 (MTZ) 53 (VAN)	1.9 (143/7507) BP not known	2.1 (152/7225) BP not known	

[a]See section Abbreviations

observed (McDonald et al. 2018; Dingsdag and Hunter 2017). For all these reasons, the major guidelines have removed MTZ as a first-line treatment option for CDI, suggesting to use it only when VAN or FDX is not available (McDonald et al. 2018; Johnson et al. 2021; Kelly et al. 2021; van Prehn et al. 2021). The use of oral vancomycin and FDX is supported by the high concentrations that these antibiotics reach in the stool, although only FDX inhibits sporulation and shows antimicrobial activity on spores, with minimal effects on gut microbiota (Krutova et al. 2022a, b).

An increase in the geometric mean of MICs for MTZ in isolates RT 027 (1.1–1.42 mg/L), RT 001/072 (0.65 mg/L), RT 106 (0.65 mg/L), RT 356 (0.61 mg/L), and in the nontoxigenic RT 010 (1.5 mg/L), compared to other RTs (0.13–0.41 mg/L) has been observed (Moura et al. 2013; Freeman et al. 2015a, b). Although the clinical relevance of strains with reduced susceptibility to MTZ is still unclear, it has been suggested a potential impact of strains RT 027 with reduced susceptibility to MTZ on the pathophysiology of recurrent CDIs (Richardson et al. 2015). In addition, strains RT 027 with this characteristic have been identified as cause of severe infections in Israel (Adler et al. 2015; Kouzegaran et al. 2016). In particular, a wide outbreak caused by a strain RT 027 with high MIC values for MTZ has been reported in Jerusalem in 2013 (Adler et al. 2015).

C. difficile colonies with increased MICs to MTZ can be isolated in the presence of subinhibitory concentrations of antibiotic (Peláez et al. 2008; Moura et al. 2013). Heteroresistance, which is the capacity of a part of bacterial population to acquire resistance and grow in presence of an antibiotic, could be considered a pre-resistance stage in *C. difficile* (Falagas et al. 2008; Peláez et al. 2008). Mean concentrations of MTZ in the feces of treated patients are not so high (from 0.8 to 24.2 μg/g) (Bolton and Culshaw 1986); therefore it is possible that the concentrations achieved in the colon may be insufficient for the treatment of infections due to strains with higher MIC values for MTZ (Brazier et al. 2001; Baines et al. 2008; Moura et al. 2013).

Vancomycin Reduced susceptibility to VAN in *C. difficile* is not largely diffused as in *Enterococci* and *Staphylococci*, although an increased number of strains with higher MICs to this antibiotic (MICs range >2–16 mg/L) have been isolated in different geographic areas (Table 1). In particular, recent reviews report the percentage of *C. difficile* strains resistant to VAN comprised between 0.6% and 4.1%, depending on the method of testing and the breakpoint used (Table 2). The clinical significance of strains with reduced susceptibility to VAN remains to be determined due to the high concentrations that this antibiotic reaches in the gastrointestinal tract (Young et al. 1985). Anyway, it is noteworthy that reduced susceptibilities to VAN and to MTZ are reported in several RTs, including RT 027, RT 001, RT 017, RT 078, and RT 356/607 (Chia et al. 2013; Goudarzi et al. 2013; Adler et al. 2015; Freeman et al. 2015a, b; Miller-Roll et al. 2016). Recently, it has been observed a statistically significant increase in resistance to VAN in *C. difficile* strains isolated post-2012 compared to those isolated pre-2012, a datum that correlated with the global increase in vancomycin usage (Saha et al. 2019). Furthermore, a recent study reports the emergence of *C. difficile* isolates with reduced susceptibility to VAN in the patient population of Houston and Nairobi, with the 26% and 67% of resistant isolates, respectively (Darkoh et al. 2022). The spread of *C. difficile* strains with these characteristics may determine challenges in the treatment of this infection, with serious public health implications, highlighting the necessity that a routinely susceptibility testing and analysis of the mechanisms of resistant of the circulating *C. difficile* strains will be expanded.

Fidaxomicin Approved for clinical use in 2011, FDX shows potent activity against *C. difficile.* FDX is suggested in initial nonsevere CDI and in the first recurrence, while both FDX and VAN represent an equal option of treatment in severe CDI (McDonald et al. 2018; van Prehn et al. 2021; Kelly et al. 2021; Johnson et al. 2021; Krutova et al. 2022a, b), although FDX resulted inferior to vancomycin for clinical response to infections due to RT 027 (Louie et al. 2011; Cornely et al. 2012). Until today, few isolates with MICs comprised from 1 to 64 mg/L have been identified (Freeman et al. 2020; Goldstein et al. 2011; Karlowsky et al. 2020; Schwanbeck et al. 2019; Peng et al. 2017); however data are limited due to the unavailability of a MIC breakpoint and commercial E-test for this antibiotic (Sholeh et al. 2020).

Rifamycins (RFs) Rifamycins (RFs), in particular rifampin (RIF), have been reported as a potential treatment for CDI (Garey et al. 2008; Basu et al. 2010). Anyway, recent data indicate that the percentage of *C. difficile* clinical isolates resistant to RIF ranged between 1.9% and 91.5% (Table 1). The emergence of strains resistant to RIF in several countries suggests that the broad-spectrum activity of this antibiotic does not always apply,\ and also implies that it could cause CDI following long-term antibiotic treatment (Kurahara et al. 2022). In fact, selective pressure after exposure to antibiotic seems to have a role in selecting *C. difficile* colonies resistant to RFs (Curry et al. 2009; Miller et al. 2011a, b). Therefore, resistant *C. difficile* strains might emerge even during therapy (Johnson et al. 2009; Carman et al. 2012). RFs are commonly used as antituberculosis (TB) agents. Interestingly, in Poland, all strains belonging to the emergent RT 046 isolated from patients affected by TB and treated with prolonged RIF therapy showed high MICs to these antibiotics (Obuch-Woszczatyński et al. 2013). Susceptibility to RIF correlated completely with susceptibility to RFX (Miller et al. 2011b). Thus, susceptibility of the rifamycin class in *C. difficile* can be assessed by testing susceptibility to RIF.

3 Multidrug Resistance in *C. difficile*

Recent studies indicate that the percentage of MDR *C. difficile* strains is variable in the different countries and may reach high levels as observed in China and Kenya, 74.5% and 85.9%, respectively (Table 1). The first European prospective survey of *C. difficile* infections in 2005 showed that 55% of resistant clinical isolates were MDR (Spigaglia et al. 2011). Data indicate that the MDR patterns mainly include resistance to CLI, FQs, ERY, and CFs (Table 1).

Table 3 Antibiotic resistance of the predominant and emergent *C. difficile* ribotypes (RTs)

RT	Toxinogenic profile[a]	Clade	Resistance reported[b]	References
001/072	*tcdA*+, *tcdB*+, *cdtA*−, *cdtB*−	1	MXF, LEV, ERY, CTX, CLI, RIF, TET, MTZ, VAN, IMP, CIP, CRO	Byun et al. (2019), Baghani et al. (2020), Barbanti and Spigaglia (2020), Lew et al. (2020), Martínez-Meléndez et al. (2020), Jon et al. (2021), Abdrabou et al. (2022)
012	*tcdA*+, *tcdB*+, *cdtA*−, *cdtB*−	1	MXF, CLI, ERY, RIF, TET, AMP	Byun et al. (2019), Lew et al. (2020), Abdrabou et al. (2022), Martínez-Meléndez et al. (2020)
014/020	*tcdA*+, *tcdB*+, *cdtA*−, *cdtB*−	1	CLI, ERY, TET; RIF, MXF, LEV, IMP, AMP, GAT	Jiménez et al. (2018), Byun et al. (2019), Kato et al. (2019), Tickler et al. (2019), Barbanti and Spigaglia (2020), Lew et al. (2020), Martínez-Meléndez et al. (2020), Jon et al. (2021), Abdrabou et al. (2022)
018	*tcdA*+, *tcdB*+, *cdtA*−, *cdtB*−	1	ERY, CLI, MXF, AMP, CTT, IPM, GAT	Byun et al. (2019), Kato et al. (2019), Tickler et al. (2019), Barbanti and Spigaglia (2020), Lew et al. (2020), Martínez-Meléndez et al. (2020), Abdrabou et al. (2022)
106	*tcdA*+, *tcdB*+, *cdtA*−, *cdtB*−	1	CLI, MXF, RIF, ERY, LEV	Jiménez et al. (2018), Tickler et al. (2019), Barbanti and Spigaglia (2020), Lew et al. (2020), Martínez-Meléndez et al. (2020), Abdrabou et al. (2022), Du et al. (2022)
027/176	*tcdA*+, *tcdB*+, *cdtA*+, *cdtB*+	2	ERY, CLI, MXF, MTZ, RIF, CIP, IMP, CRO	Tickler et al. (2019), Martínez-Meléndez et al. (2020), Jon et al. (2021), Abdrabou et al. (2022), Aptekorz et al. (2022)
244	*tcdA*+, *tcdB*+, *cdtA*+, *cdtB*+	2	NR[b]	Knight et al. (2015a, b), Martínez-Meléndez et al. (2020)
181	*tcdA*+, *tcdB*+, *cdtA*+, *cdtB*+	2	MXF, CLI , RIF	Abdrabou et al. (2022)
023	*tcdA*+, *tcdB*+, *cdtA*+, *cdtB*+	3	CIP, CLI, RIF	Martínez-Meléndez et al. (2020), Shaw et al. (2020), Abdrabou et al. (2022)
017	*tcdA*−, *tcdB*+, *cdtA*−, *cdtB*−	4	TET, CLI, RIF, MXF, ERY, AMP CTT, IPM , CRO, RFX	Jiménez et al. (2018), Byun et al. (2019), Lew et al. (2020), Martínez-Meléndez et al. (2020), Jon et al. (2021), Abdrabou et al. (2022), Imwattana et al. (2022)
078	*tcdA*+, *tcdB*+, *cdtA*+, *cdtB*+	5	DOX, TGC, TET, MXF, CLI, ERY, RIF, CHL, LNZ, CIP	Knight et al. (2019), Barbanti and Spigaglia (2020), Martínez-Meléndez et al. (2020), Jon et al. (2021), Abdrabou et al. (2022)
126	*tcdA*+, *tcdB*+, *cdtA*+, *cdtB*+	5	TET, MXF, CLI. ERY, RIF	Knight et al. (2019), Barbanti and Spigaglia (2020), Martínez-Meléndez et al. (2020), Abdrabou et al. (2022)

[a]*tcdA* and *tcdB*: genes encoding for toxins A and B, respectively; *cdtA* and *cdtB*: genes encoding for the binary toxin CDT subunits A and B, respectively
[b]See section Abbreviations

Interestingly, resistance to multiple antibiotics characterized several epidemic RTs (Table 3), such as RT 027 and the related types RT 176 and RT 181, recently emerged in Europe (Obuch-Woszczatyński et al. 2014; Krutova et al. 2015; Abdrabou et al. 2022; Aptekorz et al. 2022; Kachrimanidou et al. 2022). Multiple resistance to ERY, CLI, MXF, and RIF also characterized most of the strains belonging to RT 356/607 and RT 018, two genetically correlated types recently emerged in Italy (Spigaglia et al. 2010, 2015; Barbanti and Spigaglia 2020). Interestingly, RT 018 strains isolated in Korea and Japan show resistance only to CLI, ERY, and MXF (Kim et al. 2012; Senoh et al. 2015). The 20 years of use of RIFs in Italy (Salix Pharmaceuticals, Ltd. 10 December 2003, posting date) could explain the spread of this resistance in Italian *C. difficile* isolates. Strains RT 018 are highly virulent and transmissible, with a transmission index that has been demonstrated tenfold higher compared to that of strains RT 078 (Baldan et al. 2015). Old age (≥65 years), severe pulmonary comorbidity, previous use of FQs, and infection by RT 018 have been associated as significant risk factors for complicated infections (Bauer et al. 2011). *C. difficile* strains belonging to the RT 078 lineage, a lineage of global One-Health importance, also have a large antibiotic resistance (AMR) repertoire (Knight & Riley 2016; Knight et al. 2019). In particular, an evolutionary analysis has demonstrated that the acquisition of resistance to TET, due to the transposons Tn*6190* and Tn*6164*, has provided not only AMR but also a fitness advantage, contributing to the successful diffusion of this lineage (Knight et al. 2019). Finally, resistance to multiple antibiotics has driven the globally spread of RT 017, a type recognized as the cause of severe infection and outbreaks in Asia, Europe, and the USA (Imwattana et al. 2022).

4 *C. difficile* Antibiotic Susceptibility Methods

Susceptibility testing is usually performed by clinical microbiology laboratories to determine antimicrobial resistance profiles of *C. difficile* isolates recovered from patients, but it is also used to monitor resistance patterns of strains isolated during epidemiological studies and surveillance networks.

The most common antibiotic susceptibility methods used for *C. difficile* are the agar dilution (AD) and the epsilometer test, a commercially available gradient diffusion system for quantitative antibiotic susceptibility testing. The AD is indicated as the reference method for *C. difficile* by the CLSI (Clinical and Laboratory Standards Institute 2012). The AD assay shows some advantages for epidemiological studies because it is an accurate method, the choice of antibiotics to be tested is flexible and can be modified according to investigational necessity, and finally, it is suitable for large numbers of isolates. The disadvantages of the AD approach are the laborious, time-consuming steps required to prepare testing plates, particularly when the number of compounds to be tested is high and/or when only a limited number of strains are to be analyzed, and the need of skilled and experienced technologists to properly perform it. For these reasons, most laboratories use the epsilometer test, more flexible and simple, for routine. Although there were differences in MIC values between AD and epsilometer test, high categorical agreement between these methods has been demonstrated (Moura et al. 2013; Baines et al. 2008; Poilane et al. 2000). In addition, the epsilometer test allows analysis of susceptibility to multiple antibiotics for numerous strains at the same time. Despite these advantages, the high cost hinders the extensive use of this method in clinical laboratories and epidemiological studies.

Detection of strains with reduced susceptibility to MTZ poses problems in choosing the more proper antibiotic susceptibility method to test them. In fact, resistance to MTZ is often unstable, and laboratory manipulation of strains frequently results in MIC decrease toward susceptibility range (Peláez et al. 2008; Lynch et al. 2013). Recent studies suggest the agar incorporation (AI) as the method of choice to detect strains with reduced susceptibility to MTZ compared to the AD (Freeman et al. 2015a; Moura et al. 2013). Differences in the media used (Schaedlers broth and Wilkins-Chalgren agar for AI and Brucella broth/agar for AD) and in the pre-cultured period (24 h for AD and 48 h for AI) seem to affect MIC determination (Baines et al. 2008; Moura et al. 2013). In addition, a recent study indicates that heme is a crucial component for *C. difficile* medium-dependent reduction in MTZ susceptibility (Boekhoud et al. 2021). Furthermore, the CLSI and the EUCAST breakpoints for MTZ are not equivalent: the first is defined ≥32 mg/L and the second >2 mg/L (Clinical and Laboratory Standards Institute 2012; http://www.eucast.org/clinical_breakpoints/). Since methodological differences and different interpretation categories may cause discrepancies in results, influencing therapeutic decision and comparison of data, international committees are currently cooperating with the intention of harmonizing susceptibility testing and breakpoints for this antibiotic.

Disk diffusion testing is not recommended by CLSI for *C. difficile*, but some recent papers suggest that it could be an option for antimicrobial susceptibility testing of this pathogen. A study carried out in Denmark on 211 isolates showed that an excellent agreement was found between MIC results when the epsilometer test and disk diffusion were used to test *C. difficile* strain susceptibility to vancomycin, moxifloxacin, and metronidazole (Erikstrup et al. 2012). Furthermore, two studies, performed in Denmark and Brazil, respectively, successfully used disk diffusion to test *C. difficile* isolates with reduced susceptibility to metronidazole and vancomycin (Holt et al. 2015; Fraga et al. 2016). Despite these results, an exact zone diameter for breakpoints is still not determined either by CLSI or by EUCAST; therefore the debate about disk diffusion, as qualified antibiotic susceptibly testing method for *C. difficile*, is still open.

Although in some paper *C. difficile* MIC values have been obtained using broth microdilution (Genzel et al. 2014; Lim et al. 2016), CLSI recommends this method only to test *Bacteroides* species (Clinical Laboratory Standards Institute 2012). Furthermore, a study of Hastey et al. has demonstrated a negative bias for the broth microdilution when compared to the AD for *C. difficile* (Hastey et al. 2017). In this study, the MIC values obtained using the broth microdilution were lower than those obtained with AD. Furthermore, the reproducibility with broth microdilution was variable, probably dependent on the antibiotics tested. Therefore, in accordance with the CLSI guideline (Clinical and Laboratory Standards Institute 2012), the results indicate that the broth microdilution method is not equivalent with AD for *C. difficile* antimicrobial susceptibility testing.

The phenotypic tests are traditional methods to evaluate antibiotic susceptibility of *C. difficile*, but they need time (almost one week to get the results) and the isolation of *C. difficile* from patient stools. Since resistance to several antibiotics has been correlated with resistance genes and alteration in antibiotic targets, molecular analysis may be considered to investigate *C. difficile* resistance beside phenotypic tests. The decreased cost of these technologies will allow their introduction on a large scale as tool for infection control in the future, as suggested by very recent studies that demonstrate the importance of molecular analysis and comparative genomics in the epidemiological surveillance and evolutionary analysis of *C. difficile* (Ramírez-Vargas et al. 2017; Cairns et al. 2017; Imwattana et al. 2020a, b; Roxas et al. 2020; Xu et al. 2021). However, if silico methods can rapidly provide a large amount of data on antimicrobial resistance determinants at once, not always genotype correlate with phenotype. For example, in a recent study on *C. difficile*, RT 078 lineage in silico AMR screening completely matched (100%) with agar dilution MICs for FQs and TET while only poorly for macrolide-lincosamide-streptogramin B (MLSBs) (36%) (Knight et al. 2019).

5 *C. difficile* Mechanisms of Resistance

Several mechanisms responsible for antibiotic resistance have been identified in *C. difficile*, including chromosomal resistance genes, mobile genetic elements (MGEs), alterations in the antibiotic targets and/or in metabolic pathways, and biofilm formation (Table 4). Furthermore, recent evidences support that *C. difficile* resistance to some antibiotics may be complex and multifactorial.

5.1 Antibiotics Associated with CDI

Cephalosporins *C. difficile* is usually resistant to CFs and several studies report *C. difficile* overgrowth after CF therapy (Ambrose et al. 1985; de Lalla et al. 1989; Impallomeni et al. 1995). Although *C. difficile* is described as "constitutively resistant" to CFs, the mechanism of resistance to these antibiotics is still not completely characterized. The variable MIC values observed

Table 4 *C. difficile* antibiotic mechanisms of resistance

Antibiotics[a]	Mechanism of resistance	Genetic element	Resistance determinant	Target	References
MLS_B	23S rRNA methylases	Tn*5398*	*erm*B	23S rRNA	Farrow et al. (2001), Brouwer et al. (2011), Spigaglia et al. (2005, 2011)
		Tn*6194*	*erm*B		Wasels et al. (2013), He et al. (2010, 2013)
		Tn*6215*	*erm*B		Goh et al. (2013), Wasels et al. (2015a, b)
		Tn*6218*	*erm*AB/*cfr*		Dingle et al. (2014)
		Tn6189	*ermB*		Imwattana et al. (2020a, b)
		Mobile mosaic element	*ermG/mefA/msrD/vat*		Isidro et al. (2018)
	RNA methyltransferase	Tn*6218*	*cfr* B		Hansen and Vester (2015), Marin et al. (2015), Candela et al. (2017), Stojković et al. (2019), Chatedaki et al. (2019)
			cfr C		Chatedaki et al. (2019)
		pCd13-Lar	*cfr E*		Chatedaki et al. (2019)
TET	Ribosomal protection	Tn*5397*	*tet*M	16S rRNA	Roberts et al. (2001, 2011)
		Tn*916*	*tet*M		Spigaglia et al. (2005, 2007), Sebaihia et al. (2006), Brouwer et al. (2011, 2012)
		Tn*6164*	*tet*44		Corver et al.(2012)
		Tn6190	*tet*M		Knight et al. (2015a, b)
		Tn6235	*tet*M		Knetsch et al. (2014)
		Tn6944	*tetM, mefH*		Imwattana et al. (2020a, b, 2021a, b)
	Efflux pump		tet40		
CHL	Chloramphenicol acetyltransferase	Tn*4453*a and Tn*4453*b	*cat*D	23SrRNA	Wren et al. (1988, 1989)
FQs	Alteration molecular target			GyrA/GyrB	Ackermann et al. (2001), Dridi et al. (2002), Spigaglia et al. (2008a, b, 2011), Carman et al. (2009), Huang et al. (2009), Walkty et al. (2010), Liao et al. (2012), Mac Aogáin et al. (2015), Kuwata et al. (2015)
	Efflux pump		cdeA/ABC transporter efflux pump CD2068		Ngernsombat et al. (2017, 7: 9982), Dridi et al. (2004)
CBPs	Alteration molecular target			Pbp1, Pbp3	Imwattana et al. (2020a, b)
	Carbapenemases	Similar plasmid tig00001208_pilon	SHV-1		Imwattana et al. (2020a, b)
		Similar plasmid pAHTJR1	PER-1		Imwattana et al. (2020a, b)

(continued)

Table 4 (continued)

Antibiotics[a]	Mechanism of resistance	Genetic element	Resistance determinant	Target	References
CFs	Altered target			Pbps	Spigaglia (2016)
	Antibiotic enzymatic destruction		Putative β-lactamases		Sebaihia et al. (2006), Spigaglia (2016), Toth et al. (2018), Sandhu et al. (2019)
	Efflux pumps		cme/ABC transporter efflux pump CD2068		Lebel et al. (2004)
	Efflux pump		ABC transporter efflux pump CD2068		Ngernsombat et al. (2017)
MTZ	Metabolic pathways alterations			FeoB1/Pfo/ Xdh/IscR/ Nifj/ThiH/ GlyC	Chong et al. (2014), Moura et al. (2014), Dingsdag and Hunter (2017), Deshpande et al. (2020)
	Variations in expression		recA		
	Acquisition of plasmid		pCD-METRO		Boekhoud et al. (2020)
	Biofilm formation				Vuotto et al. (2016)
VAN	Alteration molecular target			VanS/R sensor kinase response regulator/ MurG	Leeds et al. (2014), Shen et al. (2020)
	Acquisition of plasmid	pX18-498			Pu et al. (2021)
	Biofilm formation				Dapa et al. (2013)
FDX	Alteration molecular target			RpoB/Mar	Kuehne et al. (2018), Schwanbeck et al. (2019)
RFs	Alteration molecular target			RpoB	O'Connor et al. (2008), Carman et al. (2009), Curry et al. (2009), Huang et al. (2009), Walkty et al. (2010), Miller et al. (2011a, b), Spigaglia et al. (2011), Liao et al. (2012), Pecavar et al. (2012), Cairns et al. (2017)

[a]See section Abbreviations

for the different CFs suggest that resistance may be strain-dependent. Antibiotic-degrading enzymes, β-lactamases, and modification of target sites, penicillin-binding proteins (PBPs), are the main mechanisms involved in resistance to these antibiotics, but a number of coding DNA sequences (CDSs) potentially involved in this resistance have been identified in *C. difficile* 630 genome and in other *C. difficile* strains (identity between 73% and 100%) (Sebaihia et al. 2006; Spigaglia 2016; Toth et al. 2018; Sandhu et al. 2019). Furthermore, efflux pumps, such as the ABC transporter CD2068, could also play a role in *C. difficile* resistance to CFs (Ngernsombat et al. 2017).

Macrolide-Lincosamide-Streptogramin B (MLSB) In *C. difficile*, resistance to the macrolide-lincosamide-streptogramin B (MLS_B) family is usually conferred by ribosomal methylation. Erythromycin ribosomal methylase (*erm*) genes of class B are the most widespread in *C. difficile* population, even if other *erm* genes have rarely been detected (Roberts et al. 1994; Spigaglia et al. 2005; Schmidt et al. 2007). In *C. difficile*, *erm*B is usually located on mobilizable genetic elements, and Tn*5398*, a mobilizable nonconjugative transposon (Tn) of 9.6 kb in length, is the best known among these elements (Farrow et al. 2001). Tn*5398* contains two copies of *erm*B gene, and it is able to transfer in vitro from *C. difficile* to *Staphylococcus aureus* and to *Bacillus subtilis* (Hächler et al. 1987; Mullany et al. 1995). Integration/excision functions to transfer Tn*5398* from the donor to the recipient strain are provided by other conjugative transposons present in the donor genome, because Tn*5398* does not have genes encoding a recombinase (Mullany et al. 2015). Integration into the recipient chromosome occurred either by homologous recombination or by using a site-specific recombinase of the recipient. It has been suggested that a portion of the donor genome containing Tn*5398* integrates by homologous recombination into the recipient (Wasels et al. 2015b).

The majority of *C. difficile* strains resistant to MLS_B show *erm*B-containing elements with different genetic organizations compared to Tn*5398* (Farrow et al. 2001; Spigaglia et al. 2005, 2011). Seventeen organizations (E1-E17) have been identified by a PCR-mapping method, and E4 was identified as the most frequent among European *C. difficile* clinical isolates (Spigaglia et al. 2011). Elements E4 are related to the conjugative element Tn*6194*, identified in *C. difficile* 2007855 (He et al. 2010, 2013; Wasels et al. 2013). Tn*6194* has a conjugative region related to that of Tn*916*, a large family of conjugative Tns widely spread in both Gram-positive and Gram-negative bacteria, and an accessory region related to Tn*5398*, and it is able to in vitro transfer from *C. difficile* to *Enterococcus faecalis* (Wasels et al. 2014).

Tn*6215* is a peculiar mobilizable element of about 13 kb in length found in *C. difficile* CD80 (Goh et al. 2013). Noteworthy, conjugation-like mechanism or phage ΦC2 transduction can be involved in the transfer of this element between one *C. difficile* strain and another. Furthermore, it has recently been suggested that a transformation-like mechanism can be responsible for the transfer of Tn*6215* and Tn*5398* when *C. difficile* CD13 is used as recipient strain (Wasels et al. 2015b).

Although *erm*B-containing elements have a cost on the *C. difficile* fitness in vitro (Wasels et al. 2013), these elements are common in *C. difficile* population suggesting that regardless of the burden on fitness, other factors (i.e., the capability of transfer and the intrinsic genetic characteristics of strains) are involved in their successful spread.

Resistance to both ERY and CLI or only to ERY has been observed also in *C. difficile* strains negative for *erm* genes (Spigaglia and Mastrantonio 2004; Pituch et al. 2006; Ratnayake et al. 2011; Spigaglia et al. 2011). Although alterations in the 23S ribosomal RNA (rRNA) or ribosomal proteins (L4 or L22) have been found in some of these strains, the same changes were also observed in susceptible isolates, and therefore, their role in resistance has been excluded (Spigaglia et al. 2011). Furthermore, treatment of resistant *erm*-negative strains with two pump inhibitors (reserpine and carbonyl cyanide m-chlorophenyl hydrazone—CCCP) did not determine any reduction in MICs, suggesting that resistance is not mediated by efflux mechanisms (Spigaglia et al. 2011). Other determinants could have a role in *C. difficile* resistance to MLS_B in the absence of *erm* genes. In particular, *cfrB*, *cfrC*, and *cfrE*, which encode a 23S rRNA methyltransferase and confer resistance to PhLOPSA (phenicols, lincosamides, oxazolidinones, pleuromutilins, and streptogramin A), have been found in several *C. difficile* strains resistant to linezolid and other clinically relevant antibiotics (Hansen and Vester 2015; Marin et al. 2015; Candela et al. 2017; Stojković et al. 2019). In addition, efflux pumps may also contribute in *C. difficile* resistance to the MLSB family of antibiotics, such as the multidrug transporter encoded by the *C. difficile cme* gene (Lebel et al. 2004).

Fluoroquinolones Alterations in the quinolone resistance-determining region (QRDR) of the gyrase (Gyr) A and/or GyrB are responsible for resistance to FQs in *C. difficile* (Ackermann et al. 2001, 2003; Dridi et al. 2002; Drudy et al. 2006, 2007). Several amino acid substitutions have been identified in the DNA gyrase subunits (Table 5), but the most common in *C. difficile* FQ-resistant strains is the substitution of threonine (Thr) in position 82 with isoleucine (Ile)—Thr82Ile—in GyrA (Ackermann et al. 2001; Dridi et al. 2002; Spigaglia et al. 2008b, 2011; Kuwata et al. 2015). Interestingly, Thr82Ile in GyrA has not a detectable cost on the fitness of *C. difficile* in vitro, suggesting that this substitution can be maintained in the bacterial population even in the absence of antibiotic selective pressure (Wasels et al. 2015a).

Resistant mutants to FQs can be obtained with high frequency after exposure of *C. difficile* susceptible strains to MXF and levofloxacin (LEV) (Spigaglia et al. 2009). During the early stage of treatment, the concentration of this drug in the human intestine is not inhibitory; therefore it is possible for a subpopulation of bacteria to acquire mutations conferring resistance to FQs. Also efflux pumps have been implicated in *C. difficile* resistance to fluoroquinolone, such as the ABC transporter CD2068 (Ngernsombat et al. 2017). In addition, overexpression of the *C. difficile cdeA* gene, encoding an efflux pump of the multidrug and toxic compound extrusion (MATE) subfamily, was reported to be able to induce fluoroquinolone resistance in *Escherichia coli* (Dridi et al. 2004).

Carbapenems Amino acid substitutions in either PBP1 or PBP3 have been reported in *C. difficile* strains resistant to imipenem (Imwattana et al. 2021a, b). In addition, also genes encoding carbapenemase SHV-1 and PER-1 have been identified in *C. difficile* strains carrying elements resampling a *Klebsiella pneumoniae* plasmid tig00001208_pilon and a *Acinetobacter haemolyticus* plasmid pAHTJR1, respectively (Imwattana et al. 2021a, b).

5.2 Antibiotics for CDI Treatment

Metronidazole MTZ is a nitro-aromatic pro-drug that needs the reduction of the 5-nitro group of the imidazole ring to become cytotoxic to bacterial cells (Goldman 1982). In *Helicobacter pylori* and *Bacteroides fragilis*, resistance to MTZ is usually conferred by nitroimidazole (*nim*) genes (Gal and Brazier 2004), but these genes have not been identified in *C. difficile* (Moura et al. 2014). Although it is not completely understood, data obtained in recent studies on strains RT 027 and RT 010 suggest that *C. difficile* resistance to MTZ is a multifactorial process that involves alterations in metabolic pathways, such as activity of nitroreductases, iron uptake, and DNA repair (Chong et al. 2014; Moura et al. 2014). A recent study, in which a highly mutable *C. difficile* strain was constructed through deletion of DNA mismatch repair genes, demonstrates that resistance to MTZ developed progressively, with MIC values from 2 to 16 mg/L, due to mutations in different loci (Deshpande et al. 2020). In fact, variations first appeared in *feoB1* (the main iron transporter) and then, in sequence, in *pfo* (a pyruvate–ferredoxin oxidoreductases), *xdh* (a putative xanthine dehydrogenase), and finally in *iscR* (an iron–sulfur cluster regulator); this last mutation confers higher MIC values to MTZ. Furthermore, MICs are amplified by variations in *nifj* (a pyruvate–ferredoxin/flavodoxin oxidoreductase), *xdh*, and *iscR* genes, but these mutations do not confer resistance without the loss of the *feoB1* gene, although the loss of *feob1* alone confers only low level of resistance to MTZ. In addition, variations in expression and/or increased concentrations of the DNA repair protein RecA have been reported in strains exposed to MTZ (Chong et al. 2014; Moura et al. 2014). Furthermore, *C. difficile* resistance to MTZ has been also associated with mutations in the thiamine biosynthesis (*thiH*), and glycerol-3-

Table 5 Amino acid substitutions detected in *C. difficile* isolates conferring resistance to antibiotics

Antibiotic[a]	Target	Amino acid position	Original residue[a]	Resistance substitution[a]	References
FQs	GyrA	43	Val	Asp	Carman et al. (2009)
		71	Asp	Val	Dridi et al. (2002), Walkty et al. (2010), Liao et al. (2012)
		81	Asp	Asn	Huang et al. (2009), Liao et al. (2012)
		82	Thr	Ile or Val	Ackermann et al. (2001), Dridi et al. (2002), Spigaglia et al. (2008a, b), Kuwata et al. (2015), Liao et al. (2012)
		118	Ala	Thr	Dridi et al. (2002)
		384	Ala	Asp	Mac Aogáin et al. (2015)
	GyrB	377	Arg	Gly	Liao et al. (2012)
		416	Ser	Ala	Liao et al. (2012)
		426	Asp	Asn or Val	Dridi et al. (2002), Spigaglia et al. (2008a, b), Liao et al. (2012)
		447	Arg	Lys	Walkty et al. (2010), Liao et al. (2012)
		466	Glu	Val	Liao et al. (2012)
	GyrA and GyrB	82/366	Thr/Ser	Ile/Ala	Huang et al. (2009), Kuwata et al. (2015)
		82/366 and 426	Thr/Ser and Asp	Ile/Ala and Val	Walkty et al. (2010), Kuwata et al. (2015)
		82/366 and 434	Thr/Ser and Gln	Ala/Ala and Lys	Kuwata et al. (2015)
		82/416	Thr/Ser	Ile/Ala	Spigaglia et al. (2008a, b), Liao et al. (2012)
		82/426	Thr/Asp	Ile/Asn	Walkty et al. (2010), Kuwata et al. (2015)
		82/426	Thr/Asp	Ile/Val	Spigaglia et al. (2011)
		82/426	Thr/Asp	Val/Val	Huang et al. (2009), Liao et al. (2012)
		82/444	Thr/Leu	Ile/Phe	Walkty et al. (2010)
RFs	RpoB	485	Ser	Phe	Cairns et al. (2017)
		492	Asp	Asn or Val	Pecavar et al. (2012)
		502	His	Arg or Asn or Leu or Tyr	O'Connor et al. (2008), Pecavar et al. (2012), Miller et al. (2011a, b)
		505	Arg	Lys	O'Connor et al. (2008), Curry et al. (2009), Miller et al. (2011a, b), Spigaglia et al. (2011), Pecavar et al. (2012)
		550	Ser	Phe or Tyr	Pecavar et al. (2012)
		448; 505	Ser; Arg	Thr; Lys	O'Connor et al. (2008), Curry et al. (2009)
		487; 502	Leu; His	Phe; Tyr	Pecavar et al. (2012)
		492; 505	Asp; Arg	Asn; Lys	O'Connor et al. (2008)
		498; 505	Ser; Arg	Thr; Lys	Curry et al. (2009), Miller et al. (2011a, b)
		502; 496	His; Pro	Tyr; Ser	Carman et al. (2009)
		502; 505	His; Arg	Asn; Lys	O'Connor et al. (2008), Curry et al. (2009), Miller et al. (2011a, b), Spigaglia et al. (2011), Pecavar et al. (2012)
		505; 548	Arg; Ile	Lys; Met	O'Connor et al. (2008), Curry et al. (2009), Pecavar et al. (2012)
CBPs	Pbp1	543	Leu	His	Imwattana et al. (2020a, b)
		555	Ala	Thr	Imwattana et al. (2020a, b)
	Pbp3	721	Tyr	Ser	Imwattana et al. (2020a, b)
FDX	RpoB	1073	Gln	Arg	Leeds et al. (2014)
		1143	Val	Asp	Kuehne et al. (2018), Schwanbeck et al. (2019)
		1143	Val	Gly	Kuehne et al. (2018), Schwanbeck et al. (2019)
		1143	Val	Phe	Kuehne et al. (2018), Schwanbeck et al. (2019)
	RpoC	273	Asp	Tyr	Baines and Wilcox (2015), Harnvoravongchai et al. (2017)

[a]See section Abbreviations

oxidoreductase (*glyC*) genes have also been associated with resistance in *C. difficile* (Dingsdag and Hunter 2017). Interestingly, a high-copy number 7 kb plasmid (pCD-METRO) has been found to be able to confer resistance to MTZ in *C. difficile* (Boekhoud et al. 2020). Although the introduction of the pCD-METRO plasmid into susceptible strain increased the MIC values >24-fold, the exact mechanism of resistance is still unclear (Boekhoud et al. 2020).

Finally, also biofilm formation seems to play a role in *C. difficile* MTZ resistance (Vuotto et al. 2016). It can be hypothesized that biofilm matrix can act as a protective barrier, inducing, at the same time, an alteration of the physiological state of the bacteria within the biofilm that determines a higher level of resistance to antibiotics.

Vancomycin VAN is the first-line antibiotic for moderate to severe CDI (Debast et al. 2014; Jarrad et al. 2015). This antibiotic, which consists of a glycosylated hexapeptide chain and cross-linked aromatic rings by aryl ether bonds, inhibits the biosynthesis of peptidoglycan, an essential component of the bacterial cell wall envelope, and it is poorly absorbed by the gastrointestinal tract (Perkins and Nieto 1974; Yu and Sun 2013). Regarding the mechanism of resistance, although Tn*1549*-like elements have been found in several *C. difficile* strains (Brouwer et al. 2011, 2012), these elements, different from the original Tn*1549* element described in *E. faecalis*, do not have a functional *van*B operon. Interestingly, a *van*G-like gene cluster homologous to that found in *E. faecalis* has also been described in *C. difficile*, but it seems not able to promote resistance to VAN (Ammam et al. 2012, 2013; Ramírez-Vargas et al. 2017). However, a recent study reports mutations in the *VanS/R* sensor kinase/response regulator in *C. difficile* RT 027 strains that triggers constitutive expression of the VanG, conferring VAN resistance (Shen et al. 2020). Interestingly, other van genes (*vanB*, *vanZ*) have been observed in *C. difficile* strains phenotypically susceptible, suggesting the possibility for *C. difficile* strains to transfer and acquire these genes of other vancomycin resistance genes (Knight et al. 2016; Woods et al. 2018). *C. difficile* isolates that are VAN resistant with mutations in other families of genes have been described. In particular, amino acid change proline (Pro) to leucine (Leu) in position 108 in the glycosyltransferase MurG has been obtained in vitro (Leeds et al. 2014). Since MurG is involved in the membrane-bound stage of peptidoglycan biosynthesis, this substitution may affect VAN activity. In addition, biofilm formation has been found to probably have a role in VAN resistance. In fact, *C. difficile* within biofilms resulted more resistant to high concentrations of VAN (20 mg/L), and subinhibitory and inhibitory concentrations of the antibiotic seem to induce biofilm formation (Dapa et al. 2013). Furthermore, the acquisition of plasmid pX18-498, a plasmid that probably has a role in cell wall integrity, may reduce susceptibility to VAN eightfold in *C. difficile* (Pu et al. 2021).

Fidaxomicin This antibiotic provides cure rates not inferior to VAN and is associated with a significantly lower rate of CDI recurrence caused by strains non-RT027 (Louie et al. 2011). Furthermore, it has a minimal impact on the composition of indigenous fecal microbiota, in particular on *Bacteroides* species (Tannock et al. 2010; Louie et al. 2012), with a high local concentration in the gut and feces (1225.1 μg/g after ten days of therapy) (Goldstein et al. 2012; Sears et al. 2012). Mutations in *rpoB* are responsible for resistance to FDX, but fortunately, these mutations are different from those described for resistance to RIF and are not cross-protective (O'Grady et al. 2021). *C. difficile* mutants with changes in position 1143 (Val to Asp, Val to Gly, and Val to Phe) of the *rpoB* have been reported not only with reduced susceptibility to FDX but also with reduced virulence and fitness in vitro (Kuehne et al. 2018; Schwanbeck et al. 2019). In addition, mutations in the *marR* (multidrug resistance-associated transcriptional regulator) or in the *rpoC*, with the amino acid substitution Asp to Tyr in position 273, may also increase MIC values to FDX (Baines and Wilcox 2015; Harnvoravongchai et al. 2017). Since mutations causing resistance to FDX arise in *rpo*B gene at distinct loci compared to those causing resistance to RIFs, FDX retains activity against strains resistant to RFs (Anti-Infective Drugs Advisory Committee Briefing Document, Optimer Pharmaceuticals, Inc. 2011).

Rifamycins Rifamycins (RFs), in particular RFX, have been proposed as "chaser therapy" for treatment of relapsing CDI (Iv et al. 2014), while FDX is a bactericidal new narrow-spectrum macrocyclic antibiotic that is used for the management of CDI with high risk for recurrences (Chaparro-Rojas and Mullane 2013). Both RIFs and FDX are inhibitors of bacterial transcription, but they have different RNA polymerase (RNAP) target sites. FDX binds to the "switch region" of RNAP, a target site that is adjacent to the RIF target but does not overlap (Mullane and Gorbach 2011; Srivastava et al. 2011).

Different amino acid substitutions have been identified within the β-subunit of the RNA polymerase (*rpo*B) of strains resistant to RIF (Table 5). Among the amino acid substitutions identified, the amino acid change arginine (Arg) to lysine (Lys) in position 505 is the most common, particularly in strains RT 027 (Miller et al. 2011b; Spigaglia et al. 2011; Carman et al. 2012; Pecavar et al. 2012). A rapid development of resistance to RIF has been described in *C. difficile* when exposed to this antibiotic; in particular a *C. difficile* RT 056 strain developed resistance within three days of RFX therapy, with MIC values that increased from 0.002 to >32 mg/L (Carman et al. 2012).

5.3 Other Antibiotics

Tetracycline In *C. difficile*, resistance to TET is due to *tet* genes (Table 4). The most widespread *tet* class is *tet*M, usually carried by conjugative Tn*916*-like elements (Spigaglia et al. 2005; Mullany et al. 2012; Dong et al. 2014). This family of transposon is responsible for the spread of antibiotic resistance (usually referred to TET but also to MLS_B and other antibiotics) to many important pathogens (Roberts and Mullany 2011). The best-known *C. difficile* element of this family is Tn*5397*, which is a 21 kb element able to transfer in vitro between *C. difficile* and *B. subtilis* or *E. faecalis* (Mullany et al. 1990; Jasni et al. 2010). A group II intron and a different excision/insertion module differentiate Tn*5397* from Tn*916*. In fact, Tn*5397* has a *tnd*X gene that encodes a large serine recombinase, while Tn*916* contains two genes, *xis*Tn and *int*Tn, encoding an excisionase and a tyrosine integrase (Roberts et al. 2001). Furthermore, Tn*916* inserts into multiple regions of the *C. difficile* genome (Mullany et al. 2012), while Tn*5397* inserts DNA predicted filamentation processes induced by cAMP (Fic) domain (Wang et al. 2006).

Different genetic organizations of Tn*916*-like elements and different *tet*M alleles have been identified in *C. difficile* (Spigaglia et al. 2005, 2006). In particular, the Tn*916* element detected in the clinical isolate CD1911 contains both *tet*M and *erm*B (Spigaglia et al. 2007). This element is nonconjugative and probably originated from the combination of one or more plasmids and a Tn*916*-like element.

Albeit more rarely, other *tet* genes have been identified in *C. difficile*. In particular, the co-presence of both *tet*M and *tet*W has been described in *C. difficile* isolates from humans and animals (Spigaglia et al. 2008a; Fry et al. 2012).

Interestingly, an element of 106 kb, the Tn*6164*, has been identified in *C. difficile* strain M120, a RT078 isolate (Corver et al. 2012). This transposon is composed by parts of other elements from different bacteria, particularly from *Thermoanaerobacter* sp. and *Streptococcus pneumoniae*, and it contains *tet*(44) and *ant*(6)-Ib, predicted to confer resistance to TET and streptomycin, respectively. Since strain M120 is susceptible to these antibiotics, Tn*6164* does not seem involved in resistance, but it seems to be associated with higher virulence of strains RT 078; in fact an analysis of data from patients indicates that mortality was more common in patients infected with strains RT078 containing Tn*6164* compared with those infected with strains without this element.

In addition, the efflux pump gene tet40 has been identified in *C. difficile* RT 078 resistant to TET (Dingle et al. 2019).

Chloramphenicol *C. difficile* resistance to chloramphenicol (CHL) is usually conferred by a chloramphenicol acetyltransferase encoded by a *cat*D gene (Wren et al. 1988, 1989) (Table 4). In *C. difficile*, the *cat*D gene is located on Tn*4453*a and Tn*4453*b, which are strictly related to the *Clostridium perfringens* mobilizable element Tn*4451* (Lyras et al. 1998). Recently, a conjugative element, designed Tn*6104*, has been described (Brouwer et al. 2011). This Tn contains genetic elements closely related to Tn*4453*ab and Tn*4451*, but instead of a *cat*D gene, it shows genes predicted to encode for transcriptional regulator, a two-component regulatory system, an ABC transporter, three sigma factors, and a putative toxin–antitoxin system, respectively. The role of these genes is not clear and remains to be determined.

6 Conclusions

C. difficile infection (CDI) is a growing concern for global public health. An increased CDI incidence, morbidity, and mortality have been reported in the last decades in association with the emergence and spread of *C. difficile* highly virulent types. *C. difficile* adaptive capability and genome plasticity have determined an increase of strains resistant to multiple antibiotics, and currently, high rates of MDR epidemic clinical isolates are reported. A wide range of mobile elements and alterations of antibiotic targets mediate resistance to several antibiotics, including the MLS_B family and FQs, which are significantly associated with CDI. Furthermore, a decreased susceptibility to the first-line antibiotics used for CDI therapy, in particular MTZ and VAN, and to those used for recurrences, such as RFs, may have a role in the low rate of response to treatment reported over the last years. Antibiotic resistances seem to be maintained in this pathogen regardless of the burden imposed by the acquisition of genetic elements/mutations conferring resistance and the decrease of antibiotic pressure. This feature may explain the persistence of "old" resistances and the rapid diffusion of "new" resistances in *C. difficile* population. The multifactorial nature of antibiotic resistances of *C. difficile* emphasizes the need for effective antimicrobial stewardships efficacious infection control programs, and alternative therapies for CDI. Evidences of several environmental and animal reservoirs of *C. difficile*, with the possibility of AMR gene transfer to and from other bacterial species, highlight that the public health surveillance of *C. difficile* with a One-Health perspective is critical for combating CDI.

References

Abdrabou AMM, Bischoff M, Mellmann A et al (2022) German *C. difficile* surveillance group. Implementation of a *Clostridioides difficile* sentinel surveillance system in Germany: first insights for 2019-2021. Anaerobe 77: 102548

Ackermann G, Tang YJ, Kueper R et al (2001) Resistance to moxifloxacin in toxigenic *Clostridium difficile* isolates is associated with mutations in *gyr* A. Antimicrob Agents Chemother 45:2348–2353

Ackermann G, Tang-Feldman YJ, Schaumann R et al (2003) Antecedent use of fluoroquinolones is associated with resistance to moxifloxacin in *Clostridium difficile*. Clin Microbiol Infect 9:526–530

Adler A, Miller-Roll T, Bradenstein R et al (2015) A national survey of the molecular epidemiology of *Clostridium difficile* in Israel: the dissemination of the ribotype 027 strain with reduced susceptibility to vancomycin and metronidazole. Diagn Microbiol Infect Dis 83:21–24

Aguilar-Zamora E, Weimer BC, Torres RC et al (2022) Molecular epidemiology and antimicrobial resistance of *Clostridioides difficile* in hospitalized patients from Mexico. Front Microbiol 12:787451

Ambrose NS, Johnson M, Burdon D et al (1985) The influence of single dose intravenous antibiotics on faecal flora and emergence of *Clostridium difficile*. J Antimicrob Chemother 15:319–326

Ammam F, Marvaud JC, Lambert T (2012) Distribution of the vanG-like gene cluster in *Clostridium difficile* clinical isolates. Can J Microbiol 58:547–551

Ammam F, Meziane-Cherif D, Mengin-Lecreulx D et al (2013) The functional vanGCd cluster of *Clostridium difficile* does not confer vancomycin resistance. Mol Microbiol 89:612–625

Aoki K, Takeda S, Miki T et al (2019) Antimicrobial susceptibility and molecular characterisation using whole-genome sequencing of *Clostridioides difficile* collected in 82 hospitals in Japan between 2014 and 2016. Antimicrob Agents Chemother. 63(12):e01259–e01219

Aptekorz M, Sacha K, Gofron Z et al (2022) Antibiotic resistance profile of RT 027/176 versus other *Clostridioides difficile* isolates in Silesia, Southern Poland. Pathogens 11:949

Baghani A, Mesdaghinia A, Kuijper EJ et al (2020) High prevalence of *Clostridioides difficile* PCR ribotypes 001 and 126 in Iran. Sci Rep 10:4658

Baines SD, O'Connor R, Freeman J et al (2008) Emergence of reduced susceptibility to metronidazole in *Clostridium difficile*. J Antimicrob Chemother 62: 1046–1052

Baines SD, Wilcox MH (2015) Antimicrobial resistance and reduced susceptibility in *Clostridium difficile*: potential consequences for induction, treatment, and recurrence of *C. difficile* infection. Antibiotics 4:267–298

Baldan R, Trovato A, Bianchini V et al (2015) A successful epidemic genotype: *Clostridium difficile* PCR ribotype 018. J Clin Microbiol 53:2575–2580

Barbanti F, Spigaglia P (2020) Microbiological characteristics of human and animal isolates of *Clostridioides difficile* in Italy: Results of the Istituto Superiore di Sanità in the years 2006-2016. Anaerobe 61:102136

Bartlett JG, Onderdonk AB, Cisneros RL et al (1977) Clindamycin-associated colitis due to a toxin-producing species of *Clostridium* in hamsters. J Infect Dis 136:701–705

Basu PP, Dinani A, Rayapudi K et al (2010) Rifaximin therapy for metronidazole-unresponsive *Clostridium difficile* infection: a prospective pilot trial. Therap Adv Gastroenterol 3:221–225

Bauer MP, Notermans DW, van Benthem BHB et al (2011) *Clostridium difficile* infection in Europe: a hospital-based survey. Lancet 377:63–73

Bignardi GE (1998) Risk factors for *Clostridium difficile* infection. J Hosp Infect 40:1–15

Bishop EJ, Tiruvoipati R (2022) Management of *Clostridioides difficile* infection in adults and challenges in clinical practice: review and comparison of current IDSA/SHEA, ESCMID and ASID guidelines. J Antimicrob Chemother 78:21–30

Boekhoud IM, Hornung BVH, Sevilla E et al (2020) Plasmid-mediated metronidazole resistance in *Clostridioides difficile*. Nat Commun 11:598

Boekhoud IM, Sidorov I, Nooij S et al (2021) COMBACTE-CDI Consortium. Haem is crucial for medium-dependent metronidazole resistance in clinical isolates of *Clostridioides difficile*. J Antimicrob Chemother 76:1731–1740

Bolton RP, Culshaw MA (1986) Faecal metronidazole concentrations during oral and intravenous therapy for antibiotic associated colitis due to *Clostridium difficile*. Gut 27:1169–1172

Brajerova M, Zikova J, Krutova M (2022) *Clostridioides difficile* epidemiology in the Middle and the Far East. Anaerobe 74:102542

Brazier JS, Fawley W, Freeman J et al (2001) Reduced susceptibility of *Clostridium difficile* to metronidazole. J Antimicrob Chemother 48:741–742

Brouwer MSM, Warburton PJ, Roberts AP et al (2011) Genetic organisation, mobility and predicted functions of genes on integrated, mobile genetic elements in sequenced strains of *Clostridium difficile*. PLoS ONE 6:e23014

Brouwer MSM, Roberts AP, Mullany P et al (2012) In silico analysis of sequenced strains of *Clostridium difficile* reveals a related set of conjugative transposons carrying a variety of accessory genes. Mob Genet Elements 2:8–12

Büchler AC, Rampini SK, Stelling S et al (2014) Antibiotic susceptibility of *Clostridium difficile* is similar worldwide over two decades despite widespread use of broad-spectrum antibiotics: an analysis done at the University Hospital of Zurich. BMC Infect Dis 14:607

Burckhardt F, Friedrich A, Beier D et al (2008) *Clostridium difficile* surveillance trends, Saxony, Germany. Emerg Infect Dis 4:691–692

Byun JH, Kim H, Kim JL et al (2019) A nationwide study of molecular epidemiology and antimicrobial susceptibility of *Clostridioides difficile* in South Korea. Anaerobe 60:102106

Candela T, Marvaud J-C, Nguyen TK et al (2017) A cfr-like Gene cfr(C) conferring linezolid resistance is common in *Clostridium difficile*. Int J Antimicrob Agents

Cairns MD, Preston MD, Hall CL et al (2017) Comparative genome analysis and global phylogeny of the toxin variant *Clostridium difficile* PCR ribotype 017 reveals the evolution of two independent sublineages. J Clin Microbiol 55:865–876

Carman RJ, Genheimer CW, Rafii F et al (2009) Diversity of moxifloxacin resistance during a nosocomial outbreak of a predominantly ribotype ARU 027 *Clostridium difficile* diarrhea. Anaerobe 15:244–248

Carman RJ, Boone JH, Grover H et al (2012) In vivo selection of rifamycin-resistant *Clostridium difficile* during rifaximin therapy. Antimicrob Agents Chemother 56:6019–6020

Chaparro-Rojas F, Mullane KM (2013) Emerging therapies for *Clostridium difficile* infection – focus on fidaxomicin. Infect Drug Resist 6:41–53

Chatedaki C, Voulgaridi I, Kachrimanidou M et al (2019) Antimicrobial susceptibility and mechanisms of resistance of Greek *Clostridium difficile* clinical isolates. J Glob Antimicrob Resist 16:53–58

Chia JH, Lai HC, Su LH et al (2013) Molecular epidemiology of *Clostridium difficile* at a medical center in Taiwan: persistence of genetically clustering of A−B+ isolates and increase of A+B+ isolates. PLoS ONE 8: e75471

Chong PM, Lynch T, McCorrister S et al (2014) Proteomic analysis of a NAP1 *Clostridium difficile* clinical isolate resistant to metronidazole. PLoS ONE 9:e82622

Cornely OA, Crook DW, Esposito R et al (2012) Fidaxomicin versus vancomycin for infection with

Clostridium difficile in Europe, Canada, and the USA: a double-blind, non-inferiority, randomised controlled trial. Lancet Infect Dis 12:281–289

Clements AC, Magalhães RJ, Tatem AJ et al (2010) *Clostridium difficile* PCR ribotype 027: assessing the risks of further worldwide spread. Lancet Infect Dis 10:395–404

Clinical and Laboratory Standards Institute (CLSI) (2012) Methods for antimicrobial susceptibility testing of anaerobic bacteria; Approved Standard-Eighth Edition. CLSI document M11-A8. ISBN 1-56238-789-8 (Print); ISBN 1-56238-790-1 (Electronic)

Corver J, Bakker D, Brouwer MSM et al (2012) Analysis of a *Clostridium difficile* PCR ribotype 078 100 kilobase island reveals the presence of a novel transposon, Tn*6164*. BMC Microbiol 12:130

Curry SR, Marsh JW, Shutt KA et al (2009) High frequency of rifampin resistance identified in an epidemic *Clostridium difficile* clone from a large teaching hospital. Clin Infect Dis 48:425–429

Dapa T, Leuzzi R, Ng YK et al (2013) Multiple factors modulate biofilm formation by the anaerobic pathogen *Clostridium difficile*. J Bacteriol 195:545–555

Darkoh C, Keita K, Odo C et al (2022) Emergence of clinical *Clostridioides difficile* isolates with decreased susceptibility to vancomycin. Clin Infect Dis 74:120–126

De Almeida MN, Heffernan H, Dervan A et al (2013) Severe *Clostridium difficile* infection in New Zealand associated with an emerging strain, PCR-ribotype 244. N Z Med J 126:9–14

de Lalla F, Privitera G, Ortisi G et al (1989) Third generation cephalosporins as a risk factor for *Clostridium difficile*-associated disease: a four-year survey in a general hospital. J Antimicrob Chemother 23:623–631

Debast SB, Bauer MP, Kuijper EJ (2014) European Society of Clinical Microbiology and Infectious Diseases: update of the treatment guidance document for *Clostridium difficile* infection. Clin Microbiol Infect 20:1–26

Deshpande A, Wu X, Huo W et al (2020) Chromosomal resistance to metronidazole in *Clostridioides difficile* can be mediated by epistasis between iron homeostasis and oxidoreductases. Antimicrob Agents Chemother 64:e00415–e00420

Dilnessa T, Getaneh A, Hailu W et al (2022) Prevalence and antimicrobial resistance pattern of *Clostridium difficile* among hospitalized diarrheal patients: A systematic review and meta-analysis. PLoS One 17: e0262597

Dingle KE, Elliott B, Robinson E et al (2014) Evolutionary history of the *Clostridium difficile* pathogenicity locus. Genome Biol Evol 6:36–52

Dingle KE, Didelot X, Quan TP et al (2019) A role for tetracycline selection in recent evolution of agriculture-associated *Clostridium difficile* PCR ribotype 078. mBio 10:e02790–e02718

Dingsdag SA, Hunter N (2017) Metronidazole: an update on metabolism, structure-cytotoxicity and resistance mechanisms. J Antimicrob Chemother 73:265–279

Dong D, Zhang L, Chen X et al (2013) Antimicrobial susceptibility and resistance mechanisms of clinical *Clostridium difficile* from Roberts Chinese tertiary hospital. Int J Antimicrob Agents 41:80–84

Dong D, Chen X, Jiang C et al (2014) Genetic analysis of Tn*916*-like elements conferring tetracycline resistance in clinical isolates of *Clostridium difficile*. Int J Antimicrob Agents 43:73–77

Dridi L, Tankovic J, Burghoffer B et al (2002) Gyr A and gyrB mutations are implicated in cross-resistance to ciprofloxacin and moxifloxacin in *Clostridium difficile*. Antimicrob Agents Chemother 46:3418–3421

Dridi L, Tankovic J, Petit JC (2004) CdeA of *Clostridium difficile*, a new multidrug efflux transporter of the MATE family. Microb Drug Resist 10:191–196

Drudy D, Quinn T, O'Mahony R et al (2006) High-level resistance to moxifloxacin and gatifloxacin associated with a novel mutation in gyrB in toxin-A-negative, toxin-B-positive *Clostridium difficile*. J Antimicrob Chemother 58:1264–1267

Drudy D, Kyne L, O'Mahony R et al (2007) *gyr*A mutations in fluoroquinolone-resistant *Clostridium difficile* PCR-027. Emerg Infect Dis 13:504–505

Du T, Choi KB, Silva A et al (2022) Characterization of healthcare-associated and community-associated *Clostridioides difficile* infections among adults, Canada, 2015-2019. Emerg Infect Dis 28:1128–1136

Dubberke ER, Olsen MA (2012) Burden of *Clostridium difficile* on the healthcare system. Clin Infect Dis 55 (Suppl 2):S88–S92

Eckert C, Coignard B, Hebert M et al (2013) Clinical and microbiological features of *Clostridium difficile* infections in France: The ICD-RAISIN 2009 national survey. Méd Mal Infect 43:67–74

Erikstrup LT, Danielsen TK, Hall V et al (2012) Antimicrobial susceptibility testing of *Clostridium difficile* using EUCAST epidemiological cut-off values and disk diffusion correlates. Clin Microbiol Infect 18: E266–E272

European Centre for Disease Prevention and Control (ECDC) (2013) Point prevalence survey of health care associated infections and antimicrobial use in European acute care hospitals

Falagas ME, Makris GC, Dimopoulos G et al (2008) Heteroresistance: a concern of increasing clinical significance? Clin Microbiol Infect 14:101–104

Farrow KA, Lyras D, Rood JI (2001) Genomic analysis of the erythromycin resistance element Tn*5398* from *Clostridium difficile*. Microbiology 147:2717–2728

Fraga EG, Nicodemo AC, Sampaio JL (2016) Antimicrobial susceptibility of Brazilian *Clostridium difficile* strains determined by agar dilution and disk. Braz J Infect Dis 20:476–481

Freeman J, Vernon J, Morris K et al (2015a) Pan-European longitudinal surveillance of antibiotic resistance among prevalent *Clostridium difficile* ribotypes. Clin Microbiol Infect 21:248.e9–248.e16

Freeman J, Vernon J, Vickers R et al (2015b) Susceptibility of *Clostridium difficile* isolates of varying antimicrobial resistance phenotypes to SMT19969 and 11 comparators. Antimicrob Agents Chemother 60: 689–692

Freeman J, Vernon J, Pilling S et al (2018) Pan-European Longitudinal Surveillance of Antibiotic Resistance among Prevalent Clostridium difficile Ribotypes Study Group. The ClosER study: results from a three-year pan-European longitudinal surveillance of antibiotic resistance among prevalent Clostridium difficile ribotypes, 2011-2014. Clin Microbiol Infect 24:724–731

Freeman J, Vernon J, Pilling S et al (2020) Five-year Pan-European, longitudinal surveillance of *Clostridium difficile* ribotype prevalence and antimicrobial resistance: the extended ClosER study. Eur J Clin Microbiol Infect Dis 39:169–177

Fry PR, Thakur S, Abley M et al (2012) Antimicrobial resistance, toxinotype, and genotypic profiling of *Clostridium difficile* isolates of swine origin. J Clin Microbiol 50:2366–2372

Gal M, Brazier JS (2004) Metronidazole resistance in Bacteroides spp. carrying nim genes and the selection of slow-growing metronidazole-resistant mutants. J Antimicrob Chemother 54:109–116

Garey KW, Salazar M, Shah D et al (2008) Rifamycin antibiotics for treatment of *Clostridium difficile*-associated diarrhea. Ann Pharmacother 42:827–835

Gargis AS, Karlsson M, Paulick AL et al (2022) Reference susceptibility testing and genomic surveillance of *Clostridioides difficile*, United States, 2012-2017. Clin Infect Dis 8:ciac817

Gao Q, Wu S, Huang H, Ni Y et al (2016) Toxin profiles, PCR ribotypes and resistance patterns of *Clostridium difficile*: a multicentre study in China, 2012-2013. Int J Antimicrob Agents 48:736–739

Genzel GH, Stubbings W, Stingu CS et al (2014) Activity of the investigational fluoroquinolone finafloxacin and seven other antimicrobial agents against 114 obligately anaerobic bacteria. Int J Antimicrob Agents 44:420–423

Goh S, Hussain H, Chang BJ et al (2013) Phage ϕC2 mediates transduction of Tn*6215*, encoding erythromycin resistance, between *Clostridium difficile* strains. MBio 4:e00840–e00813

Goldman P (1982) The development of 5-nitroimidazoles for the treatment and prophylaxis of anaerobic bacterial infections. J Antimicrob Chemother 10(Suppl A):23–33

Goldstein EJ, Citron DM, Sears P et al (2011) Comparative susceptibilities of fidaxomicin (OPT-80) of isolates collected at baseline, recurrence, and failure from patients in two fidaxomicin phase III trials of *C. difficile* infection. Antimicrob Agents Chemother 55:5194–5199

Goldstein EJ, Babakhani F, Citron DM (2012) Antimicrobial activities of fidaxomicin. Clin Infect Dis 55(Suppl 2):S143–S148

Gonzales-Luna AJ, Olaitan AO, Shen WJ et al (2021) Reduced susceptibility to metronidazole is associated with initial clinical failure in *Clostridioides difficile* infection. Open Forum Infect Dis 8:ofab365

Goorhuis A, Van der Kooi T, Vaessen N et al (2007) Spread and epidemiology of *Clostridium difficile* polymerase chain reaction ribotype 027/toxinotype III in The Netherlands. Clin Infect Dis 45:695–703

Goudarzi M, Goudarzi H, Alebouyeh M et al (2013) Antimicrobial susceptibility of *Clostridium difficile* clinical isolates in Iran. Iran Red Crescent Med J 15: 704–711

Gravel D, Miller M, Simor A, Taylor G et al (2009) Canadian Nosocomial Infection Surveillance Program. Health care-associated *Clostridium difficile* infection in adults admitted to acute care hospitals in Canada: a Canadian Nosocomial Infection Surveillance Program study. Clin Infect Dis 48:568–576

Hächler H, Berger-Bächi B, Kayser FH (1987) Genetic characterization of a *Clostridium difficile* erythromycin-clindamycin resistance determinant that is transferable to Staphylococcus aureus. Antimicrobial Agents Chemother 7:1039–1045

Hansen LH, Vester B (2015) A cfr-like gene from *Clostridium difficile* confers multiple antibiotic resistance by the same mechanism as the cfr gene. Antimicrob Agents Chemother 59:5841–5843

Harnvoravongchai P, Pipatthana M, Chankhamhaengdecha S, Janvilisri T (2017) Insights into drug resistance mechanisms in *Clostridium difficile*. Essays Biochem 61:81–88

Hastey CJ, Dale SE, Nary J et al (2017) Comparison of *Clostridium difficile* minimum inhibitory concentrations obtained using agar dilution vs broth microdilution methods. Anaerobe 44:73–77

He M, Sebaihia M, Lawley TD et al (2010) Evolutionary dynamics of *Clostridium difficile* over short and long time scales. PNAS 107:7527–7532

He M, Miyajima F, Roberts P, Ellison L et al (2013) Emergence and global spread of epidemic healthcare-associated *Clostridium difficile*. Nat Genet 45:109–113

Holt HM, Danielsen TK, Justesen US (2015) Routine disc diffusion antimicrobial susceptibility testing of *Clostridium difficile* and association with PCR ribotype 027. Eur J Clin Microbiol Infect Dis 34:2243–2246

Huang H, Weintraub A, Fang H et al (2009) Antimicrobial resistance in *Clostridium difficile*. Int J Antimicrob Agents 34:516–522

Impallomeni M, Galletly NP, Wort J et al (1995) Increased risk of diarrhoea caused by *Clostridium difficile* in elderly patients receiving cefotaxime. BMJ 311: 1345–1346

Imwattana K, Knight DR, Kullin B et al (2020a) Antimicrobial resistance in *Clostridium difficile* ribotype 017. Expert Rev Anti Infect Ther 18:17–25

Imwattana K, Kiratisin P, Riley TV, Knight DR (2020b) Genomic basis of antimicrobial resistance in non-toxigenic *Clostridium difficile* in Southeast Asia. Anaerobe 66:102290

Imwattana K, Putsathit P, Knight DR et al (2021a) Molecular characterization of, and antimicrobial resistance in, *Clostridioides difficile* from Thailand, 2017-2018. Microb Drug Resist 27:1505–1512

Imwattana K, Rodríguez C, Riley TV, Knight DR (2021b) A species-wide genetic atlas of antimicrobial resistance in *Clostridioides difficile*. Microb Genom 7:000696

Imwattana K, Putsathit P, Collins DA et al (2022) (2022) Global evolutionary dynamics and resistome analysis of *Clostridioides difficile* ribotype 017. Microb Genom 8:000792

Iv ECO, Iii ECO, Johnson DA (2014) Clinical update for the diagnosis and treatment of *Clostridium difficile* infection. World J Gastrointest Pharmacol Ther 5:1–26

Isidro J, Menezes J, Serrano M et al (2018) Genomic study of a *Clostridium difficile* multidrug resistant outbreak-related clone reveals novel determinants of resistance. Front Microbiol 6:2994

Jarrad AM, Karoli T, Blaskovich MAT et al (2015) *Clostridium difficile* drug pipeline: challenges in discovery and development of new agents. J Med Chem 58: 5164–5185

Jasni AS, Mullany P, Hussain H et al (2010) Demonstration of conjugative transposon (Tn*5397*)-mediated horizontal gene transfer between *Clostridium difficile* and Enterococcus faecalis. Antimicrob Agents Chemother 54:4924–4926

Jin D, Luo Y, Huang C et al (2017) Molecular epidemiology of *Clostridium difficile* infection in hospitalized patients in Eastern China. J Clin Microbiol 55:801–810

Jiménez A, Araya R, Paniagua D et al (2018) Molecular epidemiology and antimicrobial resistance of *Clostridium difficile* in a national geriatric hospital in Costa Rica. J Hosp Infect 99:475–480

Johnson S, Schriever C, Patel U et al (2009) Rifaximin redux: treatment of recurrent *Clostridium difficile* infections with rifaximin immediately post-vancomycin treatment. Anaerobe 15:290–291

Johnson S, Lavergne V, Skinner AM et al (2021) Clinical practice guideline by the Infectious Diseases Society of America (IDSA) and Society for Healthcare Epidemiology of America (SHEA): 2021 focused update guidelines on management of *Clostridioides difficile* infection in adults. Clin Infect Dis 73:e1029–e1044

Jon JV, Mark HW, Jane F et al (2021) Antimicrobial resistance progression in the United Kingdom: A temporal comparison of *Clostridioides difficile* antimicrobial susceptibilities. Anaerobe 70:102385

Kabała M, Gofron Z, Aptekorz M et al (2021) *Clostridioides difficile* ribotype 027 (RT027) outbreak investigation due to the emergence of rifampicin resistance using multilocus variable-number tandem repeat analysis (MLVA). Infect Drug Resist 17:3247–3254

Kachrimanidou M, Baktash A, Metallidis S et al (2020) An outbreak of *Clostridioides difficile* infections due to a 027-like PCR ribotype 181 in a rehabilitation centre: epidemiological and microbiological characteristics. Anaerobe 65:102252

Kachrimanidou M, Metallidis S, Tsachouridou O et al (2022) Predominance of *Clostridioides difficile* PCR ribotype 181 in northern Greece, 2016-2019. Anaerobe 76:102601

Karlowsky JA, Zhanel GG, Hammond GW et al (2012) Multidrug-resistant North American pulsotype 2 *Clostridium difficile* was the predominant toxigenic hospital-acquired strain in the province of Manitoba, Canada, in 2006-2007. J Med Microbiol 61:693–700

Karlowsky JA, Adam HJ, Baxter MR et al (2020) Antimicrobial susceptibility of *Clostridioides difficile* isolated from diarrhoeal stool specimens of Canadian patients: summary of results from the Canadian *Clostridioides difficile* (CAN-DIFF) surveillance study from 2013 to 2017. J Antimicrob Chemother 75:1824–1832

Kato H, Senoh M, Honda H et al (2019) *Clostridioides* (*Clostridium*) difficile infection burden in Japan: A multicenter prospective study. Anaerobe 60:102011

Khan R, Cheesbrough J (2003) Impact of changes in antibiotic policy on *Clostridium difficile*-associated diarrhoea (CDAD) over a five-year period in a district general hospital. J Hosp Infect 54:104–108

Kelly CR, Fischer M, Allegretti JR et al (2021) ACG clinical guidelines: prevention, diagnosis, and treatment of *Clostridioides difficile* infections. Am J Gastroenterol 116:1124–1147

Knetsch CW, Connor TR, Mutreja A et al (2014) Whole genome sequencing reveals potential spread of *Clostridium difficile* between humans and farm animals in the Netherlands, 2002 to 2011. Euro Surveill 19:20954

Kim J, Kang JO, Pai H et al (2012) Association between PCR ribotypes and antimicrobial susceptibility among *Clostridium difficile* isolates from healthcare-associated infections in South Korea. Int J Antimicrob Agents 40:24–29

Knight DR, Giglio S, Huntington PG et al (2015a) Surveillance for antimicrobial resistance in Australian isolates of *Clostridium difficile*, 2013-14. J Antimicrob Chemother 70:2992–2999

Knight DR, Elliott B, Chang BJ et al (2015b) Diversity and evolution in the genome of *Clostridium difficile*. Clin Microbiol Rev 28:721–741

Knight DR, Riley TV (2016) *Clostridium difficile* clade 5 in Australia: antimicrobial susceptibility profiling of PCR ribotypes of human and animal origin. J Antimicrob Chemother 71:2213–2217

Knight DR, Androga GO, Ballard SA et al (2016) A phenotypically silent vanB2 operon carried on a Tn1549-like element in *Clostridium difficile*. mSphere 1:e00177-00116

Knight DR, Kullin B, Androga GO et al (2019) Evolutionary and genomic insights into *Clostridioides difficile* sequence type 11: a diverse zoonotic and antimicrobial-resistant lineage of global one health importance. mBio 10:e00446-19

Kociolek LK, Gerding DN, Osmolski JR et al (2016) Differences in the molecular epidemiology and antibiotic susceptibility of *Clostridium difficile* isolates in pediatric and adult patients. Antimicrob Agents Chemother 60:4896–4900

Kouzegaran S, Ganjifard M, Tanha AS (2016) Detection, ribotyping and antimicrobial resistance properties of *Clostridium difficile* strains isolated from the cases of diarrhea. Mater Sociomed 28:324–328

Krutova M, Matejkova J, Tkadlec J et al (2015) Antibiotic profiling of *Clostridium difficile* ribotype 176 – A multidrug resistant relative to *C. difficile* ribotype 027. Anaerobe 36:88–90

Krutova M, Capek V, Nycova E et al (2020) The association of a reduced susceptibility to moxifloxacin in causative *Clostridium* (*Clostridioides*) *difficile* strain with the clinical outcome of patients. Antimicrob Resist Infect Control 9:98

Krutova M, de Meij TGJ, Fitzpatrick F et al (2022a) How to: *Clostridioides difficile* infection in children. Clin Microbiol Infect 28:1085–1090

Krutova M, Wilcox M, Kuijper E (2022b) *Clostridioides difficile* infection: are the three currently used antibiotic treatment options equal from pharmacological and microbiological points of view? Int J Infect Dis 124: 118–123

Kuehne SA, Dempster AW, Collery MM et al (2018) Characterization of the impact of rpoB mutations on the in vitro and in vivo competitive fitness of *Clostridium difficile* and susceptibility to fidaxomicin. J Antimicrob Chemother 73:973–98042

Kullin B, Wojno J, Abratt V et al (2017) Toxin A-negative toxin B-positive ribotype 017 *Clostridium difficile* is the dominant strain type in patients with diarrhoea attending tuberculosis hospitals in Cape Town, South Africa. Eur J Clin Microbiol Infect Dis 36: 163–175

Kurahara Y, Kobayashi T, Tachibana K, Tsuyuguchi K (2022) The feasibility of rifampicin Re-administration in patients with tuberculosis and *Clostridioides difficile* infection. J Infect Chemother 28:558–562

Kuwata Y, Tanimoto S, Sawabe E et al (2015) Molecular epidemiology and antimicrobial susceptibility of *Clostridium difficile* isolated from a university teaching hospital in Japan. Eur J Clin Microbiol Infect 34: 763–772

Lachowicz D, Pituch H, Obuch-Woszczatyński P (2015) Antimicrobial susceptibility patterns of *Clostridium difficile* strains belonging to different polymerase chain reaction ribotypes isolated in Poland in 2012. Anaerobe 31:37–41

Lebel S, Bouttier S, Lambert T (2004) The cme gene of *Clostridium difficile* confers multidrug resistance in *Enterococcus faecalis*. FEMS Microbiol Lett 238:93–100

Lee J-H, Lee Y, Lee K et al (2014) The changes of PCR ribotype and antimicrobial resistance of *Clostridium difficile* in a tertiary care hospital over 10 years. J Med Microbiol 63:819–823

Leeds JA, Sachdeva M, Mullin S et al (2014) In vitro selection, via serial passage, of *Clostridium difficile* mutants with reduced susceptibility to fidaxomicin or vancomycin. J Antimicrob Chemother 69:41–44

Lessa FC, Gould CV, McDonald LC (2012) Current status of *Clostridium difficile* infection epidemiology. Clin Infect Dis 55:65–70

Lessa FC, Mu Y, Bamberg WM et al (2015) Burden of *Clostridium difficile* Infection in the United States. N Engl J Med 372:825–834

Lew T, Putsathit P, Sohn KM et al (2020) Antimicrobial susceptibilities of *Clostridium difficile* isolates from 12 Asia-Pacific Countries in 2014 and 2015. Antimicrob Agents Chemother 64:e00296-20

Liao CH, Ko WC, Lu JJ et al (2012) Characterizations of clinical isolates of *Clostridium difficile* by toxin genotypes and by susceptibility to 12 antimicrobial agents, including fidaxomicin (OPT-80) and rifaximin: a multicenter study in Taiwan. Antimicrob Agents Chemother 56:3943–3949

Lim SC, Foster NF, Riley TV (2016) Susceptibility of *Clostridium difficile* to the food preservatives sodium nitrite, sodium nitrate and sodium metabisulphite. Anaerobe 37:67–71

Louie TJ, Miller MA, Mullane KM et al (2011) Fidaxomicin versus vancomycin for *Clostridium difficile* infection. N Engl J Med 364:422–431

Louie TJ, Cannon K, Byrne B et al (2012) Fidaxomicin preserves the intestinal microbiome during and after treatment of *Clostridium difficile* infection (CDI) and reduces both toxin reexpression and recurrence of CDI. Clin Infect Dis 55(Suppl 2):S132–S142

Lynch T, Chong P, Zhang J et al (2013) Characterization of a stable, metronidazole-resistant *Clostridium difficile* clinical isolate. PLoS ONE 8:e53757

Lyras D, Storie C, Huggins AS et al (1998) Chloramphenicol resistance in *Clostridium difficile* is encoded on Tn*4453* transposons that are closely related to Tn*4451* from *Clostridium* perfringens. Antimicrob Agents Chemother 42:1563–1156

Lyras D, Cooper MA (2015) *Clostridium difficile* drug pipeline: challenges in discovery and development of new agents. J Med Chem 58:5164–5185

Mac Aogáin M, Kilkenny S, Walsh C et al (2015) Identification of a novel mutation at the primary dimer interface of GyrA conferring fluoroquinolone resistance in *Clostridium difficile*. J Glob Antimicrob Resist 3:295–299

Marin M, Martin A, Alcala L et al (2015) *Clostridium difficile* isolates with high linezolid MICs harbor the multiresistance gene cfr. Antimicrob Agents Chemother 59:586–589

Martínez-Meléndez A, Morfin-Otero R, Villarreal-Treviño L et al (2020) Molecular epidemiology of predominant and emerging *Clostridioides difficile* ribotypes. J Microbiol Methods 175:105974

McDonald LC, Killgore GE, Thompson A et al (2005) An epidemic, toxin gene-variant strain of *Clostridium difficile*. N Engl J Med 353(23):24323–24341

McDonald LC, Gerding DN, Johnson S et al (2018) Clinical practice guidelines for *Clostridium difficile* infection in adults and children: 2017 update by the Infectious Diseases Society of America (IDSA) and Society for Healthcare Epidemiology of America (SHEA). Clin Infect Dis 66:e1–e4

Miller BA, Chen LF, Sexton DJ et al (2011a) Comparison of the burdens of hospital-onset, healthcare facility-associated *Clostridium difficile* infection and of healthcare-associated infection due to methicillin-resistant Staphylococcus aureus in community hospitals. Infect Control Hosp Epidemiol 32:387–390

Miller MA, Blanchette R, Spigaglia P et al (2011b) Divergent rifamycin susceptibilities of *Clostridium difficile* strains in Canada and Italy and predictive accuracy of rifampin Etest for rifamycin resistance. J Clin Microbiol 49:4319–4321

Miller-Roll T, Na'amnih W, Cohen D et al (2016) Molecular types and antimicrobial susceptibility patterns of *Clostridium difficile* isolates in different epidemiological settings in a tertiary care center in Israel. Diagn Microbiol Infect Dis 86:450–454

Moura I, Spigaglia P, Barbanti F et al (2013) Analysis of metronidazole susceptibility in different *Clostridium difficile* PCR ribotypes. J Antimicrob Chemother 68: 362–365

Moura I, Monot M, Tani C et al (2014) Multidisciplinary analysis of a nontoxigenic *Clostridium difficile* strain with stable resistance to metronidazole. Antimicrob Agents Chemother 58:4957–4960

Mullane KM, Gorbach S (2011) Fidaxomicin: first-in-class macrocyclic antibiotic. Expert Rev Anti Infect Ther 9:767–777

Mullany P, Wilks M, Lamb I et al (1990) Genetic analysis of a tetracycline resistance determinant from *Clostridium difficile* and its conjugal transfer to and from Bacillus subtilis. J Gen Microbiol 136:1343–1349

Mullany P, Wilks M, Tabaqchali S (1995) Transfer of macrolide-lincosamide-streptogramin B (MLS) resistance in *Clostridium difficile* is linked to a gene homologous with toxin A and is mediated by a conjugative transposon, Tn*5398*. J Antimicrob Chemother 2:305–315

Mullany P, Williams R, Langridge GC et al (2012) Behavior and target site selection of conjugative transposon Tn*916* in two different strains of toxigenic *Clostridium difficile*. Appl Environ Microbiol 78:2147–2153

Mullany P, Allan E, Roberts AP (2015) Mobile genetic elements in *Clostridium difficile* and their role in genome function. Res Microbiol 166:361–367

Musher DM, Aslam S, Logan N et al (2005) Relatively poor outcome after treatment of *Clostridium difficile* colitis with metronidazole. Clin Infect Dis 40:1586–1590

Mutai WC, Mureithi MW, Anzala O et al (2021) High prevalence of multidrug-resistant *Clostridioides difficile* following extensive use of antimicrobials in hospitalized patients in Kenya. Front Cell Infect Microbiol 8(10):604986

Muto CA, Pokrywka M, Shutt K et al (2005) A large outbreak of *Clostridium difficile*-associated disease with an unexpected proportion of deaths and colectomies at a teaching hospital following increased fluoroquinolone use. Infect Control Hosp Epidemiol 26:273–280

Ngernsombat C, Sreesai S, Harnvoravongchai P et al (2017) CD2068 potentially mediates multidrug efflux in *Clostridium difficile*. Sci Rep 7:9982

Norman KN, Scott HM, Harvey RB et al (2014) Comparison of antimicrobial susceptibility among *Clostridium difficile* isolated from an integrated human and swine population in Texas. Foodborne Pathog Dis 11:257–264

O'Connor JR, Galang MA, Sambol SP et al (2008) Rifampin and rifaximin resistance in clinical isolates of *Clostridium difficile*. Antimicrob Agents Chemother 52: 2813–2817

O'Grady K, Knight DR, Riley TV (2021) Antimicrobial resistance in *Clostridioides difficile*. Eur J Clin Microbiol Infect Dis 40:2459–2478

Obuch-Woszczatyński P, Dubiel G, Harmanus C et al (2013) Emergence of *Clostridium difficile* infection in tuberculosis patients due to a highly rifampicin-resistant PCR ribotype 046 clone in Poland. Eur J Clin Microbiol Infect Dis 32:1027–1103

Obuch-Woszczatyński P, Lachowicz D, Schneider A et al (2014) Occurrence of *Clostridium difficile* PCR-ribotype 027 and it's closely related PCR-ribotype 176 in hospitals in Poland in 2008-2010. Anaerobe 28:13–17

Oka K, Osaki T, Hanawa T et al (2012) Molecular and microbiological characterization of *Clostridium difficile* isolates from single, relapse, and reinfection cases. J Clin Microbiol 50:915–921

Optimer Pharmaceuticals, Inc (2011) Anti-Infective Drugs Advisory Committee Briefing Document: Dificid™ (fidaxomicin tablets) for the treatment of *Clostridium difficile* infection (CDI), also known as Clostridium difficile-associated diarrhea (CDAD), and for reducing the risk of recurrence when used for treatment of initial CDI. http://www.fda.gov/downloads/AdvisoryCommittees/CommitteesMeetingMaterials/Drugs/Anti-InfectiveDrugsAdvisoryCommittee/UCM249354.pdf

Pecavar V, Blaschitz M, Hufnagl P et al (2012) High-resolution melting analysis of the single nucleotide polymorphism hot-spot region in the *rpo*B gene as an indicator of reduced susceptibility to rifaximin in *Clostridium difficile*. J Med Microbiol 61:780–785

Peng Z, Addisu A, Alrabaa S, Sun X (2017) Antibiotic resistance and toxin production of *Clostridium difficile* isolates from the hospitalized patients in a large hospital in Florida. Front Microbiol 8:2584

Peláez T, Cercenado E, Alcalá L et al (2008) Metronidazole resistance in *Clostridium difficile* is heterogeneous. J Clin Microbiol 46:3028–3032

Pépin JL, Valiquette ME, Alary ME et al (2004) *Clostridium difficile*-associated diarrhea in a region of Quebec from 1991-2003: a changing pattern disease severity. CMAJ 17:466–472

Pépin J, Alary ME, Valiquette L et al (2005a) Increasing risk of relapse after treatment of *Clostridium difficile* Colitis in Quebec, Canada. Clin Infect Dis 40:1591–1597

Pépin JL, Valiquette ME, Clossette B (2005b) Mortality attributed to nosocomial *Clostridium difficile*-associated disease during an epidemic caused by a hypervirulent strain in Quebec. CMAJ 173:1037–1042

Perkins HR, Nieto M (1974) The chemical basis for the action of the vancomycin group of antibiotics. Ann N Y Acad Sci 235:348–363

Pituch H, Brazier JS, Obuch-Woszczatynski P et al (2006) Prevalence and association of PCR ribotypes of *Clostridium difficile* isolated from symptomatic patients from Warsaw with macrolide-lincosamide-streptogramin B (MLSB) type resistance. J Med Microbiol 55:207–213

Poilane I, Cruaud P, Torlotin JC et al (2000) Comparison of the E test to the reference agar dilution method for antibiotic susceptibility testing of *Clostridium difficile*. Clin Microbiol Infect 6:155–156

Pu M, Cho JM, Cunningham SA et al (2021) Plasmid acquisition alters vancomycin susceptibility in *Clostridioides difficile*. Gastroenterology 160:941–945

Putsathit P, Hong S, George N et al (2021) Antimicrobial resistance surveillance of *Clostridioides difficile* in Australia, 2015-18. J Antimicrob Chemother 76: 1815–1821

Ramírez-Vargas G, Quesada-Gómez C, Acuña-Amador L et al (2017) A *Clostridium difficile* lineage endemic to Costa Rican hospitals is multidrug resistant by acquisition of chromosomal mutations and novel mobile genetic elements. Antimicrob Agents Chemother 61

Ratnayake L, McEwen J, Henderson N et al (2011) Control of an outbreak of diarrhoea in a vascular surgery unit caused by a high-level clindamycin-resistant *Clostridium difficile* PCR ribotype 106. J Hosp Infect 79: 242–247

Redelings MD, Sorvillo F, Mascola L (2007) Increase in *Clostridium difficile*-related mortality rates, United States, 1999-2004. Emerg Infect Dis 13:1417–1419

Richardson C, Kim P, Lee C et al (2015) Comparison of *Clostridium difficile* isolates from individuals with recurrent and single episode of infection. Anaerobe 33:105–108

Roberts MC, McFarland LV, Mullany P et al (1994) Characterization of the genetic basis of antibiotic resistance in *Clostridium difficile*. J Antimicrob Chemother 33:419–429

Roberts AP, Johanesen PA, Lyras D et al (2001) Comparison of Tn5397 from *Clostridium difficile*, Tn*916* from Enterococcus faecalis and the CW459*tet*(M) element from *Clostridium* perfringens shows that they have similar conjugation regions but different insertion and excision modules. Microbiology 147:1243–1251

Roberts AP, Mullany P (2011) Tn*916*-like genetic elements: a diverse group of modular mobile elements conferring antibiotic resistance. FEMS Microbiol Rev 35:856–871

Rodríguez-Pardo D, Almirante B, Bartolomé RM et al (2013) Epidemiology of *Clostridium difficile* infection and risk factors for unfavorable clinical outcomes: results of a hospital-based study in Barcelona, Spain. J Clin Microbiol 51:1465–1473

Roxas BAP, Roxas JL, Claus-Walker R et al (2020) Phylogenomic analysis of *Clostridioides difficile* ribotype 106 strains reveals novel genetic islands and emergent phenotypes. Sci Rep 10:22135

Saha S, Kapoor S, Tariq R et al (2019) Increasing antibiotic resistance in *Clostridioides difficile*: a systematic review and meta-analysis. Anaerobe 58:35–46

Salix Pharmaceuticals, Ltd (2003) Salix receives FDA notification that rifaximin amendment considered a complete response. Salix Pharmaceuticals, Raleigh, NC. 10 December 2003, posting date. http://www.businesswire.com/news/home/20031210005070/en/Salix-Receives-FDA-Notification-Rifamixin-Amendment-Considered

Sandhu BK, Edwards AN, Anderson SE et al (2019) Regulation and anaerobic function of the *Clostridioides difficile* β-lactamase. Antimicrob Agents Chemother 64:e01496-19

Santos A, Isidro J, Silva C et al (2016) Molecular and epidemiologic study of *Clostridium difficile* reveals unusual heterogeneity in clinical strains circulating in different regions in Portugal. Clin Microbiol Infect 22: 695–700

Schmidt C, Löffler B, Ackermann G (2007) Antimicrobial phenotypes and molecular basis in clinical strains of *Clostridium difficile*. Diagn Microbiol Infect Dis 59:1–5

Schwanbeck J, Riedel T, Laukien F et al (2019) Characterization of a clinical *Clostridioides difficile* isolate with markedly reduced fidaxomicin susceptibility and a V1143D mutation in rpoB. J Antimicrob Chemother 74:6–10

Sears P, Crook DW, Louie TJ et al (2012) Fidaxomicin attains high fecal concentrations with minimal plasma concentrations following oral administration in patients with *Clostridium difficile* infection. Clin Infect Dis 55(Suppl. 2):S116–SS12

Sebaihia M, Wren BW, Mullany P et al (2006) The multidrug-resistant human pathogen *Clostridium difficile* has a highly mobile, mosaic genome. Nat Genet 38:779–786

Senoh M, Kato H, Fukuda T et al (2015) Predominance of PCR-ribotypes, 018 (smz) and 369 (trf) of *Clostridium difficile* in Japan: a potential relationship with other global circulating strains? J Med Microbiol 64:1226–1236

Shaw HA, Preston MD, Vendrik KEW et al (2020) The recent emergence of a highly related virulent *Clostridium difficile* clade with unique characteristics. Clin Microbiol Infect 26:492–498

Shayganmehr F-S, Alebouyeh M, Azimirad M et al (2015) Association of tcdA+/tcdB+ Clostridium *difficile* genotype with emergence of multidrug-resistant strains conferring metronidazole resistant phenotype. Iran Biomed J 19:143–148

Shen WJ, Deshpande A, Hevener KE et al (2020) Constitutive expression of the cryptic vanGCd operon promotes vancomycin resistance in *Clostridioides difficile* clinical isolates. J Antimicrob Chemother 75(4): 859–867

Sholeh M, Krutova M, Forouzesh M et al (2020) Antimicrobial resistance in *Clostridioides* (*Clostridium*) *difficile* derived from humans: a systematic review and meta-analysis. Antimicrob Resist Infect Control 9:158

Slimings C, Riley TV (2014) Antibiotics and hospital-acquired *Clostridium difficile* infection: update of systematic review and meta-analysis. J Antimicrob Chemother 69:881–891

Spigaglia P, Mastrantonio P (2004) Comparative analysis of *Clostridium difficile* clinical isolates belonging to different genetic lineages and time periods. J Med Microbiol 53:1129–1136

Spigaglia P, Carucci V, Barbanti F et al (2005) ErmB determinants and Tn*916*-like elements in clinical isolates of *Clostridium difficile*. Antimicrob Agents Chemother 49:2550–2553

Spigaglia P, Barbanti F, Mastrantonio P (2006) New variants of the *tet*(M) gene in *Clostridium difficile* clinical isolates harbouring Tn*916*-like elements. J Antimicrob Chemother 57:1205–1209

Spigaglia P, Barbanti F, Mastrantonio P (2007) Detection of a genetic linkage between genes coding for resistance to tetracycline and erythromycin in *Clostridium difficile*. Microb Drug Resist 13:90–95

Spigaglia P, Barbanti F, Mastrantonio P (2008a) Tetracycline resistance gene *tet*(W) in the pathogenic bacterium *Clostridium difficile*. Antimicrob Agents Chemother 52:770–773

Spigaglia P, Barbanti F, Mastrantonio P et al (2008b) Fluoroquinolone resistance in *Clostridium difficile* isolates from a prospective study of *C. difficile* infections in Europe. J Med Microbiol 57:784–789

Spigaglia P, Barbanti F, Louie T et al (2009) Molecular analysis of the *gyr*A and *gyr*B quinolone resistance-determining regions of fluoroquinolone-resistant *Clostridium difficile* mutants selected in vitro. Antimicrob Agents Chemother 53:2463–2468

Spigaglia P, Barbanti F, Dionisi AM et al (2010) *Clostridium difficile* isolates resistant to fluoroquinolones in Italy: emergence of PCR ribotype 018. J Clin Microbiol 48:2892–2896

Spigaglia P, Barbanti F, Mastrantonio P, European Study Group on Clostridium difficile (ESGCD) (2011) Multidrug resistance in European *Clostridium difficile* clinical isolates. J Antimicrob Chemother 66:2227–2234

Spigaglia P, Barbanti F, Morandi M et al (2015) Diagnostic testing for *Clostridium difficile* in Italian microbiological laboratories. Anaerobe 37:29–33

Spigaglia P (2016) Recent advances in the understanding of antibiotic resistance in *Clostridium difficile* infection. Ther Adv Infect Dis 3:23–42

Srivastava A, Talaue M, Liu S et al (2011) New target for inhibition of bacterial RNA polymerase: "switch region". Curr Opin Microbiol Antimicrob Genom 14: 532–543

Stojković V, Ulate MF, Hidalgo-Villeda F et al (2019) cfr (B), cfr(C), and a new cfr-like gene, cfr(E), in *Clostridium difficile* strains recovered across Latin America. Antimicrob Agents Chemother 64:e01074–e01019

Tannock GW, Munro K, Taylor C et al (2010) A new macrocyclic antibiotic, fidaxomicin (OPT-80), causes less alteration to the bowel microbiota of *Clostridium difficile*-infected patients than does vancomycin. Microbiology 156:3354–3359

Tenover FC, Tickler IA, Persing DH (2012) Antimicrobial-resistant strains of *Clostridium difficile* from North America. Antimicrob Agents Chemother 56:2929–2932

Terhes G, Maruyama A, Latkóczy K et al (2014) In vitro antibiotic susceptibility profile of *Clostridium difficile* excluding PCR ribotype 027 outbreak strain in Hungary. Anaerobe 30:41–44

Tickler IA, Obradovich AE, Goering RV et al (2019) Changes in molecular epidemiology and antimicrobial resistance profiles of *Clostridioides* (*Clostridium*) *difficile* strains in the United States between 2011 and 2017. Anaerobe 60:102050

Toth M, Stewart NK, Smith C et al (2018) Intrinsic class D b-lactamases of *Clostridium difficile*. mBio 9: e01803-18

van Prehn J, Reigadas E, Vogelzang EH et al (2021) European Society of Clinical Microbiology and Infectious Diseases: 2021 update on the treatment guidance document for *Clostridioides difficile* infection in adults. Clin Microbiol Infect 27:S1–S21

Vardakas KZ, Polyzos KA, Patouni K et al (2012) Treatment failure and recurrence of *Clostridium difficile* infection following treatment with vancomycin or metronidazole: a systematic review of the evidence. Int J Antimicrob Agents 40:1–8

Varshney JB, Very KJ, Williams JL et al (2014) Characterization of *Clostridium difficile* isolates from human fecal samples and retail meat from Pennsylvania. Foodborne Pathog Dis 11:822–829

Viprey VF, Davis GL, Benson AD et al (2022) A point-prevalence study on community and inpatient Clostridioides difficile infections (CDI): results from Combatting Bacterial Resistance in Europe CDI (COMBACTE-CDI), July to November 2018. Euro Surveill 27:2100704

Vuotto C, Moura I, Barbanti F et al (2016) Subinhibitory concentrations of metronidazole increase biofilm formation in *Clostridium difficile* strains. Pathog Dis 74: ftv114

Walkty A, Boyd DA, Gravel D et al (2010) Molecular characterization of moxifloxacin resistance from Canadian *Clostridium difficile* clinical isolates. Diagn Microbiol Infect Dis 66:419–424

Wang H, Smith MCM, Mullany P (2006) The conjugative transposon Tn*5397* has a strong preference for integration into its *Clostridium difficile* target site. J Bacteriol 188:4871–4878

Wasels F, Spigaglia P, Barbanti F et al (2013) *Clostridium difficile erm*(B)-containing elements and the burden on the in vitro fitness. J Med Microbiol 62:1461–1467

Wasels F, Monot M, Spigaglia P et al (2014) Inter- and intraspecies transfer of a *Clostridium difficile* conjugative transposon conferring resistance to MLSB. Microb Drug Resist 20:555–560

Wasels F, Kuehne SA, Cartman ST et al (2015a) Fluoroquinolone resistance does not impose a cost on the fitness of *Clostridium difficile* in vitro. Antimicrob Agents Chemother 59:1794–1796

Wasels F, Spigaglia P, Barbanti F et al (2015b) Integration of *erm*(B)-containing elements through large chromosome fragment exchange in *Clostridium difficile*. Mob Genet Elements 1:12–16

Weber I, Riera E, Déniz C et al (2013) Molecular epidemiology and resistance profiles of *Clostridium difficile* in a tertiary care hospital in Spain. Int J Med Microbiol 303:128–133

Wehrhahn MC, Keighley C, Kurtovic J et al (2019) A series of three cases of severe *Clostridium difficile* infection in Australia associated with a binary toxin producing clade 2 ribotype 251 strain. Anaerobe 55: 117–123

Wiström J, Norrby SR, Myhre EB et al (2001) Frequency of antibiotic-associated diarrhea in 2462 antibiotic-treated hospitalized patients: a prospective study. J Antimicrob Chemother 47:43–50

Woods E, Wetzel D, Mukerjee M, McBride S (2018) Examination of the *Clostridioides* (*Clostridium*) *difficile* VanZ ortholog, CD1240. Anaerobe 53:108–115

Wren BW, Mullany P, Clayton C et al (1988) Molecular cloning and genetic analysis of a chloramphenicol acetyltransferase determinant from *Clostridium difficile*. Antimicrob Agents Chemother 32:1213–1121

Wren BW, Mullany P, Clayton C et al (1989) Nucleotide sequence of a chloramphenicol acetyl transferase gene from *Clostridium difficile*. Nucleic Acids Res 17:4877

Xu X, Luo Y, Chen HG et al (2021) Genomic evolution and virulence association of *Clostridioides difficile* sequence type 37 (ribotype 017) in China. Emerg Microbes Infect 10:1331–1345

Young GP, Ward PB, Bayley N et al (1985) Antibiotic-associated colitis due to *Clostridium difficile*: double-blind comparison of vancomycin with bacitracin. Gastroenterology 89:1038–1045

Yu X, Sun D (2013) Macrocyclic drugs and synthetic methodologies toward macrocycles. Molecules 18: 6230–6268

Zar FA, Bakkanagari SR, Moorthi KM, Davis MB (2007) A comparison of vancomycin and metronidazole for the treatment of *Clostridium difficile*-associated diarrhea, stratified by disease severity. Clin Infect Dis 45: 302–307

Comparative Genomics of *Clostridioides difficile*

Sandra Janezic, Julian R. Garneau, and Marc Monot

Abstract

Clostridioides difficile, a Gram-positive spore-forming anaerobic bacterium, has rapidly emerged as the leading cause of nosocomial diarrhoea in hospitals. The availability of large numbers of genome sequences, mainly due to the use of next-generation sequencing methods, has undoubtedly shown their immense advantages in the determination of *C. difficile* population structure. The implementation of fine-scale comparative genomic approaches has paved the way for global transmission and recurrence studies, as well as more targeted studies, such as the PaLoc or CRISPR/Cas systems. In this chapter, we provide an overview of recent and significant findings on *C. difficile* using comparative genomic studies with implications for epidemiology, infection control and understanding of the evolution of *C. difficile*.

1 Introduction

Clostridioides difficile infection (CDI) is currently the most commonly occurring nosocomial diarrhoea in healthcare environments (Davies et al. 2016; Suetens et al. 2018). This major pathogen synthesizes two toxins (toxin A and toxin B) encoded in a pathogenicity locus (PaLoc), which are generally recognized as the main virulence factors. Additionally, some strains also produce a third, unrelated toxin, CDT, encoded in CdtLoc, whose role in pathogenicity has not been fully determined (Kuehne et al. 2014). Over the last decade, the incidence and severity of CDI have increased significantly, mainly owing to the emergence of new strain variants. Molecular typing methods were extensively used to understand its epidemiology, genetic diversity and evolution. The *C. difficile* population structure contains hundreds of strain types organized in phylogenetic clades (Dingle et al. 2014; Elliott et al. 2014; Janezic et al. 2016; Knight et al. 2021).

The first complete genome sequence of a *C. difficile* strain was published in 2006 (Sebaihia et al. 2006) enabling the development of comparative genomics. Initially, microarray comparative

S. Janezic (✉)
National Laboratory for Health, Environment and Food (NLZOH), Maribor, Slovenia

Faculty of Medicine, University of Maribor, Maribor, Slovenia
e-mail: sandra.janezic@nlzoh.si

J. R. Garneau
Department of Fundamental Microbiology, University of Lausanne, Lausanne, Switzerland

M. Monot
Institut Pasteur, Université Paris Cité, Plate-forme Technologique Biomics, Paris, France

P. Mastrantonio, M. Rupnik (eds.), *Updates on Clostridioides difficile in Europe*, Advances in Experimental Medicine and Biology 1435, https://doi.org/10.1007/978-3-031-42108-2_10

genome hybridizations (CGH) were used in global studies to estimate the diversity and evolution of strains (Table 1). However, many laboratories worldwide can now afford frequent and even routine sequencing of *C. difficile* strains. The availability of large numbers of genome sequences, mainly due to the use of next-generation sequencing (NGS) methods, has undoubtedly demonstrated their immense advantages in the determination of *C. difficile* population structure and evolution. The implementation of fine-scale comparative genomic approaches has paved the way for global transmission and recurrence studies, development of whole genome-based typing schemes and also more targeted studies such as organization and evolution of the PaLoc/CdtLoc or the CRISPR/Cas systems (Table 1). Whole genome sequence comparisons have also advanced our understanding of the cross-transmission of *C. difficile* between human and non-human *C. difficile* reservoirs: animals, food and the environment (discussed in a chapter "Non-human *C. difficile* Reservoirs and Sources: Animals, Food, Environment).

Here we provide an overview of the significant findings on *C. difficile* using comparative genomic studies that shed new light on the epidemiology, population structure, evolution and the clinical practice used for *C. difficile*.

2 Global Comparative Genomics

2.1 Population Structure of *C. difficile* Species

Lemee et al. (2004) conducted the first analysis of genetic relationship and population structure of *C. difficile* isolates using multilocus sequence typing (MLST). They identified 34 different MLST sequence types (MLST-ST) among 72 isolates. Phylogenetic analysis demonstrated three distinct phylogenetic clades with no specific association between a particular clade and host or geographic origin. Furthermore, they showed that the loci included in the MLST scheme were in linkage disequilibrium demonstrating a clonal population structure (i.e. mutational evolution) of *C. difficile* species (Lemee et al. 2004). The study by Griffiths et al. (2010) using a different MLST scheme (with different set of genes) confirmed the clonal population structure of *C. difficile* and identified two additional lineages, one represented by ST-22 (PCR ribotype 023, toxinotype IV) and the genetically distant outlier of ST-11 (PCR ribotype 078, toxinotype V). In 2012, Knetsch et al. (2012) described a putative sixth lineage, represented by a single sequence type (ST-122, PCR ribotype 131). However phylogenetic analysis based on core genome comparison did not confirm the topology of the tree and placed this strain as an outlier within clade 1, possibly a recombinant between clade 1 and clade 2 (Dingle et al. 2014). In a recent study, ST-122 was assigned to clade 1 (Knight et al. 2021). The population structure composed of clades was later confirmed by comparisons of whole genome sequences on more diverse collection of strains (He et al. 2010; Dingle et al. 2014; see also review papers: Knight et al. 2015; Janezic and Rupnik 2015). The high concordance of MLST and core genome phylogeny demonstrated that MLST could be used as a good proxy for whole genome comparisons (Griffiths et al. 2010; Didelot et al. 2012; Dingle et al. 2014).

Based on MLST and whole genome analysis, the *C. difficile* population can now be divided into ten distinct phylogenetic clades. Five major clades which are designated from 1 to 5 and five cryptic clades C-I, C-II, C-III, C-IV and C-V (Janezic et al. 2016; Knight et al. 2021; Williamson et al. 2022).

The population structure composed of the first six clades (1–5 and C-I) was defined mainly from isolates originating from humans and in lesser extent from animals (Dingle et al. 2014). In 2016, two new, highly divergent clades were identified among environmental (mainly soil) *C. difficile* isolates from Slovenia (Fig. 1). The new clades were designated C-II and C-III (Janezic et al. 2016). The topology of the MLST-based tree was also confirmed by whole genome comparison (core genome MLST) (Bletz et al. 2018). It was hypothesized that because of

Table 1 Early comparative genomic studies that paved the way for our understanding of the *C. difficile* epidemiology and evolution

Year	Strains	References	Topics	Summary
Hybridization: MicroArrays				
2006	8	Sebaihia et al. (2006)	Comparison	Core genome
–	75	Stabler et al. (2006)	Evolution	Phylogenomics
2009	73	Janvilisri et al. (2009)	Comparison	Core and divergence between hosts
2010	167	Scaria et al. (2010)	Comparison	Core genome
–	94	Marsden et al. (2010)	Comparison	UK and European ribotype 027
Sequencing: Sanger & NGS				
2009	2	Stabler et al. (2009)	Comparison	Historic and modern ribotype 027
2010	29	He et al. (2010)	Evolution	Short and long time scales
–	15	Scaria et al. (2010)	Comparison	Core genome
2011	14	Forgetta et al. (2011)	Comparison	Severe disease-associated genomic markers
2012	15	Eyre et al. (2012)	Transmission	WGS for outbreak detection
–	486	Didelot et al. (2012)	Transmission	Micro-evolution
2013	151	He et al. (2013)	Evolution	Emergence and global spread of ribotype 027
–	1	Eyre et al. (2013d)	Evolution	Short-term stability of a single ribotype 027
–	1223	Eyre et al. (2013b)	Transmission	Identification of diverse source of infection
–	15	Eyre et al. (2013a)	Transmission	Detection of mixed infection
–	176	Eyre et al. (2013c)	Transmission	Role of asymptomatic carriage in transmission
2014	1693	Dingle et al. (2014)	Evolution	History of the pathogenicity locus
–	48	Kurka et al. (2014)	Typing	Ribotype and MLST correlation
–	185	Eyre et al. (2014)	Antibiotics	Fidaxomicin in relapse and reinfection
–	3	Moura et al. (2014)	Antibiotics	Metronidazole resistance
–	31[a]	Moura et al. (2014), Hargreaves et al. (2014)	CRISPR	Distribution and diversity
2015	53	Mac Aogain et al. (2015)	Recurrence	Discrimination between relapses and reinfections
–	18[a]	Boudry et al. (2015)	CRISPR	Mechanistic and physiology
–	3	Monot et al. (2015)	Evolution	Model of the pathogenicity locus evolution
2016	96	Quesada-Gomez et al. (2016)	Toxins	Specificity of hypervirulent clade 2 TcdB proteins
–	5	Chowdhury et al. (2016)	Toxins	Toxin-negative strains in human and animals
–	108	Kumar et al. (2016)	Transmission	Relapse and reinfection of ribotype 027
2017	35	Sim et al. (2017)	Recurrence	Rate of relapses and reinfections
–	277	Cairns et al. (2017)	Evolution	Phylogeny of ribotype 017
–	265	Mawer et al. (2017)	Transmission	Symptomatic patients but faecal toxin negative
–	971	Eyre et al. (2017)	Transmission	WGS as hospital surveillance tools

[a]*C. difficile* phage or prophage

high abundance of isolates from these two clades in the environment and only sporadic isolation from clinical samples, these strains could represent native environmental isolates, which are not primarily associated with humans and/or animals (Janezic et al. 2016). Two additional cryptic clades, designated as clades C-IV and C-V, were described in a recent study (Williamson et al. 2022).

The most heterogeneous clade, in terms of the number of MLST-STs and PCR ribotypes, is clade 1, where more than 380 different MLST-STs are identified (data from PubMLST *C. difficile* database, accessed 14.3.2023) (Table 2). Many strains from this clade are of clinical significance, for example, PCR ribotypes 014, 002, 001, 015 and 018, which are among ten most prevalent PCR ribotypes isolated from CDI (*C. difficile* infection) patients in Europe (Davies et al. 2016). Clade 2 is comprised of 90 different MLST-STs, including ST-1 (PCR ribotype 027), a well-known epidemic strain, and two emerging

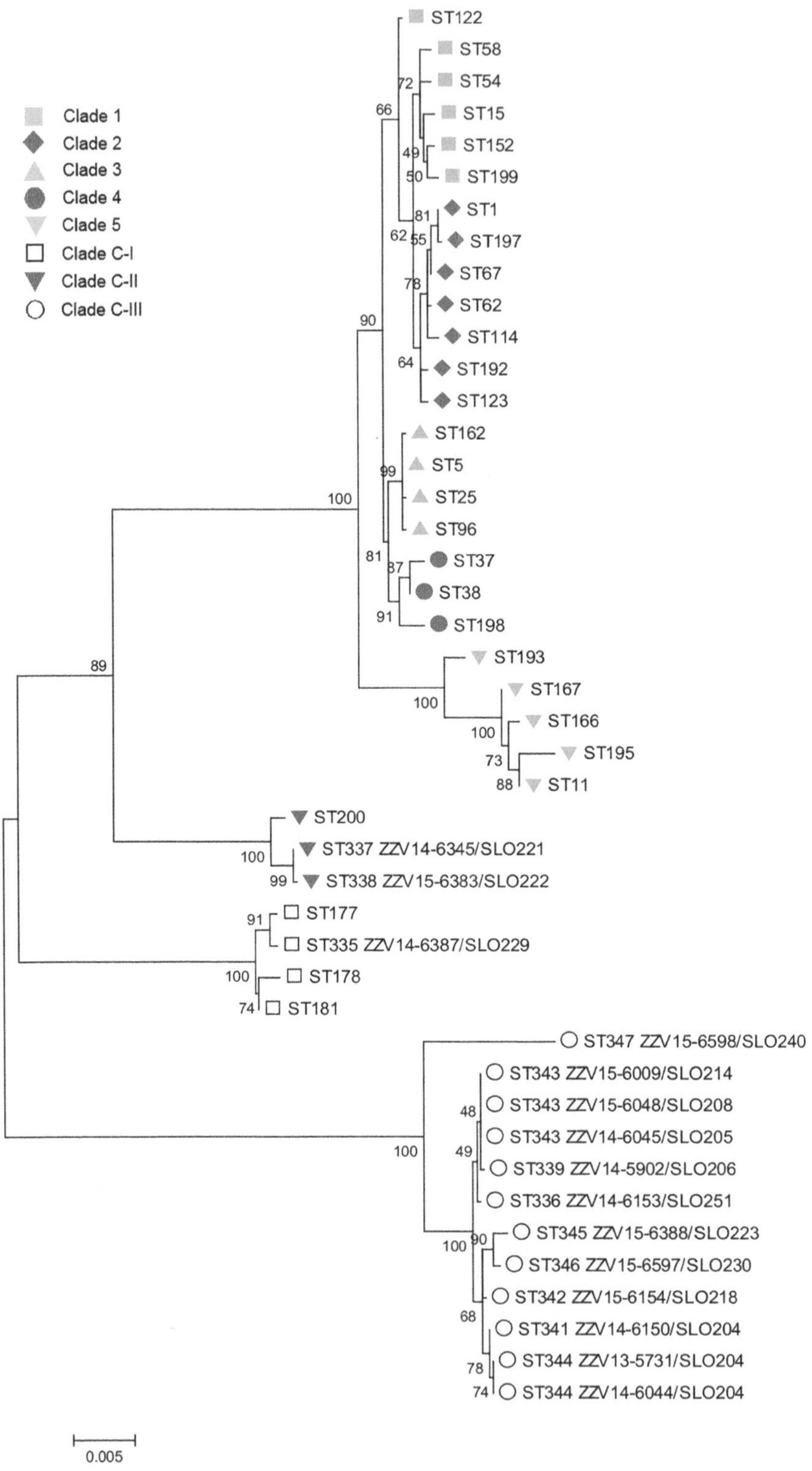

Fig. 1 Maximum likelihood phylogenetic tree showing the population structure of *C. difficile* composed of clades with tree highly divergent cryptic clades (C-I, C-II and C-III) (Reproduced from Janezic et al. 2016)

Table 2 Overview of heterogeneity within *C. difficile* clades and correlation between main MLST-ST and PCR ribotypes

Clade	Nr. MLST ST[a]	Most known PCR ribotypes/MLST-ST(s)[b]	
1	381	001	ST-3
		002	ST-8, ST-35, ST-48 and ST-146
		012	ST-54
		014	ST-2, ST-13:14, ST-49:50 and ST-132
		015	ST-10 and ST-44
		018	ST-17
2	90	027	ST-1
		176	ST-1
		244	ST-41
3	19	023	ST-5, ST-22 and ST-25
4	98	017	ST-37 and ST-86
5	35	033	ST-11
		126	ST-11
		078	ST-11
C-I[c]	33	ND	ST-177:181
C-II[c]	13	ND	ST-200 and ST-337:338
C-III[c]	23	ND	ST-336, ST-339 and ST-341:347
C-IV[c]	2	ND	ST-823 and ST-835
C-V[c]	2	ND	ST-784 and ST-786

[a]Data from PubMLST *C. difficile* database (accessed 14.3.2023)
[b]Data from Griffiths et al. (2010), Stabler et al. (2012), Knetsch et al. (2012), Dingle et al. (2014) and Williamson et al. (2022)
[c]Data from Janezic et al. (2016), Knight et al. (2021) and Williamson et al. (2022)

ribotypes 176 (ST-1) and 244 (ST-41) (Valiente et al. 2012; Lim et al. 2014). In clade 3, 19 different STs are present, and the best-known representative is PCR ribotype 023 (represented with ST-5, ST-22 and ST-25), which is also often isolated from humans in European countries (Davies et al. 2016). Clade 4 comprises 98 STs, mainly non-toxigenic strains and toxigenic strains that produce only toxin B (A−B+), the most known being PCR ribotype 017. Despite altered toxin expression, this strain is widespread, especially in Asia (Shin et al. 2008; Collins et al. 2013; Imwattana et al. 2019). One of the best-known representatives of clade 5 is PCR ribotype 078 (ST-11), a well-established strain in animal population (Jhung et al. 2008; Knight et al. 2019), which has in recent years emerged as an important strain causing human CDI, especially in the community, and is recognized as the strain with zoonotic potential (Rupnik et al. 2008; Knetsch et al. 2014; Knight et al. 2019). Although in the first studies this clade appeared to be more homogeneous, consisting primarily of ST-11 strains (Griffiths et al. 2010; Dingle et al. 2011; Knetsch et al. 2012), later studies (MLST and WGS) demonstrated the opposite as there are currently 35 STs found in clade 5 (Table 2). Besides A+B+ and non-toxigenic strains, clade 5 strains can also have atypical PaLoc organization resulting in different toxigenic profiles: A−B+ (CDT+) and A−B− (CDT+) (Elliot et al. 2009; Rupnik and Janezic 2016). To date, 33, 13, 23, 2 and 2 different MLST-STs have been found in the five cryptic clades C-I, C-II, C-III, C-IV and C-V, respectively (Table 2). These clades were primarily associated only with non-toxigenic strains; however, in a more recent publication, strains with genes homologous to *tcd*A and *tcd*B were described in all five cryptic clades; *tcd*B was found in strains from clades C-I, C-II, C-III and C-V, and *tcd*A was present only in two clades, C-II and C-IV. None of the strains from the cryptic lineages contained both toxin genes. In all but one clade (C-II), homologues to CdtLoc have been described (Monot et al. 2015; Williamson et al. 2022; Ducarmon et al. 2022). Interestingly, homologous toxin genes are not part of the typical PaLoc and CdtLoc genes, and in some isolates, the toxin genes were found on plasmids and prophages (Riedel et al. 2017; Ramírez-Vargas et al. 2018; Williamson et al. 2022).

Strains from cryptic clades are often found in the environment (Janezic et al. 2016; Williamson et al. 2022); however, there is a growing body of evidence that these strains can cause human CDI (Janezic et al. 2015; Monot et al. 2015; Ramirez-Vargas and Rodríguez 2020; Ducarmon et al. 2022).

Cryptic clades represent highly divergent lineages (Fig. 1); thus, it has been suggested that these groups of strains might represent novel species or subspecies (Dingle et al. 2014). Average nucleotide identity (ANI) analysis, which measures pairwise nucleotide identity shared between genomes and is used to assess genetic relatedness of strains, confirmed this, since ANI values for all cryptic clades were below the species threshold (isolates with ANI of at least 95% are considered to belong to the same species). Furthermore, ANI values between all five cryptic clades were below the species threshold, indicating that each clade may represent a novel species (Williamson et al. 2022; Knight et al. 2021). With ANI-based taxonomy, *C. difficile* isolates from cryptic clades were placed in the genus *Clostridioides* between *C. difficile* (clades 1–5) and *C. mangenotii* (Knight et al. 2021).

Large-scale analyses of strains from diverse sources and geographic origins also revealed that significant microdiversity exists within clades and that *C. difficile* is continuously evolving (Table 2) (Griffiths et al. 2010; Dingle et al. 2011, 2014; Knetsch et al. 2012; Janezic et al. 2016; Williamson et al. 2022; Knight et al. 2021).

2.2 Worldwide Evolution of Important *C. difficile* PCR Ribotypes

2.2.1 Epidemic *C. difficile* PCR Ribotype 027

C. difficile PCR ribotype 027 has gained much interest because of its rapid emergence worldwide in the beginning of twenty-first century. This strain has been associated with large CDI outbreaks and increased morbidity and mortality, which have first started to appear in the USA and Canada (Pepin et al. 2004; Loo et al. 2005; McDonald et al. 2005; Pepin et al. 2005). The strain was later also introduced in Europe, with the first outbreaks documented in the United Kingdom, and in following years also in continental Europe (Kuijper et al. 2008). Although the prevalence of PCR ribotype 027 declined markedly in Europe, the strain remains one of the most common strains causing CDI (Bauer et al. 2011; Davies et al. 2016). To explore the global population structure and genetic changes associated with its rapid emergence and global spread, He et al. (2013) sequenced genomes of 151 strains, representing the global population of ribotype 027 strains, collected between 1985 and 2010. They showed that ribotype 027 population consists of two genetically distinct fluoroquinolone-resistant (FQR1 and FQR2) epidemic lineages. Both lineages independently acquired the same mutation in DNA gyrase, which confers fluoroquinolone resistance, and a novel conjugative transposon (CTn5-like element, Tn6192). These were the only two genetic traits differentiating FQR1 and FQR2 lineages from the historic 027 isolates and were most likely key genetic changes associated with the rapid emergence of ribotype 027. Also, low level of genomic diversity within the core genome of the 151 PCR ribotype 027 strains analysed was demonstrated, with only 536 SNPs identified. Only two of these SNPs (limited to a single isolate) were discovered in PaLoc region of historic and epidemic isolates (He et al. 2013) which contrasts with the previously assumptions that genetic changes in PaLoc were the cause of emergence of *C. difficile* 027 (McDonald et al. 2005; Warny et al. 2005).

Although both lineages emerged in North America, they showed a different global spread and limited geographic clustering. FQR1 originated in Pittsburgh (Pennsylvania, USA) and was subsequently transmitted to Switzerland and South Korea. The FQR2 lineage which contains most epidemic strains was transmitted to continental Europe and the United Kingdom on several different occasions, and a single introduction to Australia was demonstrated. Phylogenetic analysis of UK collection of epidemic FQR2 strains further demonstrated frequent long-range transmissions within the United Kingdom, some of them associated with large-scale outbreaks that gained attention also in media (He et al. 2013).

2.2.2 Toxin Variant *C. difficile* PCR Ribotype 017

Another important strain that has gained much attention is PCR ribotype 017 (toxinotype VIII, MLST ST-37). Despite producing only one of the three *C. difficile* toxins (A−B+CDT−), PCR ribotype 017 strains are causing clinically significant infections and outbreaks worldwide (Drudy et al. 2007; Collins et al. 2013; Cairns et al. 2015; Isidro et al. 2018; Imwattana et al. 2019). Initially, ribotype 017 strains were identified in outbreaks in Asia where they were responsible for most of CDIs (Collins et al. 2013). It was hypothesized that this strain has spread from Asia throughout the world (Drudy et al. 2007; Collins et al. 2013; Cairns et al. 2015). To define the population structure and patterns of global spread, Cairns et al. (2017) conducted a phylogenetic analysis of a global collection of PCR ribotype 017 strains. Two-hundred and seventy-seven PCR ribotype 017 strains, including human, animal and environmental isolates, were obtained from all six continents, isolated between 1990 and 2013. Phylogenetic analysis based of the core SNPs demonstrated presence of two genetically diverse lineages (SL1 and SL2) which are geographically and temporally widespread. Multiple clonal expansions were observed in both lineages. Phylogeographic analysis suggested, contrary to the current Asia-origin hypothesis, that ribotype 017 originated in North America, from where the strain was first introduced to Europe and then from Europe to Asia and Australia, and then spread worldwide. However, a very recent study contrast this and places the probable origin of RT 017 in Asia/Europe (Imwattana et al. 2019), and the key genetic change responsible for global emergence of RT 017 is the acquisition of *erm*B-positive transposon Tn6194 (Imwattana et al. 2022).

2.3 *C. difficile* Transmissions and Epidemiology of Recurrent CDI

2.3.1 *C. difficile* Transmissions in the Hospital Environment

In the past, assessment of the genetic relatedness of *C. difficile* isolates has been hampered by the use of suboptimal genotyping methods that do not have sufficient discriminatory power (e.g. PCR ribotyping, MLST) to distinguish between closely related strains. Whole genome sequence analysis which enables comparison at the highest level of genetic resolution has been widely adopted for global and national *C. difficile* surveillance and has revealed some novel insights about transmission dynamics (Eyre et al. 2012, 2013a–c; Widmer et al. 2017; Kong et al. 2019; Garcia-Fernandez et al. 2019) and recurrent infections (Mac Aogain et al. 2015; Kumar et al. 2016; Sim et al. 2017; Kociolek et al. 2018).

Two different genomic approaches are normally used for typing of *C. difficile* isolates: comparison based on single nucleotide polymorphisms (SNP typing) or gene-by-gene approach that compares multiple genes across the genome (i.e. core genome (cg) or whole genome (wg) multi-locus sequence typing (cg- or wgMLST) (reviewed in Janezic and Rupnik (2019) and Schurch et al. (2018)). Three schemes with corresponding databases for cg- or wgMLST typing are available for *C. difficile* and can be used in a freely available online database, which automatically retrieves short reads from public repositories (Enterobase) (Zhou et al. 2020; Frentrup et al. 2020), or commercial software such as SeqSphere (Ridom) (Bletz et al. 2018) or BioNumerics (bioMérieux), each with its own database (Janezic and Rupnik 2019). In addition, a hash-based cgMLST was described, allowing rapid comparison of *C. difficile* genomes without the need to maintain a centralized database to assign allelic profiles (Eyre et al. 2019).

A study by Baktash et al. (2022) demonstrated that cgMLST typing could be an alternative to PCR ribotyping, which, in contrast to PCR ribotyping, can be easily standardized. A lower cgMLST threshold has been proposed to define clonal isolates, especially for PCR ribotypes with low intra-ribotype diversity (Baktash et al. 2022).

Estimating the rates at which bacterial genomes evolve (e.g. within-host diversity and short-term evolution) is critical for understanding transmission patterns (Duchene et al. 2016). For *C. difficile*, rates of short-term evolution and within-host diversity have been explored in several studies, using serial samples from patients with recurrent or ongoing CDI, and in vitro gut models of CDI. In all these studies, similar estimations of evolutionary rates were obtained, 1–2 SNPs/genome/year and within-host diversity of 0.30 SNPs/genome/year (Didelot et al. 2012; Eyre et al. 2013b, d). By using these estimations, two isolates obtained less than 124 days apart would be expected to have 0–2 SNP differences, and isolates obtained 124–364 days apart should exhibit 0–3 SNP differences (Eyre et al. 2013b). This definition of genetically related isolates (i.e. isolates that are most likely a result of direct transmission) has now been widely adopted.

It was traditionally believed that most cases of *C. difficile* infections are acquired within hospital settings, where they are being transmitted from person to person (Vonberg et al. 2008; Khanna and Pardi 2012). Eyre and colleagues (Eyre et al. 2013b) compared genomic sequences of 1223 *C. difficile* isolates and demonstrated that only 35% of cases were acquired from another known case within hospital settings and only a subset of these cases shared the same ward with at least one other case or had some sort of hospital contact, which is much lower than expected. Almost half (45%) of the isolates were genetically unrelated (≥10 SNPs) to any other previous case and could not be linked by transmission (direct or indirect), meaning that they were likely acquired from sources other than symptomatic patients. Identification of a rather diverse pool of *C. difficile* strains indicates the existence of substantial reservoirs of *C. difficile* and that transmission routes other than those due to symptomatic CDI patients should be considered (e.g. asymptomatic patients and the environment) (Eyre et al. 2013b). The same group has also described a novel approach using WGS that enables assessment of the extent of infection transmission within healthcare institutions by measuring the proportion of cases that are acquired from a previous case (i.e. linked cases) (Eyre et al. 2017).

Several studies have focused on the importance of asymptomatic colonized patients for the onward transmission of CDI showing that asymptomatic carriage is common. While 6–15% of patients are found to be colonized at admission, the onward transmission is not as frequent as in symptomatic patients (Eyre et al. 2013c; Curry et al. 2013; Kong et al. 2019; Halstead et al. 2019).

In addition, it seems that transmission from colonization (symptomatically or asymptomatically) is strain-related. More frequent onward transmission was seen for some important healthcare-associated PCR ribotypes, that is, 027, 001 and 106 in contrast to community- and livestock-associated PCR ribotypes (e.g. 014/020 and 078/126) (García-Fernández et al. 2019). The greater transmission of certain strains (e.g. PCR ribotype 027) in the hospital environment was confirmed also in the study by Martin et al. (2018). Different patterns of PCR ribotype spread across a wide collection of strains in Europe were described in a study by Eyre et al. (2018), where healthcare-associated PCR ribotypes (027 and 001/072) showed patterns of within-country and hospital clustering, consistent with local transmission, whereas for other PCR ribotypes, genetically closely related isolates were found widespread in Europe, consistent with dissemination from diverse sources.

2.3.2 *C. difficile* Recurrence: Reinfections Versus Relapses

WGS is a valuable tool to understand the epidemiology of CDI recurrences with greater accuracy, especially within the hospital settings with endemic strains (Eyre et al. 2014; Mac Aogain et al. 2015; Kumar et al. 2016). Recurrent *C. difficile* infections occur in up to 25% of patients after the first CDI episode, and discriminating between reinfections (infection with newly acquired strain) and relapses (recurrent episode due to the original strain) is important for CDI management: infection prevention and treatment, respectively (Kelly 2012).

A similar methodology that is used in transmission studies can also be applied in studies resolving the contribution of relapses and reinfections in recurrent CDI. Relapse is defined as a recurrent infection with an isolate differing by ≤2 SNPs from the isolate from initial episode and reinfection involving pairs of isolates differing in ≥10 SNPs (Eyre et al. 2014). Mac Aogain et al. (2015) applied this methodology to 19 patients with recurrent CDI to resolve the nature of the recurrences and demonstrated that majority of recurrences (16 out of 19) were due to relapse with endogenous strain. Similar findings were also found in a study by Eyre et al. (2014) that used WGS to determine whether CDI recurrences in 93 patients (28 were treated with fidaxomicin and 65 were treated with vancomycin) were due to reinfection or relapse. Overall, 79.6% (74 of 93) of the recurrent CDI cases were due to relapse. Reinfection accounted for only one-fifth of the recurrences.

2.4 Influence of SNPs on Virulence and Phenotype of CD630 Derivatives

The *C. difficile* strain CD630 was isolated in 1982 in Zurich, Switzerland, from a patient with pseudomembranous colitis (Sebaihia et al. 2006). This is the first strain of *C. difficile* whose genome has been sequenced and whose derivatives have been used as a model strain for the generation of mutants in different studies exploring the importance of *C. difficile* toxins in pathogenesis. Two groups that used isogenic mutants (in which production of one of both toxins was ablated) of erythromycin-sensitive derivatives (630E and 630Δ*erm)* from the strain CD630 obtained contradictory results on the virulence potential of toxin A (TcdA) (Collery et al. 2016). In a study by Lyras et al. (2009), the outcome was that the *tcd*B mutant, producing only toxin TcdA (A+B−), was unable to cause disease in hamster model, whereas in a study by Kuehne et al. (2010), the authors demonstrated that both toxins, TcdA and TcdB, are capable of causing disease in a hamster model. Both strains possessed the same deletion of *erm*B gene and were isolated in two different laboratories by repeated sub-culturing of strain CD630 (Collery et al. 2016). Re-sequencing of both strains revealed that both strains had several SNPs, compared to the published genome of CD630, which were most likely accumulated during sub-culturing. Genetic variations between the strains were found responsible for the phenotypic differences observed in both mutants (growth rate, motility, sporulation and virulence), explaining the different outcomes of both studies. Since 630Δ*erm* strain more closely resembles the progenitor strain, the authors concluded that this strain should be favoured over 630E and that re-sequencing of the genomes of mutant strains should become a routine practice (Collery et al. 2016).

2.5 Comparative Genomic Analysis of Non-toxigenic Strains

Comparative genomic studies have demonstrated that non-toxigenic *C. difficile* strains are represented in all clades, alongside toxigenic isolates (Dingle et al. 2014; Monot et al. 2015). Although toxin-negative *C. difficile* strains can be isolated from patients and animals with gastrointestinal diseases, they are not considered to play a role in disease (Vedantam et al. 2012). Chowdhury et al. (2016) undertook a comparative genomic analysis of five toxin-negative strains (lacking all toxin genes or their homologues described in recent years in isolates from cryptic clades) isolated from faeces from humans and animals with symptoms of gastrointestinal (GI) disease. Even though the authors stated that GI symptoms were likely due to non-toxigenic *C. difficile*, this could also be due to undetected co-infection with toxigenic *C. difficile* or to infection with a yet unknown or un-cultivable organism. Phylogenetic analysis demonstrated that all five isolates clustered with toxigenic isolates (belonging also to the same MLST-ST) and had also a similar virulence-associated gene repertoire as those found in toxigenic strains (e.g. genes required for sporulation (*spo0A*) and adhesion (*groEL*, *fliC*), genes coding for surface proteins (*slpA* and *cwp*) necessary for colonization of the gut and different serine-proteases and metalloproteases).

3 Targeted Comparative Genomics

3.1 Evolution of the *C. difficile* Pathogenicity Locus

The pathogenicity locus encodes the exotoxins TcdA and TcdB, which are the two main virulence factors involved in CDI. Bacterial strains completely lacking this genomic region are unable to cause the disease and its associated symptoms, so it appears of outmost importance to understand how this locus has been acquired and how it can evolve over time (Cohen et al. 2000). Comparative genomics which is a very powerful approach to elucidate the evolutionary history of the PaLoc has shown that this locus has undergone a very complex and intriguing eventful history (Dingle et al. 2014; Elliott et al. 2014; Monot et al. 2015; Janezic et al. 2020). However, the conclusion drawn from such analyses is likely in constant evolution as it depends on the strains available.

3.1.1 PaLoc Acquisition and Exchange

Dingle et al. (2014) have estimated that the most recent acquisition of the PaLoc would have occurred some 500 years ago. The latest exchange of the PaLoc between *C. difficile* strains has been calculated to about 300 years, and the most recent PaLoc loss from the genome would have happened in recent times (~30 years ago). Because of the very long genomic fragments concurrently swapped during these recent PaLoc losses and exchanges, host-mediated homologous recombination is thought to be the mechanism by which these recent events have arisen. These observations were made possible by plotting the distribution of indels and SNPs at the chromosomal scale and by analysing in more details the SNP plots for the regions around the PaLoc. Distinctively, specific recombination mediated by an integrase supplied in *trans* appears to be the mechanism involved in the initial PaLoc acquisition. The reason for this is the absence of recombination signatures on DNA sequences distant from the PaLoc in non-toxigenic strains (Dingle et al. 2014).

Brouwer et al. (2013) demonstrated experimentally that non-toxigenic *C. difficile* strains could be converted into toxin producers by horizontal gene transfer and genetic recombination. It is worrying to think that different versions of the PaLoc can be acquired and transferred seemingly at any time by any strain because this makes all the non-toxigenic strains possible candidates for becoming toxin producers (Brouwer et al. 2013). The possible acquisition of the PaLoc by non-toxigenic strains that already exhibit high resistance to antibiotics widely used in clinics for the treatment of CDI (e.g. strains belonging to ribotype 010 highly resistant to metronidazole (Moura et al. 2013, 2014)) is a very concerning scenario.

In almost all toxigenic *C. difficile* strains from clades 1 to 5, PaLoc and CdtLoc are chromosomally encoded, and in toxigenic strains from cryptic clades (C-I to C-V), the toxin gene homologues are always carried on extrachromosomal elements, further supporting possible mobility of both toxin loci via horizontal transfer (Dingle et al. 2014; Riedel et al. 2017; Ramírez-Vargas et al. 2018). CdtLoc was found on the prophage phiSemix9P1 (Riedel et al. 2017), and a recent study showed that strains with prophage-encoded

CdtLoc are common in cade C-III environmental isolates from Slovenia (Williamson et al. 2022). Plasmid pHSJD-312 from the clade C-I isolate was found to carry *tcd*B and the complete CdtLoc (Ramírez-Vargas et al. 2018). These putative conjugative plasmids and phages are mainly found in isolates from cryptic clades, but they do not appear to be restricted to these cryptic lineages since, in a very recent study conducted in Costa Rica, putative plasmids carrying toxin genes were found among members of *C. difficile* clades 2 and 4 (Ramirez-Vargas and Rodríguez 2020).

All of these recent findings concerning PaLoc are of great importance and can have profound repercussions on the evolution of the disease in clinics. It is highly conceivable that the events reported here and the related mechanisms might be more prevalent than first thought and may be relevant to other commensal and pathogenic bacteria as well.

3.1.2 PaLoc Organization and Evolution

The evolutionary history of the PaLoc was first studied by performing comparative genomics of *C. difficile* genomes from a collection of 1693 toxigenic and non-toxigenic strains (Dingle et al. 2014). Further studies refined the established model by adding new PaLoc variants (Elliott et al. 2014; Janezic et al. 2015; Monot et al. 2015; Ramírez-Vargas et al. 2018; Janezic et al. 2020; Williamson et al. 2022) leading to the actual known gene content organizations of the PaLoc detailed in Fig. 2a and mechanisms underlying the PaLoc variant variability.

Analysis of all known major toxinotypes, strains that reflect genetic diversity of PaLoc (Rupnik and Janezic 2016), demonstrated that PaLoc has a modular structure, composed of interspersed blocks of sequences corresponding to functional domains, with each block having a different evolutionary history and that variability in PaLoc is a result of single nucleotide substitution and recombination events that play an important role in the evolution of the PaLoc variants (Janezic et al. 2020).

Monot et al. (2015) first described two types of genomic organization of the PaLoc that each contained only one of the two toxins (A+B− and A−B+). These two "Mono-Toxins PaLocs" were located at different positions in the *C. difficile* genome far from the usual PaLoc integration site, which was not described before. Based on sequence similarity analysis, the authors detected two gene remnants of these PaLoc variants in the classical PaLoc, that is, "Bi-Toxin PaLoc". Altogether, this work supports a scenario in which the "Bi-Toxin PaLoc" was generated by a fusion of two "Mono-Toxins PaLoc" from ancestral *C. difficile* strains through multiple independent PaLoc acquisitions (Fig. 2b) (Monot et al. 2015).

The PaLoc could also be altered during evolution by insertion of mobile elements. These strains have been associated with milder clinical phenotypes, and the presence of the transposable element Tn6218 is believed to be responsible for this change in the bacterial phenotype (Dingle et al. 2014). This specific genetic region has probably undergone many different exchanges or separate acquisition events, as many accessory genes have been noticed in several variants widely spread in the *C. difficile* population. It is important to carefully study and follow this type of transposable region such as Tn6218, as it carries, among others, a set of genes providing high-level resistance to antibiotics used in clinical settings (Spigaglia et al. 2011; Kelly 2012; Deshpande et al. 2013). Elements related to Tn6218 have been found in other various genomes such as *Bifidobacterium breve*, *Ruminococcus*, *Lachnospiraceae* and *Coprobacillus* sp., suggesting that the transfer of this element between different species is also highly probable and should undoubtedly be further investigated (Dingle et al. 2014).

First identified and described by Braun et al. in 2000, IStrons represent another type of mobile genetic element that has been shown to create variations inside the *C. difficile* genome and inside the PaLoc region (Rupnik et al. 2008). It has been hypothesized that the original IStron (CdISt1-0) is the result of a fusion event between an insertion element (IS) and a group I intron, generating a novel class of chimeric ribozymes adapted to propagate in eubacterial genomes

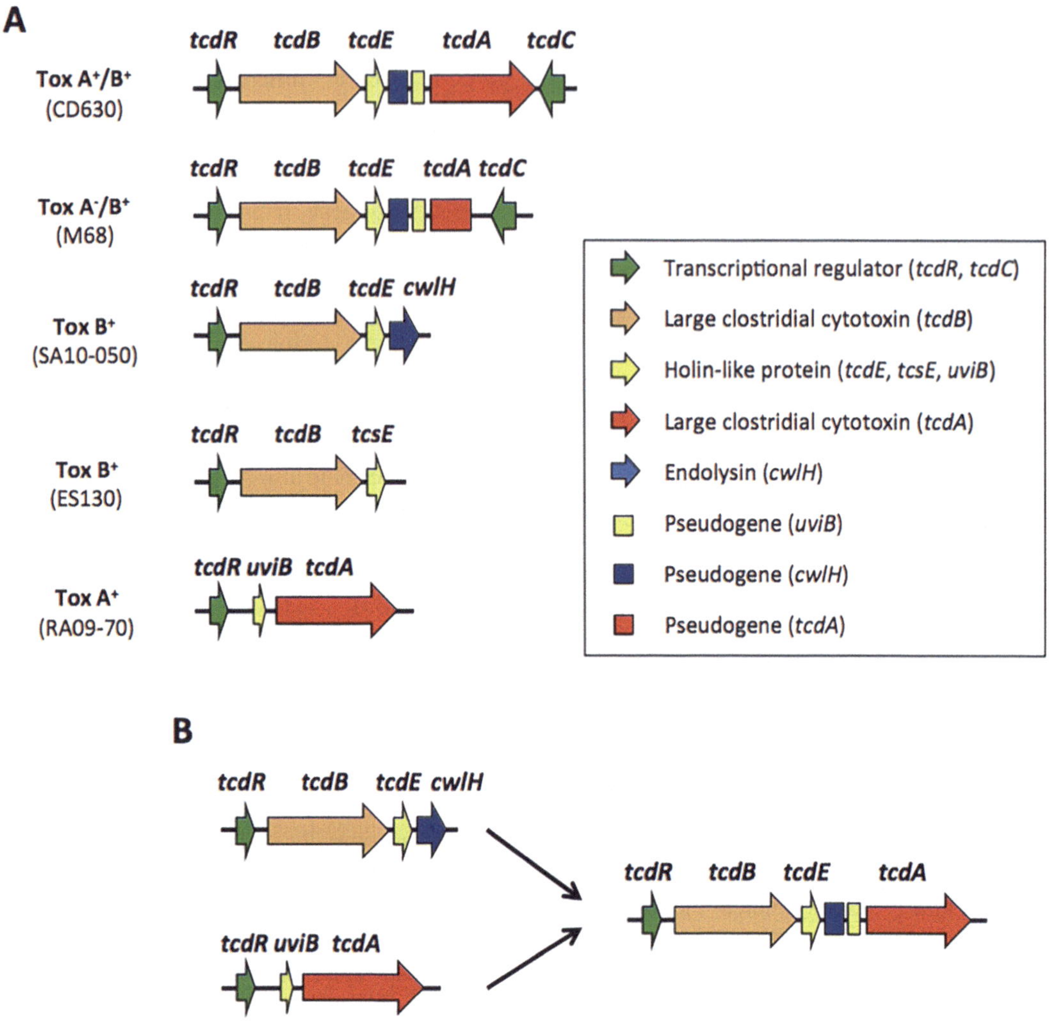

Fig. 2 PaLoc diversity and evolution. (**a**) Known types of *C. difficile* PaLoc and (**b**) model of evolution from « Mono-Toxin » to « Bi-Toxin » PaLoc [Adapted from Figure 6B and S7 of Monot et al. 2015)

(Hasselmayer et al. 2004). Widely spread in *C. difficile* genomes, four variants of IStrons have now been identified, all exhibiting a self-splicing ribozyme activity and which transposition was found to be harmless for the interrupted gene (i.e. does not affect TcdA toxin production in *C. difficile*). Braun et al. (2000) hypothesized that this particular chimeric element might be more efficient and more adapted, as the risk of mutation usually observed during transposition of an IS-elements is significantly reduced by the precise splicing activity provided by the group I intron (Braun et al. 2000).

The complex relationship between *C. difficile* and the PaLoc, and the multiple ways by which it is able to evolve, can ostensibly lead to concrete repercussions on its virulence and epidemiology. This is illustrated by the characterization of a clinical strains exhibiting new major variants of the PaLoc producing only the toxin A or toxin B or its homologues found in cryptic clades (Eckert et al. 2013; Monot et al. 2015; Janezic et al. 2015; Ramírez-Vargas et al. 2018). These types of strains are not always detected by routine diagnostic assays, and dissemination of this type of strains could lead to a problematic under-diagnostic scenarios (Monot et al. 2015; Ramírez-Vargas et al. 2018; Ducarmon et al. 2022). Toxins A and B and CDT gene homologues are indeed very divergent, with low sequence identity to PaLoc and CdtLoc genes from clades 1–5, which can result in routine toxin detection test failures (PCR and antigen-based), suggesting that toxigenic strains from cryptic clades could be much more common in the environment and in clinical samples than previously thought (Riedel et al. 2017; Ramírez-Vargas et al. 2018; Williamson et al. 2022; Ducarmon et al. 2022).

3.2 Advances in CRISPR/Cas Systems and Phage-Host Interaction

Mobile genetic elements (MGE) and especially bacteriophages are major contributors and facilitators of genetic evolution in bacteria, including *C. difficile*. It has been suggested that *C. difficile* is exhibiting a complex, highly mobile, and mosaic genome because it is striving in an environment where it is constantly being confronted to numerous interacting bacteria and phages also struggling to survive (Sebaihia et al. 2006). Therefore, *C. difficile* is incessantly incorporating favourable genetic material useful for its adaptation while simultaneously developing defence mechanisms to limit the incorporation and influence of harmful genetic material (Boudry et al. 2015). A myriad of defence mechanisms against foreign MGE and phages are now well known, but the CRISPR/Cas system has only recently been more actively explored in *C. difficile*. CRISPR/Cas systems have been defined in 3 major types (I, II and III), further divided in 12 different subtypes (Makarova et al. 2011, 2013; Maikova et al. 2018). *C. difficile* only harbours the subtype I-B, a system probably acquired by means of horizontal gene transfer (HGT) from *Archaea* (Richter et al. 2012; Peng et al. 2014).

3.2.1 CRISPR Mechanism and Physiology

The analogy between the mammalian acquired immunity and the bacterial CRISPR/Cas system is often used, since bacteria can become protected against genetically akin phages after exposition, in a fashion reminiscent of vaccination. Bacteria memorize previous unsuccessful infections by acquiring small sequences of the assailants and integrating them to its own genome, inside a specific region or array containing other similar protective sequences. Those sequences, called "spacers" in the CRISPR/Cas array system, are used by the bacteria to scan and recognize the identical or near identical sequences, called "protospacer", in the genome of a future potentially more lethal phage invader. When the sequence is recognized, a functional CRISPR system is able to neutralize the infecting agent by cutting and digesting its DNA, interrupting the infection cycle, which may also result in the acquisition of additional protective sequences.

Recently, important findings have been made for this system in *C. difficile* using comparative genomics associated with laboratory procedures, such as transcriptome sequencing (RNA-Seq) and plasmid conjugation efficiency assays (Hargreaves et al. 2014; Boudry et al. 2015). These analyses have allowed to conclude that the CRISPR/Cas system in *C. difficile* was functional and used in this species, since many genes and arrays coding for important components of the CRISPR arrays were actively transcribed. Nine different CRISPR arrays were found to be present and transcribed in the epidemic strain R20291, and reference strain 630 exhibited 12 expressed arrays (Boudry et al. 2015). Analysis of the targets for the identified spacers showed that a unique phage could be targeted by numerous different spacers, surely to increase the efficiency of phage neutralization by the system (Boudry et al. 2015). This could also be an indication that phage has the ability to evade the CRISPR system using a mutational process. Contrastively, a single spacer can have the ability to target conserved genes present in multiple related phages, thus bestowing them with an efficient and inexpensive defence against multiple potential invaders at once. Boudry et al. (2015) concluded that there is a good correlation between the real and predicted phage susceptibilities according to the spacer content of the bacterial strains and the theoretically predicted phage targets. Remarkably, the spacer sequences found in *C. difficile* strain 630 were anticipated to target all known and isolated clostridial bacteriophages. Experimentally, this strain exhibited resistance to infection by all the phages that could be tested.

The CRISPR/Cas system seems particularly active and meaningful in *C. difficile* as numerous highly active CRISPR arrays are found, which moreover greatly contrast with what is observed in other bacteria such as *E. coli* and *Streptococcus pyogenes* in which the CRISPR loci are barely expressed or even completely silent (Pougach et al. 2010; Deltcheva et al. 2011).

3.2.2 CRISPR Distribution and Diversity

Hargreaves et al. (2014) determined the distribution and diversity of the CRISPR/Cas system in *C. difficile*. To do this, they examined the relationships between spacers and 31 *C. difficile* phages and prophage genomes. The spacer content is thought to bring a good perception of the predominant and relatively recent phage predation history (Diez-Villasenor et al. 2010). However, a large number of spacers match sequences of unknown nature, possibly targeting unknown *C. difficile* phages or even non-clostridial phages. In several *C. difficile* strains, they also found CRISPR arrays inside prophage genomes, which is considered an unusual situation for this system (Hargreaves et al. 2014; Boudry et al. 2015). Those phages carried spacers that were found to match sequences of other bacteriophages. Once they have successfully integrated the bacterial genome, prophages could plausibly use those spacers in order to give them an advantage over other phages by blocking their capacity to infect the same strain (Hargreaves et al. 2014). Prophages possessing CRISPR arrays are thought to rely on the bacterial host for the proper functioning of the system, since the *cas* operon containing the set of genes necessary to process the arrays was always absent (Boudry et al. 2015).

To obtain a global view of the distribution of the CRISPR/Cas system in *C. difficile*, Boudry et al. (2015) tested the presence of *cas* operons in 2207 *C. difficile* published and available genomes. Nearly 90% of them possessed a complete *cas* operon, making the CRISPR/Cas a common system in this bacterium.

4 Conclusions

The evolution of comparative genomics of *C. difficile* strains from molecular typing and microarrays to whole genome sequencing has enabled significant improvements in the determination of the population structure and evolutionary history of *C. difficile*. Beyond a deeper understanding of the diversity of strains, WGS makes possible the emergence of new area of research such as transmission or reinfection studies.

Another aspect to be considered is the availability of massive sequence data allowing the analysis of specific loci. Owing to its importance in virulence, PaLoc has been extensively explored, and it has been concluded that this locus is constantly evolving.

This leads us to conclude that the findings of comparative genomics are highly dependent on the strains available, thus making the availability of raw data in public databases of primordial importance.

Acknowledgements JG was supported by a discovery grant from the Natural Sciences and Engineering Research Council of Canada (NSERC #341450-2010). SJ was supported by the Slovenian Research Agency (grant number J4-8224).

References

Baktash A, Corver J, Harmanus C, Smits WK, Fawley W, Wilcox MH, Kumar N, Eyre DW, Indra A, Mellmann A, Kuijper EJ (2022) Comparison of whole-genome sequence-based methods and PCR ribotyping for subtyping of Clostridioides difficile. J Clin Microbiol 60(2):e0173721. https://doi.org/10.1128/JCM.01737-21

Bauer MP, Notermans DW, van Benthem BH, Brazier JS, Wilcox MH, Rupnik M, Monnet DL, van Dissel JT, Kuijper EJ, Group ES (2011) Clostridium difficile infection in Europe: a hospital-based survey. Lancet 377(9759):63–73. https://doi.org/10.1016/S0140-6736(10)61266-4

Bletz S, Janezic S, Harmsen D, Rupnik M, Mellmann A (2018) Defining and evaluating a core genome multilocus sequence typing scheme for genome-wide typing of Clostridium difficile. J Clin Microbiol 56(6): e01987-17. https://doi.org/10.1128/JCM.01987-17

Boudry P, Semenova E, Monot M, Datsenko KA, Lopatina A, Sekulovic O, Ospina-Bedoya M, Fortier LC, Severinov K, Dupuy B, Soutourina O (2015) Function of the CRISPR-Cas system of the human pathogen Clostridium difficile. MBio 6(5):e01112-01115. https://doi.org/10.1128/mBio.01112-15

Braun V, Mehlig M, Moos M, Rupnik M, Kalt B, Mahony DE, von Eichel-Streiber C (2000) A chimeric ribozyme in Clostridium difficile combines features of group I introns and insertion elements. Mol Microbiol 36(6): 1447–1459

Brouwer MS, Roberts AP, Hussain H, Williams RJ, Allan E, Mullany P (2013) Horizontal gene transfer converts non-toxigenic Clostridium difficile strains into toxin producers. Nat Commun 4:2601. https://doi.org/10.1038/ncomms3601

Cairns MD, Preston MD, Hall CL, Gerding DN, Hawkey PM, Kato H, Kim H, Kuijper EJ, Lawley TD, Pituch H, Reid S, Kullin B, Riley TV, Solomon K, Tsai PJ, Weese JS, Stabler RA, Wren BW (2017) Comparative genome analysis and global phylogeny of the toxin variant Clostridium difficile PCR ribotype 017 reveals the evolution of two independent sublineages. J Clin Microbiol 55(3):865–876. https://doi.org/10.1128/JCM.01296-16

Cairns MD, Preston MD, Lawley TD, Clark TG, Stabler RA, Wren BW (2015) Genomic epidemiology of a protracted hospital outbreak caused by a toxin A-negative Clostridium difficile sublineage PCR ribotype 017 strain in London, England. J Clin Microbiol 53(10):3141–3147. https://doi.org/10.1128/JCM.00648-15

Chowdhury G, Joshi S, Bhattacharya S, Sekar U, Birajdar B, Bhattacharyya A, Shinoda S, Ramamurthy T (2016) Extraintestinal infections caused by non-toxigenic vibrio cholerae non-O1/non-O139. Front Microbiol 7:144. https://doi.org/10.3389/fmicb.2016.00144

Cohen SH, Tang YJ, Silva J Jr (2000) Analysis of the pathogenicity locus in Clostridium difficile strains. J Infect Dis 181(2):659–663. https://doi.org/10.1086/315248

Collery MM, Kuehne SA, McBride SM, Kelly ML, Monot M, Cockayne A, Dupuy B, Minton NP (2016) What's a SNP between friends: the influence of single nucleotide polymorphisms on virulence and phenotypes of Clostridium difficile strain 630 and derivatives. Virulence:1–15. https://doi.org/10.1080/21505594.2016.1237333

Collins DA, Hawkey PM, Riley TV (2013) Epidemiology of Clostridium difficile infection in Asia. Antimicrob Resist Infect Control 2(1):21. https://doi.org/10.1186/2047-2994-2-21

Curry RD, Muto CA, Schlackman JL, Pasculle AV, Shutt KA, Marsh JW, Harrison LH (2013) Use of multilocus variable number of tandem repeats analysis genotyping to determine the role of asymptomatic carriers in Clostridium difficile transmission. Clin Infect Dis 57(8):1094–1102. https://doi.org/10.1093/cid/cit475

Davies KA, Ashwin H, Longshaw CM, Burns DA, Davis GL, Wilcox MH, Group ES (2016) Diversity of Clostridium difficile PCR ribotypes in Europe: results from the European, multicentre, prospective, biannual, point-prevalence study of Clostridium difficile infection in hospitalised patients with diarrhoea (EUCLID), 2012 and 2013. Euro Surveill 21(29). https://doi.org/10.2807/1560-7917.ES.2016.21.29.30294

Deltcheva E, Chylinski K, Sharma CM, Gonzales K, Chao Y, Pirzada ZA, Eckert MR, Vogel J, Charpentier E (2011) CRISPR RNA maturation by trans-encoded small RNA and host factor RNase III. Nature 471(7340):602–607. https://doi.org/10.1038/nature09886

Deshpande A, Pasupuleti V, Thota P, Pant C, Rolston DD, Sferra TJ, Hernandez AV, Donskey CJ (2013) Community-associated Clostridium difficile infection and antibiotics: a meta-analysis. J Antimicrob Chemother 68(9):1951–1961. https://doi.org/10.1093/jac/dkt129

Didelot X, Eyre DW, Cule M, Ip CL, Ansari MA, Griffiths D, Vaughan A, O'Connor L, Golubchik T, Batty EM, Piazza P, Wilson DJ, Bowden R, Donnelly PJ, Dingle KE, Wilcox M, Walker AS, Crook DW, Peto TE, Harding RM (2012) Microevolutionary analysis of Clostridium difficile genomes to investigate transmission. Genome Biol 13(12):R118. https://doi.org/10.1186/gb-2012-13-12-r118

Diez-Villasenor C, Almendros C, Garcia-Martinez J, Mojica FJ (2010) Diversity of CRISPR loci in Escherichia coli. Microbiology 156(Pt 5):1351–1361. https://doi.org/10.1099/mic.0.036046-0

Dingle KE, Elliott B, Robinson E, Griffiths D, Eyre DW, Stoesser N, Vaughan A, Golubchik T, Fawley WN, Wilcox MH, Peto TE, Walker AS, Riley TV, Crook DW, Didelot X (2014) Evolutionary history of the Clostridium difficile pathogenicity locus. Genome Biol Evol 6(1):36–52. https://doi.org/10.1093/gbe/evt204

Dingle KE, Griffiths D, Didelot X, Evans J, Vaughan A, Kachrimanidou M, Stoesser N, Jolley KA, Golubchik T, Harding RM, Peto TE, Fawley W, Walker AS, Wilcox M, Crook DW (2011) Clinical Clostridium difficile: clonality and pathogenicity locus diversity. PLoS One 6(5):e19993. https://doi.org/10.1371/journal.pone.0019993

Drudy D, Harnedy N, Fanning S, Hannan M, Kyne L (2007) Emergence and control of fluoroquinolone-resistant, toxin A-negative, toxin B-positive Clostridium difficile. Infect Control Hosp Epidemiol 28(8): 932–940. https://doi.org/10.1086/519181

Duchene S, Holt KE, Weill FX, Le Hello S, Hawkey J, Edwards DJ, Fourment M, Holmes EC (2016) Genome-scale rates of evolutionary change in bacteria. Microb Genom 2(11):e000094. https://doi.org/10.1099/mgen.0.000094

Ducarmon QR, van der Bruggen T, Harmanus C, Sanders IMJG, Daenen LGM, Fluit AC, Vossen RHAM, Kloet SL, Kuijper EJ, Smits WK (2022) Clostridioides difficile infection with isolates of cryptic clade C-II: a genomic analysis of polymerase chain reaction ribotype 151. Clin Microbiol Infect. S1198-743X(22)00606-1. https://doi.org/10.1016/j.cmi.2022.12.003

Eckert C, Coignard B, Hebert M, Tarnaud C, Tessier C, Lemire A, Burghoffer B, Noel D, Barbut F, Group IC-RW (2013) Clinical and microbiological features of Clostridium difficile infections in France: the ICD-RAISIN 2009 national survey. Med Mal Infect 43(2):67–74. https://doi.org/10.1016/j.medmal.2013.01.004

Elliott B, Dingle KE, Didelot X, Crook DW, Riley TV (2014) The complexity and diversity of the Pathogenicity Locus in Clostridium difficile clade 5. Genome Biol Evol 6(12):3159–3170. https://doi.org/10.1093/gbe/evu248

Elliott B, Reed R, Chang BJ, Riley TV (2009) Bacteremia with a large clostridial toxin-negative, binary toxin-positive strain of Clostridium difficile. Anaerobe 15(6):249–251. https://doi.org/10.1016/j.anaerobe.2009.08.006

Eyre DW, Peto TEA, Crook DW, Walker AS, Wilcox MH (2019) Hash-based core genome multilocus sequence typing for Clostridium difficile. J Clin Microbiol 58(1): e01037-19. https://doi.org/10.1128/JCM.01037-19

Eyre DW, Davies KA, Davis G, Fawley WN, Dingle KE, De Maio N, Karas A, Crook DW, Peto TEA, Walker AS, Wilcox MH (2018) Two distinct patterns of Clostridium difficile diversity across Europe indicating contrasting routes of spread. Clin Infect Dis 67(7): 1035–1044. https://doi.org/10.1093/cid/ciy252

Eyre DW, Babakhani F, Griffiths D, Seddon J, Del Ojo EC, Gorbach SL, Peto TE, Crook DW, Walker AS (2014) Whole-genome sequencing demonstrates that fidaxomicin is superior to vancomycin for preventing reinfection and relapse of infection with Clostridium difficile. J Infect Dis 209(9):1446–1451. https://doi.org/10.1093/infdis/jit598

Eyre DW, Cule ML, Griffiths D, Crook DW, Peto TE, Walker AS, Wilson DJ (2013a) Detection of mixed infection from bacterial whole genome sequence data allows assessment of its role in Clostridium difficile transmission. PLoS Comput Biol 9(5):e1003059. https://doi.org/10.1371/journal.pcbi.1003059

Eyre DW, Cule ML, Wilson DJ, Griffiths D, Vaughan A, O'Connor L, Ip CL, Golubchik T, Batty EM, Finney JM, Wyllie DH, Didelot X, Piazza P, Bowden R, Dingle KE, Harding RM, Crook DW, Wilcox MH, Peto TE, Walker AS (2013b) Diverse sources of C. difficile infection identified on whole-genome sequencing. N Engl J Med 369(13):1195–1205. https://doi.org/10.1056/NEJMoa1216064

Eyre DW, Fawley WN, Rajgopal A, Settle C, Mortimer K, Goldenberg SD, Dawson S, Crook DW, Peto TEA, Walker AS, Wilcox MH (2017) Comparison of control of Clostridium difficile infection in six english hospitals using whole-genome sequencing. Clin Infect Dis. https://doi.org/10.1093/cid/cix338

Eyre DW, Golubchik T, Gordon NC, Bowden R, Piazza P, Batty EM, Ip CL, Wilson DJ, Didelot X, O'Connor L, Lay R, Buck D, Kearns AM, Shaw A, Paul J, Wilcox MH, Donnelly PJ, Peto TE, Walker AS, Crook DW (2012) A pilot study of rapid benchtop sequencing of Staphylococcus aureus and Clostridium difficile for outbreak detection and surveillance. BMJ Open 2(3). https://doi.org/10.1136/bmjopen-2012-001124

Eyre DW, Griffiths D, Vaughan A, Golubchik T, Acharya M, O'Connor L, Crook DW, Walker AS, Peto TE (2013c) Asymptomatic Clostridium difficile colonisation and onward transmission. PLoS One 8(11):e78445. https://doi.org/10.1371/journal.pone.0078445

Eyre DW, Walker AS, Freeman J, Baines SD, Fawley WN, Chilton CH, Griffiths D, Vaughan A, Crook DW, Peto TE, Wilcox MH (2013d) Short-term genome stability of serial Clostridium difficile ribotype 027 isolates in an experimental gut model and recurrent human disease. PLoS One 8(5):e63540. https://doi.org/10.1371/journal.pone.0063540

Forgetta V, Oughton MT, Marquis P, Brukner I, Blanchette R, Haub K, Magrini V, Mardis ER, Gerding DN, Loo VG, Miller MA, Mulvey MR, Rupnik M, Dascal A, Dewar K (2011) Fourteen-genome comparison identifies DNA markers for severe-disease-associated strains of Clostridium difficile. J Clin Microbiol 49(6):2230–2238. https://doi.org/10.1128/JCM.00391-11

Frentrup M, Zhou Z, Steglich M, Meier-Kolthoff JP, Göker M, Riedel T, Bunk B, Spröer C, Overmann J, Blaschitz M, Indra A, von Müller L, Kohl TA, Niemann S, Seyboldt C, Klawonn F, Kumar N, Lawley TD, García-Fernández S, Cantón R, Del Campo R, Zimmermann O, Groß U, Achtman M, Nübel U (2020) A publicly accessible database for Clostridioides difficile genome sequences supports tracing of transmission chains and epidemics. Microb Genom 6(8):mgen000410. https://doi.org/10.1099/mgen.0.000410

García-Fernández S, Frentrup M, Steglich M, Gonzaga A, Cobo M, López-Fresneña N, Cobo J, Morosini M-I, Cantón R, Del Campo R, Nübel U (2019) Whole-genome sequencing reveals nosocomial Clostridioides difficile transmission and a previously unsuspected epidemic scenario. Sci Rep 9(1):6959. https://doi.org/10.1038/s41598-019-43464-4

Griffiths D, Fawley W, Kachrimanidou M, Bowden R, Crook DW, Fung R, Golubchik T, Harding RM, Jeffery KJ, Jolley KA, Kirton R, Peto TE, Rees G, Stoesser N, Vaughan A, Walker AS, Young BC, Wilcox M, Dingle KE (2010) Multilocus sequence typing of Clostridium difficile. J Clin Microbiol 48(3):770–778. https://doi.org/10.1128/JCM.01796-09

Halstead FD, Ravi A, Thomson N, Nuur M, Hughes K, Brailey M, Oppenheim BA (2019) Whole genome sequencing of toxigenic Clostridium difficile in asymptomatic carriers: insights into possible role in transmission. J Hosp Infect 102(2):125–134. https://doi.org/10.1016/j.jhin.2018.10.012

Hargreaves KR, Flores CO, Lawley TD, Clokie MR (2014) Abundant and diverse clustered regularly interspaced short palindromic repeat spacers in Clostridium difficile strains and prophages target multiple phage types within this pathogen. MBio 5(5):e01045-01013. https://doi.org/10.1128/mBio.01045-13

Hasselmayer O, Braun V, Nitsche C, Moos M, Rupnik M, von Eichel-Streiber C (2004) Clostridium difficile IStron CdISt1: discovery of a variant encoding two complete transposase-like proteins. J Bacteriol 186(8):2508–2510

He M, Miyajima F, Roberts P, Ellison L, Pickard DJ, Martin MJ, Connor TR, Harris SR, Fairley D, Bamford KB, D'Arc S, Brazier J, Brown D, Coia JE, Douce G, Gerding D, Kim HJ, Koh TH, Kato H, Senoh M, Louie T, Michell S, Butt E, Peacock SJ, Brown NM, Riley T, Songer G, Wilcox M, Pirmohamed M, Kuijper E, Hawkey P, Wren BW, Dougan G, Parkhill J, Lawley TD (2013) Emergence and global spread of epidemic healthcare-associated Clostridium difficile. Nat Genet 45(1):109–113. https://doi.org/10.1038/ng.2478

He M, Sebaihia M, Lawley TD, Stabler RA, Dawson LF, Martin MJ, Holt KE, Seth-Smith HM, Quail MA, Rance R, Brooks K, Churcher C, Harris D, Bentley SD, Burrows C, Clark L, Corton C, Murray V, Rose G, Thurston S, van Tonder A, Walker D, Wren BW, Dougan G, Parkhill J (2010) Evolutionary dynamics of Clostridium difficile over short and long time scales. Proc Natl Acad Sci U S A 107(16):7527–7532. https://doi.org/10.1073/pnas.0914322107

Imwattana K, Putsathit P, Collins DA, Leepattarakit T, Kiratisin P, Riley TV, Knight DR (2022) Global evolutionary dynamics and resistome analysis of Clostridioides difficile ribotype 017. Microbial Genoms 8(3):000792. https://doi.org/10.1099/mgen.0.000792

Imwattana K, Knight DR, Kullin B, Collins DA, Putsathit P, Kiratisin P, Riley TV (2019) Clostridium difficile ribotype 017 – characterization, evolution and epidemiology of the dominant strain in Asia. Emerg Microbes Infect 8(1):796–807. https://doi.org/10.1080/22221751.2019.1621670

Isidro J, Menezes J, Serrano M, Borges V, Paixão P, Mimoso M, Martins F, Toscano C, Santos A, Henriques AO, Oleastro M (2018) Genomic study of a Clostridium difficile multidrug resistant outbreak-related clone reveals novel determinants of resistance. Front Microbiol 9

Janezic S, Dingle K, Alvin J, Accetto T, Didelot X, Crook DW, Lacy DB, Rupnik M (2020) Comparative genomics of Clostridioides difficile toxinotypes identifies module-based toxin gene evolution. Microb Genom 6(10):e000449. https://doi.org/10.1099/mgen.0.000449

Janezic S, Rupnik M (2019) Development and implementation of whole genome sequencing-based typing schemes for Clostridioides difficile. Front Public Health 7:309. https://doi.org/10.3389/fpubh.2019.00309

Janezic S, Marin M, Martin A, Rupnik M (2015) A new type of toxin A-negative, toxin B-positive Clostridium difficile strain lacking a complete tcdA gene. J Clin Microbiol 53(2):692–695. https://doi.org/10.1128/JCM.02211-14

Janezic S, Potocnik M, Zidaric V, Rupnik M (2016) Highly divergent Clostridium difficile strains isolated from the environment. PLoS One 11(11):e0167101. https://doi.org/10.1371/journal.pone.0167101

Janezic S, Rupnik M (2015) Genomic diversity of Clostridium difficile strains. Res Microbiol 166(4):353–360. https://doi.org/10.1016/j.resmic.2015.02.002

Janvilisri T, Scaria J, Thompson AD, Nicholson A, Limbago BM, Arroyo LG, Songer JG, Grohn YT, Chang YF (2009) Microarray identification of Clostridium difficile core components and divergent regions associated with host origin. J Bacteriol 191(12):3881–3891. https://doi.org/10.1128/JB.00222-09

Jhung MA, Thompson AD, Killgore GE, Zukowski WE, Songer G, Warny M, Johnson S, Gerding DN, McDonald LC, Limbago BM (2008) Toxinotype V Clostridium difficile in humans and food animals. Emerg Infect Dis 14(7):1039–1045. https://doi.org/10.3201/eid1407.071641

Kelly CP (2012) Can we identify patients at high risk of recurrent Clostridium difficile infection? Clin Microbiol Infect 18(Suppl 6):21–27. https://doi.org/10.1111/1469-0691.12046

Khanna S, Pardi DS (2012) Clostridium difficile infection: new insights into management. Mayo Clin Proc 87(11):1106–1117. https://doi.org/10.1016/j.mayocp.2012.07.016

Knetsch CW, Connor TR, Mutreja A, van Dorp SM, Sanders IM, Browne HP, Harris D, Lipman L, Keessen EC, Corver J, Kuijper EJ, Lawley TD (2014) Whole genome sequencing reveals potential spread of Clostridium difficile between humans and farm animals in the Netherlands, 2002 to 2011. Euro Surveill 19(45):20954. https://doi.org/10.2807/1560-7917.es2014.19.45.20954

Knetsch CW, Terveer EM, Lauber C, Gorbalenya AE, Harmanus C, Kuijper EJ, Corver J, van Leeuwen HC

(2012) Comparative analysis of an expanded Clostridium difficile reference strain collection reveals genetic diversity and evolution through six lineages. Infect Genet Evol 12(7):1577–1585. https://doi.org/10.1016/j.meegid.2012.06.003

Knight DR, Imwattana K, Kullin B, Guerrero-Araya E, Paredes-Sabja D, Didelot X, Dingle KE, Eyre DW, Rodríguez C, Riley TV (2021) Major genetic discontinuity and novel toxigenic species in Clostridioides difficile taxonomy. s 10. https://doi.org/10.7554/eLife.64325

Knight DR, Kullin B, Androga GO, Barbut F, Eckert C, Johnson S, Spigaglia P, Tateda K, Tsai P-J, Riley TV (2019) Evolutionary and genomic insights into Clostridioides difficile sequence type 11: a diverse zoonotic and antimicrobial-resistant lineage of global one health importance. mBio 10(2):e00446-19. https://doi.org/10.1128/mBio.00446-19

Knight DR, Elliott B, Chang BJ, Perkins TT, Riley TV (2015) Diversity and evolution in the genome of Clostridium difficile. Clin Microbiol Rev 28(3):721–741. https://doi.org/10.1128/CMR.00127-14

Kociolek LK, Gerding DN, Espinosa RO, Patel SJ, Shulman ST, Ozer EA (2018) Clostridium difficile whole genome sequencing reveals limited transmission among symptomatic children: a single-center analysis. Clin Infect Dis 67(2):229–234. https://doi.org/10.1093/cid/ciy060

Kong LY, Eyre DW, Corbeil J, Raymond F, Walker AS, Wilcox MH, Crook DW, Michaud S, Toye B, Frost E, Dendukuri N, Schiller I, Bourgault A-M, Dascal A, Oughton M, Longtin Y, Poirier L, Brassard P, Turgeon N, Gilca R, Loo VG (2019) Clostridium difficile: investigating transmission patterns between infected and colonized patients using whole genome sequencing. Clin Infect Dis 68(2):204–209. https://doi.org/10.1093/cid/ciy457

Kuehne SA, Collery MM, Kelly ML, Cartman ST, Cockayne A, Minton NP (2014) Importance of toxin A, toxin B, and CDT in virulence of an epidemic Clostridium difficile strain. J Infect Dis 209(1):83–86. https://doi.org/10.1093/infdis/jit426

Kuehne SA, Cartman ST, Heap JT, Kelly ML, Cockayne A, Minton NP (2010) The role of toxin A and toxin B in Clostridium difficile infection. Nature 467(7316):711–713. https://doi.org/10.1038/nature09397

Kuijper EJ, Barbut F, Brazier JS, Kleinkauf N, Eckmanns T, Lambert ML, Drudy D, Fitzpatrick F, Wiuff C, Brown DJ, Coia JE, Pituch H, Reichert P, Even J, Mossong J, Widmer AF, Olsen KE, Allerberger F, Notermans DW, Delmee M, Coignard B, Wilcox M, Patel B, Frei R, Nagy E, Bouza E, Marin M, Akerlund T, Virolainen-Julkunen-A, Lyytikainen O, Kotila S, Ingebretsen A, Smyth B, Rooney P, Poxton IR, Monnet DL (2008) Update of Clostridium difficile infection due to PCR ribotype 027 in Europe, 2008. Euro Surveill 13(31)

Kumar N, Miyajima F, He M, Roberts P, Swale A, Ellison L, Pickard D, Smith G, Molyneux R, Dougan G, Parkhill J, Wren BW, Parry CM, Pirmohamed M, Lawley TD (2016) Genome-based infection tracking reveals dynamics of Clostridium difficile transmission and disease recurrence. Clin Infect Dis 62(6):746–752. https://doi.org/10.1093/cid/civ1031

Kurka H, Ehrenreich A, Ludwig W, Monot M, Rupnik M, Barbut F, Indra A, Dupuy B, Liebl W (2014) Sequence similarity of Clostridium difficile strains by analysis of conserved genes and genome content is reflected by their ribotype affiliation. PLoS One 9(1):e86535. https://doi.org/10.1371/journal.pone.0086535

Lemee L, Dhalluin A, Pestel-Caron M, Lemeland JF, Pons JL (2004) Multilocus sequence typing analysis of human and animal Clostridium difficile isolates of various toxigenic types. J Clin Microbiol 42(6):2609–2617. https://doi.org/10.1128/JCM.42.6.2609-2617.2004

Lim SK, Stuart RL, Mackin KE, Carter GP, Kotsanas D, Francis MJ, Easton M, Dimovski K, Elliott B, Riley TV, Hogg G, Paul E, Korman TM, Seemann T, Stinear TP, Lyras D, Jenkin GA (2014) Emergence of a ribotype 244 strain of Clostridium difficile associated with severe disease and related to the epidemic ribotype 027 strain. Clin Infect Dis 58(12):1723–1730. https://doi.org/10.1093/cid/ciu203

Loo VG, Poirier L, Miller MA, Oughton M, Libman MD, Michaud S, Bourgault AM, Nguyen T, Frenette C, Kelly M, Vibien A, Brassard P, Fenn S, Dewar K, Hudson TJ, Horn R, Rene P, Monczak Y, Dascal A (2005) A predominantly clonal multi-institutional outbreak of Clostridium difficile-associated diarrhea with high morbidity and mortality. N Engl J Med 353(23):2442–2449. https://doi.org/10.1056/NEJMoa051639

Lyras D, O'Connor JR, Howarth PM, Sambol SP, Carter GP, Phumoonna T, Poon R, Adams V, Vedantam G, Johnson S, Gerding DN, Rood JI (2009) Toxin B is essential for virulence of Clostridium difficile. Nature 458(7242):1176–1179. https://doi.org/10.1038/nature07822

Mac Aogain M, Moloney G, Kilkenny S, Kelleher M, Kelleghan M, Boyle B, Rogers TR (2015) Whole-genome sequencing improves discrimination of relapse from reinfection and identifies transmission events among patients with recurrent Clostridium difficile infections. J Hosp Infect 90(2):108–116. https://doi.org/10.1016/j.jhin.2015.01.021

Maikova A, Severinov K, Soutourina O (2018) New insights into functions and possible applications of Clostridium difficile CRISPR-Cas system. Front Microbiol 9

Makarova KS, Haft DH, Barrangou R, Brouns SJ, Charpentier E, Horvath P, Moineau S, Mojica FJ, Wolf YI, Yakunin AF, van der Oost J, Koonin EV (2011) Evolution and classification of the CRISPR-Cas systems. Nat Rev Microbiol 9(6):467–477. https://doi.org/10.1038/nrmicro2577

Makarova KS, Wolf YI, Koonin EV (2013) The basic building blocks and evolution of CRISPR-CAS systems. Biochem Soc Trans 41(6):1392–1400. https://doi.org/10.1042/BST20130038

Marsden GL, Davis IJ, Wright VJ, Sebaihia M, Kuijper EJ, Minton NP (2010) Array comparative hybridisation reveals a high degree of similarity between UK and European clinical isolates of hypervirulent Clostridium difficile. BMC Genom 11:389. https://doi.org/10.1186/1471-2164-11-389

Martin JSH, Eyre DW, Fawley WN, Griffiths D, Davies K, Mawer DPC, Peto TEA, Crook DW, Walker AS, Wilcox MH (2018) Patient and strain characteristics associated with Clostridium difficile transmission and adverse outcomes. Clin Infect Dis 67(9):1379–1387. https://doi.org/10.1093/cid/ciy302

Mawer DPC, Eyre DW, Griffiths D, Fawley WN, Martin JSH, Quan TP, Peto TEA, Crook DW, Walker AS, Wilcox MH (2017) Contribution to Clostridium difficile transmission of symptomatic patients with toxigenic strains who are fecal toxin negative. Clin Infect Dis 64(9):1163–1170. https://doi.org/10.1093/cid/cix079

McDonald LC, Killgore GE, Thompson A, Owens RC Jr, Kazakova SV, Sambol SP, Johnson S, Gerding DN (2005) An epidemic, toxin gene-variant strain of Clostridium difficile. N Engl J Med 353(23):2433–2441. https://doi.org/10.1056/NEJMoa051590

Monot M, Eckert C, Lemire A, Hamiot A, Dubois T, Tessier C, Dumoulard B, Hamel B, Petit A, Lalande V, Ma L, Bouchier C, Barbut F, Dupuy B (2015) Clostridium difficile: new insights into the evolution of the pathogenicity locus. Sci Rep 5:15023. https://doi.org/10.1038/srep15023

Moura I, Monot M, Tani C, Spigaglia P, Barbanti F, Norais N, Dupuy B, Bouza E, Mastrantonio P (2014) Multidisciplinary analysis of a nontoxigenic Clostridium difficile strain with stable resistance to metronidazole. Antimicrob Agents Chemother 58(8):4957–4960. https://doi.org/10.1128/AAC.02350-14

Moura I, Spigaglia P, Barbanti F, Mastrantonio P (2013) Analysis of metronidazole susceptibility in different Clostridium difficile PCR ribotypes. J Antimicrob Chemother 68(2):362–365. https://doi.org/10.1093/jac/dks420

Peng L, Pei J, Pang H, Guo Y, Lin L, Huang R (2014) Whole genome sequencing reveals a novel CRISPR system in industrial Clostridium acetobutylicum. J Ind Microbiol Biotechnol 41(11):1677–1685. https://doi.org/10.1007/s10295-014-1507-3

Pepin J, Valiquette L, Alary ME, Villemure P, Pelletier A, Forget K, Pepin K, Chouinard D (2004) Clostridium difficile-associated diarrhea in a region of Quebec from 1991 to 2003: a changing pattern of disease severity. CMAJ 171(5):466–472. https://doi.org/10.1503/cmaj.1041104

Pepin J, Valiquette L, Cossette B (2005) Mortality attributable to nosocomial Clostridium difficile-associated disease during an epidemic caused by a hypervirulent strain in Quebec. CMAJ 173(9):1037–1042. https://doi.org/10.1503/cmaj.050978

Pougach K, Semenova E, Bogdanova E, Datsenko KA, Djordjevic M, Wanner BL, Severinov K (2010) Transcription, processing and function of CRISPR cassettes in Escherichia coli. Mol Microbiol 77(6):1367–1379. https://doi.org/10.1111/j.1365-2958.2010.07265.x

Quesada-Gomez C, Lopez-Urena D, Chumbler N, Kroh HK, Castro-Pena C, Rodriguez C, Orozco-Aguilar J, Gonzalez-Camacho S, Rucavado A, Guzman-Verri C, Lawley TD, Lacy DB, Chaves-Olarte E (2016) Analysis of TcdB proteins within the hypervirulent clade 2 reveals an impact of RhoA glucosylation on Clostridium difficile proinflammatory activities. Infect Immun 84(3):856–865. https://doi.org/10.1128/IAI.01291-15

Ramírez-Vargas G, López-Ureña D, Badilla A, Orozco-Aguilar J, Murillo T, Rojas P, Riedel T, Overmann J, González G, Chaves-Olarte E, Quesada-Gómez C, Rodríguez C (2018) Novel Clade C I Clostridium difficile strains escape diagnostic tests, differ in pathogenicity potential and carry toxins on extrachromosomal elements. Sci Rep 8:13951. https://doi.org/10.1038/s41598-018-32390-6

Ramírez-Vargas G, Rodríguez C (2020) Putative conjugative plasmids with tcdB and cdtAB genes in Clostridioides difficile. Emerg Infect Dis J 26(9) CDC. https://doi.org/10.3201/eid2609.191447

Richter H, Zoephel J, Schermuly J, Maticzka D, Backofen R, Randau L (2012) Characterization of CRISPR RNA processing in Clostridium thermocellum and Methanococcus maripaludis. Nucleic Acids Res 40(19):9887–9896. https://doi.org/10.1093/nar/gks737

Riedel T, Wittmann J, Bunk B, Schober I, Spröer C, Gronow S, Overmann J (2017) A Clostridioides difficile bacteriophage genome encodes functional binary toxin-associated genes. J Biotechnol 250:23–28. https://doi.org/10.1016/j.jbiotec.2017.02.017

Rupnik M, Janezic S (2016) An update on Clostridium difficile toxinotyping. J Clin Microbiol 54(1):13–18. https://doi.org/10.1128/JCM.02083-15

Rupnik M, Widmer A, Zimmermann O, Eckert C, Barbut F (2008) Clostridium difficile toxinotype V, ribotype 078, in animals and humans. J Clin Microbiol 46(6):2146. https://doi.org/10.1128/JCM.00598-08

Scaria J, Ponnala L, Janvilisri T, Yan W, Mueller LA, Chang YF (2010) Analysis of ultra low genome conservation in Clostridium difficile. PLoS One 5(12):e15147. https://doi.org/10.1371/journal.pone.0015147

Schürch AC, Arredondo-Alonso S, Willems RJL, Goering RV (2018) Whole genome sequencing options for bacterial strain typing and epidemiologic analysis based on single nucleotide polymorphism versus gene-by-gene-based approaches. Clin Microbiol Infect 24(4):350–354. https://doi.org/10.1016/j.cmi.2017.12.016

Sebaihia M, Wren BW, Mullany P, Fairweather NF, Minton N, Stabler R, Thomson NR, Roberts AP, Cerdeno-Tarraga AM, Wang H, Holden MT,

Wright A, Churcher C, Quail MA, Baker S, Bason N, Brooks K, Chillingworth T, Cronin A, Davis P, Dowd L, Fraser A, Feltwell T, Hance Z, Holroyd S, Jagels K, Moule S, Mungall K, Price C, Rabbinowitsch E, Sharp S, Simmonds M, Stevens K, Unwin L, Whithead S, Dupuy B, Dougan G, Barrell B, Parkhill J (2006) The multidrug-resistant human pathogen Clostridium difficile has a highly mobile, mosaic genome. Nat Genet 38(7):779–786. https://doi.org/10.1038/ng1830

Shin BM, Kuak EY, Yoo SJ, Shin WC, Yoo HM (2008) Emerging toxin A-B+ variant strain of Clostridium difficile responsible for pseudomembranous colitis at a tertiary care hospital in Korea. Diagn Microbiol Infect Dis 60(4):333–337. https://doi.org/10.1016/j.diagmicrobio.2007.10.022

Sim JH, Truong C, Minot SS, Greenfield N, Budvytiene I, Lohith A, Anikst V, Pourmand N, Banaei N (2017) Determining the cause of recurrent Clostridium difficile infection using whole genome sequencing. Diagn Microbiol Infect Dis 87(1):11–16. https://doi.org/10.1016/j.diagmicrobio.2016.09.023

Spigaglia P, Barbanti F, Mastrantonio P, European Study Group on Clostridium d (2011) Multidrug resistance in European Clostridium difficile clinical isolates. J Antimicrob Chemother 66(10):2227–2234. https://doi.org/10.1093/jac/dkr292

Stabler RA, Gerding DN, Songer JG, Drudy D, Brazier JS, Trinh HT, Witney AA, Hinds J, Wren BW (2006) Comparative phylogenomics of Clostridium difficile reveals clade specificity and microevolution of hypervirulent strains. J Bacteriol 188(20):7297–7305. https://doi.org/10.1128/JB.00664-06

Stabler RA, He M, Dawson L, Martin M, Valiente E, Corton C, Lawley TD, Sebaihia M, Quail MA, Rose G, Gerding DN, Gibert M, Popoff MR, Parkhill J, Dougan G, Wren BW (2009) Comparative genome and phenotypic analysis of Clostridium difficile 027 strains provides insight into the evolution of a hypervirulent bacterium. Genome Biol 10(9):R102. https://doi.org/10.1186/gb-2009-10-9-r102

Stabler RA, Dawson LF, Valiente E, Cairns MD, Martin MJ, Donahue EH, Riley TV, Songer JG, Kuijper EJ, Dingle KE, Wren BW (2012) Macro and micro diversity of Clostridium difficile isolates from diverse sources and geographical locations. PLoS One 7:e31559. https://doi.org/10.1371/journal.pone.0031559

Suetens C, Latour K, Kärki T, Ricchizzi E, Kinross P, Moro ML, Jans B, Hopkins S, Hansen S, Lyytikäinen O, Reilly J, Deptula A, Zingg W, Plachouras D, Monnet DL, Group the H-AIPS (2018) Prevalence of healthcare-associated infections, estimated incidence and composite antimicrobial resistance index in acute care hospitals and long-term care facilities: results from two European point prevalence surveys, 2016 to 2017. Eurosurveillance 23(46):1800516. https://doi.org/10.2807/1560-7917.ES.2018.23.46.1800516

Valiente E, Dawson LF, Cairns MD, Stabler RA, Wren BW (2012) Emergence of new PCR ribotypes from the hypervirulent Clostridium difficile 027 lineage. J Med Microbiol 61(Pt 1):49–56. https://doi.org/10.1099/jmm.0.036194-0

Vedantam G, Clark A, Chu M, McQuade R, Mallozzi M, Viswanathan VK (2012) Clostridium difficile infection: toxins and non-toxin virulence factors, and their contributions to disease establishment and host response. Gut Microbes 3(2):121–134. https://doi.org/10.4161/gmic.19399

Vonberg RP, Kuijper EJ, Wilcox MH, Barbut F, Tull P, Gastmeier P, European Cd-ICG, European Centre for Disease P, Control, van den Broek PJ, Colville A, Coignard B, Daha T, Debast S, Duerden BI, van den Hof S, van der Kooi T, Maarleveld HJ, Nagy E, Notermans DW, O'Driscoll J, Patel B, Stone S, Wiuff C (2008) Infection control measures to limit the spread of Clostridium difficile. Clin Microbiol Infect 14(Suppl 5):2–20. https://doi.org/10.1111/j.1469-0691.2008.01992.x

Warny M, Pepin J, Fang A, Killgore G, Thompson A, Brazier J, Frost E, McDonald LC (2005) Toxin production by an emerging strain of Clostridium difficile associated with outbreaks of severe disease in North America and Europe. Lancet 366(9491):1079–1084. https://doi.org/10.1016/S0140-6736(05)67420-X

Widmer AF, Frei R, Erb S, Stranden A, Kuijper EJ, Knetsch CW, Tschudin-Sutter S (2017) Transmissibility of Clostridium difficile without contact isolation: results from a prospective observational study with 451 patients. Clin Infect Dis 64(4):393–400. https://doi.org/10.1093/cid/ciw758

Williamson CHD, Stone NE, Nunnally AE, Roe CC, Vazquez AJ, Lucero SA, Hornstra H, Wagner DM, Keim P, Rupnik M, Janezic S, Sahl JW (2022) Identification of novel, cryptic Clostridioides species isolates from environmental samples collected from diverse geographical locations. Microb Genom 8(2):000742. https://doi.org/10.1099/mgen.0.000742

Zhou Z, Alikhan N-F, Mohamed K, Fan Y, Agama Study Group, Achtman M (2020) The EnteroBase user's guide, with case studies on Salmonella transmissions, Yersinia pestis phylogeny, and Escherichia core genomic diversity. Genome Res 30(1):138–152. https://doi.org/10.1101/gr.251678.119

An Updated View on the Cellular Uptake and Mode-of-Action of *Clostridioides difficile* Toxins

Panagiotis Papatheodorou, Nigel P. Minton, Klaus Aktories, and Holger Barth

Abstract

Research on the human gut pathogen *Clostridioides (C.) difficile* and its toxins continues to attract much attention as a consequence of the threat to human health posed by hypervirulent strains. Toxin A (TcdA) and Toxin B (TcdB) are the two major virulence determinants of *C. difficile*. Both are single-chain proteins with a similar multidomain architecture. Certain hypervirulent *C. difficile* strains also produce a third toxin, namely binary toxin CDT (*C. difficile transferase*). *C. difficile* toxins are the causative agents of *C. difficile*-associated diseases (CDADs), such as antibiotics-associated diarrhea and pseudomembranous colitis. For that reason, considerable efforts have been expended to unravel their molecular mode-of-action and the cellular mechanisms responsible for their uptake. Many of these studies have been conducted in European laboratories. Here, we provide an update on our previous review (Papatheodorou et al. Adv Exp Med Biol, 2018) on important advances in *C. difficile* toxins research.

P. Papatheodorou (✉) · H. Barth
Institute of Experimental and Clinical Pharmacology, Toxicology and Pharmacology of Natural Products, Ulm University Medical Center, Ulm, Germany
e-mail: panagiotis.papatheodorou@uni-ulm.de

N. P. Minton
BBSRC/EPSRC Synthetic Biology Research Centre, University of Nottingham, Nottingham, UK

K. Aktories
Institute of Experimental and Clinical Pharmacology and Toxicology, University of Freiburg, Freiburg, Germany

1 Introduction

The human gut pathogen *Clostridioides (C.) difficile* is capable of producing at least three exotoxins, namely toxin A (TcdA), toxin B (TcdB), and the binary toxin CDT (*C. difficile* transferase). The number of toxins and the quantities produced vary between different *C. difficile* strains. Certain hypervirulent strains release all three toxins during infection. Others produce strain-specific isoforms of toxins A and B (Rupnik and Janezic 2016). As shown recently by Mansfield and colleagues in a comprehensive phylogenomic analysis of more than 8000 *C. difficile* strains, toxin A genes clustered into 7 (A1–A7) and toxin B genes into 12 (B1–B12) distinct subtypes (Mansfield et al. 2020).

Toxin A and toxin B are related but they differ in structure and function from the binary toxin CDT. However, the three toxins share some fundamental similarities during the intoxication process. All three toxins are released by the bacteria and enter into host cells via receptor-mediated endocytosis. An enzymatically active portion of the toxins then escapes from acidified endosomes

P. Mastrantonio, M. Rupnik (eds.), *Updates on Clostridioides difficile in Europe*, Advances in Experimental Medicine and Biology 1435, https://doi.org/10.1007/978-3-031-42108-2_11

into the host cell cytosol in order to reach and modify its specific target proteins. In the case of toxins A and B, the enzyme portion is a glucosyltransferase that inactivates small GTPases of the Rho family. The enzyme portion of CDT is an ADP-ribosyltransferase that modifies G-actin.

In the following sections, we will provide an updated view on the current knowledge about *C. difficile* toxins' cellular uptake and mode-of-action which is fundamental for understanding their pathophysiological role in *C. difficile* infections (CDI).

2 Structure, Uptake and Mode-of-Action of *C. difficile* Toxins A and B

2.1 Modular Composition of *C. difficile* Toxins A and B

Toxins A and B are large, single-chain protein toxins that comprise several functional domains. The two toxins exhibit a high sequence homology (~50% amino acid identity) and an identical multidomain architecture indicating that a gene duplication event led to the existence of two nearly-identical toxins in *C. difficile* (von Eichel-Streiber et al. 1992a). Both toxins are also highly similar to other large clostridial toxins (LCTs; also denoted as clostridial glucosylating toxins (CGTs)), such as *Paeniclostridium sordellii* (formerly *Clostridium sordellii*) lethal toxin (TcsL) and hemorrhagic toxin, *Clostridium novyi* α-toxin (Tcnα), and *Clostridium perfringens* TpeL toxin (Voth and Ballard 2005; Aktories et al. 2017).

The large size of toxin A (2710 amino acids; 308 kDa) and toxin B (2366 amino acids; 270 kDa) led quite early to the assumption that both toxins contain several domains with specific functions during the intoxication process. Eventually, a number of fundamental findings confirmed the modular composition of toxins A and B, which is also true for the other LCTs. The modular composition of toxins A and B was initially described by Jank and Aktories in 2008 with the so-called ABCD model, where A stands for *biological activity*, B for *binding*, C for *cutting*, and D for *delivery* (Jank and Aktories 2008). The following subsections summarize the current knowledge about the functional domains of toxins A and B.

2.1.1 The CROP Domain

At first, a region consisting of series of combined, repetitive oligopeptides (CROPs) was identified and characterized in the C-terminal part of toxin A (von Eichel-Streiber and Sauerborn 1990; von Eichel-Streiber et al. 1992b). In toxin A, the CROP domain makes up nearly one-third of the complete protein and consists of 7 long repeats of 30 residues and 31 short repeats of 15–21 residues. In toxin B, the CROP domain contains 7 long repeats of 30 residues and only 21 short repeats of 20–23 residues and thus is significantly shorter than in toxin A. The number and length of the repeating CROP modules have been found to vary between toxins from different *C. difficile* isolates (Rupnik et al. 1998). Historically, the CROP domain was considered to start around residue 1849 of toxin A and residue 1852 of toxin B, respectively. However, according to more recent structural studies by Orth et al., the CROP domain starts at glycine-1832 for toxin A and at glycine-1834 for toxin B (Orth et al. 2014).

A series of studies including monoclonal antibodies or recombinant toxin fragments have provided evidence for a role of the CROP domain of toxin A in receptor binding (Frey and Wilkins 1992; Sauerborn et al. 1997; Frisch et al. 2003). In further studies, crystal structures of two C-terminal fragments (terminal 127 and 255 residues) of toxin A were obtained, thus providing new insights into the overall structure of the CROP domain (Ho et al. 2005; Greco et al. 2006). The CROP domain of toxin A adopts a solenoid-like (screw-like) fold (Ho et al. 2005; Greco et al. 2006; Jank and Aktories 2008). One of the two CROP structures was obtained by co-crystallization with the trisaccharide Galα1–3-Galβ1–4GlcNAc, which was found to interact with toxin A in earlier reports (Krivan et al. 1986; Tucker and Wilkins 1991; Greco et al. 2006). However, this carbohydrate structure is not present on human cells and thus is unlikely to be part of intestinal receptors of toxin A in humans. The carbohydrate-binding properties of the CROP domain of toxin A were also supported by a study from Dingle and colleagues (Dingle et al. 2008). Notably, the CROP domain of toxins A and B is similar to certain saccharide-binding proteins from *Streptococcus downei* and *Streptococcus mutans* (Wren 1991).

In the meantime, partial and full-length 3D structures of toxin A and toxin B have become available, revealing that the CROP domain is a rather flexible structure, which can switch between an open and closed conformation in a pH-dependent manner (Simeon et al. 2019; Chen et al. 2019a, 2022a; Aminzadeh et al. 2022).

Tam and colleagues identified a rather unexpected interaction of the CROP domain of toxin B, but not of toxin A, with intestinal bile acids. This reversible interaction led to inhibition of toxin B, by inducing major conformational changes in the toxin (Tam et al. 2020). The authors speculated that bile acids may help in the timing of the action of toxin in different areas of the gastrointestinal tract.

2.1.2 The Glucosyltransferase Domain (GTD)

In 1995, the group of Klaus Aktories (Freiburg, Germany) found that toxins A and B modify the small GTPase Rho and other members of the Rho subfamily via transfer of the glucose moiety from the co-substrate uridine diphosphate (UDP)-glucose to threonine-37 of the GTPase (Just et al. 1995a, b). Thus, it became apparent that toxins A and B are bacterial glucosyltransferases capable of inactivating small GTPases of host cells. Deletion analyses from Hofmann et al. with toxin B revealed glucosyltransferase activity in the N-terminal part of the toxin (Hofmann et al. 1997). Studies with TcsL proposed the clostridial glucosylating toxins are retaining glucosyltransferases (Vetter et al. 2000; Geyer et al. 2003). This means that the modification of the small GTPases by the toxins results in an α-anomeric configuration of the attached glucose.

In 2005, the crystal structure of the glucosyltransferase domain (GTD) of toxin B in the presence of UDP-glucose and Mn^{2+} was determined (Reinert et al. 2005). It became obvious from the 3D structure that the GTD of toxin B belongs to the glucosyltransferase type A (GT-A) family. Subsequent biochemical studies revealed important residues that are crucial for the enzymatic activity: residues 364–516 are important for substrate recognition (Hofmann et al. 1998); an essential and highly conserved DXD motif between amino acids 286 and 288 is involved in binding Mn^{2+} (Busch et al. 2000); residue tryptophan-102 is involved in UDP-glucose binding (Busch et al. 2000); isoleucine-383 and glutamine-385 are crucial residues for the co-substrate specificity (Jank et al. 2005); a four-helical-bundle subdomain at the N-terminus of the glucosyltransferase is required for the interaction with the inner plasma membrane (Geissler et al. 2010). Additional essential amino acids for substrate binding were identified by Jank and colleagues (Jank et al. 2007).

In 2012, D'Urzo and coworkers presented the crystal structure of the GTD of toxin A bound to Mn^{2+} and UDP-glucose (D'Urzo et al. 2012). In the same year, Pruitt and colleagues succeeded in solving the structure of the GTD of toxin A in the presence and absence of its co-substrate UDP-glucose (Pruitt et al. 2012). More recently, Alvin and Lacy reported new crystal structures of the GTDs of toxins A and B in complex with a non-hydrolyzable UDP-glucose analog and an *apo*-like structure of the GTD of toxin B (Alvin and Lacy 2017).

Interestingly, a GTD, highly similar to that of toxin A and toxin B, but with a unique substrate profile, was recently identified in the N-terminal part of the YGT (*Yersinia* glucosyltransferase) toxin from *Yersinia mollaretii* (Ost et al. 2020).

2.1.3 The Cysteine Protease Domain (CPD)

In 2003, Barth and colleagues showed with toxin B that only the N-terminal GTD reaches the cytosol after completion of the uptake process (Pfeifer et al. 2003). Thus, it was feasible that processing of toxins A and B is a prerequisite of the intoxication process. The cleavage site of toxin B was identified between leucine-543 and glycine-544 (Rupnik et al. 2005). Yet it was not clear whether the processing of toxins A and B occurs by a host protease or an internal domain of the toxins. Eventually, the group of Eichel-Streiber (Mainz, Germany) identified a small cytosolic compound, namely inositol hexakisphosphate (InsP6), which is capable of inducing autocatalytic processing of toxins A and B (Reineke et al. 2007). However, it was still not clear how processing occurred. This question was answered, when a cysteine protease domain (CPD), which is located adjacent to the GTD, was identified by Egerer et al. in toxins A and B (Egerer et al. 2007). A fragment of toxin B comprising only the GTD and the CPD is cleaved in the presence of InsP6, indicating that InsP6 induces autocatalytic processing of toxins A and B by activating the CPD. Lysine-600 of the CPD is essential for InsP6-binding, whereas cysteine-698, histidine-653, or aspartate-587 of toxin B represent the catalytic triad of the protease (Egerer et al. 2007, 2009).

A first 3D structure of the CPD (bound to InsP6) was provided for toxin A in 2009 by the group of Borden Lacy. The crystal structure uncovered a highly basic pocket that is required for InsP6-binding, which is separated from the active site by a beta-flap structure (Pruitt et al. 2009). Later, the 3D structure of the InsP6-bound CPD of toxin B was presented either in the absence (Shen et al. 2011) or in the presence of a specific small molecule inhibitor (Puri et al. 2010), respectively. It became apparent from these studies that InsP6-binding allosterically improves the access of the active site to its substrate. Interestingly, a structural study from Chumbler et al. revealed the requirement for zinc in the mechanism of autoprocessing of toxins A and B (Chumbler et al. 2016).

More recent studies identified another, rather unexpected function of the CPD, namely its contribution to host receptor binding (Pan et al. 2021; Chen et al. 2021; Jiang et al. 2022). By performing a mutagenesis screen, Pan and colleagues identified glycine-624 and glycine-626 of the CPD as key residues involved in receptor binding (Pan et al. 2021). However, independent structural analyses of toxin:receptor complexes by Chen et al. and Jiang et al. identified serine-573, arginine-575, and glutamate-564 as crucial residues of the CPD for receptor interaction (Chen et al. 2021; Jiang et al. 2022).

Collaborative work of the Barth and Di Masi research groups with the clinical research group of Stefano Di Bella provided evidence that human serum albumin specifically interacts with toxins A and B and prevents as self-defense mechanism the autocatalytic processing of the toxins, thus providing an explanation for the clinical correlation between CDI severity and hypoalbuminemia (Di Masi et al. 2018).

2.1.4 Delivery and Receptor-Binding Domain (DRBD)

The delivery and receptor-binding domain (DRBD) comprises a central region between the CPD and the CROP domain of toxins A and B with two independent functions. It consists of a translocation domain and a C-terminally adjacent receptor-binding domain. The designation DRBD, first coined by Chen and colleagues (Chen et al. 2019a), takes novel findings into account, which are described in the following subchapters.

2.1.4.1 Translocation Domain (TD) of DRBD

During cellular uptake, toxins A and B are trapped in endosomes and presumably form pores, which allow the translocation of the GTD into the cytosol. A relatively large region, initially delimited between the CPD and the CROP domain of toxins A and B and denoted as translocation domain (TD), was initially suggested to be involved in these processes (Dove et al. 1990; von Eichel-Streiber et al. 1992a; Barroso et al. 1994). In 2011, Genisyuerek et al. aimed to more precisely narrow down the TD of toxin B and its pore-forming region (PFR). They found that a small segment reaching from amino acid residues 830–990 of toxin B is already sufficient for pore formation, at least in artificial lipid bilayers (Genisyuerek et al. 2011). Notably, this segment is partially consistent throughout with a small globular domain first described in the three-dimensional structure of toxin A (Chumbler et al. 2016). By a series of C-terminal deletions of toxin B that were fused to the receptor-binding domain of the diphtheria toxin (DTRD), Genisyuerek and colleagues then identified that amino acids 830–1550 of the toxin are sufficient for translocation of the enzyme portion into the cytosol (Genisyuerek et al. 2011). This finding gave a first hint that the region between amino acids 1551 and 1834 (start of the CROP domain) might not be part of the hitherto defined TD.

The TD of toxins A and B, which is found in all members of the clostridial glucosylating toxins family, seems to represent a conserved protein delivery apparatus, conserved in over 700 proteins from bacteria outside of clostridia (Orrell et al. 2020). Among those proteins is also the toxin YART (*Yersinia* ADP-ribosyltransferase) from *Yersinia mollaretii*, which, astonishingly, is capable of translocating an ADP-ribosyltransferase moiety into target cells (Ost et al. 2020). Thus, it appears that the TD is quite versatile and not only restricted to the delivery of a preceding glucosyltransferase moiety into cells. Despite the current progress in solving the 3D structures of toxins A and B (Chen et al. 2019a; Aminzadeh et al. 2022), the membrane-embedded structure of the TD of toxins A and B still remains enigmatic and there is no clue about the nature of the translocation pores formed by both toxins in endosomal membranes.

2.1.5 Receptor-Binding Domain of DRBD

When it became clear that the TD of toxins A and B is much shorter than previously assumed, the question remained about the function of the undescribed toxin segment between the newly delimited TD and the CROP domain. Successive discoveries suggested that this domain is involved in binding of toxins A and B to the cell surface (Gerhard 2017).

Already in 1994, Barroso et al. tested various C-terminally truncated toxin B variants in intoxication assays and found that removal of the CROP domain did not fully diminish cytotoxicity (Barroso et al. 1994). In this study, the authors did not use purified proteins but lysates from *E. coli* that expressed the various toxin B variants. Later, Frisch et al. observed that an N-terminally extended CROP domain of toxin A competitively inhibited intoxication of cells by toxin A more efficiently than the CROP domain alone (Frisch et al. 2003). Eventually, two German laboratories from Freiburg (Aktories and Papatheodorou) and Hanover (Just and Gerhard) confirmed in 2011 with purified recombinant proteins that the CROP domain is not absolutely required for binding and uptake of toxins A and B into host cells (Olling et al. 2011; Genisyuerek et al. 2011).

The concept of CROP-independent binding and uptake of toxins A and B was further supported by the identification of the homologous TpeL toxin from *C. perfringens*, which is naturally devoid of a CROP domain (Amimoto et al. 2007). The group of Papatheodorou and Aktories substantiated that the C-terminus of TpeL represents its receptor-binding domain by identifying the low-density lipoprotein receptor-related protein 1 (LRP1) as host receptor for TpeL and by showing direct binding between the TpeL C-terminus and an extracellular portion of LRP1 (Schorch et al. 2014). In the same study, the authors also proved independent cell surface-binding of a fragment of toxin B covering residues 1349–1811, which virtually corresponds to the proposed receptor-binding domain of TpeL. Furthermore, the authors were able to competitively inhibit cell binding of CROP-deficient toxin B by co-incubation with this fragment. These data argued strongly for a two-receptor model of toxins A and B, where the toxins independently bind host receptors via the CROP domain or the newly defined receptor-binding domain. Confusingly enough, experimental data from a work by Manse and Baldwin suggested at least three independent binding sites in toxin B (Manse and Baldwin 2015). Later, Lambert and Baldwin provided additional direct evidence for dual receptor-binding sites in toxin A (Lambert and Baldwin 2016).

Eventually, the recent progress in the identification of toxin B receptors that bind to the newly defined receptor-binding domain, together with newly obtained structural information about the toxin:receptor complexes (described in a following section of this chapter), constituted the strongest evidence for the existence of additional binding sites outside of the CROP domain. However, as will be discussed in a following subsection, these discoveries also question the concept of standalone receptor-binding domains within toxins A and B.

2.1.6 Structural Information on the Modular Composition of Toxins A and B

The multidomain architecture of toxins A and B had already become evident in earlier attempts to obtain low resolution structures of the holotoxins by small-angle X-ray scattering (SAXS) and negative stain electron microscopy, respectively (Albesa-Jové et al. 2010; Pruitt et al. 2010). In 2016, the group of Borden Lacy reported the long-sought crystal structure of toxin A. Despite the fact that the structure of toxin A obtained in this study did not include the CROP domain, it showed for the first time how the other domains are organized within the holotoxin. In addition, the structure included additional domains of toxin A whose structure had not been solved so far, such as the TD and the later discovered, second receptor-binding domain (Chumbler et al. 2016). Eventually, additional holotoxin 3D structures provided in more recent years for toxin A (Aminzadeh et al. 2022) and toxin B (Chen et al. 2019a) undoubtedly confirmed the modular composition of toxins A and B and completed our understanding about the tertiary structure of the toxins and the structural organization of their domains.

2.2 Binding and Uptake of *C. difficile* Toxins A and B

2.2.1 Host Receptors of Toxins A and B

In recent years, powerful genetic screens based on the CRISPR/Cas9 system were established that led to the discovery of yet unknown host receptors of toxins A and B.

Toxin A was known to interact with cell surface carbohydrate structures (Krivan et al. 1986; Clark et al. 1987; Tucker and Wilkins 1991) and with two proteins, namely the sucrase-isomaltase (Pothoulakis et al. 1996) and the glycoprotein gp96 (Na et al. 2008). However, more recent CRISPR/Cas9 screens with toxin A identified additional receptors, such as sulfated glycosaminoglycans and the low-density lipoprotein receptor (LDLR) (Tao et al. 2019). Interestingly, sulfated glycosaminoglycans and the LDLR were also found in a CRISPR/Cas9 screen for the identification of receptors of the toxin A-related toxin Tcnα (Zhou et al. 2021).

For toxin B, several hitherto unknown receptors were identified, such as CSPG4 (chondroitin sulphate proteoglycan-4) (Yuan et al. 2015), PVRL3 (poliovirus receptor-like 3) (LaFrance et al. 2015), members of the Wnt receptor frizzled (FZD) family (i.e., FZD2) (Tao et al. 2016), and the low-density lipoprotein receptor-related protein 1 (LRP1) (Guo et al. 2022). Interestingly, the 3D structure of a toxin B:FZD2 complex revealed that an endogenous FZD-bound fatty acid functions as a co-receptor for toxin B binding (Chen et al. 2018).

Whereas LRP1 binds to the CROP domain of toxin B, PVRL3 and frizzled proteins are considered to interact with distinct regions within the receptor-binding domain of the DRBD. Interestingly, LRP1 has also been identified as a host receptor of the related toxin TpeL (Schorch et al. 2014) and was recently discussed to be involved in the cellular uptake of toxin A (Schöttelndreier et al. 2020).

Initially, it was controversially discussed whether toxin B interacts with CSPG4 via the CROP domain (Tao et al. 2016) or via the receptor-binding domain of the DRBD (Yuan et al. 2015). Later, structural analyses of toxin B:CSPG4 complexes provided evidence that not a standalone receptor-binding domain of toxin B, but rather a groove formed by the CPD, the DRBD, a hinge region, and the CROP domain binds CSPG4 (Gupta et al. 2017; Chen et al. 2021; Jiang et al. 2022). Henkel and colleagues confirmed that binding of toxin B to CSPG4 and frizzled proteins is independent and additive (Henkel et al. 2020), which was in line with a recent structural analysis of a toxin B:CSPG4:FZD2 complex obtained by others (Jiang et al. 2022).

Consecutive studies have shown that certain toxin B variants from multi-locus sequence typing (MLST) clade 2 strains do not use frizzled proteins as host receptors (Chung et al. 2018; López-Ureña et al. 2019; Pan et al. 2021). Consequently, tissue factor pathway inhibitor (TFPI) was identified as receptor for certain toxin B variants from subtypes TcdB2, TcdB4, and TcdB7 (Luo et al. 2022; Tian et al. 2022). Structural and phylogenic analysis revealed that toxin B harbors a key receptor-binding region interacting with either TFPI or frizzled proteins (Luo et al. 2022; Tian et al. 2022). Interestingly, the same binding region shows a unique sequence in the toxin B-related toxin TcsL and promotes binding to semaphorins 6A and 6B (Tian et al. 2020; Lee et al. 2020).

2.2.2 Endocytic Pathways for the Cellular Uptake of Toxins A and B

Upon binding to a cell surface receptor, toxins A and B are taken up into host cells via receptor-mediated endocytosis. For many years, the exact endocytic pathway for the uptake of toxins A and B remained unclear. At first glance, Kushnaryov and Sedmak provided evidence for endocytosis of *C. difficile* toxin A via coated pits, by visualizing colloidal gold labeled toxin A in CHO cells by electron microscopy (Kushnaryov and Sedmak 1989). In 2010, Papatheodorou et al. aimed to study the endocytic uptake of toxins A and B in more detail by the use of pharmacological and genetic inhibitors of distinct endocytic pathways (Papatheodorou et al. 2010). Their findings indicated that the endocytic uptake of toxins A and B involves a dynamin-dependent process that is mainly governed by clathrin (Papatheodorou et al. 2010). Gerhard and colleagues confirmed that clathrin and dynamin are substantially involved in endocytosis of toxin A and toxin A_{1-1874} (lacking almost the entire CROP domain). However, as inhibition or knockdown of clathrin did not completely prevent uptake of toxin A and toxin A_{1-1874}, the authors suggested alternative endocytic routes for the toxin (Gerhard et al. 2013). Indeed, Chandrasekaran and colleagues later reported that the uptake of toxin A into CaCo-2 and MEF (mouse embryonic fibroblast) cells is clathrin-independent but requires dynamin and the Fer-CIP4 homology-BAR (F-BAR) domain-containing protein PACSIN2 (Chandrasekaran et al. 2016). A recent study proposed a novel feature of toxin B, which is associated with its endocytic uptake into cells, namely the inhibition of lysosomal activity. The authors claimed that especially phagocytic cells with high endocytotic activity might be affected by this toxin effect (Klepka et al. 2022).

2.2.3 Delivery of the GTD into the Cytosol

Toxins A and B are the so-called short-trip toxins, which deliver their enzymatic portion into the cytosol directly after reaching endosomal compartments via receptor-mediated endocytosis. The translocation of the GTD across the endosomal membrane is still the least understood step of the intoxication process of toxins A and B, respectively. This is mainly due to the lack of structural information of membrane-embedded conformations of the toxins, either prior, during, or directly after the translocation event. Acidification of endosomal vesicles by vacuolar H^+-ATPases triggers conformational changes within toxins A and B, leading to the exposure of hydrophobic segments responsible for the insertion of the toxins into the endosomal membrane (Qa'dan et al. 2000; Qa'Dan et al. 2001). Low pH-dependent pore formation of toxins A and B in cellular and artificial membranes was confirmed by the Aktories group (Qa'dan et al. 2000; Barth et al. 2001; Giesemann et al. 2006). Formation of a pore in the endosomal membrane by the toxins' TD might be an essential step for the delivery of the GTD into the cytosol. It is generally assumed that toxins A and B are able to form membrane pores as monomers and independent of host cell proteins. Pore formation of toxins A and B can be forced to occur also at the plasma membrane by artificially acidifying the extracellular medium of cultured cells (Qa'dan et al. 2000; Barth et al. 2001; Giesemann et al. 2006).

Giesemann et al. could show that the efficacy of pore formation by toxins A and B was dependent on membrane cholesterol (Giesemann et al. 2006). This finding was further substantiated by the discovery of Papatheodorou and colleagues that toxins A and B require an active and functional sterol regulatory element-binding protein 2 (SREBP-2) pathway, which regulates the cholesterol content in membranes, for efficient intoxication of target cells (Papatheodorou et al. 2019). The crucial role of membrane cholesterol for toxins A and B was further corroborated by protecting cells from both toxins after preincubation with the compound U18666A, an established inhibitor of cholesterol biosynthesis and/or intracellular transport (Papatheodorou et al. 2021). Membrane cholesterol presumably assists in the correct positioning of trans-membrane segments of toxins A and B within endosomal membranes for generating a functional translocation pore. The membrane cholesterol-dependence of toxins A and B represents an "Achilles' heel" and might be useful for therapeutic targeting of both toxins.

The GTD is not required for pore formation of toxins A and B at the plasma membrane or in artificial lipid bilayers (Barth et al. 2001; Genisyuerek et al. 2011). Black lipid bilayer experiments with purified toxins revealed that the pores formed by toxins A and B are more of a temporary nature and, presumably, less stable than pores formed by other classical pore-forming bacterial toxins (Barth et al. 2001; Genisyuerek et al. 2011). According to recent data on atomic force imaging of toxin B in a supported lipid bilayer, no "pre-pore" state is assumed for membrane insertion and pore formation of toxin B (Brander et al. 2019).

Most likely, the GTD of toxins A and B needs to be unfolded during the translocation process. However, it remains an open question as to how unfolding of the GTD is initiated and whether the unfolded GTD dips into the membrane pore via its N- or C-terminus. In addition, it is not clear if the GTD translocates across the membrane pore alone or together with the adjacent CPD. Of importance is the recent finding of Steinemann and colleagues that the activity of the toxins A and B depends on the chaperonin TRiC/CCT. The chaperonin subunits CCT4/5 directly interact and facilitate refolding of the GTD after translocation into the cytosol (Steinemann et al. 2018).

The pH-driven translocation of the GTD of toxin B into the cytosol can be pharmacologically prevented and thereby cells protected from intoxication, as demonstrated by the Barth group for the approved drug bacitracin (Zhu et al. 2019).

2.3 Mode-of-Action of Toxins A and B

2.3.1 Glucosylation-Dependent Effects and Consequences

C. difficile toxins A and B were the first toxins to be shown to modify target proteins by glycosylation (Just et al. 1995a, b). Meanwhile, it is clear that this type of post-translational modification is used by many toxins to interfere with eukaryotic cell functions, including various types of large clostridial glucosylating toxins (Just et al. 1996; Jank and Aktories 2008; Jank et al. 2015a) but also toxins from *Legionella* (Belyi et al. 2006), *Photorhabdus* (Jank et al. 2013), *Yersinia* (Jank et al. 2015b; Ost et al. 2020), and *E. coli* (EPEC) species (Li et al. 2013).

Toxins A and B catalyze the glucosylation of Rho GTPases by utilizing UDP-glucose as a co-substrate (Just et al. 1995a, b). Other related clostridial glycosyltransferases (e.g., *C. novyi* α-toxin and *C. perfringens* TpeL) prefer UDP-*N*-acetylglucosamine (UDP-GlcNAc) (Selzer et al. 1996; Nagahama et al. 2011; Guttenberg et al. 2012). Primary substrates of toxins A and B are RhoA,B,C, Rac1,2, and Cdc42 but also other isoforms of the Rho family such as TC10 and RhoG are modified. Secondary substrates are also some Ras proteins like Rap1,2, Ral, and Ras (Just and Gerhard 2004; Zeiser et al. 2013). However, according to Genth and colleagues, the (H/K/N) Ras isoforms seem to be target substrates only for toxin A and not for toxin B (Genth et al. 2018).

Rho proteins are 21–25 kDa GTP-binding proteins and members of the Ras superfamily. The ~20 Rho family members are switch proteins governed by a GTPase cycle and act as master regulators of the actin cytoskeleton and of numerous cellular processes, such as cell migration, phagocytosis and intracellular traffic, cell cycle progression and apoptosis (Nobes and Hall 1994; Burridge and Wennerberg 2004; Jaffe and Hall 2005; Aktories 2011; Lemichez and Aktories 2013). Rho proteins are inactive in the GDP (guanosine diphosphate)-bound state and become activated after nucleotide exchange and GTP (guanosine triphosphate)-binding (Bishop and Hall 2000; Cherfils and Zeghouf 2013). This GDP/GTP exchange is mediated by numerous guanine nucleotide exchange factors (GEFs) (García-Mata and Burridge 2007). Active Rho proteins interact with various effector proteins to elicit cellular functions (Bishop and Hall 2000; Burridge and Wennerberg 2004). This active state is terminated by GTP hydrolysis, which is stimulated by various GTPase-activating proteins (GAPs) (Tcherkezian and Lamarche-Vane 2007; Cherfils and Zeghouf 2013). Active GTP-bound Rho proteins are cell membrane associated, which is caused by N-terminal isoprenylation. Inactive, GDP-bound Rho proteins are extracted from membranes by GDIs (guanine nucleotide dissociation inhibitors) and are in a GDI-Rho complex in the cytosol.

C. difficile toxins glucosylate Rho proteins in threonine-37, and Rac and Cdc42 in threonine-35, which is the equivalent residue (Just et al. 1995a, b). This modification blocks the signal/switch functions of Rho proteins, because they are no longer able to interact with effectors. Glucosylation inhibits the activation of Rho GTPases by GEF proteins and completely blocks the interaction with GAPs (Sehr et al. 1998). Moreover, glucosylation fixes Rho proteins in their inactive conformation (Vetter et al. 2000; Geyer et al. 2003). Additionally, it was shown that glucosylated Rho proteins remain attached to the cell membrane and are not extracted from membranes by GDI proteins (Genth et al. 1999).

Of major interest are the first-of-their-kind co-crystal structures of the glucosyltransferase of two distinct toxin B variants in complex with human Cdc42 and R-Ras or of toxin A with RhoA, which provide a better understanding of the interaction of the toxins with their substrates. Structural comparisons between the toxin-substrate co-crystals revealed both the conserved and divergent features of toxin A and toxin B in terms of substrate recognition (Liu et al. 2021; Chen et al. 2022b). A recent study from Paparella and colleagues expanded our mechanistic knowledge about the glucosylation of Rho GTPases by toxin B, which seems to involve an S_Ni mechanism with a distinct oxocarbenium phosphate ion pair transition state (Paparella et al. 2022). The S_Ni mechanism has been already suggested by Reinert et al. (Reinert et al. 2005). Knowledge of the transition state structure of the glucosyltransferase reaction of toxin B will help in designing transition state analogs as putative small molecule inhibitors against the toxin, thus providing novel antitoxin-based therapy options against CDADs.

Other options to specifically inhibit the glucosyltransferase reaction by toxin B are the compounds castanospermine (Jank et al. 2008) or, as shown recently by our group, ambroxol (Heber et al. 2021). Formerly, Giesemann and colleagues have shown that the human α-defensins HNP-1, HNP-3, and enteric HD-5 are capable of preventing Rho protein glucosylation by toxin B by direct inhibition of the GTD (Giesemann et al. 2008). However, more recent studies from our group have shown that the α-defensins inhibit toxins A and B most likely also by direct interaction resulting in aggregation (Fischer et al. 2020; Korbmacher et al. 2020; Barthold et al. 2022).

Because glucosylation of Rho proteins blocks all functions of the switch proteins, *C. difficile* toxins A and B affect numerous cellular functions. Therefore, important questions are: How is the action of the toxins related to their pathophysiological effects? What kind of actions of toxins A and B result in diarrhea, inflammation, and enterocolitis. Which are the major symptoms of *C. difficile* infection?

Cytopathological effects of toxins A and B are characterized by gross changes in cell morphology, redistribution of the actin cytoskeleton, loss of stress fibers, and retraction of the cell body with remaining irregular cell extensions, a process, which was called arborization (Ottlinger and Lin 1988; Fiorentini et al. 1990; Malorni et al.

1990; Fiorentini and Thelestam 1991). All these effects can be referred to inhibition of Rho protein functions.

Especially, glucosylation of Rac appears to be essential for the cytopathic effects of toxins A and B (Halabi-Cabezon et al. 2008). The RacQ61L mutant, which is hardly modified by the toxins, prevents cytopathic effects. The toxins alter cell–cell contacts and cell adhesion, which also depend on Rho proteins, thereby barrier functions of enterocytes are disabled (Hecht et al. 1988, 1992; Nusrat et al. 1995, 2001). The functional consequences are paracellular fluxes as a consequence of alteration of tight junctions, which depend on Rho and actin (Nusrat et al. 1995; Hirase et al. 2001).

According to Petersen et al., Rac1 glucosylation by toxin B results in Cyclin D1 suppression and arrested G1-S transition, which might delay epithelial renewal and decrease the repair capacity of the colonic epithelium (Petersen et al. 2022). In addition, Rac1 glucosylation by toxin A has been shown to inhibit the Wnt/β-catenin signaling pathway that controls target genes essential for maintaining epithelial barrier function and epithelial cell repair after injury (Martins et al. 2020). Notably, toxin B inhibits the same pathway, but by direct binding to frizzled receptors and antagonizing the binding of their ligand Wnt (Tao et al. 2016; Chen et al. 2019b). Of interest is the recent finding that the inactivation of RhoA by toxins A and B leads to decreased YAP (Yes-associated protein)/TAZ (transcriptional coactivator with PDZ-binding motif) activity in colonic epithelial cells (Song et al. 2021). YAP and TAZ are transcriptional co-activators downstream of the Hippo pathway and implicated in the intestinal regeneration (Gregorieff et al. 2015). In line with these observations, Mileto and colleagues found that toxin B damages colonic stem cells and impairs epithelial repair and recovery from CDI (Mileto et al. 2020).

Toxins A and B were shown to induce apoptosis in several types of cells (Mahida et al. 1996; Fiorentini et al. 1998; Qa'Dan et al. 2002; Brito et al. 2002). Induction of apoptosis (at least at low and moderate toxin concentrations) essentially depends on the glucosyltransferase activity of the toxins (Brito et al. 2002; Gerhard et al. 2008). For toxin B, it was shown recently that the catenin family member plakoglobin and the cell death-related chromatin factor HMBGB1 are required for inducing apoptosis (Li et al. 2022). Fettucciari and colleagues have shown that toxin B is capable of inducing apoptosis in enteric glial cells by activating three different signaling pathways mediated by caspases, calpains, and cathepsin B (Fettucciari et al. 2022).

Ng and coworkers reported that toxins A and B induce inflammasome activation in an ASC (apoptosis-associated speck-like protein)-dependent manner, thereby causing the release of IL-1β (Ng et al. 2010). More recently, the group of Feng Shao showed that Pyrin, which is encoded by the Mediterranean fever gene MEFV, acts as an intracellular "sensor" for toxin-modified RhoA-dependent inflammasome activation (Xu et al. 2014). However, it is not clear whether Pyrin is specifically relevant only for immune cells but not for epithelial cells. Pyrin associates with the ASC adaptor protein thereby activating pro-caspase 1 (Lu and Wu 2015). Caspase-1 is a key enzyme to activate IL-1β and IL-18, the final common path of inflammasome activation. Inflammasome formation appears to be regulated by phosphorylation of Pyrin and binding to 14-3-3 proteins that keeps Pyrin in an inactive state (Gao et al. 2016). Moreover, it was reported that Pyrin is phosphorylated by Rho effector protein kinase N (PKN), resulting in binding to 14-3-3 proteins and inhibition of inflammasome activation (Park et al. 2016).

Toxin-induced activation and release of IL-1β can induce release of IL-6, interferon-γ (IFN-γ), and IL-8, respectively. IL-8 is a highly potent neutrophil attractant and its release from monocytes exposed to toxin A or toxin B was described quite early (Linevsky et al. 1997). This is in line with the strong neutrophil invasion into colon mucosa that occurs during *C. difficile* infection and which is probably essentially involved in mucosal damage (Mahida et al. 1996; Linevsky et al. 1997; Steiner et al. 1997; Warny et al. 2000; Ishida et al. 2004; Jafari et al. 2013). Neutrophils, in turn, are resistant to the toxin B-catalyzed glucosylation of RhoA and Rac1 (Chaves-Cordero et al. 2022).

Surprisingly, blocking autoprocessing of toxin B, either by mutagenesis or by chemical inhibition, significantly enhances its proinflammatory activities. It was shown in an animal model that a non-cleavable mutant of toxin B was significantly more potent in inducing proinflammatory cytokines in human colonic tissues and immune cells, when compared to the wild-type toxin (Zhang et al. 2018).

An additional recent finding is of interest, where it was shown that the pyrin inflammasome triggers pyroptosis (Russo et al. 2016). Pyroptosis is featured by cell swelling followed by cell lysis with massive release of cellular content that can induce strong inflammation (Miao et al. 2010; Jorgensen and Miao 2015).

Importantly, Pruss and Sonnenburg have shown that toxins A and B-induced inflammation increases aldose reductase expression in the host. This in turn leads to the release of host-produced sorbitol from damaged tissue, which serves as a diet-independent resource for *C. difficile* during inflammation (Pruss and Sonnenburg 2021). Of note is also the finding that the *C. difficile* toxins A and B induce the production of vascular endothelial growth factor A (VEGF-A) in colonocytes, leading to increased colonic vascular permeability and promoting disease pathogenesis (Huang et al. 2019). Other studies have shown that toxins A and/or B influence activities and/or levels of host proteins such as DRA (downregulated in adenoma) or Hsp27 (heat shock factor 27) and that these changes contribute to the diarrheal phenotype of CDI (Coffing et al. 2018; Yanda et al. 2020).

2.3.2 Glucosylation-Independent Effects of Toxins A and B

While the abovementioned toxin actions depend on the glucosyltransferase activity of toxins A and B, toxin effects have been described which reportedly occur with "glucosyltransferase-dead" toxins. For example, it has been reported that toxin-induced production of reactive oxygen species (ROS) participates in enteritis and necrosis caused by *C. difficile* toxins (Qiu et al. 1999; Donald et al. 2013; Farrow et al. 2013; Wohlan et al. 2014). However, these toxin effects occurred at very high concentrations of toxins (often 100–1000 times higher than that necessary for cytopathic effects). Therefore, the pathophysiological relevance was not clear. A recent study from Stieglitz and colleagues have shown that Ras is a central upstream regulator for GTD-independent necrosis in epithelial cells induced by toxin B (Stieglitz et al. 2022).

The first *C. difficile* mutant (630 strain) producing glucosyltransferase-defective toxin B was generated by the group of Roman Melnyk in 2020. They examined the mutant in vivo in a murine and hamster model and found that the GTD activity is indispensable for disease pathogenesis (Bilverstone et al. 2020). Later, Peritore-Galve et al. assessed glucosyltransferase-dependent and independent effects of toxins A and B, by infecting mice with *C. difficile* (BI/Nap1/027 strain) toxin mutants (Peritore-Galve et al. 2022). The authors confirmed epithelial damage through a GTD-independent process, but they also found that the inactivation of the GTD activity suppressed diarrhea and deleterious immune responses.

Domain architecture, 3D structure, uptake and mode-of-action of toxins A and B are depicted in Fig. 1.

2.4 Relative Importance of Toxins A and B in *Clostridioides difficile* Infection

Historically, symptoms of CDI were mainly attributed to the action of toxin A, due to the fact that only purified toxin A but not toxin B was able to cause disease symptoms in hamsters when applied intragastrically (Lyerly et al. 1988). However, *C. difficile* strains have been isolated from symptomatic patients that produce only toxin B (Lyerly et al. 1992; Kim et al. 2012). Thus, two previous studies from the laboratories of Nigel Minton (Nottingham, UK) and Julian Rood (Melbourne, Australia) have attempted to more precisely determine in the hamster infection model the in vivo relevance of toxins A and B. To

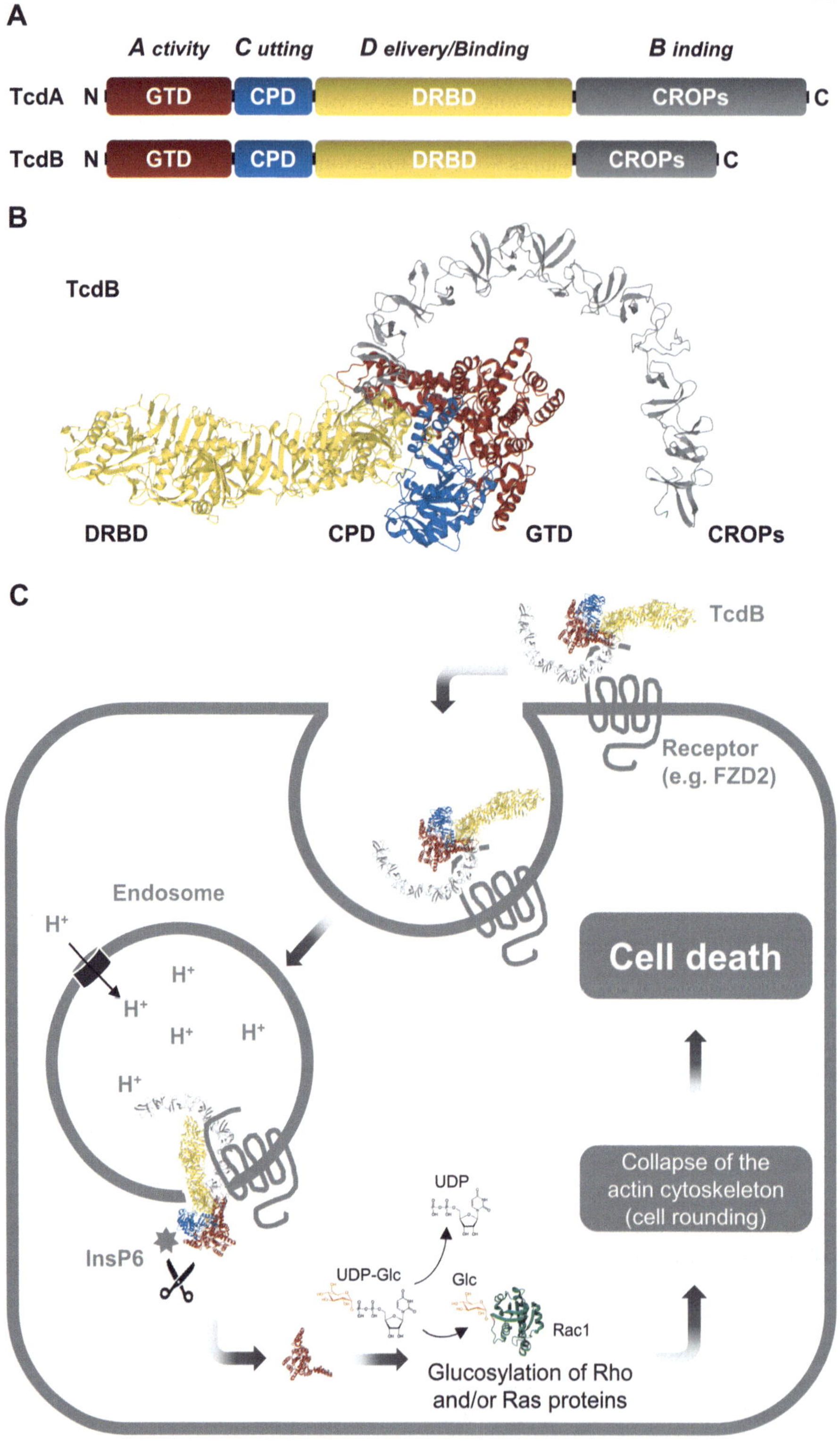

Fig. 1 Domain architecture, structure, uptake and mode-of-action of toxin A (TcdA) and toxin B (TcdB). (**a**) Domain architecture of TcdA and TcdB, (**b**) 3D structure of TcdB, and (**c**) model of the uptake and mode-of-action of TcdB is shown (details explained in the main text). GTD (red), glucosyltransferase domain; CPD (blue),

this end, both laboratories generated isogenic *C. difficile* mutants in the same strain (*C. difficile* 630) defective in the production of either toxin A or toxin B. Whereas both studies showed that toxin B alone causes disease symptoms in hamsters, contradictory results were obtained in terms of the importance of toxin A. Whereas a toxin B mutant created in the Rood group and which was capable of producing only toxin A did not cause disease in hamsters (Lyras et al. 2009), the equivalent mutant from the Minton group remained virulent (Kuehne et al. 2010). Compelling evidence has been provided recently by the Minton group that the reason for the observed contradiction resides in the use of two different erythromycin-sensitive derivatives of strain 630 for mutagenesis, which are genetically and phenotypically distinct. Unique Single Nucleotide Polymorphisms (SNPs) were identified in both strains that dramatically affected certain phenotypes, as well having marked effects on the transcriptome, which most likely impact on virulence (Collery et al. 2017). The isolation of a toxin A-positive, toxin B-negative *C. difficile* strain from a clinical case of CDI further supports the in vivo relevance of toxin A (Monot et al. 2015).

More recently, Lin et al. identified clinical pathogenic *C. difficile* strains (>5% of isolates) that produce high levels of toxin A but minimal or no toxin B, thus corroborating the significance of toxin A in CDI (Lin et al. 2020). Another group has shown for the first time that toxemia frequently occurs in patients suffering from CDI and that high serum levels especially of toxin A correlate with disease severity (Granata et al. 2021). However, although cecal histological damage was found in the hamster model after infection with a *C. difficile* strain producing only toxin A, clinical signs of infection or increased mortality were not observed (Marvaud et al. 2019). The findings from Marvaud and colleagues reveal that the hamster model might not be ideal for studying the in vivo effects of toxin A.

3 Structure, Uptake and Mode-of-Action of the Binary *C. difficile* Transferase (CDT)

3.1 Bipartite Composition of CDT

In contrast to toxins A and B, CDT is an AB-type binary toxin composed of a binding and translocation component (CDTb) and a separate enzyme component (CDTa). CDTb mediates binding to the host receptor LSR (lipolysis-stimulated lipoprotein receptor) at the cell surface, internalization of CDTa into endocytic vesicles, and pore formation in endosomes for the translocation of CDTa into the cytosol of host cells. Pore formation of CDTb is accomplished by oligomerization of CDTb into heptamers that are capable of integrating into the endosomal membrane. CDTa is an ADP-ribosyltransferase that is specific for monomeric G-actin.

3.1.1 The Binding Component of CDT

CDT is most similar to other clostridial binary toxins, such as *Clostridium perfringens* iota-toxin and *Clostridium spiroforme* toxin CST, and more distantly related to *Clostridium botulinum* C2 toxin. All those toxins are actin ADP-ribosyltransferases that resemble anthrax toxin of *Bacillus anthracis* with respect to their binding components. For instance, CDTb exhibits a 36% identity to protective antigen (PA), the binding component of anthrax toxin (Young and

Fig. 1 (continued) cysteine protease domain; DRBD (yellow), delivery and receptor-binding domain; CROPs (gray), combined repetitive oligopeptides; InsP6, inositol hexakisphosphate; UDP-Glc, UDP-glucose; Glc, glucose; UDP, uridine diphosphate; FZD2, frizzled-2. 3D structures of TcdB (PDB ID: 6OQ5; Chen et al. 2019a) and of Rac1 (PDB ID: 3TH5; Krauthammer et al. 2012) were generated with Mol* (Sehnal et al. 2021). Membrane-inserting structure of TcdB is fictitious and for representation only

Collier 2007). Much that we know about the structure-to-function relationship of CDTb was initially learned from previous extensive studies on the binding components of the anthrax toxin (PA) and, in part, the C2 toxin (C2II). From the already available structures of PA (Petosa et al. 1997; Schleberger et al. 2006; Egerer et al. 2007), it was possible to deduce that CDTb consists of four domains (I to IV) with distinct functions. Domain I at the N-terminus forms the activation domain and is followed by Domain II, which is involved in membrane insertion and pore formation. Domain III is responsible for pore formation and oligomerization. The C-terminal Domain IV corresponds to the receptor-binding domain of CDTb (Barth et al. 2004). Domain IV is highly similar among the binding components of CDT (CDTb), CST (CSTb), and iota-toxin (Ib). Interestingly, binding and enzymatic components are mutually interchangeable among CDT, CST, and iota-toxin, but not among the latter toxins and the C2 or anthrax toxin (Popoff and Boquet 1988; Considine and Simpson 1991).

CDTb is expressed as a precursor protein of 876 amino acids (~90 kDa) including an N-terminal signal peptide. Serine-type proteases activate the CDTb precursor by removal of a 20 kDa peptide from the N-terminus (Perelle et al. 1997). The activated binding component has as a size of ~75 kDa and is now able to form heptamers. It is unclear, whether the activation and oligomerization process occurs prior or after binding of the CDTb precursor to host cells (Gerding et al. 2014).

Nowadays, 3D structures of the CDTb pore are available (Anderson et al. 2020; Xu et al. 2020), also together with the enzyme component CDTa (Sheedlo et al. 2020; Kawamoto et al. 2022). The recent cryo-EM structures presented by Borden Lacy's group (Nashville, USA) strongly support a pre-pore-to-pore model for CDTb insertion and pore formation in endosomal membranes as a heptamer and reveal a previously unknown carbohydrate-binding domain (Anderson et al. 2020). This novel carbohydrate-binding domain was confirmed by Xu and colleagues and seems to include a calcium-binding site (Xu et al. 2020). Notably, in both studies, CDTb was found in a di-heptamer macromolecular assembly, stabilized by the Domain IV, but it remains an open question, whether such a di-heptamer has any biological relevance (Anderson et al. 2020; Xu et al. 2020).

In Domain IV of CDTb, which is separated by a long linker from the other domains, residues crucial for the interaction with the host receptor LSR were identified (Anderson et al. 2020), and the LSR-binding interface matches with an LSR-interacting epitope previously uncovered by our group with a transposon-based mutagenesis approach (Hemmasi et al. 2015).

One molecule CDTa is capable of interacting with the center of one CDTb heptamer already in the "preinsertion" state (Sheedlo et al. 2020), implying that endosomal acidification might stimulate CDTa translocation rather than CDTb pore formation. Recently, Kawamoto and colleagues confirmed that one CDTa molecule binds to the heptameric CDTb subunit and showed in exciting studies how CDTa binding induces the unfolding and tilting of the first N-terminal α-helix of CDTa, which is gripped by the CDTb pore after conformational changes to one of the constriction sites of the pore (Kawamoto et al. 2022).

3.1.2 The Enzyme Component of CDT

The enzyme component of CDT (CDTa) has a size of ~53 kDa and consists of 463 amino acids, including an N-terminal signal sequence of 43 amino acids, which is probably cleaved by proteolysis (Perelle et al. 1997). The mature CDTa finally has a size of ~48 kDa (420 amino acids) and is most similar to the enzyme components of iota-toxin (Ia; 84% sequence identity) and CST (CSTa; 82% sequence identity). CDTa has been crystallized and its structure revealed two domains with similar folding, which might originate from a duplication process of an ancient ADP-ribosyltransferase gene (Han et al. 1999; Sundriyal et al. 2009). Amino acids 1–215 of mature CDTa are probably involved in the interaction with CDTb, whereas amino acids 224-420 harbor the catalytically active ADP-ribosyltransferase portion. CDTa belongs to the R-S-E class of ADP-ribosyltransferases, which are characterized by the presence of a typical arginine residue (R), an STS motif (S), and an EXE motif (E) (Hottiger et al. 2010; Gerding et al. 2014). The enzyme components of the iota-toxin has been crystallized either in the presence of a stable NAD^+ analog (Tsuge et al. 2008) or in complex with actin (Tsurumura et al. 2013). Recently, NMR (nuclear magnetic resonance) assignments were reported for the CDTb-interacting and the active portion of CDTa (Roth et al. 2016a, b).

3.2 Binding and Uptake of CDT

3.2.1 The Lipolysis-Stimulated Lipoprotein Receptor

As for toxins A and B, binding to a specific structure at the cell surface of host cells is a prerequisite of the intoxication process of CDT. CDT belongs to the iota-like toxins, a subfamily of the family of clostridial, binary actin ADP-ribosylating toxins. It was already known from a previous study that iota-like toxins use a proteinaceous receptor for cell entry (Stiles et al. 2000). Eventually, in 2011, the LSR (lipolysis-stimulated lipoprotein receptor) was identified as host receptor for iota-like toxins with the help of a novel genetic screen (haploid genetic screen), which is based on the human haploid cell line Hap1 (Papatheodorou et al. 2011). Interestingly, it turned out that LSR is the host receptor also for the CDT-related *C. perfringens* iota-toxin and *C. spiroforme* toxin but not for the more distantly related *C. botulinum* C2 toxin (Papatheodorou et al. 2011, 2012). LSR acts as a lipoprotein receptor in the liver for the clearance of chylomicron remnants from the blood, but is also expressed in various other tissues, including the intestine (Yen et al. 1994, 1999; Mesli et al. 2004). Later studies identified a role of LSR in the formation of tricellular tight junctions (Masuda et al. 2011; Furuse et al. 2012; Czulkies et al. 2017). Another more recent study found that LSR is critically required for proper blood-brain barrier formation (Sohet et al. 2015). Eventually, several studies found a role of LSR in cancer progression and metastasis (Papatheodorou and Aktories 2017). As shown by us, an immunoglobulin (Ig)-like, V-type domain of LSR, present in its N-terminal, extracellular part, is bound by CDTb (Hemmasi et al. 2015).

3.2.2 Endocytic Pathways for the Cellular Uptake of CDT

Until now, the endocytic route of CDT (and other iota-like toxins) has not been entirely clarified. However, it was shown by the group of Michel Popoff (Paris, France) that dynamin, but not clathrin, is required for cellular uptake of iota-toxin (Gibert et al. 2011). In this study, colocalization of iota-toxin with the interleukin-2 receptor in endocytic vesicles was observed, indicating a similar endocytic route for both proteins (Gibert et al. 2011). The endocytic uptake of the interleukin-2 receptor is negatively regulated by RhoGDI (RhoGDP-dissociation inhibitor) (Lamaze et al. 2001). Strikingly, iota-toxin entry into Cos-1 cells was inhibited upon overexpression of RhoGDI (Gibert et al. 2011). Endocytic uptake of CDT and other iota-like toxins might involve lipid rafts, since oligomers of the binding components have been identified in detergent-resistant, cholesterol-rich membrane microdomains (Hale et al. 2004; Nagahama et al. 2004). In line with these reports, Papatheodorou and colleagues observed clustering of LSR into lipid rafts after binding of CDTb (Papatheodorou et al. 2013). LSR-clustering into lipid rafts occurred also after binding of the RBD of CDTb, which is not able to oligomerize by itself (Papatheodorou et al. 2013). Wigelsworth et al. found that the lipid rafts-protein CD44 (cluster of differentiation 44) is required for cellular uptake of CDT (Wigelsworth et al. 2012). Interestingly, CD44 was found in lipid rafts from Ib-treated Vero cells (Blonder et al. 2005). It might be possible that CD44 interacts with LSR-CDT complexes in lipid rafts, thus facilitating the endocytic uptake of the toxin.

3.2.3 Role of Chaperones During the Cellular Uptake of CDT

The delivery of CDTa into the host cell cytosol depends on CDTb, which under acidic conditions likely forms pores in endosomal membranes that serve as translocation channels for the transmembrane transport of CDTa (Roeder et al. 2014; Ernst et al. 2016). The Barth group found that pH-driven transport of CDTa across endosomal membranes requires the activities of certain host cell chaperones (Roeder et al. 2014). In vitro, CDTa directly and specifically binds to the heat shock proteins Hsp90 and Hsp70, as well as to some peptidyl-prolyl cis/trans isomerases (PPIases) of the cyclophilin (Cyp) and FK506 binding protein (FKBP) families (Kaiser et al. 2011; Ernst et al. 2015, 2017). The current model suggests that these host cell factors specifically and selectively facilitate the intracellular trans-membrane transport of ADP-ribosylating toxins by interacting with the ADP-ribosyltransferase domain of the A subunits (Ernst 2022). These findings were mainly obtained by Katharina Ernst from the laboratory of Holger Barth (Ulm, Germany) and contribute to a better understanding of the cellular uptake of CDT into human cells and to the development of novel pharmacological strategies against infections with hypervirulent, CDT-producing *C. difficile* strains (Ernst et al. 2021b).

3.3 Mode-of-Action of CDT

CDT ADP-ribosylates monomeric G-actin in arginine-177. Thus, modification of actin occurs at the same residue of actin that is also modified by other binary actin-ADP-ribosylating toxins, including *C. botulinum* C2 toxin and *C. perfringens* iota-toxin (Vandekerckhove et al. 1987, 1988). In contrast to monomeric G-actin, polymerized F-actin is not a substrate of CDT and of any other related binary toxin, because arginine-177 is not available for modification in the double helix of F-actin (Holmes et al. 1990; Margarit et al. 2006). Essential for actin functions is the ability of the microfilament protein to reversibly polymerize from G- to F-actin, a process that is tightly regulated by numerous actin binding proteins (Dominguez and Holmes 2011). Early studies obtained with *C. botulinum* C2 toxin and *C. perfringens* iota-toxin showed that modification of actin in arginine-177 inhibits actin polymerization (Aktories et al. 1986; Schering et al. 1988). This holds also true for CDT-induced ADP-ribosylation of actin. Moreover, all previous results obtained with other types of binary actin-ADP-ribosylating toxins that modify arginine-177 of actin can be reliably referred to the action of CDT. This includes the early finding that ADP-ribosylated actin binds to plus ends of F-actin filaments and acts as a capping protein to block F-actin elongation by inhibition of the binding of non-ADP-ribosylated actin (Wegner and Aktories 1988; Aktories and Wegner 1989; Weigt et al. 1989; Perieteanu et al. 2010). Also, the interaction of actin with actin binding proteins (for example, gelsolin) that is largely affected by toxin-induced ADP-ribosylation (Wille et al. 1992), is similarly relevant for CDT.

Binary toxin-induced F-actin depolymerization has typical cytotoxic effects in cell culture (Wiegers et al. 1991), resulting in rounding-up of cells and loss of cell adherence followed by apoptosis (Heine et al. 2008). Notably, not only the actin cytoskeleton but also microtubules are affected by binary actin-depolymerizing toxins. CDT and other actin-depolymerizing toxins induce long microtubule-based protrusions (Schwan et al. 2009). These cell membrane protrusions form a network of long tentacle-like structures on the surface of epithelial cells. Microtubule-based protrusions are dynamic structures. They grow and retract. CDT-induced depolymerization causes the mislocalization of capture proteins like ACF7 and Clasp2, which are involved in stabilization of growing microtubules at the actin cell cortex (Kodama et al. 2003; Drabek et al. 2006). Without appropriate capture proteins at the cell membrane, microtubule growth is no longer stopped, resulting in protrusion formation (Schwan et al. 2009). More recent studies indicate that septins, which are GTP-binding proteins that can reversibly oligomerize (Mostowy and Cossart 2012), are crucially involved in toxin-induced protrusion formation (Nölke et al. 2016). Moreover, these findings also show that septin-dependent protrusion formation is regulated by the Rho protein family member Cdc42 and its effectors Borg (binder of Rho GTPases) (Nölke et al. 2016).

CDT-induced partial depolymerization of F-actin disturbs re-cycling of vesicles at the basolateral side of epithelial cells. Thereby, the vesicles, which contain extracellular matrix (ECM) proteins like fibronectin and vitronectin, are re-routed from the basolateral side to the apical membrane, where microtubules form protrusions. Here, fibronectin and other ECM proteins are released (Schwan et al. 2014).

Recently, Schwan and colleagues identified a novel target substrate of CDT, namely the actin-related protein Arp2 (Schwan et al. 2022). This host protein is a constituent of the Arp2/3 complex, which regulates the formation of branched actin filament networks (Lai et al. 2008). ADP-ribosylation of Arp2 by CDT inhibited the function of Arp2 and thus the cellular actions of the Arp2/3 complex, a process which might contribute to the CDT-induced destabilization of the actin cytoskeleton of target cells (Schwan et al. 2022). However, it remains unclear whether Arp2-modification by CDT has any relevance in the pathophysiology of CDI.

Recently, the Barth group discovered that CDTb is cytotoxic to cultured cells in the absence of CDTa, most likely due to pore formation in the plasma membrane (Landenberger et al. 2021). Intoxication of cells by CDTb is prevented by pore blockers, such as chloroquine and derivatives and does not occur in cells lacking the LSR receptor (Ernst et al. 2021a; Landenberger et al. 2021). When Simpson and colleagues investigated the pathophysiological role of CDTb in vivo with a *C. difficile* strain only expressing CDTb, CDTb alone did not induce significant disease in mice while increased disease severity and mortality were observed in hamsters infected with the CDTb-expressing *C. difficile* strain (Simpson et al. 2023).

The current model of CDT's uptake and mode-of-action is depicted in Fig. 2.

3.4 Role of CDT During *C. difficile* Infection

Although CDT is a very potent and efficient cytotoxin, its role in *C. difficile* infection is not well understood. Only in extremely few cases, *C. difficile*-dependent enterocolitis could be traced back to CDT in the absence of *C. difficile* toxins A and B. What is then its role in disease? The group of Nigel Minton (University of Nottingham, UK) assessed the virulence of all possible combinations of isogenic *C. difficile* toxin mutants in the hamster infection model and found that CDT is a factor that increases the virulence of *C. difficile* in the presence of toxins A and B (Kuehne et al. 2014). Marvaud and colleagues observed cecal alterations in hamsters infected with a *C. difficile* strain producing only CDT (Marvaud et al. 2019). Later, Mark Wilcox's group (University of Leeds, UK) analyzed data from >1000 CDI patients and found that CDT is significantly associated with the risk of all-cause mortality (Berry et al. 2017). Eventually, William Petri's group retrospectively tested stool from 215 CDI patients and consistently observed increased disease severity and worse clinical outcomes in CDT-positive patients (Young et al. 2022).

Several mechanisms are discussed for CDT's contribution in CDI pathogenesis. CDT may increase the adherence of bacteria due to the formation of a network of microtubule-based protrusions, which facilitates adherence of *C. difficile* bacteria (Schwan et al. 2009). In the same direction points the finding that CDT causes the redistribution of fibronectin from the basolateral membrane of epithelial cells to the apical side, where it acts as a receptor for *C. difficile* (Schwan et al. 2014). CDT-induced fibronectin-redistribution was recently confirmed in a *Drosophila* line expressing the enzymatically active CDT subunit CDTa (Schwartz et al. 2020).

Moreover, it is of interest that CDT was shown to efficiently induce apoptosis of protective colonic eosinophils in a TLR2-dependent manner (Cowardin et al. 2016) and Lothar Jänsch's group

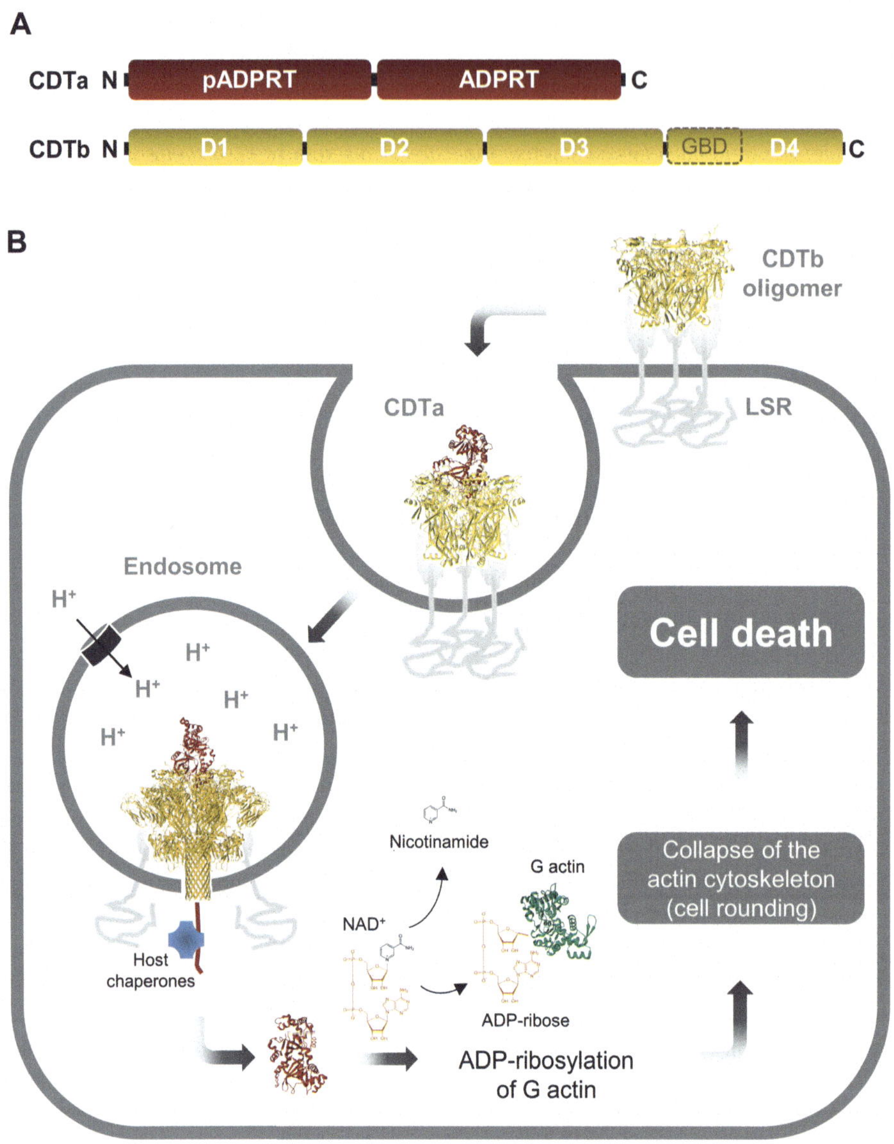

Fig. 2 Domain architecture, uptake and mode-of-action of CDT. (**a**) Domain architecture of CDTa and CDTb, and (**b**) model of the uptake and mode-of-action of CDT is shown (details explained in the main text). pADPRT, pseudo-ADP-ribosyltransferase; ADPRT, ADP-ribosyltransferase; D1, domain; D2, domain 2; D3, domain 3; D4, domain 4; GBD, glycan-binding domain; LSR, lipolysis-stimulated lipoprotein receptor. 3D structures of the CDTb oligomer in the pre-pore conformation in complex or without CDTa (PDB ID: 6V1S; Sheedlo et al. 2020), of the CDTb pore with long stem in complex with CDTa (PDB ID: 7VNN, Kawamoto et al. 2022), of CDTa (PDB ID: 2WN4; Sundriyal et al. 2009), and of G-actin (PDB ID: 2HF3; Rould et al. 2006) were generated with Mol* (Sehnal et al. 2021)

(Helmholtz Centre for Infection Research, Braunschweig, Germany) recently showed that CDT activates human mucosal-associated invariant T (MAIT) cells leading to cell degranulation of the lytic granule components perforin and granzyme B (Marquardt et al. 2021). Noteworthy, the binding component CDTb was sufficient for these effects on MAIT cells.

A phosphoproteomic analysis with epithelial Hep-2 cells revealed that nearly 1100 "phosphosites" responded to CDT treatment, but with moderate changes on proteome level (Stieglitz et al. 2021). This work also proved for the first time ADP-ribosylation of actin on arginine-177 on a proteomic level. It should be considered that actin and microtubules play a crucial role in activation of the inflammasome (Gao et al. 2016). Also, this could be an important functional connection eventually leading to increase in virulence of *C. difficile* in the presence of CDT, and toxins A and B.

4 Conclusions

It is well-accepted that *C. difficile* diseases are mainly governed by the production of protein toxins, including *C. difficile* toxins A (TcdA) and B (TcdB). More evidence has accumulated in recent years, that the third toxin, CDT, contributes to the pathophysiology of *C. difficile* infections. Therefore, recent progress in our knowledge especially about the uptake and mode of actions of these toxins is key for the understanding of the pathophysiology of *C. difficile* infections and the development of novel therapeutic strategies against the diseases caused by the pathogen. Even though great strides have been made in recent years, especially in identifying toxin receptors and solving toxin structures, many open questions remain. In respect to toxin A and toxin B, the membrane translocation of these toxins into target cells is still largely enigmatic. Moreover, *C. difficile* enterocolitis is characterized by severe inflammation and cell necrosis. The precise pathophysiological pathways caused by the toxins leading to inflammation and necrosis are still not satisfactorily understood and explained. The great success of fecal transplantation in therapy of *C. difficile* diseases indicates that the microbiome is crucially involved in the pathogenesis of *C. difficile* infections. This also indicates a pivotal role of the immune system of the host. Therefore, the actions of *C. difficile* toxins on various types of immune cells in context of intestinal tissue should be studied in detail.

References

Aktories K (2011) Bacterial protein toxins that modify host regulatory GTPases. Nat Rev Microbiol 9:487–498. https://doi.org/10.1038/NRMICRO2592

Aktories K, Wegner A (1989) ADP-ribosylation of actin by clostridial toxins. J Cell Biol 109:1385–1387. https://doi.org/10.1083/jcb.109.4.1385

Aktories K, Bärmann M, Ohishi I et al (1986) Botulinum C2 toxin ADP-ribosylates actin. Nature 322:390–392. https://doi.org/10.1038/322390a0

Aktories K, Schwan C, Jank T (2017) Clostridium difficile toxin biology. Annu Rev Microbiol 71:281–307. https://doi.org/10.1146/annurev-micro-090816-093458

Albesa-Jové D, Bertrand T, Carpenter EP et al (2010) Four distinct structural domains in Clostridium difficile toxin B visualized using SAXS. J Mol Biol 396: 1260–1270. https://doi.org/10.1016/J.JMB.2010.01.012

Alvin JW, Lacy DB (2017) Clostridium difficile toxin glucosyltransferase domains in complex with a non-hydrolyzable UDP-glucose analogue. J Struct Biol 198:203–209. https://doi.org/10.1016/j.jsb.2017.04.006

Amimoto K, Noro T, Oishi E, Shimizu M (2007) A novel toxin homologous to large clostridial cytotoxins found in culture supernatant of Clostridium perfringens type C. Microbiology 153:1198–1206. https://doi.org/10.1099/MIC.0.2006/002287-0

Aminzadeh A, Larsen CE, Boesen T, Jørgensen R (2022) High-resolution structure of native toxin A from Clostridioides difficile. EMBO Rep 23. https://doi.org/10.15252/embr.202153597

Anderson DM, Sheedlo MJ, Jensen JL, Lacy DB (2020) Structural insights into the transition of Clostridioides difficile binary toxin from prepore to pore. Nat Microbiol 5:102–107. https://doi.org/10.1038/s41564-019-0601-8

Barroso LA, Moncrief JS, Lyerly DM, Wilkins TD (1994) Mutagenesis of the Clostridium difficile toxin B gene and effect on cytotoxic activity. Microb Pathog 16: 297–303. https://doi.org/10.1006/mpat.1994.1030

Barth H, Pfeifer G, Hofmann F et al (2001) Low ph-induced formation of ion channels by Clostridium difficile toxin B in target cells. J Biol Chem 276: 10670–10676. https://doi.org/10.1074/JBC.M009445200

Barth H, Aktories K, Popoff MR, Stiles BG (2004) Binary bacterial toxins: biochemistry, biology, and applications of common Clostridium and Bacillus proteins. Microbiol Mol Biol Rev 68:373–402. https://doi.org/10.1128/MMBR.68.3.373-402.2004

Barthold L, Heber S, Schmidt CQ et al (2022) Human α-defensin-6 neutralizes Clostridioides difficile toxins TcdA and TcdB by direct binding. Int J Mol Sci 23. https://doi.org/10.3390/ijms23094509

Belyi Y, Niggeweg R, Opitz B et al (2006) Legionella pneumophila glucosyltransferase inhibits host elongation factor 1A. Proc Natl Acad Sci U S A 103:16953–16958. https://doi.org/10.1073/PNAS.0601562103

Berry CE, Davies KA, Owens DW, Wilcox MH (2017) Is there a relationship between the presence of the binary toxin genes in Clostridium difficile strains and the severity of C. difficile infection (CDI)? Eur J Clin Microbiol Infect Dis 36:2405–2415. https://doi.org/10.1007/s10096-017-3075-8

Bilverstone TW, Garland M, Cave RJ et al (2020) The glucosyltransferase activity of C. difficile toxin b is required for disease pathogenesis. PLoS Pathog 16. https://doi.org/10.1371/journal.ppat.1008852

Bishop AL, Hall A (2000) Rho GTPases and their effector proteins. Biochem J 348(Pt 2):241–255. https://doi.org/10.1042/0264-6021:3480241

Blonder J, Hale ML, Chan KC et al (2005) Quantitative profiling of the detergent-resistant membrane proteome of Iota-b toxin induced Vero cells. J Proteome Res 4: 523–531. https://doi.org/10.1021/PR049790S

Brander S, Jank T, Hugel T (2019) AFM imaging suggests receptor-free penetration of lipid bilayers by toxins. Langmuir 35:365–371. https://doi.org/10.1021/acs.langmuir.8b03146

Brito GAC, Fujji J, Carneiro-Filho BA et al (2002) Mechanism of Clostridium difficile toxin A-induced apoptosis in T84 cells. J Infect Dis 186:1438–1447. https://doi.org/10.1086/344729

Burridge K, Wennerberg K (2004) Rho and Rac take center stage. Cell 116:167–179. https://doi.org/10.1016/S0092-8674(04)00003-0

Busch C, Hofmann F, Gerhard R, Aktories K (2000) Involvement of a conserved tryptophan residue in the UDP-glucose binding of large clostridial cytotoxin glycosyltransferases. J Biol Chem 275:13228–13234. https://doi.org/10.1074/JBC.275.18.13228

Chandrasekaran R, Kenworthy AK, Lacy DB (2016) Clostridium difficile toxin A undergoes clathrin-independent, PACSIN2-dependent endocytosis. PLoS Pathog 12. https://doi.org/10.1371/JOURNAL.PPAT.1006070

Chaves-Cordero C, Quesada-Gómez C, Chaves-Olarte E, Barquero-Calvo E (2022) Human neutrophils are resistant to Clostridioides difficile toxin B. Anaerobe 74. https://doi.org/10.1016/j.anaerobe.2022.102553

Chen P, Tao L, Wang T et al (2018) Structural basis for recognition of frizzled proteins by clostridium difficile toxin B. Science 360:664–669. https://doi.org/10.1126/science.aar1999

Chen P, Lam K-h, Liu Z et al (2019a) Structure of the full-length Clostridium difficile toxin B. Nat Struct Mol Biol 26:712–719. https://doi.org/10.1038/s41594-019-0268-0

Chen P, Tao L, Liu Z et al (2019b) Structural insight into Wnt signaling inhibition by Clostridium difficile toxin B. FEBS J 286:874–881. https://doi.org/10.1111/febs.14681

Chen P, Zeng J, Liu Z et al (2021) Structural basis for CSPG4 as a receptor for TcdB and a therapeutic target in Clostridioides difficile infection. Nat Commun 12. https://doi.org/10.1038/s41467-021-23878-3

Chen B, Basak S, Chen P et al (2022a) Structure and conformational dynamics of Clostridioides difficile toxin A. Life Sci Alliance 5. https://doi.org/10.26508/LSA.202201383

Chen B, Liu Z, Perry K, Jin R (2022b) Structure of the glucosyltransferase domain of TcdA in complex with RhoA provides insights into substrate recognition. Sci Rep 12. https://doi.org/10.1038/s41598-022-12909-8

Cherfils J, Zeghouf M (2013) Regulation of small GTPases by GEFs, GAPs, and GDIs. Physiol Rev 93: 269–309. https://doi.org/10.1152/PHYSREV.00003.2012

Chumbler NM, Rutherford SA, Zhang Z et al (2016) Crystal structure of Clostridium difficile toxin A. Nat Microbiol 1. https://doi.org/10.1038/NMICROBIOL.2015.2

Chung S-Y, Schöttelndreier D, Tatge H et al (2018) The conserved Cys-2232 in Clostridioides difficile toxin B modulates receptor binding. Front Microbiol 9:2314. https://doi.org/10.3389/fmicb.2018.02314

Clark GF, Krivan HC, Wilkins TD, Smith DF (1987) Toxin A from Clostridium difficile binds to rabbit erythrocyte glycolipids with terminal Gal alpha 1-3Gal beta 1-4GlcNAc sequences. Arch Biochem Biophys 257:217–229. https://doi.org/10.1016/0003-9861(87)90561-3

Coffing H, Priyamvada S, Anbazhagan AN et al (2018) Clostridium difficile toxins a and b decrease intestinal SLC26A3 protein expression. Am J Physiol Gastrointest Liver Physiol 315:G43–G52. https://doi.org/10.1152/ajpgi.00307.2017

Collery MM, Kuehne SA, McBride SM et al (2017) What's a SNP between friends: The influence of single nucleotide polymorphisms on virulence and phenotypes of clostridium difficile strain 630 and derivatives. Virulence 8:767–781. https://doi.org/10.1080/21505594.2016.1237333

Considine RV, Simpson LL (1991) Cellular and molecular actions of binary toxins possessing ADP-ribosyltransferase activity. Toxicon 29:913–936. https://doi.org/10.1016/0041-0101(91)90076-4

Cowardin CA, Buonomo EL, Saleh MM et al (2016) The binary toxin CDT enhances Clostridium difficile virulence by suppressing protective colonic eosinophilia. Nat Microbiol 1. https://doi.org/10.1038/NMICROBIOL.2016.108

Czulkies BA, Mastroianni J, Lutz L et al (2017) Loss of LSR affects epithelial barrier integrity and tumor xenograft growth of CaCo-2 cells. Oncotarget 8:37009–37022. https://doi.org/10.18632/ONCOTARGET.10425

Di Masi A, Leboffe L, Polticelli F et al (2018) Human serum albumin is an essential component of the host defense mechanism against clostridium difficile intoxication. J Infect Dis 218:1424–1435. https://doi.org/10.1093/infdis/jiy338

Dingle T, Wee S, Mulvey GI et al (2008) Functional properties of the carboxy-terminal host cell-binding domains of the two toxins, TcdA and TcdB, expressed by Clostridium difficile. Glycobiology 18:698–706. https://doi.org/10.1093/GLYCOB/CWN048

Dominguez R, Holmes KC (2011) Actin structure and function. Annu Rev Biophys 40:169–186. https://doi.org/10.1146/ANNUREV-BIOPHYS-042910-155359

Donald RGK, Flint M, Kalyan N et al (2013) A novel approach to generate a recombinant toxoid vaccine against Clostridium difficile. Microbiology (United Kingdom) 159:1254–1266. https://doi.org/10.1099/MIC.0.066712-0

Dove CH, Wang SZ, Price SB et al (1990) Molecular characterization of the Clostridium difficile toxin A gene. Infect Immun 58:480–488. https://doi.org/10.1128/iai.58.2.480-488.1990

Drabek K, van Ham M, Stepanova T et al (2006) Role of CLASP2 in microtubule stabilization and the regulation of persistent motility. Curr Biol 16:2259–2264. https://doi.org/10.1016/J.CUB.2006.09.065

D'Urzo N, Malito E, Biancucci M et al (2012) The structure of Clostridium difficile toxin A glucosyltransferase domain bound to Mn2+ and UDP provides insights into glucosyltransferase activity and product release. FEBS J 279:3085–3097. https://doi.org/10.1111/J.1742-4658.2012.08688.X

Egerer M, Giesemann T, Jank T et al (2007) Auto-catalytic cleavage of Clostridium difficile toxins A and B depends on cysteine protease activity. J Biol Chem 282:25314–25321. https://doi.org/10.1074/JBC.M703062200

Egerer M, Giesemann T, Herrmann C, Aktorles K (2009) Autocatalytic processing of Clostridium difficile toxin B: binding of inositol hexakisphosphate. J Biol Chem 284:3389–3395. https://doi.org/10.1074/JBC.M806002200

Ernst K (2022) Requirement of peptidyl-prolyl Cis/trans isomerases and chaperones for cellular uptake of bacterial AB-type toxins. Front Cell Infect Microbiol 12. https://doi.org/10.3389/fcimb.2022.938015

Ernst K, Langer S, Kaiser E et al (2015) Cyclophilin-facilitated membrane translocation as pharmacological target to prevent intoxication of mammalian cells by binary clostridial actin ADP-ribosylated toxins. J Mol Biol 427:1224–1238. https://doi.org/10.1016/J.JMB.2014.07.013

Ernst K, Liebscher M, Mathea S et al (2016) A novel Hsp70 inhibitor prevents cell intoxication with the actin ADP-ribosylating Clostridium perfringens iota toxin. Sci Rep 6. https://doi.org/10.1038/SREP20301

Ernst K, Schmid J, Beck M et al (2017) Hsp70 facilitates trans-membrane transport of bacterial ADP-ribosylating toxins into the cytosol of mammalian cells. Sci Rep 7. https://doi.org/10.1038/s41598-017-02882-y

Ernst K, Landenberger M, Nieland J et al (2021a) Characterization and pharmacological inhibition of the pore-forming Clostridioides difficile cdtb toxin. Toxins (Basel) 13. https://doi.org/10.3390/toxins13060390

Ernst K, Sailer J, Braune M, Barth H (2021b) Intoxication of mammalian cells with binary clostridial enterotoxins is inhibited by the combination of pharmacological chaperone inhibitors. Naunyn Schmiedebergs Arch Pharmacol 394:941–954. https://doi.org/10.1007/s00210-020-02029-3

Farrow MA, Chumbler NM, Lapierre LA et al (2013) Clostridium difficile toxin B-induced necrosis is mediated by the host epithelial cell NADPH oxidase complex. Proc Natl Acad Sci U S A 110:18674–18679. https://doi.org/10.1073/PNAS.1313658110

Fettucciari K, Marguerie F, Fruganti A et al (2022) Clostridioides difficile toxin B alone and with pro-inflammatory cytokines induces apoptosis in enteric glial cells by activating three different signalling pathways mediated by caspases, calpains and cathepsin B. Cell Mol Life Sci 79. https://doi.org/10.1007/s00018-022-04459-z

Fiorentini C, Thelestam M (1991) Clostridium difficile toxin A and its effects on cells. Toxicon 29:543–567. https://doi.org/10.1016/0041-0101(91)90050-2

Fiorentini C, Malorni W, Paradisi S et al (1990) Interaction of Clostridium difficile toxin A with cultured cells: cytoskeletal changes and nuclear polarization. Infect Immun 58:2329–2336. https://doi.org/10.1128/IAI.58.7.2329-2336.1990

Fiorentini C, Fabbri A, Falzano L et al (1998) Clostridium difficile toxin B induces apoptosis in intestinal cultured cells. Infect Immun 66:2660–2665. https://doi.org/10.1128/iai.66.6.2660-2665.1998

Fischer S, Ückert AK, Landenberger M et al (2020) Human peptide α-defensin-1 interferes with Clostridioides difficile toxins TcdA, TcdB, and CDT. FASEB J 34:6244–6261. https://doi.org/10.1096/fj.201902816R

Frey SM, Wilkins TD (1992) Localization of two epitopes recognized by monoclonal antibody PCG-4 on Clostridium difficile toxin A. Infect Immun 60:2488–2492. https://doi.org/10.1128/iai.60.6.2488-2492.1992

Frisch C, Gerhard R, Aktories K et al (2003) The complete receptor-binding domain of Clostridium difficile toxin A is required for endocytosis. Biochem Biophys Res Commun 300:706–711. https://doi.org/10.1016/S0006-291X(02)02919-4

Furuse M, Oda Y, Higashi T et al (2012) Lipolysis-stimulated lipoprotein receptor: A novel membrane protein of tricellular tight junctions. Ann N Y Acad Sci 1257:54 58. https://doi.org/10.1111/J.1749-6632.2012.06486.X

Gao W, Yang J, Liu W et al (2016) Site-specific phosphorylation and microtubule dynamics control Pyrin inflammasome activation. Proc Natl Acad Sci U S A

113:E4857–E4866. https://doi.org/10.1073/PNAS.1601700113

García-Mata R, Burridge K (2007) Catching a GEF by its tail. Trends Cell Biol 17:36–43. https://doi.org/10.1016/J.TCB.2006.11.004

Geissler B, Tungekar R, Satchell KJF (2010) Identification of a conserved membrane localization domain within numerous large bacterial protein toxins. Proc Natl Acad Sci U S A 107:5581–5586. https://doi.org/10.1073/PNAS.0908700107

Genisyuerek S, Papatheodorou P, Guttenberg G et al (2011) Structural determinants for membrane insertion, pore formation and translocation of Clostridium difficile toxin B. Mol Microbiol 79:1643–1654. https://doi.org/10.1111/j.1365-2958.2011.07549.x

Genth H, Aktories K, Just I (1999) Monoglucosylation of RhoA at threonine 37 blocks cytosol-membrane cycling. J Biol Chem 274:29050–29056. https://doi.org/10.1074/JBC.274.41.29050

Genth H, Junemann J, Lämmerhirt CM et al (2018) Difference in mono-O-glucosylation of Ras subtype GTPases between toxin A and toxin B from Clostridioides difficile strain 10463 and lethal toxin from Clostridium sordellii strain 6018. Front Microbiol 9:3078. https://doi.org/10.3389/fmicb.2018.03078

Gerding DN, Johnson S, Rupnik M, Aktories K (2014) Clostridium difficile binary toxin CDT: mechanism, epidemiology, and potential clinical importance. Gut Microbes 5:15–27. https://doi.org/10.4161/GMIC.26854

Gerhard R (2017) Receptors and binding structures for clostridium difficile toxins A and B. Curr Top Microbiol Immunol 406:79–96. https://doi.org/10.1007/82_2016_17

Gerhard R, Nottrott S, Schoentaube J et al (2008) Glucosylation of Rho GTPases by Clostridium difficile toxin A triggers apoptosis in intestinal epithelial cells. J Med Microbiol 57:765–770. https://doi.org/10.1099/JMM.0.47769-0

Gerhard R, Frenzel E, Goy S, Olling A (2013) Cellular uptake of Clostridium difficile TcdA and truncated TcdA lacking the receptor binding domain. J Med Microbiol 62:1414–1422. https://doi.org/10.1099/JMM.0.057828-0

Geyer M, Wilde C, Selzer J et al (2003) Glucosylation of Ras by Clostridium sordellii lethal toxin: consequences for effector loop conformations observed by NMR spectroscopy. Biochemistry 42:11951–11959. https://doi.org/10.1021/BI034529V

Gibert M, Monier MN, Ruez R et al (2011) Endocytosis and toxicity of clostridial binary toxins depend on a clathrin-independent pathway regulated by Rho-GDI. Cell Microbiol 13:154–170. https://doi.org/10.1111/J.1462-5822.2010.01527.X

Giesemann T, Jank T, Gerhard R et al (2006) Cholesterol-dependent pore formation of Clostridium difficile toxin A. J Biol Chem 281:10808–10815. https://doi.org/10.1074/JBC.M512720200

Giesemann T, Guttenberg G, Aktories K (2008) Human alpha-defensins inhibit Clostridium difficile toxin B. Gastroenterology 134:2049–2058. https://doi.org/10.1053/J.GASTRO.2008.03.008

Granata G, Mariotti D, Ascenzi P et al (2021) High serum levels of toxin A correlate with disease severity in patients with Clostridioides difficile infection. Antibiot (Basel, Switzerland) 10. https://doi.org/10.3390/antibiotics10091093

Greco A, Ho JGS, Lin SJ et al (2006) Carbohydrate recognition by Clostridium difficile toxin A. Nat Struct Mol Biol 13:460–461. https://doi.org/10.1038/NSMB1084

Gregorieff A, Liu Y, Inanlou MR et al (2015) Yap-dependent reprogramming of Lgr5(+) stem cells drives intestinal regeneration and cancer. Nature 526: 715–718. https://doi.org/10.1038/NATURE15382

Guo S, Chen Y, Liu J et al (2022) Low-density lipoprotein receptor-related protein 1 is a CROPs-associated receptor for Clostridioides infection toxin B. Sci China Life Sci 65:107–118. https://doi.org/10.1007/s11427-021-1943-9

Gupta P, Zhang Z, Sugiman-Marangos SN et al (2017) Functional defects in Clostridium difficile TcdB toxin uptake identify CSPG4 receptor-binding determinants. J Biol Chem 292:17290–17301. https://doi.org/10.1074/jbc.M117.806687

Guttenberg G, Hornei S, Jank T et al (2012) Molecular characteristics of Clostridium perfringens TpeL toxin and consequences of mono-O-GlcNAcylation of Ras in living cells. J Biol Chem 287:24929–24940. https://doi.org/10.1074/JBC.M112.347773

Halabi-Cabezon I, Huelsenbeck J, May M et al (2008) Prevention of the cytopathic effect induced by Clostridium difficile Toxin B by active Rac1. FEBS Lett 582:3751–3756. https://doi.org/10.1016/J.FEBSLET.2008.10.003

Hale ML, Marvaud JC, Popoff MR, Stiles BG (2004) Detergent-resistant membrane microdomains facilitate Ib oligomer formation and biological activity of Clostridium perfringens iota-toxin. Infect Immun 72:2186–2193. https://doi.org/10.1128/IAI.72.4.2186-2193.2004

Han S, Craig JA, Putnam CD et al (1999) Evolution and mechanism from structures of an ADP-ribosylating toxin and NAD complex. Nat Struct Biol 6:932–936. https://doi.org/10.1038/13300

Heber S, Barthold L, Baier J et al (2021) Inhibition of Clostridioides difficile toxins TcdA and TcdB by ambroxol. Front Pharmacol 12:809595. https://doi.org/10.3389/fphar.2021.809595

Hecht G, Pothoulakis C, LaMont JT, Madara JL (1988) Clostridium difficile toxin A perturbs cytoskeletal structure and tight junction permeability of cultured human intestinal epithelial monolayers. J Clin Invest 82:1516–1524. https://doi.org/10.1172/JCI113760

Hecht G, Koutsouris A, Pothoulakis C et al (1992) Clostridium difficile toxin B disrupts the barrier function of

T84 monolayers. Gastroenterology 102:416–423. https://doi.org/10.1016/0016-5085(92)90085-D

Heine K, Pust S, Enzenmüller S, Barth H (2008) ADP-ribosylation of actin by the Clostridium botulinum C2 toxin in mammalian cells results in delayed caspase-dependent apoptotic cell death. Infect Immun 76:4600–4608. https://doi.org/10.1128/IAI.00651-08

Hemmasi S, Czulkies BA, Schorch B et al (2015) Interaction of the Clostridium difficile binary toxin CDT and its host cell receptor, lipolysis-stimulated lipoprotein receptor (LSR). J Biol Chem 290:14031–14044. https://doi.org/10.1074/JBC.M115.650523

Henkel D, Tatge H, Schöttelndreier D et al (2020) Receptor binding domains of TcdB from Clostridioides difficile for chondroitin sulfate proteoglycan-4 and frizzled proteins are functionally independent and additive. Toxins (Basel) 12. https://doi.org/10.3390/toxins12120736

Hirase T, Kawashima S, Wong EYM et al (2001) Regulation of tight junction permeability and occludin phosphorylation by RhoA-p160ROCK-dependent and -independent mechanisms. J Biol Chem 276:10423–10431. https://doi.org/10.1074/JBC.M007136200

Ho JGS, Greco A, Rupnik M, Ng KKS (2005) Crystal structure of receptor-binding C-terminal repeats from Clostridium difficile toxin A. Proc Natl Acad Sci U S A 102:18373–18378. https://doi.org/10.1073/PNAS.0506391102

Hofmann F, Busch C, Prepens U et al (1997) Localization of the glucosyltransferase activity of Clostridium difficile toxin B to the N-terminal part of the holotoxin. J Biol Chem 272:11074–11078. https://doi.org/10.1074/jbc.272.17.11074

Hofmann F, Busch C, Aktories K (1998) Chimeric clostridial cytotoxins: Identification of the N-terminal region involved in protein substrate recognition. Infect Immun 66:1076–1081. https://doi.org/10.1128/iai.66.3.1076-1081.1998

Holmes KC, Popp D, Gebhard W, Kabsch W (1990) Atomic model of the actin filament. Nature 347:44–49. https://doi.org/10.1038/347044a0

Hottiger MO, Hassa PO, Lüscher B et al (2010) Toward a unified nomenclature for mammalian ADP-ribosyltransferases. Trends Biochem Sci 35: 208–219. https://doi.org/10.1016/J.TIBS.2009.12.003

Huang J, Kelly CP, Bakirtzi K et al (2019) Clostridium difficile toxins induce VEGF-A and vascular permeability to promote disease pathogenesis. Nat Microbiol 4:269–279. https://doi.org/10.1038/s41564-018-0300-x

Ishida Y, Maegawa T, Kondo T et al (2004) Essential involvement of IFN-γ in Clostridium difficile toxin A-induced enteritis. J Immunol 172:3018–3025. https://doi.org/10.4049/JIMMUNOL.172.5.3018

Jafari NV, Kuehne SA, Bryant CE et al (2013) Clostridium difficile modulates host innate immunity via toxin-independent and dependent mechanism(s). PLoS One 8. https://doi.org/10.1371/JOURNAL.PONE.0069846

Jaffe AB, Hall A (2005) Rho GTPases: biochemistry and biology. Annu Rev Cell Dev Biol 21:247–269. https://doi.org/10.1146/ANNUREV.CELLBIO.21.020604.150721

Jank T, Aktories K (2008) Structure and mode of action of clostridial glucosylating toxins: the ABCD model. Trends Microbiol 16:222–229. https://doi.org/10.1016/J.TIM.2008.01.011

Jank T, Reinert DJ, Giesemann T et al (2005) Change of the donor substrate specificity of clostridium difficile toxin B by site-directed mutagenesis. J Biol Chem 280: 37833–37838. https://doi.org/10.1074/JBC.M506836200

Jank T, Giesemann T, Aktories K (2007) Clostridium difficile glucosyltransferase toxin B-essential amino acids for substrate binding. J Biol Chem 282:35222–35231. https://doi.org/10.1074/JBC.M703138200

Jank T, Ziegler MOP, Schulz GE, Aktories K (2008) Inhibition of the glucosyltransferase activity of clostridial Rho/Ras-glucosylating toxins by castanospermine. FEBS Lett 582:2277–2282. https://doi.org/10.1016/J.FEBSLET.2008.05.025

Jank T, Bogdanović X, Wirth C et al (2013) A bacterial toxin catalyzing tyrosine glycosylation of Rho and deamidation of G q and G i proteins. Nat Struct Mol Biol 20:1273–1280. https://doi.org/10.1038/NSMB.2688

Jank T, Belyi Y, Aktories K (2015a) Bacterial glycosyltransferase toxins. Cell Microbiol 17:1752–1765. https://doi.org/10.1111/CMI.12533

Jank T, Eckerle S, Steinemann M et al (2015b) Tyrosine glycosylation of Rho by Yersinia toxin impairs blastomere cell behaviour in zebrafish embryos. Nat Commun 6. https://doi.org/10.1038/NCOMMS8807

Jiang M, Shin J, Simeon R et al (2022) Structural dynamics of receptor recognition and pH-induced dissociation of full-length Clostridioides difficile Toxin B. PLoS Biol 20. https://doi.org/10.1371/journal.pbio.3001589

Jorgensen I, Miao EA (2015) Pyroptotic cell death defends against intracellular pathogens. Immunol Rev 265: 130–142. https://doi.org/10.1111/IMR.12287

Just I, Gerhard R (2004) Large clostridial cytotoxins. Rev Physiol Biochem Pharmacol 152:23–47. https://doi.org/10.1007/S10254-004-0033-5

Just I, Selzer J, Wilm M et al (1995a) Glucosylation of Rho proteins by Clostridium difficile toxin B. Nature 375:500–503. https://doi.org/10.1038/375500a0

Just I, Wilm M, Selzer J et al (1995b) The enterotoxin from Clostridium difficile (ToxA) monoglucosylates the Rho proteins. J Biol Chem 270:13932–13936. https://doi.org/10.1074/jbc.270.23.13932

Just I, Selzer J, Hofmann F et al (1996) Inactivation of Ras by Clostridium sordellii lethal toxin-catalyzed glucosylation. J Biol Chem 271:10149–10153. https://doi.org/10.1074/JBC.271.17.10149

Kaiser E, Kroll C, Ernst K et al (2011) Membrane translocation of binary actin-ADP-ribosylating toxins from Clostridium difficile and Clostridium perfringens Is

facilitated by Cyclophilin A and Hsp90. Infect Immun 79:3913–3921. https://doi.org/10.1128/IAI.05372-11
Kawamoto A, Yamada T, Yoshida T et al (2022) Cryo-EM structures of the translocational binary toxin complex CDTa-bound CDTb-pore from Clostridioides difficile. Nat Commun 13. https://doi.org/10.1038/s41467-022-33888-4
Kim J, Pai H, Seo MR, Kang JO (2012) Clinical and microbiologic characteristics of tcdA-negative variant clostridium difficile infections. BMC Infect Dis 12. https://doi.org/10.1186/1471-2334-12-109
Klepka C, Sandmann M, Tatge H et al (2022) Impairment of lysosomal function by Clostridioides difficile TcdB. Mol Microbiol 117:493–507. https://doi.org/10.1111/mmi.14864
Kodama A, Karakesisoglou I, Wong E et al (2003) ACF7: An essential integrator of microtubule dynamics. Cell 115:343–354. https://doi.org/10.1016/S0092-8674(03)00813-4
Korbmacher M, Fischer S, Landenberger M et al (2020) Human α-defensin-5 efficiently neutralizes Clostridioides difficile toxins TcdA, TcdB, and CDT. Front Pharmacol 11. https://doi.org/10.3389/FPHAR.2020.01204
Krauthammer M, Kong Y, Ha BH et al (2012) Exome sequencing identifies recurrent somatic RAC1 mutations in melanoma. Nat Genet 44:1006–1014. https://doi.org/10.1038/NG.2359
Krivan HC, Clark GF, Smith DF, Wilkins TD (1986) Cell surface binding site for Clostridium difficile enterotoxin: Evidence for a glycoconjugate containing the sequence Galα1-3Galβ1-4GlcNAc. Infect Immun 53: 573–581. https://doi.org/10.1128/iai.53.3.573-581.1986
Kuehne SA, Cartman ST, Heap JT et al (2010) The role of toxin A and toxin B in Clostridium difficile infection. Nature 467:711–713. https://doi.org/10.1038/NATURE09397
Kuehne SA, Collery MM, Kelly ML et al (2014) Importance of toxin a, toxin b, and cdt in virulence of an epidemic clostridium difficile strain. J Infect Dis 209: 83–86. https://doi.org/10.1093/INFDIS/JIT426
Kushnaryov VM, Sedmak JJ (1989) Effect of Clostridium difficile enterotoxin A on ultrastructure of Chinese hamster ovary cells. Infect Immun 57:3914–3921. https://doi.org/10.1128/iai.57.12.3914-3921.1989
LaFrance ME, Farrow MA, Chandrasekaran R et al (2015) Identification of an epithelial cell receptor responsible for Clostridium difficile TcdB-induced cytotoxicity. Proc Natl Acad Sci U S A 112:7073–7078. https://doi.org/10.1073/PNAS.1500791112
Lai FPL, Szczodrak M, Block J et al (2008) Arp2/3 complex interactions and actin network turnover in lamellipodia. EMBO J 27:982–992. https://doi.org/10.1038/EMBOJ.2008.34
Lamaze C, Dujeancourt A, Baba T et al (2001) Interleukin 2 receptors and detergent-resistant membrane domains define a clathrin-independent endocytic pathway. Mol Cell 7:661–671. https://doi.org/10.1016/S1097-2765(01)00212-X
Lambert GS, Baldwin MR (2016) Evidence for dual receptor-binding sites in Clostridium difficile toxin A. FEBS Lett 590:4550–4563. https://doi.org/10.1002/1873-3468.12487
Landenberger M, Nieland J, Roeder M et al (2021) The cytotoxic effect of Clostridioides difficile pore-forming toxin CDTb. Biochim Biophys Acta Biomembr 1863. https://doi.org/10.1016/j.bbamem.2021.183603
Lee H, Beilhartz GL, Kucharska I et al (2020) Recognition of semaphorin proteins by P. sordellii lethal toxin reveals principles of receptor specificity in clostridial toxins. Cell 182:345–356.e16. https://doi.org/10.1016/j.cell.2020.06.005
Lemichez E, Aktories K (2013) Hijacking of Rho GTPases during bacterial infection. Exp Cell Res 319:2329–2336. https://doi.org/10.1016/J.YEXCR.2013.04.021
Li S, Zhang L, Yao Q et al (2013) Pathogen blocks host death receptor signalling by arginine GlcNAcylation of death domains. Nature 501:242–246. https://doi.org/10.1038/NATURE12436
Li Y, Xu W, Ren Y et al (2022) Plakoglobin and high-mobility group box 1 mediate intestinal epithelial cell apoptosis induced by Clostridioides difficile TcdB. MBio 13:e0184922. https://doi.org/10.1128/mbio.01849-22
Lin Q, Pollock NR, Banz A et al (2020) Toxin A-predominant pathogenic Clostridioides difficile: A novel clinical phenotype. Clin Infect Dis 70:2628–2633. https://doi.org/10.1093/cid/ciz727
Linevsky JK, Pothoulakis C, Keates S et al (1997) IL-8 release and neutrophil activation by Clostridium difficile toxin-exposed human monocytes. Am J Physiol Gastrointest Liver Physiol 273. https://doi.org/10.1152/ajpgi.1997.273.6.g1333
Liu Z, Zhang S, Chen P et al (2021) Structural basis for selective modification of Rho and Ras GTPases by Clostridioides difficile toxin B. Sci Adv 7. https://doi.org/10.1126/sciadv.abi4582
López-Ureña D, Orozco-Aguilar J, Chaves-Madrigal Y et al (2019) Toxin b variants from clostridium difficile strains vpi 10463 and nap1/027 share similar substrate profile and cellular intoxication kinetics but use different host cell entry factors. Toxins (Basel) 11. https://doi.org/10.3390/toxins11060348
Lu A, Wu H (2015) Structural mechanisms of inflammasome assembly. FEBS J 282:435–444. https://doi.org/10.1111/FEBS.13133
Luo J, Yang Q, Zhang X et al (2022) TFPI is a colonic crypt receptor for TcdB from hypervirulent clade 2 C. difficile. Cell 185:980–994.e15. https://doi.org/10.1016/j.cell.2022.02.010
Lyerly DM, Krivan HC, Wilkins TD (1988) Clostridium difficile: its disease and toxins. Clin Microbiol Rev 1: 1–18. https://doi.org/10.1128/CMR.1.1.1
Lyerly DM, Barroso LA, Wilkins TD et al (1992) Characterization of a toxin A-negative, toxin B-positive strain of Clostridium difficile. Infect Immun 60:4633–4639. https://doi.org/10.1128/iai.60.11.4633-4639.1992
Lyras D, O'Connor JR, Howarth PM et al (2009) Toxin B is essential for virulence of Clostridium difficile.

Nature 458:1176–1179. https://doi.org/10.1038/NATURE07822

Mahida YR, Makh S, Hyde S et al (1996) Effect of Clostridium difficile toxin A on human intestinal epithelial cells: Induction of interleukin 8 production and apoptosis after cell detachment. Gut 38:337–347. https://doi.org/10.1136/gut.38.3.337

Malorni W, Fiorentini C, Paradisi S et al (1990) Surface blebbing and cytoskeletal changes induced in vitro by toxin B from Clostridium difficile: an immunochemical and ultrastructural study. Exp Mol Pathol 52:340–356. https://doi.org/10.1016/0014-4800(90)90074-N

Manse JS, Baldwin MR (2015) Binding and entry of Clostridium difficile toxin B is mediated by multiple domains. FEBS Lett 589:3945–3951. https://doi.org/10.1016/J.FEBSLET.2015.11.017

Mansfield MJ, Tremblay BJM, Zeng J et al (2020) Phylogenomics of 8,839 Clostridioides difficile genomes reveals recombination-driven evolution and diversification of toxin A and B. PLoS Pathog 16. https://doi.org/10.1371/journal.ppat.1009181

Margarit SM, Davidson W, Frego L, Stebbins CE (2006) A steric antagonism of actin polymerization by a Salmonella virulence protein. Structure 14:1219–1229. https://doi.org/10.1016/J.STR.2006.05.022

Marquardt I, Jakob J, Scheibel J et al (2021) Clostridioides difficile toxin CDT induces cytotoxic responses in human mucosal-associated invariant T (MAIT) cells. Front Microbiol 12:752549. https://doi.org/10.3389/fmicb.2021.752549

Martins CS, Costa DVS, Lima BB et al (2020) Clostridioides difficile toxin A-induced Wnt/β-catenin pathway inhibition is mediated by Rac1 glucosylation. Front Microbiol 11. https://doi.org/10.3389/FMICB.2020.01998

Marvaud J-C, Quevedo-Torres S, Eckert C et al (2019) Virulence of new variant strains of Clostridium difficile producing only toxin A or binary toxin in the hamster model. New Microbes New Infect 32: 100590. https://doi.org/10.1016/j.nmni.2019.100590

Masuda S, Oda Y, Sasaki H et al (2011) LSR defines cell corners for tricellular tight junction formation in epithelial cells. J Cell Sci 124:548–555. https://doi.org/10.1242/JCS.072058

Mesli S, Javorschi S, Bérard AM et al (2004) Distribution of the lipolysis stimulated receptor in adult and embryonic murine tissues and lethality of LSR-/- Embryos at 12.5 to 14.5 days of gestation. Eur J Biochem 271: 3103–3114. https://doi.org/10.1111/J.1432-1033.2004.04223.X

Miao EA, Leaf IA, Treuting PM et al (2010) Caspase-1-induced pyroptosis is an innate immune effector mechanism against intracellular bacteria. Nat Immunol 11: 1136–1142. https://doi.org/10.1038/NI.1960

Mileto SJ, Jardé T, Childress KO et al (2020) Clostridioides difficile infection damages colonic stem cells via TcdB, impairing epithelial repair and recovery from disease. Proc Natl Acad Sci U S A 117:8064–8073. https://doi.org/10.1073/pnas.1915255117

Monot M, Eckert C, Lemire A et al (2015) Clostridium difficile: new insights into the evolution of the pathogenicity locus. Sci Rep 5. https://doi.org/10.1038/SREP15023

Mostowy S, Cossart P (2012) Septins: The fourth component of the cytoskeleton. Nat Rev Mol Cell Biol 13: 183–194. https://doi.org/10.1038/NRM3284

Na X, Kim H, Moyer MP et al (2008) gp96 is a human colonocyte plasma membrane binding protein for Clostridium difficile toxin A. Infect Immun 76:2862–2871. https://doi.org/10.1128/IAI.00326-08

Nagahama M, Yamaguchi A, Hagiyama T et al (2004) Binding and internalization of Clostridium perfringens iota-toxin in lipid rafts. Infect Immun 72:3267–3275. https://doi.org/10.1128/IAI.72.6.3267-3275.2004

Nagahama M, Ohkubo A, Oda M et al (2011) Clostridium perfringens TpeL glycosylates the Rac and Ras subfamily proteins. Infect Immun 79:905–910. https://doi.org/10.1128/IAI.01019-10

Ng J, Hirota SA, Gross O et al (2010) Clostridium difficile toxin-induced inflammation and intestinal injury are mediated by the inflammasome. Gastroenterology 139. https://doi.org/10.1053/J.GASTRO.2010.04.005

Nobes C, Hall A (1994) Regulation and function of the Rho subfamily of small GTPases. Curr Opin Genet Dev 4:77–81. https://doi.org/10.1016/0959-437X(94)90094-9

Nölke T, Schwan C, Lehmann F et al (2016) Septins guide microtubule protrusions induced by actin-depolymerizing toxins like Clostridium difficile transferase (CDT). Proc Natl Acad Sci U S A 113:7870–7875. https://doi.org/10.1073/PNAS.1522717113

Nusrat A, Giry M, Turner JR et al (1995) Rho protein regulates tight junctions and perijunctional actin organization in polarized epithelia. Proc Natl Acad Sci U S A 92:10629–10633. https://doi.org/10.1073/pnas.92.23.10629

Nusrat A, Von Eichel-Streiber C, Turner JR et al (2001) Clostridium difficile toxins disrupt epithelial barrier function by altering membrane microdomain localization of tight junction proteins. Infect Immun 69:1329–1336. https://doi.org/10.1128/IAI.69.3.1329-1336.2001

Olling A, Goy S, Hoffmann F et al (2011) The repetitive oligopeptide sequences modulate cytopathic potency but are not crucial for cellular uptake of Clostridium difficile toxin A. PLoS One 6. https://doi.org/10.1371/JOURNAL.PONE.0017623

Orrell KE, Mansfield MJ, Doxey AC, Melnyk RA (2020) The C. difficile toxin B membrane translocation machinery is an evolutionarily conserved protein delivery apparatus. Nat Commun 11. https://doi.org/10.1038/s41467-020-14306-z

Orth P, Xiao L, Hernandez LD et al (2014) Mechanism of action and epitopes of Clostridium difficile toxin B-neutralizing antibody bezlotoxumab revealed by X-ray crystallography. J Biol Chem 289:18008–18021. https://doi.org/10.1074/JBC.M114.560748

Ost GS, Wirth C, Bogdanović X et al (2020) Inverse control of Rab proteins by Yersinia

ADP-ribosyltransferase and glycosyltransferase related to clostridial glucosylating toxins. Sci Adv 6. https://doi.org/10.1126/sciadv.aaz2094

Ottlinger ME, Lin S (1988) Clostridium difficile toxin B induces reorganization of actin, vinculin, and talin in cultured cells. Exp Cell Res 174:215–229. https://doi.org/10.1016/0014-4827(88)90156-5

Pan Z, Zhang Y, Luo J et al (2021) Functional analyses of epidemic Clostridioides difficile toxin B variants reveal their divergence in utilizing receptors and inducing pathology. PLoS Pathog 17. https://doi.org/10.1371/JOURNAL.PPAT.1009197

Paparella AS, Cahill SM, Aboulache BL, Schramm VL (2022) Clostridioides difficile TcdB toxin glucosylates Rho GTPase by an SNi mechanism and ion pair transition state. ACS Chem Biol 17:2507–2518. https://doi.org/10.1021/acschembio.2c00408

Papatheodorou P, Aktories K (2017) Receptor-binding and uptake of binary actin-ADP-ribosylating toxins. Curr Top Microbiol Immunol 406:119–133. https://doi.org/10.1007/82_2016_46

Papatheodorou P, Zamboglou C, Genisyuerek S et al (2010) Clostridial glucosylating toxins enter cells via clathrin-mediated endocytosis. PLoS One 5. https://doi.org/10.1371/JOURNAL.PONE.0010673

Papatheodorou P, Carette JE, Bell GW et al (2011) Lipolysis-stimulated lipoprotein receptor (LSR) is the host receptor for the binary toxin Clostridium difficile transferase (CDT). Proc Natl Acad Sci U S A 108: 16422–16427. https://doi.org/10.1073/PNAS.1109772108

Papatheodorou P, Wilczek C, Nölke T et al (2012) Identification of the cellular receptor of Clostridium spiroforme toxin. Infect Immun 80:1418–1423. https://doi.org/10.1128/IAI.06378-11

Papatheodorou P, Hornuss D, Nölke T et al (2013) Clostridium difficile binary toxin CDT induces clustering of the lipolysis-stimulated lipoprotein receptor into lipid rafts. MBio 4. https://doi.org/10.1128/MBIO.00244-13

Papatheodorou P, Song S, López-Ureña D et al (2019) Cytotoxicity of Clostridium difficile toxins A and B requires an active and functional SREBP-2 pathway. FASEB J 33:4883–4892. https://doi.org/10.1096/fj.201801440R

Papatheodorou P, Kindig S, Badilla-Lobo A et al (2021) The compound U18666A inhibits the intoxication of cells by Clostridioides difficile toxins TcdA and TcdB. Front Microbiol 12. https://doi.org/10.3389/FMICB.2021.784856

Park YH, Wood G, Kastner DL, Chae JJ (2016) Pyrin inflammasome activation and RhoA signaling in the autoinflammatory diseases FMF and HIDS. Nat Immunol 17:914–921. https://doi.org/10.1038/NI.3457

Perelle S, Gibert M, Bourlioux P et al (1997) Production of a complete binary toxin (actin-specific ADP-ribosyltransferase) by Clostridium difficile CD196. Infect Immun 65:1402–1407. https://doi.org/10.1128/iai.65.4.1402-1407.1997

Perieteanu AA, Visschedyk DD, Merrill AR, Dawson JF (2010) ADP-ribosylation of cross-linked actin generates barbed-end polymerization-deficient F-actin oligomers. Biochemistry 49:8944–8954. https://doi.org/10.1021/BI1008062

Peritore-Galve FC, Shupe JA, Cave RJ et al (2022) Glucosyltransferase-dependent and independent effects of Clostridioides difficile toxins during infection. PLoS Pathog 18. https://doi.org/10.1371/journal.ppat.1010323

Petersen L, Stroh S, Schöttelndreier D et al (2022) The essential role of Rac1 glucosylation in Clostridioides difficile toxin B-induced arrest of G1-S transition. Front Microbiol 13. https://doi.org/10.3389/FMICB.2022.846215

Petosa C, Collier RJ, Klimpel KR et al (1997) Crystal structure of the anthrax toxin protective antigen. Nature 385:833–838. https://doi.org/10.1038/385833A0

Pfeifer G, Schirmer J, Leemhuis J et al (2003) Cellular uptake of Clostridium difficile toxin B. Translocation of the N-terminal catalytic domain into the cytosol of eukaryotic cells. J Biol Chem 278:44535–44541. https://doi.org/10.1074/JBC.M307540200

Popoff MR, Boquet P (1988) Clostridium spiroforme toxin is a binary toxin which ADP-ribosylates cellular actin. Biochem Biophys Res Commun 152:1361–1368. https://doi.org/10.1016/S0006-291X(88)80435-2

Pothoulakis C, Gilbert RJ, Cladaras C et al (1996) Rabbit sucrase-isomaltase contains a functional intestinal receptor for Clostridium difficile toxin A. J Clin Invest 98:641–649. https://doi.org/10.1172/JCI118835

Pruitt RN, Chagot B, Cover M et al (2009) Structure-function analysis of inositol hexakisphosphate-induced autoprocessing in Clostridium difficile toxin A. J Biol Chem 284:21934–21940. https://doi.org/10.1074/JBC.M109.018929

Pruitt RN, Chambers MG, Ng KKS et al (2010) Structural organization of the functional domains of Clostridium difficile toxins A and B. Proc Natl Acad Sci U S A 107: 13467–13472. https://doi.org/10.1073/PNAS.1002199107

Pruitt RN, Chumbler NM, Rutherford SA et al (2012) Structural determinants of Clostridium difficile toxin A glucosyltransferase activity. J Biol Chem 287:8013–8020. https://doi.org/10.1074/JBC.M111.298414

Pruss KM, Sonnenburg JL (2021) C. difficile exploits a host metabolite produced during toxin-mediated disease. Nature 593:261–265. https://doi.org/10.1038/s41586-021-03502-6

Puri AW, Lupardus PJ, Deu E et al (2010) Rational design of inhibitors and activity-based probes targeting clostridium difficile virulence factor TcdB. Chem Biol 17: 1201–1211. https://doi.org/10.1016/J.CHEMBIOL.2010.09.011

Qa'Dan M, Spyres LM, Ballard JD (2000) pH-induced conformational changes in Clostridium difficile toxin

B. Infect Immun 68:2470–2474. https://doi.org/10.1128/IAI.68.5.2470-2474.2000

Qa'Dan M, Spyres LM, Ballard JD (2001) pH-enhanced cytopathic effects of Clostridium sordellii lethal toxin. Infect Immun 69:5487–5493. https://doi.org/10.1128/IAI.69.9.5487-5493.2001

Qa'Dan M, Ramsey M, Daniel J et al (2002) Clostridium difficile toxin B activates dual caspase-dependent and caspase-independent apoptosis in intoxicated cells. Cell Microbiol 4:425–434. https://doi.org/10.1046/J.1462-5822.2002.00201.X

Qiu B, Pothoulakis C, Castagliuolo I et al (1999) Participation of reactive oxygen metabolites in Clostridium difficile toxin A-induced enteritis in rats. Am J Physiol Gastrointest Liver Physiol 276. https://doi.org/10.1152/ajpgi.1999.276.2.g485

Reineke J, Tenzer S, Rupnik M et al (2007) Autocatalytic cleavage of Clostridium difficile toxin B. Nature 446: 415–419. https://doi.org/10.1038/NATURE05622

Reinert DJ, Jank T, Aktories K, Schulz GE (2005) Structural basis for the function of Clostridium difficile toxin B. J Mol Biol 351:973–981. https://doi.org/10.1016/J.JMB.2005.06.071

Roeder M, Nestorovich EM, Karginov VA et al (2014) Tailored cyclodextrin pore blocker protects mammalian cells from clostridium difficile binary toxin CDT. Toxins (Basel) 6:2097–2114. https://doi.org/10.3390/TOXINS6072097

Roth BM, Godoy-Ruiz R, Varney KM et al (2016a) 1H, 13C, and 15N resonance assignments of an enzymatically active domain from the catalytic component (CDTa, residues 216–420) of a binary toxin from Clostridium difficile. Biomol NMR Assign 10:213–217. https://doi.org/10.1007/S12104-016-9669-8

Roth BM, Varney KM, Rustandi RR, Weber DJ (2016b) 1HN, 13C, and 15N resonance assignments of the CDTb-interacting domain (CDTaBID) from the Clostridium difficile binary toxin catalytic component (CDTa, residues 1–221). Biomol NMR Assign 10: 335–339. https://doi.org/10.1007/S12104-016-9695-6

Rould MA, Wan Q, Joel PB et al (2006) Crystal structures of expressed non-polymerizable monomeric actin in the ADP and ATP states. J Biol Chem 281:31909–31919. https://doi.org/10.1074/JBC.M601973200

Rupnik M, Janezic S (2016) An update on clostridium difficile toxinotyping. J Clin Microbiol 54:13–18. https://doi.org/10.1128/JCM.02083-15

Rupnik M, Avesani V, Janc M et al (1998) A novel toxinotyping scheme and correlation of toxinotypes with serogroups of Clostridium difficile isolates. J Clin Microbiol 36:2240–2247. https://doi.org/10.1128/jcm.36.8.2240-2247.1998

Rupnik M, Pabst S, Rupnik M et al (2005) Characterization of the cleavage site and function of resulting cleavage fragments after limited proteolysis of Clostridium difficile toxin B (TcdB) by host cells. Microbiology 151:199–208. https://doi.org/10.1099/MIC.0.27474-0

Russo HM, Rathkey J, Boyd-Tressler A et al (2016) Active caspase-1 induces plasma membrane pores that precede pyroptotic lysis and are blocked by lanthanides. J Immunol 197:1353–1367. https://doi.org/10.4049/JIMMUNOL.1600699

Sauerborn M, Leukel P, Von Eichel-Streiber C (1997) The C-terminal ligand-binding domain of Clostridium difficile toxin A (TcdA) abrogates TcdA-specific binding to cells arid prevents mouse lethality. FEMS Microbiol Lett 155:45–54. https://doi.org/10.1016/S0378-1097(97)00365-0

Schering B, Bärmann M, Chhatwal GS et al (1988) ADP-ribosylation of skeletal muscle and non-muscle actin by Clostridium perfringens iota toxin. Eur J Biochem 171:225–229. https://doi.org/10.1111/j.1432-1033.1988.tb13780.x

Schleberger C, Hochmann H, Barth H et al (2006) Structure and action of the binary C2 toxin from Clostridium botulinum. J Mol Biol 364:705–715. https://doi.org/10.1016/J.JMB.2006.09.002

Schorch B, Song S, Van Diemen FR et al (2014) LRP1 is a receptor for Clostridium perfringens TpeL toxin indicating a two-receptor model of clostridial glycosylating toxins. Proc Natl Acad Sci U S A 111: 6431–6436. https://doi.org/10.1073/PNAS.1323790111

Schöttelndreier D, Langejürgen A, Lindner R, Genth H (2020) Low density lipoprotein receptor-related protein-1 (LRP1) is involved in the uptake of Clostridioides difficile toxin A and serves as an internalizing receptor. Front Cell Infect Microbiol 10. https://doi.org/10.3389/fcimb.2020.565465

Schwan C, Stecher B, Tzivelekidis T et al (2009) Clostridium difficile toxin CDT induces formation of microtubule-based protrusions and increases adherence of bacteria. PLoS Pathog 5. https://doi.org/10.1371/JOURNAL.PPAT.1000626

Schwan C, Kruppke AS, Nölke T et al (2014) Clostridium difficile toxin CDT hijacks microtubule organization and reroutes vesicle traffic to increase pathogen adherence. Proc Natl Acad Sci U S A 111:2313–2318. https://doi.org/10.1073/PNAS.1311589111

Schwan C, Lang AE, Schlosser A et al (2022) Inhibition of Arp2/3 complex after ADP-ribosylation of Arp2 by binary clostridioides toxins. Cells 11. https://doi.org/10.3390/CELLS11223661

Schwartz R, Guichard A, Franc NC et al (2020) A drosophila model for Clostridium difficile toxin CDT reveals interactions with multiple effector pathways. iScience 23:100865. https://doi.org/10.1016/j.isci.2020.100865

Sehnal D, Bittrich S, Deshpande M et al (2021) Mol* Viewer: modern web app for 3D visualization and analysis of large biomolecular structures. Nucleic Acids Res 49:W431–W437. https://doi.org/10.1093/NAR/GKAB314

Sehr P, Joseph G, Genth H et al (1998) Glucosylation and ADP ribosylation of Rho proteins: Effects on nucleotide binding, GTPase activity, and effector coupling.

Biochemistry 37:5296–5304. https://doi.org/10.1021/bi972592c

Selzer J, Hofmann F, Rex G et al (1996) Clostridium novyi α-toxin-catalyzed incorporation of GlcNAc into Rho subfamily proteins. J Biol Chem 271:25173–25177. https://doi.org/10.1074/jbc.271.41.25173

Sheedlo MJ, Anderson DM, Thomas AK, Borden Lacy D (2020) Structural elucidation of the Clostridioides difficile transferase toxin reveals a single-site binding mode for the enzyme. Proc Natl Acad Sci U S A 117: 6139–6144. https://doi.org/10.1073/pnas.1920555117

Shen A, Lupardus PJ, Gersch MM et al (2011) Defining an allosteric circuit in the cysteine protease domain of Clostridium difficile toxins. Nat Struct Mol Biol 18: 364–371. https://doi.org/10.1038/NSMB.1990

Simeon R, Jiang M, Chamoun-Emanuelli AM et al (2019) Selection and characterization of ultrahigh potency designed ankyrin repeat protein inhibitors of C. difficile toxin B. PLoS Biol 17. https://doi.org/10.1371/journal.pbio.3000311

Simpson M, Bilverstone T, Leslie J et al (2023) Clostridioides difficile binary toxin binding component increases virulence in a Hamster model. Open Forum Infect Dis 10. https://doi.org/10.1093/OFID/OFAD040

Sohet F, Lin C, Munji RN et al (2015) LSR/angulin-1 is a tricellular tight junction protein involved in blood-brain barrier formation. J Cell Biol 208:703–711. https://doi.org/10.1083/JCB.201410131

Song J, Shen X, Huang Z et al (2021) Clostridium difficile toxin A and toxin B inhibit YAP in the colonic epithelial cells. J Biochem Mol Toxicol 35. https://doi.org/10.1002/jbt.22652

Steinemann M, Schlosser A, Jank T, Aktories K (2018) The chaperonin TRiC/CCT is essential for the action of bacterial glycosylating protein toxins like Clostridium difficile toxins A and B. Proc Natl Acad Sci U S A 115: 9580–9585. https://doi.org/10.1073/pnas.1807658115

Steiner TS, Flores CA, Pizarro TT, Guerrant RL (1997) Fecal lactoferrin, interleukin-1beta, and interleukin-8 are elevated in patients with severe Clostridium difficile colitis. Clin Diagn Lab Immunol 4:719–722. https://doi.org/10.1128/CDLI.4.6.719-722.1997

Stieglitz F, Gerhard R, Pich A (2021) The binary toxin of Clostridioides difficile alters the proteome and phosphoproteome of HEp-2 cells. Front Microbiol 12:725612. https://doi.org/10.3389/fmicb.2021.725612

Stieglitz F, Gerhard R, Hönig R et al (2022) TcdB of Clostridioides difficile mediates RAS-dependent necrosis in epithelial cells. Int J Mol Sci 23. https://doi.org/10.3390/ijms23084258

Stiles BG, Hale ML, Marvaud JC, Popoff MR (2000) Clostridium perfringens iota toxin: binding studies and characterization of cell surface receptor by fluorescence-activated cytometry. Infect Immun 68: 3475–3484. https://doi.org/10.1128/IAI.68.6.3475-3484.2000

Sundriyal A, Roberts AK, Shone CC, Acharya KR (2009) Structural basis for substrate recognition in the enzymatic component of ADP-ribosyltransferase toxin CDTa from Clostridium difficile. J Biol Chem 284: 28713–28719. https://doi.org/10.1074/JBC.M109.043018

Tam J, Icho S, Utama E et al (2020) Intestinal bile acids directly modulate the structure and function of C. difficile TcdB toxin. Proc Natl Acad Sci U S A 117:6792–6800. https://doi.org/10.1073/pnas.1916965117

Tao L, Zhang J, Meraner P et al (2016) Frizzled proteins are colonic epithelial receptors for C. difficile toxin B. Nature 538:350–355. https://doi.org/10.1038/NATURE19799

Tao L, Tian S, Zhang J et al (2019) Sulfated glycosaminoglycans and low-density lipoprotein receptor contribute to Clostridium difficile toxin A entry into cells. Nat Microbiol 4:1760–1769. https://doi.org/10.1038/s41564-019-0464-z

Tcherkezian J, Lamarche-Vane N (2007) Current knowledge of the large RhoGAP family of proteins. Biol Cell 99:67–86. https://doi.org/10.1042/BC20060086

Tian S, Liu Y, Wu H et al (2020) Genome-wide CRISPR screen identifies semaphorin 6A and 6B as receptors for Paeniclostridium sordellii toxin TcsL. Cell Host Microbe 27:782–792.e7. https://doi.org/10.1016/j.chom.2020.03.007

Tian S, Xiong X, Zeng J et al (2022) Identification of TFPI as a receptor reveals recombination-driven receptor switching in Clostridioides difficile toxin B variants. Nat Commun 13. https://doi.org/10.1038/s41467-022-33964-9

Tsuge H, Nagahama M, Oda M et al (2008) Structural basis of actin recognition and arginine ADP-ribosylation by Clostridium perfringens ι-toxin. Proc Natl Acad Sci U S A 105:7399–7404. https://doi.org/10.1073/PNAS.0801215105

Tsurumura T, Tsumori Y, Qiu H et al (2013) Arginine ADP-ribosylation mechanism based on structural snapshots of iota-toxin and actin complex. Proc Natl Acad Sci U S A 110:4267–4272. https://doi.org/10.1073/PNAS.1217227110

Tucker KD, Wilkins TD (1991) Toxin A of Clostridium difficile binds to the human carbohydrate antigens I, X, and Y. Infect Immun 59:73–78. https://doi.org/10.1128/iai.59.1.73-78.1991

Vandekerckhove J, Schering B, Bärmann M, Aktories K (1987) Clostridium perfringens iota toxin ADP-ribosylates skeletal muscle actin in Arg-177. FEBS Lett 225:48–52. https://doi.org/10.1016/0014-5793(87)81129-8

Vandekerckhove J, Schering B, Barmann M, Aktories K (1988) Botulinum C2 toxin ADP-ribosylates cytoplasmic β/γ-actin in arginine 177. J Biol Chem 263:696–700

Vetter IR, Hofmann F, Wohlgemuth S et al (2000) Structural consequences of mono-glucosylation of Ha-Ras by Clostridium sordellii lethal toxin. J Mol Biol 301: 1091–1095. https://doi.org/10.1006/JMBI.2000.4045

von Eichel-Streiber C, Sauerborn M (1990) Clostridium difficile toxin A carries a C-terminal repetitive structure homologous to the carbohydrate binding region of streptococcal glycosyltransferases. Gene 96:107–113. https://doi.org/10.1016/0378-1119(90)90348-U

von Eichel-Streiber C, Laufenberg-Feldmann R, Sartingen S et al (1992a) Comparative sequence analysis of the Clostridium difficile toxins A and B. MGG Mol Gen Genet 233:260–268. https://doi.org/10.1007/BF00587587

von Eichel-Streiber C, Sauerborn M, Kuramitsu HK (1992b) Evidence for a modular structure of the homologous repetitive C-terminal carbohydrate-binding sites of Clostridium difficile toxins and Streptococcus mutans glucosyltransferases. J Bacteriol 174:6707–6710. https://doi.org/10.1128/jb.174.20.6707-6710.1992

Voth DE, Ballard JD (2005) Clostridium difficile toxins: mechanism of action and role in disease. Clin Microbiol Rev 18:247–263. https://doi.org/10.1128/CMR.18.2.247-263.2005

Warny M, Keates AC, Keates S et al (2000) p38 MAP kinase activation by Clostridium difficile toxin A mediates monocyte necrosis, IL-8 production, and enteritis. J Clin Invest 105:1147–1156. https://doi.org/10.1172/JCI7545

Wegner A, Aktories K (1988) ADP-ribosylated actin caps the barbed ends of actin filaments. J Biol Chem 263: 13739–13742. https://doi.org/10.1016/s0021-9258(18)68303-6

Weigt C, Just I, Wegner A, Aktories K (1989) Nonmuscle actin ADP-ribosylated by botulinum C2 toxin caps actin filaments. FEBS Lett 246:181–184. https://doi.org/10.1016/0014-5793(89)80279-0

Wiegers W, Just I, Muller H et al (1991) Alteration of the cytoskeleton of mammalian cells cultured in vitro by Clostridium botulinum C2 toxin and C3 ADP-ribosyltransferase. Eur J Cell Biol 54:237–245

Wigelsworth DJ, Ruthel G, Schnell L et al (2012) CD44 promotes intoxication by the clostridial iota-family toxins. PLoS One 7. https://doi.org/10.1371/JOURNAL.PONE.0051356

Wille M, Just I, Wegner A, Aktories K (1992) ADP-ribosylation of gelsolin-actin complexes by clostridial toxins. J Biol Chem 267:50–55. https://doi.org/10.1016/s0021-9258(18)48456-6

Wohlan K, Goy S, Olling A et al (2014) Pyknotic cell death induced by Clostridium difficile TcdB: chromatin condensation and nuclear blister are induced independently of the glucosyltransferase activity. Cell Microbiol 16:1678–1692. https://doi.org/10.1111/CMI.12317

Wren BW (1991) A family of clostridial and streptococcal ligand-binding proteins with conserved C-terminal repeat sequences. Mol Microbiol 5:797–803. https://doi.org/10.1111/j.1365-2958.1991.tb00752.x

Xu H, Yang J, Gao W et al (2014) Innate immune sensing of bacterial modifications of Rho GTPases by the Pyrin inflammasome. Nature 513:237–241. https://doi.org/10.1038/NATURE13449

Xu X, Godoy-Ruiz R, Adipietro KA et al (2020) Structure of the cell-binding component of the Clostridium difficile binary toxin reveals a di-heptamer macromolecular assembly. Proc Natl Acad Sci U S A 117:1049–1058. https://doi.org/10.1073/pnas.1919490117

Yanda MK, Guggino WB, Cebotaru L (2020) A new role for heat shock factor 27 in the pathophysiology of Clostridium difficile toxin B. Am J Physiol Gastrointest Liver Physiol 318:G120–G129. https://doi.org/10.1152/ajpgi.00166.2019

Yen FT, Mann CJ, Guermani LM et al (1994) Identification of a lipolysis-stimulated receptor that is distinct from the LDL receptor and the LDL receptor-related protein. Biochemistry 33:1172–1180. https://doi.org/10.1021/bi00171a017

Yen FT, Masson M, Clossais-Besnard N et al (1999) Molecular cloning of a lipolysis-stimulated remnant receptor expressed in the liver. J Biol Chem 274: 13390–13398. https://doi.org/10.1074/JBC.274.19.13390

Young JAT, Collier RJ (2007) Anthrax toxin: receptor binding, internalization, pore formation, and translocation. Annu Rev Biochem 76:243–265. https://doi.org/10.1146/ANNUREV.BIOCHEM.75.103004.142728

Young MK, Leslie JL, Madden GR et al (2022) Binary toxin expression by Clostridioides difficile is associated with worse disease. Open Forum Infect Dis 9:ofac001. https://doi.org/10.1093/ofid/ofac001

Yuan P, Zhang H, Cai C et al (2015) Chondroitin sulfate proteoglycan 4 functions as the cellular receptor for Clostridium difficile toxin B. Cell Res 25:157–168. https://doi.org/10.1038/CR.2014.169

Zeiser J, Gerhard R, Just I, Pich A (2013) Substrate specificity of clostridial glucosylating toxins and their function on colonocytes analyzed by proteomics techniques. J Proteome Res 12:1604–1618. https://doi.org/10.1021/PR300973Q

Zhang Y, Li S, Yang Z et al (2018) Cysteine protease-mediated autocleavage of Clostridium difficile toxins regulates their proinflammatory activity. Cell Mol Gastroenterol Hepatol 5:611–625. https://doi.org/10.1016/j.jcmgh.2018.01.022

Zhou Y, Li D, Luo J et al (2021) Sulfated glycosaminoglycans and low-density lipoprotein receptor mediate the cellular entry of Clostridium novyi alpha-toxin. Cell Res 31:935–938. https://doi.org/10.1038/S41422-021-00510-Z

Zhu Z, Schnell L, Müller B et al (2019) The antibiotic bacitracin protects human intestinal epithelial cells and stem cell-derived intestinal organoids from Clostridium difficile toxin TcdB. Stem Cells Int 2019. https://doi.org/10.1155/2019/4149762

Clostridioides difficile Biofilm

Claudia Vuotto, Gianfranco Donelli, Anthony Buckley, and Caroline Chilton

Abstract

Clostridioides difficile infection (CDI), previously *Clostridium difficile* infection, is a symptomatic infection of the large intestine caused by the spore-forming anaerobic, gram-positive bacterium *Clostridioides difficile*. CDI is an important healthcare-associated disease worldwide, characterized by high levels of recurrence, morbidity, and mortality. CDI is observed at a higher rate in immunocompromised patients after antimicrobial therapy, with antibiotics disrupting the commensal microbiota and promoting *C. difficile* colonization of the gastrointestinal tract.

A rise in clinical isolates resistant to multiple antibiotics and the reduced susceptibility to the most commonly used antibiotic molecules have made the treatment of CDI more complicated, allowing the persistence of *C. difficile* in the intestinal environment.

Gut colonization and biofilm formation have been suggested to contribute to the pathogenesis and persistence of *C. difficile*. In fact, biofilm growth is considered as a serious threat because of the related antimicrobial tolerance that makes antibiotic therapy often ineffective. This is the reason why the involvement of *C. difficile* biofilm in the pathogenesis and recurrence of CDI is attracting more and more interest, and the mechanisms underlying biofilm formation of *C. difficile* as well as the role of biofilm in CDI are increasingly being studied by researchers in the field.

Findings on *C. difficile* biofilm, possible implications in CDI pathogenesis and treatment, efficacy of currently available antibiotics in treating biofilm-forming *C. difficile* strains, and some antimicrobial alternatives under investigation will be discussed here.

C. Vuotto (✉) · G. Donelli
Microbial Biofilm Laboratory, IRCCS Fondazione Santa Lucia, Rome, Italy
e-mail: c.vuotto@hsantalucia.it

A. Buckley
Microbiome and Nutritional Sciences Group, School of Food Science & Nutrition, University of Leeds, Leeds, UK

C. Chilton
Healthcare Associated Infection Research Group, Section of Molecular Gastroenterology, Leeds Institute for Medical Research at St James, University of Leeds, Leeds, UK

1 Introduction

Microbial biofilms are considered as the 'true' habitat for many causative agents of infection and disease, defending microorganisms against environmental stressors, such as pH and temperature. These microbial communities growing on biotic and abiotic surfaces are embedded in a self-produced extracellular matrix containing

P. Mastrantonio, M. Rupnik (eds.), *Updates on* Clostridioides difficile in Europe, Advances in Experimental Medicine and Biology 1435, https://doi.org/10.1007/978-3-031-42108-2_12

components such as polysaccharides, proteins, signalling molecules, extracellular DNA and RNA (Di Martino 2018; Vandana and Das 2022). Biofilms offers to microorganisms an efficacious protection from antibiotics (Goldberg 2002; Hu et al. 2019) and disinfectants (Peng et al. 2002; Bridier et al. 2011; Xue et al. 2012), as well as the possibility to survive in conditions of nutrient deficiency (Koch et al. 2001). The transition from the free-living to the sessile mode of growth is dynamic, and driven by biophysical mechanisms which control the reversible and irreversible attachment to a surface, the development of a single-species community or a polymicrobial one, and then the dispersion of cells from the biofilm (Percival et al. 2015; Wang et al. 2022a, b). Bacterial biofilms are often depicted as surface-attached bacteria with a mushroom-shaped structure, but, especially in vivo, in clinical and industrial settings, and in the environment, biofilms often are also observed as non-surface-attached aggregates (Sauer et al. 2022). Biofilm formation is tightly controlled by several regulators, including quorum sensing (QS), cyclic diguanylate (c-di-GMP), small non-coding RNAs (sRNAs), and ribonucleases (Condinho et al. 2022). In particular, intracellular and intercellular communications within biofilm are supported by signals released when cell density reaches a critical level, a phenomenon known as Quorum Sensing (QS) (Lindsay and Von Holy 2006; Li and Tian 2012; Wang et al. 2022a, b). QS is a cell density-dependent gene regulation through the production of signalling molecules, termed autoinducers (AI), that activate the maturation and disassembly of the biofilm in a coordinated manner, with dispersal of microbial cells into the surrounding environment increasing the dissemination risk and the colonization of new niches (Donelli 2006). This 'lifestyle' allows pathogenic microorganisms to acquire numerous advantages in terms of survivability and spread in hostile environments (Hall-Stoodley et al. 2004; Schulze et al. 2021).

The human gut is a clear example of a rich and highly differentiated microbial ecosystem, consisting of a huge number of microbial species that play a crucial role in maintaining metabolic and immunological homeostasis (Cummings et al. 2004; Hou et al. 2022).

Although most of the studies have been conducted on faecal samples that mirror the intestinal microbiota present in the lumen, it is increasingly recognized that some microbial genera and species preferentially reside in the mucus layer and in the epithelial crypts of the intestine (Zoetendal et al. 2002). For example, Clostridium, Lactobacillus, Enterococcus, and Akkermansia have been identified as mucus-associated genera (Schiffrin and Blum 2002), coexisting in association with both the mucosal membrane and the intestinal luminal particles (Macfarlane and Dillon 2007). These mucosal communities show different fermentation profiles (Macfarlane and Macfarlane 2006) that may be important in modulating the host's immune system and contributing to some inflammatory bowel diseases (ulcerative colitis, Crohn's disease), due to their proximity to the epithelial surface (Macfarlane et al. 2011). Single-species biofilms of gut pathogens, such as *Escherichia coli*, *Salmonella*, and *Vibrio* spp., are the most studied as their extremely adhesive and invasive features can modify the dynamics of the gut and cause infections (Azriel et al. 2015; Sengupta et al. 2016; Rossi et al. 2017; Owrangi et al. 2017). Furthermore, other intestinal isolates, belonging to anaerobic species of genera Bacteroides, Clostridium, Finegoldia, Fusobacterium, Prevotella, and Veillonella have been demonstrated to be able to develop as in vitro mono-species biofilms, and to interact with each other by forming dual-species biofilms (Donelli et al. 2012).

A growing interest in the potential biofilm growth of *C. difficile* has been recorded in recent years, due to the prominence of this microorganism as an aetiologic agent of nosocomial diarrhoea worldwide.

CDI is one of the principal threats to hospitalized and immunocompromised patients, mainly following antibiotic treatment of infections. In fact, *C. difficile* colonizes 0–15% of the healthy human population (Furuya-Kanamori et al. 2015), but an antimicrobial therapy can deplete the competitor gut microbiota inducing a shift from asymptomatic colonization to CDI (Ferreyra et al. 2014; Pérez-Cobas et al. 2014). CDI is currently treated with the antibiotics fidaxomicin, vancomycin, and metronidazole (MTZ) (Nelson et al. 2017), but the accumulation of resistance mechanisms or the reduced susceptibility to antibiotics commonly used against milder cases of CDI, e.g. MTZ (Dupont 2013), may allow *C. difficile* to persist after treatment. The selective advantage for their dissemination is mainly gained through the acquisition of mobile genetic elements involved in antibiotic resistance and alterations of the antibiotic target sites (Spigaglia 2016).

Although the two main *C. difficile* virulence factors, toxin A and toxin B (Carter et al. 2012), and the actin-ADP-ribosylating toxin, play the major role in clinical manifestation of CDI, other putative virulence factors, including those that may play a role in adherence and colonization, have to be taken into account. These include the surface layer proteins (SLPs), encoded by *slpA*, involved in adherence and inflammatory stimulation, and the extracellular matrix-binding domain, the surface anchor protein needed for covalent attachment to peptidoglycan. In addition, the fimbriae, fibronectin binding proteins, flagella, the heat shock protein GroEL, and the extracellular polysaccharides must be all considered as additional factors involved in *C. difficile* pathogenesis (Awad et al. 2014; Sebaihia et al. 2006).

The importance of adhesive properties as a key virulence factor lies in the fact that adherence is the first and most essential step of the biofilm growth cycle (Römling and Balsalobre 2012; Percival et al. 2015).

In this framework, the complex multifactorial process leading to the *C. difficile* biofilm formation (Dawson et al. 2012; Đapa et al. 2013; Dapa and Unnikrishnan 2013) should be taken into due consideration when planning interventional procedures, mainly in light of the recurrent CDI in ~20% of patients (Barbut et al. 2000). In fact, it has been hypothesized that the biofilm-growing ability of *C. difficile* may play a role in the persistence and recurrence of CDI (Frost et al. 2021). Difficult-to-treat *C. difficile* strains exhibiting both antibiotic resistance and strong biofilm production are increasingly isolated; Rahmoun et al recently identified an association between reduced susceptibility to MTZ or vancomycin and biofilm formation ability (Rahmoun et al. 2021).

Thus, a better understanding of the process of *C. difficile* biofilm formation as well as its contribution to CDI recurrence could significantly improve disease prevention and treatment.

Findings on *C. difficile* biofilm, possible implications of biofilm formation in CDI pathogenesis, treatment efficacy of currently available antibiotics, and some antimicrobial alternatives under investigation will be here discussed.

2 Main Features of *C. difficile* Biofilm

The mechanisms underlying biofilm formation in *Clostridium* species, particularly *C. difficile* (Pantaléon et al. 2014), as well as the role of biofilm in CDI have not been investigated as extensively as in other bacterial species (Hall-Stoodley and Stoodley 2009).

Members of the biofilm microbiota community can impact *C. difficile* biofilm formation by acting either antagonistically or synergistically (Normington et al. 2021). However, in a eubiotic status, *C. difficile* forms part of the healthy, multispecies biofilm during asymptomatic carriage (Chilton et al. 2018), and 3–15% of healthy adults are asymptomatically colonized by *C. difficile*, with commensal species acting antagonistically and providing resistance against *C. difficile* colonization.

Studies on interbacterial interactions among commensal bacteria within the gut microbiota and *C. difficile* showed that the introduction of the pathogen resulted in increased adhesion of commensals, especially *Bacteroides* spp., and inhibition of *C. difficile* proliferation (Hassall et al. 2021). Not only do bacterial interactions provide protection from *C. difficile* colonization, interactions with yeast (*Candida* spp.) biofilms have been reported to be critical in determining the outcome of an infection. It has been suggested the reduced capacity of biofilm production by

Candida strains isolated from CDI patients might have a role in the development of *C. difficile* infection, with higher biofilm production being observed in *Candida albicans* strains isolated from healthy donors compared to that of the yeasts cultured from CDI patients (Brunetti et al. 2021).

On the contrary, some metabolic products and/or other bacterial species can act synergistically and induce *C. difficile* biofilm formation. Metabolic interactions of *C. difficile* with commensal species have been investigated by combining genome-scale metabolic reconstructions of *C. difficile*, *Bacteroides thetaiotaomicron*, *Faecalibacterium prausnitzii*, and *E. coli*. Decreased carbohydrate/increased amino acid levels and/or increased primary bile acid levels induced increases in *C. difficile* and decreases in *F. prausnitzii* abundances (Phalak and Henson 2019). In confirmation of the involvement of microbiota-generated bile acids in *C. difficile* biofilm production, and therefore in *C. difficile* persistence and risk of relapse, Dubois and colleagues reported that sub-lethal concentrations of deoxycholate stimulate biofilm formation by inducing metabolic pathways and cell envelope reorganization, and represses toxin and spore production (Dubois et al. 2019). The same research group demonstrated that extracellular pyruvate induces *C. difficile* biofilm formation in the presence of deoxycholate and, in the absence of deoxycholate, pyruvate supplementation was sufficient to induce biofilm formation, suggesting that pyruvate-induced biofilm formation might be a key process of the pathogen persistence (Tremblay et al. 2021).

Since *C. difficile* biofilm-forming ability may affect its pathogenesis and persistence when a disruption of gut microbiota occurs, research on the ability of *C. difficile* to form a biofilm has attracted considerable interest, with a number of in vitro studies being carried out in this regard. Donelli and co-workers, by using crystal violet staining and Field Emission Scanning Electron Microscopy (FESEM), first showed that a clinical isolate of *C. difficile* (CdiBs21) formed a moderate biofilm on flat bottomed plastic tissue culture plates (Donelli et al. 2012). Afterwards, the hypervirulent strain R20291 was revealed to be a strong biofilm producer, identifying a link between sporulation and biofilm formation with a biofilm reduction in a *spo0A* mutant (Dawson et al. 2012). Further analysis, by Đapa and co-workers on the massive biofilm formation of R20291 strain, confirmed the involvement of virulence-associated protein, Cwp84, flagella, and the putative quorum-sensing regulator, LuxS. In the same conditions, the strain 630 formed a weak biofilm (Đapa et al. 2013).

Biofilm formation by hypervirulent and other *C. difficile* strains showed differences in terms of ability to form weak, moderate, or strongly adherent biofilms, with the hypervirulent strains always producing greater biofilms (Hammond et al. 2014; Mathur et al. 2016; Piotrowski et al. 2017).

Regarding the adhesion properties on abiotic surfaces, strains were characterized for their ability to form a biofilm, adhesion on an inert surface, and hydrophobicity, and no correlation was highlighted between the ability of thirty-seven strains to adhere/form biofilm and their virulence, except for the incapacity of non-motile strains to form a high biofilm (Pantaléon et al. 2018).

Biofilm structure is supported by the EPS (extracellular polymeric substances) matrix, mainly composed of proteins, extracellular DNA (eDNA), and polysaccharides, that provides the scaffold by which bacteria adhere to each other and to surfaces. EPS matrix is responsible for the impenetrability of bacterial biofilms, thus contributing to the antibiotic resistance in vivo as well as to the escape from immune responses during infection. Specifically, *C. difficile* biofilm is composed of a multi-component matrix (Fig. 1) made of proteins, extracellular DNA, and polysaccharide II (PSII) (Dawson et al. 2012; Đapa et al. 2013). The latter is an antigen commonly found on the surface of all *C. difficile* species (Ganeshapillai et al. 2008) and detected in the matrix of several *C. difficile* strains (Đapa et al. 2013; Semenyuk et al. 2014). Semenyuk and colleagues found, in the *C. difficile* biofilm matrix extract and in the whole cell extracts, six proteins involved in metabolism: formate–tetrahydrofolate ligase, acetyl-CoA acetyltransferase, 2-hydroxyisocaproate

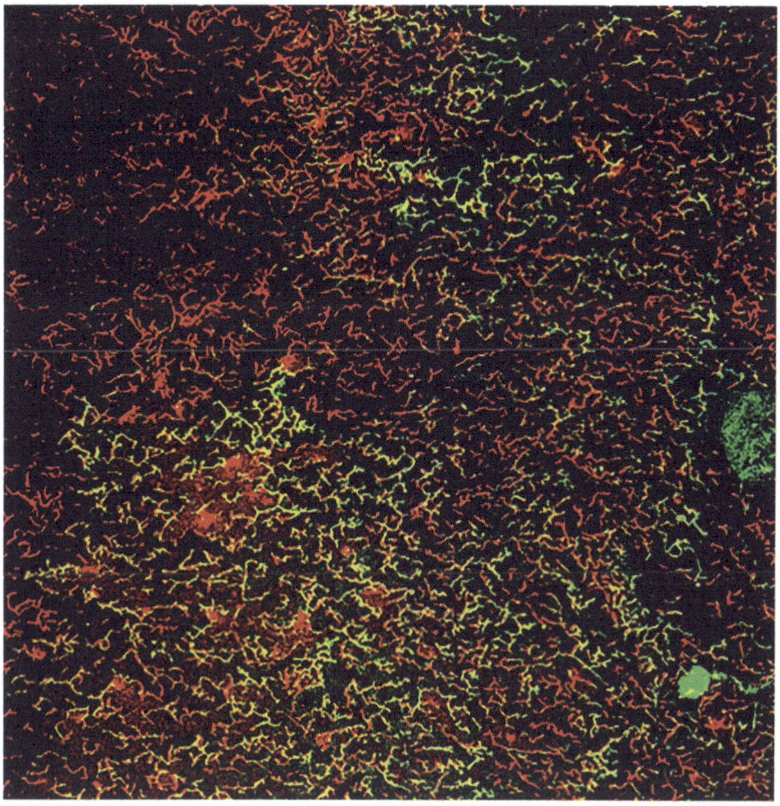

Fig. 1 Confocal laser scanning microscopy (CLSM) analysis of *C. difficile* in vitro biofilm after 48 h. The red-fluorescent propidium iodide stain labels bacteria, while the lectin Concanavalin A binds to residues of the exopolysaccharide matrix

CoA-transferase, NAD (Nicotinamide Adenine Dinucleotide)-specific glutamate dehydrogenase, 3-hydroxybutyryl-CoA dehydrogenase, and fructose-bisphosphate aldolase. On the contrary, cell wall-associated proteins were revealed in cell-surface extracts only, the matrix proteins not arising from the cell surface. These proteins, possibly originated from the cell lysis, most likely contribute in some way to biofilm formation (Semenyuk et al. 2014).

Detachment assays performed on 102 *C. difficile* isolates demonstrated that slpA and cwp84 were differentially expressed by *C. difficile* ribotype and biofilm production level, and that the highest detachment of biofilm was achieved after proteinase K treatment (>90%), concluding that proteins seem to have an important role in the biofilm's initial adherence and maturation (Martínez-Meléndez et al. 2021).

Electron micrographs show that *C. difficile* biofilms are constituted by heterogeneous populations of vegetative cells, sporulating cells, and cell debris (Donelli et al. 2012; Dawson et al. 2012).

Additionally, toxin A expression, which is linked to metabolism, has been reported to be down-regulated in biofilm (Poquet et al. 2018), while spores have reduced germination efficiency in mature biofilms, thus presumably facilitating the preservation of a dormant population ready to cause recurrent infections (Semenyuk et al. 2014). Interestingly, by indirect immunofluorescence analysis, the presence of two exosporium proteins (i.e. CdeC and the N-terminal domain of BclA1) has been detected in the spores in *C. difficile* biofilms (Pizarro-Guajardo et al. 2016a). By transmission electron microscopy, it has also been demonstrated that two exosporium morphotypes, one with a thick outermost exosporium layer and another with a thin outermost exosporium layer, were formed during biofilm development (Pizarro-Guajardo et al. 2016b; Pickering et al. 2018). Dormant spores located within biofilms were detected for the duration of the experiment within a triple-stage chemostat gut model inoculated with indigenous gut microbiota and *C. difficile* cells (Crowther et al. 2014a, b). Sessile spores, displaying increased recalcitrance to germination, may be compared to superdormant spores of *Bacillus* spp. (Ghosh et al. 2009), resulting persister cells. Spores derived by biofilm-growing *C. difficile* strains of four ribotypes (001, 020, 027, & 078) also exhibited increased thermotolerance when compared to spores collected from planktonic culture (Pickering et al. 2018).

It has been shown that eDNA is an essential component of the *C. difficile* biofilm matrix, as incubation with DNase I reduces the biofilm biomass produced (Dawson et al. 2012; Đapa et al. 2013; Semenyuk et al. 2014). One way to explain the presence of eDNA and cell debris seen within *C. difficile* biofilms could be through the differential expression of toxin–antitoxin (TA) systems. TA systems comprise a stable toxin, which is intracellular and only affects an essential cellular process, and an unstable antitoxin, which sequesters the effect of the toxin (Wen et al. 2014). The cell death and lysis caused by TA systems in a small percentage of the bacterial cells could contribute to the assembly of the matrix during biofilm formation for the 'greater good' of the population (Gil et al. 2015). The *C. difficile* genome encodes a number of putative TA systems (Gil et al. 2015) with the MazE-MazFTA system best described (Rothenbacher et al. 2012), however their contribution towards biofilm formation has not been determined.

The complex biofilm architecture of *C. difficile* strains has been analysed in different in vitro studies by FESEM (Fig. 2) and Confocal Laser Scanning Microscopy (CLSM). FESEM micrographs of *C. difficile* grown on glass coverslips revealed wide mats of rod-shaped vegetative cells, spores, and sporulating cells interconnected by a network of extracellular material constituted by cell debris and string-like material connecting the cells (Fig. 3). The appearance seems to be consistent with their being biofilms and with other SEM observations on plastics (Dawson et al. 2012; Semenyuk et al. 2014) or agar (Lipovsek et al. 2013).

CLSM analysis describes more accurately the biofilm architecture, allowing definition of the thickness and to visualize cells inside the biofilm (Fig. 4). Semenyuk and colleagues explored the evolution of biofilm structure and composition over the time, identifying, after 24 h, regions with a high concentration of apparently proliferating cells and cell debris as well as small colonies, distant from the main biofilm colony, interpreted as sites of new growth formed by cells migrated from the larger colony edge. After 3 days, together with rod-shaped cells and apparent cell debris, authors detected ovoid cells in the biofilm that were identified as spores by phase contrast microscopy. At 6 days, most of the cells in the biofilm had become spores with isolated regions of vegetative cells (Semenyuk et al. 2014). As already demonstrated for other bacterial species, the *C. difficile* biofilm thickness tends to increase every day, even if the depth varied according to the areas (Dawson et al. 2012; Maldarelli et al. 2016), and also the amount of matrix constituting biofilm increases proportionally (Dapa and Unnikrishnan 2013). In 2018, Poquet and colleagues observed in situ an intact biofilm displaying a sparse, heterogeneous, and high 3D structure made of rods and micro-aggregates, cell micro-aggregation being shown to play a major role in biofilm formation and architecture (Poquet et al. 2018).

Regarding *C. difficile* biofilm formation in vivo, clumps of *C. difficile* cells have been observed in a mouse model associated with damaged tissue (Lawley et al. 2009), while aggregation or clusters of *C. difficile* cells were observed in hamster and monoxenic mouse, respectively (Spencer et al. 2014; Soavelomandroso et al. 2017).

3 Genetic Factors Behind *C. difficile* Biofilm Formation

The formation of *C. difficile* biofilms is a multifactorial process involving many virulence-associated proteins and potentially several complex networks to regulate biofilm formation. The cell surface of *C. difficile* plays a pivotal role

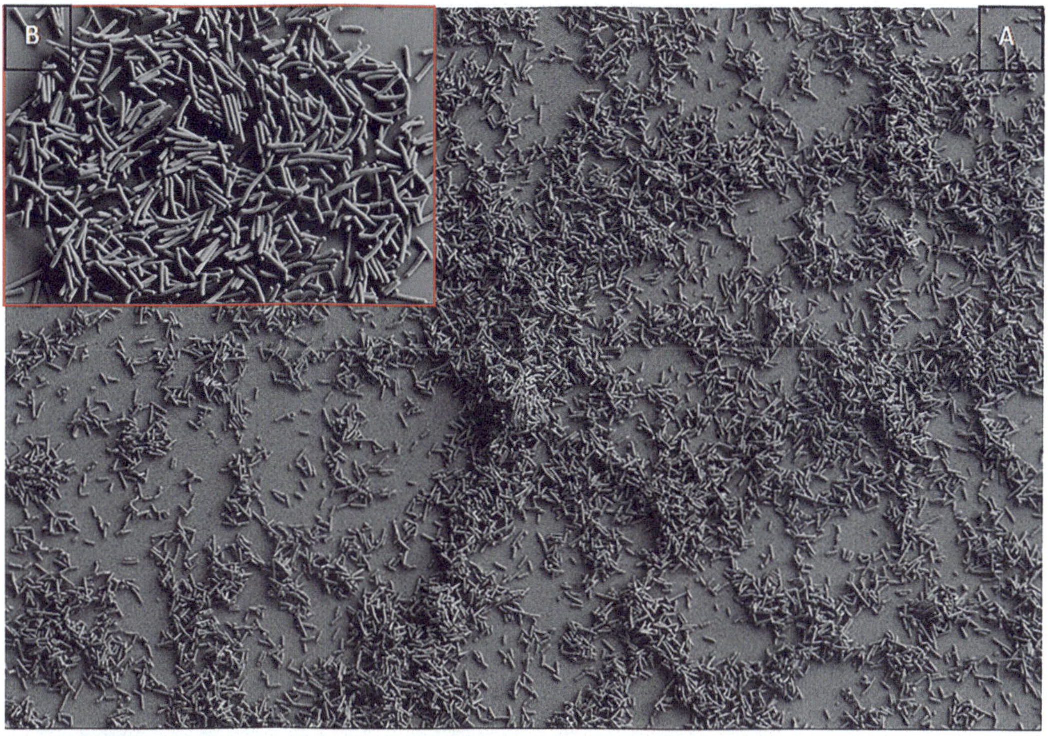

Fig. 2 FESEM analysis of *C. difficile* biofilm formed in vitro after 48 h. Biofilms micrographs were obtained at an accelerating voltage of 2 kV with magnifications of 1000× (**a**) and 5000× (**b**)

throughout the whole biofilm process, from the initial adherence of a cell to the dispersal of biofilm. Thus, structures directly involved in biofilm formation have been identified by investigating proteins and macromolecules present on the cell surface. Flagella, Type IV pili

Fig. 3 FESEM analysis of *C. difficile* biofilm formed on glass coverslips after 5 days; mushroom-like structures formed by rod-shaped vegetative cells, spores, sporulating cells, and cell debris. Biofilms micrographs were obtained at an accelerating voltage of 2 kV with magnifications of 5000×

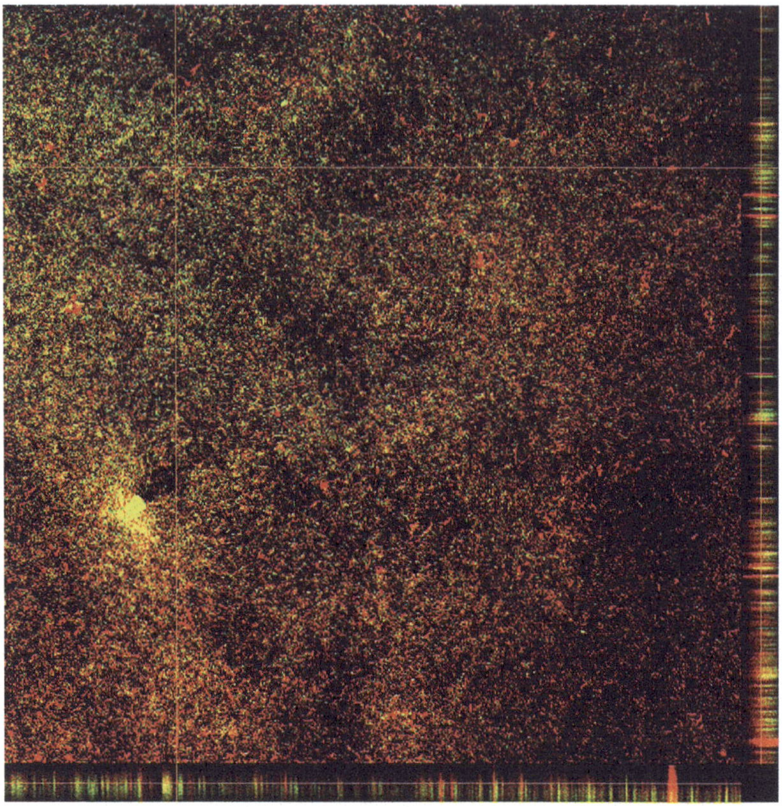

Fig. 4 Three-dimensional CLSM image of *C. difficile* biofilm grown in vitro for 5 days. The red-fluorescent propidium iodide stain labels bacteria, while the lectin Concanavalin A binds to residues of the exopolysaccharides

(T4P), and the S-layers are all implicated in *C. difficile* biofilm formation.

In the closely related bacterium, *Clostridium perfringens*, T4P plays an important role in twitching motility, biofilm formation, and disease pathogenesis (Varga et al. 2006). The T4P filament in *Clostridium* spp. is typically made up of a major pilin subunit, PilA, and minor pilin subunits, PilJ and PilW, with further genes putatively involved in the retraction of the pilus to provide the twitching motility (Varga et al. 2006; Piepenbrink et al. 2014; Piepenbrink et al. 2015; Melville and Craig 2013). T4P were once thought only to be present in Gram-negative bacteria, but Varga et al. (2006) first identified several putative pilin genes within the genome of *C. difficile* strain 630, and Goulding et al. (2009) used immunogold labelling to show that pili structures are present on the cell surface of *C. difficile* during infection in hamsters. Analysis of pilin gene transcripts from in vitro *C. difficile* biofilm cultures shows an up-regulation of *pilA1* transcripts compared to planktonic cultures (Maldarelli et al. 2016), which is even more prominent in *C. difficile* strain R20291 compared to strain 630 (Purcell et al. 2016). The importance of T4P in *C. perfringens* can be seen in mutants that are defective in T4P formation as these mutants display abnormal biofilm formation compared to the wild-type strain (Varga et al. 2008). In *C. difficile*, mutants that have a disrupted *pilA1* gene lack T4P structures on the cell surface under laboratory conditions (Bordeleau et al. 2015). Interestingly, T4P play an important role in the early stages of *C. difficile* biofilm formation, as mutants with a *pilA1* disruption show a reduced biofilm biomass compared to wild-type (Maldarelli et al. 2016; Purcell et al. 2016). Additionally, *pilJ* and *pilW* null mutants also displayed reduced biofilm formation, where PilJ and PilW were found to recognize and bind extracellular DNA found in biofilms (Ronish et al. 2022); extracellular DNA forms part of the extracellular matrix (Dawson et al. 2021). However, T4P seem to play little role in the maturation of a biofilm, as these mutants showed no difference in biofilm biomass compared to wild-type when grown over 7 days (Maldarelli et al. 2016). Up to nine putative pilin-like proteins are encoded on the *C. difficile* genome (Melville and Craig 2013; Maldarelli et al. 2014), three of these being designated as major pilin subunits (*pilA1-3*). The biological function for each of these pilin genes remains unclear, with current hypotheses suggesting T4P made from these different pilin subunits could perform different functions, or pilin switching could be a mechanism for immune evasion, or, as many of these are not located in T4P operons, these could be non-functional. In other bacteria, T4P play a pivotal role in biofilm formation and disease pathogenesis; T4P are essential for passage of *Neisseria meningitidis* to cross the blood-brain barrier (Nassif et al. 1994), whilst T4P-mediated motility is important for *Pseudomonas aeruginosa* early biofilm development (Klausen et al. 2003). The role of T4P during in vivo *C. difficile* colonization and persistence remains to be investigated.

Recent work on how T4P is regulated in *C. difficile* has identified the bacterial secondary messenger molecule Bis-(3′-5′)-cyclic dimeric guanosine monophosphate (c-di-GMP) as a key component to the regulatory pathway. In Gram-negative bacteria, c-di-GMP modulates virulence attributes, such as biofilm formation in *Vibrio cholerae* (Tischler and Camilli 2005) and *P. aeruginosa* (Kulasakara et al. 2006), decreased flagella-mediated motility in *E. coli*, and cell differentiation in *Caulobacter crescentus* (Aldridge et al. 2003). Two enzymes, diguanylate cyclases (DGCs) and phosphodiesterases (PDEs) that either synthesize or degrade c-di-GMP (Römling and Amikam 2006), tightly control the intracellular levels of c-di-GMP. *C. difficile* is unusual among Gram-positive organisms by the number of DGCs & PDEs encoded on the genome; strain 630 has 37 putative c-di-GMP metabolizing enzymes. Ectopic expression of 31 of these enzymes in the surrogate organism, *V. cholerae*, confirmed these genes as either having DGC or PDE activity (Bordeleau et al. 2011). Interestingly, heterologous and homologous expression of *C. difficile* 630 CD1420 (*dccA*) in either *V. cholerae* or *C. difficile*, respectively, increased cellular levels of c-di-GMP and induced biofilm formation (Bordeleau et al. 2011; Purcell et al. 2012). Through overexpression of *dccA*, high intracellular levels of c-di-GMP resulted in increased expression of the genes in the T4P operon and a greater number of pili observed on the cell surface (Bordeleau et al. 2015). High intracellular levels of c-di-GMP also regulate the cell surface attachment of two proteins, CD2831 and CD3246, which both proteins contribute towards early-stage biofilm formation (Dawson et al. 2021); CD2831 is a collagen-binding protein important for colonization (Arato et al. 2019). In other bacteria, c-di-GMP controls the transcription and translation of many genes by direct binding to c-di-GMP riboswitches (Sudarsan et al. 2008). Riboswitches are mRNA molecules that bind small molecules (such as c-di-GMP) resulting in the transcription of downstream genes (Winkler and Breaker 2005). In this way, the same small molecule can coordinate multiple genetic pathways. RNA-seq experiments first identified a Type II c-di-GMP riboswitch located upstream of the start of the *C. difficile* major T4P operon (*pilA1*) (Soutourina et al. 2013), called Cdi2_4, which is switched 'ON' via a conformational change upon binding c-di-GMP to the riboswitch to relieve a predicted Rho-independent transcription terminator (Bordeleau et al. 2015). Between different strains of *C. difficile* there appears to be subtle variations in *pilA1* expression patterns during biofilm formation with strains 630 and R20291 (Purcell et al. 2016), which could be due to differences in the total c-di-GMP levels.

Research on other regulatory proteins within *C. difficile* suggests that its pathogenesis is intimately linked to the metabolic state of the bacterium (Bouillaut et al. 2015). CodY is a pleotropic regulator involved in the adaptive response of Gram-positive bacteria to low nutrient levels, and in *C. difficile*, an estimated 52 genes are directly regulated by CodY (Dineen et al. 2010; Bouillaut et al. 2015). One of these genes is *pdcA* (*CD1515*), which is a PDE enzyme that affects the regulation of flagella biosynthesis by influencing c-di-GMP levels (Purcell et al. 2012, 2017). Thus, through this regulatory pathway, *C. difficile* biofilm formation is connected to the nutrient availability of the bacterium. C-di-GMP acts as a signalling molecule coordinating the transition from a planktonic, motile lifestyle to a sessile, biofilm lifestyle in many bacterial pathogens. In *C. difficile*, high c-di-GMP levels directly repress the major flagella operon *flgB* through a Type I c-di-GMP riboswitch, Cdi1_3, located 496 bp upstream of the *flgB* start codon (Sudarsan et al. 2008; Soutourina et al. 2013). Through the two types of riboswitches, one family of signalling molecules can regulate the expression of T4P and flagella biosynthesis during *C. difficile* biofilm formation.

A decrease in flagella transcripts would indicate a limited role for flagella during biofilm formation, and targeted disruption of *fliC* gene in strains 630 or R20291 had no effect on biofilm formation compared to the wild-type strains (Faulds-Pain et al. 2014; Valiente et al. 2016). However, one report has shown a *fliC* mutant had significantly less biofilm biomass compared to wild-type (Ðapa et al. 2013). The genetic organization of the *C. difficile* flagella operon can be split into three parts, however the F2 locus is the most divergent between the genomes of different *C. difficile* strains (Stabler et al. 2009; Stevenson et al. 2015). The F2 locus encodes genes involved in glycosylation of the flagella with sugar moieties, and the disruption of these genes resulted in the production of flagella on the cell surface even though most of these mutants were non-motile (Twine et al. 2009; Faulds-Pain et al. 2014; Valiente et al. 2016). Interestingly, these mutants produced more biofilm biomass compared to the wild-type strain (Faulds-Pain et al. 2014; Valiente et al. 2016). In the closely related bacterium, *Bacillus subtilis*, inhibition of flagella rotation acts as a mechanical trigger to activate the DegS-DegU two-component signal transduction system, which regulates biofilm formation and matrix production (Cairns et al. 2013, 2014). Although no DegS/DegU homologues have been identified in *C. difficile*, this could be why these flagellate, non-motile mutants produced more biofilm biomass, although more work is needed to understand the regulatory mechanisms behind this phenotype.

Using riboswitches is one-way *C. difficile* regulates the change from motility to biofilm, however other regulatory RNA molecules appear to play a role in biofilm formation. Small non-coding RNAs (sRNAs) act by base pairing with their target mRNAs, leading to modulation of mRNA stability or translation (Chao and Vogel 2010; Soutourina 2017). Buckley et al. (2021) identified 25 different sRNAs that were specifically expressed under biofilm conditions compared with planktonic growth. Some sRNAs require an RNA chaperone protein called Hfq to help the base pair binding of the sRNA and mRNA molecules. In other bacteria, mutating Hfq has pleotropic effects on cell physiology, ranging from increased sensitivity to external stresses (detergents, iron limitation, and oxidative stress), to increased biofilm formation, or reduced virulence (Chao and Vogel 2010). Boudry et al. (2014) used a knockdown approach to decrease Hfq protein levels fivefold compared to wild-type to determine its contribution towards cell physiology. Using this approach, the authors observed an increase in biofilm formation in the Hfq depleted strain, indicating that sRNAs play a role in negatively regulating biofilm formation. Alongside this, the authors observed a decrease in flagella present on the cell surface and increased expression of cell wall/membrane proteins, all of which could have contributed to the increase in biofilm formation (Boudry et al. 2014). The extent of sRNA binding to Hfq has been studied by Fuchs et al. (2021), where 26 different sRNAs were bound to Hfq in co-immunoprecipitation experiments, and these sRNA-Hfq complexes extended the half-life of the sRNA.

Another cell surface organelle that has been implicated in *C. difficile* biofilm formation is the S-layer. The *C. difficile* S-layer (Cerquetti et al. 2000) is a two-dimensional paracrystalline protein array coating the cell and is made up of SlpA subunits that are post-translational cleaved by another protein called Cwp84 (de la Riva et al. 2011; Fagan and Fairweather 2014). The S-layer harbours up to 28 different cell wall proteins that are anchored to the cell wall by CWB2 protein domains (Fagan and Fairweather 2014; Willing et al. 2015). Disruption of *cwp84* results in uncleaved SlpA in the cell wall, which in turn results in aberrant retention of other cell wall proteins at the cell surface (Kirby et al. 2009; de la Riva et al. 2011). The effect of *cwp84* disruption on biofilm formation was dependent on the strain background. In strain R20291 a *cwp84* mutant showed reduced biofilm formation (Đapa et al. 2013), whereas in strain 630 this mutant showed an increase in biofilm formation (Pantaléon et al. 2015). As these strains encode a different array of proteins that are predicted to associate with the S-layer (Biazzo et al. 2013), an immature S-layer may contain different surface-associated proteins between the two strains. Whether the S-layer per se is involved in biofilm formation or if this effect is due to the proteins associated with the S-layer remains uncertain. Other surface-associated proteins have been found to influence biofilm formation. A mutated, a putative lipoprotein, CD1687, had reduced biofilm formation (Dubois et al. (2019), and a serine/threonine kinase null mutant (PrkC, involved in cell wall homeostasis) formed more biofilm compared with wild type (Cuenot et al. 2019). Although it is not immediately clear how these proteins can influence biofilm formation.

Regarding quorum sensing, bacteria detect a threshold level of autoinducer (AI) molecules and activate a signal cascade that leads to altered gene expression. The AI-2 molecule is synthesized by LuxS and produced by Gram-positive and Gram-negative bacteria. Due to the number of bacteria that can produce and detect AI-2 molecules, this quorum signalling mechanism is thought to function as an intra- and interspecies communication molecule. *C. difficile* encodes a *luxS* homologue and produces a chemically active AI-2 molecule that can induce homologous and heterologous gene expression (Carter et al. 2005;Lee and Song 2005). Biofilm formation in a *C. difficile luxS* mutant was severely diminished compared to wild-type strain, where not even a bacterial monolayer was able to form (Đapa et al. 2013; Slater et al. 2019). The regulatory pathway behind AI-2 induced biofilm formation is currently unknown. In the gut mucosa, *C. difficile* interacts with members of the sessile community (Lawley et al. 2009; Buckley et al. 2011; Donelli et al. 2012; Crowther et al. 2014a, b; Semenyuk et al. 2015), where such interspecies signalling could play an important role for disease progression. Indeed, Slater et al. (2019) showed a LuxS-dependent inhibition of *C. difficile* biofilm formation when co-cultured with *Bacteroides fragilis*.

Based on current research, it seems clear that the genetic regulation behind *C. difficile* biofilm formation is extremely complex and several different global regulators that link various metabolic pathways influence it. The *C. difficile* sporulation master regulator, *spo0A*, besides coordinating sporulation by undergoing post-translational phosphorylation (Spo0A-P) in order to activate the sigma factor cascade (Pettit et al. 2014; Al-Hinai et al. 2015), also plays a role in biofilm formation. Disrupting the *spo0A* gene resulted in a reduced biofilm phenotype that could be restored by complementation (Dawson et al. 2012; Đapa et al. 2013). In *Bacillus* spp., the intracellular concentration of Spo0A-P is critical to determining if the cell proceeds down either the sporulation pathway (high Spo0A-P levels), or biofilm pathway (low Spo0A-P levels) (Mhatre et al. 2014). During the early stages of biofilm formation Spo0A-P induces the expression of *sinI*, which inhibits a protein that represses the biofilm matrix genes, SinR (Vlamakis et al. 2013; Cairns et al. 2014). *C. difficile* encodes homologues of *sinI* and *sinR* (Edwards et al. 2014), however their role in regulating biofilms, and the regulon of SinR, is unknown and deserves further investigation. Other global regulators have a role in biofilm formation. Deletion of *ccpA* (glucose and carbon regulator) and *codY* (nutritional regulator) reduced in vitro biofilm formation (Dubois et al. 2019), phase variation of the CmrRST signal transduction system affected biofilm formation (Garrett et al. 2022), and deletion of the flagella assembly factor (FliW) increased biofilm formation (Zhu et al. 2021). The intricate role of global regulators appears to confirm that there is a link between biofilm formation and the metabolic status of the cells. In studies by Poquet et al. (2018), Brauer et al. (2021), Buckley et al. (2021), and Tremblay et al. (2021), the changes to the metabolic landscape as a *C. difficile* cell transitions from a planktonic, motile lifestyle into a biofilm have been detailed. These changes occur at the cell surface, in a reduction of nutrient transporters and an increase in oligo-peptide transporters, and a shift in carbon metabolism, with upregulations in the Wood–Ljungdahl pathway, to generate important central metabolites such as acetyl-CoA.

It has also been demonstrated that biofilm formation in *C. difficile* may be stress-inducible; exposure of cells to sub-inhibitory concentrations of antibiotics, such as MTZ and vancomycin, induces biofilm formation (Đapa et al. 2013; Vuotto et al. 2016). In other bacteria, this stress-induced biofilm formation is induced by the SOS regulatory network in response to DNA damage, through activation of the transcriptional repressor, LexA, by the recombinase protein, RecA (Butala et al. 2009). Mutation of *lexA* in *C. difficile* caused pleotropic effects to the cell: elongated cell morphology, decreased sporulation & motility, and increased biofilm formation (Walter et al. 2015). In silico analysis of predicted LexA binding sites within the *C. difficile* genome suggests LexA could regulate up to 29 loci (Walter et al. 2014). How this regulatory pathway contributes to *C. difficile* biofilm formation is unclear and warrants further investigation.

4 In Vitro and In Vivo Models to Study the Interactions of Sessile Microorganisms

The mammalian intestinal mucosa is home to a complex mixture of microbial communities, which can aggregate to form mats or biofilm structures over epithelial cells and play a crucial role in host health and disease (Motta et al. 2021). *C. difficile* cells can associate with these microbial communities during CDI (Lawley et al. 2009; Goulding et al. 2009; Spencer et al. 2014; Semenyuk et al. 2015). Since understanding the interplay between *C. difficile* and the sessile community could be key to designing defined microbial treatments for recurrent CDI, several in vitro and in vivo models have been developed to study these kinds of interactions. In vitro models allow researchers to manipulate and control certain factors and/or conditions, thus providing a valuable tool for biofilm research. Systems can be defined as ‘closed’ or ‘open’. Closed (or static) biofilm models, such as the popular microtitre tray-based models, are based on batch culture, in which there is limited nutrient availability and aeration, as well as a build-up of metabolic products. Open (or dynamic) biofilm model systems are based on continuous flow models, whereby fresh media replace metabolic products and waste constantly. Whilst open biofilm models may be better able to simulate sheer forces and flow, they often require more technical expertise and complex equipment than closed systems, and so are less amenable to high throughput workloads.

The use of microtitre trays is one of the simplest methods used to investigate biofilm formation. A 24 well format has been consistently successful in growing *C. difficile* in both mono- and poly-microbial species biofilm and has been used to determine some of the genetic mechanisms behind *C. difficile* biofilm formation, as well as its interaction with other gut microbiota. Donelli et al. (2012) found that several gastrointestinal residing bacteria were able to cooperatively form a biofilm when co-cultured together and highlighted a positive interaction between *C. difficile* and *Finegoldia magna*. Similar positive interactions were described by Dubois et al, who grew mono- and dual-species biofilms with different permutations of bile salts and demonstrated that *Clostridium scindens* enhances biofilm formation of *C. difficile* by converting cholate into deoxycholate (Dubois et al. 2019), and by Smith et al, who demonstrated that *E. faecalis* readily forms biofilms with *C. difficile*, enhancing *C. difficile* survival during vancomycin treatment (Smith et al. 2022). Conversely, mixed biofilms formed with *E. coli* did not impact *C. difficile* survival following vancomycin. Antagonistic effects of probiotic organisms including *Lactobacillus rhamnosus*,

Bifidobacterium longum, and *B. breve* have also recently been described (Normington et al. 2021).

Regarding the open (or dynamic) biofilm model systems, Poquet et al grew *C. difficile* mono-species biofilms in continuous flow glass microfermentors (based on those reported by Ghigo (2001) for investigation of *E coli* biofilms) to investigate gene expression and biofilm architecture, reporting that gene expression is widely reprogrammed, particularly with respect to cell surface properties and metabolism, and likely controlled to some extent by CD2214 and CD2215 regulators (Poquet et al. 2018). More gut-reflective chemostat systems seeded with pooled human faeces have investigated *C. difficile* interactions with gut organisms. Single stage microfermentors (known as MinBioReactor Arrays (MBRAs)) have successfully been used to simulate *C. difficile* growth and toxin production (Robinson et al. 2014) and investigate colonization resistance (Mahnic et al. 2020). *C. difficile* has been described in multi-species biofilms formed on mucus coated coverslips within these fermentors (Engevik et al. 2021) and interactions between *C. difficile* and microbiota species have been investigated; *Fusobacterium nucleatum* has been shown to bind *C. difficile* and promote biofilm formation in mucus (Engevik et al. 2021).

Crowther et al. (2014a, b) developed a modified version of the continuous triple chemostat system (Macfarlane et al. 1998) to monitor the sessile populations, by using glass rods suspended from the lid. During simulated CDI, the authors observed consistent sessile populations comprised of *Bacteroides* spp., *Bifidobacteria* spp., *Lactobacillus* spp., and *Enterococcus* spp. on different rods sampled at the same time (Crowther et al. 2014a, b). Normington and colleagues recently used bacterial 16S rRNA sequencing analysis to further identify the composition of the sessile community from these rods (Normington et al. 2021). *C. difficile* spores were shown to associate with and persist within the biofilm, as it has been visualized microscopically in Fig. 5. Both sessile spore and vegetative populations were isolated during the CDI phase (Crowther et al. 2014a, b) and transfer of biofilm containing *C. difficile* spores into a *C. difficile* negative model with disrupted microbiota resulted in simulated CDI (Normington et al. 2021), providing further evidence that biofilm may provide a reservoir for *C. difficile* spores, facilitating recurrent infections. In vivo models of CDI have been used to specifically identify the bacterial populations associated with the mucus layer during disease. Confocal laser scanning microscopy and immunochemistry analyses demonstrated that *C. difficile* bacteria were distributed heterogeneously over the intestinal tissue during monoculture in germ free mice (Soavelomandroso et al. 2017), with highest bioburdens in caecum and colon. Bacteria were localized both inside and outside the mucus layer, mostly encased in 3-D structures overlaying the mucus layer. Using paraffin embedded sections, to preserve the mucus layer, and fluorescent in situ hybridization (FISH), Semenyuk et al. (2015) identified *C. difficile* vegetative cells within the outer mucus layer. Microbial taxonomy analysis from 16S rRNA sequences recognized other bacterial genera residing within the mucus layer, from several families belonging to Bacteroidetes and Firmicutes (*Lactobacillaceae*, *Lachnospiraceae*, and *Clostridium* cluster XVII and XIV). Those microbial species that directly interact with *C. difficile* in vivo are still unknown, however Smith et al recently performed FISH during CDI in mice and described the co-localization of Enterococci and *C. difficile* in biofilm-like mucosal structures (Smith et al. 2022). Interestingly, during the early phase of CDI an increase in *Enterbacteriaceae* was observed within the mucosal populations (Semenyuk et al. 2015). Whether such an increase enhances *C. difficile* recruitment into the biofilm, or allows extra-intestinal invasion (Ng et al. 2013), as seen by (Goulding et al. 2009; Lawley et al. 2009), remains to be determined.

Human intestinal enteroids have been developed (Chang-Graham et al. 2019; McCracken et al. 2011) enabling biofilm formation on clinically relevant differentiated human intestinal cells to be monitored. Such systems have been used to study *C. difficile* growth in both mono- and co-culture. Leslie et al. performed the first

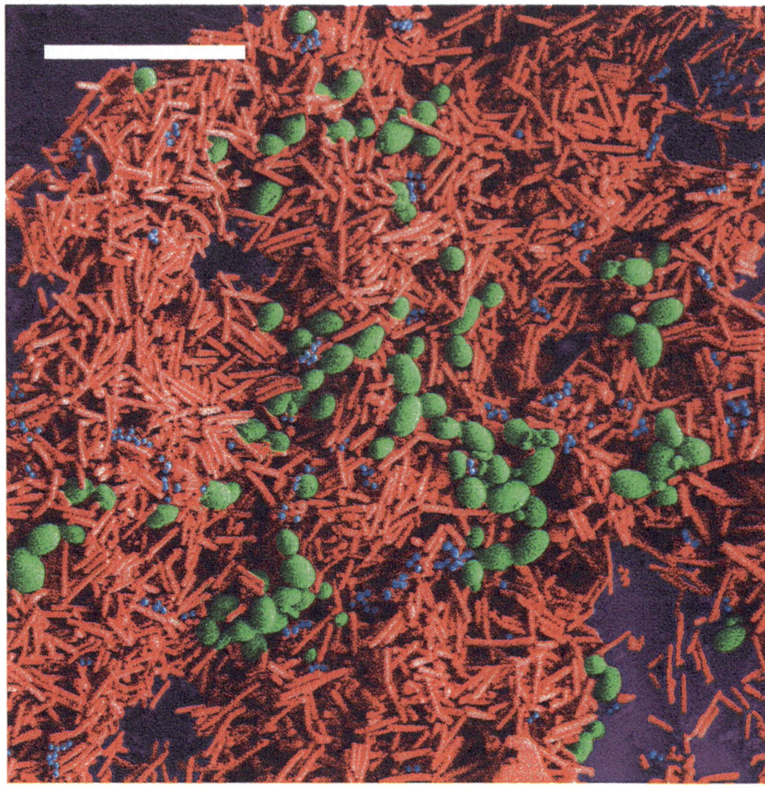

Fig. 5 Scanning electron micrograph of an in vitro polymicrobial biofilm. Using the method outlined in Vuotto et al. (2015), a biofilm containing *C. difficile* (red cells), *Candida* spp. (green cells), and *Staphylococcus* spp. (blue cells) was grown anaerobically for 3 days. White scale bar indicates 20 μm. SEM image taken from Normington C. et al., 2017. Influence of gut microflora on *C. difficile* biofilm formation. In Microbiology Society Annual Conference. p. P418

experiments to demonstrate *C. difficile* can persist in the lumen of human intestinal organoids for up to 12 h and cause loss of barrier function in a toxin-dependent manner, confirming previously identified toxin mechanisms (Leslie et al. 2015). Interactions between toxins and epithelial cells have been further investigated in similar human intestinal colonoids derived from human jejunal tissue (Engevik et al. 2020). Importantly, in work identifying the interactions between *C. difficile* and *F. nucleatum* leading to increased biofilm formation in mucus, human colonoids were used to confirm findings from biofilms grown within bioreactors (Engevik et al. 2021). The complex mucus layers in the colonoid model facilitated increased understanding for the role of mucus in *C. difficile* infection. As interest in gastrointestinal biofilm in health and disease increases, more technological advances in intestinal biofilm models are likely to materialize, providing opportunity to the research community to further the understanding of *C. difficile* biofilm formation and growth in the clinical setting, and the importance of biofilm in the *C. difficile* infection process.

5 Effects of Antibiotics on *C. difficile* Biofilm

Biofilm formation has been demonstrated to be an important factor enhancing antimicrobial resistance (Ciofu et al. 2017). In fact, during infection the biofilm mode of growth protects cells from antibiotic treatment, being 10–1000-fold more resistant to killing by bactericidal antimicrobials, compared to logarithmic-phase planktonic cells, and therefore exhibiting increased antibiotic tolerance (Mah and O'Toole 2001; Spoering and Lewis 2001; Hoiby et al. 2010).

Several mechanisms can contribute to antibiotic tolerance when cells grow in biofilm; including the biofilm matrix, acting as a physical barrier that affects penetration of antimicrobial agents

(Flemming and Wingender 2010), the presence of persister cells (Shah et al. 2006), and the genetic mutations occurring within bacteria in biofilm (Tyerman et al. 2013).

Tolerance mechanisms have been proposed in *C. difficile* biofilm (Đapa et al. 2013), so the effect of antibiotics most commonly used to treat CDI, such as MTZ and vancomycin (Peng et al. 2017), has been assessed against biofilm-growing cells and preformed biofilms.

Semenyuk and colleagues determined that 630 and VPI 10463 *C. difficile* cells grown as biofilm for 20 h had greater resistance to MTZ than planktonic cells, with 1 μg/ml of antibiotic inhibiting liquid cell growth by about 100-fold and 100 μg/ml causing only about a 10-fold reduction in the sessile cells. These data demonstrated that biofilms conferred a 100-fold increase in MTZ resistance (Semenyuk et al. 2014).

In addition to being ineffective to counteract in vitro *C. difficile* biofilm, it has been demonstrated that, at sub-inhibitory concentrations, MTZ even enhances biofilm formation in specific cases. In particular, three non-toxigenic clinical strains belonging to Polymerase Chain Reaction (PCR) -ribotype 010 with different MTZ susceptibility profiles exhibited variation in biofilm-forming ability. In presence of MTZ, a susceptible strain and a strain with reduced susceptibility revealed a significant increase in biofilm biomass, due to a more abundant EPS matrix production, while the biofilm-forming ability of the stable-resistant strain was not affected by the antibiotic pressure (Vuotto et al. 2016). This study highlights the possibility that the exposure of *C. difficile* to low concentrations of antibiotic present in the gut at the beginning or end of antibiotic therapy for CDI could serve as stress signal and, thus, stimulate biofilm production, with severe clinical implications for the treatment failure and recurrence of CDI. When similar experiments were carried out by using *B. fragilis*, opposed results were obtained. In fact, sub-inhibitory concentrations of MTZ were able to inhibit biofilm formation (Silva et al. 2014). Confirmation of enhanced biofilm production in sub-inhibitory concentrations of MTZ comes from Doan and colleagues. Additionally, they also observed an inhibition of motility and downregulation of flagellar genes for CD17-146 *C. difficile* strain with reduced susceptibility to MTZ, as well as elongated morphology, increased adherence to intestinal Caco-2/TC7 cells, and colonization in monoxenic and conventional mouse models (Doan et al. 2022). Vancomycin, compared to MTZ, demonstrates a higher clinical cure rate in adults with severe CDI and a similar clinical cure rate in moderate CDI cases, thus becoming the recommended therapy for more severe cases (Ofosu 2016). However, regarding its ability to act against mature biofilms, a number of papers have been published on staphylococcal species (Meeker et al. 2016; Ozturk et al. 2016; Hashem et al. 2017; Jimi et al. 2017) but limited and not encouraging data are so far available for *C. difficile*.

Đapa and co-workers first analysed the influence of vancomycin on biofilms of a *C. difficile* strain belonging to the PCR-ribotype 027, by examining the effects of different concentrations of antibiotic. High concentrations of vancomycin (20 μg/mL) failed to kill bacteria within biofilms while sub-inhibitory and inhibitory concentrations of vancomycin (0.25 μg/mL and 0.5 μg/mL, respectively) induced *C. difficile* biofilm formation. This suggests that increased antibiotic resistance in *C. difficile* may be mediated by the thick biofilm matrix and/or by the physiological state of bacteria within biofilms (Đapa et al. 2013). These results were corroborated by Mathur et al. (2016), who observed low efficacy of vancomycin against various PCR ribotypes.

Using a triple-stage human gut model, Crowther and colleagues simulated CDI and determined the effect of vancomycin on the motile and sessile *C. difficile* populations. Vancomycin exposure reduced the *C. difficile* planktonic populations to below the limit of detection, however the sessile populations were unaffected. This could be due to the levels of vancomycin that were detected within the biofilms [mean 40.4 mg/L (range 38.7–43.4 mg/L)] compared to those (54.7 mg/L) of the vessel lumen (Crowther et al. 2014a, b). A reduced level of vancomycin within the biofilm could prevent a critical level of vancomycin from being achieved. A differential response of sessile bacteria to antimicrobial administration was clearly observed, with *C. difficile* spores being largely unresponsive either to clindamycin instillation or perturbation of gut microbiota (Crowther et al. 2014a, b).

In vitro assays with sub-MICs of vancomycin did not affect *C. difficile* biofilm formation nor affect planktonic growth (Hamada et al. 2020).

The effect of tigecycline, teicoplanin, rifampicin, and nitazoxanide was also evaluated on the biofilm of five different *C. difficile* strains, noting that the sensitivities of these biofilms to different antimicrobials were strain-dependent, regardless of the produced biomass (Mathur et al. 2016).

More recently, fidaxomicin has been evaluated against *C. difficile* biofilm. James et al. demonstrated that fidaxomicin was significantly more effective than vancomycin or MTZ ($P < 0.001$) at killing vegetative cells in established biofilms, as well as more effective than MTZ at reducing viable spore counts in biofilms ($P < 0.05$). Fluorescently labelled fidaxomicin penetrated *C. difficile* biofilms in 1 h, and after 24 h treatment, fidaxomicin completely disrupted the biofilm structure (James et al. 2018).

Sub-minimum inhibitory concentrations of fidaxomicin were also tested against *C. difficile* UK027 biofilm, with biofilm inhibition and planktonic growth delay being observed (Hamada et al. 2020).

This finding reinforces the importance of antibiotic susceptibility testing against *C. difficile* biofilms, mainly in recurrent infections that may be induced by a strain that is both antibiotic tolerant and biofilm producing.

6 Alternatives to Counteract Biofilm-Growing *C. difficile*

Antibiotic administration, although carried out at higher doses over a prolonged period, often fails to counteract biofilm-related infections. In addition, antibiotic overuse and misuse are key factors contributing to the global increase of antibiotic resistance. Alternative therapeutic agents with antibacterial properties that prevent, disrupt, weaken, or kill the microbial community within a biofilm, are becoming increasingly attractive. In particular, anti-biofilm compounds: (1) may prevent biofilm formation by killing planktonic cells or blocking bacterial adhesion; (2) may counteract mature biofilms by destabilizing the matrix or by making the microbial cells susceptible to antimicrobial and/or host defence mechanisms; (3) may undo virulence factors involved in biofilm formation or may affect quorum sensing; (4) may have a bactericidal effect on biofilm-growing cells (Roy et al. 2017).

Efforts to fight these microbial communities include the use of different compounds, alone or in combination, to target different phases of biofilm, drug repurposing, peptides, nanomaterials, and medical devices coatings refractory to microbial adhesion or functionalized with anti-biofilm compounds (Ribeiro et al. 2016).

The conventional antibiotics used in CDI therapy are often unsuccessful and recurrent infections may occur, perhaps due to its ability to grow as a biofilm thus impairing antimicrobial activity. Different approaches, which are an alternative to the use of antibiotics, have been proposed to decrease *C. difficile* biofilm formation or disrupt mature biofilm.

Among the huge number of antimicrobial compounds today at our disposal, relatively few have been tested so far against *C. difficile* biofilm, and some of them are natural products. The first one tested was Manuka honey; its anti-biofilm properties on other species being already demonstrated (Badet and Quero 2011). Biofilms formed by two *C. difficile* strains, a ribotype 027 strain and a ribotype 106 strain, were used to test the effect of Manuka honey at varying concentrations of 1–50% (w/v). A dose-dependent response was observed for both test strains, with the optimum Manuka honey activity obtained at 40–50% (v/v) (Hammond et al. 2014). Consistent results were also obtained by evaluating its efficacy on clinical *C. difficile* strains belonging to four prominent PCR ribotypes (R017, R023, R027, and R046) (Piotrowski et al. 2017). Other essential oils have also been tested for their activity against

C. difficile biofilm. Brazilian Red Propolis (BRP) has been demonstrated to be effective against four strains of *C. difficile* in reducing ($P < 0.05$) the biomass, matrix proteins, and matrix carbohydrates of growing biofilms when used at MIC (625 mg/mL), while, at 8xMIC, was able to reduce ($P < 0.05$) the biomass and matrix proteins of mature biofilms (Costa et al. 2021). Essential oils of wild oregano, black pepper, and garlic reduced in vitro biofilm production of *C. difficile* clinical isolates belonging to six different PCR ribotypes, with the best activity of oregano oil (Aleksić et al. 2022). A component of coconut oil, i.e. the lauric acid, exhibits potent antimicrobial activities against multiple toxigenic *C. difficile* isolates, due to reactive oxygen species (ROS) generation and cell membrane damage, also considerably reduced both biofilm formation in vitro and in vivo (mouse infection model), and preformed biofilms in a dose-dependent manner (Yang et al. 2018). The antimicrobial agent thuricin CD, a sactibiotic produced by a bacterial strain derived from a human faecal sample, was also assessed against biofilms of R027, Liv022 R106, and DPC6350, alone or in combination with some antibiotics commonly used to treat CDI. Results underlined the effectiveness of thuricin CD against all the tested strains and its ability to significantly potentiate the efficacy of the antibiotics rifampicin, tigecycline, vancomycin, and teicoplanin against R027 biofilms (Mathur et al. 2016).

The ability of probiotics and/or prebiotics to interfere with adhesion and biofilm-forming abilities of *C. difficile* strains have been also investigated. The probiotic yeast Saccharomyces boulardii CNCM I-745 is able to inhibit of in vitro *C. difficile* biofilm formation by modification of extrapolymeric matrix components, in the dual-species biofilm (Lacotte et al. 2022). About prebiotics, it has been demonstrated that, among all the tested carbohydrates (cellobiose, fructooligosaccharides, inulin, mannose, and raffinose), fructooligosaccharides and mannose significantly decreased adhesion ($P < 0.001$) of *C. difficile* strains on non-mucous secreting HT-29, mucous secreting HT-29 MXT, and CCD 841 CoN cells lines. Sub-inhibitory concentrations (1%) of fructooligosaccharides and mannose increased *C. difficile* biofilm formation (Piotrowski et al. 2019). Combination of prebiotic (fructooligosaccharides) and bacterial strain (*Bacteroides* sp.) also decreased the adhesion of *C. difficile* (Piotrowski et al. 2022).

Interestingly, Rashid and colleagues assessed the in vitro effectiveness of hospital disinfectants against *C. difficile* spores embedded within biofilms and observed that not one of seven hospital disinfectants were able to completely eliminate cells from the 72 or 120 h biofilms of 5 *C. difficile* strains. Clorox, orthophthalaldehyde (OPA), and Virex were the most effective at killing *C. difficile* spores and Clorox and Virex were most effective in reducing biomass (Rashid et al. 2019).

More innovative proposals to avoid treatment failure and recurrent CDI infection have been sought through the use of bacteriophages and photodynamic therapy.

It has been demonstrated that some bacteriophages have good activity against biofilms of different species by invading it and significantly reducing the viable numbers of cells. Accordingly, bacteriophages appear to be a highly promising therapeutic option for eradicating CDI by replacing antibiotics or supplementing them (Azeredo and Sutherland 2008). Nale and colleagues evaluated the impact of a four-phage cocktail on *C. difficile* ribotype 014/020 biofilm, in vitro alone or in combination with vancomycin treatment in *Galleria mellonella* larva CDI model. Phages were able to prevent in vitro biofilm formation, to penetrate established biofilms, and also to reduce colonization and/or prevent disease in the *Galleria mellonella* model, when used alone or in combination with vancomycin (Nale et al. 2016).

Photodynamic therapy, more frequently applied to determine its usefulness to treat periodontal (Sculean et al. 2015) and wound (Percival et al. 2014) infections, has also been tested against planktonic and sessile-growing *C. difficile* strains. This approach exploits the ability of light-activated photosensitizers (PS) to produce reactive oxygen species (ROS) lethal to cells. Three of thirteen PS screened were able to kill 99.9% of the tested *C. difficile* strains both in planktonic and biofilm states, after exposure to red laser light (0.2 J/cm^2) (De Sordi et al. 2015). Although PS are an interesting perspective for biofilm eradication, as they work by producing free radical species, their use in the human gastrointestinal tract remains limited without further development of the technology.

Discoveries of alternative *C. difficile* treatments include rhodanine derivatives (AbdelKhalek et al. 2016) and acyldepsipeptides (Gil and Paredes-Sabja 2016) that exhibit in vitro activity against planktonic populations, while their efficacy against the sessile populations remains to be evaluated.

In addition to the antimicrobial compounds already tested and the other approaches mentioned above, further possibilities to interfere with *C. difficile* biofilm could presumably come from the discovery of novel compounds that bind c-di-GMP riboswitches (Furukawa et al. 2012), from the use of DNase as enhancer of the effect of MTZ (Machado et al. 2015), or by employing specific QS inhibitors able to interfere with biofilm maturation (Đapa et al. 2013).

7 Conclusions

Biofilms are the most representative form of bacterial growth in the large intestine, with biofilm formation being known to influence the ability of pathogens to colonize and establish during infection.

Clinically relevant strains of *C. difficile* have been proven to be able to form biofilms in vitro that appear as complex cellular processes involving an array of different regulating proteins, intracellular chemical signals, and effector proteins, all having a role in different aspects of bacterial physiology. It is clear that *C. difficile* cells entering into a biofilm lifestyle undergo changes to the cell surface and cellular metabolism to compensate. This picture is still incomplete and the details of the precise function and regulation of each of these proteins/pathways remain to be studied. During CDI, multi-species biofilms are important, but little is known on the contribution of these microbes to promote or inhibit *C. difficile* biofilm formation, and this is true on host factors that can affect biofilm formation. There are still fundamental questions on *C. difficile* in vivo biofilm formation that need answers, like the contribution of *C. difficile* towards the production of the extracellular matrix (ECM) and the heterogeneity of cell physiology within mucosal biofilms. This intertwinement is likely to allow an accurate modulation of the differentiation pathways for motility, biofilm formation, or sporulation at a spatiotemporal manner.

C. difficile biofilm formation in vivo has been demonstrated, although not studied in great depth. However, understanding gleaned from mature biofilm in vitro suggests biofilm formation may be critical to in vivo colonization by *C. difficile* and subsequent CDI development. Along with the formation of spores, biofilm formation by *C. difficile* could hypothetically explain the occurrence of recurrent infections supporting a potential infection model involving the colonization of the colon by *C. difficile* through the formation of microcolonies or biofilms, followed by toxin production. This in vivo biofilm mode of growth possibly protects the bacterium from the cellular immune responses triggered by the toxins and from the antibiotic treatment. In the light of the above, a deeper knowledge of the factors involved in the *C. difficile* biofilm development during infection might provide an advanced understanding of the role of biofilm in CDI.

References

AbdelKhalek A, Ashby CR Jr, Patel BA, Talele TT, Seleem MN (2016) In vitro antibacterial activity of rhodanine derivatives against pathogenic clinical isolates. PLoS ONE 11(10):e0164227

Aldridge P, Paul R, Goymer P et al (2003) Role of the GGDEF regulator PleD in polar development of *Caulobacter crescentus*. Mol Microbiol 47:1695–1708

Aleksić A, Stojanović-Radić Z, Harmanus C et al (2022) In vitro anti-clostridial action and potential of the spice herbs essential oils to prevent biofilm formation of hypervirulent *Clostridioides difficile* strains isolated from hospitalized patients with CDI. Anaerobe 76: 102604

Al-Hinai MA, Jones SW, Papoutsakis ET (2015) The Clostridium sporulation programs: diversity and preservation of endospore differentiation. Microbiol Mol Biol Rev 79:19–37

Arato V, Gasperini G, Giusti F et al (2019) Dual role of the colonization factor CD2831 in *Clostridium difficile* pathogenesis. Sci Rep 9:5554

Awad MM, Johanesen PA, Carter GP et al (2014) *Clostridium difficile* virulence factors: Insights into an anaerobic spore-forming pathogen. Gut Microbes 5: 579–593

Azeredo J, Sutherland IW (2008) The use of phages for the removal of infectious biofilms. Curr Pharm Biotechnol 9:261–266

Azriel S, Goren A, Rahav G et al (2015) The stringent response regulator DksA is required for *Salmonella enteric Serovar Typhimurium* growth in minimal medium, motility, biofilm formation, and intestinal colonization. Infect Immun 84:375–384

Badet C, Quero F (2011) The *in vitro* effect of manuka honeys on growth and adherence of oral bacteria. Anaerobe 17:19–22

Barbut F, Richard A, Hamadi K et al (2000) Epidemiology of recurrences or reinfections of *Clostridium difficile*-associated diarrhea. J Clin Microbiol 38:2386–2388

Biazzo M, Cioncada R, Fiaschi L et al (2013) Diversity of cwp loci in clinical isolates of *Clostridium difficile*. J Med Microbiol 62:1444–1452

Bordeleau E, Fortier LC, Malouin F et al (2011) c-di-GMP turn-over in *Clostridium difficile* is controlled by a plethora of diguanylatecyclases and phosphodiesterases. PLoS Genet 7:e1002039

Bordeleau E, Purcell EB, Lafontaine DA et al (2015) Cyclic di-GMP riboswitch-regulated type IV pili contribute to aggregation of *Clostridium difficile*. J Bacteriol 197:819–832

Boudry P, Gracia C, Monot M et al (2014) Pleiotropic role of the RNA chaperone protein Hfq in the human pathogen *Clostridium difficile*. J Bacteriol 196:3234–3248

Bouillaut L, Dubois T, Sonenshein AL et al (2015) Integration of metabolism and virulence in *Clostridium difficile*. Res Microbiol 166:375–383

Brauer M, Lassek C, Hinze C et al (2021) What's a biofilm?—how the choice of the biofilm model impacts the protein inventory of *Clostridioides difficile*. Front Microbiol 12:682111

Bridier A, Briandet R, Thomas V, Dubois-Brissonnet F (2011) Resistance of bacterial biofilms to disinfectants: a review. Biofouling 27:1017–1032

Brunetti G, Giuliani A, Navazio AS et al (2021) Candida gut colonization, yeast species distribution, and biofilm production in *Clostridioides difficile* infected patients: a comparison between three populations in two different time periods. Braz J Microbiol 52:1845–1852

Buckley AM, Spencer J, Candlish D et al (2011) Infection of hamsters with the UK *Clostridium difficile* ribotype 027 outbreak strain R20291. J Med Microbiol 60: 1174–1180

Butala M, Zgur-Bertok D, Busby SJ (2009) The bacterial LexA transcriptional repressor. Cell Mol Life Sci 66 (1):82–93. https://doi.org/10.1007/s00018-008-8378-6

Buckley AM, Ewin D, Moura IB et al (2021) Insights into the regulatory mechanisms of *Clostridioides difficile* biofilm formation. bioRxiv. https://doi.org/10.1101/2021.02.19.431970

Cairns LS, Marlow VL, Bissett E et al (2013) A mechanical signal transmitted by the flagellum controls signalling in *Bacillus subtilis*. Mol Microbiol 90:6–21

Cairns LS, Hobley L, Stanley-Wall NR (2014) Biofilm formation by *Bacillus subtilis*: New insights into regulatory strategies and assembly mechanisms. Mol Microbiol 93:587–598

Carter GP, Purdy D, Williams P et al (2005) Quorum sensing in *Clostridium difficile*: Analysis of a luxS-type signalling system. J Med Microbiol 54:119–127

Carter GP, Rood JI, Lyras D (2012) The role of toxin A and toxin B in the virulence of *Clostridium difficile*. Trends Microbiol 20:21–29

Cerquetti M, Molinari A, Sebastianelli A et al (2000) Characterization of surface layer proteins from different *Clostridium difficile* clinical isolates. Microb Pathog 28:363–372

Chang-Graham AL, Danhof HA, Engevik MA et al (2019) Human intestinal enteroids with inducible neurogenin-3 expression as a novel model of gut hormone secretion. Cell Mol Gastroenterol Hepatol 8:209–229

Chao Y, Vogel J (2010) The role of Hfq in bacterial pathogens. Cur Opin Microbiol 13:24–33

Chilton CH, Pickering DS, Freeman J (2018) Microbiologic factors affecting *Clostridium difficile* recurrence. Clin Microbiol Infect 24:476–482

Ciofu O, Rojo-Molinero E, Macià MD et al (2017) Antibiotic treatment of biofilm infections. APMIS 125: 304–319

Condinho M, Carvalho B, Cruz A et al (2022) The role of RNA regulators, quorum sensing and c-di-GMP in bacterial biofilm formation. FEBS Open Bio. https://doi.org/10.1002/2211-5463.13389

Costa CL, Azevedo CP, Quesada-Gómez C et al (2021) Inhibitory effect of Brazilian red propolis on planktonic and biofilm forms of *Clostridioides difficile*. Anaerobe 69:102322

Crowther GS, Chilton CH, Todhunter SL et al (2014a) Comparison of planktonic and biofilm-associated communities of *Clostridium difficile* and indigenous gut microbiota in a triple-stage chemostat gut model. J Antimicrob Chemother 69:2137–2147

Crowther GS, Chilton CH, Todhunter SL et al (2014b) Development and validation of a chemostat gut model to study both planktonic and biofilm modes of growth of *Clostridium difficile* and human microbiota. PLoS One 9:e88396

Cuenot E, Garcia-Garcia T, Douche T et al (2019) The Ser/Thr kinase PrkC participates in cell wall homeostasis and antimicrobial resistance in *Clostridium difficile*. Infect Immun 87:e00005–e00019

Cummings JH, Antoine JM, Azpiroz F et al (2004) PASSCLAIM: gut health and immunity. Eur J Nutr 43:II118–II173

Dapa T, Unnikrishnan M (2013) Biofilm formation by *Clostridium difficile*. Gut Microbes 4:397–402

Ðapa T, Leuzzi R, Baban ST et al (2013) Multiple factors modulate biofilm formation by the anaerobic pathogen *Clostridium difficile*. J Bacteriol 195:545–555

Dawson LF, Valiente E, Faulds-Pain A et al (2012) Characterisation of *Clostridium difficile* biofilm formation, a role for Spo0A. PLoS One 7:e50527

Dawson LF, Peltier J, Hall CL et al (2021) Extracellular DNA, cell surface proteins and c-di-GMP promote biofilm formation in *Clostridioides difficile*. Sci Rep 11:3244

de la Riva L, Willing SE, Tate EW et al (2011) Roles of cysteine proteases Cwp84 and Cwp13 in biogenesis of the cell wall of *Clostridium difficile*. J Bacteriol 193: 3276–3285

De Sordi L, Butt MA, Pye H et al (2015) Development of Photodynamic Antimicrobial Chemotherapy (PACT) for *Clostridium difficile*. PLoS One 10:e0135039

Di Martino P (2018) Extracellular polymeric substances, a key element in understanding biofilm phenotype. AIMS Microbiol 4:274–288

Dineen SS, McBride SM, Sonenshein AL (2010) Integration of metabolism and virulence by *Clostridium difficile* CodY. J Bacteriol 192:5350–5362

Doan TH, Bernet-Camard MF, Hoÿs S et al (2022) Impact of subinhibitory concentrations of metronidazole on morphology, motility, biofilm formation and colonization of *Clostridioides difficile*. Antibiotics (Basel) 11: 624

Donelli G (2006) Vascular catheter-related infection and sepsis. Surg Infect (Larchmt) 7:S25–S27

Donelli G, Vuotto C, Cardines R et al (2012) Biofilm-growing intestinal anaerobic bacteria. FEMS Immunol Med Microbiol 65:318–325

Dubois T, Tremblay YDN, Hamiot A et al (2019) A microbiota-generated bile salt induces biofilm formation in *Clostridium difficile*. NPJ Biofilms Microbiomes 5:14

Dupont HL (2013) Diagnosis and management of *Clostridium difficile* infection. Clin Gastroenterol Hepatol 11:1216–1223

Edwards AN, Nawrocki KL, McBride SM (2014) Conserved oligopeptide permeases modulate sporula tion initiation in *Clostridium difficile*. Infect Immun 82: 4276–4291

Engevik MA, Danhof HA, Chang-Graham AL et al (2020) Human intestinal enteroids as a model of *Clostridioides difficile*-induced enteritis. Am J Physiol Gastrointest Liver Physiol 318:G870–G888

Engevik MA, Danhof HA, Auchtung J et al (2021) *Fusobacterium nucleatum* adheres to *Clostridioides difficile* via the RadD adhesin to enhance biofilm formation in intestinal mucus. Gastroenterology 160: 1301–1314.e8

Fagan RP, Fairweather NF (2014) Biogenesis and functions of bacterial S-layers. Nat rev Microbiol 12: 211–222

Faulds-Pain A, Twine SM, Vinogradov E et al (2014) The post-translational modification of the *Clostridium difficile* flagellin affects motility, cell surface properties and virulence. Mol Microbiol 94:272–289

Ferreyra JA, Wu KJ, Hryckowian AJ et al (2014) Gut microbiota-produced succinate promotes *C. Difficile* infection after antibiotic treatment or motility disturbance. Cell Host Microbe 16:770–777

Flemming HC, Wingender J (2010) The biofilm matrix. Nat Rev Microbiol 8:623–633

Frost LR, Cheng JKJ, Unnikrishnan M (2021) *Clostridioides difficile* biofilms: a mechanism of persistence in the gut? PLoS Pathog 17:e1009348

Fuchs M, Lamm-Schmidt V, Sulzer J et al (2021) An RNA-centric global view of *Clostridioides difficile* reveals broad activity of Hfq in a clinically important gram-positive bacterium. Proc Natl Acad Sci U S A 118:e2103579118

Furukawa K, Gu H, Sudarsan N et al (2012) Identification of ligand analogues that control c-di-GMP riboswitches. ACS Chem Biol 7:1436–1443

Furuya-Kanamori L, Marquess J, Yakob L et al (2015) Asymptomatic *Clostridium Difficile* colonization: epidemiology and clinical implications. BMC Infect Dis 15:516

Ganeshapillai J, Vinogradov E, Rousseau J et al (2008) *Clostridium difficile* cell-surface polysaccharides composed of pentaglycosyl and hexaglycosyl phosphate repeating units. Carbohydr Res 343:703e10

Garrett EM, Mehra A, Sekulovic O et al (2022) Multiple regulatory mechanisms control the production of CmrRST, an atypical signal transduction system in *Clostridioides difficile*. mBio 13:e0296921

Ghigo JM (2001) Natural conjugative plasmids induce bacterial biofilm development. Nature 412:442–445

Ghosh S, Zhang P, Li YQ et al (2009) Superdormant spores of Bacillus species have elevated wet-heat resistance and temperature requirements for heat activation. J Bacteriol 191:5584–5591

Gil F, Paredes-Sabja D (2016) Acyldepsipeptide antibiotics as a potential therapeutic agent against *Clostridium difficile* recurrent infections. Future Microbiol 11:1179–1189

Gil F, Pizarro-Guajardo M, Álvarez R (2015) *Clostridium difficile* recurrent infection: possible implication of TA systems. Future Microbiol 10:1649–1657

Goldberg J (2002) Biofilms and antibiotic resistance: a genetic linkage. Trends Microbiol 10:264

Goulding D, Thompson H, Emerson J et al (2009) Distinctive profiles of infection and pathology in hamsters infected with *Clostridium difficile* strains 630 and B1. Infect Immun 77:5478–5485

Hall-Stoodley L, Stoodley P (2009) Evolving concepts in biofilm infections. Cell Microbiol 11:1034–1043

Hall-Stoodley L, Costerton JW, Stoodley P (2004) Bacterial biofilms: from the natural environment to infectious diseases. Nat Rev Microbiol 2:95–108

Hamada M, Yamaguchi T, Ishii Y et al (2020) Inhibitory effect of fidaxomicin on biofilm formation in *Clostridioides difficile*. J Infect Chemother 26:685–692

Hammond EN, Donkor ES, Brown CA (2014) Biofilm formation of *Clostridium difficile* and susceptibility to Manuka honey. BMC Complement Altern Med 14:329

Hashem AA, Abd El Fadeal NM et al (2017) *In vitro* activities of vancomycin and linezolid against biofilm-producing methicillin-resistant staphylococci species isolated from catheter-related bloodstream infections from an Egyptian tertiary hospital. J Med Microbiol 66:744–752

Hassall J, Cheng JKJ, Unnikrishnan M (2021) Dissecting individual interactions between pathogenic and commensal bacteria within a multispecies gut microbial community. mSphere 6(2):e00013-21. https://doi.org/10.1128/mSphere.00013-21

Hoiby N, Bjarnsholt T, Givskov M et al (2010) Antibiotic resistance of bacterial biofilms. Int J Antimicrob Agents 35:322–332

Hou K, Wu ZX, Chen XY et al (2022) Microbiota in health and diseases. Signal Transduct Target Ther 7:135

Hu X, Kang F, Yang B et al (2019) Extracellular polymeric substances acting as a permeable barrier hinder the lateral transfer of antibiotic resistance genes. Front Microbiol 10:736

James GA, Chesnel L, Boegli L et al (2018) Analysis of *Clostridium difficile* biofilms: imaging and antimicrobial treatment. J Antimicrob Chemother 73:102–108

Jimi S, Miyazaki M, Takata T et al (2017) Increased drug resistance of meticillin-resistant *Staphylococcus aureus* biofilms formed on a mouse dermal chip model. J Med Microbiol 66:542–550

Kirby JM, Ahern H, Roberts AK et al (2009) Cwp84, a surface-associated cysteine protease, plays a role in the maturation of the surface layer of *Clostridium difficile*. J Biol Chem 284:34666–34673

Klausen M, Aaes-Jørgensen A, Molin S et al (2003) Involvement of bacterial migration in the development of complex multicellular structures in *Pseudomonas aeruginosa* biofilms. Mol Microbiol 50:61–68

Koch B, Worm J, Jensen LE et al (2001) Carbon limitation induces s-dependent gene expression in *Pseudomonas fluorescens* in soil. Appl Environ Microbiol 67:3363–3370

Kulasakara H, Lee V, Brencic A et al (2006) Analysis of *Pseudomonas aeruginosa* diguanylatecyclases and phosphodiesterases reveals a role for bis-(3′-5′)-cyclic-GMP in virulence. Proc Natl Acad Sci USA 103:2839–2844

Lacotte PA, Simons A, Bouttier S et al (2022) Inhibition of *in vitro Clostridioides difficile* biofilm formation by the probiotic yeast *Saccharomyces boulardii* CNCM I-745 through modification of the extracellular matrix composition. Microorganisms 10:1082

Lawley TD, Clare S, Walker AW et al (2009) Antibiotic treatment of *Clostridium difficile* carrier mice triggers a supershedder state, spore-mediated transmission, and severe disease in immunocompromised hosts. Infect Immun 77:3661–3669

Lee ASY, Song KP (2005) LuxS/autoinducer-2 quorum sensing molecule regulates transcriptional virulence gene expression in *Clostridium difficile*. Biochem Biophys Res Commun 335:659–666

Leslie JL, Huang S, Opp JS et al (2015) Persistence and toxin production by *Clostridium difficile* within human intestinal organoids result in disruption of epithelial paracellular barrier function. Infect Immun 83:138–145

Li YH, Tian X (2012) Quorum sensing and bacterial social interactions in biofilms. Sensors (Basel) 12:2519–2538

Lindsay D, von Holy A (2006) Bacterial biofilms within the clinical setting: what healthcare professionals should know. J Hosp Infect 64:313–325

Lipovsek S, Leitinger G, Rupnik M (2013) Ultrastructure of *Clostridium difficile* colonies. Anaerobe 24:66e70

Macfarlane S, Dillon JF (2007) Microbial biofilms in the human gastrointestinal tract. J Appl Microbiol 102:1187–1196

Macfarlane S, Macfarlane GT (2006) Composition and metabolic activities of bacterial biofilms colonizing food residues in the human gut. Appl Environ Microbiol 72:6204–6211

Macfarlane GT, Macfarlane S, Gibson GR (1998) Validation of a three-stage compound continuous culture system for investigating the effect of retention time on the ecology and metabolism of bacteria in the human colon. Microb Ecol 35:180–187

Macfarlane S, Bahrami B, Macfarlane GT (2011) Mucosal biofilm communities in the human intestinal tract. Adv Appl Microbiol 75:111–143

Machado D, Castro J, Palmeira-de-Oliveira A et al (2015) Bacterial vaginosis biofilms: challenges to current therapies and emerging solutions. Front Microbiol 6:152

Mah TF, O'Toole GA (2001) Mechanisms of biofilm resistance to antimicrobial agents. Trends Microbiol 9:34–39

Mahnic A, Auchtung JM, Poklar Ulrih N et al (2020) Microbiota in vitro modulated with polyphenols shows decreased colonization resistance against *Clostridioides difficile* but can neutralize cytotoxicity. Sci Rep 10:8358

Maldarelli GA, De Masi L, von Rosenvinge EC et al (2014) Identification, immunogenicity and cross-reactivity of Type IV pilin and pilin-like proteins from *Clostridium difficile*. Pathog Dis 71:302–314

Maldarelli GA, Piepenbrink KH, Scott AJ et al (2016) Type IV pili promote early biofilm formation by *Clostridium difficile*. Pathog Dis 74:ftw061

Martínez-Meléndez A, Morfin-Otero R, Villarreal-Treviño L et al (2021) Analysis of biofilm production and expression of adhesion structures of circulating *Clostridioides difficile* strains from Mexico. Enferm Infecc Microbiol Clin (Engl Ed). https://doi.org/10.1016/j.eimc.2021.01.017

Mathur H, Rea MC, Cotter PD et al (2016) The efficacy of thuricin CD, tigecycline, vancomycin, teicoplanin, rifampicin and nitazoxanide, independently and in paired combinations against *Clostridium difficile* biofilms and planktonic cells. Gut Pathog 8:20

Mccracken KW, Howell JC, Wells JM, Spence JR (2011) Generating human intestinal tissue from pluripotent stem cells *in vitro*. Nat Protoc 6:1920–1928

Meeker DG, Beenken KE, Mills WB et al (2016) Evaluation of antibiotics active against methicillin-resistant

Staphylococcus aureus based on activity in an established biofilm. Antimicrob Agents Chemother 60:5688–5694
Melville S, Craig L (2013) Type IV pili in gram-positive bacteria. Microbiol Mol Biol Rev 77:323–341
Mhatre E, Monterrosa RG, Kovács ÁT (2014) From environmental signals to regulators: modulation of biofilm development in Gram-positive bacteria. J Basic Microbiol 54:616–632
Motta JP, Wallace JL, Buret AG et al (2021) Gastrointestinal biofilms in health and disease. Nat Rev Gastroenterol Hepatol 18:314–334
Nale JY, Chutia M, Carr P et al (2016) 'Get in Early'; biofilm and wax moth (Galleria mellonella) models reveal new insights into the therapeutic potential of *Clostridium difficile* bacteriophages. Front Microbiol 7:1383
Nassif X, Beretti JL, Lowy J et al (1994) Roles of pilin and PilC in adhesion of Neisseria meningitidis to human epithelial and endothelial cells. Proc Natl Acad Sci U S A. 91:3769–3773
Nelson RL, Suda KJ, Evans CT (2017) Antibiotic treatment for *Clostridium difficile*-associated diarrhoea in adults. Cochrane Database Syst Rev 3:CD004610
Ng KM, Ferreyra JA, Higginbottom SK et al (2013) Microbiota-liberated host sugars facilitate post-antibiotic expansion of enteric pathogens. Nature 502: 96–99
Normington C, Moura IB, Bryant JA et al (2021) Biofilms harbour *Clostridioides difficile*, serving as a reservoir for recurrent infection. npj Biofilms Microbiomes 7:16
Ofosu A (2016) *Clostridium difficile* infection: a review of current and emerging therapies. Ann Gastroenterol 29: 147–154
Owrangi B, Masters N, Vollmerhausen TL et al (2017) Comparison between virulence characteristics of dominant and non-dominant *Escherichia coli* strains of the gut and their interaction with Caco-2 cells. Microb Pathog 105:171 176
Ozturk B, Gunay N, Ertugrul BM et al (2016) Effects of vancomycin, daptomycin, and tigecycline on coagulase-negative staphylococcus biofilm and bacterial viability within biofilm: an in vitro biofilm model. Can J Microbiol 62:735–743
Pantaléon V, Bouttier S, Soavelomandroso AP et al (2014) Biofilms of *Clostridium species*. Anaerobe 30:193–198
Pantaléon V, Soavelomandroso AP, Bouttier S et al (2015) The *Clostridium difficile* protease Cwp84 modulates both biofilm formation and cell-surface properties. PLoS ONE 10:1–20
Pantaléon V, Monot M, Eckert C et al (2018) *Clostridium difficile* forms variable biofilms on abiotic surface. Anaerobe 53:34–37
Peng JS, Tsai WC, Chou CC (2002) Inactivation and removal of *Bacillus cereus* by sanitizer and detergent. Int J Food Microbiol 77:11–18
Peng Z, Jin D, Kim HB et al (2017) Update on antimicrobial resistance in *Clostridium difficile*: resistance mechanisms and antimicrobial susceptibility testing. J Clin Microbiol 55:1998–2008
Percival SL, Suleman L, Francolini I et al (2014) The effectiveness of photodynamic therapy on planktonic cells and biofilms and its role in wound healing. Future Microbiol 9:1083–1094
Percival SL, Suleman L, Vuotto C et al (2015) Healthcare-associated infections, medical devices and biofilms: risk, tolerance and control. J Med Microbiol 64:323–334
Pérez-Cobas AE, Artacho A, Ott SJ et al (2014) Structural and functional changes in the gut microbiota associated to *Clostridium Difficile* infection. Front Microbiol 5: 220–234
Pettit LJ, Browne HP, Yu L et al (2014) Functional genomics reveals that *Clostridium difficile* Spo0A coordinates sporulation, virulence and metabolism. BMC genomics 15:160
Phalak P, Henson MA (2019) Metabolic modeling of *Clostridium difficile* associated dysbiosis of the gut microbiota. Processes 7:97
Pickering DS, Wilcox MH, Chilton CH (2018) Biofilm-derived spores of *Clostridioides* (*Clostridium*) *difficile* exhibit increased thermotolerance compared to planktonic spores. Anaerobe 54:169–171
Piepenbrink KH, Maldarelli GA, de la Peña CF et al (2014) Structure of *Clostridium difficile* PilJ exhibits unprecedented divergence from known Type IV pilins. J Biol Chem 289:4334–4345
Piepenbrink KH, Maldarelli GA, Martinez de la Peña CF et al (2015) Structural and evolutionary analyses show unique stabilization strategies in the Type IV pili of *Clostridium difficile*. Structure 23:385–396
Piotrowski M, Karpiński P, Pituch H, van Belkum A, Obuch-Woszczatyński P (2017) Antimicrobial effects of Manuka honey on in vitro biofilm formation by *Clostridium difficile*. Eur J Clin Microbiol Infect Dis. https://doi.org/10.1007/s10096-017-2980-1
Piotrowski M, Wultańska D, Obuch-Woszczatyński P, Pituch H (2019) Fructooligosaccharides and mannose affect *Clostridium difficile* adhesion and biofilm formation in a concentration-dependent manner. Eur J Clin Microbiol Infect Dis 38:1975–1984
Piotrowski M, Wultańska D, Pituch H (2022) Effect of prebiotics on Bacteroides sp. adhesion and biofilm formation and synbiotic effect on *Clostridioides difficile*. Future Microbiol 17:363–375
Pizarro-Guajardo M, Calderón-Romero P, Castro-Córdova P et al (2016a) Ultrastructural variability of the exosporium layer of *Clostridium difficile* spores. Appl Environ Microbiol 82:2202–2209
Pizarro-Guajardo M, Calderón-Romero P, Paredes-Sabja D (2016b) Ultrastructure variability of the exosporium layer of *Clostridium difficile* spores from sporulating cultures and biofilms. Appl Environ Microbiol 82: 5892–5898
Poquet I, Saujet L, Canette A et al (2018) *Clostridium difficile* biofilm: remodeling metabolism and cell

surface to build a sparse and heterogeneously aggregated architecture. Front Microbiol 9:2084

Purcell EB, McKee RW, McBride SM et al (2012) Cyclic diguanylate inversely regulates motility and aggregation in *Clostridium difficile*. J Bacteriol 194:3307–3316

Purcell EB, McKee RW, Bordeleau E et al (2016) Regulation of Type IV pili contributes to surface behaviours of historical and epidemic strains of *Clostridium difficile*. J Bacteriol 198:565–577

Purcell EB, McKee RW, Courson DS et al (2017) A nutrient-regulated cyclic diguanylate phosphodiesterase controls *Clostridium difficile* biofilm and toxin production during stationary phase. Infect Immun 85(9):e00347-17

Rahmoun LA, Azrad M, Peretz A (2021) Antibiotic resistance and biofilm production capacity in *Clostridioides difficile*. Front Cell Infect Microbiol 11:683464

Rashid T, Haghighi F, Hasan I et al (2019) Activity of hospital disinfectants against vegetative cells and spores of *Clostridioides difficile* embedded in biofilms. Antimicrob Agents Chemother 64:e01031-19

Ribeiro SM, Felício MR, Boas EV et al (2016) New frontiers for anti-biofilm drug development. Pharmacol Ther 160:133–144

Robinson CD, Auchtung JM, Collins J, Britton RA (2014) Epidemic *Clostridium difficile* strains demonstrate increased competitive fitness compared to nonepidemic isolates. Infect Immun 82:2815–2825

Römling U, Amikam D (2006) Cyclic di-GMP as a second messenger. Cur Opin Microbiol 9:218–228

Römling U, Balsalobre C (2012) Biofilm infections, their resilience to therapy and innovative treatment strategies. J Intern Med 272(6):541–561. https://doi.org/10.1111/joim.12004

Ronish LA, Sidner B, Yu Y, Piepenbrink KH (2022) Recognition of extracellular DNA by type IV pili promotes biofilm formation by *Clostridioides difficile*. J Biol Chem 298:102449

Rossi E, Cimdins A, Lüthje P et al (2017) "It's a gut feeling"—*Escherichia coli* biofilm formation in the gastrointestinal tract environment. Crit Rev Microbiol 9:1–30

Rothenbacher FP, Suzuki M, Hurley JM et al (2012) *Clostridium difficile* MazF toxin exhibits selective, not global, mRNA cleavage. J Bacteriol 194:3464–3474

Roy R, Tiwari M, Donelli G et al (2017) Strategies for combating bacterial biofilms: A focus on anti-biofilm agents and their mechanisms of action. Virulence. https://doi.org/10.1080/21505594.2017.1313372

Sauer K, Stoodley P, Goeres DM et al (2022) The biofilm life cycle: expanding the conceptual model of biofilm formation. Nat Rev Microbiol 20:608–620

Schiffrin EJ, Blum S (2002) Interactions between the microbiota and the intestinal mucosa. Eur J Clin Nutr 56(Suppl 3):S60–S64

Schulze A, Mitterer F, Pombo JP, Schild S (2021) Biofilms by bacterial human pathogens: Clinical relevance—development, composition and regulation—therapeutical strategies. Microb Cell 8:28–56

Sculean A, Aoki A, Romanos G et al (2015) Is photodynamic therapy an effective treatment for periodontal and peri-implant infections? Dent Clin North Am 59: 831–858

Sebaihia M, Wren BW, Mullany P et al (2006) The multidrug resistant pathogen *Clostridium difficile* has a highly mobile mosaic genome. Nat Genet 38:779–786

Semenyuk EG, Laning ML, Foley J et al (2014) Spore formation and toxin production in *Clostridium difficile* biofilms. PLoS One 9:e87757

Semenyuk EG, Poroyko VA, Johnston PF et al (2015) Analysis of bacterial communities during *Clostridium difficile* infection in the mouse. Infect Immun 83:4383–4391

Sengupta C, Mukherjee O, Chowdhury R (2016) Adherence to intestinal cells promotes biofilm formation in *Vibrio cholerae*. J Infect Dis 214:1571–1578

Shah D, Zhang Z, Khodursky A et al (2006) Persisters: a distinct physiological state of *E. coli*. BMC Microbiol 6:53

Silva JO, Martins Reis AC, Quesada-Gómez C et al (2014) *In vitro* effect of antibiotics on biofilm formation by *Bacteroides fragilis* group strains isolated from intestinal microbiota of dogs and their antimicrobial susceptibility. Anaerobe 28:24–28

Slater RT, Frost LR, Jossi SE et al (2019) *Clostridioides difficile* LuxS mediates inter-bacterial interactions within biofilms. Sci Rep 9:9903

Smith AB, Jenior ML, Keenan O et al (2022) Enterococci enhance *Clostridioides difficile* pathogenesis. Nature 611:780–786

Soavelomandroso AP, Gaudin F, Hoys S et al (2017) Biofilm structures in a mono-associated mouse model of *Clostridium difficile* infection. Front Microbiol 8: 2086

Soutourina O (2017) RNA-based control mechanisms of *Clostridium difficile*. Curr Opin Microbiol 36:62–68

Soutourina OA, Monot M, Boudry P et al (2013) Genome-wide identification of regulatory RNAs in the human pathogen *Clostridium difficile*. PLoS Genet 9: e1003493

Spencer J, Leuzzi R, Buckley A et al (2014) Vaccination against *Clostridium difficile* using toxin fragments: Observations and analysis in animal models. Gut Microbes 5:23–22

Spigaglia P (2016) Recent advances in the understanding of antibiotic resistance in *Clostridium difficile* infection. Ther Adv Infect Dis 3:23–42

Spoering AL, Lewis K (2001) Biofilms and planktonic cells of *Pseudomonas aeruginosa* have similar resistance to killing by antimicrobials. J Bacteriol 183: 6746–6751

Stabler RA, He M, Dawson L, Martin M, Valiente E, Corton C, Lawley TD, Sebaihia M, Quail MA, Rose G, Gerding DN, Gibert M, Popoff MR, Parkhill J, Dougan G, Wren BW (2009) Comparative genome and phenotypic analysis of Clostridium difficile 027 strains provides insight into the evolution of a hypervirulent bacterium. Genome Biol 10(9):R102. https://doi.org/10.1186/gb-2009-10-9-r102

Stevenson E, Minton NP, Kuehne SA (2015) The role of flagella in *Clostridium difficile* pathogenicity. Trends Microbiol 23:1–8

Sudarsan N, Lee ER, Weinberg Z et al (2008) Riboswitches in eubacteria sense the second messenger cyclic di-GMP. Science 321:411–413

Tischler AD, Camilli A (2005) Cyclicdiguanylate regulates *Vibrio cholera* virulence gene expression. Infect Immun 73:5873–5882

Tremblay YDN, Durand BAR, Hamiot A et al (2021) Metabolic adaption to extracellular pyruvate triggers biofilm formation in *Clostridioides difficile*. ISME J 15:3623–3635

Twine SM, Reid CW, Aubry A et al (2009) Motility and flagellar glycosylation in *Clostridium difficile*. J Bacteriol 191:7050–7062

Tyerman JG, Ponciano JM, Joyce P et al (2013) The evolution of antibiotic susceptibility and resistance during the formation of *Escherichia coli* biofilms in the absence of antibiotics. BMC Evol Biol 13:22

Valiente E, Bouché L, Hitchen P et al (2016) Role of glycosyltransferases modifying type B flagellin of emerging hypervirulent *Clostridium difficile* lineages and their impact on motility and biofilm formation. J Biol Chem 291:25450–25461

Vandana, Das S (2022) Genetic regulation, biosynthesis and applications of extracellular polysaccharides of the biofilm matrix of bacteria. Carbohydr Polym 291: 119536

Varga JJ, Nguyen V, O'Brien DK et al (2006) Type IV pili-dependent gliding motility in the Gram-positive pathogen *Clostridium perfringens* and other Clostridia. Mol Microbiol 62:680–694

Varga JJ, Therit B, Melville SB (2008) Type IV pili and the CcpA protein are needed for maximal biofilm formation by the gram-positive anaerobic pathogen *Clostridium perfringens*. Infect Immun 76:4944–4951

Vlamakis H, Chai Y, Beauregard P et al (2013) Sticking together: building a biofilm the *Bacillus subtilis* way. Nat Rev Microbiol 11:157–168

Vuotto C, Moura I, Barbanti F et al (2016) Sub-inhibitory concentrations of metronidazole increase biofilm formation in *Clostridium difficile* strains. Pathog Dis 74:ftv114

Walter BM, Rupnik M, Hodnik V et al (2014) The LexA regulated genes of the *Clostridium difficile*. BMC Microbiol 14:88

Walter BM, Cartman ST, Minton NP et al (2015) The SOS response master regulator LexA is associated with sporulation, motility and biofilm formation in *Clostridium difficile*. PLoS One 10:1–17

Wang X, Blumenfeld R, Feng XQ, Weitz DA (2022a) 'Phase transitions' in bacteria—from structural transitions in free living bacteria to phenotypic transitions in bacteria within biofilms. Phys Life Rev 43:98–138

Wang Y, Bian Z, Wang Y (2022b) Biofilm formation and inhibition mediated by bacterial quorum sensing. Appl Microbiol Biotechnol 106:6365–6381

Wen Y, Behiels E, Devreese B (2014) Toxin-antitoxin systems: their role in persistence, biofilm formation, and pathogenicity. Pathog Dis 70:240–249

Willing SE, Candela T, Shaw HA et al (2015) *Clostridium difficile* surface proteins are anchored to the cell wall using CWB2 motifs that recognise the anionic polymer PSII. Mol Microbiol 96:596–608

Winkler WC, Breaker RR (2005) Regulation of bacterial gene expression by riboswitches. Ann Rev Microbiol 59:487–517

Xue Z, Sendamangalam VR, Gruden CL, Seo Y (2012) Multiple roles of extracellular polymeric substances on resistance of biofilm and detached clusters. Environ Sci Technol 46:13212–13219

Yang HT, Chen JW, Rathod J et al (2018) Lauric acid is an inhibitor of *Clostridium difficile* growth in vitro and reduces inflammation in a mouse infection model. Front Microbiol 8:2635

Zhu D, Wang S, Sun X (2021) FliW and CsrA govern flagellin (FliC) synthesis and play pleiotropic roles in virulence and physiology of *Clostridioides difficile* R20291. Front Microbiol 12:735616

Zoetendal EG, von Wright A, Vilpponen-Salmela T et al (2002) Mucosa-associated bacteria in the human gastrointestinal tract are uniformly distributed along the colon and differ from the community recovered from faeces. Appl Environ Microbiol 68:3401–3407

Clostridioides difficile Sporulation

Mónica Serrano, Diogo Martins, and Adriano O. Henriques

Abstract

Some members of the Firmicutes phylum, including many members of the human gut microbiota, are able to differentiate a dormant and highly resistant cell type, the endospore (hereinafter spore for simplicity). Spore-formers can colonize virtually any habitat and, because of their resistance to a wide variety of physical and chemical insults, spores can remain viable in the environment for long periods of time. In the anaerobic enteric pathogen *Clostridioides difficile* the aetiologic agent is the oxygen-resistant spore, while the toxins produced by actively growing cells are the main cause of the disease symptoms. Here, we review the regulatory circuits that govern entry into sporulation. We also cover the role of spores in the infectious cycle of *C. difficile* in relation to spore structure and function and the main control points along spore morphogenesis.

M. Serrano (✉) · D. Martins · A. O. Henriques
Instituto de Tecnologia Química e Biológica António Xavier, Oeiras, Portugal
e-mail: serrano@itqb.unl.pt

1 Importance of Spores in the *Clostridioides difficile* Life Cycle

To cause disease, *C. difficile*-ingested spores must germinate in the host gastrointestinal tract to return to vegetative cell growth (through spore germination and cell outgrowth) and subsequent toxin production (Fig. 1a). During germination the spore loses its resistance properties. Bile salts play a significant role in spore germination and outgrowth, and this explains, in part, the importance of the host microbiota and its associated metabolome in *C. difficile* infection (CDI) (Smits et al. 2016; Sorg and Sonenshein 2008). Ingested spores survive through the gastro-intestinal tract, reaching the small intestine, where they can germinate in response to bile salts such as cholate, and to amino acids, such as glycine and histidine that act as co-germinants (Koenigsknecht et al. 2015; Sorg and Sonenshein 2008; Wheeldon et al. 2011; Wilson 1983). The germinated spores pass through the ileum into the large intestine, an anaerobic environment where vegetative growth can take place. In a healthy host, however, the normal microbiota, including species such as *Clostridium scindens* that produce a 7α-dehydroxylating activity, metabolizes the cholate derivatives into secondary bile salts, such as deoxycholate and lithocholate that act as growth inhibitors of *C. difficile*, preventing colonization by this organism (Buffie et al. 2015; Giel et al.

P. Mastrantonio, M. Rupnik (eds.), *Updates on Clostridioides difficile in Europe*, Advances in Experimental Medicine and Biology 1435, https://doi.org/10.1007/978-3-031-42108-2_13

2010; Sorg and Sonenshein 2008; Theriot and Young 2015). After antibiotic treatment, the commensal gut microbiota is disturbed and the organisms that convert primary into secondary bile salts are reduced. Thus, the level of growth inhibitory secondary bile salts in the large intestine is reduced, and growth of *C. difficile* is enhanced, leading to host colonization (Sorg and Sonenshein 2008; Theriot and Young 2015). Most of the *C. difficile* vegetative cells are flagellated. Flagella confer motility to the cells but are also important to sense surfaces before adherence and colonization (Aubry et al. 2012; Baban et al. 2013; Faulds-Pain et al. 2014; Stevenson et al. 2015). Once the cell finds a surface, it starts dividing and form cell clusters or microcolonies. In mice, the bacteria persist as microcolonies and biofilm-like structures at the surface of the intestinal mucosa during months (Lawley et al. 2012). Evidence suggests that spores and toxins are produced within these biofilm-like structures (Dawson et al. 2012; Dubois et al. 2019; Semenyuk et al. 2014; Thapa et al. 2013). Eventually the toxins, which are the main *C. difficile* virulence factors, will damage the epithelial barrier, leading to infection and disease (Abt et al. 2016; Kordus et al. 2022; Smits et al. 2016). In vivo, after ingestion of spores by antibiotic-treated mice, vegetative growth is detected 6 h thereafter, while it takes 24 h for spores and toxins to be detected, and disease symptoms develop by hour 30. At this time, spores reach close to 20% of the viable *C. difficile* cell counts (Koenigsknecht et al. 2015). This indicates that spores and toxin, the main transmission and virulence factors, are produced concomitantly. To disseminate, *C. difficile* has to produce spores in numbers higher than those that were ingested. Premature and/or robust spore production may decrease toxin production and vice versa if the two processes are largely antithetical and restricted to sub-populations of toxin or spore producers. The balance between toxinogenesis and sporogenesis appears critical for the outcome of the infectious cycle. Previous studies revealed that toxin and sporulation partially overlap in that a sub-population of about 20% of the sporulating express the TcdA-encoding gene in the mother cell (Donnelly et al. 2022; Ransom et al. 2018).

Spores are also linked to persistence and chronic infection due to resistance to antibiotics (Deakin et al. 2012; Lawley et al. 2012). It was shown recently that *C. difficile* spores can be internalized by intestinal epithelial cells (Castro-Córdova et al. 2021, 2023). Toxin production by the vegetative cells was shown to contribute to spore adherence and internalization by the epithelial cells. Once inside the cells, the spore remains dormant even in a dysbiosis environment enriched in bile salts that trigger cell proliferation. Spores may be released from the cell during the normal renewal of the intestinal epithelium and contribute to recurrence of CDI (Fig. 1a).

In toto, spore formation is an important factor in the resistance of *C. difficile* to antibiotics, persistence in the host causing high rates of disease recurrence, and in the environment, leading to host to host dissemination.

2 Making a Dormant Cell Inside Another Cell

Sporulation is a developmental process that proceeds through a series of well-defined morphological stages that culminate in the lysis of the mother cell and release of a spore (Fig. 1b). The main morphological stages of sporulation and the overall architecture of the mature spore are well conserved among endospore formers (Fimlaid et al. 2013; Henriques and Moran 2007; Pereira et al. 2013) (Figs. 1b and 2). Soon after the cells, which in most spore-forming organisms are rod-shaped, embark into sporulation, an asymmetric, polar, division produces a smaller forespore, the future spore, and a larger mother cell.

Following asymmetric division, the mother cell starts to engulf the forespore eventually releasing it as a free protoplast, surrounded by the mother cell cytoplasm and isolated from the external medium (Fig. 1b). The engulfed forespore is separated from the mother cell cytoplasm by two membranes that derive from the septal membranes. The membrane that faces the

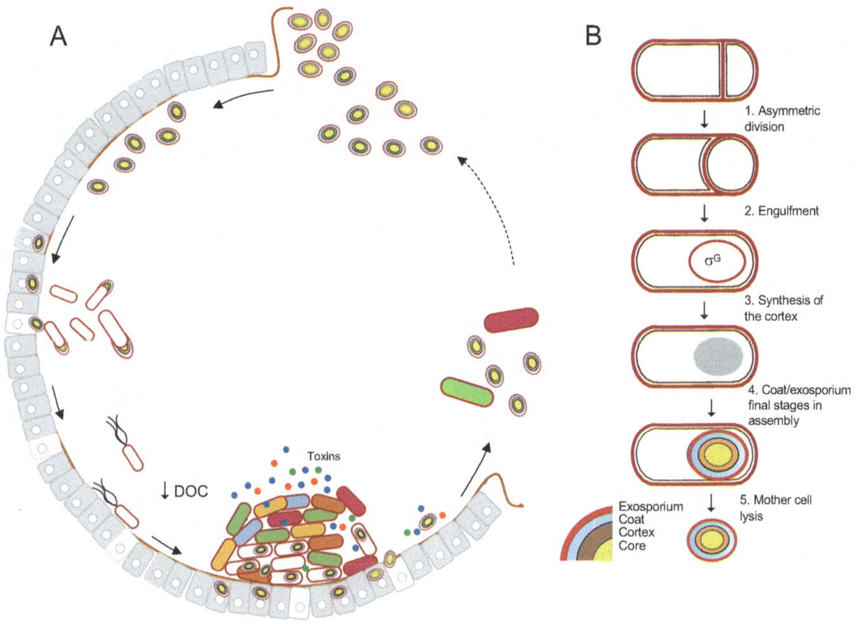

Fig. 1 Schematic view of the *Clostridioides difficile* life cycle. (**a**) An obligate anaerobe, *C. difficile* is usually found outside the host in the form of dormant spores. Infection starts with the ingestion of spores. Once in the small intestine, spores are exposed to bile salts and germinate. While primary bile salts, such as cholate (CA), induce spore germination and promote vegetive growth, secondary bile salts, such as deoxycholate (DOC), more abundant in the large intestine, inhibit vegetative growth. Bile salts levels are influenced by the commensal gut microbiota. *C. scindens*, for instance, encodes a 7α-dehydroxylating activity which converts CA into DOC. After antibiotic treatment the commensal gut microbiota is disturbed and the representation of species capable of 7α-dehydroxylation is reduced. Thus, growth is enhanced in the large intestine, leading to host colonization. The vegetative cells are motile and formation of flagellum was shown to be important for infection. Once the cell finds a surface, it can divide and form cell clusters or microcolonies. Evidence suggests that toxins and spores are produced within the biofilm. Toxins contribute for spore adherence and internalization by the intestinal epithelial cells. Spore may be released from the cell during the normal renewal of the intestinal epithelium. Shedding of the spores to the environment will allow the infection of new hosts, while spores that remain inside the host can be the cause of disease recurrence. (**b**) Sequence of the morphological events leading to spore differentiation: (1) asymmetric division of the sporangium; (2) intermediate stage in the process of engulfment of the forespore (the future spore) by the larger mother cell; (3) engulfment completion, isolating the forespore from the surrounding medium; (4) synthesis of the spore surface layers, the spore cortex and the coat and exosporium. (5) Finally, upon lysis of the mother cell the spore is released to the environment. The various layers detected in mature spores are indicated

mother cell cytoplasm is called the outer membrane and the membrane that contacts the forespore cytoplasm is called the inner membrane. This topology creates an extracellular compartment, the interspace, between the two forespore membranes of opposite polarity. The engulfed forespore is then surrounded by two peptidoglycan (PG) layers, the germ cell wall and the cortex, and two proteinaceous layers, the coat and the exosporium. At this point, the spore develops full resistance to physical (UV and temperature) and chemical (organic solvents and oxidative compounds) agents (reviewed by Setlow and Christie 2023). At the end of the differentiation process the mother cell lyses releasing the spore into the environment (Fig. 1b).

The innermost compartment in mature spores is the core where the chromosome is located. The high concentration of dipicolinic acid (DPA) that is chelated with calcium (Ca-DPA), the low water content, and the saturation of DNA with α/β-type small acid-soluble spore proteins (SASPs) give

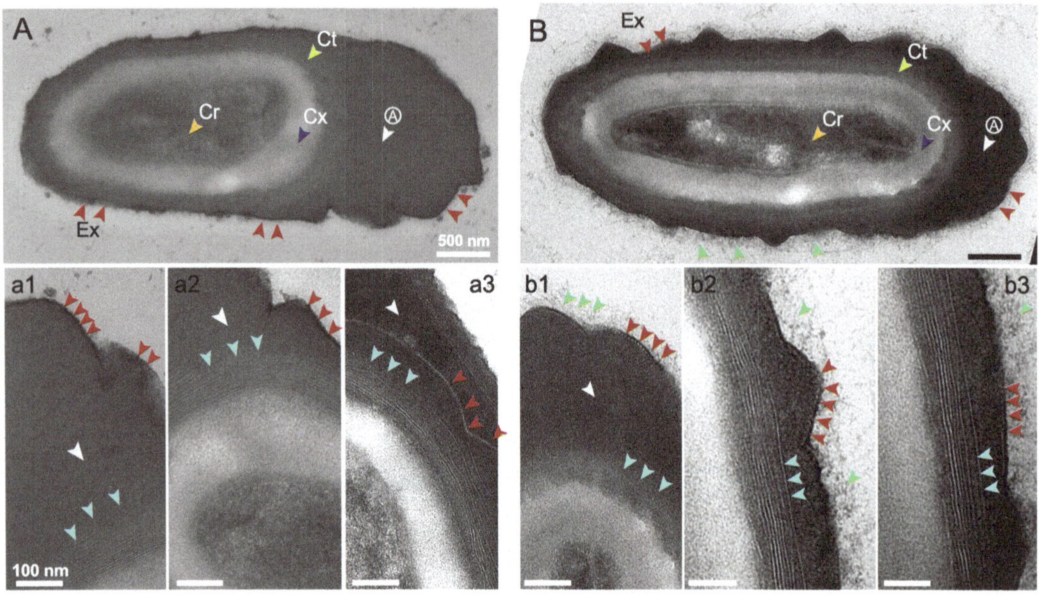

Fig. 2 Spore structure. (**a**) The top panel shows transmission electron microscopy (TEM) images of a wild-type spore of *C. difficile* spore produced by strain 630Δ*erm*. The image corresponds to a longitudinal view and shows the spore body and the appendage region (A). Other spore features are: the core, Cr; the cortex, Cx; the coat, Ct; the exosporium, Ex. The bottom panels show higher magnification images of details of the spore on the top panel (**a1** and **a2**) and a detail of the spore body/appendage region for a spore still inside the mother cell (**a3**). In a1 to a3, the blue arrowheads point to the lamellar region that forms the spore coat. The brown arrowheads in **a1** and **a2** point to a segmented layer at the edge of the exosporium; in **a3,** the brown arrowheads point to a structure in the mother cell cytoplasm that has the appearance of the segmented layer at the edge of the spore highlighted in **a1** and **a2**. (**b**) The top panels shows a TEM image of a longitudinal section of a spore produced by an epidemic strain of ribotype 126. The different spore structures are labelled as in **a**, but note that the electrodense layer (brown arrowheads) that encircles the lamellar coat (yellow arrowhead) is much thicker that in 630Δ*erm* spores and has a bumpy appearance. As for the spore in **a**, this layer is continuous with the spore appendage. The green arrowheads point to material that projects from the edge of the spore and may correspond to glycosylated proteins. In **a** and **b**, the electrodense layer is normally interpreted as being the exosporium. In an alternative view (see text for a discussion) the exosporium is the thin segmented layer at the edge of the spore, closely connected to the underlying electrodense layer which in turn may correspond to an outer coat. The scale bars represent 500 nm for the top panels and 100 nm for the bottom panels

the core its resistance properties (Nerber and Sorg 2021; Setlow 2006, 2014). The inner forespore membrane has very low permeability to small molecules, including water (Cowan et al. 2004). The composition of the primordial germ cell wall is similar to that of the cell wall of the vegetative cell, and it will latter become the cell wall of the outgrowing bacterium (Atrih et al. 1998; Meador-Parton and Popham 2000). The spore cortex PG is essential for the maintenance of the dehydrated state of the spore core and spore mineralization, which in turn are essential for heat resistance (Setlow 2014). While the germ cell wall is assembled from forespore precursors by a machinery inserted in the forespore inner membrane (McPherson et al. 2001), synthesis of the spore cortex is a function of the mother cell and takes place across the forespore outer membrane (Alabdali et al. 2021; Diaz et al. 2018; Vasudevan et al. 2007). The spore cortex is covered by several layers composed mainly of proteins that together form the spore surface layers. In *B. subtilis*, the spore surface layers are formed by a coat and a more external crust (Driks and Eichenberger 2016; Henriques and Moran 2007). The coat is organized in two main layers: an inner lamellar coat layer and an electrodense outer coat layer and is composed of dozens of proteins that provide mechanical integrity and contribute to spore resistance against small toxic molecules, protects the cortex against PG-breaking enzymes such as host produced lysozyme, and modulates the interaction of spores with germinants (McKenney et al. 2013). The crust is glycosylated and closely associated to the spore outer coat (Bartels et al. 2019; Shuster et al. 2019). In pathogens such as *B. cereus* and *B. anthracis*, the coat, which is not so clearly differentiated into an inner and an outer layer, is surrounded by an additional layer, called the exosporium, which has a balloon like appearance and is separated from the underlying coat by an interspace of unknown composition (Ball et al. 2008; Ohye and Murrell 1973; Stewart 2015). The exosporium has a basal layer from which long radial projections of glycosylated collagen-like proteins, such as BclA, emanate. BclA mediates an interaction with receptors at the surface of eukaryotic cells, which in the case of *B. anthracis* is important for spore internalization (Sylvestre et al. 2002).

The core, the cortex, a lamellar coat, an electrodense exosporium, and a prominent polar appendage are the standard descriptions of a *C. difficile* spore, based on transmission electron microscopy (TEM) of thin sections (Antunes et al. 2018; Calderón-Romero et al. 2018; Paredes-Sabja et al. 2014; Pizarro-Guajardo et al. 2020) (Fig. 2). An alternative view of the structural organization of the spore surface is that the electrodense layer corresponds to an outer coat, whereas the exosporium is a thin layer at the edge of the spore. This layer is often seen in electron micrographs of spores regardless of whether the electrodense layer is thin and relatively smooth as is the case for spores of the commonly used strain 630*Δerm* or is thick and with bumpy appearance as in spores of epidemics strains (Fig. 2). This alternative view of the structural organization of the spore surface layers is discussed in Sect. 4. The exosporium is permeable to germinants but excludes lytic enzymes; it is the first line of contact of the spore with host cells, the immune system, and with the environment, and influences adhesion to cells and abiotic surfaces (Paredes-Sabja et al. 2014; Stewart 2015).

3 The Genetic Regulation of Sporulation in *C. difficile*

3.1 Regulatory Networks, an Overview

Entry into sporulation is governed by the response regulator Spo0A in all sporeformers in which its function was experimentally assessed (Abecasis et al. 2013; Galperin et al. 2012; Traag et al. 2013a). While Spo0A is universally conserved among sporeformers, the presence of a *spo0A*-encoding gene in the genome does not by itself identifies an organism as a sporeformer (Galperin 2013; Galperin et al. 2012). In *B. subtilis*, Spo0A is auto-regulatory, phosphorylated through a phosphorelay and subject to just-in-time regulation (Chastanet and Losick 2011; Grossman 1995; Jiang et al. 2000; Sonenshein 2000). As described below, the regulation of Spo0A production and activation in *C. difficile* differs significantly (Lee et al. 2022). Following formation of the forespore and the mother cell, cell type-specific gene expression is controlled by a cascade of RNA polymerase sigma factors in the order σ^F, σ^E, σ^G, and σ^K (Haldenwang 1995; Higgins and Dworkin 2012; Hilbert and Piggot 2004; Saujet et al. 2014). Studies in *B. subtilis* have shown that the cascade is hierarchical, in that the activity of a sigma factor is dependent on the activity of the prior sigma in the cascade, and it involves cell–cell signalling pathways that connect the two lines of gene expression (De Hoon et al. 2010; Higgins and Dworkin 2012; Hilbert and Piggot 2004). As such, the activation of the consecutive sigma factors alternates between the two cells. The cell–cell signalling pathways operate at critical stages in morphogenesis, coupling the activation of each sigma factor to the successful completion of key intermediate structures. These morphological checkpoints maintain the successive waves of gene expression in register with progression through the morphological stages of sporulation (De Hoon et al. 2010; Higgins and Dworkin 2012; Hilbert and Piggot 2004). In *B. subtilis*, σ^F is activated in the forespore soon after asymmetric division and is transiently active in this compartment (Margolis et al. 1991); soon after and in a σ^F-dependent manner σ^E is activated in the mother cell (Driks and Losick 1991; Hofmeister et al. 1995). Both σ^F and σ^E thus control early stages of morphogenesis, that is, prior to engulfment completion; σ^G and σ^K, in turn, are active mainly during late stages, following engulfment completion. σ^E is required for the activity of σ^G and the activity of σ^K depends on that of σ^G (Camp and Losick 2009; Cutting et al. 1991; Doan et al. 2009).

We review below the main mechanisms involved in the activation of Spo0A and the four cell type-specific sigma factors highlighting the main players and the deviations to the *B. subtilis* paradigm that are specific to *C. difficile*. One general observation is that the morphological checkpoints appear degenerate for σ^F and σ^E and may not be in place for σ^G and σ^K. The significance of this observation is still unclear, as most of the mechanisms enforcing morphological coupling that have been found in *B. subtilis* are predicted to be present in an ancestral of all sporeformers (Abecasis et al. 2013; De Hoon et al. 2010; Galperin 2013; Galperin et al. 2012; Ramos-Silva et al. 2019). In this chapter, when experimental evidence is still missing for *C. difficile* but the regulatory proteins are conserved, we describe the mode of action of these proteins during sporulation in *B. subtilis*. The information available for *B. subtilis* is also used to give context to the work in *C. difficile*.

The total number of genes involved in sporulation in *C. difficile*, including those found by genome wide screens (RNAseq and random mutagenesis), is of at least 1000, i.e., about 25% of the genome (Dembek et al. 2015; Fimlaid et al. 2013; Pereira et al. 2013; Pettit et al. 2014). Remarkably, nearly all of the genes found by RNAseq as Spo0A-controlled were also found in a TnSeq screen as required for sporulation (Dembek et al. 2015; Fimlaid et al. 2013; Pettit et al. 2014). Thus, nearly all of the Spo0A-controlled functions are important for spore formation.

3.2 Sporulation Initiation

3.2.1 Regulatory Events in the Cell Entering Sporulation

In *C. difficile*, the regulatory networks that control sporulation initiation are still not well understood. *spo0A* transcription is under the control of σ^A and σ^H, the vegetative, housekeeping sigma factor, and a stationary RNA polymerase sigma factor, respectively (Saujet et al. 2011). A *sigH* mutant does not form spores and is blocked at the onset of sporulation, prior to asymmetric division (Saujet et al. 2011). This phenotype is a consequence not only of the positive regulation of *spo0A* by σ^H but also of genes involved in chromosome segregation (see below). *spo0A* is also subject to positive autoregulation (Underwood et al. 2009) (Fig. 3).

Spo0A, as other response regulator proteins, consists of a receiver domain and a DNA binding domain and is activated by phosphorylation of the receiver domain in a conserved aspartate residue (D56 in the *C. difficile* protein) (DiCandia et al. 2022). Once phosphorylated, a conformational change results in homodimerization and binding to specific sequences, the Spo0A boxes, in the regulatory regions of its target genes; depending on the location of the boxes relative to the core promoter, Spo0A can act as a positive or a negative transcriptional regulator (Asayama et al. 1995; DiCandia et al. 2022; Lewis et al. 2002). In *B. subtilis*, low-threshold genes, i.e., those with high affinity Spo0A boxes, are activated before high-threshold genes, i.e., with low affinity Spo0A boxes (Fujita et al. 2005). This allows the sequential deployment of several adaptive responses as cells enter stationary, before sporulation (the key genes having low affinity boxes) is triggered (Fujita and Losick 2005). This is possible because at least in laboratory (domesticated) strains, phosphorylation of Spo0A increases gradually through signal integration mediated by a phosphorelay (Fujita and Losick 2005). Spo0A is also auto-regulatory and at least in laboratory

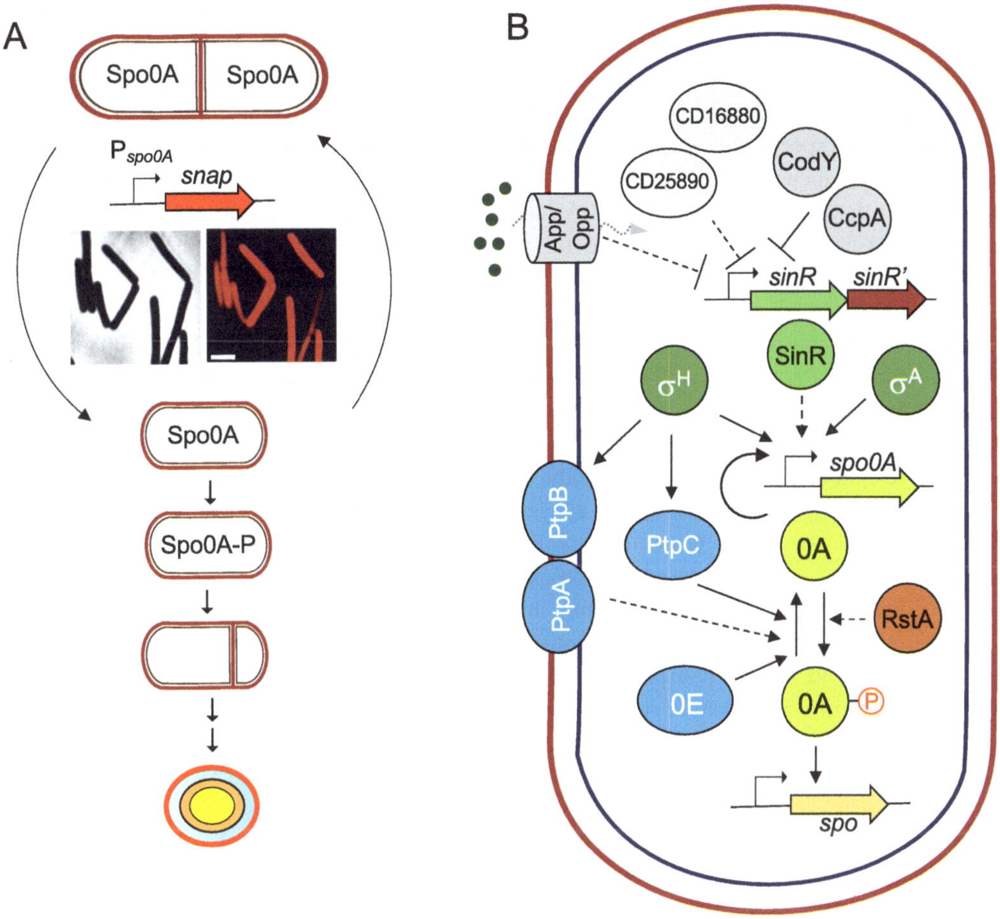

Fig. 3 Activation of Spo0A via phosphorylation. (**a**) During vegetative growth cells grow by elongating along their long axis and divide at midcell to produce two identical daughter cells. As seen by fluorescence microscopy of *C. difficile* cells carrying a P*spoOA*⁻ *SNAP*^{Cd} transcriptional fusion, *spo0A* is expressed in all cells during growth. In response to as yet unidentified signals, Spo0A is activated by phosphorylation triggering the expression of the genes coding for the first cell type-specific regulators of sporulation, as well as the genes required for the asymmetric division of the cell. Scale bar, 1 μm. (**b**) Expression of *spo0A* is under the control of σ^A and σ^H and indirectly, is positively regulated by SinR, a helix-turn-helix-type transcriptional regulator. Expression of the *sinRR′* operon is inhibited by two global regulators, that respond to nutrient availability, CodY and CcpA and indirectly by the oligopeptide permeases, Opp and App, the putative ribonuclease CD25890 and the response regulator CD16880. Levels of Spo0A-P are positively regulated by RstA, a DNA binding protein of the RRNPP family, that may respond to environmental signals. PtpC and Spo0E act as phosphatases that directly dephosphorylate Spo0A-P, while PtpA and PtpB are orphan histidine kinases acting indirectly to decrease the levels of Spo0A-P. Solid lines and dashed lines indicate direct and indirect regulatory networks, respectively

(domesticated) strains, a bistable switch is created which restricts sporulation to a fraction of the population (Chastanet and Losick 2011; Veening et al. 2005). As discussed later, in gut isolates, this bistable switch is not seen, and sporulation is induced during growth (Serra et al. 2014). In *C. difficile*, *spo0A* is constitutively expressed, and the levels of phosphorylated Spo0A correlate with the level of protein accumulation, suggesting that phosphorylation is not a limiting step for entry into sporulation (Martins et al. 2021; Rosenbusch et al. 2012) (Fig. 3a). As such, and in contrast, to what is seen in *B. subtilis*, higher accumulation of phosphorylated Spo0A leads to increased levels of sporulation (Dembek et al. 2017; Fujita and Losick 2005; Kempher et al. 2022; Martins et al. 2021).

Response regulators are usually phosphorylated by specific histidine kinases that sense environmental cues. In *C. difficile*, no histidine kinase(s) that directly activates Spo0A has been identified and there is no evidence for a phosphorelay involved in Spo0A activation, as in *B. subtilis* (Grossman 1995; Sonenshein 2000). An indirect positive regulator of Spo0A phosphorylation is RstA (Edwards et al. 2016, 2019). RstA is a DNA binding protein of the RRNPP family, which includes proteins known to respond to quorum sensing signals (Neiditch et al. 2017). It is possible that by an as yet unknown mechanism RstA promotes Spo0A phosphorylation in response to environment signals (Edwards et al. 2022) (Fig. 3b). In *C. difficile*, perhaps because *spo0A* is constitutively expressed and Spo0A phosphorylation is not a limiting step for sporulation, several direct and indirect negative regulators of *spo0A* expression and Spo0A phosphorylation have been identified; these are thought to control the proper timing of sporulation initiation (Fig. 3b).

The *C. difficile* genome codes for five orphan histidine kinases (CD14920, CD24920, CD15790, CD19490, and CD13520) with similarity to the *B. subtilis* sporulation-associated kinases. However, deletion of CD19490 and CD13520 (CprK) has no impact on sporulation, and a strain with a deletion in CD14920 (PtpA), CD24920 (PtpB), or CD15790 (PtpC) has a hypersporulation phenotype (Edwards et al. 2022; Suárez et al. 2013). In PtpA and PtpC, the conserved histidine residue responsible for the transfer of the phosphate group to the response regulator, is important for the function of these kinases (Edwards et al. 2022). In the case of PtpC, direct phosphorylation of and interaction with Spo0A was shown in vitro (Underwood et al. 2009). However, the hypersporulation phenotype observed in vivo does not support a direct role for this kinase in the phosphorylation of Spo0A (Edwards et al. 2022). In fact, many histidine kinases are also phosphatases, removing the phosphate from the response regulator aspartate residues, this may be the case of PtpC, in vivo. PtpA may also act directly as a Spo0A phosphatase or alternatively is an indirect regulator of Spo0A phosphorylation. PtpB function does not require the conserved histidine residue but it was shown to act in the same pathway of PtpA. The three orphan histidine kinase shown to play a role in sporulation initiation act as repressors (Edwards et al. 2022) (Fig. 3b).

A protein, Spo0E (CD32710), belonging to a family of aspartyl-phosphate phosphatases, that in *B. subtilis* specifically dephosphorylates Spo0A, is also present in *C. difficile* (Ohlsen et al. 1994; Perego and Hoch 1991). In *C. difficile* a *spo0E* mutant shows a hypersporulating phenotype. Due to this phenotype and since Spo0A and Spo0E directly interact, Spo0E is likely to directly dephosphorylate Spo0A in vivo, negatively regulating entry into sporulation (Lee et al. 2022) (Fig. 3b).

Other negative regulators act indirectly at the level of *spo0A* expression (Fig. 3b). Two global transcription regulators, that respond to nutrient availability, CodY and CcpA, affect the initiation of sporulation (Antunes et al. 2012; Dineen et al. 2010; Nawrocki et al. 2016). Also, the oligopeptide permeases, Opp and App, the putative ribonuclease CD25890, and the response regulator CD16880 are negative regulators of sporulation initiation (Edwards et al. 2014; Kempher et al. 2022; Martins et al. 2021). Deletions of the negative regulators described above lead to increased expression of the *sinRR′* operon, coding for two HTH-type transcription regulators (Fig. 3b). In *C. difficile*, SinR has a positive effect on sporulation, mainly due to an increase in the expression of *spo0A* by an unknown mechanism (Girinathan et al. 2018). In contrast, SinR′ is a negative regulator of sporulation (Girinathan et al. 2018). As an operon, *sinRR′* expression leads to greater abundance of SinR relative to SinR′, resulting in a positive effect on sporulation (Ciftci et al. 2019). Deletion of the *sinRR′* operon, however, results in an asporogenous phenotype, with the cells blocked at the onset of sporulation (Girinathan et al. 2018). In this mutant, the cellular levels of the second messenger signalling molecule c-di-GMP increase (Girinathan et al. 2018; Poquet et al. 2018). High levels of c-di-GMP inhibit sporulation initiation whereas the effect of low levels of c-di-GMP on sporulation is strain dependent (Dhungel and Govind 2021; Edwards et al. 2021). It seems likely that a series of regulatory pathways that control sporulation initiation exert their control, directly or indirectly, at the level of expression of the *sinRR′* operon (Fig. 3b).

Noteworthy, many of the regulators that control sporulation initiation also participate in the regulation of toxin production, such as CcpA, CodY, Spo0A, σ^H, RstA, Spo0E, and PtpA (Antunes et al. 2012; Childress et al. 2016; Dineen et al. 2010; Edwards et al. 2019; Lee et al. 2022; Pettit et al. 2014; Saujet et al. 2011). The intertwining of the regulatory circuits that control toxinogenesis and sporogenesis suggests that the balance between these two processes and/or the extent of expression of the toxin-encoding genes by sporulating cells is critical during the infectious cycle (Donnelly et al. 2022; Ransom et al. 2018; see also above).

3.2.2 Morphological Changes in Cells Entering Sporulation

In *C. difficile*, Spo0A controls a total of approximately 300 genes, many of them overlapping with the σ^H regulon (Fimlaid et al. 2013; Pettit et al. 2014). This may be a consequence of a direct positive regulation of Spo0A over the *sigH* gene. The Spo0A regulon includes genes involved in biofilm formation (Dawson et al. 2012), swimming motility (Pettit et al. 2014), toxin production (Underwood et al. 2009), and sporulation. Among the latter are the genes required for the asymmetric division of the cell entering sporulation and the remodelling of the chromosome to be segregated into the forespore (Fimlaid et al. 2013; Pettit et al. 2014). During the vegetative cycle of growth and division, the chromosome is replicated at midcell and the newly replicated *oriC* regions are rapidly segregated to positions at ¼ and ¾ of the cell length. In *B. subtilis*, the nucleoid assumes a compact topology in which DNA is excluded from the polar regions of the cell (reviewed by (Adams et al. 2014; Errington and Wu 2017)). Division close to the cell poles would result in the formation of anucleate cells but is prevented by a division inhibitor complex, MinCD. In contrast to Gram-negative bacteria in which the MinCD inhibitor oscillates and the MinE protein serves as topological indicator keeping the inhibitor more frequently close to the cell poles, in *B. subtilis* there is no MinE protein and the MinCD complex does not oscillate. Aberrant division is prevented through the action of two proteins: the tropomyosin-like DivIVA protein which intrinsically recognizes membranes with negative curvature and MinJ, a transmembrane protein. DivIVA and MinJ assemble as double rings flanking the division septum, early in division. MinCD is recruited to these rings and becomes physically separated from FtsZ during division while preventing aberrant formation of Z-rings close to sites of active division (Eswaramoorthy et al. 2011). Following division, DivIVA rings collapse with patches at the cell poles (Eswaramoorthy et al. 2011). These patches are present at the poles of cells entering sporulation and serve as a static attractor that interacts with and keeps the MinCD proteins at the poles (Cha and Stewart 1997; Edwards and Errington 1997; Lenarcic et al. 2009; Marston et al. 1998; Eswaramoorthy et al. 2011). MinJ is absent in *C. difficile* but DivIVA is present (CD26190). *C. difficile* is an exception among Gram-positive bacteria, in that it encodes a MinE homologue, and the three proteins, MinC, MinD, and MinE oscillate when produced in *B. subtilis* and this oscillation interferes with sporulation (Makroczyová et al. 2016). Also, the *C. difficile* DivIVA and MinD protein interact directly in a yeast two-hybrid system and in pull down assays (Valenčíková et al. 2018). The functional analysis of *divIVA* and the *minCDE* operon in *C. difficile* has not been reported. The MinCDJ system of *B. subtilis* acts in conjunction with nucleoid occlusion, the main effector of which is the Noc protein (Wu and Errington 2004), CD36730 in *C. difficile*. Noc is a CTPase that binds to 16 bp Noc-binding sites and then spreads along the DNA to form nucleoprotein complexes and binds to the membrane with both activities dependent on CTP binding (Jalal et al. 2021).

Recruitment of the DNA to the membrane prevents formation of FtsZ rings over the mass of the nucleoid (Adams et al. 2014; Jalal et al. 2021). Noc works in concert with the Min system to promote division at midcell (Rodrigues and Harry 2012; Wu and Errington 2004).

Cells entering sporulation stop growing and go through a last round of chromosome replication. Thus, they have two complete copies of the chromosome. In *B. subtilis*, binding of the RacA protein (produced under Spo0A control) helps organizing the nucleoid into an elongated structure known as the axial filament which extends from pole to pole of the cell (Ben-Yehuda et al. 2003). In *B. subtilis*, RacA also preferentially binds to a centromere-like element close to *oriC* that helps anchoring the chromosome to the cell pole; binding of centromere-RacA to the cell pole is DivIVA-dependent and displaces the division inhibitor complex (Ben-Yehuda et al. 2005; Lenarcic et al. 2009; Wu and Errington 2003). Another DNA-binding protein produced under Spo0A-control, RefZ, binds to DNA motifs around *oriC* and is required for precise septum positioning during asymmetric division; RefZ and Noc act synergistically to promote proper asymmetric division, and many cells of a *noc*/*refZ* double mutant fail to divide asymmetrically and to activate σ^F (see below), or show extra, misplaced, or aberrant septa, and a strong block in sporulation (Brown et al. 2019; Miller and Herman 2022). The axial filament configuration of the nucleoid has not been described for *C. difficile* and no RacA homologue is found in this organism. Furthermore, no RefZ homologue is present in *C. difficile*. How the activity of the oscillating MinCD is counteracted at the onset of sporulation in *C. difficile* and how the chromosome is oriented for segregation into the forespore are unknown.

In *B. subtilis*, two other proteins that are under Spo0A control are FtsZ and SpoIIE. The production of FtsZ is enhanced under Spo0A control in cells entering sporulation and this is important for the re-localization of FtsZ via helical intermediates to sites of division close to the cell poles (Ben-Yehuda and Losick 2002). SpoIIE localizes to these sites, forming what is known as an E ring, and helps stabilizing the Z rings close to the cell poles (see more on SpoIIE below (Ben-Yehuda and Losick 2002). The localization of FtsZ and SpoIIE in *C. difficile* cells entering sporulation has not been reported.

An asymmetric division then divides the developing cell into a small forespore, the future spore, and a larger mother cell. At this stage, the two cells lie side by side and both are in contact with the external medium. Asymmetric division traps about 30% of the chromosome destined to the forespore into this cell. The remaining part of this chromosome is translocated into the forespore by an ATP-dependent DNA translocase, SpoIIIE, itself required to transport parts of the chromosome that may be entrapped by the forming division septum during growth (reviewed by Chan et al. 2022). Both SpoIIIE and a paralog, SftA, contribute to this activity during growth. During sporulation, SpoIIIE assembles at the centre of the septal plate and forms two coaxial paired channels that each translocates one arm of the chromosome (Burton et al. 2007). It takes about 20 min for the complete translocation of the chromosome, the two arms at the same velocity (Burton et al. 2007). The initial trapping of the *oriC* proximal region in the forespore generates a state of transient genetic asymmetry, which contributes to the establishment of cell type-specific gene expression (see the σ^F checkpoint). In *C. difficile*, CD13240 is likely to code for a SpoIIIE homologue, since this gene, as in *B. subtilis*, is downstream of the gene coding for a TepA homologue, a ClpP-like protease involved in the degradation of the SASPs during spore outgrowth (Traag et al. 2013b). The function of the *C. difficile spoIIIE* gene has not been analysed but its requirement for sporulation is established (Dembek et al. 2015).

Also, under Spo0A control are the genes coding for the first cell type-specific sigma factors of sporulation, σ^F and σ^E (Fimlaid et al. 2013). Thus, pre-divisional cell expresses pools of σ^F and σ^E but only after the formation of the asymmetric septum is complete does σ^F and σ^E become active in the forespore and the mother cell, respectively.

3.3 Cell Type-Specific Gene Expression

3.3.1 The σ^F Checkpoint

As mentioned above, σ^F is synthesized prior to the formation of the sporulation septum, but σ^F-dependent gene expression is only detected in the forespore after asymmetric division (Pereira et al. 2013) (Fig. 4). *sigF*, that codes for σ^F, is the third gene of a tricistronic operon expressed under the control of σ^H and Spo0A (Fimlaid et al. 2013; Pettit et al. 2014; Saujet et al. 2011). The operon is conserved in *B. subtilis* and *C. difficile* (Abecasis et al. 2013; Galperin et al. 2012). Although this has not been experimentally evaluated, we presume that the function of *spoIIAA* and *spoIIAB* (the first two genes of the operon) is conserved in *C. difficile* (Fig. 4). In *B. subtilis*, SpoIIAA and SpoIIAB are both required for the forespore-specific activation of σ^F after asymmetric septation. SpoIIAB is an anti-sigma factor that binds to σ^F and holds it inactive in the predivisional cell and in the mother cell after asymmetric division (Duncan and Losick 1993; Gholamhoseinian and Piggot 1989; Min et al. 1993; Schmidt et al. 1990). SpoIIAA is an anti-anti-sigma factor, which can bind to and counteract SpoIIAB, releasing active σ^F (Carniol et al. 2004). SpoIIAB is also a serine protein kinase that can inactivate SpoIIAA by phosphorylation (Min et al. 1993) (Fig. 4c).

SpoIIE (CD34900) is a third protein required for forespore-specific activation of σ^F (Arigoni et al. 1996). SpoIIE is also conserved in *B. subtilis* and *C. difficile* and is produced in the predivisional cell under the control of Spo0A (see above; Dembek et al. 2015; Fimlaid et al. 2013; Pettit et al. 2014). In *B. subtilis*, SpoIIE is targeted to degradation by the membrane-embedded protease FtsH, which is also found in *C. difficile* (Bradshaw and Losick 2015). SpoIIE is a membrane serine phosphatase that initially localizes at the emergent polar septum. SpoIIE may be enriched in the forespore in part because its encoding gene is close to *oriC* (McBride et al. 2005). Moreover, after division completion SpoIIE relocalizes to the forespore pole where oligomerization protects it from degradation and leads to activation of the phosphatase activity (Bradshaw and Losick 2015). Desphosphorylation of SpoIIAA by the SpoIIE phosphatase leads to binding of SpoIIAA to SpoIIAB and, consequently, activation of σ^F in the forespore (Fig. 4c). Neither the cytosolic N-terminal tag that in *B. subtilis* targets SpoIIE (residues 11–37) to degradation is present in *C. difficile* SpoIIE, which has a smaller cytoplasmatic N-terminal domain, nor the genomic synteny between *ftsH* and *spoIIE* is conserved (Bradshaw and Losick 2015). Hence, compartmentalization of σ^F activity may be less stringent in *C. difficile*; in fact, uncompartmentalized σ^F-dependent gene expression was observed in approximately 10% of the *C. difficile* sporulating cells (Donnelly et al. 2022).

In *B. subtilis*, the chromosomal location of the *spoIIA* operon is also important for the compartmentalization of σ^F activity. During asymmetric division 30% of the chromosome destined to the forespore is initially trapped in this cell, with 70% of the chromosome remaining transiently in the mother cell; since the *spoIIA* operon remains in the mother cell, *spoIIAB* is initially excluded from the forespore (Dworkin and Losick 2001). In *C. difficile*, the *spoIIA* operon is localized at 79° and is also transiently located in the mother cell side of the septum. There is also evidence of concomitant degradation of SpoIIAB in the forespore (Pan et al. 2001). The transient genetic asymmetry imposed by the mechanism of forespore chromosome segregation, together with the degradation of SpoIIAB in the forespore after asymmetric division but prior to translocation of the *spoIIAB* gene both contribute to tipping the balance towards σ^F activation. It was also previously noted that the proteins bound to the chromosome are stripped off by SpoIIIE during translocation of the chromosome to the forespore. This includes the stripping of RNA polymerase associated with other sigma factors, including σ^E, or transcription factors such as Spo0A, facilitating binding of σ^F to RNA polymerase and the start a new programme of gene expression (Marquis et al. 2008).

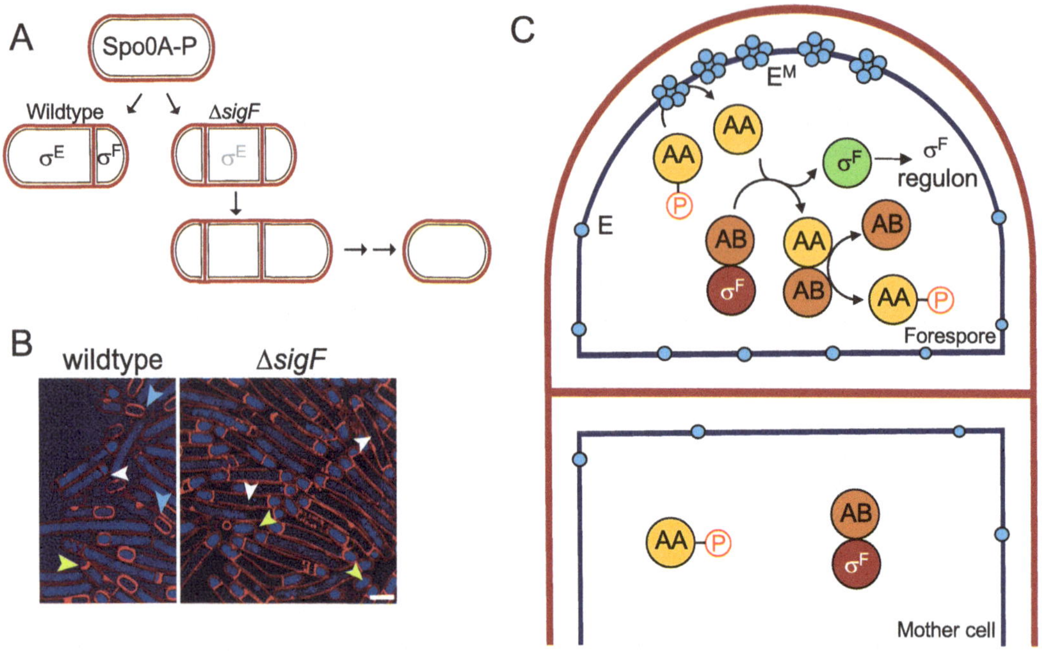

Fig. 4 The σ^F checkpoint. (**a**) At the onset of sporulation, in a wild-type strain, a single septum forms near one pole of the sporangium, creating a small cell, the forespore, and a larger cell, the mother cell. At this point, the first sporulation specific sigma factors become active, σ^F in the forespore and σ^E in the mother cell. In cells carrying a deletion of the *sigF* gene, σ^E is only activated in a small fraction of cells and a second polar septum is formed, leading to an abortive disporic phenotype, as the mother cell is anucleate. If enough nutrients are present, however, this sporangium may reinitiate longitudinal growth, with the polar forespores giving rise to the small cells that are observed in this mutant. (**b**) Fluorescence microscopy of wild-type and *sigF* mutant cells stained with the membrane dye FM4-64 and the DNA stain Hoechst. Sporulation in *C. difficile* is heterogeneous. In the wild-type cells at different stages of sporulation are observed: at asymmetric division (white arrowheads); at intermediate stage in the process of engulfment (yellow arrowheads); and with engulfment completed (note that Hoechst is excluded from the forespore; blue arrowheads). In contrast, most of the *sigF* cells show a disporic phenotype, with DNA in both prespores and a central mother cell which lacks DNA and eventually lyses (white arrowheads). Small cells characteristic of this mutant are indicated by the yellow arrowhead. The direction of chromosome translocation into the forespore by the SpoIIIE complex is not represented for simplicity. Scale bar, 1 μm. (**c**) Pathway governing the activation of σ^F, involving the SpoIIAA (AA), SpoIIAB (AB), and SpoIIE (E) proteins. σ^F is kept inactive in a complex with SpoIIAB (AB) in the pre-divisional cell and in the mother cell after asymmetric division. SpoIIE is a membrane-bound serine phosphatase that initially localizes at the polar septum. After division completion, SpoIIE is degraded in the mother cell side and relocalizes to the forespore pole where oligomerization protects it from degradation and potentiates its phosphatase activity. Desphosphorylation of SpoIIAA-P by the SpoIIE phosphatase in the forespore leads to binding of SpoIIAA to SpoIIAB, releasing σ^F from inhibition

The *sigF* mutant presents an abortive disporic phenotype and occasionally forms multiple septa close to the cell pole ((Fimlaid et al. 2013; Pereira et al. 2013); Fig. 4a, b; see also below). Also, characteristic of this mutant is the production of small cells that may originate from both polar forespores when the central mother-cell compartment, which lacks a chromosome, undergoes lysis (Fimlaid et al. 2013; Pereira et al. 2013) (Fig. 4a, b). At least in some sporulation media used in *C. difficile*, enough nutrients may be present for these cells to resume vegetative growth, as observed for the *sigF* mutant of *B. subtilis* when transferred to rich medium (Dworkin and Losick 2005) (Fig. 4a, b).

Once active σ^F transcribes 55 genes, such as *sigG* coding for the late σ^G forespore-specific sigma factor, *spoIIQ* and *spoIIP* both involved in engulfment (and in case of *spoIIQ* also in the control of late cell specific gene expression) and *spoIIR*, involved in the activation of σ^E in the mother cell (Fimlaid et al. 2013; Pereira et al. 2013).

3.3.2 The σ^E Checkpoint

As is the case for σ^F, σ^F is produced in the pre-divisional cell (Fig. 5). The *sigE* gene, which codes for σ^E, is expressed under the control of σ^A and Spo0A (Fimlaid et al. 2013; Pettit et al. 2014). σ^E is synthesized as an inactive proprotein precursor named pro-σ^E. The pro-sequence at the N-terminal of σ^E directs the sigma factor to the membrane and has to be proteolytically removed for the sigma factor to interact with core RNAP and direct gene expression. The first gene of the operon, *spoIIGA*, codes for an aspartic membrane protease, shown in *B. subtilis* to be responsible for the removal of the pro-sequence. SpoIIGA has a membrane-embedded N-terminal domain, and a C-terminal domain, facing the cytosol, that dimerize (reviewed by Hilbert and Piggot 2004; Imamura et al. 2008). The *spoIIG* operon is away from the *oriC* region, and this position may contribute to favour the accumulation of pro-σ^E and SpoIIGA on the mother cell side of the asymmetric septum (McBride et al. 2005). This is important because SpoIIGA is only active after receiving a signal from the forespore (Fig. 5c). The signal involves two secretory proteins, SpoIIR and SpoIIT, produced in the forespore under the control of σ^F (Hofmeister et al. 1995; Meeske et al. 2016). In *E. coli*, however, SpoIIR is sufficient to activate SpoIIGA, consistent with the relative mild effect that disruption of *spoIIT* has on sporulation in *B. subtilis* (Meeske et al. 2016) and with its absence from *C. difficile*. SpoIIR interacts with the N-terminal domain of SpoIIGA on the *trans* side of the membrane and this causes a conformational change that leads to the formation of active protease dimers of the C-terminal part (Imamura et al. 2008). Thus, σ^E activation is coupled to polar septation in an indirect manner which depends on the forespore-specific synthesis of two (in *B. subtilis*) or just one signalling proteins (in *C. difficile*). In the absence of σ^E activity, a second septum forms at the distal pole, creating a sporangium with two forespores, the abortive disporic phenotype also seen for *sigF* mutants, each with a copy of the genome and a central mother-cell which lacks a chromosome (the latter translocated by SpoIIIE into the second forespore) (Fig. 5a, b). Studies in *B. subtilis* have shown that formation of the second polar septum is prevented by a complex of three proteins (SpoIID, SpoIIM, and SpoIIP, also called the DMP machine, below).

As the *spoIIA* operon, *spoIIR* is also subject to "chromosomal positional regulation". *spoIIR* is located at 349° in the *C. difficile* chromosome, near *oriC*, within the 30% of the chromosome initially trapped inside the forespore after asymmetric division. Therefore, *spoIIR* is immediately transcribed in the forespore once σ^F becomes active (Dworkin and Losick 2001). In *B. subtilis*, placing *spoIIR* away from the origin-proximal part of the chromosome substantially delays *spoIIR* expression with a concomitant failure to activate σ^E in a sub-population of cells in time to prevent formation of the second polar septum; this leads to an abortive disporic phenotype and impaired sporulation (Dworkin and Losick 2001). It should be noted that if the control over the initiation of chromosome replication that functions at the onset of sporulation is relaxed, the mother cell may end up with more than one chromosome and some of the diasporic cells may complete sporulation (Eldar et al. 2009). In contrast to *B. subtilis*, in which *spoIIR* expression is only dependent on σ^F, *spoIIR* is, in *C. difficile*, expressed at low levels in the forespore in the absence of σ^F, in a Spo0A-dependent manner (Pereira et al. 2013). Therefore, in a sub-population of cells, σ^E is activated in the absence of σ^F (Fimlaid et al. 2013; Pereira et al. 2013; Saujet et al. 2013). We presume that in some cells, transient σ^E activity is detected in the central cell of the disporic sporangium (Pereira et al. 2013), before the translocation of

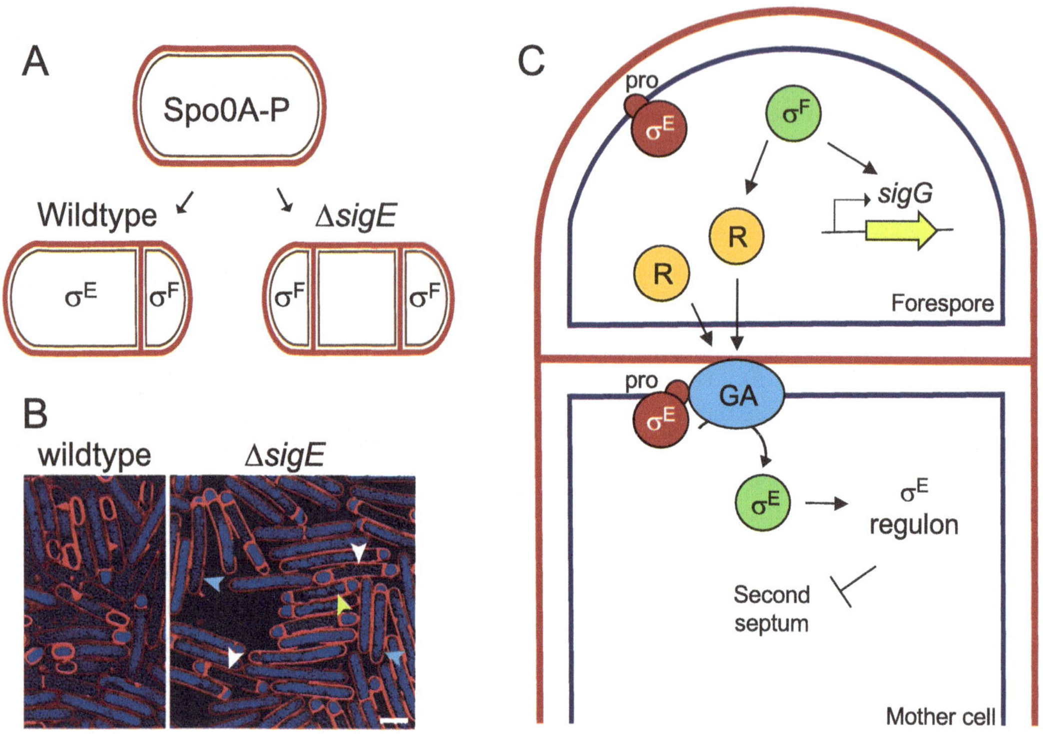

Fig. 5 σ^E checkpoint. (**a**) Since σ^E activity is required to inhibit the formation of a second polar septum, cells carrying a deletion in the *sigE* gene have a disporic phenotype. In these cells, however, σ^F becomes active in the forespore, and therefore longitudinal growth of this cell is inhibited. (**b**) Fluorescence microscopy of wild-type and *sigE* mutant cells stained with the membrane dye FM4-64 and the DNA stain Hoechst. Most of the *sigE* cells show a disporic phenotype, with DNA in both forespores and a central mother cell compartment which lacks DNA (white arrowheads). Indicated are also cells that have not translocated the chromosome to the second forespore compartment (yellow arrowheads) and cells that have not completed the formation of the second septum (blue arrowheads). Scale bar, 1 μm. (**c**) Cell–cell signal transduction pathway governing the activation of σ^E. The signalling protein SpoIIR (R) is synthetized in the forespore, mainly under the control of σ^F. SpoIIR is then secreted to the space between the two cellular compartments where it activates SpoIIGA (GA), a membrane protease responsible for converting pro-σ^E to active σ^E

the terminal proximal region of the chromosome to the second forespore. Another population may activate σ^E on time to inhibit formation of the second septum.

Soon after the sporulation septum is formed, the mother cell begins to engulf the forespore. As shown first in *B. subtilis*, the SpoIID, SpoIIM, and SpoIIP proteins, collectively designated the DMP machine, and all produced under σ^E control (Eichenberger et al. 2004; Steil et al. 2005; but note that *spoIIP* is also transcribed in the forespore from the promoter of the upstream *gpr* gene), have a central role in this process (reviewed by Khanna et al. 2020). SpoIIM is a polytopic membrane protein that in *B. subtilis* serves as a platform to bring SpoIID and SpoIIP into a complex, all three proteins anchored to the forespore outer membrane (reviewed by Khanna et al. 2020). SpoIIP is an endopeptidase that cleaves the crosslinks between the stem peptides, and also an amidase that removes the peptide stems from the glycan chains; SpoIID is a lytic transglycosylase that acts on the products generated by SpoIIP (Khanna et al. 2020). The DMP complex localizes to the leading edges of the engulfment membranes and is thought to remodel PG during engulfment (Kelly and Salgado 2019; Khanna et al. 2020). PG synthesis is required for membrane migration and for the localization of the DMP complex at the leading edges of the engulfing membrane (Ojkic et al. 2016). Importantly, proteins involved in PG synthesis, including PonA (a high molecular weight (HMW) class A bifunctional transglycosylase/transpeptidase) and all the HMW class B penicillin binding proteins (PBPs) and low molecular weight (LMW) PBPs that were tested track the leading edge of the engulfing membrane from the forespore side (Ojkic et al. 2016). Membrane migration thus appears to require the PBP-dependent PG synthesis in the forespore and the PG-degrading DMP complex acting in the mother cell (Ojkic et al. 2016). The function of the DMP complex, but not the requirement of PG synthesis in the forespore, has been studied in *C. difficile*.

Although conserved among sporeformers, *spoIIM* is dispensable for engulfment in *C. difficile* where SpoIID and SpoIIP interact directly in the absence of SpoIIM (Dembek et al. 2018). Also, in contrast to *B. subtilis*, SpoIIP is produced under the control of σ^F, and has a signal peptide, suggesting it is secreted across the forespore inner membrane. Thus, the SpoIIP catalytic domain could be released in the intermembrane space interact with the membrane tethered SpoIID (Kelly and Salgado 2019). In addition to cleavage of the signal peptide, the 38 kDa SpoIIP protein is further processed by an as yet unknown protease(s), into two isoforms, one of which, of around 30 kDa is not seen in *sigE* or *sigG* mutants (Ribis et al. 2017). This suggests that perhaps SpoIIP is inactivated towards the end of the engulfment process (reviewed by Kelly and Salgado 2019). In *B. subtilis*, as in *C. difficile*, *spoIIP* mutants are blocked soon after formation of the asymmetric septum. In *B. subtilis*, *spoIIP* mutants form bulges which result from uncoupled synthesis and degradation of PG and continued forespore chromosome translocation (reviewed by Khanna et al. 2020). In *C. difficile*, *spoIIP* mutants do not form bulges, suggesting some degree of functional redundancy at this stage. Redundancy may come from the SpoIIQ-SpoIIIAH transmembrane complex. SpoIIIAH (σ^E-dependent) and SpoIIQ (σ^F-dependent) interact through their extracytoplasmic domains in the intermembrane space forming a stable complex that has a zipper-like function, keeping the two forespore membranes close during engulfment (reviewed by Khanna et al. 2020; see also the section on late gene expression). In *C. difficile*, the SpoIIIAH-SpoIIQ zipper may actually compensate for impaired activity of the DMP machinery (Ribis et al. 2017; reviewed by Kelly and Salgado 2019). In *C. difficile*, *spoIIP* mutants show a bearding phenotype in which coat material does not adhere to the forespore surface, but rather extends into the mother cell cytoplasm (Ribis et al. 2017). This phenotype is also seen, to a lesser extent, for a *spoIID* mutant, but a *spoIID/spoIIQ* mutant displays it to the level of the *spoIIP* mutant (Ribis et al. 2017). While strengthening the idea that the DMP machinery and the SpoIIQ-SpoIIIAH complex are at least partially redundant, this observation also suggests that the two complexes have a role in the localization of the coat/exosporium proteins (see Sect. 4).

At the end of the engulfment process, the FisB protein (CD07820 in *C. difficile*), produced under the control of σ^E, mediates membrane fission, releasing the forespore in the mother cell cytoplasm (Doan et al. 2013; Landajuela et al. 2021, 2022). Interactions with acidic lipids in the membrane and homo-oligomerization of FisB are essential for its localization at fission sites, which in turn is required for membrane fission (Doan et al. 2013; Landajuela et al. 2021). The SpoIIIE-mediated translocation of the chromosome inflates the forespore (Lopez-Garrido et al. 2018) and membrane fission is energized by the mechanical energy that is produced through forespore inflation; thus, in addition to the localization and homo-oligomerization of FisB at fission sites, forespore inflation is also a necessary condition for efficient membrane fission (Landajuela et al. 2021, 2022). The membrane fission step that terminates the engulfment sequence may be conserved in *C. difficile*, as *CD07820* is also under σ^E control and the gene was found to be required for sporulation (Dembek et al. 2015).

Engulfment completion leads, in *B. subtilis*, to σ^K activation through a σ^G-dependent signal, as detailed in the following sections. σ^K then drives assembly of the cortex and coat/crust layers. Most of the proteins and precursors pool required for synthesis of the spore cortex are produced under σ^E control but formation of this structure requires σ^K (Vasudevan et al. 2007). In contrast, *C. difficile sigK* mutants still form the cortex layer (Pereira et al. 2013). The high molecular weight penicillin-binding protein SpoVD is a transpeptidase shown in *B. subtilis* to interact with a membrane-embedded transglycosylase of the shape, elongation, division and sporulation (SEDS) family, coded for by the σ^E-controlled *spoVE* gene (Henriques et al. 1998; Meeske et al. 2016; Daniel et al. 1994; Fay et al. 2010; Sjodt et al. 2020). This interaction is essential to build the spore cortex. In *C. difficile*, the *spoVE* and *spoVD* genes are conserved and the requirement for SpoVD for spore cortex synthesis was demonstrated (Alabdali et al. 2021). Cephamycin antibiotics block sporulation in *C. difficile* by targeting sporulation-specific penicillin binding proteins, including SpoVD (Sacco et al. 2022). The functional characterization of *spoVE* has not been reported.

The σ^E regulon of *C. difficile* includes at least 297 genes (Fimlaid et al. 2013; Saujet et al. 2013). In addition, to those already mentioned, regulon includes the *spoIIIA* operon, involved in the engulfment process and in cell–cell communication (see channel, below), genes required for the activation of the late mother cell-specific regulator σ^K, *sigK* and *CD1234*, genes involved in cortex synthesis such as the *yqfCD* operon, *spoVD* and *spoVQ*, spore coat and exosporium assembly, as *spoIVA*, *cotL*, *sipL*, *cdeM*, and *cdeC* (Alabdali et al. 2021; Alves Feliciano et al. 2019; Antunes et al. 2018; Calderón-Romero et al. 2018; Fimlaid et al. 2015b; Putnam et al. 2013; Serrano et al. 2016a, b; Touchette et al. 2021). σ^E also regulates the expression of at least three mother cell-specific transcriptional regulators. CD02640, of the *nrdR* family of transcriptional repressors, is in an operon (CD02639–CD02641) with *ylmC*, which is part of the sporulation genomic signature (Abecasis et al. 2013). In *B. subtilis*, *ylmC* has a paralog, *ymxH* and the double mutant shows a strong block in sporulation, after asymmetric division (Abecasis et al. 2013). *ylmC* has no paralogs in *C. difficile*; its function is not known but it codes for a putative transporter, which suggests a role in mother cell-forespore communication. The second regulatory protein is coded for by CD06290 and belongs to the catabolite repressor protein, Crp, family; proteins of this family are involved in the coordination between internal cAMP levels and expression of genes involved in sugar metabolism (Deutscher 2008). In fact, in *B. subtilis* as in *C. difficile*, the σ^E regulon includes a large group of metabolic genes, coding for oxireductases, peptidases, and nutrient transporters, that are thought to be important to maintain and adjust metabolic activity in the mother cell and enabling this cell to nurture the forespore (Fimlaid et al. 2013; Saujet et al. 2013, 2014). Some of these genes may be under the control of additional transcription factors that may not be sporulation-specific but rather co-opted to the process.

The third transcriptional regulator, SpoIIID, is mother cell-specific in both *B. subtilis* and in *C. difficile*. SpoIIID is a winged helix-turn-helix transcription factor (Chen et al. 2014). In *B. subtilis*, SpoIIID represses some σ^E-controlled genes, which are thus transiently expressed in the mother cell; it also activates transcription of others, such as the *sigK* gene, which is delayed relative to a first wave of σ^E-dependent but SpoIIID-independent genes (see also below, section on the control of σ^K activation) (Eichenberger et al. 2004). In *C. difficile*, SpoIIID inhibits the transcription of 30% of the σ^E-dependent genes (the *spoIIIA* operon, *spoIVA* and *sipL* are examples) and acts as an activator of *sigK*, *CD1234* (coding for a recombination directionality factor, as detailed below) and *cdeC* transcription (see Sect. 4) (Fig. 6). Indirectly, it controls all other members of the σ^K-regulon (Pishdadian et al. 2015).

3.3.3 Late Forespore Gene Expression

In contrast to *B. subtilis*, in *C. difficile* the activity of the late sporulation sigma factors, σ^G and σ^K, is not strictly coupled to engulfment completion (Pereira et al. 2013). *sigG* is transcribed from two promoters. The first promoter is located upstream of the *spoIIG* operon, with which *sigG* is co-transcribed. Therefore, transcription of the *sigG* gene begins in the pre-divisional cell under the control of σ^A and Spo0A. However, this poly-cistronic mRNA does not lead to σ^G accumulation and/or activation, at least in a wild-type background (Fimlaid et al. 2013). As in *B. subtilis*, the long mRNA that originates from the *spoIIG* promoter is predicted to form a stem-loop structure in the region just upstream of *sigG*. This structure may sequester the ribosome binding site of the *sigG* mRNA preventing its translation (Masuda et al. 1988) (Fig. 6c), at least under certain conditions. The second promoter is located in the *sigE-sigG* intergenic region (P*sigG*) and is transcribed specifically in the forespore under the control of σ^F (Pereira et al. 2013). The additional levels of regulation that decrease the potential for σ^G accumulation from transcripts originating at P*sigG* in *B. subtilis*, such as the use of a GTG start codon and the RBS hairpin structure that blocks the ribosome binding site, are not present in *C. difficile* (Mearls et al. 2018). Therefore, in *C. difficile*, σ^G is produced soon after asymmetric division. Once produced, σ^G is active and able to utilize its own promoter creating a positive autoregulatory loop that may be responsible for the increased σ^G-activity observed after engulfment completion (Pereira et al. 2013). Unexpectedly, σ^G accumulates in a σ^F mutant, but is not active (Fimlaid et al. 2013). It was suggested that σ^G may be produced in the forespore from the long mRNA that originates at the *spoIIGA* promoter, but that an unknown σ^F-dependent gene is required for its activation or that an inhibitor is depleted or otherwise antagonized. This inhibitor could be the SpoIIAB anti-σ factor, which is present in the forespore at least soon after asymmetric division. In *B. subtilis*, SpoIIAB is able to bind to both σ^F and σ^G (Serrano et al. 2004). In *B. subtilis*, however, $\sigma^{F.}$ drives production of a second anti-σ factor, CsfB, that specifically binds to σ^G, while σ^F is resistant to it, preventing its interaction with core RNA polymerase until, at the completion of engulfment, CsfB becomes non-functional (Chary et al. 2007; Karmazyn-Campelli et al. 2008; Rhayat et al. 2009; Serrano et al. 2011). Although *csfB* entered the sporulation gene set prior to the divergence of the Clostridia from other (aerobic) sporeformers, the gene is not found in *C. difficile* (Ramos-Silva et al. 2019). An alternative possibility to explain the apparent lack of σ^G activity in a σ^F mutant in *C. difficile*, is that translation of the long mRNA may take place in the central mother-cell of the disporic $\sigma^{F.}$ mutant, which lacks a chromosome, and therefore σ^G activity is not detected.

In *B. subtilis,* gene expression in the engulfed forespore depends on the assembly of a complex that is thought to maintain metabolic potential in the forespore when after engulfment completion this cell it becomes isolated from the external medium (Camp and Losick 2009; Doan et al. 2009). In agreement with engulfment being a distinctive feature of (endo)sporulation, the proteins that form this complex are part of a genomic signature for sporulation (Abecasis et al. 2013; Galperin et al. 2012). Studies in *B. subtilis* have shown that this complex is assembled from the products of nine genes expressed in the mother cell under σ^E control, the *spoIIIA* octacistronic operon and *gerM*, and two genes expressed in the forespore under σ^F-control, *spoIIQ* and *spoIIIL* (Blaylock et al. 2004; Camp and Losick 2009; Doan et al. 2009; Meisner et al. 2008; Rodrigues et al. 2016). Several of the *spoIIIA*-encoded proteins show structural similarity to components of type II, III, and IV secretion systems of Gram-negative bacteria, suggesting a hybrid secretion system specialized for cell–cell communication during sporulation (reviewed recently by Morlot and Rodrigues 2018). The first gene in the *spoIIIA* operon codes for a traffic ATPase, suggesting that it may energize transport from the mother cell into the forespore and accordingly, a single amino acid substitution in the ATP-binding motif of both the *B. subtilis* and *C. difficile* proteins causes the forespore membranes to collapse and impairs the activity of σ^G in this cell (Doan et al. 2009; Fimlaid et al. 2015b). In *C. difficile*, as in *B. subtilis*, the SpoIIIAH and SpoIIQ proteins interact directly

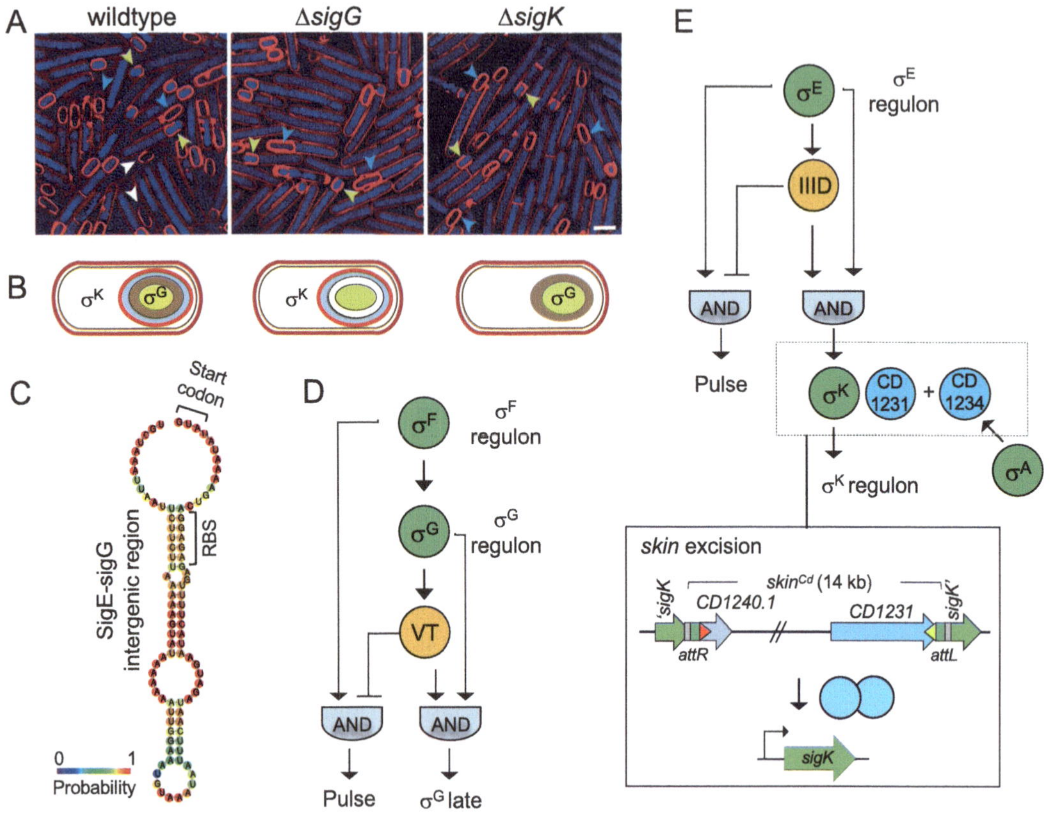

Fig. 6 Control of late gene expression during sporulation. (**a**) Fluorescence microscopy of wild-type, *sigG* and *sigK* mutant cells stained with the membrane dye FM4-64 and the DNA stain Hoechst. Cells at different stages of sporulation are indicated: at intermediate stage in the process of engulfment of the forespore by the larger mother cell (yellow arrowheads); with engulfment completed (Hoechst is excluded from the forespore; blue arrowheads) and free spores released after mother cell lysis (white arrows). Scale bar, 1 μm. (**b**) In the lower panel is a schematic representation of the stages at which *sigG* and *sigK* mutants are blocked (refer to Fig. 1b). In the *sigG* mutant at least some cells complete the engulfment process. The cortex layer is absent but there is accumulation of coat material around the forespore, indicating σ^k activity in the mother cell. A *sigK* mutant completes the engulfment process, synthesizes the cortex but not the coat and exosporium. In this mutant the forespore is often positioned slightly tilted relative to the longitudinal axis of the sporangium. (**c**) The *sigG* RBS-hairpin structure predicted to form in the mRNA originating from the *spoIIG* promoter. The *sigG* RBS, and ATG start codon are indicated. The figure was generated by ViennaRNA using the sequence of the intergenic region between *sigE* and *sigG*. (**d**) Gene expression in the forespore is governed by a regulatory cascade involving activation and repression of gene expression. σ^F turns on expression of the *sigG* gene, which codes for σ^G. σ^G turns on a regulon that includes the DNA binding protein SpoVT. In a type II coherent feed-forward loop (FFL) with AND gate logic, σ^G and SpoVT act together to switch on target genes, such as *sspA* and *sspB*. σ^F and SpoVT also form an incoherent FFL in which SpoVT acts as a repressor of a subset of σ^F-dependent genes which are expressed as a pulse. Although not represented some genes are solely under the control of σ^F and σ^G. (**e**) Organization of the mother cell transcriptional network. In a coherent FFL with AND gate logic, σ^E drives production of SpoIIID, and together both activate transcription of genes involved in σ^K production (*CD1234* and *sigK*) and *cdeC* (not represented for simplicity), delaying *skin*^{Cd} excision (which additionally requires the CD1231 recombinase) and the main period of *sigK* transcription to post-engulfment sporangia. SpoIIID also represses transcription from σ^E-dependent genes; the resulting incoherent FFL with AND gate logic results in a pulse of gene expression. Some genes are solely under the control of σ^E. The box is the schematic representation of *skin*^{Cd} excision and σ^K activation

in the intermembrane space and form a channel between two cells (Camp and Losick 2009; Fimlaid et al. 2015b; Meisner et al. 2008; Serrano et al. 2016a). SpoIIIAH and SpoIIQ, as well as SpoIIIAG and GerM form oligomeric ring and modelling studies suggested that stacked SpoIIIAH and SpoIIQ rings present a central conduit (Levdikov et al. 2012; Meisner et al. 2012; Trouve et al. 2018). SpoIIQ has a LytM domain, found in metalloendopeptidases involved in cell wall remodelling, but lacks two of the active site histidine residues including one involved in the coordination of Zn^{2+} (Levdikov et al. 2012; Meisner et al. 2012; Meisner and Moran 2011). The *C. difficile* counterpart has an intact active site suggesting that it could participate in PG remodelling during engulfment and in line with the idea that the DMP machine and the SpoIIIAH-SpoIIQ complex are partially redundant in this organism (Ribis et al. 2018; see above). Replacement of the histidine involved in Zn^{2+}-coordination, however, caused a much less pronounced phenotype than deletion of *spoIIQ*, and so the putative endopeptidase activity of SpoIIQ does not seem to make a critical contribution for engulfment in *C. difficile* (Fimlaid et al. 2015b; Serrano et al. 2016a). Both *spoIIIAH* and *spoIIQ* are required for the formation of heat-resistant spores; most cells of the *spoIIQ* and *spoIIIAH* mutants of *C. difficile* are blocked during the engulfment process but some are able to reach late stages in sporulation (Fimlaid et al. 2015b; Serrano et al. 2016a). Since σ^G is produced early in the forespore, σ^G activity is detected in *spoIIQ* or *spoIIIAH* sporangia (Fimlaid et al. 2015b; Serrano et al. 2016a). Due to the collapse of the forespore membranes in these mutants, however, the forespore specificity of σ^G activity is lost in some cells (Serrano et al. 2016a). Still, in the few cells of the *spoIIIAH* mutant in which engulfment was completed the activity of σ^G was strongly reduced (Serrano et al. 2016a), indicating that in *C. difficile* the channel may also be important for the intercellular communication between the mother cell and the forespore that allow the exchange of small molecules needed to sustain biosynthetic activity in the latter. What goes through the channel is presently not known. Recent studies, however, show that in *B. subtilis* the two cells undergo a profound metabolic remodelling, with the forespore becoming dependent on the mother cell for synthesis of metabolites required for protein synthesis; arginine was shown to be transported between the two cells (Riley et al. 2021). Given the differences in the metabolic landscape of *C. difficile*, it is even less clear what metabolites may the mother cell may convey to the forespore.

Once active σ^G transcribes a set of at least 60 genes, several of which overlap with the σ^F regulon (Fimlaid et al. 2013; Saujet et al. 2013). The σ^G regulon includes many genes that are required to prepare the forespore for dormancy, including the *spoVA* operon involved in dipicolinic acid (DPA) uptake during spore formation and later in its release during germination and the *sspA* and *sspB* genes, coding for SASPs (Baloh and Sorg 2021; Nerber and Sorg 2021). The SASPs are a family of small proteins, found in all sporeformers, highly similar at the sequence level but with a range of functions from resistance to chemicals to resistance to UV radiation (reviewed by Setlow and Christie 2023). *C. difficile* has two main SASP proteins, SspA and SspB, of the α/β type. SspA is central to spore UV resistance, with a contribution from SspB. The SASPs also make a small contribution to resistance to nitrous acid. The combined deletion of *sspA* and *sspB* prevented spore formation which led to the suggestion that the SASPs regulate/serve as a checkpoint for spore formation (Nerber and Sorg 2021). Also, under σ^F/σ^G control is *spoVT*, a transcriptional factor that downregulates at least two σ^F-dependent genes, *spoIIR* and *gpr* (Saujet et al. 2013) (Fig. 6d). Thus, repression by SpoVT creates a type II incoherent feed-forward loop that generates a pulse in the expression of these σ^F genes (a similar function is performed by *B. subtilis* RsfA (Wang et al. 2006), which is absent from *C. difficile*). SpoVT also turns on at least two σ^G-dependent genes, *sspA* and *sspB*, creating a type II coherent feed-forward loop that results in their delayed expression (Saujet et al. 2013). A *spoVT* mutant produces "sporelets" that are phase dark, a signal that the cortex is absent or reduced and no heat resistant spores (Saujet et al. 2013). The double *sspA/sspB* mutant also produces phase dark spores but it is unclear whether the phenotype of the *spoVT* mutant results from reduced or lack of expression of the *sspA* and *sspB* genes. In any event, since cortex synthesis is a function of the mother cell (see Sect. 3.3.2), this suggests that a crucial step in formation of the cortex PG requires a signal from the forespore which is under SpoVT control. This signal has not been identified.

SpoVT is the only ancillary DNA-binding protein known to be produced specifically in the forespore. Although not a DNA-binding protein, another possible regulatory protein, coded for by *CD25990*, is also under σ^G control (Saujet et al. 2013). *CD25990* is an ortholog of *B. subtilis ylyA* which codes for an RNA polymerase binding protein that together with σ^G activates the transcription of a set of genes involved in spore germination, including *gerB* (not found in *C. difficile*), *spoVA* and *sspB* (Traag et al. 2013b). Whether this function is maintained in *C. difficile* is currently unknown.

A *sigG* mutant is blocked soon after completion of the engulfment process (Fig. 6a, b). The cortex layer is absent in this mutant but there is accumulation of some coat/exosporium material around the forespore (Fimlaid et al. 2013; Pereira et al. 2013). The major period of coat assembly is known to take place in the mother cell at late stages of sporulation and relies mainly on the activity of σ^K (see below). Therefore, the morphological phenotype of a *sigG* mutant indicates that $\sigma^{K.}$ becomes active in the mother cell independently of late gene expression in the forespore.

3.3.4 Late Mother Cell Gene Expression

sigK, coding for σ^K, is interrupted by a 14.6 kb element, termed *skin* (*s*igma *K* *in*tervening), which resembles a prophage (Serrano et al. 2016b). σ^K, the late mother cell-specific sigma factor, is regulated at the level of the reconstitution and expression of the *sigK* gene (Fig. 6e). Excision of the prophage is required to join together the two coding parts of the *sigK* gene. The site-specific recombinase, CD12310, involved in *skin* excision, is synthesized in the vegetative cell under the control of σ^A and is sufficient for the integration of an engineered short *skin* element into the *sigK* gene in vitro (Serrano et al. 2016b). However, *skin* excision only occurs in the mother cell since a directionality factor, CD12340, required together with the recombinase for the excision reaction, is produced under the control of σ^E and SpoIIID, i.e, in the mother cell (Fig. 6e). This restricts excision to the mother cell and ensures that a complete copy of the genome remains in the forespore. Moreover, the requirement for SpoIIID for CD12340 transcription creates a type II coherent feed-forward loop that causes a delay in the production of the directionality factor and hence on *skin* excision. After reconstitution, the *sigK* gene is expressed under the control of σ^E and SpoIIID (again a coherent feed-forward loop), and later under the control of σ^K itself (Serrano et al. 2016b) (Fig. 6e). Thus, *sigK* transcription is confined to the mother cell and delayed, possibly coinciding with engulfment completion. In contrast to pro-σ^E and pro-σ^K of *B. subtilis*, σ^K is produced in *C. difficile* without an inhibitory pro-sequence. In *B. subtilis*, cleavage of the pro-sequence relies on the activity of a intramembrane cleaving metalloprotease, SpoIVFB, located in the forespore outer membrane; SpoIVFB is kept in an inactive complex by two other proteins, SpoIVFA and BofA. The Site-1 SpoIVB protease, made under σ^G control and secreted to the intermembrane space, leads to the activation of SpoIVFB and the production of σ^K (reviewed by Sun et al. 2021). The inhibitory proteins SpoIVFA and BofA occupy the active site cleft of SpoIVFB (Olenic et al. 2022). SpoIVB cleavage of SpoIVFA in the intermembrane space triggers a conformational change in SpoIVFB, which transits from a closed to an open state allowing pro-σ^K to reach the active site region (Ramírez-Guadiana et al. 2018).

The SpoIVFA, SpoIVFB, and BofA proteins are not found in *C. difficile*, as could be expected given the lack of a pro-sequence in σ^K. Paradoxically, *C. difficile* has two paralogs of the SpoIVB protease of as yet unknown function. Their study could illuminate a longstanding mystery of sporulation in *B. subtilis*: mutations that bypass the need for forespore signalling in the activation of pro-σ^K, including removal of the pro-sequence from pro-σ^K, do not bypass a requirement for SpoIVB for sporulation (Oke et al. 1997; Wakeley et al. 2000). This indicates that SpoIVB has another function in sporulation, one that has remained elusive. In any case, of the three levels of control over the time of production of active σ^K in *B. subtilis*, only two are found in *C. difficile*: one is the time of CD12340 transcription and *skin* excision and the other is the time of *sigK* transcription.

Since σ^E is activated in a sub-population of cells in the *sigF* mutant, *sigK* is expressed in these cells (Fimlaid et al. 2013; Pereira et al. 2013; Saujet et al. 2013). However, many σ^K-dependent genes are downregulated in the *sigF* mutant. Interestingly, single-cell analysis shows σ^K activity only in the cells that activate σ^E just on time to inhibit the formation of the second septum (Pereira et al. 2013). In these cells, a copy of the chromosome is kept in the mother cell enabling σ^K-dependent gene expression. Since the reconstitution the *sigK* gene is also σ^E-dependent this may cause a delay in the accumulation of σ^K and cells with the disporic phenotype, will not have chromosomal DNA on the mother cell by the time σ^K accumulates and hence its activity is not detected.

σ^K activity is reduced in cells of the *spoIIIAH* and *spoIIQ* mutants, during and after engulfment of the forespore by the mother cell (Serrano et al. 2016a). This has not been observed in *B. subtilis* and suggests an expanded role of the channel in *C. difficile* in which the SpoIIIAH-SpoIIQ channel is not only required to maintain late gene expression in the forespore but also in the mother cell. Since the direction of transport through the channel is expected to be from the mother cell to the forespore (Camp and Losick 2009; Doan et al. 2009; Meisner et al. 2008), a possible explanation is that in the absence of the channel an inhibitory metabolite accumulates in the mother cell that inhibits σ^K.

A Δ*skin* strain produces spores but shows premature σ^K activity affecting the assembly of the coat and exosporium spore surface layers (Serrano et al. 2016b). An interesting possibility is that the recombinase CD12310 may function independently of CD12340 in vegetative cells causing the permanent elimination of the *skin* element in some cells; in fact there are epidemic strains of RT033, RT017, and RT127 that lack this element (Alves et al. 2022). It is possible that these *skin*-less epidemic strains produce spores with altered outermost spore layers, which are the first line of contact of the spore with host cells and the immune system, and therefore some diversity may increase the chances to escape the immune system.

4 Assembly of the Spore Surface Layers

4.1 Early Events

Studies in *B. subtilis* have shown that assembly of the spore coat and crust begins soon after the activation of σ^E, when the mother cell begins to engulf the forespore (McKenney and Eichenberger 2012). In *B. subtilis*, the early events in coat/crust assembly are governed by a group of the so-called morphogenetic proteins, a term first used to designate proteins that control the localization of several others but not the transcription of their encoding genes (Zheng et al. 1988). Two morphogenetic proteins that prepare the surface of the spore for receiving the coat are SpoVM and SpoIVA; both proteins are conserved in *C. difficile*. The first is a 26 amino acid long peptide that recognizes positive membrane curvature (Kim et al. 2017; Ramamurthi et al. 2006, 2009). The second is an ATPase (Ramamurthi and Losick 2008). The two are interdependent for assembly in *B. subtilis*, but in *C. difficile*, SpoVM is largely dispensable for coat/exosporium assembly (Ribis et al. 2017). As shown in *B. subtilis*, SpoIVA forms cables around the surface of the forespore, in an ATP-dependent manner, and is required for the localization of all known coat/crust proteins (Ramamurthi and Losick 2008). In *spoIVA* mutants, in both *B. subtilis* and *C. difficile*, the coat proteins are produced but accumulate as long swirls of partially structured material in the mother cell cytoplasm (Putnam et al. 2013; Roels et al. 1992; Stevens et al. 1992). In *B. subtilis*, *spoIVA* mutants do not form the cortex, in contrast in *C. difficile*, the cortex is formed although it appears thinner and may present other defects (Putnam et al. 2013).

In *B. subtilis*, SpoIVA recruits the proteins that serve as the main organizers (sometimes termed hubs) of the inner coat, outer coat, and crust sub-layers. The proteins that compose these layers, in turn, are targeted to the spore surface presumably by interacting with the corresponding hubs at various times during sporulation, and according to different kinetics classes, in part determined by their position on the transcriptional cascade (McKenney and Eichenberger 2012). In a second stage, the coat/crust proteins migrate around the spore, in a process termed encasement that requires a third morphogenetic protein, SpoVID. SpoVID has a N-terminal "morphogenetic" domain, responsible for the encasement function which is thought to directly interact with its client proteins (Wang et al. 2009). Noteworthy, no proteins are known in *C. difficile*, which clearly function as hubs for coat and exosporium assembly, and the timeline of production of the coat and exosporium components and their genetic dependencies for localization has not been systematically studied. This N-terminal domain is common to *Spo*VID, *C*otE (the outer coat hub in *B. subtilis*), and *S*ipL, a SpoVID homologue from *C. difficile*, and is termed SPOCS (Delerue et al. 2022; Putnam et al. 2013). Following the SPOCS domain, in both SipL and SpoVID, is a central region most likely disordered and a C-terminal LysM domain (Delerue et al. 2022; Putnam et al. 2013; Wang et al. 2009). LysM domains are PG-binding modules but also mediate protein–protein interactions as shown recently. For example, the *B. subtilis* inner coat hub SafA has a LysM domain that interacts at an early stage in assembly with SpoVID and is required for encasement by SafA and the inner coat proteins; later in sporulation, the LysM domain of SafA is involved in binding of the protein to the cortex PG (Pereira et al. 2019). Like SpoVID, SipL has a C-terminal LysM domain, which is involved in an interaction with SpoIVA that is essential for assembly of the coat and exosporium (Touchette et al. 2019) (Fig. 7a). The "bearding" phenotype of a *sipL* mutant suggests that the encasement function is conserved in *C. difficile* (Putnam et al. 2013).

4.2 Cortex/Coat/Exosporium Connections

The LysM domain of SpoVID does not bind PG, unlike that of SafA and other extracellular LysM domains; rather, as mentioned above, it mediates an interaction with SpoIVA (Delerue et al. 2022). Remarkably, the LysM domain of SpoVID binds to lipid II in the membrane (Delerue et al. 2022). A model has been proposed in which defects in coat assembly detected at the level of SpoIVA, expose the LysM domain of SpoVID, which by sequestering lipid II blocks cortex synthesis. SpoVID thereby establishes a checkpoint linking the morphogenesis of the cortex and coat layers (Delerue et al. 2022). It is not known whether the LysM domain of SipL binds lipid II and has a role similar to that of *B. subtilis* SpoVID.

An interesting insertional mutant has been described in *C. difficile* that forms spores completely devoid of both a visible coat and exosporium as assessed by TEM, and thus, with an exposed cortex (Alves Feliciano et al. 2019). The mutation disrupts the CD10650 gene, renamed *cotL* (Alves Feliciano et al. 2019). Several proteins were strongly reduced or missing from coat/exosporium extracts of the *cotL* mutant, but the expression of the corresponding σ^E- and σ^K-controlled genes was not affected, on which basis CotL was designated a morphogenetic protein (Alves Feliciano et al. 2019). A CotL-SNAP^{Cd} fusion was produced throughout sporulation, in the mother cell, from tandem σ^E- and σ^K-dependent promoters, and the fusion protein initially targeted to curved septa, eventually formed a shell around the entire spore (Alves Feliciano et al. 2019). Although no TEM images of sporulating cells have been published, studies with translational SNAP^{Cd} fusions show that the localization of several σ^E- and σ^K-dependent was impaired. In particular, in the *cotL* mutant the σ^K-

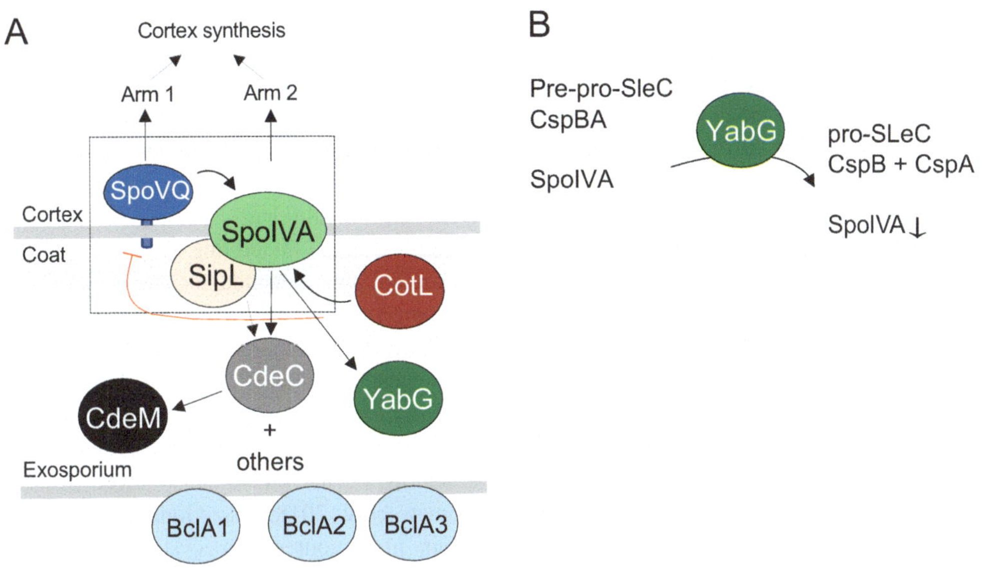

Fig. 7 Assembly of the surface layers. (**a**) The machinery involved in the initial stages of coat assembly and a model for the assembly of the coat and exosporium layers of the spore. CotL is required for the proper localization of SpoIVA, which in turn may recruit SipL. Both SpoIVA and SipL recruit downstream proteins including CdeC that possibly recruits CdeM and also possibly, the YabG protease. SpoVQ is a membrane protein and its soluble domain may reside in the cortex region. SpoIVA, SipL, and SpoVQ interact directly and form a complex close to the forespore inner membrane. It is hypothesized here that this complex controls two arms in cortex synthesis. Accordingly, in the absence of SpoIVA, SipL, or SpoVQ cortex synthesis is also impaired but in *spoVQ*/*spoIVA* or *spoVQ*/*sipL* double mutants, no cortex is formed. As CotL is also required for the localization of SpoIVA, and as such, affects the SpoIVA/SipL cortex synthesis arm and since *cotL* mutants form what appears to be a normal cortex, it seems possible that CotL normally represses the activity of the SpoVQ arm (red line). The SpoVQ arm then compensates for the impaired SpoIVA/SipL function in a *cotL* mutant. The figure also represents the relative position of the collagen-like BclA proteins, but note that the set of these proteins varies among strains (Pizarro-Guajardo et al. 2014). (**b**) Role of the YabG protease in coat assembly and the processing of the CspBA and pre-pro-SleC proteins, involved in germination. The levels of extractable SpoIVA increase in a *yabG* mutant, suggesting that SpoIVA is a YabG substrate. It is not known whether YabG localizes to the coat/exosporium, cortex, or both layers of the spore

dependent protein CdeC accumulated as dots at the spore surface and throughout the mother cell cytoplasm and never encircled the forespore (Alves Feliciano et al. 2019). Importantly, SpoIVA-SNAP^{Cd} localized to the septum during the initial stages of engulfment but later it also accumulated as dots at the spore surface and throughout the mother cell cytoplasm. This seems to indicate that CotL is involved in the persistent localization and/or polymerization of SpoIVA. If so, then CotL would be at the top of the hierarchy for coat/exosporium assembly. *cotL* and *spoIVA* mutants differ, in that the cortex is formed in the latter but is reduced in the former (Alves Feliciano et al. 2019; Putnam et al. 2013). Formation of the cortex, therefore, does not seem to require proper localization of SpoIVA. It is not known, however, whether sufficient SpoIVA to allow cortex synthesis localizes around the spore in the *cotL* mutant. An alternative model that is also consistent with the data is that CotL normally antagonizes a factor required for cortex assembly, and SpoIVA in turn antagonizes this activity of CotL. This model does not exclude a more direct role of SpoIVA in cortex formation (Fig. 7a).

SpoVQ is a mother cell-specific bitopic transmembrane protein recently identified as interacting directly with SpoIVA and SipL

(Touchette et al. 2021). A *spoVQ* mutant is severely impaired in the formation of heat resistant spores and shows a much reduced cortex layer and high rates of spontaneous spore germination (Touchette et al. 2021). While *spoIVA* or *sipL* mutant shows relatively mild defects in cortex synthesis, in *spoVQ* and *spoIVA* or *sipL* double mutants synthesis of the cortex is completely blocked (Touchette et al. 2021). One interpretation is that cortex synthesis can proceed or be activated via a *spoVQ*-dependent pathway, more important, and a *spoIVA*/*sipL*-dependent pathway; CotL could function upstream of the SpoIVA/SipL arm, but it is not known whether *cotL* is required for the localization of SpoVQ (Fig. 7a). SpoVQ localizes around the forespore presumably at the outer membrane; this localization is largely independent of *spoIVA* or *sipL*, although in the absence of SpoIVA the signal from a SpoVQ-mCherry fusion increased (Touchette et al. 2021). In turn, localization of mCherry-SpoIVA was not affected in *spoVQ* sporangia, but the localization of a mCherry fusion to the C-terminus of SipL was impaired in the absence of SpoVQ (Touchette et al. 2021). Since the C-terminus of SipL carries the LysM domain involved in an interaction with SpoIVA, the observation seems to indicate that the interaction with SpoIVA becomes more important in the absence of SpoVQ. Whether the soluble domain of SpoVQ lies in the mother cell cytoplasm or in the intermembrane space is unclear. Touchette and co-authors raise the possibility that the soluble domain somehow control synthesis of the cortex in the interspace (Touchette et al. 2021). Then, the question of how SpoIVA and SipL interact with SpoVQ arises. Since the C-terminal mCherry fusion to SipL disrupts localization of the protein in a *spoVQ* mutant, the authors speculate that the LysM domain of SipL could reach the cortex and proposed that SipL could help "staple" the coat and cortex layers by analogy with the role of the LysM-containing SafA protein of *B. subtilis* (Pereira et al. 2019; Touchette et al. 2021).

4.3 The Mature Spore

The accepted description of a *C. difficile* spore, as imaged by thin sectioning TEM, includes a lamellar coat with no clear distinction between inner/outer layers, reminiscent of *B. cereus/B. anthracis* and an electrodense exosporium closely connected to the underlying coat; the spore has a prominent polar appendage (Antunes et al. 2018; Calderón-Romero et al. 2018; Paredes-Sabja et al. 2014; Pizarro-Guajardo et al. 2020) (Fig. 2). Cysteine-rich proteins that have been extracted from spores, such as CdeC and CdeM (see below), were later shown to be required for the assembly of the electrodense layer (Calderón-Romero et al. 2018; Díaz-González et al. 2015). Their high content on cysteine invokes the exosporium proteins of *B. cereus*/*B. anthracis*, which self-assemble into honeycomb-like 2D lattices and of the crust proteins of *B. subtilis*, coded for by the *cotVWXYZ* cluster (Ball et al. 2008; Jiang et al. 2015). The exosporium is formed by a thin basal layer from which radial projections of the glycosylated collagen-like BclA protein protrude (reviewed by Stewart 2015). The crust of *B. subtilis* spores also consists of a thin surface layer, clearly seen when the structural organization of the outer coat is perturbed, as in *cotG* mutants (Freitas et al. 2020) and is also glycosylated. The crust layer of *B. subtilis* spores thus seems analogous to the exosporium of *B. cereus*/*B. anthracis*. The difference is that the exosporium is a balloon-like layer separated from the coat by an interspace, while the crust is closely adherent to the coat. An alternative view of the structural organization of the *C. difficile* spores is that the exosporium may be the thin layer observed at the edge of the electrodense outer coat by analogy with the crust of *B. subtilis* (Fig. 2). TEM images suggest that the coat and the electrodense region (that corresponds to the exosporium in the accepted interpretation) both have a lamellar structure; the outermost part of it is covered by an electrodense layer, dependent on assembly of CdeM (Antunes et al. 2018). Thus, the coat would have a common lamellar stratum, different from the exosporium.

C. difficile spores have a long appendage at one of its poles, the length of which varies greatly among strains (Antunes et al. 2018; Pizarro-Guajardo et al. 2020). This structure is continuous with the rest of the coat and exosporium; mechanical treatment of the spores with glass beads in the presence of reducing agents releases a shell with the shape of the spore plus the appendage (Antunes et al. 2018). TEM images suggest that the lamellar structure of the coat also forms the bulk of the appendage; this is clearly seen at the transition zone between the spore body and the appendage. Again, electrodensity of the appendage would be given by CdeM. The appendage thus seems to be an extension of the coat layers and it could be surrounded, like the rest of the spore, by a thin, closely connected exosporium. The definition of the layers could be functional or based on the presence of cysteine-rich proteins, which is the accepted view. No mutant has been described which maintains an exosporium layer but lacks the morphological features of the coat, and so evidence for a different interpretation is lacking.

Formation of a robust appendage is seen for only a fraction of the wild type spores, but how this morphological bifurcation comes about is not understood. A protein that is central to appendage formation is CdeM; without it, the appendage is short and disorganized and the spores loose rigidity as shown by atomic force microscopy (Antunes et al. 2018). However, a bifurcation into spores with and without an appendage is still seen, indicating that other proteins participate in the formation of this structure. Wild type spores with a short appendage or no appendage germinate faster than those with a more robust appendage and this may be because they are thinner and more permeable, as also proposed for *cdeM* spores (Antunes et al. 2018). However, in either the wild type or the *cdeM* mutant, the spores with a short appendage or no appendage are more impaired in germination induced by taurocholate in a rich medium, than their counterparts, in the wild type or the mutant, with a visible appendage (Antunes et al. 2018). Since the levels of mature SleC, and also of CspA, CspB, and CspC, all of which are required for germination as detailed below, do not vary between the two populations, the appendage may serve a structural role in germination (Antunes et al. 2018).

Several proteins were identified in the coat and exosporium layers through proteomics of coat/exosporium extracts prepared from purified spore suspensions (Abhyankar et al. 2019; Lawley et al. 2009). Several of these proteins are produced under σ^E or σ^K control, but others are not sporulation-specific. Some are predicted to be enzymes with roles in spore protection, such as a rubrerythrin (CD28450, σ^G-controlled), manganese catalases (CotD, σ^K controlled, and CotG), and a superoxide dismutase (SodA, σ^E-controlled) (Lawley et al. 2009; reviewed by (Paredes-Sabja et al. 2022). Others modulate the adherence of spores to host cells. Among these are three collagen-like glycoproteins BclA1, BclA2, and BclA3 (all σ^K-controlled; Paredes-Sabja et al. 2022).These proteins may influence the assembly of the exosporium layer in strain 630 (Phetcharaburanin et al. 2014). Epidemic strains, such as R20291 of ribotype 027, have a thick exosporium, with a bump-like appearance and hair-like projections in sharp contrast to the smooth appearance of the spore surface in strain 630 (Pizarro-Guajardo et al. 2020) (Fig. 2). Both BclA2 and BclA3 are present in these projections, but deletion of *bclA3* is sufficient to abolish their formation (Pizarro-Guajardo et al. 2014). Importantly, BclA3 is involved in the interaction of spores with host cells and spore internalization (Castro-Córdova et al. 2021). Other proteins that may be involved in spore–host cell interactions are a putative haemagglutinin/adhesin (CD05140), SlpA, produced by vegetative cells, and a major component of the S-layer (Lawley et al. 2009), and CotE, a bi-functional enzyme with a peroxiredoxin N-terminal domain and a C-terminal chitinase domain (Permpoonpattana et al. 2011, 2013). CotE is essential for the binding of spores to the mucus layer, and its mucin-degrading activity, which is associated with the chitinase domain, facilitates the interaction of spores with the colonic mucosa, colonization, and virulence (Hong et al. 2017).

Proteins thought to be specifically associated with the electrodense exosporium include the cysteine-rich proteins CdeA, CdeC, and CdeM (Antunes et al. 2018; Barra-Carrasco et al. 2013; Calderón-Romero et al. 2018; Díaz-González et al. 2015). *cdeC* and *cdeM* mutants, which have a misassembled coat and exosporium, show altered colonization and virulence, in keeping with the importance of the exosporium in the interaction of spores with host cells (Antunes et al. 2018; Calderón-Romero et al. 2018; Castro-Córdova et al. 2021).

The ability of spores to adhere to intestinal epithelial cells is also important because a fraction of the spores is internalized in a E-cadherin-dependent manner, contributing to the recurrence of the disease (Castro-Córdova et al. 2023). Internalization of the *C. difficile* spores was only studied in the epidemic strains R20291 (of ribotype 27, see above) and 630Δ*erm* (RT012) (Castro-Córdova et al. 2021, 2023), two strains that have been widely used for *C. difficile* studies in the laboratory. It would be interesting to extend these studies to strains of other ribotypes, as at least one study has shown that spores produced by epidemic strains show greater adherence to epithelial cells than spores from non-epidemic strains (Vitucci et al. 2020). As we wrote above, the exosporium is the first line of contact of the spore with the host epithelial cells and the host immune system. The exosporium layer, i.e., the electrodense layer of spores in the accepted view, shows a great heterogeneity between strain, ribotypes and even intra-strain (Pizarro-Guajardo et al. 2016, 2020). How this variability arises is unclear. It could be due to stochastic fluctuations in gene expression during spore formation, and/or due to genetic variation among strains. An example of the former is the absence of the *skin* element in some epidemic strains which may lead to premature activity of σ^K and several alterations in the composition and structure of the exosporium (Serrano et al. 2016b; see above). The impact of spore heterogeneity in adherence and internalization is yet to be studied.

4.4 Germination

Proteins that are specifically involved in the germination of *C. difficile* spores include CspBA, CspC, and pre-pro-SleC (reviewed by Baloh and Sorg 2021). Interdomain processing of CspBA releases CspB, whereas cleavage of pre-pro-SleC produces pro-SleC (Kevorkian et al. 2016; Shrestha et al. 2019). CspB is a subtilisin-like serine protease involved in the activation, together with the germinant receptor CspC, of pro-SleC (Adams et al. 2013; Francis et al. 2015). SleC, in turn, is a cortex hydrolase essential for spore germination in response to bile salts (Adams et al. 2013; Bhattacharjee et al. 2016; Francis et al. 2013; Kochan et al. 2018). These proteins may all be localized mainly to the cortex (Baloh et al. 2022) and a model proposes that they form a germinosome complex (Kochan et al. 2018).

A cysteine (thiol) protease coded by the *yabG* gene intervenes in processing of several spore proteins (Fig. 7b). The *yabG* gene is the only in the σ^K regulon that contributes to a genomic signature for sporulation (Abecasis et al. 2013; Shrestha et al. 2019). This reflects not only the diversity in the composition of the surface layers but also the importance of YabG in the assembly of the spore surface layers. In *B. subtilis*, YabG associates with the spore coat and is required for the cleavage of at least six coat proteins which facilitates their subsequent cross-linking by a transglutaminase (Kuwana et al. 2006, 2007; Takamatsu et al. 2000). None of the transglutaminase-dependent proteins (nor the transglutaminase itself) are conserved in *C. difficile*. In fact, SpoIVA may be the only YabG substrate common to the two organisms ((Kevorkian et al. 2016); see also below). Two other proteins that are *C. difficile*-specific YabG substrates are CspBA and pre-pro-SleC (Kevorkian et al. 2016; Shrestha et al. 2019). Spores produced by a *yabG* mutant have higher levels of extractable SpoIVA and accumulate CspBA and pre-pro-SleC (Kevorkian et al. 2016). While disruption of *yabG* causes a reduction in the rate of spore germination, the insertional mutation, as well as missense mutations in the gene, also renders spores independent of co-germinants for the taurocholate-induced germination (Shrestha et al. 2019). However, while spores of *yabG* point mutants are still stimulated by co-germinants, spores of a *yabG* insertional mutant are not; this was interpreted as showing that YabG is involved in the processing of a protein required for co-germinant recognition (Shrestha et al. 2019). (The germination proteins may form a germinosome complex, possibly in localized in the cortex (see above) but it is unclear whether they also localize to the coat/exosporium (Fimlaid et al. 2015a; Shrestha et al. 2019) (Fig. 7b). Processing of CspBA and pre-proSleC by YabG could presumably establish the final localization of these proteins. The role of *yabG* in the processing of proteins involved in coat assembly and of CspBA and pre-pro-SleC shows a connection, as yet poorly understood, between the formation of the coat/exosporium and the assembly of the germinosome.

When activated, SleC begins to hydrolyze the cortex PG. The cortex is a modified form of PG that contains muramic acid residues in the δ-lactam form. This modification, a hallmark of spores, creates the substrate for SleC allowing it to hydrolyze the cortex leaving intact the primordial germ cell wall (Popham et al. 1996; Popham and Bernhards 2015). Synthesis of muramic-δ-lactam requires two enzymes in *B. subtilis*: CwlD is a muramoyl-L-alanine amidase that first removes a peptide side chain from the N-acetylmuramic; the second step involves deacetylation of muramic acid and lactam ring formation and requires the PdaA deacetylase (Gilmore et al. 2004). Expression of *cwlD* occurs both in the forespore and in the mother cell (under σ^E control) but may be required only in the former; expression of *pdaA* is mother cell-specific, σ^E (Gilmore et al. 2004). In *C. difficile*, synthesis of muramic-δ-lactam requires PdaA and CwlD and an additional lipoprotein, GerS, specific to members of the Peptostreptococcaceae family (Diaz et al. 2018). A *gerS* mutant fails to degrade the cortex, and thus has a severe germination defect, even though in the mutant mature SleC is found at wild-type levels (Fimlaid et al. 2015a). GerS is a lipoprotein secreted into the intermembrane space; activation of SleC requires secretion of GerS but not lipidation (Alves Feliciano et al. 2021; Diaz et al. 2018). Importantly, the mutant is less virulent in an hamster model of infection which is probably caused by its impairment in germination (Diaz et al. 2018; Fimlaid et al. 2015a).

5 Conservation and Diversity of the Sporulation Programme

After formation of the forespore and the mother cell, the sporulation pathway can be described as three main levels: (1) the presence and sequential activation of the four cell type-specific sigma factors and checkpoint regulation; (2) the type II feed-forward loops, coherent and incoherent and other network motifs that ensure fidelity of the morphogenetic process; (3) structural genes and other effectors (De Hoon et al. 2010). The sigma factors and their sequential activation alternating between the forespore and the mother cell are strongly conserved features among sporeformers, but there are exceptions. In *C. perfringens*, *C. acetobutylicum*, and *C. botulinum*, for instance, σ^K is required for an early stage of sporulation, prior to asymmetric division (reviewed by Al-Hinai et al. 2015). Also, the late checkpoints of *B. subtilis* are less conserved as exemplified by the somewhat degenerated morphological checkpoints of *C. difficile*. The feed-forward loops are also less conserved. The SpoIIID ancillary factor is conserved between *B. subtilis* and *C. difficile* and divides the σ^E regulon into early (SpoIIID-independent genes), late (SpoIIID-dependent genes), and genes that are expressed as a pulse (repressed by SpoIIID). But other ancillary factors known for *B. subtilis*, such as RsfA and GerE, which work in a similar way but with σ^F and σ^K, are missing in *C. difficile*. Together with the morphological checkpoints, the feed-forward loops fine-tune waves of gene expression within the sigma-specified modules and are also thought to contribute to the fidelity of spore morphogenesis. Although it was initially proposed that the "older" Clostridia could have a simpler morphogenetic programme, the ancestral sporeformer on the basis of the Firmicutes possessed most of the control mechanisms found in *B. subtilis* but absent in *C. difficile* (Ramos-Silva et al. 2019). Another possibility is that a less tightly controlled morphogenetic programme results in structural variations of the resulting spores that may be advantageous for a pathogen.

The last level in the functional and evolutionary hierarchy of the sporulation network comprises the genes that code for the machinery involved in building the spore, including those that code for the structural components of structures such as the coat and exosporium. This is the least conserved level (Abecasis et al. 2013; De Hoon et al. 2010; Galperin et al. 2012; Ramos-Silva et al. 2019). Interestingly, the evolutionary history of *B. subtilis* and related organisms indicates that many sporulation genes have appeared the novo, while in *C. difficile* horizontal gene transfer has been a major driving force in evolution; the gut ecosystem may be an environment that favours this process (Ramos-Silva et al. 2019). Significantly, examples of genes acquired by *C. difficile* through horizontal gene transfer include *cotF*, coding for a putative ferrodoxin, and *cdeC* and *cdeA*, coding for two of the cysteine-rich proteins of the coat and exosporium layers (Ramos-Silva et al. 2019).

6 Gut Feelings

6.1 The Specificity of Sporulation in the Gut

A large number of species of the gut microbiota in healthy humans are anaerobic sporeformers (Browne et al. 2016). The variety of sporeformers found in the gut suggests that sporulation is a cornerstone activity of the gut microbiota. Some of these organisms may sporulate, or at least form resistant cells, in new ways. For instance, in some strains of *Ruminococcus* σ^G appears to be missing, yet the organism was isolated after treatments that select for spores (Browne et al. 2016; Ramos-Silva et al. 2019). Interesting deviations to standard pathway of sporulation have been described for gut sporeformers. For instance, *Epulopiscium* (a symbiont of the surgeonfish) and *Metabacterium polyspora* (a guinea pig symbiont) produce multiple intracellular offspring inside the mother cell and use sporulation as a form of propagation in the gut (Angert and Losick 1998). Also, in *B. subtilis* isolates from the gut of various species including humans, sporulation is triggered during growth and cultures reach higher spore titres than laboratory strains (Serra et al. 2014; Tam et al. 2006). Remarkably, these strains loose the "sporulation during growth" characteristic, as well as other social traits such as biofilm formation, in a very short period of time under laboratory conditions, by acquiring single mutations that reduce sporulation kinetics (Barreto et al. 2020). In some cases, sporulation ability is lost altogether (Maughan et al. 2007). Clearly, in the absence of environmental pressure to maintain sporulation, energy consuming and complex developmental processes may be rapidly attenuated. "Attenuation" upon domestication has also been shown for pathogens such as *Escherichia coli*, *Bartonella henselae*, *Salmonella*, *S. aureus*, among others (Arvand et al. 2006; Davidson et al. 2008; Liu et al. 2017; Somerville et al. 2002).

The phenotypical characterization of *C. difficile* epidemics strains should be done taking extreme care with sub-culturing. It was shown previously that the two *C. difficile* strains more widely used in the laboratory, 630Δ*erm* (RT012) and R20291 (RT027), accumulate mutations during culturing in the lab (Collery et al. 2017; Monteford et al. 2021). This observation suggests that these strains have become "domesticated", losing or attenuating the ability of their wild ancestors to carry out behaviours such as sporulation and biofilm formation.

6.2 The Quality Versus Quantity Trade-Off

Epidemics strains of *C. difficile* have variable sporulation kinetics under laboratory conditions (Barbanti and Spigaglia 2016; Hong et al. 2019;

Zidaric et al. 2012; Zidaric and Rupnik 2016). Whether early sporulation or maximum sporulation capacity has more impact in *C. difficile* transmission is not known. It was suggested that earlier sporulation could be important for survival in a non-gut environment where oxygen is present (Zidaric and Rupnik 2016). In any case, the timing of sporulation may also be a variable to take into account in interpreting the success of epidemic strains. In *B. subtilis* the timing of sporulation controls spore germination (Mutlu et al. 2018). A trade-off between spore quantity and quality was observed, in that spores that form early in a culture respond more efficiently to germinants, whereas a delay in spore formation affects the rate of germination (Mutlu et al. 2018). Moreover, different trade-offs are adopted by strains living in different ecological niches. For example, while a *B. subtilis* gut isolate sporulates faster generating high-quality spores (spores that germinate efficiently), a soil isolate produces more spores but at later stages of growth (Mutlu et al. 2020). In *C. difficile* a possible trade-off between spore quantity and quality was never tested, but different strains may have adopted different strategies, some producing more spores which would favour transmission; others produce a lower number of spores that could germinate faster in the gut under the right conditions, enabling earlier colonization (Carlson et al. 2015).

6.3 The Colonization Versus Dissemination Trade-Off

Even in the absence of antibiotic pressure, *C. difficile* microcolonies persisting at the surface of the intestinal mucosa must overcome challenges such as competition for nutrients and the host defence mechanisms (Lawley et al. 2012). These challenges may lead to the emergence of mutants better adapted to in vivo conditions. Most within-host adaptation in common human pathogens occurs within antigenic loci, enabling bacteria to escape the immune system. There are also examples of recurrent evolution in quorum sensing genes such as *S. aureus* and *Pseudomonas aeruginosa* (Lelong et al. 2011; Suligoy et al. 2018). In these cases, adaptation to the host is usually concomitant with a decrease in virulence. Large-scale genomic analysis has shown that host adaptation of some lineages of gut Firmicutes is associated with sporulation loss (Browne et al. 2021). Former-spore-formers evolve to colonize to high cell numbers enabling direct transmission. In contrast, spore-formers colonize to less extent but the shedding of spores through faecal material allows the transmission through the environment at long distance and during long temporal intervals (Browne et al. 2021). In an obligate anaerobe gut pathogen the loss of sporulation would not be anticipated given its importance in the lifecycle, however colonization by *C. difficile* for a long time may impact on the spore yield produced by the bacteria. Longitudinal studies of within-host evolution of *C. difficile* complemented with whole genomic sequence and phenotypical characterization would shed some light about the importance of sporulation in host adaptation. In a few studies in which isolates from relapse cases were analysed, an improvement of the efficiency of sporulation was never observed. In fact, in at least two cases with strains from RT082 and RT017, a significant decrease in sporulation efficiency was observed ((Oka et al. 2012; Plaza-Garrido et al. 2015) and our unpublished results). Interestingly, in one study, isolates from relapse cases show higher rate of germination in the absence of germinants (Oka et al. 2012). These results may indicate that once again the spore surface layers are altered in these spores. Variations in the exosporium may also play a role in the evasion of the host immune system and may reflect within-host adaptation.

6.4 Sporulation in Epidemics Strains

According to recent reports, the most predominant ribotypes found in Europe are RT027, RT014-20, RT001, and RT078 (Freeman et al. 2020; Mengoli et al. 2022; Zhao et al. 2021). Ribotypes 014 and 020 are so similar that there is almost no literature that distinguishes between the two. They produce both the TcdA and TcdB toxins (Knight et al. 2016). Ribotype 078 is predominant among livestock and is often correlated with a possible zoonotic origin for some *C. difficile* strains. The 078 ribotype affects a younger population and is more associated with community-acquired (CA)-CDI. This ribotype produces TcdA, TcdB, and CDT toxins. RT001 is also one of the RTs that are frequently recovered from patients in Europe, and its association with severe CDI has been reported. The majority of the strains belonging to this type produces both TcdA and TcdB toxins. Another ribotype which prevalence has raised in Europe is RT017. Recently, Portugal reported a prevalence superior to 10% (Imwattana et al. 2019; Isidro et al. 2018). Together with RT027, all these ribotypes have in common a multidrug-resistant profile.

Strains from RT027 were responsible for one of the most *C. difficile* threatening that emerged in 2003–2004 in Canada and caused an outbreak of severe CDI cases in more than 30 hospitals (Kuijper et al. 2007; Pépin et al. 2004). Despite being isolated for the first time in North America, this ribotype has spread to and is still the predominant one in Europe (Freeman et al. 2020; Mengoli et al. 2022; Zhao et al. 2021). Since its emergence it became the major focus of *C. difficile* studies and the strain R20291 has been used as the type strain for this ribotype (Vitucci et al. 2020). Although several studies in vitro report that they produce large amounts of spores, going even further as to proclaim that they produce more than non-RT027 strains, and in turn are more virulent, others demonstrate that there is no association between sporulation characteristics and ribotypes (Åkerlund et al. 2008; Burns et al. 2010, 2011). RT017 and RT027 strains are usually resistant to fluoroquinolones, third-generation cephalosporins, clindamycin, and rifampicin (Freeman et al. 2020; Isidro et al. 2018). Since these sudden outbreaks occurred after the administration of antibiotics that belong to the fluoroquinolones group (gatifloxacin or moxifloxacin), it is believed that the use of these large-spectrum antibiotics selected bacterial resistance (Razavi et al. 2007). Due to this specificity, a mere change in antibiotic therapy would help to decrease dissemination of these ribotypes.

The success of a ribotype is linked not only to its ability to cause disease but also to be disseminated between hosts. In case of *C. difficile* dissemination relies mainly on spores and therefore there was an attempt to correlate the success of specific ribotypes to an increase in sporulation frequency (Smits 2013). However, in vitro, a similar or even larger variation between strains of the same ribotype was observed than between different ribotypes (Burns et al. 2010, 2011). Also, comparation between epidemic and non-epidemic strains has shown that although in the animal model the epidemic strains were more virulent than non-epidemic strain, the in vitro phenotypes, such as spore production, were not predictive of their virulence in vivo (Vitucci et al. 2020). These results suggest more than spore yield, evaluation of other properties linked to sporulation, such as the ability of different strains to persist, to cause relapsing disease and host-to-host transmission in vivo, may be more predictive of its epidemic success (Deakin et al. 2012).

Acknowledgements This work was also supported by "Fundação para a Ciência e a Tecnologia" (FCT), Portugal through grants PTDC/BIA-MIC/32542/2017 to AOH and PTDC/BIA-MIC/29293/2017 to MS, by the European Union Marie Sklodowska Curie Innovative Training Networks (contract number 642068) to AOH. The work was also funded by project LISBOA-01-0145-FEDER-007660 ("Microbiologia Molecular, Estrutural e Celular") funded by FEDER funds through COMPETE2020—"-Programa Operacional Competitividade e Internacionalização", and by project PPBI—Portuguese Platform of BioImaging (PPBI-POCI-01-0145-FEDER-022122) co-funded by national funds from OE—"-Orçamento de Estado" and by European funds from FEDER—"Fundo Europeu de Desenvolvimento Regional". DM is the recipient of a PhD fellowship (PD/BD/143148/2019) within the scope of the PhD programme INTERFACE funded by FCT.

References

Abecasis AB, Serrano M, Alves R, Quintais L, Pereira-Leal JB, Henriques AO (2013) A genomic signature and the identification of new sporulation genes. J Bacteriol 195:2101–2115. https://doi.org/10.1128/JB.02110-12

Abhyankar WR, Zheng L, Brul S, de Koster CG, de Koning LJ (2019) Vegetative cell and spore proteomes of Clostridioides difficile show finite differences and reveal potential protein markers. J Proteome Res 18: 3967–3976. https://doi.org/10.1021/acs.jproteome.9b00413

Abt MC, McKenney PT, Pamer EG (2016) Clostridium difficile colitis: pathogenesis and host defence. Nat Rev Microbiol 14:609–620. https://doi.org/10.1038/nrmicro.2016.108

Adams CM, Eckenroth BE, Putnam EE, Doublié S, Shen A (2013) Structural and functional analysis of the CspB protease required for Clostridium spore germination. PLoS Pathog 9:e1003165

Adams DW, Wu LJ, Errington J (2014) Cell cycle regulation by the bacterial nucleoid. Curr Opin Microbiol 22: 94–101. https://doi.org/10.1016/j.mib.2014.09.020

Åkerlund T, Persson I, Unemo M, Norén T, Svenungsson B, Wullt M, Burman LG (2008) Increased sporulation rate of epidemic Clostridium difficile type 027/NAP1. J Clin Microbiol 46:1530–1533. https://doi.org/10.1128/JCM.01964-07

Alabdali YAJ, Oatley P, Kirk JA, Fagan RP (2021) A cortex-specific penicillin-binding protein contributes to heat resistance in Clostridioides difficile spores. Anaerobe 70:102379. https://doi.org/10.1016/j.anaerobe.2021.102379

Al-Hinai MA, Jones SW, Papoutsakis ET (2015) The Clostridium sporulation programs: diversity and preservation of endospore differentiation. Microbiol Mol Biol Rev 79:19-LP-37. https://doi.org/10.1128/MMBR.00025-14

Alves F, Nunes A, Castro R, Sequeira A, Moreira O, Matias R, Rodrigues JC, Silveira L, Gomes JP, Oleastro M (2022) Assessment of the transmission dynamics of Clostridioides difficile in a farm environment reveals the presence of a new toxigenic strain connected to swine production. Front Microbiol 13: 858310. https://doi.org/10.3389/fmicb.2022.858310

Alves Feliciano C, Douché T, Giai Gianetto Q, Matondo M, Martin-Verstraete I, Dupuy B (2019) CotL, a new morphogenetic spore coat protein of Clostridium difficile. Environ Microbiol 21:984–1003. https://doi.org/10.1111/1462-2920.14505

Alves Feliciano C, Eckenroth BE, Diaz OR, Doublié S, Shen A (2021) A lipoprotein allosterically activates the CwlD amidase during Clostridioides difficile spore formation. PLoS Genet 17:e1009791. https://doi.org/10.1371/journal.pgen.1009791

Angert ER, Losick RM (1998) Propagation by sporulation in the guinea pig symbiont Metabacterium polyspora. Proc Natl Acad Sci U S A 95:10218–10223. https://doi.org/10.1073/pnas.95.17.10218

Antunes A, Camiade E, Monot M, Courtois E, Barbut F, Sernova NV, Rodionov DA, Martin-Verstraete I, Dupuy B (2012) Global transcriptional control by glucose and carbon regulator CcpA in Clostridium difficile. Nucleic Acids Res 40:10701–10718. https://doi.org/10.1093/nar/gks864

Antunes W, Pereira FC, Feliciano C, Saujet L, dos Vultos T, Couture-Tosi E, Pechine S, Bruxelle J-F, Janoir C, Melo L, Brito P, Martin-Verstraete I, Serrano M, Dupuy B, Henriques AO (2018) Structure and assembly of a *Clostridioides difficile* spore polar appendage. bioRxiv. https://doi.org/10.1101/468637

Arigoni F, Duncan L, Alper S, Losick R, Stragier P (1996) SpoIIE governs the phosphorylation state of a protein regulating transcription factor sigma F during sporulation in Bacillus subtilis. Proc Natl Acad Sci U S A 93: 3238–3242. https://doi.org/10.1073/pnas.93.8.3238

Arvand M, Schubert H, Viezens J (2006) Emergence of distinct genetic variants in the population of primary Bartonella henselae isolates. Microbes Infect 8:1315–1320. https://doi.org/10.1016/j.micinf.2005.12.015

Asayama M, Yamamoto A, Kobayashi Y (1995) Dimer form of phosphorylated Spo0A, a transcriptional regulator, stimulates the spo0F transcription at the initiation of sporulation in Bacillus subtilis. J Mol Biol 250:11–23. https://doi.org/10.1006/jmbi.1995.0354

Atrih A, Zöllner P, Allmaier G, Williamson MP, Foster SJ (1998) Peptidoglycan structural dynamics during germination of Bacillus subtilis 168 endospores. J Bacteriol 180:4603–4612. https://doi.org/10.1128/JB.180.17.4603-4612.1998

Aubry A, Hussack G, Chen W, KuoLee R, Twine SM, Fulton KM, Foote S, Carrillo CD, Tanha J, Logan SM (2012) Modulation of toxin production by the flagellar regulon in Clostridium difficile. Infect Immun 80: 3521–3532. https://doi.org/10.1128/IAI.00224-12

Baban ST, Kuehne SA, Barketi-Klai A, Cartman ST, Kelly ML, Hardie KR, Kansau I, Collignon A, Minton NP (2013) The role of flagella in Clostridium difficile pathogenesis: comparison between a non-epidemic and an epidemic strain. PLoS One 8:e73026. https://doi.org/10.1371/journal.pone.0073026

Ball DA, Taylor R, Todd SJ, Redmond C, Couture-Tosi E, Sylvestre P, Moir A, Bullough PA (2008) Structure of the exosporium and sublayers of spores of the Bacillus cereus family revealed by electron crystallography. Mol Microbiol 68:947–958. https://doi.org/10.1111/j.1365-2958.2008.06206.x

Baloh M, Sorg JA (2021) Clostridioides difficile SpoVAD and SpoVAE interact and are required for dipicolinic acid uptake into spores. J Bacteriol 203:e0039421. https://doi.org/10.1128/JB.00394-21

Baloh M, Nerber HN, Sorg JA (2022) Imaging Clostridioides difficile spore germination and germination proteins ABSTRACT. J Bacteriol 204(7). https://doi.org/10.1128/jb.00210-22

Barbanti F, Spigaglia P (2016) Characterization of Clostridium difficile PCR-ribotype 018: A problematic emerging type. Anaerobe 42:123–129. https://doi.org/10.1016/j.anaerobe.2016.10.003

Barra-Carrasco J, Olguín-Araneda V, Plaza-Garrido Á, Miranda-Cárdenas C, Cofré-Araneda G, Pizarro-Guajardo M, Sarker MR, Paredes-Sabja D (2013) The Clostridium difficile exosporium cysteine (CdeC)-rich protein is required for exosporium morphogenesis and coat assembly. J Bacteriol 195:3863–3875

Barreto HC, Cordeiro TN, Henriques AO, Gordo I (2020) Rampant loss of social traits during domestication of a Bacillus subtilis natural isolate. Sci Rep 10:18886. https://doi.org/10.1038/s41598-020-76017-1

Bartels J, Blüher A, López Castellanos S, Richter M, Günther M, Mascher T (2019) The Bacillus subtilis endospore crust: protein interaction network, architecture and glycosylation state of a potential glycoprotein layer. Mol Microbiol 112:1576–1592. https://doi.org/10.1111/mmi.14381

Ben-Yehuda S, Losick R (2002) Asymmetric cell division in B. subtilis involves a spiral-like intermediate of the cytokinetic protein FtsZ. Cell 109:257–266. https://doi.org/10.1016/s0092-8674(02)00698-0

Ben-Yehuda S, Rudner DZ, Losick R (2003) RacA, a bacterial protein that anchors chromosomes to the cell poles. Science 299:532–536. https://doi.org/10.1126/science.1079914

Ben-Yehuda S, Fujita M, Liu XS, Gorbatyuk B, Skoko D, Yan J, Marko JF, Liu JS, Eichenberger P, Rudner DZ, Losick R (2005) Defining a centromere-like element in Bacillus subtilis by Identifying the binding sites for the chromosome-anchoring protein RacA. Mol Cell 17:773–782. https://doi.org/10.1016/j.molcel.2005.02.023

Bhattacharjee D, McAllister KN, Sorg JA (2016) Germinants and their receptors in Clostridia. J Bacteriol 198:2767–2775. https://doi.org/10.1128/JB.00405-16

Blaylock B, Jiang X, Rubio A, Moran CP, Pogliano K (2004) Zipper-like interaction between proteins in adjacent daughter cells mediates protein localization. Genes Dev 18:2916–2928. https://doi.org/10.1101/gad.1252704

Bradshaw N, Losick R (2015) Asymmetric division triggers cell-specific gene expression through coupled capture and stabilization of a phosphatase. eLife 4:e08145. https://doi.org/10.7554/eLife.08145

Brown EE, Miller AK, Krieger IV, Otto RM, Sacchettini JC, Herman JK (2019) A DNA-binding protein tunes septum placement during Bacillus subtilis sporulation. J Bacteriol 201

Browne HP, Forster SC, Anonye BO, Kumar N, Neville BA, Stares MD, Goulding D, Lawley TD (2016) Culturing of 'unculturable' human microbiota reveals novel taxa and extensive sporulation. Nature 533:543–546. https://doi.org/10.1038/nature17645

Browne HP, Almeida A, Kumar N, Vervier K, Adoum AT, Viciani E, Dawson NJR, Forster SC, Cormie C, Goulding D, Lawley TD (2021) Host adaptation in gut Firmicutes is associated with sporulation loss and altered transmission cycle. Genome Biol 22:204. https://doi.org/10.1186/s13059-021-02428-6

Buffie CG, Bucci V, Stein RR, McKenney PT, Ling L, Gobourne A, No D, Liu H, Kinnebrew M, Viale A, Littmann E, van den Brink MRM, Jenq RR, Taur Y, Sander C, Cross J, Toussaint NC, Xavier JB, Pamer EG (2015) Precision microbiome reconstitution restores bile acid mediated resistance to Clostridium difficile. Nature 517:205–208. https://doi.org/10.1038/nature13828

Burns DA, Heap JT, Minton NP (2010) The diverse sporulation characteristics of *Clostridium difficile* clinical isolates are not associated with type. Anaerobe 16:618–622. https://doi.org/10.1016/j.anaerobe.2010.10.001

Burns DA, Heeg D, Cartman ST, Minton NP (2011) Reconsidering the sporulation characteristics of hypervirulent Clostridium difficile BI/NAP1/027. PLoS One 6:e24894. https://doi.org/10.1371/journal.pone.0024894

Burton BM, Marquis KA, Sullivan NL, Rapoport TA, Rudner DZ (2007) The ATPase SpoIIIE transports DNA across fused septal membranes during sporulation in Bacillus subtilis. Cell 131:1301–1312. https://doi.org/10.1016/j.cell.2007.11.009

Calderón-Romero P, Castro-Córdova P, Reyes-Ramírez R, Milano-Céspedes M, Guerrero-Araya E, Pizarro-Guajardo M, Olguín-Araneda V, Gil F, Paredes-Sabja D (2018) Clostridium difficile exosporium cysteine-rich proteins are essential for the morphogenesis of the exosporium layer, spore resistance, and affect C. difficile pathogenesis. PLoS Pathog 14:e1007199

Camp AH, Losick R (2009) A feeding tube model for activation of a cell-specific transcription factor during sporulation in Bacillus subtilis. Genes Dev 23:1014–1024. https://doi.org/10.1101/gad.1781709

Carlson PE, Kaiser AM, McColm SA, Bauer JM, Young VB, Aronoff DM, Hanna PC (2015) Variation in germination of Clostridium difficile clinical isolates correlates to disease severity. Anaerobe 33:64–70. https://doi.org/10.1016/j.anaerobe.2015.02.003

Carniol K, Eichenberger P, Losick R (2004) A threshold mechanism governing activation of the developmental regulatory protein sigma F in Bacillus subtilis. J Biol Chem 279:14860–14870. https://doi.org/10.1074/jbc.M314274200

Castro-Córdova P, Mora-Uribe P, Reyes-Ramírez R, Cofré-Araneda G, Orozco-Aguilar J, Brito-Silva C, Mendoza-León MJ, Kuehne SA, Minton NP, Pizarro-Guajardo M, Paredes-Sabja D (2021) Entry of spores into intestinal epithelial cells contributes to recurrence of Clostridioides difficile infection. Nat Commun 12:1140. https://doi.org/10.1038/s41467-021-21355-5

Castro-Córdova P, Otto-Medina M, Montes-Bravo N, Brito-Silva C, Lacy DB, Paredes-Sabja D (2023) Redistribution of the novel Clostridioides difficile spore adherence receptor E-cadherin by TcdA and TcdB increases spore binding to adherens junctions. Infect Immun 91:e0047622. https://doi.org/10.1128/iai.00476-22

Cha JH, Stewart GC (1997) The divIVA minicell locus of Bacillus subtilis. J Bacteriol 179:1671–1683. https://doi.org/10.1128/jb.179.5.1671-1683.1997

Chan H, Mohamed AMT, Grainge I, Rodrigues CDA (2022) FtsK and SpoIIIE, coordinators of chromosome segregation and envelope remodeling in bacteria. Trends Microbiol 30:480–494. https://doi.org/10.1016/j.tim.2021.10.002

Chary VK, Xenopoulos P, Piggot PJ (2007) Expression of the sigmaF-directed csfB locus prevents premature appearance of sigmaG activity during sporulation of Bacillus subtilis. J Bacteriol 189:8754–8757. https://doi.org/10.1128/JB.01265-07

Chastanet A, Losick R (2011) Just-in-time control of Spo0A synthesis in Bacillus subtilis by multiple regulatory mechanisms. J Bacteriol 193:6366–6374. https://doi.org/10.1128/JB.06057-11

Chen B, Himes P, Liu Y, Zhang Y, Lu Z, Liu A, Yan H, Kroos L (2014) Structure of bacterial transcription factor SpoIIID and evidence for a novel mode of DNA binding. J Bacteriol 196:2131–2142. https://doi.org/10.1128/JB.01486-13

Childress KO, Edwards AN, Nawrocki KL, Anderson SE, Woods EC, McBride SM (2016) The phosphotransfer protein CD1492 represses sporulation initiation in Clostridium difficile. Infection Immun 84:3434–3444. https://doi.org/10.1128/IAI.00735-16

Ciftci Y, Girinathan BP, Dhungel BA, Hasan MK, Govind R (2019) Clostridioides difficile SinR′ regulates toxin, sporulation and motility through protein-protein interaction with SinR. Anaerobe 59:1–7. https://doi.org/10.1016/j.anaerobe.2019.05.002

Collery MM, Kuehne SA, McBride SM, Kelly ML, Monot M, Cockayne A, Dupuy B, Minton NP (2017) What's a SNP between friends: The influence of single nucleotide polymorphisms on virulence and phenotypes of *Clostridium difficile* strain 630 and derivatives. Virulence 8:767–781. https://doi.org/10.1080/21505594.2016.1237333

Cowan AE, Olivastro EM, Koppel DE, Loshon CA, Setlow B, Setlow P (2004) Lipids in the inner membrane of dormant spores of Bacillus species are largely immobile. Proc Natl Acad Sci USA 101:7733–7738. https://doi.org/10.1073/pnas.0306859101

Cutting S, Driks A, Schmidt R, Kunkel B, Losick R (1991) Forespore-specific transcription of a gene in the signal transduction pathway that governs Pro-sigma K processing in Bacillus subtilis. Genes Dev 5:456–466. https://doi.org/10.1101/gad.5.3.456

Daniel RA, Drake S, Buchanan CE, Scholle R, Errington J (1994) The Bacillus subtilis spoVD gene encodes a mother-cell-specific penicillin-binding protein required for spore morphogenesis. J Mol Biol 235:209–220. https://doi.org/10.1016/S0022-2836(05)80027-0

Davidson CJ, White AP, Surette MG (2008) Evolutionary loss of the rdar morphotype in Salmonella as a result of high mutation rates during laboratory passage. ISME J 2:293–307. https://doi.org/10.1038/ismej.2008.4

Dawson LF, Valiente E, Faulds-Pain A, Donahue EH, Wren BW (2012) Characterisation of Clostridium difficile biofilm formation, a role for Spo0A. PLoS One 7: e50527. https://doi.org/10.1371/journal.pone.0050527

Deakin LJ, Clare S, Fagan RP, Dawson LF, Pickard DJ, West MR, Wren BW, Fairweather NF, Dougan G, Lawley TD (2012) The Clostridium difficile spo0A gene is a persistence and transmission factor. Infect Immun 80:2704–2711. https://doi.org/10.1128/IAI.00147-12

De Hoon MJL, Eichenberger P, Vitkup D (2010) Hierarchical evolution of the bacterial sporulation network. Curr Biol 20:R735–R745. https://doi.org/10.1016/j.cub.2010.06.031

Delerue T, Anantharaman V, Gilmore MC, Popham DL, Cava F, Aravind L, Ramamurthi KS (2022) Bacterial developmental checkpoint that directly monitors cell surface morphogenesis. Dev Cell 57:344–360.e6. https://doi.org/10.1016/j.devcel.2021.12.021

Dembek M, Barquist L, Boinett CJ, Cain AK, Mayho M, Lawley TD, Fairweather NF, Fagan RP (2015) High-throughput analysis of gene essentiality and sporulation in Clostridium difficile. mBio 6:e02383-14. https://doi.org/10.1128/mBio.02383-14

Dembek M, Willing SE, Hong HA, Hosseini S, Salgado PS, Cutting SM (2017) Inducible expression of spo0A as a universal tool for studying sporulation in Clostridium difficile. Front Microbiol 8:1793. https://doi.org/10.3389/fmicb.2017.01793

Dembek M, Kelly A, Barwinska-Sendra A, Tarrant E, Stanley WA, Vollmer D, Biboy J, Gray J, Vollmer W, Salgado PS (2018) Peptidoglycan degradation machinery in Clostridium difficile forespore engulfment. Mol Microbiol 110:390–410. https://doi.org/10.1111/mmi.14091

Deutscher J (2008) The mechanisms of carbon catabolite repression in bacteria. Curr Opin Microbiol 11:87–93. https://doi.org/10.1016/j.mib.2008.02.007

Dhungel BA, Govind R (2021) Phase-variable expression of *pdcB* , a phosphodiesterase, influences sporulation in *Clostridioides difficile*. Mol Microbiol 116:1347–1360. https://doi.org/10.1111/mmi.14828

Diaz OR, Sayer CV, Popham DL, Shen A (2018) Clostridium difficile lipoprotein GerS is required for cortex modification and thus spore germination. mSphere 3: e00205-18. https://doi.org/10.1128/mSphere.00205-18

Díaz-González F, Milano M, Olguin-Araneda V, Pizarro-Cerda J, Castro-Córdova P, Tzeng S-C, Maier CS, Sarker MR, Paredes-Sabja D (2015) Protein composition of the outermost exosporium-like layer of Clostridium difficile 630 spores. J Proteomics 123:1–13. https://doi.org/10.1016/j.jprot.2015.03.035

DiCandia MA, Edwards AN, Jones JB, Swaim GL, Mills BD, McBride SM (2022) Identification of functional Spo0A residues critical for sporulation in Clostridioides difficile. J Mol Biol 434:167641. https://doi.org/10.1016/j.jmb.2022.167641

Dineen SS, McBride SM, Sonenshein AL (2010) Integration of metabolism and virulence by Clostridium difficile CodY. J Bacteriol 192:5350–5362. https://doi.org/10.1128/JB.00341-10

Doan T, Morlot C, Meisner J, Serrano M, Henriques AO, Moran CP Jr, Rudner DZ (2009) Novel secretion

apparatus maintains spore integrity and developmental gene expression in Bacillus subtilis. PLoS Genet 5: e1000566. https://doi.org/10.1371/journal.pgen.1000566
Doan T, Coleman J, Marquis KA, Meeske AJ, Burton BM, Karatekin E, Rudner DZ (2013) FisB mediates membrane fission during sporulation in Bacillus subtilis. Genes Dev 27:322–334. https://doi.org/10.1101/gad.209049.112
Donnelly ML, Shrestha S, Ribis JW, Kuhn P, Krasilnikov M, Alves Feliciano C, Shen A (2022) Development of a dual-fluorescent-reporter system in Clostridioides difficile reveals a division of labor between virulence and transmission gene expression. mSphere 7:e0013222. https://doi.org/10.1128/msphere.00132-22
Driks A, Eichenberger P (2016) The spore coat. Microbiol Spectr 4. https://doi.org/10.1128/microbiolspec.TBS-0023-2016
Driks A, Losick R (1991) Compartmentalized expression of a gene under the control of sporulation transcription factor sigma E in Bacillus subtilis. Proc Natl Acad Sci U S A 88:9934–9938. https://doi.org/10.1073/pnas.88.22.9934
Dubois T, Tremblay YDN, Hamiot A, Martin-Verstraete I, Deschamps J, Monot M, Briandet R, Dupuy B (2019) A microbiota-generated bile salt induces biofilm formation in Clostridium difficile. NPJ Biofilms Microbiomes 5:14. https://doi.org/10.1038/s41522-019-0087-4
Duncan L, Losick R (1993) SpoIIAB is an anti-sigma factor that binds to and inhibits transcription by regulatory protein sigma F from Bacillus subtilis. Proc Natl Acad Sci U S A 90:2325–2329. https://doi.org/10.1073/pnas.90.6.2325
Dworkin J, Losick R (2001) Differential gene expression governed by chromosomal spatial asymmetry. Cell 107:339–346. https://doi.org/10.1016/s0092-8674(01)00528-1
Dworkin J, Losick R (2005) Developmental commitment in a bacterium. Cell 121:401–409. https://doi.org/10.1016/j.cell.2005.02.032
Edwards DH, Errington J (1997) The Bacillus subtilis DivIVA protein targets to the division septum and controls the site specificity of cell division. Mol Microbiol 24:905–915. https://doi.org/10.1046/j.1365-2958.1997.3811764.x
Edwards AN, Nawrocki KL, McBride SM (2014) Conserved oligopeptide permeases modulate sporulation initiation in Clostridium difficile. Infect Immun 82:4276–4291. https://doi.org/10.1128/IAI.02323-14
Edwards AN, Tamayo R, McBride SM (2016) A novel regulator controls Clostridium difficile sporulation, motility and toxin production. Mol Microbiol 100: 954–971. https://doi.org/10.1111/mmi.13361
Edwards AN, Anjuwon-Foster BR, McBride SM (2019) RstA is a major regulator of Clostridioides difficile toxin production and motility. mBio 10. https://doi.org/10.1128/mBio.01991-18
Edwards AN, Willams CL, Pareek N, McBride SM, Tamayo R (2021) c-di-GMP inhibits early sporulation in Clostridioides difficile. mSphere 6
Edwards AN, Wetzel D, DiCandia MA, McBride SM (2022) Three orphan histidine kinases inhibit Clostridioides difficile sporulation. J Bacteriol 204: e00106-22. https://doi.org/10.1128/jb.00106-22
Eichenberger P, Fujita M, Jensen ST, Conlon EM, Rudner DZ, Wang ST, Ferguson C, Haga K, Sato T, Liu JS, Losick R (2004) The program of gene transcription for a single differentiating cell type during sporulation in Bacillus subtilis. PLoS Biol 2:e328. https://doi.org/10.1371/journal.pbio.0020328
Eldar A, Chary VK, Xenopoulos P, Fontes ME, Losón OC, Dworkin J, Piggot PJ, Elowitz MB (2009) Partial penetrance facilitates developmental evolution in bacteria. Nature 460:510–514. https://doi.org/10.1038/nature08150
Errington J, Wu LJ (2017) Cell cycle machinery in Bacillus subtilis. Subcell Biochem 84:67–101. https://doi.org/10.1007/978-3-319-53047-5_3
Eswaramoorthy P, Erb ML, Gregory JA, Silverman J, Pogliano K, Pogliano J, Ramamurthi KS (2011) Cellular architecture mediates DivIVA ultrastructure and regulates min activity in Bacillus subtilis. mBio 2: e00257-11. https://doi.org/10.1128/mBio.00257-11
Faulds-Pain A, Twine SM, Vinogradov E, Strong PCR, Dell A, Buckley AM, Douce GR, Valiente E, Logan SM, Wren BW (2014) The post-translational modification of the *Clostridium difficile* flagellin affects motility, cell surface properties and virulence. Mol Microbiol 94:272–289. https://doi.org/10.1111/mmi.12755
Fay A, Meyer P, Dworkin J (2010) Interactions between late-acting proteins required for peptidoglycan synthesis during sporulation. J Mol Biol 399:547–561. https://doi.org/10.1016/j.jmb.2010.04.036
Fimlaid KA, Bond JP, Schutz KC, Putnam EE, Leung JM, Lawley TD, Shen A (2013) Global analysis of the sporulation pathway of Clostridium difficile. PLoS Genet 9:e1003660. https://doi.org/10.1371/journal.pgen.1003660
Fimlaid KA, Jensen O, Donnelly ML, Francis MB, Sorg JA, Shen A (2015a) Identification of a novel lipoprotein regulator of Clostridium difficile spore germination. PLoS Pathog 11:e1005239
Fimlaid KA, Jensen O, Donnelly ML, Siegrist MS, Shen A (2015b) Regulation of Clostridium difficile spore formation by the SpoIIQ and SpoIIIA proteins. PLoS Genet 11:e1005562. https://doi.org/10.1371/journal.pgen.1005562
Francis MB, Allen CA, Shrestha R, Sorg JA (2013) Bile acid recognition by the Clostridium difficile germinant receptor, CspC, is important for establishing infection. PLoS Pathog 9:e1003356. https://doi.org/10.1371/journal.ppat.1003356
Francis MB, Allen CA, Sorg JA (2015) Spore cortex hydrolysis precedes dipicolinic acid release during

Clostridium difficile spore germination. J Bacteriol 197:2276–2283. https://doi.org/10.1128/JB.02575-14

Freeman J, Vernon J, Pilling S, Morris K, Nicolson S, Shearman S, Clark E, Palacios-Fabrega JA, Wilcox M, Pan-European Longitudinal Surveillance of Antibiotic Resistance among Prevalent Clostridium difficile Ribotypes' Study Group (2020) Five-year Pan-European, longitudinal surveillance of Clostridium difficile ribotype prevalence and antimicrobial resistance: the extended ClosER study. Eur J Clin Microbiol Infect Dis 39:169–177. https://doi.org/10.1007/s10096-019-03708-7

Freitas C, Plannic J, Isticato R, Pelosi A, Zilhão R, Serrano M, Baccigalupi L, Ricca E, Elsholz AKW, Losick R, Henriques AO (2020) A protein phosphorylation module patterns the Bacillus subtilis spore outer coat. Mol Microbiol 114:934–951. https://doi.org/10.1111/mmi.14562

Fujita M, Losick R (2005) Evidence that entry into sporulation in Bacillus subtilis is governed by a gradual increase in the level and activity of the master regulator Spo0A. Genes Dev 19:2236–2244. https://doi.org/10.1101/gad.1335705

Fujita M, González-Pastor JE, Losick R (2005) High- and low-threshold genes in the Spo0A regulon of *Bacillus subtilis*. J Bacteriol 187:1357–1368. https://doi.org/10.1128/JB.187.4.1357-1368.2005

Galperin MY (2013) Genome diversity of spore-forming firmicutes. Microbiol Spectr 1. https://doi.org/10.1128/microbiolspectrum.TBS-0015-2012

Galperin MY, Mekhedov SL, Puigbo P, Smirnov S, Wolf YI, Rigden DJ (2012) Genomic determinants of sporulation in Bacilli and Clostridia: towards the minimal set of sporulation-specific genes. Environ Microbiol 14: 2870–2890. https://doi.org/10.1111/j.1462-2920.2012.02841.x

Gholamhoseinian A, Piggot PJ (1989) Timing of spoII gene expression relative to septum formation during sporulation of Bacillus subtilis. J Bacteriol 171:5747–5749. https://doi.org/10.1128/jb.171.10.5747-5749.1989

Giel JL, Sorg JA, Sonenshein AL, Zhu J (2010) Metabolism of bile salts in mice influences spore germination in Clostridium difficile. PLoS One 5:e8740. https://doi.org/10.1371/journal.pone.0008740

Gilmore ME, Bandyopadhyay D, Dean AM, Linnstaedt SD, Popham DL (2004) Production of muramic delta-lactam in Bacillus subtilis spore peptidoglycan. J Bacteriol 186:80–89. https://doi.org/10.1128/JB.186.1.80-89.2004

Girinathan BP, Ou J, Dupuy B, Govind R (2018) Pleiotropic roles of Clostridium difficile sin locus. PLoS Pathog 14:e1006940. https://doi.org/10.1371/journal.ppat.1006940

Grossman AD (1995) Genetic networks controlling the initiation of sporulation and the development of genetic competence in Bacillus subtilis. Annu Rev Genet 29:477–508. https://doi.org/10.1146/annurev.ge.29.120195.002401

Haldenwang WG (1995) The sigma factors of Bacillus subtilis. Microbiol Rev 59:1–30

Henriques AO, Moran CP (2007) Structure, assembly, and function of the spore surface layers. Annu Rev Microbiol 61:555–588. https://doi.org/10.1146/annurev.micro.61.080706.093224

Henriques AO, Glaser P, Piggot PJ, Moran CP Jr (1998) Control of cell shape and elongation by the rodA gene in Bacillus subtilis. Mol Microbiol 28:235–247

Higgins D, Dworkin J (2012) Recent progress in Bacillus subtilis sporulation. FEMS Microbiol Rev 36:131–148. https://doi.org/10.1111/j.1574-6976.2011.00310.x

Hilbert DW, Piggot PJ (2004) Compartmentalization of gene expression during Bacillus subtilis spore formation. Microbiol Mol Biol Rev MMBR 68:234–262. https://doi.org/10.1128/MMBR.68.2.234-262.2004

Hofmeister AEM, Londono-Vallejo A, Harry E, Stragier P, Losick R (1995) Extracellular signal protein triggering the proteolytic activation of a developmental transcription factor in B. subtilis. Cell 83:219–226. https://doi.org/10.1016/0092-8674(95)90163-9

Hong HA, Ferreira WT, Hosseini S, Anwar S, Hitri K, Wilkinson AJ, Vahjen W, Zentek J, Soloviev M, Cutting SM (2017) The spore coat protein CotE facilitates host colonization by Clostridium difficile. J Infect Dis 216:1452–1459. https://doi.org/10.1093/infdis/jix488

Hong S, Knight DR, Chang B, Carman RJ, Riley TV (2019) Phenotypic characterisation of Clostridium difficile PCR ribotype 251, an emerging multi-locus sequence type clade 2 strain in Australia. Anaerobe 60:102066. https://doi.org/10.1016/j.anaerobe.2019.06.019

Imamura D, Zhou R, Feig M, Kroos L (2008) Evidence that the Bacillus subtilis SpoIIGA protein is a novel type of signal-transducing aspartic protease. J Biol Chem 283:15287–15299. https://doi.org/10.1074/jbc.M708962200

Imwattana K, Knight DR, Kullin B, Collins DA, Putsathit P, Kiratisin P, Riley TV (2019) Clostridium difficile ribotype 017—characterization, evolution and epidemiology of the dominant strain in Asia. Emerg Microbes Infect 8:796–807. https://doi.org/10.1080/22221751.2019.1621670

Isidro J, Santos A, Nunes A, Borges V, Silva C, Vieira L, Mendes AL, Serrano M, Henriques AO, Gomes JP, Oleastro M (2018) Imipenem resistance in *Clostridium difficile* Ribotype 017, Portugal. Emerg Infect Dis 24: 741–745. https://doi.org/10.3201/eid2404.170095

Jalal ASB, Tran NT, Wu LJ, Ramakrishnan K, Rejzek M, Gobbato G, Stevenson CEM, Lawson DM, Errington J, Le TBK (2021) CTP regulates membrane-binding activity of the nucleoid occlusion protein Noc. Mol Cell 81:3623–3636.e6. https://doi.org/10.1016/j.molcel.2021.06.025

Jiang M, Shao W, Perego M, Hoch JA (2000) Multiple histidine kinases regulate entry into stationary phase and sporulation in Bacillus subtilis. Mol Microbiol 38:

535–542. https://doi.org/10.1046/j.1365-2958.2000.02148.x

Jiang S, Wan Q, Krajcikova D, Tang J, Tzokov SB, Barak I, Bullough PA (2015) Diverse supramolecular structures formed by self-assembling proteins of the Bacillus subtilis spore coat. Mol Microbiol 97:347–359. https://doi.org/10.1111/mmi.13030

Karmazyn-Campelli C, Rhayat L, Carballido-López R, Duperrier S, Frandsen N, Stragier P (2008) How the early sporulation sigma factor sigmaF delays the switch to late development in Bacillus subtilis. Mol Microbiol 67:1169–1180. https://doi.org/10.1111/j.1365-2958.2008.06121.x

Kelly A, Salgado PS (2019) The engulfasome in C. difficile: Variations on protein machineries. Anaerobe 60. https://doi.org/10.1016/j.anaerobe.2019.102091

Kempher ML, Morris SC, Shadid TM, Menon SK, Ballard JD, West AH (2022) Response regulator CD1688 is a negative modulator of sporulation in Clostridioides difficile. J Bacteriol 204:e00130-22. https://doi.org/10.1128/jb.00130-22

Kevorkian Y, Shirley DJ, Shen A (2016) Regulation of Clostridium difficile spore germination by the CspA pseudoprotease domain. Biochimie 122:243–254. https://doi.org/10.1016/j.biochi.2015.07.023

Khanna K, Lopez-Garrido J, Pogliano K (2020) Shaping an endospore: architectural transformations during Bacillus subtilis sporulation. Annu Rev Microbiol 74:361–386. https://doi.org/10.1146/annurev-micro-022520-074650

Kim EY, Tyndall ER, Huang KC, Tian F, Ramamurthi KS (2017) Dash-and-recruit mechanism drives membrane curvature recognition by the small bacterial protein SpoVM. Cell Syst 5:518–526.e3. https://doi.org/10.1016/j.cels.2017.10.004

Knight DR, Squire MM, Collins DA, Riley TV (2016) Genome analysis of Clostridium difficile PCR ribotype 014 lineage in Australian pigs and humans reveals a diverse genetic repertoire and signatures of long-range interspecies transmission. Front Microbiol 7:2138. https://doi.org/10.3389/fmicb.2016.02138

Kochan TJ, Shoshiev MS, Hastie JL, Somers MJ, Plotnick YM, Gutierrez-Munoz DF, Foss ED, Schubert AM, Smith AD, Zimmerman SK, Carlson PE, Hanna PC (2018) Germinant synergy facilitates clostridium difficile spore germination under physiological conditions. mSphere 3:1–13. https://doi.org/10.1128/mSphere.00335-18

Koenigsknecht MJ, Theriot CM, Bergin IL, Schumacher CA, Schloss PD, Young VB (2015) Dynamics and establishment of Clostridium difficile infection in the murine gastrointestinal tract. Infect Immun 83:934–941. https://doi.org/10.1128/IAI.02768-14

Kordus SL, Thomas AK, Lacy DB (2022) Clostridioides difficile toxins: mechanisms of action and antitoxin therapeutics. Nat Rev Microbiol 20:285–298. https://doi.org/10.1038/s41579-021-00660-2

Kuijper EJ, Coignard B, Brazier JS, Suetens C, Drudy D, Wiuff C, Pituch H, Reichert P, Schneider F, Widmer AF, Olsen KE, Allerberger F, Notermans DW, Barbut F, Delmée M, Wilcox M, Pearson A, Patel BC, Brown DJ, Frei R, Akerlund T, Poxton IR, Tüll P (2007) Update of Clostridium difficile-associated disease due to PCR ribotype 027 in Europe. Euro Surveill 12:E1–E2. https://doi.org/10.2807/esm.12.06.00714-en

Kuwana R, Okuda N, Takamatsu H, Watabe K (2006) Modification of GerQ reveals a functional relationship between Tgl and YabG in the coat of *Bacillus subtilis* spores. J Biochem 139:887–901. https://doi.org/10.1093/jb/mvj096

Kuwana R, Takamatsu H, Watabe K (2007) Expression, localization and modification of YxeE spore coat protein in Bacillus subtilis. J Biochem 142:681–689. https://doi.org/10.1093/jb/mvm179

Landajuela A, Braun M, Rodrigues CDA, Martínez-Calvo A, Doan T, Horenkamp F, Andronicos A, Shteyn V, Williams ND, Lin C, Wingreen NS, Rudner DZ, Karatekin E (2021) FisB relies on homo-oligomerization and lipid binding to catalyze membrane fission in bacteria. PLoS Biol 19:e3001314. https://doi.org/10.1371/journal.pbio.3001314

Landajuela A, Braun M, Martínez-Calvo A, Rodrigues CDA, Gomis Perez C, Doan T, Rudner DZ, Wingreen NS, Karatekin E (2022) Membrane fission during bacterial spore development requires cellular inflation driven by DNA translocation. Curr Biol 32:4186–4200.e8. https://doi.org/10.1016/j.cub.2022.08.014

Lawley TD, Croucher NJ, Yu L, Clare S, Sebaihia M, Goulding D, Pickard DJ, Parkhill J, Choudhary J, Dougan G (2009) Proteomic and genomic characterization of highly infectious Clostridium difficile 630 spores. J Bacteriol 191:5377–5386. https://doi.org/10.1128/JB.00597-09

Lawley TD, Clare S, Walker AW, Stares MD, Connor TR, Raisen C, Goulding D, Rad R, Schreiber F, Brandt C, Deakin LJ, Pickard DJ, Duncan SH, Flint HJ, Clark TG, Parkhill J, Dougan G (2012) Targeted restoration of the intestinal microbiota with a simple, defined bacteriotherapy resolves relapsing Clostridium difficile disease in mice. PLoS Pathog 8:e1002995. https://doi.org/10.1371/journal.ppat.1002995

Lee CD, Rizvi A, Edwards AN, DiCandia MA, Vargas Cuebas GG, Monteiro MP, McBride SM (2022) Genetic mechanisms governing sporulation initiation in Clostridioides difficile. Curr Opin Microbiol 66:32–38. https://doi.org/10.1016/j.mib.2021.12.001

Lelong E, Marchetti A, Simon M, Burns JL, van Delden C, Köhler T, Cosson P (2011) Evolution of Pseudomonas aeruginosa virulence in infected patients revealed in a Dictyostelium discoideum host model. Clin Microbiol Infect 17:1415–1420. https://doi.org/10.1111/j.1469-0691.2010.03431.x

Lenarcic R, Halbedel S, Visser L, Shaw M, Wu LJ, Errington J, Marenduzzo D, Hamoen LW (2009) Localisation of DivIVA by targeting to negatively curved membranes. EMBO J 28:2272–2282. https://doi.org/10.1038/emboj.2009.129

Levdikov VM, Blagova EV, McFeat A, Fogg MJ, Wilson KS, Wilkinson AJ (2012) Structure of components of an intercellular channel complex in sporulating Bacillus subtilis. Proc Natl Acad Sci U S A 109:5441–5445. https://doi.org/10.1073/pnas.1120087109

Lewis RJ, Scott DJ, Brannigan JA, Ladds JC, Cervin MA, Spiegelman GB, Hoggett JG, Barák I, Wilkinson AJ (2002) Dimer formation and transcription activation in the sporulation response regulator Spo0A. J Mol Biol 316:235–245. https://doi.org/10.1006/jmbi.2001.5331

Liu B, Eydallin G, Maharjan RP, Feng L, Wang L, Ferenci T (2017) Natural Escherichia coli isolates rapidly acquire genetic changes upon laboratory domestication. Microbiology 163:22–30. https://doi.org/10.1099/mic.0.000405

Lopez-Garrido J, Ojkic N, Khanna K, Wagner FR, Villa E, Endres RG, Pogliano K (2018) Chromosome translocation inflates Bacillus forespores and impacts cellular morphology. Cell 172:758–770.e14. https://doi.org/10.1016/j.cell.2018.01.027

Makroczyová J, Jamroškovič J, Krascsenitsová E, Labajová N, Barák I (2016) Oscillating behavior of Clostridium difficile Min proteins in Bacillus subtilis. MicrobiologyOpen 5:387–401. https://doi.org/10.1002/mbo3.337

Margolis P, Driks A, Losick R (1991) Establishment of cell type by compartmentalized activation of a transcription factor. Science 254:562–565. https://doi.org/10.1126/science.1948031

Marquis KA, Burton BM, Nollmann M, Ptacin JL, Bustamante C, Ben-Yehuda S, Rudner DZ (2008) SpoIIIE strips proteins off the DNA during chromosome translocation. Genes Dev 22:1786–1795. https://doi.org/10.1101/gad.1684008

Marston AL, Thomaides HB, Edwards DH, Sharpe ME, Errington J (1998) Polar localization of the MinD protein of Bacillus subtilis and its role in selection of the mid-cell division site. Genes Dev 12:3419–3430. https://doi.org/10.1101/gad.12.21.3419

Martins D, DiCandia MA, Mendes AL, Wetzel D, McBride SM, Henriques AO, Serrano M (2021) CD25890, a conserved protein that modulates sporulation initiation in Clostridioides difficile. Sci Rep 11: 7887. https://doi.org/10.1038/s41598-021-86878-9

Masuda ES, Anaguchi H, Yamada K, Kobayashi Y (1988) Two developmental genes encoding sigma factor homologs are arranged in tandem in Bacillus subtilis. Proc Natl Acad Sci U S A 85:7637–7641. https://doi.org/10.1073/pnas.85.20.7637

Maughan H, Masel J, Birky CW, Nicholson WL (2007) The roles of mutation accumulation and selection in loss of sporulation in experimental populations of *Bacillus subtilis*. Genetics 177:937–948. https://doi.org/10.1534/genetics.107.075663

McBride SM, Rubio A, Wang L, Haldenwang WG (2005) Contributions of protein structure and gene position to the compartmentalization of the regulatory proteins sigma(E) and SpoIIE in sporulating Bacillus subtilis. Mol Microbiol 57:434–451. https://doi.org/10.1111/j.1365-2958.2005.04712.x

McKenney PT, Eichenberger P (2012) Dynamics of spore coat morphogenesis in Bacillus subtilis. Mol Microbiol 83:245–260. https://doi.org/10.1111/j.1365-2958.2011.07936.x

McKenney PT, Driks A, Eichenberger P (2013) The Bacillus subtilis endospore: assembly and functions of the multilayered coat. Nat Rev Microbiol 11:33–44. https://doi.org/10.1038/nrmicro2921

McPherson DC, Driks A, Popham DL (2001) Two class a high-molecular-weight penicillin-binding proteins of Bacillus subtilis play redundant roles in sporulation. J Bacteriol 183:6046–6053. https://doi.org/10.1128/JB.183.20.6046-6053.2001

Meador-Parton J, Popham DL (2000) Structural analysis of Bacillus subtilis spore peptidoglycan during sporulation. J Bacteriol 182:4491–4499. https://doi.org/10.1128/JB.182.16.4491-4499.2000

Mearls EB, Jackter J, Colquhoun JM, Farmer V, Matthews AJ, Murphy LS, Fenton C, Camp AH (2018) Transcription and translation of the sigG gene is tuned for proper execution of the switch from early to late gene expression in the developing Bacillus subtilis spore. PLoS Genet 14:e1007350. https://doi.org/10.1371/journal.pgen.1007350

Meeske AJ, Rodrigues CDA, Brady J, Lim HC, Bernhardt TG, Rudner DZ (2016) High-throughput genetic screens identify a large and diverse collection of new sporulation genes in Bacillus subtilis. PLoS Biol 14: e1002341. https://doi.org/10.1371/journal.pbio.1002341

Meisner J, Moran CP (2011) A LytM domain dictates the localization of proteins to the mother cell-forespore interface during bacterial endospore formation. J Bacteriol 193:591–598. https://doi.org/10.1128/JB.01270-10

Meisner J, Wang X, Serrano M, Henriques AO, Moran CP (2008) A channel connecting the mother cell and forespore during bacterial endospore formation. Proc Natl Acad Sci U S A 105:15100–15105. https://doi.org/10.1073/pnas.0806301105

Meisner J, Maehigashi T, André I, Dunham CM, Moran CP (2012) Structure of the basal components of a bacterial transporter. Proc Natl Acad Sci U S A 109: 5446–5451. https://doi.org/10.1073/pnas.1120113109

Mengoli M, Barone M, Fabbrini M, D'Amico F, Brigidi P, Turroni S (2022) Make it less difficile: understanding genetic evolution and global spread of Clostridioides difficile. Genes 13:2200. https://doi.org/10.3390/genes13122200

Miller AK, Herman JK (2022) RefZ and Noc act synthetically to prevent aberrant divisions during Bacillus subtilis sporulation. J Bacteriol 204:e0002322. https://doi.org/10.1128/jb.00023-22

Min KT, Hilditch CM, Diederich B, Errington J, Yudkin MD (1993) Sigma F, the first compartment-specific transcription factor of B. subtilis, is regulated by an anti-sigma factor that is also a protein kinase. Cell 74: 735–742. https://doi.org/10.1016/0092-8674(93)90520-z

Monteford J, Bilverstone TW, Ingle P, Philip S, Kuehne SA, Minton NP (2021) What's a SNP between friends: the lineage of Clostridioides difficile R20291 can effect research outcomes. Anaerobe 71:102422. https://doi.org/10.1016/j.anaerobe.2021.102422

Morlot C, Rodrigues CDA (2018) The new kid on the block: a specialized secretion system during bacterial sporulation. Trends Microbiol 26:663–676. https://doi.org/10.1016/j.tim.2018.01.001

Mutlu A, Trauth S, Ziesack M, Nagler K, Bergeest J-P, Rohr K, Becker N, Höfer T, Bischofs IB (2018) Phenotypic memory in Bacillus subtilis links dormancy entry and exit by a spore quantity-quality tradeoff. Nat Commun 9:69. https://doi.org/10.1038/s41467-017-02477-1

Mutlu A, Kaspar C, Becker N, Bischofs IB (2020) A spore quality–quantity tradeoff favors diverse sporulation strategies in Bacillus subtilis. ISME J 14:2703–2714. https://doi.org/10.1038/s41396-020-0721-4

Nawrocki KL, Edwards AN, Daou N, Bouillaut L, McBride SM (2016) CodY-dependent regulation of sporulation in Clostridium difficile. J Bacteriol 198: 2113–2130. https://doi.org/10.1128/JB.00220-16

Neiditch MB, Capodagli GC, Prehna G, Federle MJ (2017) Genetic and structural analyses of RRNPP intercellular peptide signaling of gram-positive bacteria. Annu Rev Genet 51:311–333. https://doi.org/10.1146/annurev-genet-120116-023507

Nerber HN, Sorg JA (2021) The small acid-soluble proteins of Clostridioides difficile are important for UV resistance and serve as a check point for sporulation. PLoS Pathog 17:e1009516. https://doi.org/10.1371/journal.ppat.1009516

Ohlsen KL, Grimsley JK, Hoch JA (1994) Deactivation of the sporulation transcription factor Spo0A by the Spo0E protein phosphatase. Proc Natl Acad Sci U S A 91:1756–1760. https://doi.org/10.1073/pnas.91.5.1756

Ohye DF, Murrell WG (1973) Exosporium and spore coat formation in Bacillus cereus T. J Bacteriol 115:1179–1190. https://doi.org/10.1128/jb.115.3.1179-1190.1973

Ojkic N, López-Garrido J, Pogliano K, Endres RG (2016) Cell-wall remodeling drives engulfment during Bacillus subtilis sporulation. eLife 5:e18657. https://doi.org/10.7554/eLife.18657

Oka K, Osaki T, Hanawa T, Kurata S, Okazaki M, Manzoku T, Takahashi M, Tanaka M, Taguchi H, Watanabe T, Inamatsu T, Kamiya S (2012) Molecular and microbiological characterization of Clostridium difficile isolates from single, relapse, and reinfection cases. J Clin Microbiol 50:915–921. https://doi.org/10.1128/JCM.05588-11

Oke V, Shchepetov M, Cutting S (1997) SpoIVB has two distinct functions during spore formation in Bacillus subtilis. Mol Microbiol 23:223–230. https://doi.org/10.1046/j.1365-2958.1997.2091573.x

Olenic S, Heo L, Feig M, Kroos L (2022) Inhibitory proteins block substrate access by occupying the active site cleft of Bacillus subtilis intramembrane protease SpoIVFB. Elife 11:e74275. https://doi.org/10.7554/eLife.74275

Pan Q, Garsin DA, Losick R (2001) Self-reinforcing activation of a cell-specific transcription factor by proteolysis of an anti-sigma factor in B. subtilis. Mol Cell 8: 873–883. https://doi.org/10.1016/s1097-2765(01)00362-8

Paredes-Sabja D, Shen A, Sorg JA (2014) Clostridium difficile spore biology : sporulation , germination , and spore structural proteins. Trends Microbiol 22: 406–416

Paredes-Sabja D, Cid-Rojas F, Pizarro-Guajardo M (2022) Assembly of the exosporium layer in Clostridioides difficile spores. Curr Opin Microbiol 67:102137. https://doi.org/10.1016/j.mib.2022.01.008

Pépin J, Valiquette L, Alary M-E, Villemure P, Pelletier A, Forget K, Pépin K, Chouinard D (2004) Clostridium difficile-associated diarrhea in a region of Quebec from 1991 to 2003: a changing pattern of disease severity. CMAJ 171:466–472. https://doi.org/10.1503/cmaj.1041104

Perego M, Hoch JA (1991) Negative regulation of Bacillus subtilis sporulation by the spo0E gene product. J Bacteriol 173:2514–2520. https://doi.org/10.1128/jb.173.8.2514-2520.1991

Pereira FC, Saujet L, Tomé AR, Serrano M, Monot M, Couture-Tosi E, Martin-Verstraete I, Dupuy B, Henriques AO (2013) The spore differentiation pathway in the enteric pathogen Clostridium difficile. PLoS Genet 9:e1003782. https://doi.org/10.1371/journal.pgen.1003782

Pereira FC, Nunes F, Cruz F, Fernandes C, Isidro AL, Lousa D, Soares CM, Moran CP, Henriques AO, Serrano M (2019) A LysM domain intervenes in sequential protein-protein and protein-peptidoglycan interactions important for spore coat assembly in Bacillus subtilis. J Bacteriol 201. https://doi.org/10.1128/JB.00642-18

Permpoonpattana P, Tolls EH, Nadem R, Tan S, Brisson A, Cutting SM (2011) Surface layers of Clostridium difficile endospores. J Bacteriol 193:6461–6470. https://doi.org/10.1128/JB.05182-11

Permpoonpattana P, Phetcharaburanin J, Mikelsone A, Dembek M, Tan S, Brisson MC, La Ragione R, Brisson AR, Fairweather N, Hong HA, Cutting SM (2013) Functional characterization of Clostridium difficile spore coat proteins. J Bacteriol 195:1492–1503. https://doi.org/10.1128/JB.02104-12

Pettit LJ, Browne HP, Yu L, Smits WK, Fagan RP, Barquist L, Martin MJ, Goulding D, Duncan SH, Flint HJ, Dougan G, Choudhary JS, Lawley TD (2014) Functional genomics reveals that Clostridium difficileSpo0A coordinates sporulation, virulence and metabolism. BMC Genomics 15:160. https://doi.org/10.1186/1471-2164-15-160

Phetcharaburanin J, Hong HA, Colenutt C, Bianconi I, Sempere L, Permpoonpattana P, Smith K, Dembek M, Tan S, Brisson M-C, Brisson AR, Fairweather NF, Cutting SM (2014) The spore-associated protein BclA1 affects the susceptibility of animals to colonization and infection by Clostridium difficile. Mol Microbiol 92:1025–1038. https://doi.org/10.1111/mmi.12611

Pishdadian K, Fimlaid KA, Shen A (2015) SpoIIID-mediated regulation of sigK function during Clostridium difficile sporulation. Mol Microbiol 95:189–208. https://doi.org/10.1111/mmi.12856

Pizarro-Guajardo M, Olguín-Araneda V, Barra-Carrasco J, Brito-Silva C, Sarker MR, Paredes-Sabja D (2014) Characterization of the collagen-like exosporium protein, BclA1, of Clostridium difficile spores. Anaerobe 25:18–30. https://doi.org/10.1016/j.anaerobe.2013.11.003

Pizarro-Guajardo M, Calderón-Romero P, Paredes-Sabja D (2016) Ultrastructure variability of the exosporium layer of Clostridium difficile spores from sporulating cultures and biofilms. Appl Environ Microbiol 82:5892–5898

Pizarro-Guajardo M, Calderón P, Romero-Rodriguez A, Paredes-Sabja D (2020) Characterization of exosporium layer variability of Clostridioides difficile spores in the epidemically relevant strain R20291. Front Microbiol 11:1345

Plaza-Garrido Á, Miranda-Cárdenas C, Castro-Córdova P, Olguín-Araneda V, Cofré-Araneda G, Hernández-Rocha C, Carman R, Ibáñez P, Fawley WN, Wilcox MH, Gil F, Calderón IL, Fuentes JA, Guzmán-Durán AM, Alvarez-Lobos M, Paredes-Sabja D (2015) Outcome of relapsing Clostridium difficile infections do not correlate with virulence-, spore- and vegetative cell-associated phenotypes. Anaerobe 36:30–38. https://doi.org/10.1016/j.anaerobe.2015.09.005

Popham DL, Bernhards CB (2015) Spore peptidoglycan. Microbiol Spectr 3. https://doi.org/10.1128/microbiolspec.TBS-0005-2012

Popham DL, Helin J, Costello CE, Setlow P (1996) Muramic lactam in peptidoglycan of Bacillus subtilis spores is required for spore outgrowth but not for spore dehydration or heat resistance. Proc Natl Acad Sci USA 93:15405–15410. https://doi.org/10.1073/pnas.93.26.15405

Poquet I, Saujet L, Canette A, Monot M, Mihajlovic J, Ghigo J-M, Soutourina O, Briandet R, Martin-Verstraete I, Dupuy B (2018) Clostridium difficile biofilm: remodeling metabolism and cell surface to build a sparse and heterogeneously aggregated architecture. Front Microbiol 9:2084. https://doi.org/10.3389/fmicb.2018.02084

Putnam EE, Nock AM, Lawley TD, Shen A (2013) SpoIVA and SipL are Clostridium difficile spore morphogenetic proteins. J Bacteriol 195:1214–1225

Ramamurthi KS, Losick R (2008) ATP-driven self-assembly of a morphogenetic protein in Bacillus subtilis. Mol Cell 31:406–414. https://doi.org/10.1016/j.molcel.2008.05.030

Ramamurthi KS, Clapham KR, Losick R (2006) Peptide anchoring spore coat assembly to the outer forespore membrane in Bacillus subtilis. Mol Microbiol 62:1547–1557. https://doi.org/10.1111/j.1365-2958.2006.05468.x

Ramamurthi KS, Lecuyer S, Stone HA, Losick R (2009) Geometric cue for protein localization in a bacterium. Science 323:1354–1357. https://doi.org/10.1126/science.1169218

Ramírez-Guadiana FH, Rodrigues CDA, Marquis KA, Campo N, Barajas-Ornelas RDC, Brock K, Marks DS, Kruse AC, Rudner DZ (2018) Evidence that regulation of intramembrane proteolysis is mediated by substrate gating during sporulation in Bacillus subtilis. PLoS Genet 14:e1007753. https://doi.org/10.1371/journal.pgen.1007753

Ramos-Silva P, Serrano M, Henriques AO (2019) From root to tips: sporulation evolution and specialization in *Bacillus subtilis* and the intestinal pathogen *Clostridioides difficile*. Mol Biol Evol 36:2714–2736. https://doi.org/10.1093/molbev/msz175

Ransom EM, Kaus GM, Tran PM, Ellermeier CD, Weiss DS (2018) Multiple factors contribute to bimodal toxin gene expression in Clostridioides (Clostridium) difficile. Mol Microbiol 110:533–549. https://doi.org/10.1111/mmi.14107

Razavi B, Apisarnthanarak A, Mundy LM (2007) Clostridium difficile: emergence of hypervirulence and fluoroquinolone resistance. Infection 35:300–307. https://doi.org/10.1007/s15010-007-6113-0

Rhayat L, Duperrier S, Carballido-López R, Pellegrini O, Stragier P (2009) Genetic dissection of an inhibitor of the sporulation sigma factor σG. J Mol Biol 390:835–844. https://doi.org/10.1016/j.jmb.2009.05.073

Ribis JW, Ravichandran P, Putnam EE, Pishdadian K, Shen A (2017) The conserved spore coat protein SpoVM is largely dispensable in Clostridium difficile spore formation. mBio 2:1–16. https://doi.org/10.1128/mSphere.00315-17

Ribis JW, Fimlaid KA, Shen A (2018) Differential requirements for conserved peptidoglycan remodeling enzymes during Clostridioides difficile spore formation. Mol Microbiol 110:370–389. https://doi.org/10.1111/mmi.14090

Riley EP, Lopez-Garrido J, Sugie J, Liu RB, Pogliano K (2021) Metabolic differentiation and intercellular nurturing underpin bacterial endospore formation. Sci Adv 7:eabd6385. https://doi.org/10.1126/sciadv.abd6385

Rodrigues CDA, Harry EJ (2012) The Min system and nucleoid occlusion are not required for identifying the division site in Bacillus subtilis but ensure its efficient utilization. PLoS Genet 8:e1002561. https://doi.org/10.1371/journal.pgen.1002561

Rodrigues CDA, Ramírez-Guadiana FH, Meeske AJ, Wang X, Rudner DZ (2016) GerM is required to assemble the basal platform of the SpoIIIA-SpoIIQ transenvelope complex during sporulation in *Bacillus subtilis*: GerM is required to assemble the SpoIIIA-SpoIIQ complex. Mol Microbiol 102:260–273. https://doi.org/10.1111/mmi.13457

Roels S, Driks A, Losick R (1992) Characterization of spoIVA, a sporulation gene involved in coat morphogenesis in Bacillus subtilis. J Bacteriol 174:575–585. https://doi.org/10.1128/jb.174.2.575-585.1992

Rosenbusch KE, Bakker D, Kuijper EJ, Smits WK (2012) C. difficile 630Δerm Spo0A regulates sporulation, but does not contribute to toxin production, by direct high-affinity binding to target DNA. PLoS One 7:e48608. https://doi.org/10.1371/journal.pone.0048608

Sacco MD, Wang S, Adapa SR, Zhang X, Lewandowski EM, Gongora MV, Keramisanou D, Atlas ZD, Townsend JA, Gatdula JR, Morgan RT, Hammond LR, Marty MT, Wang J, Eswara PJ, Gelis I, Jiang RHY, Sun X, Chen Y (2022) A unique class of Zn2+-binding serine-based PBPs underlies cephalosporin resistance and sporogenesis in Clostridioides difficile. Nat Commun 13:4370. https://doi.org/10.1038/s41467-022-32086-6

Saujet L, Monot M, Dupuy B, Soutourina O, Martin-Verstraete I (2011) The key sigma factor of transition phase, sigh, controls sporulation, metabolism, and virulence factor expression in Clostridium difficile. J Bacteriol 193:3186–3196. https://doi.org/10.1128/JB.00272-11

Saujet L, Pereira FC, Serrano M, Soutourina O, Monot M, Shelyakin PV, Gelfand MS, Dupuy B, Henriques AO, Martin-Verstraete I (2013) Genome-wide analysis of cell type-specific gene transcription during spore formation in Clostridium difficile. PLoS Genet 9: e1003756. https://doi.org/10.1371/journal.pgen.1003756

Saujet L, Pereira FC, Henriques AO, Martin-Verstraete I (2014) The regulatory network controlling spore formation in Clostridium difficile. FEMS Microbiol Lett 358:1–10

Schmidt R, Margolis P, Duncan L, Coppolecchia R, Moran CP, Losick R (1990) Control of developmental transcription factor sigma F by sporulation regulatory proteins SpoIIAA and SpoIIAB in Bacillus subtilis. Proc Natl Acad Sci U S A 87:9221–9225. https://doi.org/10.1073/pnas.87.23.9221

Semenyuk EG, Laning ML, Foley J, Johnston PF, Knight KL, Gerding DN, Driks A (2014) Spore formation and toxin production in Clostridium difficile biofilms. PLoS One 9:e87757. https://doi.org/10.1371/journal.pone.0087757

Serra CR, Earl AM, Barbosa TM, Kolter R, Henriques AO (2014) Sporulation during growth in a gut isolate of Bacillus subtilis. J Bacteriol 196:4184–4196. https://doi.org/10.1128/JB.01993-14

Serrano M, Neves A, Soares CM, Moran CP, Henriques AO (2004) Role of the anti-sigma factor SpoIIAB in regulation of σG during Bacillus subtilis sporulation. J Bacteriol 186:4000–4013. https://doi.org/10.1128/JB.186.12.4000-4013.2004

Serrano M, Real G, Santos J, Carneiro J, Moran CP Jr, Henriques AO (2011) A negative feedback loop that limits the ectopic activation of a cell type-specific sporulation sigma factor of Bacillus subtilis. PLOS Genetics 7:e1002220

Serrano M, Crawshaw AD, Dembek M, Monteiro JM, Pereira FC, Pinho MG, Fairweather NF, Salgado PS, Henriques AO (2016a) The SpoIIQ-SpoIIIAH complex of Clostridium difficile controls forespore engulfment and late stages of gene expression and spore morphogenesis. Mol Microbiol 100:204–228. https://doi.org/10.1111/mmi.13311

Serrano M, Kint N, Pereira FC, Saujet L, Boudry P, Dupuy B, Henriques AO, Martin-Verstraete I (2016b) A recombination directionality factor controls the cell type-specific activation of σK and the fidelity of spore development in Clostridium difficile. PLOS Genet 12: e1006312

Setlow P (2006) Spores of Bacillus subtilis: their resistance to and killing by radiation, heat and chemicals. J Appl Microbiol 101:514–525. https://doi.org/10.1111/j.1365-2672.2005.02736.x

Setlow P (2014) Spore resistance properties. Microbiol Spectr 2:1–14. https://doi.org/10.1128/microbiolspec

Setlow P, Christie G (2023) New thoughts on an old topic: secrets of bacterial spore resistance slowly being revealed. Microbiol Mol Biol Rev. https://doi.org/10.1128/mmbr.00080-22

Shrestha R, Cochran AM, Sorg JA (2019) The requirement for co-germinants during Clostridium difficile spore germination is influenced by mutations in yabG and cspA. PLoS Pathog 15:e1007681. https://doi.org/10.1371/journal.ppat.1007681

Shuster B, Khemmani M, Abe K, Huang X, Nakaya Y, Maryn N, Buttar S, Gonzalez AN, Driks A, Sato T, Eichenberger P (2019) Contributions of crust proteins to spore surface properties in Bacillus subtilis: Crust morphology and spore hydrophobicity. Mol Microbiol 111:825–843. https://doi.org/10.1111/mmi.14194

Sjodt M, Rohs PDA, Gilman MSA, Erlandson SC, Zheng S, Green AG, Brock KP, Taguchi A, Kahne D, Walker S, Marks DS, Rudner DZ, Bernhardt TG, Kruse AC (2020) Structural coordination of polymerization and crosslinking by a SEDS–bPBP peptidoglycan synthase complex. Nat Microbiol 5:813–820

Smits WK (2013) Hype or hypervirulence: A reflection on problematic *C. difficile* strains. Virulence 4:592–596. https://doi.org/10.4161/viru.26297

Smits WK, Lyras D, Lacy DB, Wilcox MH, Kuijper EJ (2016) Clostridium difficile infection. Nat Rev Dis Primers 2:16020. https://doi.org/10.1038/nrdp.2016.20

Somerville GA, Beres SB, Fitzgerald JR, DeLeo FR, Cole RL, Hoff JS, Musser JM (2002) In vitro serial passage of Staphylococcus aureus: changes in physiology, virulence factor production, and agr nucleotide sequence.

J Bacteriol 184:1430–1437. https://doi.org/10.1128/JB.184.5.1430-1437.2002

Sonenshein AL (2000) Control of sporulation initiation in Bacillus subtilis. Curr Opin Microbiol 3:561–566. https://doi.org/10.1016/s1369-5274(00)00141-7

Sorg JA, Sonenshein AL (2008) Bile salts and glycine as cogerminants for Clostridium difficile spores. J Bacteriol 190:2505–2512. https://doi.org/10.1128/JB.01765-07

Steil L, Serrano M, Henriques AO, Völker U (2005) Genome-wide analysis of temporally regulated and compartment-specific gene expression in sporulating cells of Bacillus subtilis. Microbiology (Reading) 151:399–420. https://doi.org/10.1099/mic.0.27493-0

Stevens CM, Daniel R, Illing N, Errington J (1992) Characterization of a sporulation gene, spoIVA, involved in spore coat morphogenesis in Bacillus subtilis. J Bacteriol 174:586–594. https://doi.org/10.1128/jb.174.2.586-594.1992

Stevenson E, Minton NP, Kuehne SA (2015) The role of flagella in Clostridium difficile pathogenicity. Trends Microbiol 23:275–282. https://doi.org/10.1016/j.tim.2015.01.004

Stewart GC (2015) The exosporium layer of bacterial spores: a connection to the environment and the infected host. Microbiol Mol Biol Rev MMBR 79:437–457. https://doi.org/10.1128/MMBR.00050-15

Suárez JM, Edwards AN, McBride SM (2013) The Clostridium difficile cpr locus is regulated by a noncontiguous two-component system in response to type A and B lantibiotics. J Bacteriol 195:2621–2631. https://doi.org/10.1128/JB.00166-13

Suligoy CM, Lattar SM, Noto Llana M, González CD, Alvarez LP, Robinson DA, Gómez MI, Buzzola FR, Sordelli DO (2018) Mutation of Agr is associated with the adaptation of Staphylococcus aureus to the host during chronic osteomyelitis. Front Cell Infect Microbiol 8:18. https://doi.org/10.3389/fcimb.2018.00018

Sun G, Yang M, Jiang L, Huang M (2021) Regulation of pro-σK activation: a key checkpoint in Bacillus subtilis sporulation. Environ Microbiol 23:2366–2373. https://doi.org/10.1111/1462-2920.15415

Sylvestre P, Couture-Tosi E, Mock M (2002) A collagen-like surface glycoprotein is a structural component of the Bacillus anthracis exosporium. Mol Microbiol 45:169–178. https://doi.org/10.1046/j.1365-2958.2000.03000.x

Takamatsu H, Kodama T, Imamura A, Asai KEI, Kobayashi K (2000) The Bacillus subtilis yabG gene is transcribed by SigK RNA polymerase during sporulation, and yabG mutant spores have altered coat protein composition. J Bacteriol 182:1883–1888

Tam NKM, Uyen NQ, Hong HA, Duc LH, Hoa TT, Serra CR, Henriques AO, Cutting SM (2006) The intestinal life cycle of Bacillus subtilis and close relatives. J Bacteriol 188:2692–2700. https://doi.org/10.1128/JB.188.7.2692-2700.2006

Thapa T, Leuzzi R, Ng YK, Baban ST, Adamo R, Kuehne SA, Scarselli M, Minton NP, Serruto D, Unnikrishnan M (2013) Multiple factors modulate biofilm formation by the anaerobic pathogen Clostridium difficile. J Bacteriol 195:545–555. https://doi.org/10.1128/JB.01980-12

Theriot CM, Young VB (2015) Interactions between the gastrointestinal microbiome and Clostridium difficile. Annu Rev Microbiol 69:445–461. https://doi.org/10.1146/annurev-micro-091014-104115

Touchette MH, de la Puebla HB, Ravichandran P, Shen A (2019) SpoIVA-SipL complex formation is essential for Clostridioides difficile spore assembly. J Bacteriol. https://doi.org/10.1128/JB.00042-19

Touchette MH, Benito de la Puebla H, Alves Feliciano C, Tanenbaum B, Schenone M, Carr SA, Shen A (2021) Identification of a novel regulator of Clostridioides difficile cortex formation. mSphere 6:e0021121. https://doi.org/10.1128/mSphere.00211-21

Traag BA, Pugliese A, Eisen JA, Losick R (2013a) Gene conservation among endospore-forming bacteria reveals additional sporulation genes in Bacillus subtilis. J Bacteriol 195:253–260. https://doi.org/10.1128/JB.01778-12

Traag BA, Ramirez-Peralta A, Wang Erickson AF, Setlow P, Losick R (2013b) A novel RNA polymerase-binding protein controlling genes involved in spore germination in *Bacillus subtilis*: A novel regulatory protein for germination genes. Mol Microbiol 89:113–122. https://doi.org/10.1111/mmi.12262

Trouve J, Mohamed A, Leisico F, Contreras-Martel C, Liu B, Mas C, Rudner DZ, Rodrigues CDA, Morlot C (2018) Structural characterization of the sporulation protein GerM from Bacillus subtilis. J Struct Biol 204:481–490. https://doi.org/10.1016/j.jsb.2018.09.010

Underwood S, Guan S, Vijayasubhash V, Baines SD, Graham L, Lewis RJ, Wilcox MH, Stephenson K (2009) Characterization of the sporulation initiation pathway of Clostridium difficile and its role in toxin production. J Bacteriol 191:7296–7305. https://doi.org/10.1128/JB.00882-09

Valenčíková R, Krascsenitsová E, Labajová N, Makroczyová J, Barák I (2018) Clostridial DivIVA and MinD interact in the absence of MinJ. Anaerobe 50:22–31. https://doi.org/10.1016/j.anaerobe.2018.01.013

Vasudevan P, Weaver A, Reichert ED, Linnstaedt SD, Popham DL (2007) Spore cortex formation in Bacillus subtilis is regulated by accumulation of peptidoglycan precursors under the control of sigma K. Mol Microbiol 65:1582–1594. https://doi.org/10.1111/j.1365-2958.2007.05896.x

Veening J-W, Hamoen LW, Kuipers OP (2005) Phosphatases modulate the bistable sporulation gene expression pattern in Bacillus subtilis. Mol Microbiol 56:1481–1494. https://doi.org/10.1111/j.1365-2958.2005.04659.x

Vitucci JC, Pulse M, Tabor-Simecka L, Simecka J (2020) Epidemic ribotypes of Clostridium (now

Clostridioides) difficile are likely to be more virulent than non-epidemic ribotypes in animal models. BMC Microbiol 20:27. https://doi.org/10.1186/s12866-020-1710-5

Wakeley PR, Dorazi R, Hoa NT, Bowyer JR, Cutting SM (2000) Proteolysis of SpolVB is a critical determinant in signalling of Pro-sigmaK processing in Bacillus subtilis. Mol Microbiol 36:1336–1348. https://doi.org/10.1046/j.1365-2958.2000.01946.x

Wang ST, Setlow B, Conlon EM, Lyon JL, Imamura D, Sato T, Setlow P, Losick R, Eichenberger P (2006) The forespore line of gene expression in Bacillus subtilis. J Mol Biol 358:16–37. https://doi.org/10.1016/j.jmb.2006.01.059

Wang KH, Isidro AL, Domingues L, Eskandarian HA, McKenney PT, Drew K, Grabowski P, Chua M-H, Barry SN, Guan M, Bonneau R, Henriques AO, Eichenberger P (2009) The coat morphogenetic protein SpoVID is necessary for spore encasement in Bacillus subtilis. Mol Microbiol 74:634–649. https://doi.org/10.1111/j.1365-2958.2009.06886.x

Wheeldon LJ, Worthington T, Lambert PA (2011) Histidine acts as a co-germinant with glycine and taurocholate for Clostridium difficile spores. J Appl Microbiol 110:987–994. https://doi.org/10.1111/j.1365-2672.2011.04953.x

Wilson KH (1983) Efficiency of various bile salt preparations for stimulation of Clostridium difficile spore germination. J Clin Microbiol 18:1017–1019. https://doi.org/10.1128/jcm.18.4.1017-1019.1983

Wu LJ, Errington J (2003) RacA and the Soj-Spo0J system combine to effect polar chromosome segregation in sporulating Bacillus subtilis. Mol Microbiol 49:1463–1475. https://doi.org/10.1046/j.1365-2958.2003.03643.x

Wu LJ, Errington J (2004) Coordination of cell division and chromosome segregation by a nucleoid occlusion protein in Bacillus subtilis. Cell 117:915–925. https://doi.org/10.1016/j.cell.2004.06.002

Zhao H, Nickle DC, Zeng Z, Law PYT, Wilcox MH, Chen L, Peng Y, Meng J, Deng Z, Albright A, Zhong H, Xu X, Zhu S, Shen J, Blanchard RL, Dorr MB, Shaw PM, Li J (2021) Global landscape of Clostridioides difficile phylogeography, antibiotic susceptibility, and toxin polymorphisms by post-hoc whole-genome sequencing from the MODIFY I/II studies. Infect Dis Ther 10:853–870. https://doi.org/10.1007/s40121-021-00426-6

Zheng LB, Donovan WP, Fitz-James PC, Losick R (1988) Gene encoding a morphogenic protein required in the assembly of the outer coat of the Bacillus subtilis endospore. Genes Dev 2:1047–1054. https://doi.org/10.1101/gad.2.8.1047

Zidaric V, Rupnik M (2016) Sporulation properties and antimicrobial susceptibility in endemic and rare Clostridium difficile PCR ribotypes. Anaerobe 39:183–188. https://doi.org/10.1016/j.anaerobe.2016.04.010

Zidaric V, Pardon B, dos Vultos T, Deprez P, Brouwer MSM, Roberts AP, Henriques AO, Rupnik M (2012) Different antibiotic resistance and sporulation properties within multiclonal Clostridium difficile PCR Ribotypes 078, 126, and 033 in a single calf farm. Appl Environ Microbiol 78:8515–8522. https://doi.org/10.1128/AEM.02185-12

Membrane Vesicles of *Clostridioides difficile* and Other Clostridial Species

Shan Goh and Jameel Inal

Abstract

Membrane vesicles are secreted by growing bacterial cells and are important components of the bacterial secretome, with a role in delivering effector molecules that ultimately enable bacterial survival. Membrane vesicles of *Clostridioides difficile* likely contribute to pathogenicity and is a new area of research on which there is currently very limited information. This chapter summarizes the current knowledge on membrane vesicle formation, content, methods of characterization and functions in Clostridia and model Gram-positive species.

1 Introduction to Membrane Vesicles

Extracellular vesicles (EVs) are natural lipid-lined nanoparticulate products of actively growing eukaryotes and prokaryotes. EVs produced by Gram-negative bacteria are known as outer membrane vesicles (OMVs), while EVs from Gram-positive bacteria are commonly known as membrane vesicles (MVs). There is a greater understanding of OMVs, which were described first; however, the number of studies on MVs is increasing because of the diverse roles that they play in bacterial fitness and pathogenesis, and their many potential applications in medicine and biotechnology (Brown et al. 2015; Kaparakis-Liaskos and Kufer 2020; Bali et al. 2022). For example, studies of the MVs of Gram-positive bacteria such as *Staphylococcus aureus*, *Bacillus anthracis*, *Bacillus subtilis*, and *Clostridium perfringens* indicate functions in biofilm formation (He et al. 2017), survival against host cell killing and antibiotic susceptibility (Andreoni et al. 2019; Askarian et al. 2018), toxin secretion (Rivera et al. 2010; Jiang et al. 2014), phage infection (Tzipilevich et al. 2017), and immunogenicity (Jiang et al. 2014; Bitto et al. 2021a). There are many more studies which only describe MV contents (Lee et al. 2009; Kim et al. 2019; Afonina et al. 2021; Ichikawa et al. 2021; Bitto et al. 2021a); however, some of the components described could potentially be involved in horizontal gene transfer, antimicrobial resistance transfer, and quorum sensing.

In *Clostridioides difficile*, potential functions of MVs are secretion of toxins, horizontal gene transfer, phage infection, immunogenicity, and biofilm formation. As there have been only two

S. Goh (✉)
Department of Clinical, Pharmaceutical and Biological Sciences, School of Life and Medical Sciences, University of Hertfordshire, Hatfield, UK
e-mail: s.goh5@herts.ac.uk

J. Inal
Department of Clinical, Pharmaceutical and Biological Sciences, School of Life and Medical Sciences, University of Hertfordshire, Hatfield, UK

School of Human Sciences, London Metropolitan University, London, UK
e-mail: j.inal@londonmet.ac.uk

P. Mastrantonio, M. Rupnik (eds.), *Updates on Clostridioides difficile in Europe*, Advances in Experimental Medicine and Biology 1435, https://doi.org/10.1007/978-3-031-42108-2_14

studies of *C. difficile* MVs (by Nicholas et al. 2017; Lopes et al. 2019), the rationales for these will be presented in the context of current knowledge on MVs from Clostridial species (Table 1). Two studies published in the 1980s in *Clostridium acetobutylicum* (Driessen et al. 1988) and *Clostridium thermoautotrophicum* (Hugenholtz et al. 1987) are recorded in Table 1, however the MVs in these studies were artificially created by osmotic lysis for the purpose of investigating transport of molecules across cell membranes. As such, these two reports will be excluded from further discussion. Knowledge on MVs from model Gram-positive species and OMVs from Gram-negative species will be drawn upon where appropriate, as they are more extensively studied.

2 Vesiculogenesis

2.1 Mechanisms in Gram-Positive Bacteria

MV biogenesis has been studied in model Gram-positive organisms but not in *C. difficile* and Clostridia. MV formation in *S. aureus* and *S. epidermidis* was recently examined using super-resolution stochastic optical reconstruction microscopy (STORM) in addition to TEM and SEM (Jeong et al. 2022). MVs were formed either through membrane blebbing or explosive cell lysis, both of which were first described for OMVs (Schwechheimer and Kuehn 2015; Turnbull et al. 2016). A new observation for membrane blebbing was that MV precursors were seen in the periplasmic space of *S. aureus* with intact peptidoglycan. Another recent study on *Lacticaseibacillus casei* observed MVs in the periplasmic space (da Silva Barreira et al. 2022). Jeong et al. (2022) found that MV precursors were released either from localized cell wall lysis or less commonly after coating with peptidoglycan. The former could be due to cell wall restructuring facilitated by modifying enzymes (e.g., hydrolases) during cell division. Explosive cell lysis results from an expanded periplasmic space that ruptures the cell wall leading to release of cell debris and MV formation external to the cell (Jeong et al. 2022). MVs that had formed in different ways had different sizes and varied in surface components. Although MV biogenesis has not been investigated in Clostridia, it is likely that MVs from Clostridial species, including *C. difficile*, are formed in a similar way.

2.2 Regulation of Vesiculogenesis

Regulation of MV formation is less well understood. In the Gram-positive bacteria *Streptococcus pyogenes*, strain to strain variation in MV production was associated with a two-component system control of virulence regulator sensor operon (*covRS*) (Resch et al. 2016). In *Clostridium perfringens,* deletion of the sporulation master regulator gene, *spo0A,* and orphan sensor histidine kinase genes *CPE1316* or *reeS* that phosphorylate SpoA, significantly reduced MV production. However, deletion of *sigF*, which is essential for sporulation did not affect MV production (Obana et al. 2017). This suggests that phosphorylation of Spo0A rather than sporulation itself is important for MV formation. In *C. difficile*, phosphorylated Spo0A has a global effect in addition to sporulation and has been shown to indirectly control toxin production, cell envelope structure, flagellar formation, and butyrate production (Pettit et al. 2014). Spo0A was shown to negatively regulate the *sinRI* operon, which regulates sporulation, toxin production, motility, and biofilm formation (Dhungel Babita and Govind 2020). Hence, Spo0A may influence vesiculogenesis in *C. difficile*.

2.3 Phage and Antibiotics

In *B. subtilis*, *S. aureus*, and *L. casei*, prophage induction and endolysin activity aided MV release (Toyofuku et al. 2017; Andreoni et al. 2019; da Silva Barreira et al. 2022), indicating that prophages can regulate MV formation. However, in *Enterococcus faecalis* phage did not appear to be involved in MV production (Afonina et al. 2021). In *C. difficile* (Nicholas et al. 2017)

Table 1 Summary of reports on naturally secreted MVs of Clostridial species, unless stated otherwise

Clostridial species	Strains	MV size range (nm) and growth phase	Purification[a]	Physical characterization techniques[b]	Cargo/uptake by mammalian cells	MV function	Reference
C. difficile	VPI 10463	17–100, Late log	0.22 μm UF No DG	TEM, sodium dodecyl-sulphate polyacrylamide gel electrophoresis (SDS-PAGE), bicinchoninic acid (BCA) assay	Proteomic analysis found 262 proteins (185 cytoplasm, 50 membrane, and 27 extracellular)	≥5 μg/mL MV increased expression of IL-1b, IL-6, IL-8, and MCP-1 genes in CaCo-2 cells; ≥ 20 μg/mL for visible cytotoxicity of CaCo-2 cells	Nicholas et al. (2017)
C. difficile	R20291	20–400, stationary	No filtration UF DG-O	TEM, SEM, DLS, SDS-PAGE	Toxin A detected by Western blot, proteomic analysis found 194 proteins (cytoplasmic such as tcdA, tcdB, RNA polymerase, flagellin, ribosomal proteins, Spo0A, SecA)	Not reported	Lopes et al. (2019)
C. perfringens	Strain 13	~50–550, log to stationary (6, 12, 18, 24 h)	0.45μm UC DG-O	TEM, SDS-PAGE, BCA assay	Peptidoglycan and membrane proteins, phospholipase C toxin negative by Western blot	30 and 300 ng/mL MVs increased IL-6 and TNF-α production in J774.1 mouse macrophage-like cells. 300 ng of MV increased expression of *tlr2*, *il-6*, and *tnf-α*. No cytotoxic effect	Obana et al. (2017)
C. perfringens	CP4 mainly. Other strains CP273, CP404, CP888, CP971, JGS4143	20–400 Stationary	0.45 μm UC DG-O	TEM, SDS-PAGE, lipophilic fluorophore dialkylcarbocyanine iodide (DiI) staining, BCA assay	Alpha and NetB toxin negative by Western blot. DNA from MVs were PCR positive for 16S rRNA, *plc*, *pfoA* genes. Proteomic analysis found 432 proteins (24% membrane, 18% cytoplasm, 9% intracellular organelle, 8% ribosome, 10% plasma membrane) and Beta2 toxin	5, 10 μg/ml of CP4 MVs induced granulocyte-colony stimulating factor (G-CSF), tumor necrosis factor alpha (TNF-a), interleukin-6 (IL-6) production in RAW264.7 cells. Mice inoculated with 10 μg/mL CP4 MV also produced more	Jiang et al. (2014)

(continued)

Table 1 (continued)

Clostridial species	Strains	MV size range (nm) and growth phase	Purification[a]	Physical characterization techniques[b]	Cargo/uptake by mammalian cells	MV function	Reference
						G-CSF, TNF-a, IL-6, but MVs did not protect mice from *C. perfringens* challenge. No cytotoxic effect on RAW264.7 cells	
C. botulinum	Group I type A 62A, type A 7I03-H, and type B Okra, Group II type E strain Iwanai	~80–300, except type E 50–500, Stationary (4–5 days)	0.45 μm UC DG-O	TEM, NTA, SDS-PAGE, BCA assay	Uptake in RAW264.7 cells was via actin polymerization-dependent endocytosis pathway. determined by fluorescence microscopy of fluorescein isothiocyanate (FITC)-stained MVs	10 μg/mL MV increased expression of IL-6, TNF, IL-1b in RAW264.7 mouse macrophage-like cell line. No cytotoxic effect	Kobayashi et al. (2022)
C. sporogenes	ATCC7955	~80–400, stationary (4–5 days)					
C. scindens	ATCC35704, VPI12708	~60–350, stationary (4–5 days)					
C. butyricum	Probiotic strain from Ace Bio Product Co., Ltd.	Median of 152, Stationary (24 h)	0.22μm UC No DG	AFM, DLS, lipophilic fluorophore PKH67 staining, Bradford assay	Peptidoglycan content was 0.03 ng/μg MVs by silkworm larvae plasma absorbance assay. MV uptake by DC2.4 and RAW264.7 cells was by macropinocytosis and clathrin-mediated endocytosis, determined by fluorescence microscopy	0.1 μg MV induced TNF-a, IL-6, TGFb-1 expression, and production in RAW264.7 cells	Morishita et al. (2021)
C. butyricum	MIYAIRI 588 (CBM588) from Miyarisan Pharmaceutical	30–200, not stated	0.22 μm UF DG-S	TEM, NTA, Dil staining	Not reported	15 μg MV daily reduced ulcerative colitis in C57BL/6 mice compared to untreated group. MV	Liang et al. (2022)

	Co. Ltd (Tokyo, Japan)					upregulated expression genes for MUC2, ZO-1, Arg1, IL-10, and polarized M2-type macrophages in vivo	
C. thermocellum	DSM 1313	36–300, mostly 50–100, not stated	0.22 μm UF No DG	TEM, Bradford assay	Cellulosomes on the surface of MVs	Not applicable	Ichikawa et al. (2019)
C. thermocellum	DSM 1313	Not stated	0.22 μm UF No DG	TEM	5 major compounds in surfactin-treated MVs by MS/MS, one of which was cis-2-decenoic acid, known as a diffusible signal factor (DSF) that can interact with diverse organisms	Lysed MV promoted growth of *C. thermocellum* in vitro, DSF may have broad-range communication function	Ichikawa et al. (2021)
C. acetobutylicum	ATCC824	Not stated, Exponential phase (MVs generated by osmotic lysis)	Not done	Not done	Phospholipid composition ~60–65% phosphatidylethanolamine and mono-N methylated phosphatidylethanolamine, 20% phosphatidylglycerol, 15% cardiolipin	This study fused cytochrome c oxidase into MVs to study amino acid transport across fused membrane. Membranes were fused by freeze-thaw sonication	Driessen et al. (1988)
C. thermoautotrophicum	701/5	Not stated, Exponential phase (MVs generated by osmotic lysis)	Not done	Not done	Carbon monoxide dehydrogenase and methylene tetrahydrofolate dehydrogenase activity investigated	Carbon monoxide dehydrogenase of MV mediated redox of b-type cytochromes and other electron carriers	Hugenholtz et al. (1987)

[a]Major purification steps involve MV separation from cells by filtration through either 0.22 or 0.45 μm filters, MV concentration by either ultrafiltration (UF) or ultracentrifugation (UC), and MV purification by density gradient (DG) using either OptiPrep™ (O) or sucrose (S) if carried out

[b]Physical characterization includes methods to visualize MV and determine sizes using transmission electron microscopy (TEM), scanning electron microscopy (SEM), or atomic force microscopy (AFM); methods to determine particle size by dynamic light scattering (DLS) and nanoparticle tracking analysis (NTA); methods to quantify particles by NTA or fluorescence staining using lipophilic dyes; methods to quantify or analyse protein in MVs by BCA assay, Bradford assay, and SDS-PAGE

and other Gram-positive bacteria, proteomic analysis of MV contents has found phage structural proteins as major products (Resch et al. 2016; Afonina et al. 2021; Champagne-Jorgensen et al. 2021; da Silva Barreira et al. 2022). Whole phage virions were observed within *B. subtilis* MVs (Toyofuku et al. 2017), phage DNA was detected in *Lacticaseibacillus rhamnosus* MVs (Champagne-Jorgensen et al. 2021), and phage RNA was detected in *S. pyogenes* (Resch et al. 2016). Hence, there is ample data indicating phage can be involved in MV formation.

Antibiotics can affect MV formation. This has not only been observed in OMVs (Bos et al. 2021) but also for MVs of *Enterococcus faecium* (Kim et al. 2019) and *S. aureus* (Andreoni et al. 2019). This may be the combined effect of antibiotic activity on the cell wall (for cell-wall targeting antibiotics) and antibiotic-induced stress responses leading to prophage-induced MV formation and regulation of genes involved in MV production.

3 MV Contents

3.1 Variability in Content

MVs serve as a secretory system and are selectively packaged (meaning that only certain proteins produced within bacterial cells are found in MVs) and the composition of MVs can be altered. For example, *Streptococcus mutans srtA*-deficient mutants produced as many MVs as wild-type cells but with differing protein profiles (Liao et al. 2014). Consistent with selective packaging, MV content is dependent on growth phase (Obana et al. 2017; Zavan et al. 2019; Jeong et al. 2022), growth state (i.e. planktonic or sessile) (Grande et al. 2017), growth medium (Askarian et al. 2018; Kim et al. 2019), and different mechanisms of biogenesis (Jeong et al. 2022), all of which lead to different biological functions of MVs.

3.2 Virulence Factors

MVs of many Gram-positive species contain virulence factors, such as anthrax toxins in *B. anthracis* (Rivera et al. 2010), alpha-toxin, enterotoxin B, and virulence-associated factors in *S. aureus* MVs (Thay et al. 2013; Jeong et al. 2022; Askarian et al. 2018). Toxins were not found in *C. difficile* VPI 10463 MVs (Nicholas et al. 2017), but TcdA was apparently detected in *C. difficile* R20291 MVs by western blot (Lopes et al. 2019). MV cargo is dependent on strain, growth conditions, and phase, and these differences between the two studies could explain these observations (Table 1). Phospholipase C and beta2 toxins were found in *C. perfringens* MVs (Jiang et al. 2014; Obana et al. 2017), but neurotoxins were not detected in *C. botulinum* nor *C. sporogenes* MVs (Kobayashi et al. 2022). The other Clostridia in Table 1 are not known to be toxigenic.

3.3 Fitness Factors

MVs can also contain enzymes that improve fitness. For instance, *S. aureus* MVs were found to contain beta lactamase which mediated ampicillin-resistance of other bacterial species (Lee et al. 2013). *C. difficile* MVs contained VanZ (Nicholas et al. 2017) which mediates teicoplanin resistance (Woods et al. 2018; Sur et al. 2022). *C. thermocellum* MVs contained cellulosomes which degrade cellulose (Ichikawa et al. 2019), enabling the bacterium to degrade plant matter and survive in diverse environments. Cellulosomes are valuable for bioprocessing and creating biofuels from plant biomass.

3.4 Nucleic Acids

C. perfringens MVs contained 16S rRNA, alpha-toxin gene *plc*, and perfringolysin O gene *pfoA* (Jiang et al. 2014). *S. mutans* MVs contained eDNA (non-selectively packaged) which aided early stage of biofilm formation (Liao et al. 2014). *Lactobacillus reuteri* MVs from planktonic cells and biofilm cells were compared for eDNA abundance, and MVs from biofilm cells were found to contain significantly more eDNA, perhaps having a role in biofilm maintenance (Grande et al. 2017). *S. pyogenes* MVs contained

RNA which were in different relative proportions to RNA in the cytoplasm, and this was hypothesized to enable rapid adaptive cell responses to environmental changes (Resch et al. 2016).

3.5 Cell Envelope

MVs contain cell envelope components such as cell membrane, peptidoglycan, and surface layer (S layer) proteins. *C. difficile* has an S layer (Fagan and Fairweather 2014), which was found to be incorporated into MVs. SlpA, Cwp8, Cwp10, Cwp66 were found in the proteome of *C. difficile* VPI 10463 MVs (Nicholas et al. 2017). Other Gram-positive species which have an S layer and are known to form MVs are *C. thermocellum* (Ichikawa et al. 2019), *C. botulinum* (Kobayashi et al. 2022), and *B. anthracis* (Rivera et al. 2010), although S layer presence in MVs was not mentioned in those studies. S layer was reported to be observed morphologically on OMVs of the Gram-negative bacterium *Campylobacter fetus* (Farace et al. 2022). New models of vesiculogenesis should take into consideration the S layer. Peptidoglycan was found to be a major component in MVs of *C. perfringens* strain 13 (Obana et al. 2017) and *C. butyricum* (Morishita et al. 2021). However, in MVs of *C. difficile* (Nicholas et al. 2017) and *C. perfringens* strain CP4 (Jiang et al. 2014) there were proportionally more membrane proteins than peptidoglycan.

4 MV Preparation and Quantification

4.1 General Method of Separation, Concentration, and Purification

MV extraction methods from Clostridia are very similar to other bacteria and involve filtration of bacterial culture supernatant, then ultrafiltration or ultracentrifugation to concentrate MVs in the supernatant. Finally, MVs are washed and purified, most commonly through a density gradient such as OptiPrep™ (Table 1). To obtain purified MVs from Clostridial species, usually 0.4–1 L of bacterial culture is used. It is widely acknowledged by EV researchers that non-vesicular entities (e.g. phage, lipoproteins, extracellular proteins) cannot be fully separated from EVs using the methods described above (Théry et al. 2018). To assign effects to EVs, demonstrating a lack of effect in the remaining EV-depleted sample should be considered (Théry et al. 2018). Otherwise, a combination of purification methods and developing EV-selective methods (Nakao et al. 2014) to achieve as high a purity of EVs as possible should be employed.

The choice of materials or methods for EV separation, concentration, and purification has been shown to significantly impact their characterization (Vergauwen et al. 2017; Théry et al. 2018; Bitto et al. 2021b; Steć et al. 2022). The International Society for Extracellular Vesicles (ISEV) has published Minimal Information for Studies of Extracellular Vesicles (MISEV 2018) (Théry et al. 2018), some of which is relevant to bacterial EVs. Similar guidelines for a standardized approach to preparing and analysing bacterial EVs should be considered (Bitto et al. 2021b).

4.2 Common Quantification Methods

MVs can be quantified by: (1) protein content of vesicles, using BCA, Bradford, or Qubit assays; (2) particle number, using light scattering techniques such as nanoparticle tracking analysis (NTA); or (3) fluorescence labelling with lipophilic dyes followed by detection by flow cytometry or NTA. Although quantification by protein content is the most used method of MV quantification in bacterial EV studies (including for Clostridial species, see Table 1), perhaps because of convenience, this method can overestimate co-purified protein contaminants, or underestimate MV protein content if detergent was not used to disrupt MVs to release protein prior to quantification, or if the sample was not purified. Importantly, protein content will not correlate with MV particle number, potentially affecting experimental findings (Théry et al. 2018; Bitto et al. 2021b; Steć et al. 2022).

A lipid dye FM1-43 has been validated for MV protein mass quantification using the Bradford assay and applied on several Gram-positive species (Resch et al. 2016; Toyofuku et al. 2017; Andreoni et al. 2019). FM1-43 was used to track MV release in *E. coli* (Bos et al. 2021). We have developed a fluorescence-based semi-quantitative assay using FM1-43 for *C. difficile* MVs validated against NTA particle counts (unpublished). Compared to NTA, this method requires less sophisticated equipment, is medium-throughput, is quicker and easier to obtain a measurement (Szatanek et al. 2017), and has lower background. However, it is semi-quantification and requires particle counting to generate a standard curve at the start.

4.3 Clostridial MV Characteristics

The most common parameters of MV characterization are particle size distribution, morphology by TEM, and protein content. Amount and size of bacterial MVs, and protein quantity of MV are species and strain dependent (Bitto et al. 2021b; Kobayashi et al. 2022), as well as growth phase and growth medium dependent (Askarian et al. 2018; Jeong et al. 2022). MVs of *C. difficile* VPI10463 and R20291 were 20–400 nm, possibly with larger MVs forming in stationary phase of R20291 compared with log phase of VPI 10463 growth. MVs produced at stationary phase of other Clostridia have a similar size range (Table 1). In *C. perfringens*, MVs produced at different time points were found to vary in yield, size, and composition, whereas MVs from stationary phase cultures were larger than those from log phase cultures (Obana et al. 2017).

C. difficile MVs appear spherical with a lipid bilayer of uniform thickness (Nicholas et al. 2017), similar to MVs from *C. thermocellum* (Ichikawa et al. 2019). However, samples consisting of some MVs with a thicker lipid bilayer were observed in *C. perfringens* (Jiang et al. 2014; Obana et al. 2017), *C. botulinum*, *C. sporogenes*, *C. scindens* (Kobayashi et al. 2022), and *C. butyricum* (Liang et al. 2022). The significance of a thicker membrane is unknown; it could be a staining artefact or associated with spontaneously self-assembled vesicles from bacterial debris external to bacterial cells (Huang et al. 2017). For protein content, see Sects. 3.1–3.5.

5 Potential Functions of *C. difficile* MVs

5.1 Toxin Secretion

While there is evidence of toxin secretion mechanisms in *C. difficile* being mediated by TcdE (Govind and Dupuy 2012; Govind et al. 2015) and Cwp19 (Wydau-Dematteis et al. 2018), the absence of these proteins did not always abolish toxin secretion (Olling et al. 2012; Wydau-Dematteis et al. 2018), indicating the existence of additional secretory pathways. The first study on *C. difficile* MVs did not find toxins as a constituent (Nicholas et al. 2017), however it is known that MV cargo is dependent on growth phase and conditions (perhaps affecting MV biogenesis pathways). The presence of toxins in *C. difficile* MVs should be examined at varying growth phases and conditions.

5.2 Horizontal Gene Transfer

A DNase-sensitive mechanism for DNA transfer requiring the presence of viable donors has been demonstrated in *C. difficile* (Khodadoost et al. 2017). This suggests that MVs may be involved in a transformation-like DNA transfer known as "vesicle-mediated gene transfer" or VMGT (Fulsundar et al. 2014). VMGT has been demonstrated in Gram-negative bacteria. OMVs can carry plasmid DNA capable of gene transfer within (*Acinetobacter baumannii* is one example (Rumbo et al. 2011)) and between species (Fulsundar et al. 2014), where interspecies transfer was not entirely dependent on the relatedness of OMV donor and recipient (Tran and Boedicker

2017). OMVs were actively associated, or loaded, with DNA in growing bacterial cells (Tran and Boedicker 2017) and DNA could be carried externally and internally, hence DNase treatment reduced transfer frequency (Bitto et al. 2017). In Gram-positive bacteria such as *S. aureus* (Bitto et al. 2021a) and *S. mutans* (Liao et al. 2014), MVs were shown to carry DNA externally and internally. Two mechanisms of VMGT have been proposed based on OMV-mediated transfer of plasmid borne *bla* in *Acinetobacter baylyi*: either dependent on competence factor uptake of DNA external/internal of the MV, or internalization of MV and release of vesicular DNA (Fulsundar et al. 2014). This is an area that requires more investigation, particularly for Gram positives.

MVs were also shown to facilitate phage transduction in *B. subtilis* (Tzipilevich et al. 2017). MVs derived from a phage-susceptible *B. subtilis* strain could transfer the SPP1 phage receptor YueB, enabling SPP1 transducing particles to infect and transfer a plasmid to normally phage-resistant cells (Tzipilevich et al. 2017). This mechanism is described further in the next section.

5.3 Phage Infection

OMVs are known to carry phage receptors which enable phage adsorption and DNA ejection (e.g., *Salmonella typhimurium* OMVs and P22 phage) (Manning and Kuehn 2011; Stephan et al. 2020; Bali et al. 2022). Recently, in the Gram-positive organism *B. subtilis*, MVs carrying phage receptors could transfer them to the cell surface of phage-resistant cells, resulting in the sensitization of normally phage-resistant cells to phage infection (Tzipilevich et al. 2017). This phenomenon would lead to increased dissemination of phages in a cell population and could be a naturally occurring mechanism for phage expansion of host range in some species but not others (Augustyniak et al. 2022). On the other hand, OMVs carrying phage receptors have been shown to act as decoys to protect bacteria against phage infection (Reyes-Robles et al. 2018; Stephan et al. 2020; Augustyniak et al. 2022). *C. difficile* phages identified so far have relatively narrow host ranges, in part defined by phage receptor-binding proteins recognizing a cell surface layer protein, SlpA, as a phage receptor on bacterial cells (Royer et al. 2022). Proteomics analysis of *C. difficile* MVs by Nicholas et al., 2017 found SlpA, suggesting *C. difficile* MVs may be capable of transferring phage receptors or acting as decoys for phage infection.

5.4 Immunogenicity

MVs are known to contain cell wall, cell membrane, and cytoplasmic content which are immunogenic. Nicholas et al. (2017) reported that *C. difficile* MVs from a toxigenic strain stimulated expression of pro-inflammatory cytokine genes for IL-1β, IL-6, IL-8, and MCP-1 in CaCo-2 cells, an intestinal cell line. Although toxins were not detected in MVs, CaCo-2 cells treated with >1 μg/mL MVs led to cytotoxicity (Table 1). MVs of other Clostridia were also shown to be immunogenic. *C. perfringens* MVs induced secretion of pro-inflammatory cytokines IL-6 and TNF-α in J774.1 cells, a macrophage cell line, via TLR2 signalling (Obana et al. 2017). Similar observations, in addition to G-CSF production, were reported by Jiang et al. (2014) in RAW264.7 cells, a macrophage cell line, and sera of MV-immunized mice. *C. botulinum*, *C. sporogenes*, and *C. scindens* MVs induced expression of inflammatory cytokine genes for IL-1-β, IL-6, and TNF in CaCo-2; IL-6, CXCL2, and CCL2 in CMT-3 (an intestinal cell line); and IL-6, IL-8, and CCL2 in RAW264.7 cell lines (Kobayashi et al. 2022). *C. butyricum* induced expression of inflammatory cytokine genes for IL-6, TNF-α, and TGFβ-1 in RAW264.7 cells (Morishita et al. 2021). *C. butyricum* was shown to upregulate gene expression of gut barrier-related proteins (MUC2, ZO-1, Arg1, IL-10) in mice and induced polarization of anti-inflammatory M2-type macrophages (Liang et al. 2022).

Exploring the use of MVs as vaccines for protection against *C. perfringens* infection has been disappointing; MVs were found to not protect mice from *C. perfringens* challenge (Jiang et al. 2014). However, use of MVs as disease intervention agents has shown promise; *C. butyricum* MVs protected mice from ulcerative colitis through M2 macrophage transformation and gut microbiome modulation (Liang et al. 2022). Vaccines being developed for *C. difficile* have focused on antigenic toxin components (Nibbering et al. 2021; Razim et al. 2021). Since colonization of non-toxigenic *C. difficile* leads to immuno-protection against toxigenic strains, vaccine development could also include proteins other than toxins, such as cell wall and flagellar proteins (Nibbering et al. 2021). In this regard, MVs from non-toxigenic *C. difficile* could be effective vaccines if their contents can be controlled to exclude cytotoxic agents.

5.5 Biofilm Formation

MVs were found to be involved in biofilm formation in *S. aureus*, where the presence of MVs derived from vancomycin cultured methicillin-resistant *S. aureus* (MRSA) significantly increased cell attachment and aggregation (He et al. 2017). MVs can be produced by cells within a biofilm, as demonstrated in *L. reuteri*. MVs from planktonic cells (pMV) or biofilms (bMV) were compared and bMVs contained more eDNA, but less protein compared with pMVs (Grande et al. 2017). This may indicate an important role for vesiculated eDNA in maintaining biofilms. Biofilm formation by *C. difficile* depends on multiple factors, as reviewed by Frost et al. (2021). Spo0A is important for biofilm formation in *C. difficile* (Dawson et al. 2012), similar to *C. perfringens* (Huang et al. 2004) which also depends on Spo0A for efficient MV formation (Obana et al. 2017). It may be possible that both MV and biofilm production in *C. difficile* are regulated by Spo0A and that MVs contribute to biofilm formation as seen in *S. aureus*. Quorum sensing (QS) also contributes to biofilm formation in *C. difficile*. A mutant unable to synthesize the QS signalling molecule AI-2 produced significantly less biofilm containing less eDNA and had reduced expression of prophage genes (Slater et al. 2019). The possible relationship between AI-2 and prophage induction may extend to MV production, since prophage induction can increase vesiculogenesis (see Sect. 5.3) a proportion of which may contain vesiculated eDNA as mentioned in Sects. 3.4 and 5.2.

6 Concluding Remarks

Membrane vesicles are increasingly recognized as playing important roles in bacterial survival. Research in MVs is catching up to OMVs, and MVs in *C. difficile* deserve to be investigated because they potentially have multiple functions and could interact broadly within and between species. Although this chapter proposed many concepts based on related Clostridia because of the currently very limited knowledge on *C. difficile* MVs, many findings on MVs so far apply to both Gram-positive and Gram-negative bacteria, hence some aspects of MVs could be conserved in prokaryotes. On the other hand, it is worth noting that while *C. difficile* was classified as a Clostridia, phylogenetically it is distantly related to *C. botulinum, C. sporogenes, C. scindens, C. perfringens, C. butyricum* and recently re-classified as Clostridioides in its own cluster (XIa) (Lawson et al. 2016; Cruz-Morales et al. 2019). Therefore, some differences between MVs from *C. difficile* and those of other Clostridial species could be expected.

References

Afonina I, Tien B, Nair Z et al (2021) The composition and function of *Enterococcus faecalis* membrane vesicles. microLife 2:uqab002. https://doi.org/10.1093/femsml/uqab002

Andreoni F, Toyofuku M, Menzi C et al (2019) Antibiotics stimulate formation of vesicles in *Staphylococcus aureus* in both phage-dependent and -independent fashions and via different routes. Antimicrob Agents Chemother 63:e01439-01418. https://doi.org/10.1128/AAC.01439-18

Askarian F, Lapek JD Jr, Dongre M et al (2018) Staphylococcus aureus membrane-derived vesicles promote bacterial virulence and confer protective immunity in murine infection models. Front Microbiol 9:262. https://doi.org/10.3389/fmicb.2018.00262

Augustyniak D, Olszak T, Drulis-Kawa Z (2022) Outer membrane vesicles (OMVs) of *Pseudomonas aeruginosa* provide passive resistance but not sensitization to LPS-specific phages. Viruses 14(1):121

Bali K, Lu Z, McCoy R et al (2022) Multiparametric sensing of outer membrane vesicle-derived supported lipid bilayers demonstrates the specificity of bacteriophage interactions. bioRxiv. https://doi.org/10.1101/2022.12.13.520201

Bitto NJ, Chapman R, Pidot S et al (2017) Bacterial membrane vesicles transport their DNA cargo into host cells. Sci Rep 7(1):7072. https://doi.org/10.1038/s41598-017-07288-4

Bitto NJ, Cheng L, Johnston EL et al (2021a) Staphylococcus aureus membrane vesicles contain immunostimulatory DNA, RNA and peptidoglycan that activate innate immune receptors and induce autophagy. J Extracell Vesicles 10(6):e12080. https://doi.org/10.1002/jev2.12080

Bitto NJ, Zavan L, Johnston EL et al (2021b) Considerations for the analysis of bacterial membrane vesicles: methods of vesicle production and quantification can influence biological and experimental outcomes. Microbiol Spectr 9(3):e0127321. https://doi.org/10.1128/Spectrum.01273-21

Bos J, Cisneros LH, Mazel D (2021) Real-time tracking of bacterial membrane vesicles reveals enhanced membrane traffic upon antibiotic exposure. Sci Adv 7(4). https://doi.org/10.1126/sciadv.abd1033

Brown L, Wolf JM, Prados-Rosales R et al (2015) Through the wall: extracellular vesicles in Gram-positive bacteria, mycobacteria and fungi. Nat Rev Microbiol 13(10):620–630. https://doi.org/10.1038/nrmicro3480

Champagne-Jorgensen K, Jose TA, Stanisz AM et al (2021) Bacterial membrane vesicles and phages in blood after consumption of *Lacticaseibacillus rhamnosus* JB-1. Gut microbes 13(1):1993583. https://doi.org/10.1080/19490976.2021.1993583

Cruz-Morales P, Orellana CA, Moutafis G et al (2019) Revisiting the evolution and taxonomy of Clostridia, a phylogenomic update. Genome Biol Evol 11(7): 2035–2044. https://doi.org/10.1093/gbe/evz096

da Silva Barreira D, Lapaquette P, Novion Ducassou J et al (2022) Spontaneous prophage induction contributes to the production of membrane vesicles by the Gram-positive bacterium *Lacticaseibacillus casei* BL23. MBio 13(5):e02375-02322. https://doi.org/10.1128/mbio.02375-22

Dawson LF, Valiente E, Faulds-Pain A et al (2012) Characterisation of *Clostridium difficile* biofilm formation, a role for Spo0A. PLoS One 7(12):e50527. https://doi.org/10.1371/journal.pone.0050527

Dhungel Babita A, Govind R (2020) Spo0A suppresses sin locus expression in *Clostridioides difficile*. mSphere 5 (6):e00963-00920. https://doi.org/10.1128/mSphere.00963-20

Driessen AJ, Ubbink-Kok T, Konings WN (1988) Amino acid transport by membrane vesicles of an obligate anaerobic bacterium, *Clostridium acetobutylicum*. J Bacteriol 170(2):817–820. https://doi.org/10.1128/jb.170.2.817-820.1988

Fagan RP, Fairweather NF (2014) Biogenesis and functions of bacterial S-layers. Nat Rev Microbiol 12(3):211–222. https://doi.org/10.1038/nrmicro3213

Farace P, Cravero S, Taibo C et al (2022) Campylobacter fetus releases S-layered and immunoreactive outer membrane vesicles. Rev Argent Microbiol 54(2): 74–80. https://doi.org/10.1016/j.ram.2021.06.001

Frost LR, Cheng JKJ, Unnikrishnan M (2021) *Clostridioides difficile* biofilms: A mechanism of persistence in the gut? PLoS Pathog 17(3):e1009348. https://doi.org/10.1371/journal.ppat.1009348

Fulsundar S, Harms K, Flaten GE et al (2014) Gene transfer potential of outer membrane vesicles of *Acinetobacter baylyi* and effects of stress on vesiculation. Appl Environ Microbiol 80(11):3469–3483. https://doi.org/10.1128/AEM.04248-13

Govind R, Dupuy B (2012) Secretion of *Clostridium difficile* toxins A and B requires the holin-like protein TcdE. PLoS Pathog 8(6):e1002727. https://doi.org/10.1371/journal.ppat.1002727

Govind R, Fitzwater L, Nichols R (2015) Observations on the role of TcdE isoforms in *Clostridium difficile* toxin secretion. J Bacteriol 197(15):2600–2609. https://doi.org/10.1128/jb.00224-15

Grande R, Celia C, Mincione G et al (2017) Detection and physicochemical characterization of membrane vesicles (MVs) of *Lactobacillus reuteri* DSM 17938. Front Microbiol 8. https://doi.org/10.3389/fmicb.2017.01040

He X, Yuan F, Lu F et al (2017) Vancomycin-induced biofilm formation by methicillin-resistant *Staphylococcus aureus* is associated with the secretion of membrane vesicles. Microb Pathog 110:225–231. https://doi.org/10.1016/j.micpath.2017.07.004

Huang IH, Waters M, Grau RR et al (2004) Disruption of the gene (*spo0A*) encoding sporulation transcription factor blocks endospore formation and enterotoxin production in enterotoxigenic *Clostridium perfringens* type A. FEMS Microbiol Lett 233(2):233–240. https://doi.org/10.1016/j.femsle.2004.02.014

Huang C, Quinn D, Sadovsky Y et al (2017) Formation and size distribution of self-assembled vesicles. Proc Natl Acad Sci U S A 114(11):2910–2915. https://doi.org/10.1073/pnas.1702065114

Hugenholtz J, Ivey DM, Ljungdahl LG (1987) Carbon monoxide-driven electron transport in *Clostridium thermoautotrophicum* membranes. J Bacteriol 169(12):5845–5847. https://doi.org/10.1128/jb.169.12.5845-5847.1987

Ichikawa S, Ogawa S, Nishida A et al (2019) Cellulosomes localise on the surface of membrane vesicles from the cellulolytic bacterium *Clostridium thermocellum*. FEMS Microbiol Lett 366(12). https://doi.org/10.1093/femsle/fnz145

Ichikawa S, Tsuge Y, Karita S (2021) Metabolome analysis of constituents in membrane vesicles for *Clostridium thermocellum* growth stimulation. Microorganisms 9(3):593
Jeong D, Kim MJ, Park Y et al (2022) Visualizing extracellular vesicle biogenesis in gram-positive bacteria using super-resolution microscopy. BMC Biol 20(1): 270. https://doi.org/10.1186/s12915-022-01472-3
Jiang Y, Kong Q, Roland KL et al (2014) Membrane vesicles of *Clostridium perfringens* type A strains induce innate and adaptive immunity. Int J Med Microbiol 304(3):431–443. https://doi.org/10.1016/j.ijmm.2014.02.006
Kaparakis-Liaskos M, Kufer TA (2020) Bacterial membrane vesicles: biogenesis, functions and applications. Springer, Cham. https://doi.org/10.1007/978-3-030-36331-4
Khodadoost L, Hussain H, Mullany P (2017) Plasmids can transfer to *Clostridium difficile* CD37 and 630Δerm both by a DNase resistant conjugation-like mechanism and a DNase sensitive mechanism. FEMS Microbiol Lett 364(21). https://doi.org/10.1093/femsle/fnx208
Kim MH, Kim SY, Son JH et al (2019) Production of membrane vesicles by *Enterococcus faecium* cultured with or without subinhibitory concentrations of antibiotics and their pathological effects on epithelial cells. Front Cell Infect Microbiol 9. https://doi.org/10.3389/fcimb.2019.00295
Kobayashi N, Abe K, Akagi S et al (2022) Membrane vesicles derived from *Clostridium botulinum* and related Clostridial species induce innate immune responses via MyD88/TRIF signaling in vitro. Front Microbiol 13:720308
Lawson PA, Citron DM, Tyrrell KL et al (2016) Reclassification of *Clostridium difficile* as *Clostridioides difficile* (Hall and O'Toole 1935) Prévot 1938. Anaerobe 40:95–99. https://doi.org/10.1016/j.anaerobe.2016.06.008
Lee EY, Choi DY, Kim DK et al (2009) Gram-positive bacteria produce membrane vesicles: proteomics-based characterization of *Staphylococcus aureus*-derived membrane vesicles. Proteomics 9(24):5425–5436. https://doi.org/10.1002/pmic.200900338
Lee J, Lee E-Y, Kim S-H et al (2013) *Staphylococcus aureus* extracellular vesicles carry biologically active β-lactamase. Antimicrob Agents Chemother 57(6): 2589–2595. https://doi.org/10.1128/AAC.00522-12
Liang L, Yang C, Liu L et al (2022) Commensal bacteria-derived extracellular vesicles suppress ulcerative colitis through regulating the macrophages polarization and remodeling the gut microbiota. Microb Cell Fact 21(1):88. https://doi.org/10.1186/s12934-022-01812-6
Liao S, Klein MI, Heim KP et al (2014) *Streptococcus mutans* extracellular DNA is upregulated during growth in biofilms, actively released via membrane vesicles, and influenced by components of the protein secretion machinery. J Bacteriol 196(13):2355–2366. https://doi.org/10.1128/jb.01493-14
Lopes AS, Boente RF, Silva RC et al (2019) Characterization of membrane vesicles of *Clostridioides difficile*. In: 30 degree Congresso Brasileiro de Microbiologia Brazil. Sociedade Brasileira de Microbiologia
Manning AJ, Kuehn MJ (2011) Contribution of bacterial outer membrane vesicles to innate bacterial defense. BMC Microbiol 11(1):258. https://doi.org/10.1186/1471-2180-11-258
Morishita M, Horita M, Higuchi A et al (2021) Characterizing different probiotic-derived extracellular vesicles as a novel adjuvant for immunotherapy. Mol Pharm 18(3):1080–1092. https://doi.org/10.1021/acs.molpharmaceut.0c01011
Nakao R, Kikushima K, Higuchi H et al (2014) A novel approach for purification and selective capture of membrane vesicles of the periodontopathic bacterium, *Porphyromonas gingivalis*: membrane vesicles bind to magnetic beads coated with epoxy groups in a noncovalent, species-specific manner. PLoS One 9(5):e95137. https://doi.org/10.1371/journal.pone.0095137
Nibbering B, Gerding DN, Kuijper EJ et al (2021) Host immune responses to *Clostridioides difficile*: toxins and beyond. Front Microbiol 12:804949
Nicholas A, Jeon H, Selasi GN et al (2017) *Clostridium difficile*-derived membrane vesicles induce the expression of pro-inflammatory cytokine genes and cytotoxicity in colonic epithelial cells in vitro. Microb Pathog 107:6–11. https://doi.org/10.1016/j.micpath.2017.03.006
Obana N, Nakao R, Nagayama K et al (2017) Immunoactive Clostridial membrane vesicle production is regulated by a sporulation factor. Infect Immun 85(5):e00096-00017. https://doi.org/10.1128/IAI.00096-17
Olling A, Seehase S, Minton NP et al (2012) Release of TcdA and TcdB from *Clostridium difficile* cdi 630 is not affected by functional inactivation of the *tcdE* gene. Microb Pathog 52(1):92–100. https://doi.org/10.1016/j.micpath.2011.10.009
Pettit LJ, Browne HP, Yu L et al (2014) Functional genomics reveals that *Clostridium difficile* Spo0A coordinates sporulation, virulence and metabolism. BMC Genomics 15(1):160. https://doi.org/10.1186/1471-2164-15-160
Razim A, Pacyga K, Naporowski P et al (2021) Identification of linear epitopes on the flagellar proteins of *Clostridioides difficile*. Sci Rep 11(1):9940. https://doi.org/10.1038/s41598-021-89488-7
Resch U, Tsatsaronis James A, Le Rhun A et al (2016) A two-component regulatory system impacts extracellular membrane-derived vesicle production in Group A Streptococcus. mBio 7(6):e00207-00216. https://doi.org/10.1128/mBio.00207-16
Reyes-Robles T, Dillard Rebecca S, Cairns Lynne S et al (2018) *Vibrio cholerae* outer membrane vesicles inhibit bacteriophage infection. J Bacteriol 200(15): e00792-00717. https://doi.org/10.1128/JB.00792-17

Rivera J, Cordero RJB, Nakouzi AS et al (2010) *Bacillus anthracis* produces membrane-derived vesicles containing biologically active toxins. Proc Natl Acad Sci USA 107(44):19002–19007. https://doi.org/10.1073/pnas.1008843107

Royer ALM, Umansky AA, Allen M-M et al (2022) The *Clostridioides difficile* S-layer protein A (SlpA) serves as a general phage receptor. bioRxiv. https://doi.org/10.1101/2022.09.19.508581

Rumbo C, Fernández-Moreira E, Merino M et al (2011) Horizontal transfer of the OXA-24 carbapenemase gene via outer membrane vesicles: a new mechanism of dissemination of carbapenem resistance genes in Acinetobacter baumannii. Antimicrob Agents Chemother 55(7):3084–3090. https://doi.org/10.1128/aac.00929-10

Schwechheimer C, Kuehn MJ (2015) Outer-membrane vesicles from Gram-negative bacteria: biogenesis and functions. Nat Rev Microbiol 13(10):605–619. https://doi.org/10.1038/nrmicro3525

Slater RT, Frost LR, Jossi SE et al (2019) *Clostridioides difficile* LuxS mediates inter-bacterial interactions within biofilms. Sci Rep 9(1):9903. https://doi.org/10.1038/s41598-019-46143-6

Steć A, Jońca J, Waleron K et al (2022) Quality control of bacterial extracellular vesicles with total protein content assay, nanoparticles tracking analysis, and capillary electrophoresis. Int J Mol Sci 23(8):4347

Stephan MS, Broeker NK, Saragliadis A et al (2020) *In vitro* analysis of O-antigen-specific bacteriophage P22 inactivation by Salmonella outer membrane vesicles. Front Microbiol 11. https://doi.org/10.3389/fmicb.2020.510638

Sur VP, Mazumdar A, Vimberg V et al (2022) Specific inhibition of VanZ-mediated resistance to lipoglycopeptide antibiotics. Int J Mol Sci 23(1):97

Szatanek R, Baj-Krzyworzeka M, Zimoch J et al (2017) The methods of choice for extracellular vesicles (EVs) characterization. Int J Mol Sci 18(6):1153

Thay B, Wai SN, Oscarsson J (2013) *Staphylococcus aureus* α-toxin-dependent induction of host cell death by membrane-derived vesicles. PLoS One 8(1): e54661. https://doi.org/10.1371/journal.pone.0054661

Théry C, Witwer KW, Aikawa E et al (2018) Minimal information for studies of extracellular vesicles 2018 (MISEV2018): a position statement of the International Society for Extracellular Vesicles and update of the MISEV2014 guidelines. J Extracell Vesicles 7(1): 1535750. https://doi.org/10.1080/20013078.2018.1535750

Toyofuku M, Cárcamo-Oyarce G, Yamamoto T et al (2017) Prophage-triggered membrane vesicle formation through peptidoglycan damage in *Bacillus subtilis*. Nat Commun 8(1):481

Tran F, Boedicker JQ (2017) Genetic cargo and bacterial species set the rate of vesicle-mediated horizontal gene transfer. Sci Rep 7(1):8813. https://doi.org/10.1038/s41598-017-07447-7

Turnbull L, Toyofuku M, Hynen AL et al (2016) Explosive cell lysis as a mechanism for the biogenesis of bacterial membrane vesicles and biofilms. Nat Commun 7(1):11220. https://doi.org/10.1038/ncomms11220

Tzipilevich E, Habusha M, Ben-Yehuda S (2017) Acquisition of phage sensitivity by bacteria through exchange of phage receptors. Cell 168(1–2):186–199.e112. https://doi.org/10.1016/j.cell.2016.12.003

Vergauwen G, Dhondt B, Van Deun J et al (2017) Confounding factors of ultrafiltration and protein analysis in extracellular vesicle research. Sci Rep 7(1): 2704. https://doi.org/10.1038/s41598-017-02599-y

Woods EC, Wetzel D, Mukerjee M et al (2018) Examination of the *Clostridioides (Clostridium) difficile* VanZ ortholog, CD1240. Anaerobe 53:108–115. https://doi.org/10.1016/j.anaerobe.2018.06.013

Wydau-Dematteis S, El Meouche I, Courtin P et al (2018) Cwp19 Is a novel lytic transglycosylase involved in stationary-phase autolysis resulting in toxin release in Clostridium difficile. mBio 9(3). https://doi.org/10.1128/mBio.00648-18

Zavan L, Bitto NJ, Johnston EL et al (2019) Helicobacter pylori growth stage determines the size, protein composition, and preferential cargo packaging of outer membrane vesicles. Proteomics 19(1–2):1800209. https://doi.org/10.1002/pmic.201800209

Non-human *Clostridioides difficile* Reservoirs and Sources: Animals, Food, Environment

Cristina Rodriguez-Diaz, Christian Seyboldt, and Maja Rupnik

Abstract

Clostridioides difficile is ubiquitous and is found in humans, animals and in variety of environments. The substantial overlap of ribotypes between all three main reservoirs suggests the extensive transmissions. Here we give the overview of European studies investigating farm, companion and wild animals, food and environments including water, soil, sediment, wastewater treatment plants, biogas plants, air, and households. Studies in Europe are more numerous especially in last couple of years, but are still fragmented in terms of countries, animal species, or type of environment covered. Soil seem to be the habitat of divergent unusual lineages of *C. difficile*. But the most important aspect of animals and environment is their role in *C. difficile* transmissions and their potential as a source for human infection is discussed.

C. Rodriguez-Diaz (✉)
Instituto de Investigación Biomédica de Málaga y Plataforma de Nanomedicina-IBIMA Plataforma BIONAND, UGC de Aparato Digestivo, Hospital Universitario Virgen de la Victoria, Málaga, Spain

Laboratory of Food Microbiology, Fundamental and Applied Research for Animals and Health (FARAH), Department of Food Sciences, Faculty of Veterinary Medicine, University of Liège, Liège, Belgium

C. Seyboldt
Institute of Bacterial Infections and Zoonoses, Friedrich-Loeffler-Institut, Federal Research Institute for Animal Health, Jena, Germany

M. Rupnik
National Laboratory for Health, Environment and Food, NLZOH, Maribor, Slovenia

University of Maribor, Faculty of Medicine, Maribor, Slovenia

1 Introduction

Clostridioides (Clostridium) difficile is regarded mainly as an important human pathogen. Because it can colonize his natural niche, the gut, only in the absence of established gut microbiota, it seem that his natural multiplying hosts are young animals and children. As an anaerobic spore-forming bacterium, it will be transmitted from the gut into different environments. *C. difficile* is hence ubiquitous and can be found in humans, animals, and the environment with a great variety of transmission routes between them.

Several reviews suggest a common reservoir of the bacterium in the environment, food, and animals. In addition, the latest genomic sequencing techniques have revealed cross-transmission of *C. difficile* between animals and humans (Rodriguez et al. 2016; Rupnik 2007, 2010; Weese 2010; Otten et al. 2010; Hensgens et al. 2012; Rodriguez-Palacios et al. 2013; Warriner et al. 2016; Lim et al. 2020; Rivas et al. 2020; Weese 2020). Here we give the overview of studies performed to date in Europe.

P. Mastrantonio, M. Rupnik (eds.), *Updates on Clostridioides difficile in Europe*, Advances in Experimental Medicine and Biology 1435, https://doi.org/10.1007/978-3-031-42108-2_15

2 *C. difficile* in Farm Animals: European Studies

Looking back to the early research on *C. difficile*, the presence of these bacteria in farm animals first gained attention in the 1970s. The first reference in the literature describing *C. difficile* in farm animals (rabbit, horse, and cow dung) and in the environment (hay, sand, and river mud) in Europe dates from 1974 (Hafiz 1974). Thereafter, other authors in different European geographic areas also confirmed the presence of *C. difficile* and infection in hares (France) (Dabard et al. 1979), pigs (UK) (Lysons et al. 1980; Jones and Hunter 1983), goats (UK) (Hunter et al. 1981; Borriello et al. 1983), ducks, geese, rabbits, and chickens (UK) (Borriello et al. 1983). The first report of *C. difficile* in cattle in Europe was published in 2008 in which bacterial toxins were found in biological samples from calves (Pirs et al. 2008).

Over the last 20 years, several studies have investigated not only the presence and the prevalence of *C. difficile* in different farm animal species but also the pathogenic potential of the bacterium in these animals. In addition to the interest in *C. difficile* as an infectious agent in livestock animals and the economic losses that it can generate, the main objective of research groups worldwide has been to demonstrate the existence of an animal reservoir and to elucidate the relationships between potential reservoirs and *C. difficile* infection in humans, through the genetic similarities between strains. Hence, many studies also report the potential for zoonotic spread (Table 1).

2.1 *C. difficile* in Pigs and Cattle

Pigs are the farm animals that have been most commonly studied in Europe in the context of infection by *C. difficile*, followed by cattle (Fig. 1). In cattle, the described prevalence (up to 33%) is much lower than that in pigs (up to 96%) and studies have reported between 90 and 100% toxigenic strains circulating in both types of animal farms. In cattle, several studies have addressed the possibility of age and breeding effect on *C. difficile* colonization in animals and therefore different types of production systems have been investigated, including production farms, fattening farms, or dairy farms (Koene et al. 2012; Romano et al. 2012a; Zidaric et al. 2012; Rodriguez et al. 2017). A recent study also suggests that the presence of *C. difficile* PCR ribotype 033 on different farms studied may be a direct result of inter-farm trade of calves (Bandelj et al. 2018). However, in pigs, these possible differences between types of breed have not been addressed in the literature. Only two studies report the prevalence of *C. difficile* on free-range pigs, but the results of the study revealed the *C. difficile* prevalence in this population similar to the prevalence found in intensively raised animals (Álvarez-Pérez et al. 2013, 2018).

2.2 *C. difficile* in Other Less Commonly Studied Farm Animals in Europe

Poultry seem to be a natural host as colonized birds are asymptomatic, the prevalence in young animals is very high, and the diversity of ribotypes within a farm is very high. Still, not many studies in Europe have explored this species on farms. Also, goats and sheep were only recently studied in respect to *C. difficile*. A mean prevalence of 8.6% was reported in sheep, 5.8% in goats, and 33.1% in poultry (Table 1).

As interest has increased regarding the possible zoonotic transmission of *C. difficile* in recent years, new studies have investigated the prevalence and epidemiology of the bacterium in animal production types that are less commonly addressed than cattle, pigs, or poultry. An investigation conducted in Italy reported a *C. difficile* prevalence of 3% for rabbits raised in industrial holdings for food production (Drigo et al. 2015).

Table 1 Overview of recent European studies on *C. difficile* in animals

Species	References	Reported prevalence and the most prevalent ribotypes
Pigs	Pirs et al. (2008); Avbersek et al. (2009); Álvarez-Pérez et al. (2009); Indra et al. (2009); Hoffer et al. (2010); Hopman et al. (2011); Keessen et al. (2011b); Koene et al. (2012); Rodriguez et al. (2012); Álvarez-Pérez et al. (2013); Rodriguez et al. (2013); Schneeberg et al. (2013a); Noren et al. (2014); Stein et al. (2017); Krutova et al. (2018); Álvarez-Pérez et al. (2018); Barbanti and Spigaglia (2020)	22.6–96% (neonates) 0–36% (adults); 002, 005, 011, 014/020, 013, 015, 023, 029, 033, 035, 045, 046, 050, 066, 078, 126, 150, 193, 569
Cattle	Pirs et al. (2008); Avbersek et al. (2009); Hoffer et al. (2010); Koene et al. (2012); Rodriguez et al. (2012); Romano et al. (2012a); Zidaric et al. (2012); Rodriguez et al. (2013); Schneeberg et al. (2013a); Rodriguez et al. (2017); Bandelj et al. (2018); Romano et al. (2018); Barbanti and Spigaglia (2020); Marcos et al. (2021); Redding et al. (2021); Abay et al. (2022)	1.8–30.4% (neonates) 0–11% (adults) 002, 003, 012, 014, 015, 020, 029, 033, 038, 045, 066, 070, 077, 078, 081, 126, 137
Goat and ship	Koene et al. (2012); Romano et al. (2012a); Avbersek et al. (2014); Barbanti and Spigaglia (2020)	Goats 0–10.1% 001, 010, 014, 020, 045, 066 Sheep 0–18.2% 015, 056, 061, 097, 614
Poultry	Zidaric et al. (2008); Indra et al. (2009); Koene et al. (2012)	0–100% 001, 010, 014, 023, 446
Horses	Avbersek et al. (2009); Ossiprandi et al. (2010); Koene et al. (2012); Rodriguez et al. (2014a); Rodriguez et al. (2015); Kecerova et al. (2019); Schoster et al. (2019)	0–1.5% in healthy, non-hospitalized horses 3.7–33.3% 003, 005, 006, 009, 010, 012, 014, 023, 033, 035, 039, 042, 045, 046, 051, 078, 081, 126, AI-78, PR17515
Cats	Koene et al. (2012); Schneeberg et al. (2012); Álvarez-Pérez et al. (2017); Rabold et al. (2018); Alves et al. (2023)	0–16.4% 001, 009, 010, 014/020, 039, 045, 106
Dogs	Schneeberg et al. (2012); Koene et al. (2012); Wetterwik et al. (2013); Pirs et al. (2013); Álvarez-Pérez et al. (2015, 2017); Orden et al. (2017a); Spigaglia et al. (2015); Rabold et al. (2018); Janezic et al. (2018); Andrés-Lasheras et al. (2018); Rodriguez et al. (2019a); Barbanti and Spigaglia (2020); Tramuta et al. (2021); Albuquerque et al. (2021); Bjöersdorff et al. (2021); Rodríguez-Pallares et al. (2022); Finsterwalder et al. (2022); Alves et al. (2023)	0–100% (neonates) 3.4–26% (adults) 009, 010, 012, 014, 015, 018, 014/020, 020, 023, 026, 027, 031, 033, 039, 045, 056, 078, 106, 107, 123, 154, 213, 358, 430, 449, 739, 106, 107, 154, 213, 430
Rabbits (farm)	Drigo et al. (2015); Barbanti and Spigaglia (2020)	3% 002, 003, 012, 014, 017, 020, 078, 084, 205, 569, 592
Wild animals	Burt et al. (2012); Bandelj et al. (2016); Andrés-Lasheras et al. (2017); Burt et al. (2018); Krijger et al. (2019); Darwich et al. (2021); Zlender et al. (2022)	0–100% 010, 002, 005, 013, 014/020, 015, 029, 035, 056, 057, 058, 073, 078, 033, 045, 062, 087, 126, 258, 454

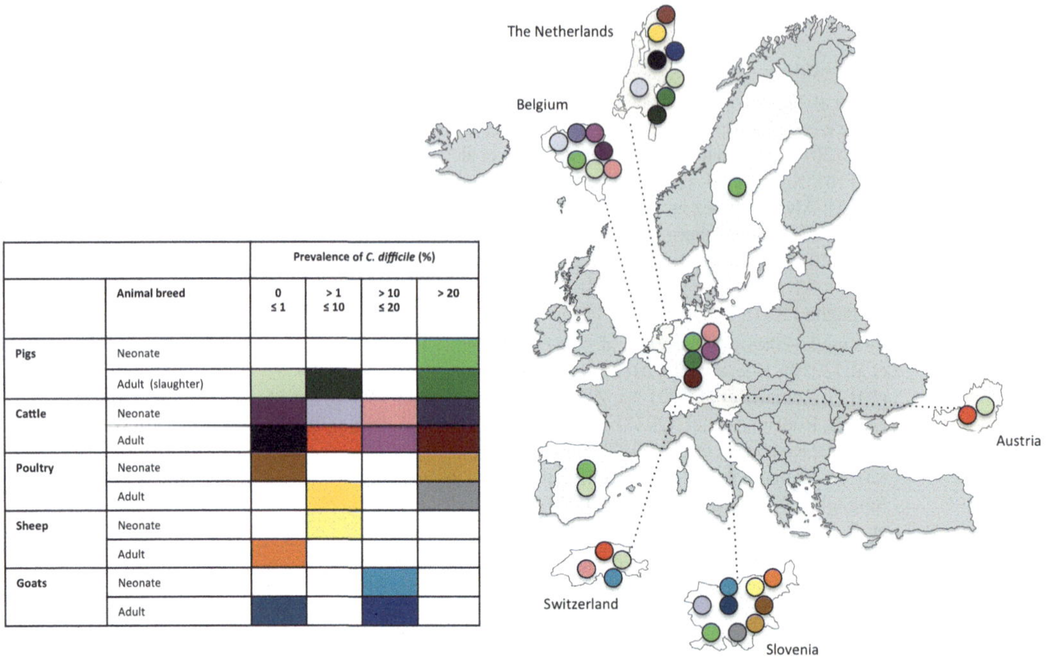

Fig. 1 Prevalence of *C. difficile* in farm animals in Europe

2.3 Factors Associated with *C. difficile* Colonization in Farm Animals

Several factors, including animal species, age, microbiota, breeding effect, and seasonality have been associated with *C. difficile* colonization in farm animals (Fig. 2) and likely apply also for other animals. It is possible that *C. difficile* is better adapted to some animal hosts than to others. The reported prevalence varies strongly between different species and studies (Rodriguez et al. 2016; Table 1). Also, laboratory diagnosis of *C. difficile* infection in animals and the performance of commercially available methods may vary depending on the animal species (Carvalho et al. 2022).

Age is the best studied among factors associated with *C. difficile* carriage in farm animals. All of the studies conducted in various European countries (Álvarez-Pérez et al. 2009; Schneeberg et al. 2013a; Bandelj et al. 2018) have shown high colonization rates in newborn animals that are either considerably reduced or eliminated in adult animals. In pig production, a *C. difficile* prevalence of 77% of piglet litter samples and 21% of sow samples was reported (Stein et al. 2017). This reduction in infection prevalence with age has two important consequences. First, the risk of foodborne transmission from contaminated animal products during harvest is greatly reduced. Second, Clostridioides difficile infection (CDI) in adult animals is very rare; therefore, *C. difficile* is currently not considered a common health problem in adult farm animals.

Regarding gut microbiota composition, in Europe, some studies have evaluated changes in the intestinal microbiota with *C. difficile* colonization in poultry (Skraban et al. 2013), calves (Redding et al. 2021), and pigs (Proctor et al. 2021). In poultry, differences in the presence of *Enterococcus cecorum*, *Lactobacillus gallinarum*, *Moniliella* sp., and *Trichosporon asahii* were detected among *C. difficile*-positive and *C. difficile*-negative animals. Interestingly, *Acidaminococcus intestini*, identified for the first time as a part of the poultry intestinal microbiota

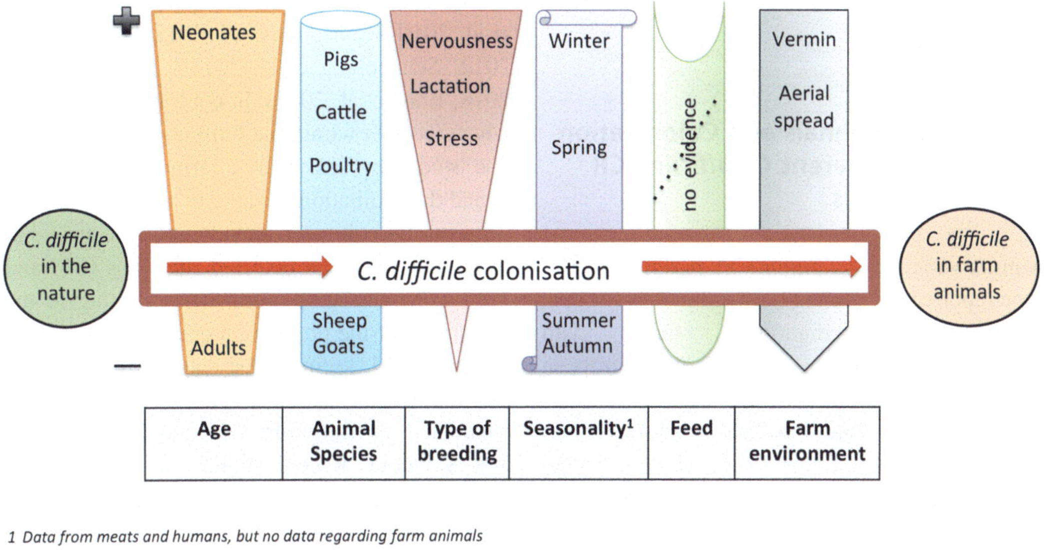

Fig. 2 Factors associated with the presence of *C. difficile* in livestock animals in Europe

in this study, was detected in high abundance in animals not colonized by *C. difficile*. In dairy calves, positive animals showed increased levels of *Ruminococcus*, *Lachnoclostridium*, *Butyricicoccus*, and *Clostridium sensu stricto 2* compared to *C. difficile*-negative animals. In pigs, the *Bacteroides*, *Fusobacterium*, *Enterobacteriaceae*, and *Sutterella* groups were dominant in younger animals, and their abundance decreased with age. *Prevotella* was the dominant group in older piglets, which is negatively associated with the abundance of *C. difficile* in young piglets. Further studies may lead to the identification of several bacterial populations that can potentially protect hosts from CDI.

2.4 Infection vs. Carriage of *C. difficile* in Farm Animals

In farms, *C. difficile* shows a similar prevalence among animals with or without diarrhoea (Pirs et al. 2008; Álvarez-Pérez et al. 2009; Koene et al. 2012; Schneeberg et al. 2013a; Rodriguez et al. 2017; Stein et al. 2017; Bandelj et al. 2018; Mertens et al. 2022), which may indicate that the bacterium is not the main causal agent of disease, but instead, an opportunistic pathogen that worsens the clinical status and outcome of affected animals. In a recent study in Spain, more than 80% of faecal samples obtained from diarrhoeic piglets showed mixed infections, including *Clostridium perfringens* (*C. perfringens*), *C. difficile*, species A rotavirus, species C rotavirus, and porcine epidemic diarrhoea virus (Monteagudo et al. 2022). In piglets, *C. difficile* causes important economic losses in farms due to both diarrhoea and premature death as well as delays in growth and reduced weight gain (Songer 2000; Squire and Riley 2013). There are a few reports of *C. difficile* infection in pigs in Europe, including one study that reported an outbreak in periparturient sows in a large outdoor production unit in Croatia (Kiss and Bilkei 2005) and one case-report study of typhlocolitis and diarrhoea in piglets in Ireland (McElroy et al. 2016). In calves and poultry, *C. difficile* has also been proposed as a possible cause of diarrhoea, enteritis, and death (Hammitt et al. 2008; Cooper et al. 2013), although there is no evidence of outbreaks due to the bacterium in these animal species. A review of these data indicates that the incidence, clinical relevance, and pathogenesis of

CDI in farm animals in Europe have not yet been elucidated.

2.5 Farm Animals and Colonization with Different *C. difficile* PCR Ribotypes

A great variety of *C. difficile* PCR ribotypes has been reported in different farm animals in Europe. Comparative international study with 12 participating European and non-European countries that included 112 strains from 13 species including farm animals has distributed strains into 50 PCR ribotypes. Some ribotypes were found across all tested species (014, 078) while some others are more likely to be associated with a given animal species (033 with cattle) (Janezic et al. 2012).

An interesting aspect is also ribotype variability within the farm. At pig farms a single PCR ribotype will be present. In cattle the variability will be greater although the number of detected types is still modest. In contrast, in poultry and rabbit farms the reported variability is very high and from 12 to 16 PCR ribotypes are found per single farm (Zidaric et al. 2008; Drigo et al. 2015).

PCR ribotype 078 is the only one that has been repeatedly reported in swine throughout different European countries and is described in several studies as the dominant type irrespective of age or diarrhoeal status (Koene et al. 2012; Rodriguez et al. 2012; Schneeberg et al. 2013a; McElroy et al. 2016; Stein et al. 2017; Krutova et al. 2018; Moloney et al. 2021). The remaining PCR ribotypes isolated from pig farms constitute a long list and include ribotypes 002, 011, 014, 015, 023, 033, 045, 126, 150, and 193; however, they have only been reported in specific studies (Avbersek et al. 2009; Hopman et al. 2011; Keessen et al. 2011b; Koene et al. 2012; Rodriguez et al. 2012; Schneeberg et al. 2013a; Noren et al. 2014; McElroy et al. 2016; Stein et al. 2017; Krutova et al. 2018).

In cattle, an even greater variety of PCR ribotypes has been isolated. PCR ribotype 078 has also been commonly detected in cattle farms in different countries in Europe (Hoffer et al. 2010; Rodriguez et al. 2012; Zidaric et al. 2012; Schneeberg et al. 2013b; Romano et al. 2018; Blasi et al. 2021). In contrast to pig farms, where isolates within the farm are clonal, at least one study on veal calves farm did not detect clonal dissemination (Zidaric et al. 2012). Calves were mostly colonized already upon the arrival to farm and two of all detected ribotypes (078 and 126) were persisting from the beginning to the last stages of the production cycle. Another PCR ribotype, 033, seems to be cattle-associated and has been described in five different studies conducted in Belgium, Germany, Switzerland, and Slovenia. Recent studies on family dairy farms revealed that the prevalence of *C. difficile* ribotype 033 increased linearly with the number of calves, with a close genetic relationship between farms (Bandelj et al. 2018), and that this ribotype together with ribotype 126 is more prevalent in cattle farms using digestate as a product of biogas plants (Masarikova et al. 2020). Other PCR ribotypes frequently associated with these animals are types 012 and 002, which were described in Belgium, the Netherlands, and Slovenia (Avbersek et al. 2009; Koene et al. 2012; Rodriguez et al. 2012; Zidaric et al. 2012). Other types like 015 and 020 were also isolated in specific studies (Rodriguez et al. 2017). The percentage of toxigenic strains in cattle varies between 70 and 100%, but no association between diarrhoeal status and colonization with specific PCR ribotypes has been established.

For other small ruminants such as goats and sheep, as well as poultry or rabbits, the presence of specific PCR ribotypes has not been widely described in part because there are only a few studies in Europe describing the presence of *C. difficile* in these animal species, and the few available studies describe a large variety composed of different types, and in other cases the studies have not carried out ribotyping characterization (Zidaric et al. 2008; Indra et al. 2009; Koene et al. 2012; Romano et al. 2012a; Avbersek et al. 2014; Candel-Pérez et al. 2021; Marcos et al. 2021) A recent study in Italy identified PCR ribotype 614 in sheep and various PCR ribotypes, such as 003, 014, and 078, among others, in rabbits (Barbanti and Spigaglia 2020) (Table 1).

2.6 Antimicrobial Susceptibility of *C. difficile* Isolates Isolated from Farm Animals

Drug resistance in *C. difficile* strains is usually associated with specific antibiotics, especially quinolones, erythromycin, and clindamycin, and with specific PCR ribotypes. In pig and cattle production, different studies have reported resistances to fluoroquinolones, ciprofloxacin, and erythromycin, especially among isolates of PCR ribotype 078 (Keessen et al. 2013; Pelaez et al. 2013), but also among PCR ribotypes 012 and 033 (Bandelj et al. 2017). Barbanti and Spigaglia (2020) reported the presence of multidrug resistant strains (to erythromycin, clindamycin, and moxifloxacin/rifampicin) in pigs and rabbits. In pork and cattle industry, the use of fluoroquinolones has also been related with the isolation of multiple antibiotic-resistant strains (Zidaric et al. 2012).

For *C. difficile* isolates from small ruminants, the limited available data in the literature reported antibiotic susceptibility to vancomycin, metronidazole, and moxifloxacin of all isolates obtained from goats and sheep and a possible relationship between PCR ribotype 045 and resistance to fluoroquinolones, beta-lactams, lincosamides, and macrolides (Avbersek et al. 2014).

Susceptibility to several other drugs, including antibiotics typically used for the treatment of CDI in humans like metronidazole, vancomycin or rifampicin, completely inhibited *C. difficile* growth (Pirs et al. 2013), which reflects no major differences in antibiotic susceptibilities between animal and human strains. In a previous study comparing human and animal isolates, the prevalence of multidrug resistant isolates, especially to erythromycin, clindamycin, and metronidazole, was found to be higher in clinical isolates (73%) than in animal isolates (30%). Resistance to erythromycin, clindamycin, or moxifloxacin was the most frequent among the animal isolates, while only 10% and 1.6% of these animal isolates showed resistance to metronidazole and rifampicin, respectively (Barbanti and Spigaglia 2020).

3 *C. difficile* in Companion Animals in Europe

Dogs and cats are the most studied companion animals. Taking the European studies involving dogs and cats together, the overall prevalence for *C. difficile* in cats is slightly lower than in dogs, but studies including cats are scarce.

In eight European studies including cats from veterinary clinics or shelters, the *C. difficile* prevalence ranged from 0 to 30% (2%, Al Saif and Brazier 1996; 15.7%, Koene et al. 2012; 3.7%, Schneeberg et al. 2012; 8%, Weber et al. 1989; 2.5%, Rabold et al. 2018; 16.4%, Alves et al. 2023) (Table 1). Both studies marking the prevalence borders included only a small number of 37 and 20 cats, respectively (Álvarez-Pérez et al. 2017 ; Borriello et al. 1983). A larger study on cats living in households yielded a prevalence of 2.5% (10 of 403) while another study in a more clinical setting yielded a prevalence of 16.4% (23 of 140) (Rabold et al. 2018; Alves et al. 2023).

More information is available in respect to dogs in Europe. The reported prevalence rates in the different studies range from 1.45% in dogs of a control group (1 of 74) up to 100% in puppies of one litter at certain time-points (Perrin et al. 1993; Álvarez-Pérez et al. 2015). Other reports describe *C. difficile* carriage rates of 3.4%–26% for dogs in different study settings (Table 1). A Germany study investigated 437 dogs in household settings and detected a carriage rate of 3.4% (15 of 437) (Rabold et al. 2018). A positivity rate of the same range 4.9% (11 of 225) was reported from Denmark where dog faecal deposits in public gardens were collected (Bjöersdorff et al. 2021). A Portuguese study with sampling from veterinary clinics and collected laboratory samples reported a prevalence of 26% (87 of 335) (Alves et al. 2023). A canine case-control study at a referral veterinary hospital in Scotland revealed

18.7% (61 of 327) (Albuquerque et al. 2021). Interestingly not only faecal samples were investigated; 24% (6 of 25) dog paws in household setting in Slovenia (Janezic et al. 2018) and nasal discharge from 4 (19%) dogs in Belgium (Rodriguez et al. 2019a) were positive for *C. difficile* reflecting the extraintestinal and environmental presence.

15 European studies reported PCR ribotypes in dogs and only five considered cats. Ribotypes 009, 010, 014/020, 039, and 106 are common in dogs and cats across Europe. The most frequently reported ribotypes in cats are 010, 039 or 039/2, 014 or 014/020 and 106 (Koene et al. 2012; Schneeberg et al. 2012; Álvarez-Pérez et al. 2017; Rabold et al. 2018; Alves et al. 2023). The most frequently described ribotypes in dogs are 009, 010, 012, 014, 014/020, 020, 023, 039, 056, 078, 106 (Table 1).

Factors most likely associated with *C. difficile* colonization in dogs and cats are age, enteric disease, antibiotic treatment, and hospitalization.

A plausible association of age and carriage rate in dogs (puppies and older animals) was reported. In puppies high prevalence up to 100% was noted in the time from 2 to 6 weeks after birth. The carriage rate in puppies markedly decreased with age and reached 3.1 and 0% at the end of the observation time (Perrin et al. 1993; Álvarez-Pérez et al. 2015). Additionally, Álvarez-Pérez et al. (2017) reported that carriage was significantly linked with age over 7 years investigating 105 dogs from 17 veterinary clinics. Rabold et al. (2018) recognized an association of *C. difficile* detection and treatment with antibiotics or proton pump inhibitors in small companion animals. Additionally, dogs and cats tended to be *C. difficile*-positive more often when the owner suffered from a chronic disease or diarrhoea (Rabold et al. 2018). A study conducted at a referral veterinary hospital in Scotland also found antibiotic treatment to be a risk factor for *C. difficile* carriage increasing with the length of treatment (Albuquerque et al. 2021), while other investigations could not find an association with antibiotic administration (Finsterwalder et al. 2022; Alves et al. 2023).

Despite some case reports of *C. difficile* infection in dogs and cats, an association with diarrhoea was not obvious in a number of studies. Regarding the available data from Europe, it seems that *C. difficile* does not cause disease in dogs and cats beyond single cases as similar percentages are isolated from symptomatic and healthy animals and no statistical correlation was detectable (Weber et al. 1989; Wetterwik et al. 2013; Duijvestijn et al. 2016; Albuquerque et al. 2021; Finsterwalder et al. 2022; Alves et al. 2023). Interestingly some studies with sampling scenarios involving veterinary clinics or hospitals showed higher prevalence (Albuquerque et al. 2021; Finsterwalder et al. 2022; Alves et al. 2023) than household or public park sampling scenarios (Rabold et al. 2018; Bjöersdorff et al. 2021). However, dogs and cats can harbour *C. difficile* strains with virulence potential (Table 1) and with exception of the longitudinal studies conducted in puppies the duration of *C. difficile* shedding was scarcely addressed. It is not clear whether a *C. difficile* carriage can be a result of a longer lasting colonization or is just connected with a short transient passage. Recently interspecies transmission of toxigenic *C. difficile* was reported involving a 10-month-old infant and the family dog, both with diarrhoea and without other diagnosis. The dog was reported with recurrent diarrhoea indicating a longer lasting carriage or infection (Rodríguez-Pallares et al. 2022).

In respect to antibiotic resistance, metronidazole-resistant *C. difficile* strains were isolated from dogs with recorded application of metronidazole (Wetterwik et al. 2013; Orden et al. 2017a) or suspected metronidazole treatment as it is commonly used for *Giardia* spp. infections in Italian dogs (Spigaglia et al. 2015). Metronidazole resistant isolates were also observed in Austria, Italy, Spain, and Portugal (Andrés-Lasheras et al. 2018; Barbanti and Spigaglia 2020; Finsterwalder et al. 2022; Alves et al. 2023). Recently, research on metronidazole resistance discovered a plasmid-mediated metronidazole resistance in European RT010 from humans and animals and RT020 strains from humans.

Resistance to clindamycin, erythromycin, and moxifloxacin is frequently detected while tetracycline and rifampicin resistance is rarely reported. Multidrug resistant isolates (MDR) isolates are not very frequent but geographically widespread, the resistance pattern clindamycin, erythromycin, and metronidazole was repeatedly noticed in dogs (Andrés-Lasheras et al. 2018; Barbanti and Spigaglia 2020; Bjöersdorff et al. 2021; Finsterwalder et al. 2022; Alves et al. 2023).

4 *C. difficile* in Horses in Europe

In contrast to other companion animals, horses are reported to develop *C. difficile* enteric disease. Foals and adult horses could be affected and outbreaks as well as sporadic cases were described. Antibiotic treatment and hospitalization have been depicted as important risk factors. *C. difficile* rates in horses with enteric disease were 5–63% in different studies. Healthy horses may harbour *C. difficile* as well; reported prevalence was ranging between 0 and 10% (reviewed in Diab et al. 2013). More recent European studies reported 0 and 1.5% in healthy and non-hospitalized horses, respectively (Kecerova et al. 2019; Schoster et al. 2019). Horses with colic and horses with diarrhoea had prevalence rates of 19% (cumulative, in three samplings) and 6.6%, respectively (Schoster et al. 2019). In a group of hospitalized horses, prevalence was 21.3% (Kecerova et al. 2019). A Swedish study found higher carriage rates of 29% in healthy foals younger than 14 days. Additionally, soil samples from stud farms contained *C. difficile* more frequent than soil samples from farms with mature horses. It was concluded that strains from the environment and healthy foals can serve as reservoir (Baverud et al. 2003). European studies report *C. difficile* in horses from Czechia, Switzerland, Slovenia, Italy, the Netherlands, and Belgium with carriage rates from 0 to 33.3% (Table 1) showing a remarkably high diversity of detected ribotypes (Avbersek et al. 2009; Koene et al. 2012; Ossiprandi et al. 2010; Rodriguez et al. 2014a, 2015; Kecerova et al. 2019; Schoster et al. 2019). Only three of these studies contain information on antibiotic resistance. In the first study conducted in Sweden, the resistance of 52 strains isolated from horses and their close environments was investigated for 10 different antibiotics. All of these strains were resistant to trimethoprim/sulphamethoxazole and bacitracin, but susceptible to metronidazole and fusidic acid. A total of 14 *C. difficile* strains, all of them isolated from hospitalized horses, were resistant to erythromycin and rifampicin (Baverud et al. 2003). As all of these strains were isolated from horses previously treated with erythromycin alone or in combination with rifampicin, authors suggest that erythromycin treatment probably selects the spread of this resistant pattern (Baverud et al. 2004). In a further study conducted in Belgium, antibiotic resistance was tested from ten strains isolated from hospitalized horses. All isolates displayed resistance to clindamycin and ceftiofur. Ceftiofur is one of the most commonly used antibiotics in the equine clinic (Rodriguez et al. 2014a). A Czech study investigated 18 isolates, whereof all were resistant to enrofloxacin, eight were resistant to tetracycline, five to clindamycin, and one to erythromycin and clindamycin (Kecerova et al. 2019).

5 *C. difficile* in Wild Animals in Europe

Limited data are available in Europe regarding the presence of *C. difficile* in wild animals outside of their direct or indirect relationships with livestock. In Slovenia, a study found *C. difficile* in barn swallows in an area identified as a barn swallow congregation point during the autumn migration of the species across Europe. The authors found an overall prevalence of 4% (4.6% (7/152) in juvenile birds and 0/23 in adults). PCR ribotypes 078, 002, and 014 were identified among a large variety of new types. The conclusions of this study focus on the possible role of barn swallows in the national and international dissemination of the bacterium (Bandelj et al. 2014). Another study also conducted in Slovenia investigated the carriage of *C. difficile* in migrating passerine birds by sampling cloacal specimens from animals during migration (Bandelj et al. 2011). However, in this study, none of the samples yielded a positive result for the presence of the bacterium. In the same country, a recent study described a *C. difficile* prevalence of 18% (4/22) in captative wild animals, including Eurasian collared dove, Tawny owl, Eurasian eagle-owl, and black stork (Zlender et al. 2022).

In Spain, the faecal shedding of *C. difficile* by 40 zoo animal species was investigated (Álvarez-Pérez et al. 2014). The bacterium was found with an infection prevalence of 3.5% in samples from the chimpanzee (*Pan troglodytes troglodytes*), dwarf goat (*Capra hircus*), Iberian ibex (*Capra pyrenaica hispanica*), and plains zebra. All isolates displayed resistance to the fluoroquinolones ciprofloxacin, enrofloxacin, and levofloxacin and belonged to PCR ribotypes 078, 039, and 110. The distribution of these PCR ribotypes typically found in farm or companion animals and humans may be explained by the close contact of zoo animals with humans and their environment as well as by continuous contact between these animals and droppings of other wild animals such as birds, which may aid in the dissemination of these common *C. difficile* strains. Also, in Spain, *C. difficile* was detected in two wild boars (prevalence of 1%) foraging in urban and peri-urban areas (Darwich et al. 2021).

In a clinical case study conducted in a zoo in Denmark, *C. difficile* was reported as a cause of Asian elephant enterocolitis. Molecular differences between the isolates obtained from three different elephants were not detected; thus, it was suggested that the same clone caused the outbreak. The origin of the contamination was not elucidated. The elephants were fed large quantities of broccoli, and authors hypothesized that sulforaphane, which is present in this vegetable, could have caused dysbiosis and subsequently led to CDI (Bojesen et al. 2006). However, because the same clone was present in all of the affected elephants, it is also possible that the broccoli itself was contaminated with toxigenic *C. difficile*; therefore, the broccoli could have been the source of contamination.

C. difficile was also investigated in zooplankton populations and associated environments at five sampling stations in the Gulf of Naples, Italy. The bacterium was detected in zooplankton samples but not in marine sediments. Many types were characterized including PCR ribotypes 009 and 066. These results demonstrated for the first time that *C. difficile* is also well adapted to aquatic marine populations that were not previously studied, which suggests that the bacterium could be transmitted through the ingestion of raw or undercooked seafood (Pasquale et al. 2011).

6 Transmissions Between Animals and Environment

Clostridium difficile colonizes the intestinal tract of animals, which then excrete the bacterial spores in the faeces. In this way, animals can serve as source of environmental contamination or as vectors in direct and indirect transmission. Environmental contamination will include manure and farm waste recycling (as fertilizers or biogas substrates), soil contamination (pastures), water contamination, or aerial contamination and some examples will be described in Sect. 7.

To assess the direct or indirect transmission of *C. difficile* by vermin in pig farms, samples of house mice, drain flies, lesser houseflies, yellow mealworms, house sparrows, and bird droppings were investigated. *C. difficile* prevalence ranging between 4 and 100% was reported, and PCR ribotype 078 was identified in each type of sampling. The authors concluded that vermin could be important sources of *C. difficile* contamination in farms (Burt et al. 2012). Similarly, a recent study conducted in north-eastern Spain reported the presence of *C. difficile* in pest species including rodents and pigeons in pig farms and the associated environment. Most of the characterized isolates were identified as the susceptible metronidazole and vancomycin strains, PCR ribotypes 078 and 126, which were also isolated from pigs. This study also confirmed the cross-transmission of bacterium between wild animals and production animals in farms, although the impact of this phenomenon on the epidemiology of *C. difficile* was not well established (Andrés-Lasheras et al. 2017). *C. difficile* was also detected in flies at dairy farms (Bandelj et al. 2016). In the Netherlands, a recent study reported the presence of *C. difficile* in rodents and insectivores in 3.2% of 347 animals tested, with a total of 13 different PCR ribotypes identified (Krijger et al. 2019). Another study also conducted in the Netherlands reported that house mice carried *C. difficile* with a prevalence of 35%. The authors also found that more than one third of the positive mice were colonized with *C. difficile* ribotypes associated with human infection (Burt et al. 2018).

In respect of dogs and cats and their role in transmission of *C. difficile* between companion animals and environment in Europe, nearly nothing is known, but two studies comprise interesting information. Occurrence of the same strain (Multi-locus variable number tandem repeat analysis (MLVA) and ribotype) in dogs and a cat indicating direct or indirect transmission was described in animal shelters in Germany (Schneeberg et al. 2012). Orden et al. (2017b) investigated recreational sandboxes for children and dogs within the Madrid region (Spain). Two of the most frequent ribotypes (009 and 106) were also reported in independent study in Madrid dogs (Álvarez-Pérez et al. 2017). A recent study also investigated the prevalence of *C. difficile* on shoe soles of veterinarians, veterinary support staff, and veterinary students at the Veterinary Faculty Campus. The prevalence found ranged from 86.7% in samples from veterinarians and 100% in samples from support staff and students. PCR ribotype 010 was the most prevalent while other common types found were identified as ribotypes 010 and 014/020. In the study, the authors highlighted the role of students' shoes as potential vectors for the spread of the bacterium (Wojtacka et al. 2021).

7 *C. difficile* in Food in Europe

Foodborne zoonotic pathogens are transmitted via the consumption of contaminated food and drinking water. The possible foodborne transmission of *C. difficile* was reported for the first time in 1983 in Europe (Borriello et al. 1983). However, currently, the importance of *C. difficile* as a zoonotic disease remains largely unknown.

Food contamination routes can be various. Apparently healthy animals can carry *C. difficile* spores through the slaughter stage and introduce a potential risk of meat contamination during processing. Vegetables would be contaminated by manure spread or irrigation with contaminated water. Root vegetables could carry *C. difficile* spores often present in soil irrespective of fertilizing.

7.1 Detection of Contaminated Meats in Retail Markets

The evidence that carcass contamination occurs inside the slaughterhouse reinforces the hypothesis of the potential risk of foodborne infections linked to the ingestion of foods contaminated with *C. difficile* spores. A recent study in Turkey reported a high prevalence of the bacterium in cattle (33.6% (83/247)) and sheep (25.3% (78/308)) carcass samples (Hampikyan et al. 2018). In Europe, meats have been found contaminated with *C. difficile* with a frequency ranging from 2.3 to 7.5%, and the main PCR ribotypes identified were 078, 001, 012, 014, 015, 045, 053, 078, and 087 (Bouttier et al. 2010; Jobstl et al. 2010; De Boer et al. 2009; Rodriguez et al. 2014b; Tkalec et al. 2020) (Table 2). Nevertheless, other surveys have failed to find *C. difficile* in meat samples (Indra et al. 2009; Hoffer et al. 2010; De Boer et al. 2009). Some recent studies have isolated the bacterium in edible chicken giblets, gizzard samples, liver, and other chicken meats at slaughterhouse (Candel-Pérez et al. 2021). Similarly, a national food surveillance for *C. difficile* in Slovenia detected the presence of the bacteria in beef, pork, and poultry, with a prevalence ranging from 3.8 to 5% (Tkalec et al. 2020). The reason for the lower variety of PCR ribotypes in meat samples is not clear considering the high variety of types found in farm animal faecal samples. One

Table 2 Overview of recent European studies on *C. difficile* in foods

Food	References	Reported prevalence and detected ribotypes
Meats	Indra et al. (2009); Von Abercron et al. (2009); Bouttier et al. (2010); De Boer et al. (2009); Hoffer et al. (2010); Jobstl et al. (2010); Rodriguez et al. (2014b); Tkalec et al. (2020); Candel-Pérez et al. (2021); Heise et al. (2021)	0–15.8% 001, 002, 003, 005, 012, 014/020, 045, 053, 071, 078, 087
Seafood	Pasquale et al. (2011, 2012); Agnoletti et al. (2019); Tkalec et al. (2020)	5.9–75% 001, 002, 003, 005, 010, 012, 014, 018, 020, 045, 046, 049, 066, 070, 078, 081, 087, 106, 220, 404, 422, 449, 569, 614, 651
Vegetables	Eckert et al. (2013); Tkalec et al. (2019, 2020, 2022); Scholtzek et al. (2022)	1.9–26.7% 001/072, 002, 003, 005, 009, 010, 011/049, 012, 014/020, 015, 018, 023, 024, 027, 029, 032, 053, 056, 070, 077, 078, 081, 085, 106, 126, 127, 128, 131, 150, 174, 204, 207, 244, 255, 276, 394, 500, 625, 864, 912, 913, 914, 915, 916, 917, 918, 919

possible explanation is that there are differences in the sporulation frequencies and susceptibilities to external agents among the different PCR ribotypes (Zidaric et al. 2012). This feature may contribute to the survival of only some PCR ribotypes to the final stages of the meat supply chain (i.e. distribution in retail markets). Furthermore, it is noteworthy that animals may not be the sole origin of *C. difficile* contamination via meat and that other sources could involve contamination during processing or in retail markets.

7.2 *C. difficile* in Foods Other than Meats in Europe

In Europe, only a couple of studies have addressed the presence of *C. difficile* in foods other than meat, such as seafood and vegetables. The prevalence reported for seafood ranges from 5.9% to more than 50% of samples showing positive results (Pasquale et al. 2011; Pasquale et al. 2012; Agnoletti et al. 2019; Tkalec et al. 2020); while the prevalence described for vegetables is slightly lower, ranging between 1.9 and 26.7% (Eckert et al. 2013; Tkalec et al. 2019, 2020; Scholtzek et al. 2022). A recent study in Slovenia points to potatoes as the vegetable most frequently contaminated by *C. difficile* (prevalence of 28%), followed by ginger (prevalence of 6.7%) and leaf vegetables (prevalence of 9.4%) (Tkalec et al. 2019). Also, in Germany, *C. difficile* was found in potatoes and salads with a prevalence of 26.7% and 1.9%, respectively (Scholtzek et al. 2022). A large study on *C. difficile* in potatoes in 12 European countries found a prevalence of 22.4% (33/147) and identified a total of 38 different ribotypes (Tkalec et al. 2022). Furthermore, several PCR ribotypes have been detected in these types of samples including PCR ribotypes 011/049, 014/020, 078, 001, and 015, among others, and most of these PCR ribotypes have also been associated with CDI in humans in European hospitals (Bauer et al. 2011; Agnoletti et al. 2019).

8 Studies on *C. difficile* in Environment in European Countries

Although the first large study including samples from non-hospital environment was done in Europe, the reports on *C. difficile* in environmental sources in European countries were scarce. However, in recent 5 years, the number of environmental studies increased and they often include also comparisons with animal or clinically relevant strains on genomic level (Table 3). Tested environments include water, soil, wastewater treatment plants (WWTP), biogas plants, air, sediment, manure, silage/hay,

Table 3 Overview of studies on *C. difficile* in environment in different European countries

Environmental sample type	Country	Positivity rate	CFU (if available)	Strain characterization	Reference
WWTP—inlet, sewage, effluent	Italy	Positivity <100%		–	Romanazzi et al. (2016)
WWTP—inlet and effluent	Switzerland	18/18		RT	Romano et al. (2012b)
WWTP—inflow	Germany	Unspecified		RT, WGS	Numberger et al. (2019)
WWTP—effluent	Slovenia	12/12		RT	Steyer et al. (2015)
WWTP effluent	Czech Republic	2/2		RT, AMR, MLVA	Cizek et al. (2022)
WWTP—diverse	Germany	12/16; 75%		AMR	Blau and Gallert (2023)
WWTP	Finland	1/1		RT	Kotila et al. (2013)
WWTP	UK	20 WWTPs		WGS	Moradigaravand et al. (2018)
Water—swimming pool	UK	4/8; 25%	1–3 CFU/100 ml	RT[a]	Al Saif and Brazier (1996)
Water—seawater	Italy	2/5; 40%		RT	Pasquale et al. (2011)
Water—seawater	UK	7/15; 46.7%	3–6 CFU/100 ml	RT[a]	Al Saif and Brazier (1996)
Water—seawater	UK	0/4		RT, AMR	Hargreaves et al. (2013)
Water—river ($n = 4$)	UK	14/16; 87.5%	1–5 CFU/100 ml	RT[a]	Al Saif and Brazier (1996)
Water—river ($n = 2$)	Czech Republic	5/12; 41.7%		RT, AMR, MLVA	Cizek et al. (2022)
Water—river ($n = 25$)	Slovenia	42/69; 60.9%		RT	Zidaric et al. (2010)
Water—puddles	Slovenia	15/104; 14.4%		RT, AMR	Janezic et al. (2016)
Water—lake	UK	7/15; 46.7%	1–5 CFU/100 ml	RT[a]	Al Saif and Brazier (1996)
Water—lake	Czech Republic	1/2		RT, AMR, MLVA	Cizek et al. (2022)
Water—inland drainage	UK	7/26; 27%		RT[a]	Al Saif and Brazier (1996)
Water—foam	UK	1/1		RT, AMR	Hargreaves et al. (2013)
Water at farms	Ireland	5/30; 17% bovine 2/30; 7% ovine 9/30; 30% broiler		–	Marcos et al. (2021)
Water—drinking bowls at dairy farm	Slovenia	3/80; 3.75%			Bandelj et al. (2016)
Tap water	Finland	1 positive/unspecified total number	28 CFU/100 ml	RT	Kotila et al. (2013)
Tap water	UK	1/18; 5.5%	1–3 CFU/100 ml	RT[a]	Al Saif and Brazier (1996)
Surfaces at public places	Sweden	0/95		AMR	Baverud et al. (2003)
Soil—spinach fields	Ireland	6/60; 10%		RT, AMR, WGS	Marcos et al. (2022)
Soil at farms	Ireland	15/30; 50% bovine 12/30; 40% ovine 13/30: 43% broiler		–	Marcos et al. (2021)

(continued)

Table 3 (continued)

Environmental sample type	Country	Positivity rate	CFU (if available)	Strain characterization	Reference
Soil (seasonality)	Belgium	45/112; 40.2% high in winter		RT, AMR	Rodriguez et al. (2019b)
Soil (farms)	Slovenia	28/80; 35%		RT	Bandelj et al. (2016)
Soil—fertilized (long-term study)	Germany	8/8		(RT, AMR)[c], WGS	Frentrup et al. (2021)
Soil—domestic garden	Slovenia	3/10; 30%		RT	Janezic et al. (2020)
Soil	Slovenia	28/78; 36.7%		RT, AMR	Janezic et al. (2016)
Soil	UK	22/104; 21.2%		RT[a]	Al Saif and Brazier (1996)
Soil	Sweden	25/598, 4%		AMR	Baverud et al. (2003)
Soil	Germany	3/3		AMR	Blau and Gallert (2023)
Sediments estuarine in 2009	UK	11/18; 61.1% (2009) 13/21; 61.9% (2010)		RT, AMR	Hargreaves et al. (2013)
Sediments	Italy	0/5		na	Pasquale et al. (2011)
Sediment	Germany	1/1		RT, WGS	Numberger et al. (2019)
Sandboxes—for dogs or children	Spain	21/40; 52.5%		RT, AMR	Orden et al. (2017b)
Households	UK	550 samples; 2.2% positive		RT[a]	Al Saif and Brazier (1996)
Households	Slovenia	19/44; 43% shoes 6/21; 28% slippers		RT, WGS	Janezic et al. (2018)
Farm—silage/hay	Slovenia	3/80; 3.75%		RT	Bandelj et al. (2016)
Farm—manure; dairy farms	Slovenia	23/80; 28.7%		RT	Bandelj et al. (2016)
Farm—chicken manure	Germany	3/3		(RT,AMR)[c], WGS	Frentrup et al. (2021)
Environmental samples[b]	Italy	na		RT, MLVA	Romano et al. (2018)
Compost—organic garbage pile	Slovenia	1/1		RT, AMR	Janezic et al. (2016)
Compost	Slovenia	9/15; 60%		RT	Janezic et al. (2020)
Biogas plants ($n = 8$)	Germany	69/154; 44.8%		–	Froschle et al. (2015)
Air—farm associated	Netherlands	Inside pig farm Air at exhausters Air at 20 m distance 2/4 positive	2–625 CFU/m^3 6–120 CFU/m^3	RT	Keessen et al. (2011a, b)
Air—dust during manure application	Germany	1		(RT, AMR)[c], WGS	Frentrup et al. (2021)

WWTP waste water treatment plant, *ABR* antibiotic resistance, *RT* PCR-Ribotype, *WGS* Whole genome sequencing, *AMR* Antimicrobial resistance

[a]Typing published in separate publication (Al-Saif et al. 1998)

[b]Samples from previous studies (WWTP, sewage sludge, seawater, freshwater)

[c]RT reported based on WGS cluster previous associations with ribotypes; AMR not found in genome sequences

sandboxes, surfaces in public places, and households.

Unsurprisingly, WWTPs seem to be the environment with very high positivity rate and *C. difficile* is often detected in all tested samples either from inlet water, sewage, or effluent (Kotila et al. 2013; Steyer et al. 2015; Romano et al. 2012b; Moradigaravand et al. 2018; Cizek et al. 2022). A single study, using non-culturing method, reported positivity rate lower than 100% (Romanazzi et al. 2016). Another report from Germany also had positivity rate lower than 100% and in this case *C. difficile* was detected in all WWTPs associated samples except in effluent (Blau and Gallert 2023).

Rivers and sediments also have variable proportions of *C. difficile*-positive samples, from 41.7 to 87.5% in river samples and from none to 61.9% in sediment samples (Table 3) (Zidaric et al. 2010; Hargreaves et al. 2013; Numberger et al. 2019; Cizek et al. 2022).

Prevalence of *C. difficile* seems to be somewhat lower in soil. Most studies on different soil types (farm associated, domestic gardens, fields, populated areas) reported positivity rates between 30 and 50% (Janezic et al. 2016; Rodriguez et al. 2019b; Janezic et al. 2020; Marcos et al. 2021) but this can depend on soil type (Table 3). As an example, the overall prevalence in more than 500 soil samples in Sweden was 4%. While soil from public environments (parks, playgrounds, gardens, cultivated fields) showed the 4% positivity, samples from pastures and paddocks in stables with only mature horses were positive only in 1% and in stud farms at 11% (Baverud et al. 2003). Spores were detected significantly more often during winter soil sampling than during the summer sampling (Rodriguez et al. 2019b). Importantly, a long-term *C. difficile* persistence of almost 3 years in a single field after manure application was described (Frentrup et al. 2021).

Sandboxes, here specified as environments different than soil, showed slightly different positivity rate if they were used by children (9 positive of 20) or designated for dogs (12 positive of 20) (Orden et al. 2017b).

Another example of unequal distribution within the given environment are biogas plants. In Germany, eight plants with different substrate use (single predominate substrate which was either grass silage or cattle manure) were sampled (Froschle et al. 2015). *C. difficile* that was most frequently detected of all clostridia tested (44.8% of samples), followed by *C. novyi* (3.9% of samples); other tested species were not detected (*C. botulinum, C. chauvoei, C. haemolyticum, C. septicum*). Animal substrates were more likely to contain *C. difficile* than plant substrates (10/17; 58.8% vs. 2/44; 4.5%). Because all settings use mixed substrates (animal and plant, with predominance of one) the positivity of digested sludge was 22 of 42 samples (52.4%) and in digestion products 35 of 51 samples (68.6%).

Two European studies have detected *C. difficile* in air. A single study has investigated airborne spore transmission within and around a pig production farm with known high *C. difficile* prevalence (Keessen et al. 2011a). *C. difficile* was detected in all farm units except in the pregnant sow unit. The detected airborne *C. difficile* colony counts ranged from 2 to 625 CFU/m^3. At farrowing unit pens with piglets of different age were sampled and the *C. difficile* spores detected in the air decreased with piglet age being highest in pens with neonatal and up to 2 weeks old piglets. Air exhausts at roofs of four different units resulted in spore counts from 6 to 120 CFU/m^3, two of four air samples at 20 m distance downwind were positive while air samples up to 140 m distance were all negative. Frentrup et al. (2021) sampled the air during the manure application on the field and detected *C. difficile* at the distance of 20 m from the tractor, but not at 50 m or 100 m.

Strain typing was done in most of the studies (Table 3). Variety of detected ribotypes within a single environment is very large, but PCR ribotypes detected almost in every study were 014 and 010. Soil, in particular in rural but not urban areas, was shown to be natural environment for very distinctive and divergent lineages of *C. difficile* strains (Janezic et al. 2016). These divergent strains from cryptic clades CI-III most likely represent individual species (Knight et al. 2021). They can possess atypical toxin genes for toxin A or B and plasmid encoded binary toxin (Riedel et al. 2017; Ramírez-Vargas et al. 2018; Williamson et al. 2022). Occasionally they are detected also in patients (Janezic et al. 2015; Ducarmon et al. 2022).

Antibiotic resistance was tested in several studies (Table 3) and mainly to only few selected antibiotics. Environmental isolates are resistant to similar antibiotics as human isolates. Interestingly, nontoxic environmental strains could be more resistant than toxigenic environmental strains (Janezic et al. 2016).

9 Importance of Animals, Food, and Environment for Human Infection

The transmission of *C. difficile* from animal and environmental source occurs via the faecal-oral route through either direct or indirect contact with contaminated surfaces (e.g. water, foods, or faeces) or when spores are ingested. Furthermore, close contact with colonized animals may also be involved in the epidemiology of *C. difficile* in humans. Potential of airborne transmissions from farms and during manure application was shown (Keessen et al. 2011a; Frentrup et al. 2021). Another interesting option for spore transmissions between settings are shoes. In the households, a higher proportion of shoes in comparison to dog paws was positive on *C. difficile* spores (Janezic et al. 2018). Potato as one of the mostly eaten vegetable in Europe was shown to be often contaminated with *C. difficile* and is probably an example how spores are transmitted transnationally (Tkalec et al. 2020, 2022).

A certain proportion of *C. difficile* strains is very likely constantly transmitted between humans, animals, and the environment as partial overlap of ribotypes isolated from humans to those found in food, animals, or environment is well documented. A comparison of PCR ribotypes isolated in a single country during 3 year period from humans, animals, and environment showed that 11 of total 90 PCR ribotypes were shared between all three reservoirs (Janezic et al. 2012). Strains within a given ribotype still represent very heterogeneous group and whole genome sequence level is needed for identity confirmation. This was initially done in two studies, one on ribotype 078 strains in Netherlands and other on ribotype 014 strains in Australia (Knight et al. 2016; Knetsch et al. 2014). Although in both studies, identity between pig and human strains was proven, the proportion of such shared strains within the studied ribotype was very low. The recent *C. difficile* studies on animal and environmental strains often include also whole genome sequence comparisons and have confirmed also shared sequence types (STs) between humans, animals, and environment (Table 3).

To date, no direct infection originating from food, animal, or environmental source was described. Single study in Finland aimed at linking environmental samples from sewage and tap water to a large gastroenteritis outbreak associated with sewage contaminated drinking water (Kotila et al. 2013). Authors claimed to report for the first time that 'waterborne transmission of *C. difficile* spores was possible and a potential cause of CDI during outbreak'. However, only limited number of samples was obtained either from environment or from patients (9 strains from 19 CDI patients). Only one patient and one tap water isolate showed same PCR ribotype (014). As this is the one of the most prevalent PCR ribotypes in humans, some animals, and most environments, only whole genome sequencing could confirm the true association and identity of both strains.

Impact and prevention of *C. difficile* foodborne transmission is an emerging issue in *C. difficile* field. The verified presence of *C. difficile* in food begets the question about the risks for consumers. If the gut microbiota is normal, intestinal colonization may be transient (i.e. in the sense that shedding can result from short-term successful bacterial colonization or from intestinal passage of the ingested dormant spores) and can occur without associated pathology. Even if the spore numbers in foods are typically low, ingestion of a small dose in combination with an altered gut microbiota may be able to trigger infection.

The spores of *C. difficile* are heat resistant and can survive gentle cooking of foods (70 °C) but cannot survive the same range of high temperatures as the spores of other clostridial species (Rodriguez-Palacios and Lejeune 2011). Therefore, thermal treatment (85 °C for 10 min) may be the best strategy for reducing the risk of foodborne transmission. Furthermore, thermal treatment is an easy household practice that should be emphasized because it is also useful for eliminating other pathogens present in foods. Under this scenario, special attention must be given to the presence of *C. difficile* in raw foods consumed directly (e.g. raw meats or fish consumed without thermal treatment), biological products (e.g. fruits or vegetables, normally grown with the help of organic fertilizers), or traditional food products in developing countries which are sometimes prepared without the appropriate hygienic procedures. In these cases, the prevalence and counts of spores may have greater importance than is currently recognized and may present an important potential risk of foodborne infection, especially in populations with gastrointestinal perturbations.

Conclusions *C. difficile* reservoirs other than humans and hospitals are becoming increasingly recognized. Following the results of numerous studies in recent years on the niche and transmission of *C. difficile* between humans, animals, the environment and food, the bacterium is widespread in the environment, animals, and foods and should now be considered as a zoonotic pathogen. In addition, new genomic sequencing technologies have revealed the presence of clones or identical strains of *C. difficile* that cluster in the same lineage in the different niches discussed in this chapter. Therefore, a comprehensive 'One Health' approach is needed in future surveillance and control studies of *C. difficile* infections.

References

Abay S, Ahmed EF, Aydin F et al (2022) Presence of *Clostridioides difficile* in cattle feces, carcasses, and slaughterhouses: molecular characterization and antibacterial susceptibility of the recovered isolates. Anaerobe 75:102575

Agnoletti F, Arcangeli G, Barbanti F et al (2019) Survey, characterization and antimicrobial susceptibility of *Clostridium difficile* from marine bivalve shellfish of North Adriatic Sea. Int J Food Microbiol 298:74–80

Albuquerque C, Pagnossin D, Landsgaard K, Simpson J, Brown D, Irvine J, Candlish D, Ridyard AE, Douce G, Millins C (2021) The duration of antibiotic treatment is associated with carriage of toxigenic and non-toxigenic strains of *Clostridioides difficile* in dogs. PLoS One 16(5):e0245949. https://doi.org/10.1371/journal.pone.0245949

Al Saif N, Brazier JS (1996) The distribution of *Clostridium difficile* in the environment of South Wales. J Med Microbiol 45:133–137

Al-Saif NM, O'Neill GL, Magee JT et al (1998) PCR-ribotyping and pyrolysis mass spectrometry fingerprinting of environmental and hospital isolates of *Clostridium difficile*. J Med Microbiol 47:117–1121

Álvarez-Pérez S, Blanco JL, Bouza E et al (2009) Prevalence of *Clostridium difficile* in diarrhoeic and non-diarrhoeic piglets. Vet Microbiol 137:302–305

Álvarez-Pérez S, Blanco JL, Pelaez T et al (2013) High prevalence of the epidemic *Clostridium difficile* PCR ribotype 078 in Iberian free-range pigs. Res Vet Sci 95:358–361

Álvarez-Pérez S, Blanco JL, Martinez-Nevado E et al (2014) Shedding of *Clostridium difficile* PCR-ribotype 078 by zoo animals, and report of an unstable metronidazole-resistant isolate from a zebra foal (Equus quagga burchellii). Vet Microbiol 169:218–222

Álvarez-Pérez S, Blanco JL, Peláez T et al (2015) Faecal shedding of antimicrobial-resistant *Clostridium difficile* strains by dogs. J Small Anim Pract 56:190–195

Álvarez-Pérez S, Blanco JL, Harmanus C et al (2017) Prevalence and characteristics of *Clostridium perfringens* and *Clostridium difficile* in dogs and cats attended in diverse veterinary clinics from the Madrid region. Anaerobe 48:47–55

Álvarez-Pérez S, Blanco JL, Astorga RJ et al (2018) Distribution and tracking of *Clostridium difficile* and *Clostridium perfringens* in a free-range pig abattoir and processing plant. Food Res Int 113:456–464

Alves F, Castro R, Pinto M, Nunes A, Pomba C, Oliveira M, Silveira L, Gomes JP, Oleastro M (2023) Molecular epidemiology of *Clostridioides difficile* in companion animals: Genetic overlap with human strains and public health concerns. Front Public Health 10:1070258. https://doi.org/10.3389/fpubh.2022.1070258

Andrés-Lasheras S, Bolea R, Mainar-Jaime RC et al (2017) Presence of *Clostridium difficile* in pig faecal samples and wild animal species associated with pig farms. J Appl Microbiol 122:462–472

Andrés-Lasheras S, Martín-Burriel I, Mainar-Jaime RC, Morales M, Kuijper E, Blanco JL, Chirino-Trejo M, Bolea R (2018) Preliminary studies on isolates of

Clostridium difficile from dogs and exotic pets. BMC Vet Res 14(1):77. https://doi.org/10.1186/s12917-018-1402-7

Avbersek J, Janezic S, Pate M et al (2009) Diversity of *Clostridium difficile* in pigs and other animals in Slovenia. Anaerobe 15:252–255

Avbersek J, Pirs T, Pate M et al (2014) *Clostridium difficile* in goats and sheep in Slovenia: characterisation of strains and evidence of age-related shedding. Anaerobe 15:252–255

Bandelj P, Trilar T, Raenik J et al (2011) Zero prevalence of *Clostridium difficile* in wild passerine birds in Europe. FEMS Microbiol Lett 321:183–185

Bandelj P, Trilar T, Blagus R et al (2014) Prevalence and molecular characterization of *Clostridium difficile* isolated from European Barn Swallows (Hirundo rustica) during migration. BMC Vet Res 10:40

Bandelj P, Blagus R, Briski F et al (2016) Identification of risk factors influencing *Clostridium difficile* prevalence in middle-size dairy farms. Vet Res 47:41

Bandelj P, Golob M, Ocepek M et al (2017) Antimicrobial susceptibility patterns of *Clostridium difficile* isolates from family dairy farms. Zoonoses Public Health 64: 213–221

Bandelj P, Harmanus C, Blagus R et al (2018) Quantification of *Clostridioides (Clostridium) difficile* in feces of calves of different age and determination of predominant *Clostridioides difficile* ribotype 033 relatedness and transmission between family dairy farms using multilocus-variable-number tandem-repeat analysis. BMC Vet Res 14(1):298

Barbanti F, Spigaglia P (2020) Microbial characteristics of human and animal isolates of *Clostridioides difficile* in Italy : Results of the Istituto Superiore di Sanità in the years 2006-2016. Amaerobe 61:102136

Bauer MP, Notermans DW, van Benthem BH et al (2011) *Clostridium difficile* infection in Europe: a hospital based survey. Lancet 377:63–73

Baverud V, Gustafsson A, Franklin A et al (2003) *Clostridium difficile*: prevalence in horses and environment, and antimicrobial susceptibility. Equine Vet J 35:465–471

Baverud V, Gustafsson A, Franklin A et al (2004) *Clostridium difficile* diarrhea: infection control in horses. Vet Clin N Am Equine Pract 20:615–630

Bjöersdorff OG, Lindberg S, Kiil K, Persson S, Guardabassi L, Damborg P (2021) Dogs are carriers of *Clostridioides difficile* lineages associated with human community-acquired infections. Anaerobe 67: 102317. https://doi.org/10.1016/j.anaerobe.2020.102317

Blasi F, Lovito C, Albini E et al (2021) *Clostridioides difficile* in calves in central Italy : prevalence, molecular typing, antimicrobial susceptibility and associations with antibiotic administration. Animals (Basel) 11(2): 515

Blau K, Gallert C (2023) Prevalence, antimicrobial resistance and toxin-encoding genes of *Clostridioides difficile* from environmental sources contaminated by feces. Antibiotics (Basel, Switzerland) 12(1):162. https://doi.org/10.3390/antibiotics12010162

Bojesen AM, Olsen KE, Bectelsen MF (2006) Fatal enterocolitis in Asian elephants (Elephas maximus) caused by *Clostridium difficile*. Vet Microbiol 116: 329–335

Borriello SP, Honour P, Turner T et al (1983) Household pets as a potential reservoir for *Clostridium difficile* infection. J Clin Pathol 36:84–87

Bouttier S, Barc MC, Felix B et al (2010) *Clostridium difficile* in ground meat, France. Emerg Infect Dis 16: 733–735

Burt SA, Siemeling L, Kuijper EJ et al (2012) Vermin on pig farms are vectors of *Clostridium difficile* PCR-ribotypes 078 and 045. Vet Microbiol 160:256–258

Burt SA, Meijer K, Burggraaff P et al (2018) Wild mice in and around the city of Utrecht, the Netherlands, are carriers of *Clostridium difficile* but not ESBL-producing *Enterobacteriaceae, Salmonella spp.* or MRSA. Lett Appl Microbiol 67(5):513–549

Candel-Pérez C, Santaella-Pascual J, Ros-Berruezo G et al (2021) Occurrence of *Clostridioides (Clostridium) difficile* in poultry giblets at slaughter and in retail pork and poultry meat in Southeastern Spain. J Food Prot 84(2):310–314

Carvalho GM, Ramos CP, Lobato FCF et al (2022) Laboratory diagnosis of *Clostridioides (Clostridium) difficile* infection in domestic animals : a short review. Anaerobe 75:102574

Cizek A, Masarikova M, Mares J, Brajerova M, Krutova M (2022) Detection of plasmid-mediated resistance to metronidazole in Clostridioides difficile from river water. Microbiology spectrum 10(4):e0080622. https://doi.org/10.1128/spectrum.00806-22

Cooper KK, Songer JG, Uzal FA (2013) Diagnosing clostridial enteric disease in poultry. J Vet Diagn Invest 25: 314–327

Dabard J, Dubos F, Martinet L et al (1979) Experimental reproduction of neonatal diarrhea in young gnotobiotic hares simultaneously associated with *Clostridium difficile* and other *Clostridium* strains. Infect Immun 24:7–11

Darwich L, Seminati C, Lopez-Olvera JR et al (2021) Detection of Beta-Lactam-Resistant Escherichia coli and toxigenic *Clostridium difficile* strains in wild boars foraging in an anthropization gradient. Animal (Basel) 11(6):1585

De Boer E, Zwartkruis-Nahuis A, Heuvelink A et al (2009) *Clostridium difficile* PCR-ribotype 078 toxinotype V found in diarrhoeal pigs identical to isolates from affected humans. Environ Microbiol 144:561–511

Diab SS, Songer G, Uzal FA (2013) *Clostridium difficile* infection in horses: a review. Vet Microbiol 167:42–49

Drigo I, Mazzolini E, Bacchin C et al (2015) Molecular characterization and antimicrobial susceptibility of *Clostridium difficile* isolated from rabbits raised for meat production. Vet Microbiol 181:303–307

Ducarmon QR, van der Bruggen T, Harmanus C, Sanders IMJG, Daenen LGM, Fluit AC, Vossen RHAM, Kloet SL, Kuijper EJ, Smits WK (2022) *Clostridioides difficile* infection with isolates of cryptic clade C-II: a genomic analysis of polymerase chain reaction ribotype 151. Clin Microbiol Infect. Advance online publication. https://doi.org/10.1016/j.cmi.2022.12.003

Duijvestijn M, Mughini-Gras L, Schuurman N, Schijf W, Wagenaar JA, Egberink H (2016) Enteropathogen infections in canine puppies: (Co-)occurrence, clinical relevance and risk factors. Vet Microbiol 15:115–122. https://doi.org/10.1016/j.vetmic.2016.09.006

Eckert C, Burghoffer B, Barbut F et al (2013) Contamination of ready to eat raw vegetables with *Clostridium difficile* in France. J Med Microbiol 62:1435–1438

Finsterwalder SK, Loncaric I, Cabal A, Szostak MP, Barf LM, Marz M, Allerberger F, Burgener IA, Tichy A, Feßler AT, Schwarz S, Monecke S, Ehricht R, Ruppitsch W, Spergser J, Künzel F (2022) Dogs as carriers of virulent and resistant genotypes of *Clostridioides difficile*. Zoonoses Public Health 69(6): 673–681. https://doi.org/10.1111/zph.1295

Frentrup M, Thiel N, Junker V, Behrens W, Münch S, Siller P, Kabelitz T, Faust M, Indra A, Baumgartner S, Schepanski K, Amon T, Roesler U, Funk R, Nübel U (2021) Agricultural fertilization with poultry manure results in persistent environmental contamination with the pathogen *Clostridioides difficile*. Environ Microbiol 23(12):7591–7602. https://doi.org/10.1111/1462-2920.15601

Froschle B, Messelhäusser U, Höller C et al (2015) Fate of *Clostridium botulinum* and incidence of pathogenic clostridia in biogas processes. J Appl Microbiol 119: 936–947

Hafiz S (1974) *Clostridium difficile* and its toxins. Ph.D. Thesis. Department of Microbiology, University of Leeds, UK

Hammitt MC, Bueschel DM, Keel MK et al (2008) A possible role for *Clostridium difficile* in the etiology of calf enteritis. Vet Microbiol 127:343–352

Hampikyan H, Bingol EB, Muratoglu K et al (2018) The prevalence of *C. difficile* in cattle and sheep carcasses and the antibiotic susceptibility of isolates. Meat Sci 139:120–124

Hargreaves KR, Colvin HV, Patel KV et al (2013) Genetically diverse *Clostridium difficile* strains harboring abundant prophages in an estuarine environment. Appl Environ Microbiol 79:6236–6243

Heise J, Witt P, Maneck C et al (2021) Prevalence and phylogenetic relationship of *Clostridioides difficile* strains in fresh poultry meat samples processed in different cutting plants. Int J Food Microbiol 339: 109032

Hensgens MP, Keessen EC, Squire MM et al (2012) *Clostridium difficile* infection in the community: a zoonotic disease? Clin Microbiol Infect 18:635–645

Hoffer E, Haechler H, Frei R et al (2010) Low occurrence of *Clostridium difficile* in faecal samples of healthy calves and pigs at slaughter and in minced meat in Switzerland. J Food Prot 73:973–975

Hopman NEM, Oorburg D, Sanders I et al (2011) High occurrence of various *Clostridium difficile* PCR-ribotypes in pigs arriving at the slaughterhouse. Vet Q 31:179–181

Hunter D, Bellhouse R, Baker K (1981) *Clostridium difficile* isolated from a goat. Vet Rec 109:291–292

Indra A, Lassing H, Baliko N et al (2009) *Clostridium difficile*: a new zoonotic agent? Wein Klin Wochensr 121:91–95

Janezic S, Ocepek M, Zidaric V et al (2012) *Clostridium difficile* genotypes other than ribotype 078 that are prevalent among human, animal and environmental isolates. BMC Microbiol 12:48

Janezic S, Marín M, Martín A, Rupnik M (2015) A new type of toxin A-negative, toxin B-positive *Clostridium difficile* strain lacking a complete tcdA gene. J Clin Microbiol 53(2):692–695. https://doi.org/10.1128/JCM.02211-14

Janezic S, Potocnik M, Zidaric V et al (2016) Highly divergent *Clostridium difficile* strains isolated from the environment. PLoS One 11:e0167101

Janezic S, Mlakar S, Rupnik M (2018) Dissemination of *Clostridium difficile* spores between environment and households: Dog paws and shoes. Zoonoses Public Health. 65(6):669–674. https://doi.org/10.1111/zph.12475

Janezic S, Smrke J, Rupnik M (2020) Isolation of *Clostridioides difficile* from different outdoor sites in the domestic environment. Anaerobe 62:102183. https://doi.org/10.1016/j.anaerobe.2020.102183

Jobstl M, Heuberger S, Indra A et al (2010) *Clostridium difficile* in raw products of animal origin. Int J Food Microbiol 138:172–175

Jones MA, Hunter D (1983) Isolation of *Clostridium difficile* from pigs. Vet Rec 112:253

Kecerova Z, Cizek A, Nyc O, Krutova M (2019) *Clostridium difficile* isolates derived from Czech horses are resistant to enrofloxacin; cluster to clades 1 and 5 and ribotype 033 predominates. Anaerobe 56:17–21. https://doi.org/10.1016/j.anaerobe.2019.01.005

Keessen EC, Donswijk CJ, Hol SP et al (2011a) Aerial dissemination of *Clostridium difficile* on a pig farm and its environment. Environ Res 111:1027–1032

Keessen EC, van den Berkt AJ, Haasjes NH et al (2011b) The relation between farm specific factors and prevalence of *C. difficile* in slaughter pigs. Vet Microbiol 154:130–134

Keessen EC, Hensgens MP, Spigaglia P et al (2013) Antimicrobial susceptibility profiles of human and piglet *Clostridium difficile* PCR-ribotype 078. Antimicrob Resist Infect Control 2:14

Kiss D, Bilkei G (2005) A new periparturient disease in Eastern Europe, *Clostridium difficile* causes postparturient sow losses. Theriogenology 63:17–23

Knetsch CW, Connor TR, Mutreja A et al (2014) Whole genome sequencing reveals potential spread of *Clostridium difficile* between humans and farm animals in the Netherlands, 2002 to 2011. Euro Surveill 19:20954

Knight DR, Squire MM, Collins DA et al (2016) Genome analysis of *Clostridium difficile* PCR ribotype

014 lineage in Australian pigs and humans reveals a diverse genetic repertoire and signatures of long-range interspecies transmission. Front Microbiol 7:2138

Knight DR, Imwattana K, Kullin B, Guerrero-Araya E, Paredes-Sabja D, Didelot X, Dingle KE, Eyre DW, Rodríguez C, Riley TV (2021) Major genetic discontinuity and novel toxigenic species in *Clostridioides difficile* taxonomy. eLife 10:e64325. https://doi.org/10.7554/eLife.64325

Koene MGJ, Mevius D, Wagenaar JA et al (2012) *Clostridium difficile* in Dutch animals: their presence, characteristics and similarities with human isolates. Clin Microbiol Infect 18:778–784

Kotila SM, Pitkänen T, Brazier J et al (2013) *Clostridium difficile* contamination of public tap water distribution system during a waterborne outbreak in Finland. Scand J Public Health 41:541–545

Krijger IM, Meerburg BG, Harmanus C et al (2019) *Clostridium difficile* in wild rodents and insectivores in the Netherlands. Lett Appl Microbiol 69(1):35–40

Krutova M, Zouharova M, Matejkova J et al (2018) The emergence of *Clostridium difficile* PCR-ribotype 078 in piglets in the Czech Republic clusters with *Clostridium difficile* PCR ribotype 078 isolates from Germany, Japan and Taiwan. Int J Med Microbiol 308(7):770–775

Lim SC, Knight DR, Riley TV (2020) *Clostridium difficile* and one health. Clin Microbiol Infect 26(7):857–863

Lysons RJ, Hall GA, Lemcke RM et al (1980) Studies of organisms possibly implicated in swine dysentery. In: Proceedings of the 6th International Pig Veterinary Society

Marcos P, Whyte P, Rogers T et al (2021) The prevalence of *Clostridioides difficile* on farms, in abattoirs and in retail foods in Ireland. Food Microbiol 98:10378. https://doi.org/10.1016/j.fm.2021.103781

Marcos P, Whyte P, Burgess C, Ekhlas D, Bolton D (2022) Detection and genomic characterisation of Clostridioides difficile from spinach fields. Pathogens (Basel, Switzerland) 11(11):1310. https://doi.org/10.3390/pathogens11111310

Masarikova M, Simkova I, Plesko M et al (2020) The colonisation of calves in Czech large-scale dairy farms by clonally-related *Clostridioides difficile* of the sequence type 11 represented by ribotypes 033 and 126. Microorganisms 8(6):901

McElroy MC, Hill M, Moloney G et al (2016) Typhlocolitis associated with *C. difficile* PCR-ribotypes 078 and 110 in neonatal piglets from a commercial Irish pig herd. Ir Vet J 69:10

Mertens N, TheuB T, Köchling M et al (2022) Pathogens detected in 205 German farms with porcine neonatal diarrhea in 2017. Vet Sci 9(2):44

Moloney G, Eyre DW, Aogáin MM et al (2021) Human and porcine transmission of *Clostridioides difficile* ribotype 078, Europe. Emerg Infect Dis 27(9): 2294–2300

Monteagudo LV, Benito AA, Lázaro-Gaspar S et al (2022) Occurrence of rotavirus A genotypes and other enteric pathogens in diarrheic suckling piglets from Spanish swine farms. Animals (Basel) 12(3):251

Moradigaravand D, Gouliouris T, Ludden C, Reuter S, Jamrozy D, Blane B, Naydenova P, Judge K, Aliyu SH, Hadjirin NF, Holmes MA, Török E, Brown NM, Parkhill J, Peacock S (2018) Genomic survey of *Clostridium difficile* reservoirs in the East of England implicates environmental contamination of wastewater treatment plants by clinical lineages. Microb Genom 4(3):e000162. https://doi.org/10.1099/mgen.0.000162

Noren T, Johansson K, Unemo M (2014) *Clostridium difficile* PCR-ribotype 046 is common among neonatal pigs and humans in Sweden. Clin Microbiol Infect 20: O2–O6

Numberger D, Riedel T, McEwen G, Nübel U, Frentrup M, Schober I, Bunk B, Spröer C, Overmann J, Grossart HP, Greenwood AD (2019) Genomic analysis of three *Clostridioides difficile* isolates from urban water sources. Anaerobe 56:22–26. https://doi.org/10.1016/j.anaerobe.2019.01.002

Orden C, Blanco JL, Álvarez-Pérez S et al (2017a) Isolation of *Clostridium difficile* from dogs with digestive disorders, including stable metronidazole-resistant strains. Anaerobe 43:78–81

Orden C, Neila C, Blanco JL et al (2017b) Recreational sandboxes for children and dogs can be a source of epidemic ribotypes of *Clostridium difficile*. Zoonoses Public Health 65(1):88–95

Ossiprandi MC, Buttrini M, Bottarelli E et al (2010) Preliminary molecular analysis of *Clostridium difficile* isolates from healthy horses in northern Italy. Comp Immunol Microbiol Infect Dis 33:e25–e29

Otten AM, Reid-Smith RJ, Fazil A et al (2010) Disease transmission model for community-associated *Clostridium difficile* infection. Epidemiol Infect 138:907–914

Pasquale V, Romano VJ, Rupnik M et al (2011) Isolation and characterization of *Clostridium difficile* from shellfish and marine environments. Folia Microbiol (Praha) 56:431–437

Pasquale V, Romano VJ, Rupnik M et al (2012) Occurrence of toxigenic *Clostridium difficile* in edible bivalve molluscs. Food Microbiol 31:309–312

Pelaez T, Alcala L, Blanco JL et al (2013) Characterization of swine isolates of *Clostridium difficile* in Spain: a potential source of epidemic multidrug resistant strains? Anaerobe 22:45–49

Perrin J, Buogo C, Gallusser A et al (1993) Intestinal carriage of *Clostridium difficile* in neonate dogs. Zentralbl Veterinarmed B 40:222–226

Pirs T, Ocepek M, Rupnik M (2008) Isolation of *Clostridium difficile* from food animals in Slovenia. J Med Microbiol 57:790–792

Pirs T, Avbersek J, Zdouc I et al (2013) Antimicrobial susceptibility of animal and human isolates of *C. difficile* by broth microdilution. J Med Microbiol 62:1478–1485

Proctor A, Cornick NA, Wang C et al (2021) Neonatal piglets are protected from *Clostridioides difficile* infection by age-dependent increase in intestinal microbial diversity. Microbiol Spectr 9(2)

Rabold D, Espelage W, Abu Sin M, Eckmanns T, Schneeberg A, Neubauer H, Möbius N, Hille K, Wieler LH, Seyboldt C, Lübke-Becker A (2018) The zoonotic potential of *Clostridium difficile* from small companion animals and their owners. PLoS One 13(2):e0193411. https://doi.org/10.1371/journal.pone.0193411

Ramírez-Vargas G, López-Ureña D, Badilla A, Orozco-Aguilar J, Murillo T, Rojas P, Riedel T, Overmann J, González G, Chaves-Olarte E, Quesada-Gómez C, Rodríguez C (2018) Novel Clade C-I *Clostridium difficile* strains escape diagnostic tests, differ in pathogenicity potential and carry toxins on extrachromosomal elements. Sci Rep 8(1):13951. https://doi.org/10.1038/s41598-018-32390-6

Redding L, Huang E, Ryave J et al (2021) *Clostridioides difficile* on dairy farms and potential risk to dairy farm workers. Anaerobe 69:102353

Riedel T, Wittmann J, Bunk B, Schober I, Spröer C, Gronow S, Overmann J (2017) A *Clostridioides difficile* bacteriophage genome encodes functional binary toxin-associated genes. J Biotechnol 250:23–28. https://doi.org/10.1016/j.jbiotec.2017.02.017

Rivas L, Dupont PY, Gilpin BJ et al (2020) Isolation and characterization of *Clostridium difficile* from a small survey of wastewater, food and animals in New Zealand. Lett Appl Microbiol 70(1):29–35

Rodriguez C, Taminiau B, Van Broeck J et al (2012) *Clostridium difficile* in young farm animals and slaughter animals in Belgium. Anaerobe 18:621–625

Rodriguez C, Avesani V, Van Broeck J et al (2013) Presence of *Clostridium difficile* in pigs and cattle intestinal contents and carcass contamination at slaughterhouse in Belgium. Int J Food Microbiol 166:256–262

Rodriguez C, Taminiau B, Brévers B et al (2014a) Carriage and acquisition rates of *Clostridium difficile* in hospitalized horses, including molecular characterization, multilocus sequence typing and antimicrobial susceptibility of bacterial isolates. Vet Microbiol 172: 309–317

Rodriguez C, Taminiau B, Avesani V et al (2014b) Multilocus sequence typing analysis and antibiotic resistance of *Clostridium difficile* strains, including molecular characterization, multilocus sequence typing and antimicrobial susceptibility of bacterial isolates. Vet Microbiol 172:309–317

Rodriguez C, Taminiau B, Brévers B et al (2015) Faecal microbiota characterisation of horses using 16 rdna barcoded pyrosequencing, and carriage rate of clostridium difficile at hospital admission. BMC Microbiol 15: 181

Rodriguez C, Taminiau B, Van Broeck J et al (2016) *Clostridium difficile* in food and animals: a comprehensive review. Adv Exp Med Biol 4:65–92

Rodriguez C, Hakimi DE, Vanleyssem R et al (2017) *Clostridium difficile* in beef cattle farms, farmers and their environment : assessing the spread of the bacterium. Vet Microbiol 210:183–187

Rodriguez C, Taminiau B, Bouchafa L, Romijn S, Van Broeck J, Delmée M, Clercx C, Daube G (2019a) *Clostridium difficile* beyond stools: dog nasal discharge as a possible new vector of bacterial transmission. Heliyon 5(5):e01629. https://doi.org/10.1016/j.heliyon.2019.e01629

Rodriguez C, Bouchafa L, Soumillion K, Ngyuvula E, Taminiau B, Van Broeck J, Delmée M, Daube G (2019b) Seasonality of *Clostridium difficile* in the natural environment. Transbound Emerg Dis 66(6): 2440–2449. https://doi.org/10.1111/tbed.13301

Rodriguez-Palacios A, Lejeune JT (2011) Moist-heat resistance, spore aging, and superdormancy in *Clostridium difficile*. Appl Environ Microbiol 77:3085–3091

Rodriguez-Palacios A, Borgmann S, Kline TR et al (2013) *Clostridium difficile* in foods and animals: history and measures to reduce exposure. Anim Health Res Rev 14:11–29

Rodríguez-Pallares S, Fernández-Palacios P, Jurado-Tarifa E, Arroyo F, Rodríguez-Iglesias MA, Galán-Sánchez F (2022) Transmission of toxigenic *Clostridioides difficile* between a pet dog with diarrhea and a 10-month-old infant. Anaerobe 74:102519. https://doi.org/10.1016/j.anaerobe.2022.102519

Romanazzi V, Bonetta S, Fornasero S et al (2016) Assessing Methanobrevibacter smithii and *Clostridium difficile* as not conventional faecal indicators in effluents of a wastewater treatment plant integrated with sludge anaerobic digestion. J Environ Manag 184:170–177

Romano V, Albanese F, Dumontet S et al (2012a) Prevalence and genotypic characterization of *Clostridium difficile* from ruminants in Switzerland. Zoonoses Public Health 59:545–548

Romano V, Pasquale V, Krovacekb K et al (2012b) Toxigenic *Clostridium difficile* PCR ribotypes from wastewater treatment plants in Southern Switzerland. Appl Environ Microbiol 78:6643–6646

Romano V, Pasqualea V, Lemee L et al (2018) *Clostridioides difficile* in the environment, food, animals and humans in southern Italy : Occurrence and genetic relatedness. Comp Immunol Microbiol Infect Dis 59:41–46. https://doi.org/10.1016/j.cimid.2018.08.006

Rupnik M (2007) Is *Clostridium difficile*-associated infection a potentially zoonotic and foodborne disease? Clin Microbiol Infect 13:457–459

Rupnik M (2010) *Clostridium difficile*: (re)emergence of zoonotic potential. Clin Infect Dis 51:583–584

Schneeberg A, Rupnik M, Neubauer H et al (2012) Prevalence and distribution of *Clostridium difficile* PCR ribotypes in cats and dogs from animal shelters in Thuringia, Germany. Anaerobe 18:484–488

Schneeberg A, Neubauer H, Schmoock G et al (2013a) *Clostridium difficile* genotypes in piglet populations in Germany. J Clin Microbiol 51:3796–3803

Schneeberg A, Neubauer H, Schomoock G et al (2013b) Presence of *Clostridium difficile* PCR ribotype clusters related to 033, 078 and 045 in diarrhoeic calves in

Germany. J Med Microbiol 62:1190–1198. Congress 1980, Copenhagen, p 231
Scholtzek AD, Heise J, Witt P et al (2022) Contamination of home-grown and retail vegetables with *Clostridioides difficile*. Anaerobe 74:102512
Schoster A, Kunz T, Lauper M, Graubner C, Schmitt S, Weese JS (2019) Prevalence of *Clostridium difficile* and *Clostridium perfringens* in Swiss horses with and without gastrointestinal disease and microbiota composition in relation to Clostridium difficile shedding. Vet Microbiol 239:108433. https://doi.org/10.1016/j.vetmic.2019.108433
Skraban J, Dzeroski S, Zenko B et al (2013) Changes of poultry faecal microbiota associated with *Clostridium difficile* colonisation. Vet Microbiol 165:416–424
Songer JG (2000) Infection of neonatal swine with *Clostridium difficile*. J Swine Health Prod 4:185–189
Spigaglia P, Drigo I, Barbanti F et al (2015) Antibiotic resistance patterns and PCR-ribotyping of *Clostridium difficile* strains isolated from swine and dogs in Italy. Anaerobe 31:42–46
Squire MM, Riley TV (2013) *Clostridium difficile* infection in human and piglets: a "One health" opportunity. Curr Top Microbiol Immunol 365:299–314
Stein K, Egan S, Lynch H et al (2017) PCR-distribution of *Clostridium difficile* in Irish pigs. Anaerobe 48:237–241
Steyer A, Gutiérrez-Aguirre I, Rački N et al (2015) The detection rate of enteric viruses and *Clostridium difficile* in a waste water treatment plant effluent. Food Environ Virol 7:164–172
Tkalec V, Janezic S, Skok B et al (2019) High Clostridium difficile contamination rates of domestic and imported potatoes compared to some other vegetables in Slovenia. Food Microbiol 78:194–200
Tkalec V, Jamnikar-Ciglenecki U, Rupnik M et al (2020) *Clostridioides difficile* in national food surveillance, Slovenia, 2015 to 2017. Euro Surveill 25(16):1900479
Tkalec V, Viprey V, Davis G et al (2022) *Clostridioides difficile* positivity rate and PCR ribotype distribution on retail potatoes in 12 European countries, January to June 2018. Euro Surveill 27(15):2100417
Tramuta C, Spigaglia P, Barbanti F, Bianchi DM, Boteva C, Di Blasio A, Zoppi S, Zaccaria T, Proroga YTR, Chiavacci L, Dondo A, Decastelli L (2021) Comparison of *Clostridioides difficile* strains from animals and humans: First results after introduction of C. difficile molecular typing and characterization at the Istituto Zooprofilattico Sperimentale of Piemonte, Liguria e Valle d'Aosta, Italy. Comp Immunol Microbiol Infect Dis 75:101623. https://doi.org/10.1016/j.cimid.2021.101623
Von Abercron SMM, Karlsson F, Wigh GT et al (2009) Low occurrence of *Clostridium difficile* in retail ground meat in Sweden. J Food Prot 72:1732–1734
Warriner K, Xu C, Habash M et al (2016) Dissemination of *Clostridium difficile* in food and the environment: significant sources of *C. difficile* community acquired infection? J Appl Microbiol 122:542–553
Weber A, Kroth P, Heil G (1989) The occurrence of *Clostridium difficile* in fecal samples of dogs and cats. Zentralbl Veterinarmed B 36:568–576
Weese JS (2010) *Clostridium difficile* in food—innocent bystander or serious threat? Clin Microbiol Infect 16: 3–10
Weese JS (2020) *Clostridium (Clostridioides) difficile* in animals. J Vet Diagn Invest 32(2):213–221
Wetterwik KJ, Trowald-Wigh G, Fernström LL et al (2013) *Clostridium difficile* in faeces from healthy dogs and dogs with diarrhea. Acta Vet Scand 55:23
Williamson CHD, Stone NE, Nunnally AE, Roe CC, Vazquez AJ, Lucero SA, Hornstra H, Wagner DM, Keim P, Rupnik M, Janezic S, Sahl JW (2022) Identification of novel, cryptic Clostridioides species isolates from environmental samples collected from diverse geographical locations. Microb Genom 8(2):000742. https://doi.org/10.1099/mgen.0.000742
Wojtacka J, Wysoc B, Kocuvan A et al (2021) High contamination rates of shoes of veterinarians, veterinary support staff and veterinary students with *Clostridioides difficile* spores. Transbound Emerg Dis 00:1–9
Zidaric V, Zemljic M, Janezic S et al (2008) High diversity of *Clostridium difficile* genotypes isolated from a single poultry farm producing replacement laying hens. Anaerobe 14:325–327
Zidaric V, Beigot S, Lapajne S et al (2010) The occurrence and high diversity of *Clostridium difficile* genotypes in rivers. Anaerobe 16(4):371–375
Zidaric V, Pardon B, Dos Vultos T et al (2012) Different antibiotic resistance and sporulation properties within multiclonal *Clostridium difficile* PCR ribotypes 078, 126, and 033 in a single calf farm. Appl Environ Microbiol 78:8515–8522
Zlender T, Golob Z, Rupnik M (2022) Low *Clostridioides difficile* positivity rate in wild animal shelter in Slovenia. Anaerobe 77:102643

The ESCMID *Study Group for Clostridioides difficile*: History, Role, and Perspectives

John E. Coia, Ed J. Kuijper, and Fidelma Fitzpatrick

Abstract

Clostridioides difficile (*C. difficile*) is a major nosocomial pathogen but is also increasingly recognised as an important diarrhoeal pathogen in the community, not always associated with antibiotics. The European Society of Clinical Microbiology and Infectious Diseases (ESCMID) Study Group for *C. difficile* (ESGCD) is a group of clinicians, scientists, and others from many European countries and further afield, who share a common interest in *C. difficile*. The aims of the Study Group are centred around raising the profile of *C. difficile* infection (CDI) in humans and animals, fostering collaboration amongst centres in different European countries and providing a forum for discussing and disseminating information. One of the principal aims of the Study Group is to raise awareness of *C. difficile* infections in Europe. ESGCD has a particular interest in the development and dissemination of European guidance on prevention, diagnosis, and treatment of CDI. This chapter will discuss the organisation of ESGCD within the ESCMID Study Group structure, the origins of the Study Group, the aims, and objectives of the group, and will highlight some of the past and present activities of ESGCD in relation to these.

J. E. Coia (✉)
Institute for Regional Health Research (IRS), University of Southern Denmark (SDU), Esbjerg, Denmark

ESCMID Study Group for *C. difficile* (ESGCD), Basel, Switzerland

ESCMID Study Group for Nosocomial Infections (ESGNI), Basel, Switzerland
e-mail: jcoia@health.sdu.dk

E. J. Kuijper
ESCMID Study Group for *C. difficile* (ESGCD), Basel, Switzerland

Department of Medical Microbiology, Center of Infectious Diseases, Leiden University Medical Center, Leiden, The Netherlands

ESCMID Study Group for Host and Microbiota Interaction (ESGHAMI), Basel, Switzerland
e-mail: e.j.kuijper@lumc.nl

F. Fitzpatrick
ESCMID Study Group for *C. difficile* (ESGCD), Basel, Switzerland

ESCMID Study Group for Host and Microbiota Interaction (ESGHAMI), Basel, Switzerland

Department of Clinical Microbiology, The Royal College of Surgeons in Ireland and Beaumont Hospital, Dublin, Ireland
e-mail: fidelmafitzpatrick@rcsi.ie

1 Introduction

Clostridioides difficile is a major nosocomial pathogen but is also increasingly recognised as an important diarrhoeal pathogen in the community, not always associated with antibiotics. The European Society of Clinical Microbiology and

P. Mastrantonio, M. Rupnik (eds.), *Updates on Clostridioides difficile in Europe*, Advances in Experimental Medicine and Biology 1435, https://doi.org/10.1007/978-3-031-42108-2_16

Infectious Diseases (ESCMID) Study Group for *Clostridioides difficile* (ESGCD) is a group of clinicians and scientists from many European countries and further afield, who share a common interest in *C. difficile*. In 2022, feedback from study group members and attendees at the European Congress of Clinical Microbiology and Infectious Diseases (ECCMID) 2022 after one of the ESGCD sessions (*Involving Patients in Research for Better Outcomes*) indicated the clear advantages of patient representation for ESGCD and *C. difficile* education and research. Hence, ESGCD was the first ESCMID study group to invite a patient representative onto the executive committee in late 2022 (ESGCD 2022).

The aims of the Study Group are centred around raising the profile of CDI in humans and animals, fostering collaboration amongst centres in different European countries and providing a forum for discussing and disseminating information. One of the principal aims of the Study Group is to raise awareness of *C. difficile* infections in European hospitals. ESGCD has a particular interest in the development and dissemination of European guidance on prevention, diagnosis, and treatment of CDI. This chapter will discuss the organisation of ESGCD within the ESCMID Study Group structure, the origins of the Study Group, the aims and objectives of the group, and will highlight some of the past and present activities of ESGCD in relation to these.

ESCMID and ESGCD

The organisation that we know today as ESCMID was originally founded in 1983 as the European Society of Clinical Microbiology (ESCM), with an initial membership of 41 people. In 1990, with the approval of 83% of the membership, the name of the society was formally changed to the European Society of Clinical Microbiology and Infectious Diseases (ESCMID). By this time, the membership had grown to 971 (Phillips 2008). In the intervening years, ESCMID has flourished to become Europe's leading society for clinical microbiology and infectious diseases with members from all European countries and all continents, and with more than 9000 individual and 40,000 affiliated members around the world (ESCMID 2022a). The Society's annual scientific meeting, the European Congress of Clinical Microbiology and Infectious Diseases (ECCMID), is now regarded as one of the premier meetings in the field and attracts over 10,000 delegates (ESCMID 2022b).

ESCMID is a non-profit organisation whose mission is to improve the diagnosis, treatment, and prevention of infection-related diseases. This is achieved by promoting and supporting research, education, training, and good medical practice. The promotion of research as a core activity of the Society has been a feature virtually since the inception of ESCM. The idea of Study Groups and Working Parties as a means to support this key aim arose from Katherine Dornbusch's proposal in 1985 that the Society should associate itself with her existing international study of antibiotic resistance. The concept of such Study Groups was strongly supported by Jacques Acar during his presidency of the Society (Phillips 2008). The success of the Study Group approach is exemplified by the European Study Group on Antibiotic Breakpoints (ESGAB), which was established in 1988, and would subsequently go on to become the European Committee on Antimicrobial Susceptibility Testing (EUCAST) in 1997. The work of this group has been a major driver of standardisation and harmonisation of clinically-focussed antimicrobial sensitivity testing in Europe. A key difference between Working Parties and Study Groups was that the former were expected to have a limited single-objective-based lifetime, while the latter would be semi-permanent as long as the topic remained of significant relevance. Thus, the Study Groups' main objectives were to bring together human and veterinary researchers, both from academia and industry, to collaborate in multi-centre studies, to address scientific issues in position papers or practice guidance, and to mount educational meetings. This is still reflected in the Study Group statutes, which state that "The Study Group shall devote itself to the promotion of research and education in diagnosis and therapy in its defined field(s) of expertise" (ESCMID 2020).

At the time of writing, ESCMID supports 32 Study Groups engaged in advancing scientific knowledge and/or disseminating professional guidelines in the field of clinical microbiology and infectious diseases (ESCMID 2022c). The Study Groups are overseen by the Scientific Affairs Subcommittee of ESCMID, and their performance is annually evaluated against a number of criteria to ensure that the required standards of scientific and professional outputs are maintained. ESGCD has consistently been one of the top performing Study Groups. In the 5-year period 2017–2021, ESGCD presented 11 symposia at ECCMID meetings. In the period 2017–2022, the study group and its members published 41 articles (including several medical guidelines as described below), and supported 10 research projects, organised 3 educational webinars, and supported 13 scientific meetings outwith ECCMID.

2 The History and Origins of ESGCD

The aetiological role of *C. difficile* in pseudomembranous colitis (PMC) has been known for 40 years (Larson et al. 1978). However, it was the emergence and rapid spread in North America and Europe of the hypervirulent PCR ribotype 027 strain (Warny et al. 2005) at the dawning of the new millennium (Honda and Dubberke 2014), which was the catalyst for a resurgence of interest in CDI. ESGCD played an important role in the recognition of PCR ribotype 027 in Europe, since Canadian researchers presented their data at the 10th ECCMID in Stockholm (2000) to ESGCD members and subsequently sent strains to the UK Anaerobic Reference Laboratory under the directorship of Dr John Brazier, who subsequently supported other European laboratories to recognise this new emerging type. This emergence of a new hypervirulent type was a stimulus for a group of scientists and clinicians with existing research and clinical interest in *Clostridium difficile* to expand the activities of the Study Group under the auspices of ESCMID. In close collaboration with the European Centre for Disease Prevention and Control (ECDC), a "white paper" was made by ESGCD to advise on diagnostics, typing, and surveillance of new emerging types (Kuijper et al. 2006)

The inaugural gathering of the nascent ESGCD was held on Sunday 28th May 2000 during the 10th ECCMID in the International Fairs Building in Stockholm, Sweden. At this time, temporary officers of ESGCD were elected, with Dr JS Brazier (UK) as the first Chairperson, and with Dr M Claros (Germany) and Professor M. Delmeé (Belgium) acting as interim Secretary and Treasurer, respectively. The original statutes of the ESGCD, which were tabled at the inaugural meeting, are shown in Fig. 1.

The first full meeting of ESGCD took place on 3 April 2001 at the 11th ECCMID in Istanbul. At this meeting, Jon Brazier was confirmed as Chairperson of the group, with Michel Delmée assuming the role of treasurer, and Dr Maja Rupnik (Slovenia) taking on the position of secretary of the group. The original aims of the Study Group were to:

- Establish the extent and prevalence of nosocomial infections with *C. difficile* in hospitals across Europe
- Compare the types of *C. difficile* in circulation in European hospitals by molecular and phenotypic methods
- Undertake a survey of *C. difficile* in animals
- Foster collaboration between participating centres worldwide on human CDI and animal CDI
- Investigate the feasibility of adopting a common typing method based on PCR ribotyping using defined primers and standardised methods
- Provide surveillance on the antimicrobial susceptibility of strains of *C. difficile* in circulation in European hospitals
- Foster links with respective national authorities on collection of surveillance data on *C. difficile* infections
- Assemble European guidelines on prevention, diagnosis, treatment, and surveillance of

STATUTES OF THE EUROPEAN STUDY GROUP ON *CLOSTRIDIUM DIFFICILE* OF ESCMID

1. The study shall carry the name "European Study Group on *Clostridium difficile* of the ESCMID". (ESGCD)
The registered office of the network shall be located in the city of the active President.

2. Aims of the study group:
The study group shall endeavour:
a) to establish the extent and prevalence of nosocomial infections with *C. difficile* in hospitals across Europe.
b) to compare the types of *C. difficile* in circulation in European hospitals by molecular and phenotypic methods
c) to foster collaboration between participating centres

d) to investigate the feasibility of adopting a common typing method in participating centres based on PCR ribotyping using defined primers and standardised methods.
e) to provide surveillance on the antimicrobial susceptibility of strains of *C. difficile* in circulation in European hospitals
f) to foster links with respective national authorities on collection of surveillance data on *C. difficile* infections
g) to assemble European guidelines on prevention, diagnosis, treatment and surveillance of *C. difficile* infections.
h) to harmonise methodologies relating to *C. difficile* - benefits in (*g*) above.

3. Membership.
Full members with the right to elect, or be eligible to the Executive Committee must be ESCMID members in good standing. Other interested parties can participate to the ESGCD activities as associate members.
The ESGCD shall consist of interested parties who can apply to the Executive Committee for membership.

4. Executive Committee.
The first General Assembly will appoint officers of the Executive to consist of a Chairman, Secretary and Treasurer. Additional members may be elected to ensure better representation from other countries. Members of the Executive are elected for a period of two years and elections will be by ballot of ESGCD members. The number of terms of office are limited to three. The Executive will meet at regular intervals (as and when necessary) and is responsible for:

a) co-ordinating the activities of ESGCD
b) dissemination of information
c) raising funds for the project
d) organising meetings
e) receiving applications for membership
f) other activities deemed appropriate

5. Activities of ESGCD.
A general meeting will be held at each ECCMID. All elections for the Executive will take place at these meetings. A scientific symposium will take place at each ECCMID to present the findings of ESGCD and the official language will be English.

6. Publications.
Publications regarding aim (g), the guidelines/position paper should be published in the ESCMID journal, *Clinical Microbiology and Infection*, through joint submission to the CMI editor and ESCMID Scientific Officer

Work plan for 2000-2001
Gain approval for statutes of ESGCD
Hold inaugural membership meeting at Stockholm ECCMID
Elect Executive Committee at Stockholm ECCMID
Plan work strategy to achieve aims of ESGCD
Initiate possible exchange of strains for typing comparison
Discuss results of questionnaire issued to volunteer participants
Discuss assessment of *C. difficile* infections in member states
Discuss grant application sources

Composition of Temporary Committee:
Dr. J.S. Brazier, Cardiff,
Prof. I.R. Poxton, Edinburgh
Prof. J.C. Petit, Paris
Prof.M. Delmee, Brussels

The list of other potentially interested parties include:
Prof. Nord, Stockholm, Sweden
Dr. Claros, Leipzig, Germany
Dr. Jousimies-Somer, Helsinki, Finland
Dr.Picazo, Madrid, Spain
Prof. Botta, Udine. Italy
Dr. Doyle, Dublin. Ireland

Tentative Programme for Istanbul 2001 (April 1st – 4th)
Report of Executive Committee
Results of typing exchange
Set up of European *C. difficile* Culture Collection.

Anticipated Sources of Funding
ESCMID start-up support grant, EU grant programmes.

Email: Brazier@cardiff.ac.uk
Eudiff2.doc

Fig. 1 Original statutes, workplan, and composition of the committee of ESGCD

C. difficile infections, and by this process to harmonise methodologies relating to CDI

- Collaborate with commercial entities developing treatments for CDI (vaccines, new antibiotics, immunotherapies)

In the intervening 22 years since those initial meetings in Stockholm and Istanbul, ESGCD has continued to hold an annual business meeting at each successive ECCMID. The members of the Executive committee also met every 6 months in Brussels or Edinburgh to discuss the progress of ESGCD activities. There have been a further five Chairpersons of ESGCD, and the full list of these, and the current executive committee, is provided in Tables 1 and 2, respectively. The current executive committee meets online every 6–8 weeks.

Veterinary and human clinical microbiologists, molecular microbiologists, antimicrobial pharmacists, and infectious diseases specialists have always participated in ESGCD. The Group has been a prolific and consistent contributor to the scientific programme of the ECCMID meetings through a wide range of symposia, workshops, presentations, and posters. ESGCD has also contributed to many of the international ClostPath meetings and to all the International *C. difficile* Symposia (ICDS; www.icds.si) in Slovenia, both by financial support and scientific presentations. However, most importantly, over this time the Study Group has evolved and grown to become a hub for research, and for the development and promulgation of standards for surveillance, diagnostics, infection prevention and control, and therapeutics for CDI. Although the focus has been on Europe, the fostering of collaborations with colleagues outwith Europe, particularly in North America, has ensured that the activities of ESGCD have helped to shape and

Table 1 List of previous ESGCD Chairpersons

Time period[a]	ESGCD chairperson
2000–2005	Jon Brazier, Cardiff, UK
2005–2008	Ian Poxton, Edinburgh, UK
2008–2016	Ed Kuijper, Leiden, Netherlands
2016–2018	John Coia, Glasgow, UK
2018–2019	Sarah Tschudin-Sutter, Basel, Switzerland

[a]Handover of chairpersonship happens at ECCMID

influence the understanding and management of CDI globally.

3 Activities and Achievements of ESGCD

The aims of the Study Group are addressed through the support and promotion of a range of activities by Study Group members, often in collaboration with other groups and institutions. These activities include:

- Scientific and clinical projects and publications initiated by the Study Group
- Scientific and clinical projects and publications initiated in collaboration with other groups and institutions, including industrial partners
- Scientific and clinical projects and publications to which Study Group members have contributed, or which have benefitted from the professional support of ESGCD
- Proposals for scientific & educational sessions at ECCMID, or under the auspices of ESCMID
- Presentations at ECCMID and other scientific meetings
- ECCMID postgraduate workshops
- Promotion and support of scientific meetings and workshops & educational activities outside of ESCMID
- Regular publication of Study Group Newsletters
- Regular educational webinars, also with other Study Groups
- Development of and participation in a range of online activities
- Funded research projects

Rather than provide an exhaustive list of all the activities and outputs of ESGCD, the remainder of this section will focus on the achievements in three key domains which are central to the aims and objectives of the Study Group. These are laboratory investigation of CDI (including diagnosis and typing), epidemiology and surveillance of CDI in Europe, and management of CDI (including infection prevention and control, and treatment). Activities in each of these areas have provided the basis for, and encouraged the development of, the collaboration amongst key stakeholders (individual clinical and research groups, organisations, and institutions) at the national and international level. A common cross-cutting element of this approach has been the role of ESGCD in development and promotion of comprehensive, evidence-based guidance in each of these areas. A key overall achievement

Table 2 Current ESGCD executive committee 2020–present

Role	Name
Chairperson	Fidelma Fitzpatrick, Dublin, Ireland
Secretary	Marcela Krutova, Prague, Czech Republic
Treasurer	Frédéric Barbut, Paris, France
Science Officer	Benoit Guery, Lausanne, Switzerland
Education Officer	Kerrie Davies, Leeds, UK
Ad hoc: Patient Representative	John Heritage, Leeds, UK

of these activities is that CDI is now recognised as a very significant clinical disease entity that requires to be controlled and managed in its own right, rather than being viewed as a troublesome complication of other medical interventions.

3.1 Laboratory Investigation of CDI

Accurate diagnosis is a cornerstone of any laboratory-based surveillance system. Moreover, even where there is the laboratory capability to undertake accurate diagnostic testing, the comparability of resulting surveillance data is crucially dependent upon the criteria employed for sampling and testing. The absence of specific guidelines which would help to facilitate reliable diagnosis and the accurate comparison of the incidence and the epidemiology of CDI from one hospital to another or from one country to another, was a key early concern of ESGCD. This was reflected in the minutes of the first meeting of the Study Group, where it was noted that a survey of diagnostic methods and testing protocols for CDI in Europe should be undertaken. This was one of the first major activities of ESGCD, and established a baseline measurement of the marked discrepancies between laboratories and between countries regarding the criteria by which *C. difficile* was investigated for, and the methods and strategies that were used for the diagnosis of CDI (Barbut et al. 2003). This lack of specific guidance was addressed by the publication in 2009 of ESCMID recommendations for the diagnosis of CDI developed by ESGCD (Crobach et al. 2009). A review of this guidance, with evaluation of the current evidence, led to the publication of updated ESGCD guidelines in 2016 (Crobach et al. 2016). In 2018, this diagnostic algorithm protocol was slightly adapted and recommended by ECDC for surveillance studies (ECDC 2018).

As has already been noted above, the emergence of CDI as a major pathogen in the early part of this century was associated with particular strains of *C. difficile*, and our current understanding of the epidemiology of CDI and ability to investigate and control outbreaks of infection with this organism remains reliant upon the development and availability of robust typing methodologies. The importance of typing in elucidating the emergence and spread of novel subtypes was highlighted in a review in 2006 (Kuijper et al. 2006), which summarised the outputs of a series of meetings organised by the ECDC with experts in the field of CDI, including ESGCD and the United States Centers for Disease Control and Prevention (CDC). ESGCD has played an important role in promoting the development, standardisation, and adoption of molecular subtyping (particularly PCR ribotyping) of *C. difficile* in Europe. In order to obtain an overview of the phenotypic and genotypic features of clinical isolates of *C. difficile*, during 2005, the Study Group undertook a 2-month prospective study of *C. difficile* infections (CDI) in 38 hospitals from 14 different European countries (Barbut et al. 2007).

In recent years, the emergence of whole-genome sequence-based (WGS) methodologies has had a profound impact on genotypic typing of a range of microorganisms, including *C. difficile.* This has already provided important insights into the emergence and spread of hypervirulent strains (He et al. 2013). This technology is increasingly being developed and applied in the routine characterisation and surveillance of *C. difficile* (Janezic and Rupnik 2019), and ESGCD members have been actively involved in these developments (Baktash et al. 2022). Applying WGS on "hypervirulent" strains from various countries revealed close similarities (clade 2) of clearly distinct PCR ribotypes, providing further support for implementing WGS as a routine typing method. For instance, ESGCD members from Greece investigated a large outbreak of *C. difficile* ribotype 181 which was very similar to RT 027 and has now spread to other countries (Kachrimanidou et al. 2020).

Measures to develop and promulgate standardised typing methodologies for *C. difficile* have been closely linked to activities to develop surveillance of CDI in Europe, and are considered in the next section.

3.2 Epidemiology and Surveillance of CDI in Europe

Following the recognition of the arrival of the new hypervirulent *C. difficile* strain, PCR ribotype 027, in 2005 in Europe, ESGCD contacted ECDC and a range of stakeholders and partners to consider how recognition and awareness of CDI could be increased, and how surveillance in Europe could be improved. As part of this, the background review document on CDI (Kuijper et al. 2006), which has been referred to above was produced. This initiative was also the catalyst for the first pan-European surveillance study, the European Clostridium Infection Survey (ECDIS), supported by ECDC. This was performed in 2008–09 and was subsequently published in the Lancet (Bauer et al. 2011). Based on the results of the ECDIS study, it was decided to provide support for further capacity building for surveillance of CDI across Europe. This resulted in the ECDIS-net project (ECDIS-net 2017), in which ESGCD and its members played a key role. ECDIS-net comprised a consortium of experts in the field of CDI including microbiologists, epidemiologists, and molecular biologists, who were all in close contact with or were part of their respective National Institutes of Health, and who were active in surveillance studies of *C. difficile*. The project aimed to enhance laboratory diagnostic capacity, standardise approaches and build capacity for molecular subtyping (particularly PCR ribotyping), and to develop a European CDI surveillance protocol. As part of this work, surveys of diagnostic and typing capacity (van Dorp et al. 2016b), and of CDI surveillance systems (Kola et al. 2016), in Europe were undertaken. Following the development of the surveillance protocol, a pilot study of standardised surveillance of *Clostridium difficile* infection in European acute care hospitals was undertaken (van Dorp et al. 2016a). The protocol developed now forms the basis of the ECDC protocol for surveillance of CDI in Europe (ECDC 2017). ESGCD has subsequently partnered with ECDC in a joint project consortium on Microbiological support to European surveillance of CDI, and ESGCD has continued to provide input into the surveillance protocol as it has continued to develop. A minireview on "How to: Surveillance of *Clostridium difficile* infections" was published in 2018 in the journal Clinical Microbiology and Infection (CMI) (Krutova et al. 2018). This article formed part of a themed issue of CMI on various aspects of CDI produced by ESGCD members (Coia 2018). More detailed protocols for CDI surveillance are available at the ECDC site (ECDC 2019). The surveillance of CDI is the subject of a separate chapter in this volume.

3.3 Management of CDI

As part of its activities ESGCD has also been active in initiatives to improve the management of CDI in Europe. Again, in keeping with the original aims and objectives of the Study Group, a particular focus has been on the development and promotion of evidence-based guidance. A number of group members were involved in the production of Infection control measures to limit the spread of *C. difficile* produced on behalf of the European *C. difficile* Infection Control Group and the ECDC which were published in 2008 (Vonberg et al. 2008). This evidence-based guidance has recently been reviewed as part of the current activities of ESGCD (see below). Guidance for treatment of CDI was developed and published by Study Group members in 2009 as the ESCMID treatment guidance document for CDI (Bauer et al. 2009). An evidence-based update of this guidance was published in 2014 by Debast et al. (2014). This clinical guideline has subsequently been updated again in 2021 following a major evidence-based review, with important changes to the guidance (van Prehn et al. 2021).

4 Current Activities of ESGCD

As can be seen from the most recent annual report (ESGCD 2021), ESGCD continues to be one of the most active ESCMID Study Groups. Six publications were authored by, or had significant contributions to or support from, ESGCD members. These comprised production of revised European treatment guidance document for *C. difficile* infection in adults (van Prehn et al. 2021) and two further publications in CMI on prophylactic interventions for prevention of *C. difficile* infection (Reigadas et al. 2021) and the need for a holistic view on management of *C. difficile* infection (Wingen-Heimann et al. 2021). The article on prophylactic interventions was chosen by CMI Editor-in-chief Leonard Leibovici as one of the top articles of 2021 (CMI 2022). There were also articles on detection of metronidazole resistance in clinical isolates of *C. difficile* (Boekhoud et al. 2021), establishment of a French surveillance system for CDI (Khanafer et al. 2021), and a retrospective study of mortality following CDI in Europe (Czepiel et al. 2021). In 2022, the publications authored by, or supported by the activities of ESGCD members have included a review of the diagnosis and management of paediatric CDI (Krutova et al. 2022a), the management of CDI in patients receiving concomitant antibiotics for other infections (Fitzpatrick et al. 2022), pharmacological and microbiological equivalence of antibiotics used for treatment of CDI (Krutova et al. 2022b), molecular epidemiology and antimicrobial resistance patterns of *C. difficile* isolates in Algerian hospitals (Boudjelal et al. 2022), and differences in motility among different *C. difficile* ribotypes (Karpiński et al. 2022).

ESGCD was also a supporting organiser of three separate virtual events in 2021. These were the "World Microbe Forum", 20–24 June 2021, the "III International Latin American, *Clostridioides difficile* Symposium", 4–6 October 2021, and "International Conference on the Molecular Biology and Pathogenesis of the Clostridia (Clostpath)", 13–17 September 2021.

At ECCMID 2022 in Lisbon, an educational workshop on "Advances in *C. difficile* infection outbreak prevention and control" was jointly organised by ESGCD and the ESCMID Study Group for Nosocomial Infections (ESGNI), and a symposium on "*C. difficile* infection: importance of learning from patients experience" focusing on the impact of CDI from the viewpoint of patients was held. In addition, an oral session "Secrets of *Clostridioides difficile*" presented a series of selected abstracts dealing with various topical aspects of CDI.

In addition to the frequent newsletters updating ESGCD members on the activities of the group (ESCMID 2023b), ESGCD has been active in developing a series of informative and educational scientific webinars either alone or in combination with other study groups. Webinar topics have included "Recurrent CDI", "Treatment and Management of Severe CDI", "*C. difficile* typing: from bench to bedside", "Beyond FMT" (in conjunction with the ESCMID Study Group for host and microbiota interaction (ESGHAMI)), and "Clinical and laboratory-themed cases in CDI". ESGCD continues to seek effective and innovative ways to support not only clinicians and scientists with active interest in this important pathogen, but also to represent the viewpoint of those who have experienced CDI. To help achieve this goal, ESGCD has recently welcomed a patient representative to join the Executive Committee and is the first ESCMID study group to have included patient representation on the executive group.

CDI was the first infectious disease for which microbiota intervention treatments were successfully developed and subsequently recommended in various international guidelines. In 2018, a new study group (ESGHAMI) was formed by ESGCD members to intensify research and interventions of the gut microbiome in other diseases than CDI (ESCMID 2023a). ESGHAMI has still close collaborations with ESGCD and organises webinars and sessions at ECCMID on behalf of both study groups.

In addition to these activities, ESGCD members have continued to be active in a variety of projects and initiatives to further our understanding and management of the challenges posed by CDI. These have included recently the "Combatting Bacterial Resistance in Europe—*Clostridioides difficile* infections" (COMBACTE-CDI) project. This activity has been reflected in a number of recent publications on various aspects of CDI including testing for metronidazole resistance, contamination of retail potatoes, a point-prevalence survey of community and hospital CDI in Europe, a survey of surveillance, management, and testing practices for CDI in Europe, and a pan-European study of cost and resource utilisation for CDI (Boekhoud et al. 2021; Tkalec et al. 2022; Viprey et al. 2022, 2023; Wingen-Heimann et al. 2022).

5 Perspectives and the Future

Since its establishment at the start of the current millennium, ESGCD has been one of the most consistently active and productive ESCMID Study Groups and has achieved considerable success in attaining the aims and objectives outlined in the original statutes. This success has been generated by a combination of approaches reflecting not only research projects and other activities undertaken by members of ESGCD itself, but also through wider collaborations. These partnerships have served to add further value to the activities of ESGCD, and have encompassed specific formal research projects in combination with other national and international partners, e.g. the European Centre for Disease Prevention and Control (ECDC), as well as less structured ad hoc interactions between groups of individual scientists and clinicians. Importantly, industrial partners also approached ESGCD for advice and participation in their projects, using the knowledge and experiences of ESGCD members. All these activities have been underpinned by a common underlying goal of mitigating the impact of CDI, and a significant proportion of the material in the subsequent chapters of this book reflect the outcomes of some of this work.

There can be little doubt that considerable progress has been made in understanding the epidemiology of CDI in Europe, and in developing comprehensive guidance for the surveillance, diagnosis, prevention, control, and management of this major nosocomial pathogen. However, despite these successes, CDI remains a very significant infection challenge in many parts of Europe and beyond. CDI is also an important disease in animals and the emergence of Type 078 in human CDI coincided with the finding of this type in diarrhoeal piglets. CDI fits in a "One Health" approach, since whole-genome sequencing has revealed genetic identity between human and animal isolates for at least two ribotypes (078 (Knetsch et al. 2014); 014 (Knight et al. 2016)). Many unresolved issues remain, and even as our knowledge advances, fresh questions arise. How can surveillance of CDI be extended in resource-poor settings? What are the virulence mechanisms of the "hypervirulent" strains? What is the precise role of the intestinal microbiota in defence against CDI? How can we prevent primary and/or recurrent CDI? What are the relative contributions of different control measures in prevention of nosocomially-acquired CDI? What is the role of asymptomatic carriage? How can diagnostic testing be improved and simplified? What is the best approach to deal with recurrent disease or severe disease? What fresh insights will the application of whole-genome sequencing, which has already challenged our existing paradigm of CDI, bring to our understanding and management of CDI? What is the variety of reservoirs contributing to hospital and community CDI? Clearly the work of, and need for, the activities of ESGCD is far from complete.

Acknowledgements The authors wish to thank Jon Brazier, Michel Delmeé, Ian Poxton, Henri Saenz, Patrick Kudyba, and the current ESGCD executive committee for their assistance in the development of this manuscript. They also particularly wish to thank ESCMID for their ongoing support of, and commitment to, the work of ESGCD.

References

Baktash A, Corver J, Harmanus C, Smits WK, Fawley W, Wilcox MH, Kumar N, Eyre DW, Indra A, Mellmann A, Kuijper EJ (2022) Comparison of whole-genome sequence-based methods and PCR ribotyping for subtyping of Clostridioides difficile. J Clin Microbiol 60(2):e0173721. https://doi.org/10.1128/jcm.01737-21

Barbut F, Delmee M, Brazier JS, Petit JC, Poxton IR, Rupnik M, Lalande V, Schneider C, Mastrantonio P, Alonso R, Kuipjer E, Tvede M (2003) A European survey of diagnostic methods and testing protocols

for Clostridium difficile. Clin Microbiol Infect 9(10): 989–996. http://www.clinicalmicrobiologyandinfection.com/article/S1198-743X(14)63055-X/pdf
Barbut F, Mastrantonio P, Delmee M, Brazier J, Kuijper E, Poxton I (2007) Prospective study of Clostridium difficile infections in Europe with phenotypic and genotypic characterisation of the isolates. Clin Microbiol Infect 13(11):1048–1057. https://doi.org/10.1111/j.1469-0691.2007.01824.x
Bauer MP, Kuijper EJ, van Dissel JT (2009) European Society of Clinical Microbiology and Infectious Diseases (ESCMID): treatment guidance document for Clostridium difficile infection (CDI). Clin Microbiol Infect 15(12):1067–1079. https://doi.org/10.1111/j.1469-0691.2009.03099.x
Bauer MP, Notermans DW, van Benthem BH, Brazier JS, Wilcox MH, Rupnik M, Monnet DL, van Dissel JT, Kuijper EJ (2011) Clostridium difficile infection in Europe: a hospital-based survey. Lancet 377(9759): 63–73. https://doi.org/10.1016/s0140-6736(10)61266-4
Boekhoud IM, Sidorov I, Nooij S, Harmanus C, Bos-Sanders I, Viprey V, Spittal W, Clark E, Davies K, Freeman J, Kuijper EJ, Smits WK (2021) Haem is crucial for medium-dependent metronidazole resistance in clinical isolates of Clostridioides difficile. J Antimicrob Chemother 76(7):1731–1740. https://doi.org/10.1093/jac/dkab097
Boudjelal Y, Krutova M, Djebbar A, Sebaihia M, Bekara MEA, Rouabhia S, Couturier J, Syed-Zaidi R, Barbut F (2022) Molecular epidemiology and antimicrobial resistance patterns of Clostridioides difficile isolates in Algerian hospitals. J Infect Dev Ctries 16(6): 1055–1063. https://doi.org/10.3855/jidc.16056
CMI (2022) Editor's top articles from 2021. Retrieved February 2nd, 2023 from https://www.journals.elsevier.com/clinical-microbiology-and-infection/editors-choice/editor-s-top-articles-from-2021
Coia J (2018) C. difficile infection—can we do better? Clin Microbiol Infect 24(5):450–451. https://doi.org/10.1016/j.cmi.2017.12.009
Crobach MJ, Dekkers OM, Wilcox MH, Kuijper EJ (2009) European Society of Clinical Microbiology and Infectious Diseases (ESCMID): data review and recommendations for diagnosing Clostridium difficile-infection (CDI). Clin Microbiol Infect 15(12):1053–1066. https://doi.org/10.1111/j.1469-0691.2009.03098.x
Crobach MJ, Planche T, Eckert C, Barbut F, Terveer EM, Dekkers OM, Wilcox MH, Kuijper EJ (2016) European Society of Clinical Microbiology and Infectious Diseases: update of the diagnostic guidance document for Clostridium difficile infection. Clin Microbiol Infect 22(Suppl 4):S63–S81. https://doi.org/10.1016/j.cmi.2016.03.010
Czepiel J, Krutova M, Mizrahi A, Khanafer N, Enoch DA, Patyi M, Deptuła A, Agodi A, Nuvials X, Pituch H, Wójcik-Bugajska M, Filipczak-Bryniarska I, Brzozowski B, Krzanowski M, Konturek K, Fedewicz M, Michalak M, Monpierre L, Vanhems P et al (2021) Mortality following clostridioides difficile infection in Europe: a retrospective multicenter case-control study. Antibiotics (Basel) 10(3). https://doi.org/10.3390/antibiotics10030299
Debast SB, Bauer MP, Kuijper EJ (2014) European Society of Clinical Microbiology and Infectious Diseases: update of the treatment guidance document for Clostridium difficile infection. Clin Microbiol Infect 20 (Suppl 2):1–26. https://doi.org/10.1111/1469-0691.12418
ECDC (2017) ECDC surveillance protocol for CDI. Retrieved August 4th, 2017 from https://ecdc.europa.eu/en/clostridium-difficile-infections/surveillance-and-disease-data
ECDC (2018) Laboratory procedures for diagnosis and typing of human C. difficile infection. Retrieved February 6th, 2023 from https://www.ecdc.europa.eu/en/publications-data/laboratory-procedures-diagnosis-and-typing-human-clostridium-difficile-infection
ECDC (2019) European surveillance of Clostridioides (Clostridium) difficile infections surveillance protocol version 2.4. Retrieved February 2nd, 2023 from https://www.ecdc.europa.eu/sites/default/files/documents/clostridium-difficile-infections-EU-surveillance-protocol-vers2.4.pdf
ECDIS-net (2017) ECDIS-net home page. Retrieved August 4th, 2017 from http://www.ecdisnet.eu/
ESCMID (2020) ESCMID study group statutes. https://www.escmid.org/research_projects/study_groups/sg_statutes/
ESCMID (2022a) About ECCMID. ESCMID. Retrieved December 2nd, 2022 from https://www.escmid.org/dates_events/eccmid/about_the_congress
ESCMID (2022b) About ESCMID. Retrieved December 2nd, 2002 from https://www.escmid.org/membership-organisation/about-escmid
ESCMID (2022c) ESCMID Study Groups. https://www.escmid.org/research_projects/study_groups/
ESCMID (2023a) ESCMID Study Group for Host and Microbiota Interaction—ESGHAMI. Retrieved February 6th, 2023 from https://www.escmid.org/research_projects/study_groups/study_groups_g_n/host_and_microbiota_interaction
ESCMID (2023b) ESGCD Newsletters. Retrieved February 2nd, 2023 from https://www.escmid.org/research_projects/study_groups/study_groups_a_f/clostridioides_difficile/newsletters
ESGCD (2021) Annual report 2021. Retrieved February 2nd, 2021 from https://www.escmid.org/fileadmin/src/media/PDFs/3Research_Projects/ESGCD/Annual_Report_ESGCD_2021.pdf
ESGCD (2022) ESGCD Newsletter Winter 2022. Retrieved February 7th, 2023 from https://www.escmid.org/fileadmin/src/media/PDFs/3Research_Projects/ESGCD/ESGCD_Newsletterwinter_2022_final.pdf
Fitzpatrick F, Safdar N, van Prehn J, Tschudin-Sutter S (2022) How can patients with Clostridioides difficile

infection on concomitant antibiotic treatment be best managed? Lancet Infect Dis 22(11):e336–e340. https://doi.org/10.1016/s1473-3099(22)00274-2

He M, Miyajima F, Roberts P, Ellison L, Pickard DJ, Martin MJ, Connor TR, Harris SR, Fairley D, Bamford KB, D'Arc S, Brazier J, Brown D, Coia JE, Douce G, Gerding D, Kim HJ, Koh TH, Kato H et al (2013) Emergence and global spread of epidemic healthcare-associated Clostridium difficile. Nat Genet 45(1): 109–113. https://doi.org/10.1038/ng.2478

Honda H, Dubberke ER (2014) The changing epidemiology of Clostridium difficile infection. Curr Opin Gastroenterol 30(1):54–62. https://doi.org/10.1097/MOG.0000000000000018

Janezic S, Rupnik M (2019) Development and Implementation of whole genome sequencing-based typing schemes for Clostridioides difficile [mini review]. Front Public Health 7. https://doi.org/10.3389/fpubh.2019.00309

Kachrimanidou M, Baktash A, Metallidis S, Tsachouridou O, Netsika F, Dimoglou D, Kassomenaki A, Mouza E, Haritonidou M, Kuijper E (2020) An outbreak of Clostridioides difficile infections due to a 027-like PCR ribotype 181 in a rehabilitation centre: Epidemiological and microbiological characteristics. Anaerobe 65:102252. https://doi.org/10.1016/j.anaerobe.2020.102252

Karpiński P, Wultańska D, Piotrowski M, Brajerova M, Mikucka A, Pituch H, Krutova M (2022) Motility and the genotype diversity of the flagellin genes fliC and fliD among Clostridioides difficile ribotypes. Anaerobe 73:102476. https://doi.org/10.1016/j.anaerobe.2021.102476

Khanafer N, Hemmendinger A, Guery B, Vachée A, Rogues AM, Gravet A, Boutoille D, Vanjak D, Barbut F, Vanhems P (2021) Establishment of a French surveillance system of Clostridioides difficile infection: Comparison of patient's characteristics with other national and European data. Anaerobe 69: 102329. https://doi.org/10.1016/j.anaerobe.2021.102329

Knetsch CW, Connor TR, Mutreja A, van Dorp SM, Sanders IM, Browne HP, Harris D, Lipman L, Keessen EC, Corver J, Kuijper EJ, Lawley TD (2014) Whole genome sequencing reveals potential spread of Clostridium difficile between humans and farm animals in the Netherlands, 2002 to 2011. Euro Surveill 19(45): 20954

Knight DR, Squire MM, Collins DA, Riley TV (2016) Genome analysis of Clostridium difficile PCR ribotype 014 lineage in Australian pigs and humans reveals a diverse genetic repertoire and signatures of long-range interspecies transmission. Front Microbiol 7. https://doi.org/10.3389/fmicb.2016.02138

Kola A, Wiuff C, Akerlund T, van Benthem BH, Coignard B, Lyytikainen O, Weitzel-Kage D, Suetens C, Wilcox MH, Kuijper EJ, Gastmeier P (2016) Survey of Clostridium difficile infection surveillance systems in Europe, 2011. Euro Surveill 21(29). https://doi.org/10.2807/1560-7917.es.2016.21.29.30291

Krutova M, Kinross P, Barbut F, Hajdu A, Wilcox MH, Kuijper EJ (2018) How to: surveillance of Clostridium difficile infections. Clin Microbiol Infect 24(5): 469–475. https://doi.org/10.1016/j.cmi.2017.12.008

Krutova M, de Meij TGJ, Fitzpatrick F, Drew RJ, Wilcox MH, Kuijper EJ (2022a) How to: Clostridioides difficile infection in children. Clin Microbiol Infect 28(8): 1085–1090. https://doi.org/10.1016/j.cmi.2022.03.001

Krutova M, Wilcox M, Kuijper E (2022b) Clostridioides difficile infection: are the three currently used antibiotic treatment options equal from pharmacological and microbiological points of view? Int J Infect Dis 124: 118–123. https://doi.org/10.1016/j.ijid.2022.09.013

Kuijper EJ, Coignard B, Tull P, ESCMID Study Group for Clostridium difficile; EU Member States; European Centre for Disease Prevention and Control (2006) Emergence of Clostridium difficile-associated disease in North America and Europe. Clin Microbiol Infect 12 (Suppl 6):2–18. https://doi.org/10.1111/j.1469-0691.2006.01580.x

Larson HE, Price AB, Honour P, Borriello SP (1978) Clostridium difficile and the aetiology of pseudomembranous colitis. Lancet 1(8073): 1063–1066. https://www.ncbi.nlm.nih.gov/pubmed/77366, http://www.thelancet.com/journals/lancet/article/PIIS0140-6736(78)90912-1/abstract

Phillips I (2008) ESCMID news supplement to issue number 1/2008. https://www.escmid.org/fileadmin/src/media/PDFs/2News_Discussions/1News_Media/escmid_news/ESCMID_News20Supplement-History_120dpi_080502.pdf

Reigadas E, van Prehn J, Falcone M, Fitzpatrick F, Vehreschild M, Kuijper EJ, Bouza E (2021) How to: prophylactic interventions for prevention of Clostridioides difficile infection. Clin Microbiol Infect 27(12):1777–1783. https://doi.org/10.1016/j.cmi.2021.06.037

Tkalec V, Viprey V, Davis G, Janezic S, Sente B, Devos N, Wilcox M, Davies K, Rupnik M (2022) Clostridioides difficile positivity rate and PCR ribotype distribution on retail potatoes in 12 European countries, January to June 2018. Euro Surveill 27(15). https://doi.org/10.2807/1560-7917.Es.2022.27.15.2100417

van Dorp SM, Kinross P, Gastmeier P, Behnke M, Kola A, Delmee M, Pavelkovich A, Mentula S, Barbut F, Hajdu A, Ingebretsen A, Pituch H, Macovei IS, Jovanovic M, Wiuff C, Schmid D, Olsen KE, Wilcox MH, Suetens C, Kuijper EJ (2016a) Standardised surveillance of Clostridium difficile infection in European acute care hospitals: a pilot study, 2013. Euro Surveill 21(29). https://doi.org/10.2807/1560-7917.es.2016.21.29.30293

van Dorp SM, Notermans DW, Alblas J, Gastmeier P, Mentula S, Nagy E, Spigaglia P, Ivanova K, Fitzpatrick F, Barbut F, Morris T, Wilcox MH, Kinross P, Suetens C, Kuijper EJ (2016b) Survey of diagnostic and typing capacity for Clostridium difficile infection in Europe, 2011 and 2014. Euro Surveill 21(29). https://doi.org/10.2807/1560-7917.es.2016.21.29.30292

van Prehn J, Reigadas E, Vogelzang EH, Bouza E, Hristea A, Guery B, Krutova M, Noren T, Allerberger F, Coia JE, Goorhuis A, van Rossen TM, Ooijevaar RE, Burns K, Scharvik Olesen BR, Tschudin-Sutter S, Wilcox MH, Vehreschild M, Fitzpatrick F et al (2021) European Society of Clinical Microbiology and Infectious Diseases: 2021 update on the treatment guidance document for Clostridioides difficile infection in adults. Clin Microbiol Infect 27 (Suppl 2):S1–S21. https://doi.org/10.1016/j.cmi.2021.09.038

Viprey VF, Davis GL, Benson AD, Ewin D, Spittal W, Vernon JJ, Rupnik M, Banz A, Allantaz F, Cleuziat P, Wilcox MH, Davies KA (2022) A point-prevalence study on community and inpatient Clostridioides difficile infections (CDI): results from Combatting Bacterial Resistance in Europe CDI (COMBACTE-CDI), July to November 2018. Euro Surveill 27(26). https://doi.org/10.2807/1560-7917.Es.2022.27.26.2100704

Viprey VF, Granata G, Vendrik KEW, Davis GL, Petrosillo N, Kuijper EJ, Vilken T, Lammens C, Schotsman JJ, Benson AD, Cataldo MA, van der Kooi TII, Wilcox MH, Davies KA (2023) European survey on the current surveillance practices, management guidelines, treatment pathways and heterogeneity of testing of Clostridioides difficile, 2018-2019: results from The Combatting Bacterial Resistance in Europe CDI (COMBACTE-CDI). J Hosp Infect 131:213–220. https://doi.org/10.1016/j.jhin.2022.11.011

Vonberg RP, Kuijper EJ, Wilcox MH, Barbut F, Tull P, Gastmeier P, European C difficile-Infection Control Group; European Centre for Disease Prevention and Control (ECDC), van den Broek PJ, Colville A, Coignard B, Daha T, Debast S, Duerden BI, van den Hof S, van der Kooi T, Maarleveld HJ, Nagy E et al (2008) Infection control measures to limit the spread of Clostridium difficile. Clin Microbiol Infect 14(Suppl 5):2–20. https://doi.org/10.1111/j.1469-0691.2008.01992.x

Warny M, Pepin J, Fang A, Killgore G, Thompson A, Brazier J, Frost E, McDonald LC (2005) Toxin production by an emerging strain of Clostridium difficile associated with outbreaks of severe disease in North America and Europe. Lancet 366(9491):1079–1084. https://doi.org/10.1016/S0140-6736(05)67420-X

Wingen-Heimann SM, van Prehn J, Kuijper EJ, Vehreschild M (2021) The need for a holistic view on management of Clostridioides difficile infection. Clin Microbiol Infect 27(10):1383–1385. https://doi.org/10.1016/j.cmi.2021.07.014

Wingen-Heimann SM, Davies K, Viprey VF, Davis G, Wilcox MH, Vehreschild M, Lurienne L, Bandinelli PA, Cornely OA, Vilken T, Hopff SM, Vehreschild JJ (2022) Clostridioides difficile infection (CDI): A pan-European multi-center cost and resource utilization study, results from the Combatting Bacterial Resistance in Europe CDI (COMBACTE-CDI). Clin Microbiol Infect. https://doi.org/10.1016/j.cmi.2022.12.019

SPRINGER NATURE

GPSR Compliance

The European Union's (EU) General Product Safety Regulation (GPSR) is a set of rules that requires consumer products to be safe and our obligations to ensure this.

If you have any concerns about our products, you can contact us on ProductSafety@springernature.com

In case Publisher is established outside the EU, the EU authorized representative is:

Springer Nature Customer Service Center GmbH
Europaplatz 3
69115 Heidelberg, Germany

The manufacturer's authorised representative in the EU is Springer Nature Customer Service Centre GmbH, Europaplatz 3, 69115 Heidelberg, Germany. If you have any concerns regarding our products, please contact ProductSafety@springernature.com

Printed and bound by CPI Group (UK) Ltd, Croydon, CR0 4YY
01/10/2026
02225896-0020